SHOCK-WAVE
AND
HIGH-STRAIN-RATE
PHENOMENA
IN MATERIALS

T0172590

MECHANICAL ENGINEERING

A Series of Textbooks and Reference Books

Editor

L. L. Faulkner

Columbus Division, Battelle Memorial Institute
and Department of Mechanical Engineering
The Ohio State University
Columbus, Ohio

1. *Spring Designer's Handbook,* Harold Carlson
2. *Computer-Aided Graphics and Design,* Daniel L. Ryan
3. *Lubrication Fundamentals,* J. George Wills
4. *Solar Engineering for Domestic Buildings,* William A. Himmelman
5. *Applied Engineering Mechanics: Statics and Dynamics,* G. Boothroyd and C. Poli
6. *Centrifugal Pump Clinic,* Igor J. Karassik
7. *Computer-Aided Kinetics for Machine Design,* Daniel L. Ryan
8. *Plastics Products Design Handbook, Part A: Materials and Components; Part B: Processes and Design for Processes,* edited by Edward Miller
9. *Turbomachinery: Basic Theory and Applications,* Earl Logan, Jr.
10. *Vibrations of Shells and Plates,* Werner Soedel
11. *Flat and Corrugated Diaphragm Design Handbook,* Mario Di Giovanni
12. *Practical Stress Analysis in Engineering Design,* Alexander Blake
13. *An Introduction to the Design and Behavior of Bolted Joints,* John H. Bickford
14. *Optimal Engineering Design: Principles and Applications,* James N. Siddall
15. *Spring Manufacturing Handbook,* Harold Carlson
16. *Industrial Noise Control: Fundamentals and Applications,* edited by Lewis H. Bell
17. *Gears and Their Vibration: A Basic Approach to Understanding Gear Noise,* J. Derek Smith
18. *Chains for Power Transmission and Material Handling: Design and Applications Handbook,* American Chain Association
19. *Corrosion and Corrosion Protection Handbook,* edited by Philip A. Schweitzer

Additional Volumes in Preparation

Mechanical Engineering Software

SHOCK-WAVE AND HIGH-STRAIN-RATE PHENOMENA IN MATERIALS

EDITED BY

MARC A. MEYERS
University of California, San Diego
La Jolla, California

LAWRENCE E. MURR
University of Texas at El Paso
El Paso, Texas

KARL P. STAUDHAMMER
Los Alamos National Laboratory
Los Alamos, New Mexico

CRC Press
Taylor & Francis Group
Boca Raton London New York

CRC Press is an imprint of the
Taylor & Francis Group, an **informa** business

CRC Press
Taylor & Francis Group
6000 Broken Sound Parkway NW, Suite 300
Boca Raton, FL 33487-2742

First issued in paperback 2019

© 1992 by Taylor & Francis Group, LLC
CRC Press is an imprint of Taylor & Francis Group, an Informa business

No claim to original U.S. Government works

ISBN-13: 978-0-8247-8579-6 (hbk)
ISBN-13: 978-0-367-40279-2 (pbk)

This book contains information obtained from authentic and highly regarded sources. Reasonable efforts have been made to publish reliable data and information, but the author and publisher cannot assume responsibility for the validity of all materials or the consequences of their use. The authors and publishers have attempted to trace the copyright holders of all material reproduced in this publication and apologize to copyright holders if permission to publish in this form has not been obtained. If any copyright material has not been acknowledged please write and let us know so we may rectify in any future reprint.

Except as permitted under U.S. Copyright Law, no part of this book may be reprinted, reproduced, transmitted, or utilized in any form by any electronic, mechanical, or other means, now known or hereafter invented, including photocopying, microfilming, and recording, or in any information storage or retrieval system, without written permission from the publishers.

For permission to photocopy or use material electronically from this work, please access www.copyright.com (http://www.copyright.com/) or contact the Copyright Clearance Center, Inc. (CCC), 222 Rosewood Drive, Danvers, MA 01923, 978-750-8400. CCC is a not-for-profit organization that provides licenses and registration for a variety of users. For organizations that have been granted a photocopy license by the CCC, a separate system of payment has been arranged.

Trademark Notice: Product or corporate names may be trademarks or registered trademarks, and are used only for identification and explanation without intent to infringe.

Library of Congress Cataloging-in-Publication Data

Shock–wave and high-strain-rate phenomena in materials / edited by
  Marc A. Meyers, Lawrence E. Murr, Karl P. Staudhammer.
        p.   cm. – (mechanical engineering)
    ISBN 0-8247-8579-7
    1. Deformations (Mechanics)   2. Shock (Mechanics)   I. Meyers,
  Marc A. II. Murr, Lawrence E.   III. Staudhammer, Karl P.
  IV. Series : Mechanical engineering (Marcel Dekker, Inc.)
  TA417.6.S46   1992
  620.1'126–dc20   92-3830
                                                                       CIP

**Visit the Taylor & Francis Web site at
http://www.taylorandfrancis.com**

**and the CRC Press Web site at
http://www.crcpress.com**

Foreword John S. Rinehart Award

This award is being established to recognize outstanding effort and creative work in the science and technology of dynamic processes in materials. This encompasses the processes by which materials are welded, formed, compacted, and synthesized, as well as dynamic deformation, fracture, and the extreme shock loading effects. The award is named after a true pioneer who witnessed and actively contributed to the field for over forty years.

This award will be given every five years, at the occasion of the EXPLOMET conferences. The selection of the first two awards, announced in August 1990, was made by a committee composed of the EXPLOMET chairmen and Dr. J. S. Rinehart. In subsequent years, the awardees will chair the committee for future awards. A permanent committee is in such a way established to select the nominees. In selecting the individuals, special attention will be given to the balance between fundamental science and technological implementation.

John S. Rinehart has not only witnessed, but actively taken part in the development of the field of dynamic deformation. He has dedicated his life to the study of stress waves in solids; the results of these investigations have been published in over 130 technical articles and three books, two of them co-authored by John Pearson. *Behavior of Metals Under Impulsive Loads, Explosive Working of Metals* and *Stress Transients in Solids*, have been the *vade mecum* of all scientists and engineers throughout the world working in the field. The simple, no-nonsense, yet fundamentally correct approach used by Dr. Rinehart combines the rigorousness of the physicist with the practicality of the engineer. His fifty-year career has been divided between government and university, and he has frequently served as a consultant to industry. He has occupied many positions of high responsibility throughout his career. Director of Research and Development for the U.S. Coast and Geodetic Survey, Director of the Mining Research Laboratory of the Colorado School of Mines, which he founded, Assistant Director of the Smithsonian Astrophysical Observatory, Head of the Mechanics Branch at the Naval Ordnance Test Station, China Lake, Professor of Mechanical Engineering at the University of Colorado. Dr. Rinehart was associated with Dr. E. J. Workman's Ordnance Research Group before this activity became a division of the New Mexico Institute of Mining and Technology in the early 1950s.

John S. Rinehart (center) personally gave the awards to the co-recipients Andrey Deribas (left) and Mark Wilkins (right).

INSCRIPTIONS

Andrey A. Deribas, co-recipient of the 1990 John S. Rinehart Award for seminal contributions to the theory of explosive welding, for the first experiments of shock synthesis and for leadership in the technological implementation of explosive fabrication. Mark L. Wilkins, co-recipient of the 1990 John S. Rinehart Award for seminal contributions to the development of hydrocodes, for their application to a multitude of dynamic problems and for leadership in the technological implementation of shock compaction.

ANDREY A. DERIBAS

A.A. Deribas was born in 1931 in Moscow. He graduated from Lomonosov University in Moscow in 1953. His field of specialization is mechanics of continuous media. He initiated his research in the field of physics of explosions in 1956. He joined the Siberian Branch of the U.S.S.R. Academy of Sciences from the very beginning with his advisor, Academician Laverentiev, one of its founders. The first scientific results in the field of the explosive hardening of metals were obtained by him in 1960. Investigations on the explosive welding of metals have been carried out since 1961. His first results in the field of explosive compaction of powders and explosive synthesis of new materials took place in 1963. He was the head of the Laboratory of the Institute of Hydrodynamics in Novosibirsk. Research on the localization of explosion was initiated there and the metallic explosive chambers were created. He is the author of over 100 scientific papers and 25 inventions, and there have been two editions of his monograph "Physics of Explosive Welding and Hardening of Metals." Since 1976 he has been the Head of the Special Design Office of High-Rate Hydrodynamics in Novosibirsk. He has a Doctor of Science degree and is a Professor at the Electro-Technological School in Novosibirsk. He is a corresponding member of the U.S.S.R. Academy of Sciences. He was awarded the Lenin Prize for Science, the Prize of Council of Ministry of the U.S.S.R. for Science, and many other awards.

MARK L. WILKINS

Mark Wilkins joined the Lawrence Livermore National Laboratory (LLNL) in 1952, the year it was founded. He developed some of the major computer simulation programs used in the design of nuclear arms. He pioneered the application of large computers to simulate material behavior in the engineering application of materials. The numerical techniques are in current use world-wide. He has been a guest lecturer at some of the leading universities and laboratories in the United States, Europe, and Asia. During the early days of the space program he worked on modeling the effects of micrometeorite impacts. During the period 1967 to 1970, he led a research project sponsored by the Defense Advanced Research Projects Agency (DARPA) to develop a fundamental understanding of penetration mechanics. In 1973 he founded a new division in the Physics Department at LLNL for experimental and theoretical research on the behavior of materials. He has published over 70 scientific papers on modeling the behavior of materials and the simulation of physical phenomena.

Preface

This book contains the proceedings of EXPLOMET 90, the third of the EXPLOMET series. This quinquennial frequency is well suited for a realistic appraisal of progress in the field. Shock wave and high-strain-rate phenomena in materials are a vast subject that has, since World War II, evolved into a coherent body of knowledge with foundations in the basic sciences of physics, chemistry, and materials science, inputs from a variety of disciplines, and broad technological applications.

The expansion of this field and its redirection can be gauged by the evolution of participation and principal themes since the inception of EXPLOMET, in 1980. Concomitantly, the emphasis of the principal research thrusts has shifted throughout the years. A constant effort throughout this period has been the fundamental study of materials response under shock loading conditions.

This book is divided into ten sections in which the chapters have been organized in a logical sequence. Section I deals with high-strain-rate deformation, while Section II covers shock and combustion synthesis. Dynamic compaction is described in Section III. Section IV (shaped charge phenomena) presents a detailed and unique coverage of this important area. Shear localization (shear bands) is the subject of Section V; dynamic fracture is presented in Section VI. Another novel area of research, shock phenomena and superconductivity, is covered in Section VII. Section VIII deals with progress in shock waves and shock loading, while Section IX introduces a third novel topic of great importance: shock and dynamic phenomena in ceramics. Finally, the traditional topics of explosive welding and metal working are given in Section X.

The contents of this book, its organizational structure, and the emphasis on materials effects are intended to make it a useful research tool for practicing scientists and engineers, as well as a teaching tool in specialized curricula dealing with dynamic effects in materials. The contributors to this book represent thirteen countries in the world; over 40 percent of the contributors are from outside the United States. Therefore, these proceedings represent a global and up-to-date appraisal of this field.

The International Conference on the Materials Effects of Shock-Wave and High-Strain-Rate Phenomena, held at the University of California, San Diego in La Jolla, California, in August 1990 was sponsored by the U.S. Army Research Office, Materials Science Division (under contract ARO DAAL-03-90-G-0068), Los Alamos National Laboratory, Center of Excellence for Advanced Materials, and University of Texas at El Paso. We gratefully acknowledge this support. The chapters composing this book required retyping and editing to ensure a reasonably coherent format. Because of the extensive retyping it is likely that we may have missed typographical errors in our review process. We will assume responsibility for these remaining errors. We would like to thank Megan Harris and Faye Ekberg (University of Texas at El Paso), Debra Vigil and Carol Cole (Los Alamos National Laboratory), and Kay Baylor (UCSD) for their competent typing.

Marc A. Meyers
Lawrence E. Murr
Karl P. Staudhammer

Contents

SECTION II: SHOCK AND COMBUSTION SYNTHESIS

SECTION III: DYNAMIC CONSOLIDATION

SECTION IV: SHAPED CHARGE PHENOMENA

SECTION V: SHEAR LOCALIZATION

SECTION VIII: SHOCK WAVES AND SHOCK LOADING

SECTION X: EXPLOSIVE WELDING AND METAL WORKING

SHOCK-WAVE
AND
HIGH-STRAIN-RATE
PHENOMENA
IN MATERIALS

Section I
High-Strain-Rate Deformation

1

Dynamic Deformation and Failure

SIA NEMAT-NASSER

Center of Excellence for Advanced Materials
Department of Applied Mechanics and Engineering Sciences
University of California, San Diego
La Jolla, CA, USA, 92093

Dynamic deformation, flow, and fracture are integral parts of all dynamic processes in materials. At one extreme is ductile plastic flow with associated failure modes by shear banding or void nucleation, growth, and coalescence. At the other end are brittle fracturing, microcracking, and pulverization. A systematic scientific study of these phenomena, with a view toward a constitutive description and computational simulation of the involved processes, has been the focus of research at UCSD's Center of Excellence for Advanced Materials, over the last few years. This research has included the development of state-of-the-art dynamic probes for recovery experiments at strain rates ranging from quasi-static to $10^6 / s$, or greater, with associated time- and spatially-resolved data acquisition facilities and techniques. This has been paralleled by systematic material characterization and observation, and consequent physically-based micromechanical modeling with computational simulation of the dynamic response and failure modes. In the course of this effort, a number of novel experimental techniques have been developed, which provide powerful means of exploring and quantifying microstructural evolution in a variety of materials, from single and polycrystalline metals to ceramics and ceramic composites.

This paper briefly reviews some of these experimental techniques and the associated diagnostics, focusing attention on novel Hopkinson bar techniques for recovery experiments. Application of these techniques to study, for example, void collapse in fcc and bcc metals, to probe material response at extremely high strains and strain rates, the phenomenon of adiabatic shear banding, and microcracking and phase transformation, are presented elsewhere in these proceedings.

I. INTRODUCTION

Dynamic processes refer to high-rate events generally associated with rapid deposition and/or transmission of energy through shock or other modes of wave motion. They can be used: (1) to develop fundamental understanding of material response under extreme pressures, temperatures, and deformation and reaction rates; (2) to create new materials with unique microstructures and, hence, properties; and (3) to monitor dynamic events in materials and to probe and characterize the corresponding inelastic flow, fracture, and failure modes. These are necessary ingredients for developing physically-based mathematical and computational models for predicting synthesis, processing, structural changes, and response and failure modes over a broad range of pressures, temperatures, and deformation rates. At one extreme is ductile plastic flow with associated failure modes by shear banding or void nucleation, growth, and coalescence. At the other end is brittle fracturing, microcracking, and pulverization; see Fig. 1.

This conference addresses three main topics in dynamic processes in materials, i.e., dynamic synthesis and processing; dynamic deformation and fracture; and dynamic probing into materials; see Fig. 2. My comments are focused on *dynamic deformation and failure modes*. My basic aim is to outline some recent innovations, specifically developed to explore certain fundamental ingredients in the dynamic deformation and failure modes of materials. These are:

* Relation between microstructure and thermomechanical properties;

* Quantitative evaluation of response and failure modes;

* Physically-based constitutive models.

Central to a program which effectively implements these ingredients into an integrated research plan, are *dynamic recovery experiments*. In such experiments the microstructural evolution is related to the loading history, as well as to the thermomechanical response. Based on microstructural characterization, micromechanical models are developed and used to construct physically-based constitutive relations with quantitatively predictive capabilities. The constitutive models are then embedded through constitutive algorithms, into large-scale dynamic computer programs, to simulate the material response, the response of the structural elements made of these materials, as well as the synthesis and processing that lead to the creation of such materials. Figure 3a provides an outline emphasizing the key role that recovery experiments play in this sequence of events.

At the University of California, San Diego's (UCSD's) Center of Excellence for Advanced Materials (CEAM), we have made a number of key innovations over the past four years, which specifically focus on addressing the elements in Fig. 3a. While it will not be possible to examine all aspects of this activity, I will focus attention on some innovations in recovery experiments, particularly using the Hopkinson bar technique, and then comment on a breakthrough in computational algorithms for large-strain large-strain-rate rate-dependent, as well as rate-independent plastic flow of materials. Before doing this, I would like to point out that, in addition to the Hopkinson bar technique, we have made interesting innovations in the use of plate impact experiments for strain rates exceeding

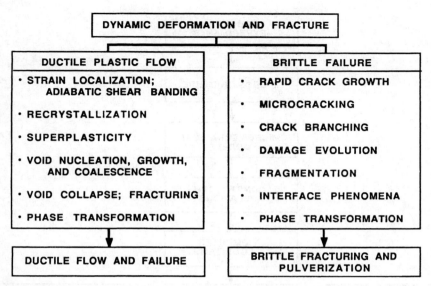

FIG. 1 *Major issues in dynamic deformation and failure modes of ductile and brittle materials.*

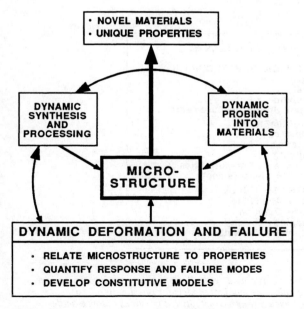

FIG. 2 *Dynamic processing in materials includes dynamic synthesis and processing, dynamic deformation and failure, and dynamic probing into materials: This conference addresses all three topics.*

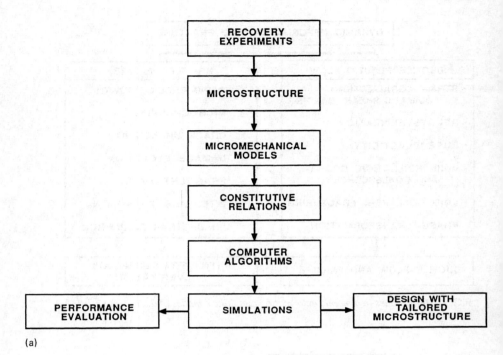

(a)

(b)

FIG. 3 *(a) Dynamic recovery experiments play a central role in relating microstructure to dynamic properties, in developing micromechanical models for physically-based constitutive relations which can be used through constitutive algorithms in large-scale computer codes to simulate response and failure modes, to design microstructure for desired performance, and to evaluate materials and structural performance ; (b) UCSD's dynamic recovery techniques.*

$10^5/s$, and in the use of laser-induced stress pulses with extremely short--of the order of nanoseconds-- durations and high amplitudes. Figure 3b outlines these, and brief accounts can be found in Nemat-Nasser *et al.* [1], and Nemat-Nasser [2].

As illustrations of the application of these techniques, we note here our observations on plastic flow of fcc and bcc single and polycrystals, both at extremely high strains and strain rates with concomitant shear localization and fracturing, as well as phase transformation, dynamic fracturing, and microcracking of ceramics containing partially stabilized zirconia precipitates; see Nemat-Nasser and Chang [3], Rogers and Nemat-Nasser [4], Ramesh *et al.* [5], and Beatty *et al.* [6].

For ductile metals, the mechanism of void collapse has been used to probe material response at extremely high strains and strain rates, using the Hopkinson bar technique; see Chang and Nemat-Nasser [7]. It is observed that materials behave in a unique manner which has not been explored in these regimes. For example, when the void collapses under pure compression at high strain rates, dynamic recrystallization can take place in an extremely short period of time, and ductile materials such as pure single-crystal copper, can fracture in a seemingly brittle manner, with microcracks extending *normal to the applied compression* (a seemingly paradoxical result), into the newly formed crystals. This result brings into focus the importance of strain and strain rate histories in the mechanical response of crystalline solids. Because of this and related issues, innovative Hopkinson bar techniques developed at UCSD, which allow the subjecting of specimens to well-defined stress pulses for recovery experiments, are of central importance. For example, one is able to subject the sample to a single tension pulse, a single compression pulse, a compression pulse followed by a tension pulse, or any desired combination, with complete trapping for recovery analysis and characterization. The techniques also permit the study of microcracking in ceramics, with full recovery.

II.   NOVEL TECHNIQUES FOR HOPKINSON BAR RECOVERY EXPERIMENTS

A. COMPRESSION EXPERIMENT

The split Hopkinson bar dynamic compression testing technique was invented by Kolsky in 1949, following the pioneering work of John and Bertram Hopkinson [8] [9] [10], and Davis [11].

In this approach, the dynamic stress-strain relation in uniaxial compression of a material is obtained by sandwiching a small sample between two elastic bars of common cross-sectional area and elasticity, called the *incident* bar and the *transmission* bar, respectively. An elastic stress pulse is imparted into the incident bar by striking it with a striker bar of given cross-sectional area and elasticity. By ensuring that all three bars remain elastic, plastic deformation is induced in the (usually) ductile sample. The stress in the sample is obtained by measuring the transmitted pulse, and the strain of the sample is calculated from the pulse reflecting off the sample back into the incident bar, where it is measured by means of a strain gauge attached to this bar; see Fig. 4.

This classical Hopkinson bar technique is *not* suitable for recovery tests, since the reflected pulse in the incident bar is again and again reflected back into this bar, subjecting the sample to repeated compression loads. To remedy this major stumbling block in recovery experiments with the split Hopkinson bar, a novel fixture has been developed at UCSD's CEAM, which generates in the incident bar a *compression pulse followed by a tension pulse.* In this manner, once the tensile pulse, which tails the compression, reaches the interface between the sample and the incident bar, the sample is softly recovered, having been subjected to a known compressive pulse. As is discussed below, the shape and amplitude of this compressive pulse can be controlled, and hence the sample can be subjected to a pre-assigned stress history in this experiment.

UCSD's loading fixture for the *stress reversal Hopkinson bar* consists of an incident bar with a *transfer flange* at its loading end, and an *incident tube* resting at the one end against the transfer flange, and at the other end, against a *reaction mass,* as shown in Fig. 5a. The incident bar, the incident tube, and the striker are of the same material (maraging steel) and cross-sectional area, i.e., they have a common impedance. The striker and the incident tube have the same length. When the striker bar impacts the transfer flange of the incident bar, the same axial compression is generated in the incident bar, incident tube, and the striker. The pulse in the incident bar travels toward the sample at the longitudinal elastic wave velocity C_0, whereas the compression in the incident tube reflects back as *compression,* once it reaches the interface with the reaction mass. This compression travels back and loads the incident bar in tension, through the transfer flange. This tensile loading takes place at exactly the instant when the release tensile pulse, which has been reflected off the free end of the striker, reaches the interface with the transfer flange. The striker and the transfer flange begin to move at a third of the impact velocity opposite the impact direction, for a short time, until the striker separats from the transfer flange, bouncing back at a third of its initial impact speed. Figure 5b shows a typical stress pulse generated by this technique.

B. PULSE SHAPING

If the length of the sample is denoted by l, then it is easy to show that the strain rate in the sample is given by $\dot\varepsilon = -2\dfrac{C_0}{l}\,\varepsilon_r$, where ε_r is the strain reflected off the sample into the incident bar. A *constant* strain rate is attained by imparting a *rectangular pulse* to the incident bar.

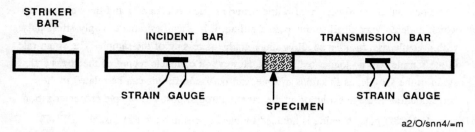

FIG. 4 *Classical compression Hopkinson bar: An elastic compressive stress pulse is imparted into the incident bar by striking it with a striker bar. The sample is subjected to repeated loading as the pulse reflects back and forth along the two bars.*

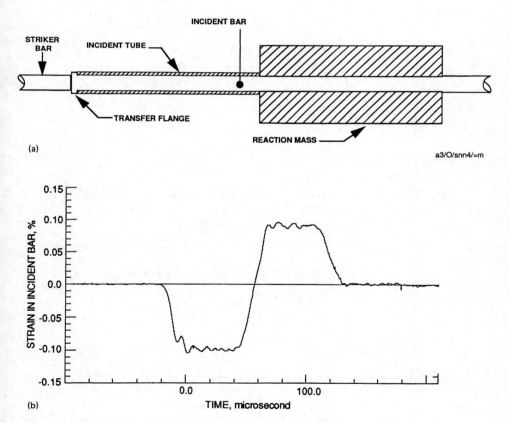

FIG. 5 *UCSD's stress reversal Hopkinson bar technique: (a) the loading fixture; (b) a typical stress pulse generated by this technique.*

For application to very hard brittle materials such as ceramics and their composites which undergo very small strains before failing, it is often desirable to apply stress pulses with a gentle rise, in order to allow more gradual stressing of the sample. The strain rate in the sample will no longer be constant. However, a complete record of the stress and strain in the sample, as functions of time, can be obtained; this can be related to the corresponding damage evolution in the sample through post-test sample characterization.

At UCSD, pulse shaping is attained by placing a suitable metal (usually OFHC) cushion between the striker and the transfer flange, attached to the latter. A detailed analysis of the plastic deformation of this kind of cushions has been given by Nemat-Nasser *et al.* [12]. Depending on the size of the cushion relative to the bars, and the velocity and the length of the striker, different pulse shapes can be generated. Figures 6a and b

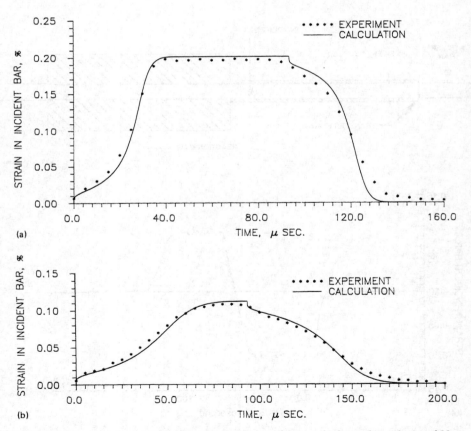

(a)

(b)

FIG. 6 *Pulse shaping: (a) for striker velocity of about 20 m/s; (b) for striker velocity of 11 m/s.*

show two extreme cases obtained using a copper cushion of 0.19" (4.8 mm) and 0.020" (0.51 mm) initial diameter and thickness, and a 3/4" (19 mm) striker of 9" (228.6 mm) length. Figure 6a is for a striker velocity of 19.72 m/sec, and Fig. 6b is for 11.02 m/sec. The theoretical prediction is based on incompressible, axisymmetric, rate-independent plastic flow of the cushion whose axial true stress, σ, and engineering (nominal) strain, ε, are related by a simple power-law, $\sigma = \sigma_0 \varepsilon^n$, where, for OFHC, direct experiments suggest $\sigma_0 \approx 570$ MPa and $n=1/5$.

An important point to bear in mind, in relation to dynamic testing of very hard and brittle samples, is that *the sample tends to indent the bars*, and, therefore, the reflected strain in the incident bar, ε_r, is *not* a measure of the strain rate in the sample. At UCSD, we attach strain gauges to the sample and directly measure the axial as well as

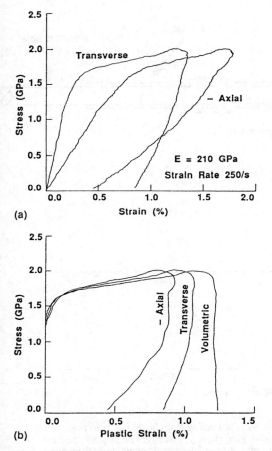

(a)

(b)

FIG. 7 *A typical result for Mg-PSZ; see Rogers and Nemat-Nasser (1990).*

the lateral strains of the sample as functions of time. A typical result for Mg-PSZ is shown in Fig. 7; see Rogers and Nemat-Nasser [4].

C. TENSION EXPERIMENT

The classical tension split Hopkinson bar has been used to obtain stress-strain relations of samples in uniaxial stress, to *failure*; see Harding *et al.* [13] and Lindholm [14]. UCSD's novel technique allows for recovery experiments by trapping the compression pulse which reflects off the sample. The design of this apparatus is sketched in Fig. 8a. The loading fixture consists of a tubular striker riding on the incident bar which passes through a gas gun and terminates with a transfer flange at one end and the sample at the other end. A precision gap separates the transfer flange from a *momentum trap* bar. This gap is set such that when the striker has imparted to the incident bar the entire tensile pulse, the gap is

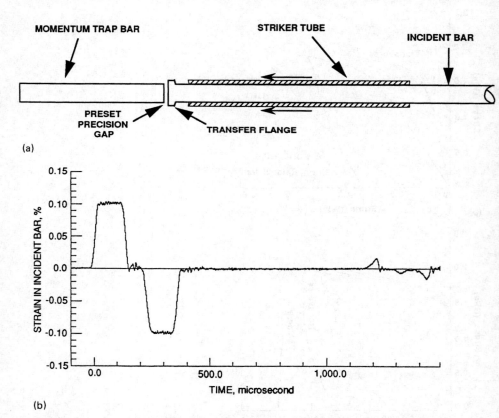

FIG. 8 *UCSD's momentum trapping technique for tension Hopkinson bar: (a) the loading fixture with momentum trap bar and precision gap; (b) a typical stress pulse generated by this technique (note that the pulse reflected off the sample is almost completely trapped).*

closed. Upon reflection as compression off the sample interface, this reflected pulse is then transmitted into the momentum trap bar and is trapped there. An example is shown in Fig. 8b.

In closing this section, we point out that it is also possible to perform *recovery* experiments with the sample having been subjected to a compression pulse *and* a tension pulse, using a modified version of the stress reversal Hopkinson technique. This requires redesigning the end of the transmission bar in contact with the sample such that, after the sample is subjected to compression and tensile pulses, it is pulled off the transmission bar; see Nemat-Nasser *et al.* [12] for details. This apparatus then allows study of the Bauschinger effect under dynamic loading.

III. MODELING AND COMPUTATION

A major component of UCSD's program is the development of physically-based constitutive models capable of predicting dynamic deformation, prior to and beyond failure. For dynamic flow of metals, this modeling includes effects such as strain rate, thermal softening, and workhardening due to plastic strain accumulation, and microstructural effects such as texture, crystal structure, precipitates, and voids or inclusions. In addition, through a series of careful experiments, attempts are made to include the *strain-rate history* effects which, up to now, have not been included in constitutive models used in large-scale codes. Recent experiments by Rashid [15] at UCSD have shown that the dislocation structures in single-crystal pure copper critically depend on the *strain-rate history*. Parallel with the above experiments and modeling, and as an important part of our program, has been the development of efficient and robust computer constitutive algorithms for implementation in dynamic computer codes.

Here we summarize the results of a recent breakthrough in explicit constitutive computational algorithms for finite-element calculations of large-deformation rate-independent and rate-dependent elastoplasticity, using a simple example based on the J_2 plasticity theory with isotropic hardening. The new algorithm provides a direct, explicit, and always nearly exact estimate of all stress components, and any internal variable that may be involved, for any prescribed deformation (or time) increment (large or small) in *one single step,* or in any desired number of substeps. The algorithm can accommodate nonsmooth yield surfaces (for rate-independent materials). This generalization is discussed elsewhere; Nemat-Nasser [2]. Here *as an illustration* we consider a *simple case* where the plastic part, D^p, of the deformation rate, D is given by

$$D^p = \dot{\gamma}\mu,\tag{1}$$

where

$$\mu = \tau'/(\sqrt{2}\tau), \qquad \tau = (\tfrac{1}{2}\tau':\tau')^{\frac{1}{2}}, \qquad \tau' = \tau - \frac{1}{3}I\,tr\,\tau.\tag{2}$$

Clearly

$$\boldsymbol{\mu}:\boldsymbol{\mu} = 1, \qquad \boldsymbol{\mu}:\dot{\boldsymbol{\mu}} = 0, \qquad \dot{\gamma} = \boldsymbol{\mu}:\boldsymbol{D}^p. \tag{3}$$

The essential physics is embedded in the quantity $\dot{\gamma}$. As *an illustration*, consider the following power law

$$\dot{\gamma} = \dot{\gamma}_0 \left[\frac{\tau}{\tau_r}\right]^m, \tag{4}$$

where $\dot{\gamma}_0$ and τ_r are the reference strain rate and the associated reference flow stress. When m is very large, this model simulates rate-independent plasticity. For moderate values of m, the rate effect becomes dominant. For the rate-independent case, we consider the simple example of the yield condition,

$$f \equiv \tau - F(\gamma; ...). \tag{5}$$

In (4), τ_r, and in (5), F, represent the resistance of the material to plastic flow. They thus embody workhardening, temperature softening, and all related microstructural evolutions which have preceded the current state.

Models used for plasticity computations have generally been based on constitutive relations which do *not* include the effects of *strain-rate history* on the plastic flow of the material. In the present illustration, this means that, e.g., τ_r is regarded a function of only the accumulated plastic strain,

$$\gamma = \int_0^t \dot{\gamma}\, dt, \tag{6}$$

as well as the temperature, but not the strain-rate history. Although it is now generally accepted that hardening depends not only on the total plastic deformation, but also on the *deformation-rate history* experienced by the material, for simplicity we will not address this issue here.

To be specific, we consider the rate-independent model first, and then comment on the rate-dependent case. In either case, we write the deformation tensor as

$$\boldsymbol{D} = \boldsymbol{D}^e + \boldsymbol{D}^p, \qquad \text{or} \quad D_{ij} = D_{ij}^e + D_{ij}^p, \tag{7}$$

where the elastic deformation rate tensor, $\boldsymbol{D}^e$, relates to an objective stress rate, $\overset{\circ}{\tau}$, through the current elasticity tensor, $\boldsymbol{C}$, by

$$\overset{\circ}{\boldsymbol{\tau}} = \boldsymbol{C}:\boldsymbol{D}^e, \qquad \text{or} \quad \overset{\circ}{\tau}_{ij} = C_{ijkl} D_{kl}^e. \tag{8}$$

In (8), $\overset{\circ}{\tau}$ is defined by

$$\overset{\circ}{\tau} = \dot{\tau} - \mathbf{\Omega}\tau + \tau\mathbf{\Omega}, \quad \text{or} \quad \overset{\circ}{\tau}_{ij} = \dot{\tau}_{ij} - \Omega_{ik}\tau_{kj} + \tau_{ik}\Omega_{kj}, \tag{9}$$

where $\mathbf{\Omega}$ is an appropriate spin; see Nemat-Nasser [2] [16] [17]. While a proper choice of the objective stress is important -- especially in kinematic hardening -- it has no bearing on the new algorithm.

Let the current stress state be on the yield surface. Define a hardening parameter by $H = dF/d\gamma = \dot{\tau}/\dot{\gamma}$, and from (1), (7), and (8), obtain

$$\overset{\circ}{\tau} = C : (D - \dot{\gamma}\mu). \tag{10}$$

Then, calculating $\mu : \overset{\circ}{\tau}$ from this equation, we arrive at

$$\dot{\tau}(\hat{t})/A + \dot{\gamma}(\hat{t}) = d(\hat{t}), \quad t < \hat{t} \le t + \Delta t, \tag{11}$$

where

$$\tau(\hat{t}) = F(\gamma(\hat{t})), \quad \gamma(\hat{t}) = \int_0^{\hat{t}} \dot{\gamma}(\theta) d\theta,$$

$$d = (\mu : C : D)/(\mu : C : \mu), \quad A = (\mu : C : \mu)/\sqrt{2}; \tag{12}$$

for isotropic C, $A = \sqrt{2}G$ (where G is the shear modulus) and $d = \mu : D$.

Assuming continuing plastic or neutral loading, i.e. $d(t) \ge 0$, we integrate (11) over the time increment to obtain

$$\tau(t + \Delta t) - \tau(t) + A\,\Delta\gamma = A\,d^*\,\Delta t, \tag{13}$$

where d^* is an *estimate* of the average value of $d(\hat{t})$ over the considered time increment; see Nemat-Nasser [2] for details. The main step in our algorithm is to tentatively assign the deviatoric part of the *total deformation rate tensor*, i.e. D', *over the entire time increment* to be only due to plastic flow, i.e., set $D^p = \dot{\gamma}(\hat{t})\mu(\hat{t}) \approx D'(\hat{t})$, $t \le \hat{t} \le t + \Delta t$, and then seek to correct the error that this assignment has introduced. With this assignment, the yield condition at $t + \Delta t$ is approximated by

$$\tau_A(t + \Delta t) = F(\gamma(t) + d^*\Delta t) \tag{14}$$

which includes the error

$$\Delta_e \tau = \tau_A(t + \Delta t) - \tau(t + \Delta t). \tag{15}$$

The error in the function $\dot{\gamma}(\hat{t})$ over the considered time interval is

$$\dot{\gamma}_{er}(\hat{t}) = d(\hat{t}) - \dot{\gamma}(\hat{t}), \qquad t \leq \hat{t} \leq t + \Delta t . \tag{16}$$

Then the error in the value of γ at $t + \Delta t$, denoted by $\Delta_e \gamma$, is given by

$$\Delta_e \gamma = \int_t^{t+\Delta t} \dot{\gamma}_{er}(\theta) \, d\theta = \dot{\gamma}_{er}^* \Delta t , \tag{17}$$

where $\dot{\gamma}_{er}^*$ is the mean value of $\dot{\gamma}_{er}(\hat{t})$ over the time interval t to $t + \Delta t$.

From (13) to (17) we have

$$\tau(t + \Delta t) - \tau(t) = A \, \dot{\gamma}_{er}^* \Delta t$$

$$= \tau_A(t + \Delta t) - \tau(t) - \Delta_e \tau \tag{18}$$

which is exact. We estimate $\Delta_e \tau$, by

$$\Delta_e \tau \approx H \Delta_e \gamma \approx H \dot{\gamma}_{er}^* \Delta t , \tag{19}$$

and arrive at

$$\tau(t + \Delta t) = \frac{A \, \tau_A(t + \Delta t) + H \, \tau(t)}{A + H} , \tag{20}$$

$$\dot{\gamma}_{er}^* = \frac{\tau_A(t + \Delta t) - \tau(t)}{(A + H) \Delta t} , \tag{21}$$

$$\Delta \gamma = (d^* - \dot{\gamma}_{er}^*) \Delta t . \tag{22}$$

Equation (20) is a near-exact estimate of the yield surface after the incremental loading associated with the prescribed deformation rate tensor D, over time increment Δt.

For elastic-viscoplastic constitutive models, with *no yield surface*, we examine an example where the effective plastic strain rate $\dot{\gamma}$ is related to the effective stress τ by (4), with the following flow stress:

$$\tau_r = \tau_0 (1 + \frac{\gamma}{\gamma_0})^N, \tag{23}$$

where γ_0 is the reference strain, and N is a material parameter. We follow essentially the same procedure, but replace the hardening parameter H by η which is defined by

$$\eta = \frac{\tau(t)}{\sqrt{2}\,G}\left[\frac{1}{m\,d\,\Delta t} + \frac{N}{\gamma_0 + \gamma(t)}\right]. \tag{24}$$

Then,

$$\tau_A(t + \Delta t) = \tau_{rA}(t + \Delta t)\left[\frac{d}{\dot{\gamma}_0}\right]^{1/m}, \quad \tau_{rA}(t + \Delta t) \equiv \tau_0\left[1 + \frac{\gamma(t) + d\,\Delta t}{\gamma_0}\right]^N. \tag{25}$$

Nemat-Nasser and Chung [18] have compared the results of this algorithm with the explicit effective tangent moduli method proposed by Peirce *et al.* [19]. It is shown that the new algorithm predicts the near-exact solution in a single step, while the effective tangent moduli method may require thousands of steps for similar accuracy. In Fig. 9 we show a typical result based on the following values of the constitutive parameters: $\gamma_0/\sqrt{3} = \tau_0/E = 6.25 \times 10^{-3}$ (E = Young's modulus, with Poisson's ratio $\nu = 0.3$) and $N = 0.08$. The dimension of stress is arbitrary. As is seen, the size of the timestep for the new algorithm is immaterial: *the new algorithm is explicit, very accurate, and always stable, independent of the size of the time increment.*

For the rate-independent case, we consider $F(\gamma) \equiv \tau_r$, where τ_r is given by (23), with $N = 0.2$. Figure 10 shows the corresponding stress-strain curve, using the new algorithm.

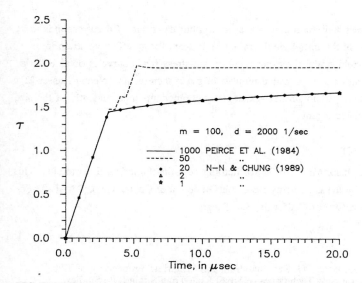

FIG. 9 *Performance demonstration of the new elastic-viscoplastic algorithm for power-law model (23) with m = 100 and flow stress (4); constitutive parameters are given after (25). The algorithm yields nearly exact results for any number of timesteps -- even one -- over the entire (unrealistically large) time increment.*

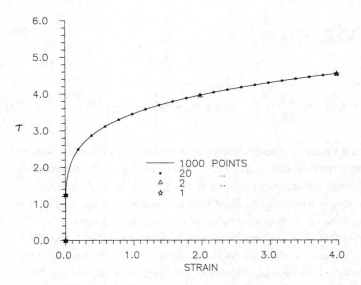

FIG. 10 *Performance demonstration of the new elastoplastic algorithm for power-law model (23). The algorithm yields nearly exact results for one, two, or any number of strain increments; strain of 4 corresponds to 640 times the initial yield strain, and the unit of stress is arbitrary.*

In this figure, the stress and strain units are arbitrary, but the strain of 4 corresponds to 640 γ_0, i.e. 640 units of the initial yield strain. As is seen, for an effective strain increment 640 times the yield strain, the exact point on the stress-strain curve is obtained in a *single strain increment* or in any desired number of strain increments. Nemat-Nasser [2] has extended this method to obtain near-exact stress components independently of the size of the time (or strain) increment.

ACKNOWLEDGMENT

The author wishes to thank Mr. Y.-F. Li for helping with the numerical illustrations. This work was supported by the U.S. Army Research Office under Contract No. DAAL-03-86-K-0169 with the University of California, San Diego.

REFERENCES

1. S. Nemat-Nasser, J. Isaacs, G. Ravichandran, and J. Starrett, Proceeding of TTCP TTP-1 Workshop on New Techniques of Small Scale High Strain Rate Studies, Australia (1989).

2. S. Nemat-Nasser, *Int'l J JSME,* Series 1, Japan Society of Mechanical Engineering, to appear (1991).

3. S. Nemat-Nasser and S.-N. Chang, *'90 Explomet Proceedings* (1991).

4. W. P. Rogers and S. Nemat-Nasser, *J. Amer. Cer. Soc.,* 73(1), pp. 136-39 (1990).

5. K. T. Ramesh, B. Altman, G. Ravichandran, and S. Nemat-Nasser, *Advances in Fracture Research,* ICF -7, pp. 811-817 (1989).

6. J. H. Beatty, L. W. Meyer, M. A. Meyers, and S. Nemat-Nasser, *'90 Explomet Proceedings* (1991).

7. S.-N. Chang and S. Nemat-Nasser, *'90 Explomet Proceedings* (1991).

8. B. Hopkinson, *Proc. Roy. Soc.* A, 74, p. 498 (1905).

9. B. Hopkinson, *Phil. Trans.* A, 213, p. 437 (1914).

10. J. Hopkinson, *Proc. Manch. Lit. Phil. Soc.,* 11, p. 40 (1872).

11. R. M. Davis, *Phil. Trans.* A, 2450, p. 3875 (1948).

12. S. Nemat-Nasser, J. Isaacs, and J. Starrett, in preparation (1991).

13. J. Harding, E. D. Wood, and J. D. Cambell, *J. Mech. Eng. Sci.,* 2, p. 88 (1960).

14. U. S. Lindholm, *J. Mech. Phys. Solids,* 12, p. 317 (1964).

15. M. Rashid, Ph.D. Thesis, University of California, San Diego (1990).

16. S. Nemat-Nasser, ASME *J. Appl. Mech.,* 50, pp. 1114-1126 (1983).

17. S. Nemat-Nasser, *Meccanica,* to appear (1990).

18. S. Nemat-Nasser and D.-T. Chung, *Comput. Methods in Appl. Mechanics and Eng.,* to appear (1991).

19. D. Pierce, C. F. Shih, and A. Needleman, *Comput. Struct.,* 18, 875-887 (1984).

2

Mechanical Behavior of Composite Materials Under Impact Loading

J. HARDING

Department of Engineering Science
University of Oxford
Parks Road, Oxford OX1-3PJ, U.K.

The problems of characterising the mechanical behaviour of composite materials under impact loading are discussed. Techniques for determining such behaviour for tensile, compressive and shear loading are described and some qualitative conclusions are drawn from the results obtained.

I. INTRODUCTION

The rate-dependence of the mechanical properties of metallic type materials may be determined using standard designs of test-piece. The macroscopic response of the bulk material, for well defined loading systems, may be related to data obtained in such tests and analytical functions, i.e. constitutive relationships, may be derived. These can then be used to describe the mechanical behaviour of structural components of different geometrical shapes and under more complex loading systems. For composite materials, however, where there are two or more phases present, several complicating factors arise. For example, the overall rate dependence of the composite will depend, to a greater or lesser extent, on the rate dependence of each of these various phases. Also of importance will be the reinforcement configuration, e.g. unidirectional, cross-ply or woven, and the type and direction of loading, e.g. tensile, shear or compressive. Thus for a unidirectionally-reinforced composite under tensile loading in the fibre direction, where fibre fracture is likely to be the process controlling the failure of the composite, the fibre properties may be expected to determine the rate dependent behaviour.   In contrast, for woven reinforced material loaded in compression in one of the principal directions of reinforcement, where a fibre buckling process is likely to control failure, the properties of the matrix will have a greater influence on the overall rate dependence.

A further complicating factor, however, for all multi-phase materials, is the presence of interfaces and hence the possibility of additional failure processes associated with the interface, for example, by deplying or interlaminar shear mechanisms.   It becomes necessary, therefore, to consider whether the critical conditions under which such processes occur are also rate dependent.

In the light of these various factors it is clear that a full characterisation of the mechanical behaviour of composite materials at impact rates of straining is likely to require the use of a wider range of test configurations and test-piece designs than are commonly encountered in the testing of simpler single-phase metallic materials.   The present review describes several of the techniques which have been used and discusses the results which have been obtained.   Although in some of these tests it is difficult to distinguish between the various failure processes and their individual dependence on the rate of loading, so that a detailed interpretation of the results may not be possible, some general conclusions may be drawn.

II. REVIEW OF EXPERIMENTAL TECHNIQUES

Problems of data interpretation arise in all testing techniques used at impact rates of loading.   For homogeneous isotropic metallic materials, however, these problems are least severe when the testing technique is based on the split Hopkinson pressure bar principle.   For this reason all the experimental methods to be described here will also be based on this principle, although it is realised that for anisotropic multi-phase materials these difficulties may be greatly increased.   Tensile, compressive and shear loading configurations will be considered, in each case with a design of test-piece intended to study, as far as possible, a single failure initiation process, although subsequent propagation of failure is likely to involve a complex interaction between several different failure processes.

In all Hopkinson-bar tests overall specimen dimensions need to be small, so as to minimise radial inertia and wave propagation effects within the specimen, while care has to be taken in designing the method of load transfer between the specimen and the loading bars so as to avoid the introduction of a region of significant impedance mismatch which could introduce stress wave reflections and thus invalidate the Hopkinson-bar analysis. When composite specimens are to be tested the need for small overall specimen dimensions may conflict with the requirement for a specimen which is large relative to the scale of the reinforcement while the anisotropic nature of the composite material can complicate the design of the specimen/loading bar interface.

A. TENSILE TESTING TECHNIQUES

A tensile version of the Hopkinson-bar apparatus for use with composite specimens is shown schematically in fig. 1. A cylindrical projectile impacts the loading block and causes an elastic tensile loading wave to propagate along the loading bar towards the specimen and output bar. The thin strip specimen is waisted in the thickness direction and has a very slow taper so as to minimise stress concentrations due to free edge effects. The state of stress within the specimen has been investigated using a two-dimensional finite element analysis. It shows the biggest stress concentrations to be at the specimen/loading bar interface and to be significantly smaller than the controlling tensile stresses in the specimen gauge region which is in a state of uniform tension.

The specimen is fixed with epoxy adhesive into parallel-sided slots in the loading bars. Strain gauge signals from two stations on the input bar and one on the output bar allow the full dynamic stress-strain curve to be derived using the standard Hopkinson-bar analysis. However, since most composites fail at low or very low strains and may show a significant rate dependence of the initial elastic deformation, for the determination of which the Hopkinson-bar analysis is not very accurate, it is usual to make, in addition, a direct determination of the specimen elongation using a fourth set of strain gauges, attached to the specimen itself. A range of fibre reinforced polymeric materials have been tested in this way and the results compared with those obtained at lower rates of strain. Some of these results are summarised below and some tentative conclusions drawn.

Initial tests on a unidirectionally-reinforced carbon/epoxy material [1], loaded in the reinforcement direction, at mean strain rates from ~0.0001/s to ~450/s, showed no effect of strain rate, see fig. 2a, on the tensile modulus, tensile strength or strain to failure; nor was there any effect of strain rate on the fracture appearance. Such behaviour is consistent with the conclusion that the tensile properties of the composite are entirely controlled by the carbon fibres the behaviour of which is entirely independent of strain rate. In contrast similar tests [2] on a plain coarse-weave carbon/epoxy material when loaded in a principal reinforcing direction, see fig. 2b, showed a small effect of strain rate on the initial tensile modulus and a more significant effect on the tensile strength and the elongation to failure. Here, however, the woven reinforcement geometry is likely to result in a stronger interaction with the matrix so that the rate dependent properties of the matrix play a more important role in the deformation process.

The rate dependence is much more marked when glass reinforcing fibres are used. Tensile stress-strain curves for a plain fine-weave glass/epoxy material [3] are shown in fig. 2c. The initial modulus and the tensile strength both increase very significantly with strain rate while, in contrast with the behaviour shown by the woven carbon/epoxy material, the overall elongation also increases with rate of loading. Part of this increased rate sensitivity may be due to a greater interaction with the matrix when a fine weave reinforcement is used. More important, however, is likely to be the rate dependent behaviour of the glass fibres themselves, the strength of which is expected to increase quite markedly at impact rates [4]. A direct confirmation of this in tensile tests on unidirectional glass/epoxy composites loaded in the reinforcement direction, however, did not prove possible. Although under quasi-static loading a tensile failure was obtained in the central gauge section of the specimen, under impact loading failure was invariably by the pull-out of glass-fibres from the matrix in the grip-regions of the specimen, see fig. 3, [5]. This implies that the ratio of the tensile strength of the glass fibres to the interfacial shear strength between the glass fibres and the epoxy matrix increases with increasing strain rate. This could be due either to an increase in the former or to a decrease in the latter.

Further evidence for the relative importance of the tensile to the shear strength in composite materials was obtained in tensile tests on some coarse satin-weave glass/polyester specimens, loaded in a principal reinforcement direction [6]. Here

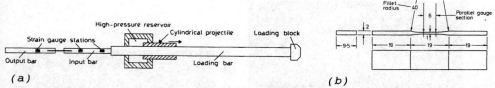

FIG. 1 *Tensile version of split Hopkinson bar (a) general assembly (schematic) (b) specimen design (dimensions in mm).*

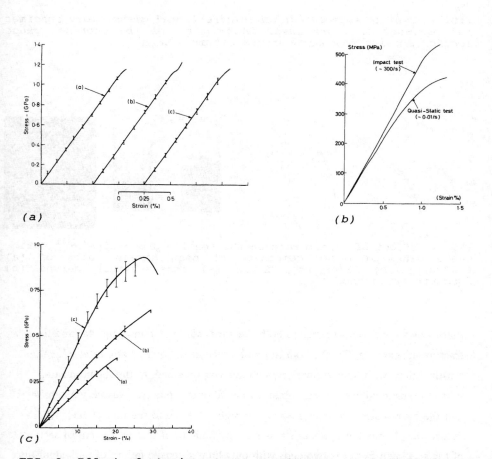

FIG. 2 *Effect of strain rate on tensile stress-strain curves for composite materials (a) unidirectional carbon/epoxy [mean strain rate (/s): a) - 0.0001; b) - 10; c) - 450;] (b) woven carbon/epoxy (c) woven glass/epoxy [mean strain rate (/s): a) - 0.0001; b) - 10; c) - 900;].*

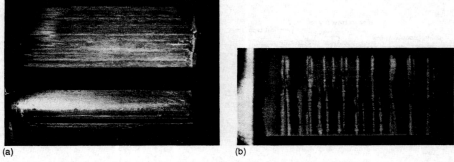

FIG. 3 *Tensile impact test on unidirectional glass/epoxy specimen (a) pull-out of fibre layer in grip region (b) unbroken gauge section showing debonding around matrix cracks.*

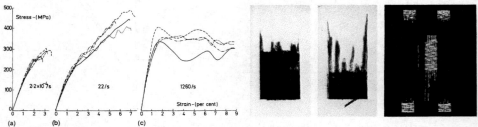

FIG. 4 Effect of strain rate on the tensile properties of a satin-weave glass/polyester composite at mean strain rates of (a) 0.0022/s; (b) 22/s; (c) 1260/s (a) stress-strain curves (b) fracture appearance.

there was a continuous change in both the stress-strain response, fig. 4a, and the fracture appearance, fig. 4b, from those observed at the quasi-static rate, where a tensile failure with limited fibre tow pull-out was obtained, to the medium rate, where a tensile failure was still obtained but fibre tow pull-out was very extensive and the overall strain to failure was quite high, ~7%, up to the impact rate, where failure was dominated by shear stresses giving pull-out of the whole central section of the specimen from the two ends with only limited tensile failure of individual fibre tows and an overall strain to failure of the order of 13%.

In the light of these results it is necessary to devise a technique for determining the effect of strain rate specifically on the shear strength of the composite specimen, i.e. both the interlaminar shear strength between adjacent reinforcing plies and the interfacial shear strength between fibre tows and the matrix.

B.   SHEAR TESTING TECHNIQUES

Several techniques for determining the rate dependence of the interlaminar shear strength in composite materials have been devised.   In two of these [7,8], both based on the torsional Hopkinson-bar, a very significant increase in shear strength with strain rate was observed for both woven and cross-ply glass/epoxy specimens. In contrast, a study of both the interlaminar and the transverse shear strength of a plain-weave carbon/epoxy laminate [9], using a test based on the double-notch shear version of the split Hopkinson-bar apparatus, showed no significant rate dependence. Unlike the previously described tensile tests, however, in none of these various shear tests was the specimen subjected to a well-defined stress system.

This problem has been tackled in a more recently developed test [10] which uses the "double-lap" shear specimen shown in fig. 5a.   In this specimen, which has to be especially laid-up and cannot be cut from existing laminates, failure occurs on pre-determined interlaminar planes.   The strain distribution along one of these planes, as derived from a two-dimensional finite element analysis, gives the results shown in fig. 5b.   It is clear that the shear strain on the interlaminar failure plane is very far from uniform.   This is a problem common to most designs of shear specimen.   It means that, although we may be able to determine the effect of strain rate on the critical load at which interlaminar shear failure occurs in the double-lap specimen, the corresponding shear stresses as estimated from the area of the interlaminar fail-ure planes will only be representative values.

With this proviso it may be reported that all such measurements so far made, for a satin weave carbon/epoxy, a plain weave carbon/epoxy, a unidirectionally rein-forced carbon/epoxy, a plain weave glass/epoxy and at the interface between plies of plain weave glass and a plain weave carbon in a hybrid carbon/glass/epoxy lay-up, showed a significant increase in the interlaminar shear strength at impact rates of loading.   While these results clearly establish the general trend, the wide variations

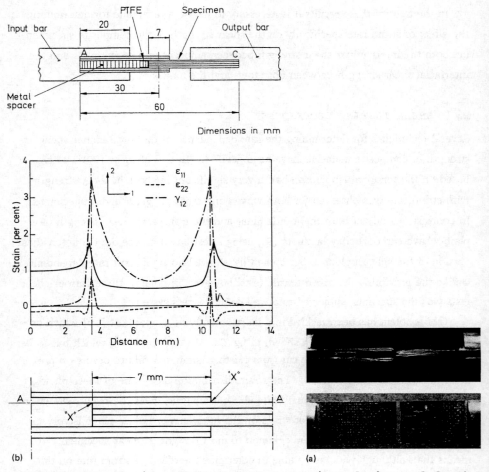

FIG. 5 *Interlaminar shear test (a) specimen design and fracture appearance (b) strain distribution on failure plane.*

in shear stress and shear strain along the interlaminar plane at failure make it difficult to determine from these tests a critical value of interlaminar shear stress for use in modelling damage accumulation processes in composites under impact loading.

C.   COMPRESSIVE TESTING TECHNIQUES

The question of specimen design arises again when compressive impact testing of composite materials is considered.   The standard design of specimen for use with the

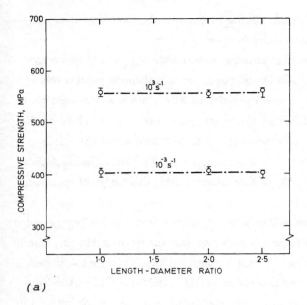

(a)

(b)

FIG. 6 *Effect of strain rate and specimen geometry on compressive behavior of woven glass/epoxy (a) ultimate compressive strength (b) quasi-static failure mode.*

compression Hopkinson-bar is a short cylinder of diameter slightly less than that of the loading bars. Most of the work on composite materials in the compression SHPB [11,12] has in fact used this design of specimen even though it is far from ideal for this type of material. In tests on a woven glass/epoxy laminate loaded in a principal reinforcing direction for specimens with different length to diameter ratio Parry [13] showed a significant increase in the ultimate compressive strength under impact loading, see fig. 6a. In the quasi-static tests failure was by a combination of shear and longitudinal splitting, see fig. 6b, with the possibility that failure was initiated at the ends of the specimen. In the impact tests the specimens completely disintegrated preventing any conclusions from being drawn.

Some evidence that the initiation of compressive failure in woven glass/epoxy composites, at both quasi-static and impact rates, was by a shearing process was obtained in tests on cylindrical specimens reinforced with a single woven glass ply on a diametral plane [14]. The first sign of damage was a sudden drop in load on the stress-strain curve, see fig. 7a, corresponding to a shear failure across the axially-aligned fibre tow near the centre of the specimen, see fig. 7b. Nevertheless doubts clearly remain regarding the validity of data obtained using this design of specimen in the Hopkinson bar test.

However, since the specimen will only be subjected to transient loading, provided the loading bars are well aligned there is no reason why the waisted thin-strip tensile specimen of fig. 1b should not also be used in compression. It is close to the design recommended for the quasi-static compression testing of unidirectional carbon/epoxy specimens [15] and has the major advantage that failure is unlikely to be initiated by end effects at the specimen/loading bar interfaces.

Using this specimen design very marked increases in the compressive strength and the strain to failure under impact loading have been found [16] for both woven carbon and woven glass/epoxy specimens loaded in a principal reinforcing direction, see fig. 8. Failure initiates by shear across the central parallel region of the specimen on a plane inclined at ~45° to both the loading and the thickness direct-ions, see fig. 9 for the glass/epoxy specimen. The damage zone is more extensive in the impacted specimens, corresponding to the much greater strain to failure.

III. DISCUSSION

Although considerable data are now becoming available on the impact mechanical response of fibre-reinforced composites, most of which show that there are quite significant effects which need to be taken into account, no clear picture is yet emerging on which to base a general approach, e.g. the development of some form of "constitutive relationship", which might be used to describe this behaviour. This may be for many reasons but perhaps primarily because the deformation of fibre re-inforced polymers is essentially a damage accumulation process involving a large

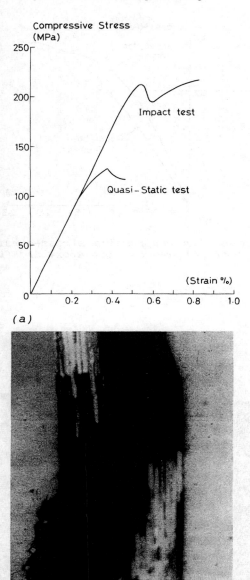

(a)

(b)

FIG. 7   *Initial compressive failure in single-layer reinforced glass/epoxy specimens (a) stress-strain curves (b) shear failure across axially-aligned roving.*

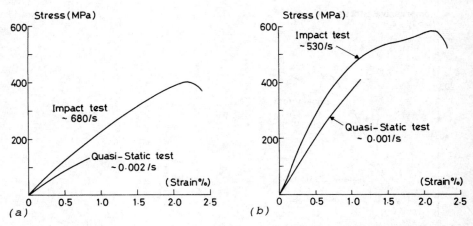

FIG. 8 *Effect of strain rate on the compressive stress-strain curves for thin strip specimens of (a) woven carbon/epoxy and (b) woven glass/epoxy.*

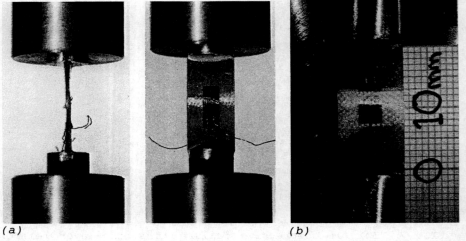

FIG. 9 *Effect of strain rate on compressive failure mode in thin strips specimens of woven glass/epoxy (a) quasi-static (b) impact*

number of different possible damage mechanisms. The testing techniques described above make some attempt to isolate and study some of these mechanisms. In other cases, e.g. in compression, the complex interaction between the fibres and the matrix make a detailed interpretation of the test data extremely difficult.

An attempt has been made [17] to apply finite element methods to the modelling of the tensile impact response of hybrid woven carbon/glass/epoxy laminates. Failure is assumed to initiate at an arbitrarily chosen site by the tensile fracture of an axially-aligned carbon tow and to be followed by limited delamination on adjacent interlaminar planes. The results obtained showed qualitative agreement with experiment but the technique cannot yet give a quantitative prediction of the hybrid impact behaviour.

IV. CONCLUSIONS

Testing techniques have been developed for studying the impact response of fibre-reinforced composites and for obtaining reliable data on their mechanical behaviour at high rates of strain. Care is required, however, in evaluating the data obtained if true "material properties" are to be derived and more work is needed if particular damage processes are to be isolated and their individual rate dependence determined.

V. ACKNOWLEDGMENT

Grateful acknowledgment is made to the Air Force Office of Scientific Research (AFSC), United States Air Force, from whom financial support was received, under Grant No. AFOSR-87-0129, during which much of the work reviewed above was performed.

REFERENCES

1. J. Harding and L. M. Welsh, in Proc ICCM IV, Fourth Int. Conf. on Composite Materials, Japan Soc. for Composite Materials, Tokyo, 1982, 845-852.

2. J. Harding, K. Saka and M. E. C. Taylor, Oxford University Engineering Laboratory Report No. 1654/1986

3. J. Harding and L. M. Welsh, J. Mater. Sci., 18, 1810-1826 (1983).

4.   A. Rotem and J. M. Lifshitz, in Proc. 26th. Annual Tech. Conf. SPI, Reinforced
      Plastics/Composites Division, Society of Plastics Industry, New York, 1971,
      paper 10-G

5.   L. M. Welsh and J. Harding, in Proc ICCM V, Fifth Int. Conf. on Composite
      Materials, TMS-AIME, 1985, 1517-1531.

6.   L. M. Welsh and J. Harding, in Proc DYMAT 85, Mechanical and Physical
      Behaviour of Materials under Dynamic Loading, Jour. de Physique, Colloque C5,
      1985, 405-414.

7.   T. Parry and J. Harding, Colloque Int. du CNRS No. 319, Plastic Behaviour of
      Anistropic Solids, CNRS, Paris 1988, 271-288.

8.   C. Y. Chiem and Z. G. Liu, in Proc IMPACT 87, Impact Loading and Dynamic
      Behaviour of Materials, DGM Informationsgesellschafft mbH, Oberursel, 1988, 2,
      579-586.

9.   S. M. Werner and C. K. H. Dharan, J. Comp. Mater., 20, 365-374 (1986)

10.   J. Harding, Y. L. Li, K. Saka and M. E. C. Taylor, in Proc. 4th. Oxford Int. Conf.
      on Mech. Properties of materials at High Rates of Strain, Inst. of Phys. Conf.
      Series No. 102, Inst. of Physics, London and Bristol, 1989, 403-410.

11.   L. J. Griffiths and D. J. Martin, J. Phys. D. Appl. Phys., 7, 2329-2341 (1974)

12.   R. L. Sierakowski, G. E. Nevill, C. A. Ross and E. R. Jones, J. Comp. Mater., 5,
      362-377 (1971)

13.   T. Parry, Oxford University Engineering Laboratory (unpublished work)

14.   Y. Bai and J. Harding, in Proc. Int. Conf. on Structural Impact and Crashworthi-
      ness, Elsevier Applied Science, London and New York, 1984, 2, 482-493

15.   P. D. Ewins, RAE Technical Report No. 71217, 1971.

16.   S. Shah, R. K. Y. Li and J. Harding, Oxford University Engineering Laboratory
      Report No. OUEL 1730/1988.

17.   Y. L. Li, C. Ruiz and J. Harding, to appear in Composites Science and
      Technology.

3

New Directions in Research on Dynamic Deformation of Materials

GEORGE MAYER

Institute for Defense Analyses
Science and Technology Division
Alexandria, Virginia 22311, U.S.A

Progress in the development of new approaches to the analysis and experimental studies of the deformation, failure, and processing of structural materials under high loading rates has been reviewed. Advances in elucidation of the response of metals, ceramics, and polymeric materials to high loading rates, including the field of synthesis and processing, and directions of future work, will be considered.

I.   INTRODUCTION

Studies of the dynamic loading and response of materials have multiplied during the past ten years, as exemplified by the host of papers presented at conferences and symposia which have been held on a regular basis (see for example, references 1-10). Because the area of dynamic loading is so broad, and due to the limitations of space in the proceedings and time at the present conference (and because of the limitations in my own knowledge), many topics, such as the treatment of detonations, shock phenomena relating to the earth, and hypervelocity studies, will be mentioned either not at all, or in passing. Suffice to say that significant attention is being devoted to these other important areas, and that they have been addressed regularly at the meetings already mentioned.

The two main areas of application discussed here which are reflected in the research activities, progress, and future directions of the field are armaments (armor, warheads, etc.) and synthesis and processing of materials. Hopefully, this bias can be overlooked, and more general extensions of the activities to be described will be seen by the reader.

35

```
┌─────────────────────────────────────────────┐
│           AREAS OF APPLICATION              │
├─────────────────────────────────────────────┤
│  • ARMOR PROTECTION                         │
│  • PENETRATION PHENOMENA                    │
│  • SHOCK WAVE EFFECTS ON STRUCTURES         │
│  • FRAGMENTATION                            │
│  • SPALLATION                              │
│  • SHAPED CHARGE JET FORMATION              │
└─────────────────────────────────────────────┘
```

FIG. 1   *Applications areas in armaments for dynamic loading phenomena*

In the decades of the Cold War, the U.S. Department of Defense became increasingly concerned about both the numbers and effectiveness of the conventional weaponry of the Soviet Union and their allies in the Warsaw Pact nations. The approach that was taken was to meet numbers of tanks and guns with smaller numbers of highly effective new weapons. On the research front of high loading rate phenomena, before embarking on new materials and other development programs, the state of understanding and areas of ignorance had to be established. The applications for new armaments were important, as is indicated in Figure 1. Thus, in 1977, the DoD asked the National Materials Advisory Board to study this problem, and a committee (the so-called Herrmann Committee) was established for this purpose. Their report "Materials Response to Ultra-High Loading Rates" (11) is still widely referenced, and the study was used as a guide by many DoD agencies for the support of research during the past decade.

With regard to shock synthesis and processing, the interests on the part of research sponsors stemmed from the possibilities of creating new materials thereto unknown, strengthening and hardening of materials by novel means, new joining methods, and less expensive routes to fabrication of advanced materials, such as intermetallics and ceramics.

II.   REVIEW OF PROGRESS AND PROSPECTS

The broad impact of the Herrmann report and its recommendations can be realized from the progress which has been made in the ten years since the publication of the report. For example, one of the major issues at the time was the separation of the numerical modelling community from the sector dealing with dynamic material property measurements, and those involved with test firings. Much broader cooperation exists now, as seen by the development of more physically realistic constitutive models which interface well with complex hydrocodes. Some examples are discussed later in this paper. Such cooperative development has extended successfully into ceramic armor materials, with the Ceramics Modelling Working Group, organized in 1988 by G.E. Cort of the Los Alamos National Laboratory, and comprised of representatives of all three communities.

In 1980, a high priority for research was the elucidation of mechanisms of dynamic failure by brittle crack propagation, ductile void growth, and adiabatic shear banding, and the subsequent incorporation of these failure modes, through models, into hydrocodes. As an example of progress, there has been intense activity, both theoretical and experimental, into shear banding. Four widely different studies are described in references (12-15). For example, a model of shear banding was developed employing the concept of the energy dissipated by a moving dislocation (12). From direct observations of shear localization with high-speed photography, the energy dissipated in a band and the resulting temperature rise were estimated (13). Over the years, better resolution and sensitivity have been brought to the measurement of shear band temperatures, the most recent being done with an elaborate array of infrared detectors (16).

The development of new dynamic mechanical property tests for materials under well characterized and controllable loading conditions was also listed as a need in 1980. Many novel testing methods and equipment have been developed since that time, including a pressure-shear plate impact test (17) for shear strain rates greater than $10^5 s^{-1}$ and controllable levels of nearly hydrostatic pressure, a high-speed torsional testing machine (18), and a novel Taylor anvil impact test performed with an instrumented compression Hopkinson bar (19). On the subject of detection methods, a white light speckle method with high speed photography (21) and moire photography (21) have been employed to assess the dynamic displacement fields and dynamic stress intensity factors of fast cracks. A major new diagnostic tool for probing the internal events during projectile/target impact in thick sections of steels, ceramics, etc. has been successfully developed and applied at the Los Alamos National Laboratory (22). The system, called PHERMEX, is an acronym for pulsed high energy (30 MeV) radiographic machine emitting x-rays.

Studies of the characterization of dynamic brittle fracture in both metals and ceramics based upon a description of the nucleation, growth, intersection, and coalescence of cracks have only begun e.g. reference 23. Existing models, such as the SRI NAG/FRAG flaw nucleation and growth models (24) should be explored and possibly extended to include crack intersections and coalescence to failure.

Another subject which required attention within the dynamic loading of materials was the development of more detailed description of material behavior at ultrahigh rates of strain. The controlling micromechanisms and ranges where they dominate, should be identified. For this problem, it would seem ideal to develop a form of failure mechanism diagram, after Ashby's scheme (25). Figure 2 shows two examples developed for an alumina ceramic deformed in tension (a) and compression (b) The differences in the maps illustrate (over a range of fairly slow-strain rate) which failure mechanisms dominate and where they prevail. The information can be used in two ways: first, by specifying the stress, temperature, strain-rate, and load state (in these two cases, tension or compression) the map will provide

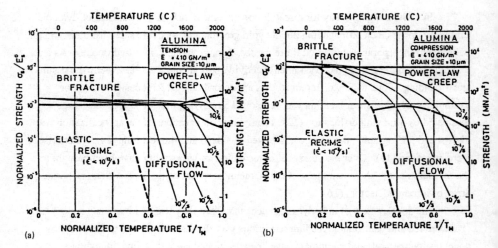

FIG. 2   *Failure-mechanism diagrams for a ceramic (alumina) loaded in (a) tension, and (b) compression. (from Cellular Solids Structure and Properties, 1988, by L.J. Gibson and M.F. Ashby, Courtesy of Pergamon Press)*

information of use for developing constitutive models. In addition, to avoid a particular failure mode, there may be flexibility in changing the grain size or other microstructural feature, altering the load or temperature, or limiting the strain-rate that the material experiences.

The foregoing are examples of topics within the broad field which have received concerted attention in the past decade in dynamic loading of materials. In order to systematically sort out our understanding in more comprehensive fashion, Figure 3 lists some important research areas which should receive continuing and new emphasis. Much progress has already been made on sorting out the roles of microstructure and defects and some of these results will be reported at this meeting (26, 27).

There has been recent interest in the behavior of ceramic materials under dynamic loading, e.g., on the role of pre-existing crack networks, their growth, and coalescence, and the behavior of pulverized and rubbelized ceramic and glassy materials under projectile impact. Figure 4 indicates topics which need additional attention.

Two subjects of major activity in dynamic loading deserve special mention. These are modeling and code development, and synthesis and processing of materials. The following sections address these topics.

RESEARCH AREAS FOR ADDED EMPHASIS
• DEVELOPMENT OF DATA BASES OF DYNAMIC PROPERTIES OF MATERIALS UNDER WELL-DESCRIBED, REPRODUCIBLE CONDITIONS
• CONTINUING STUDIES OF THE ROLES OF MICROSTRUCTURE AND DEFECTS IN DYNAMIC DEFORMATION AND FRACTURE
• CONSTRUCTION OF ASHBY DEFORMATION AND FRACTURE MAPS OVER A WIDE RANGE OF LOADING RATE AND TEMPERATURE REGIMES
• DETAILED STUDIES OF YIELD, DEFORMATION AND FAILURE AS A FUNCTION OF LOADING RATE FOR ADVANCED MATERIALS
• EXPANDED DIAGNOSTIC SPECTRUM OF IN-SITU DETECTION, WITH INCREASED SPEED, RESOLUTION, AND SENSITIVITY OF MEASUREMENT
• CONTINUING STUDIES OF THE ROLES OF MICROSTRUCTURE AND DEFECTS IN DYNAMIC DEFORMATION AND FRACTURE

FIG. 3   *Research areas for continuing and new emphasis in dynamic loading*

DYNAMIC LOADING ISSUES IN CERAMICS
• KINETICS OF CRACK NUCLEATION, GROWTH, AND INTERSECTION
• MECHANISMS OF CRACKING UNDER HIGH DYNAMIC COMPRESSIVE STRESSES
• DEVELOPMENT OF DAMAGE MODELS
• CONSTITUTIVE MODELLING OF PULVERIZED, CONFINED CERAMICS
• ENERGY DISSIPATION PHENOMENA
• EFFECTS OF PHASE TRANSFORMATIONS ON ENERGY DISSIPATION
• PROCESS ZONES IN CERAMICS
• METHODS FOR MEASURING DYNAMIC FRACTURE TOUGHNESS

FIG. 4   *Research issues in dynamic loading of ceramic materials*

A .   Materials Models and Code Developments

Wider use of hydrocodes by the shock wave community has taken place over the past two decades because of the complexity of dynamic events, levels of pressure and temperature achieved, the novel new materials which are employed in the systems being used, and finally, because of the high and still escalating costs of full-scale testing in areas such as ordnance. The numerical simulations which undergird these codes have been facilitated by new supercomputers and by advanced methods, such as parallel processing. An excellent review of hydrocode concepts has been given by Anderson (28) and earlier by Zukas and colleagues (29, 30). The proliferation of codes has been a mixed blessing. On the one hand, some problems of increasing complexity under extreme conditions are being successfully addressed by individual codes; on the other hand, little compatibility exists between codes, and it is very costly and often restrictive (need for supercomputers, or restricted by classification) to maintain all of the codes in use today. Examples of some of the codes which have been popular are listed in Figure 5. The pros and cons of Lagrangian and Eulerian codes have been examined at length. For example, Lagrangian codes are generally more computationally efficient, need a smaller number of zones than Eulerian codes for equivalent accuracy, and avoid mixed material cell computations. The Lagrangian calculations allow the behavior at material interfaces (e.g., opening of voids) to be computed employing the concept of slidelines (31) and the Lagrangian approach allows superior treatment of material behavior (constitutive relations, strain hardening, etc.). On the negative side, large grid distortions create great problems for these codes, and users have often shifted to Eulerian codes for large deformation problems where there is extensive local flow, in high velocity impact regimes, for the collapse of shaped charges, in turbulent flows, etc. In recent years, codes have been adapted to account for intense localized failure modes, such as adiabatic shear and erosion (32). With the aid of such new capabilities, some Lagrangian codes (e.g., EPIC) can extend to treat "selected" large deformation problems. A general problem (more prevalent with Eulerian codes) is the need for large computer memories and long processing times, increasing the expenses of calculation.

EXAMPLES OF HYDROCODES	
Lagrangian (Finite Element)	Eulerian (Finite Difference)
HEMP	HULL
DYNA	JOY
PRONTO	CTH
EPIC	MESA

FIG. 5   *Some examples of hydrocodes used in numerical simulations*

Much progress has been made in developing better constitutive models for use in hydrocodes. The Johnson-Cook model (33) was widely accepted and updates for use in EPIC are periodically made. Other models which take into account specific materials parameters, such as microstructure and dislocation behavior are the Zerilli-Armstrong model (34) and the Mechanical Threshold Stress or MTS model of Follansbee (35). For brittle materials, Sandia National Laboratories have developed a mesocrack continuum damage model (36), based on the notion that many brittle materials contain pre-existing microcrack networks. The initiation, growth, and interaction of such cracks contribute to the nonlinear response of these materials and the latter are not predictable by classical fracture mechanics theories. This model has been useful in predicting the dynamic response of quasi-brittle materials under tensile loads.

More recently, significant advances have been made in developing a new generation of codes which combine the favorable aspects of Lagrangian calculations with limited Eulerian features for avoidance of mesh distortion (37). The techniques are Arbitrary Lagrangian Eulerian (ALE) and Free Lagrange. The ALE method has recently been extended to allow multimaterial zones, as well as nonoscillatory second-order-accurate advection routines, which are necessary for accurate computations. The key to making the ALE technique efficient was the development of automatic criteria (as measured by grid distortion) to switch a cell from Lagrangian to Eulerian.

Reducing the number of working hydrocodes seems to be a truly formidable task, in view of the range of problems addressed by these codes, from determining the response of structures to blast waves, to penetration , and to retorting of oil shale, etc. The trends to combine the best elements of different codes, such as ALE, and attempts to combine two and three-dimensional codes into one that treats both dimensions e.g., a new EPIC code (38), are promising directions. Figure 6 lists some continuing needs in the hydrocode arena. Although the list appears formidable, progress is being made on a number of fronts. For example, the advent of larger and faster computers has allowed three-dimensional versions of some hydrocodes to be formulated (39). Also, among new ventures into data bases, what appears to be a comprehensive effort is on-going at Sandia (40).

B . *Shock Synthesis and Processing of Materials*

The area of shock processing of materials (used in the broadest sense) has seen concerted interest during the past three decades. The broadening of the field is reflected from the book on Explosive Working of Metals by Rinehart and Pearson (41) to the fairly recent volume, Shock Waves for Industrial Applications, which was brought together by Murr in 1988 (8).

The early efforts in processing (Figure 7) employing shock waves were focused on explosive forming, often of large shapes, such as hemispherical sections or cones, on

HYDROCODE ADVANCES - SELECTED NEEDS
• Extend ALE to Three Dimensions
• Generalize Adaptive Mesh Refinement
• Narrow the Gap Between Practical Hydrocodes and Physically-based Models
• Anisotropic Materials Response
• Fracture Initiation Criteria (Ductile and Brittle Materials)
• Mechanism(s) and Models of Softening from Damage Accumulation (Ductile and Brittle Materials)
• Mechanisms and Models of Fracture Propagation (Ductile and Brittle Materials)
• Comprehensive Materials Property Data

FIG. 6   *Some future needs for the advancement of hydrocodes*

EARLY AREAS OF ACTIVITY
• Explosive Forming
• Explosive Welding
• Explosive Hardening
• Explosive Cladding

FIG. 7   *Early areas of activity in shock processing of metals*

FIG. 8   *Specimen of titanium carbide fabricated by combustion synthesis and dynamic compaction (Courtesy of M.A. Meyers)*

surface hardening, and on explosive welding (42) and bonding or cladding. Interests in shock synthesis of diamond were spurred by the work of DeCarli and Jamieson (43) in 1961. DeCarli's patent (44) was followed by more patents and industrial applications of shock-synthesized diamond by the DuPont Company (45). Somewhat after his success with diamond, DeCarli demonstrated the successful shock synthesis of cubic BN from the hexagonal phase (46). Successful shock syntheses of hard materials and other materials of technological interest (e.g., intermetallics) have proceeded strongly in the U.S.S.R., Japan, and the U.S.A. during the past three decades, have been reported at the prior EXPLOMET meeting (4) and will be discussed at the present conference.

Dynamic processing and synthesis of materials have been of strong interest in the U.S.A., especially during the past decade, and this has been reflected in two studies of the National Materials Advisory Board (47, 48). Shock consolidation has been of interest for the densification of materials which are normally difficult to sinter, to avoid grain growth, and to seek a cost-effective industrial production method. More recently, dynamic consolidation has been combined with combustion synthesis (or self-sustaining, high-temperature synthesis) to yield ceramic and ceramic composite compacts close to full density (49, 50). Figure 8 shows a recent product which was dynamically compacted along with the combustion synthesis step. Figure 9 lists the more recent areas of activity in shock synthesis and processing.

Temperature predictions for shock processing have been made using hydrocodes (51) and thermal analysis models (52). Although careful and systematic approaches have been taken to explain the generation of dislocations and other defects during shock loading (53, 54), questions of defect nucleation mechanisms remain, and will be addressed at the present meeting (55). Figure 10 lists a series of areas that deserve future study. For example, transient and/or intermediate states that are generated during shock loading need to be treated both theoretically and experimentally. On a related topic, in an earlier work (56), diffusion coefficients under shock loading were reported to be 10^2-10^3 higher than normal, but no theoretical model was offered. The problem of mass transport over relatively large distances in short times has been difficult to explain. In a different vein, electrical phenomena have been reported in association with shock-related processing and fracture (57). It has also been reported that shock activation of catalyst materials can increase the reactivity of the catalysts by three orders of magnitude (48). This may represent an important practical effect, if it can be retained for a reasonable period of time.

III.   FUTURE DIRECTIONS

The areas of on-going activity and need for increased understanding represent a base of opportunity for a rapidly widening field of research. In the future, more complex materials, such

RECENT AREAS OF ACTIVITY
• Powder Compaction
• Phase Transformations
• New Compound Synthesis
• Combustion Synthesis (SHS)
• Chemical Decomposition
• Polymerization and Cross-Linking
• Shock Modification and Activation

FIG. 9   *Recent activities in shock synthesis and processing of materials*

AREAS FOR FUTURE STUDY	
• Nucleation of Defects	• Nonequilibrium Temperatures in Shock Front
• Roles of Defects as Precursors	• Measurement of Temperature and Shear Stress in Dynamics Compression
• Void Collapse in Porous Materials	• Continued Emphasis on Tailoring of Shock Profiles and Containment Designs for Fracture Avoidance
• Transient or Intermediate States	
• Diffusion Phenomena	
• Electrical Phenomena Associated with Fracture	

FIG. 10   *Subjects in shock synthesis and processing which require new emphasis*

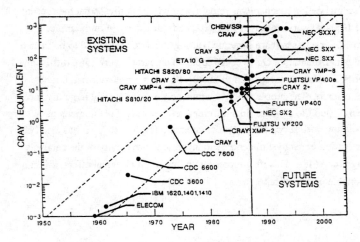

FIG. 11 *Trends in supercomputing capabilities (Courtesy of National Materials Advisory Board)*

as composites (metallic, ceramic, and organic), laminates, intermetallics, and hybrids will constitute special challenges for the high loading rate community.

Hypervelocity impact will receive new attention. Systems such as electromagnetic launchers (58) and multistage gas dynamic launchers (59) bring new experimental capabilities to this arena, which have broad applications from armament, to geosciences, and to space.

Finally, new computing capabilities on the horizon will enable computations that are impossible or impractical with machines that are available today (60) (Figure 11).

REFERENCES

1. *Shock Waves and High-Strain-Rate Phenomena in Metals*, M.A. Meyers and L.E. Murr (eds.) Plenum (1981).

2. *Shock Waves in Condensed Matter - 1983*, J.R. Asay, R.A. Graham, and G.K. Straub (eds.) North-Holland (1984).

3. *Proceedings of the Third Conference on the Mechanical Properties of Materials at High Rates of Strain 1984*, J. Harding (ed.) Institute of Physics (1984).

4. *Metallurgical Applications of Shock-Wave and High-Strain-Rate Phenomena*, L.E. Murr, K.P. Staudhammer, and M.A. Meyers (eds.) Marcel Dekker (1986).

5. *Shock Waves in Condensed Matter - 1985*, Y.M. Gupta (ed.) Plenum (1986).

6. *Hypervelocity Impact-Proceedings of the 1986 Symposium*, C.E. Anderson, Jr. (ed.) Pergamon (1987).

7. *Shock Waves in Condensed Matter 1987*, S.C. Schmidt and N.C. Holmes (eds.) North-Holland (1988).

8. *Shock Waves for Industrial Applications*, L.E. Murr (ed.) Noyes (1988).

9. *Proceedings of the International Conference on Mechanical and Physical Behavior of Materials under Dynamic Loading*, Journal de Physique, Tome 49, Colloque C3, Supplement No. 9 (Sept.. 1988).

10. *Shock Compression of Condensed Matter - 1989*, S.C. Schmidt, J.N. Johnson, and L.W. Davison (eds.) North-Holland (1990).

11. W. Herrmann, Chairman, Committee on Materials Response to Ultra-High Loading Rates, *Materials Response to Ultrahigh Loading Rates, NMAB-356*, National Academy of Sciences (1980).

12. C.S. Coffey, "Shear Band Formation and Localised Heating by Rapidly Moving Dislocations," *Mechanical Properties at High Rates of Strain 1984*, The Institute of Physics 519 (1984).

13. J.H. Giovanola, "Adiabatic Shear Banding Under Pure Shear Loading Part I: Direct Observation of Strain Localization and Energy Dissipation Measurements," *Mechanics of Materials 7*, 59 (1988).

14.   T.W. Wright and S.C. Chou "Army Studies of Large Deformation Plasticity: *Proceedings of DoD/DARPA Coordination Meeting on Advanced Armor/Anti-Armor Materials and Advanced Computational Methods*, IDA D-576, G. Mayer (ed.) Institute for Defense Analyses, 1025 (March 1989).

15.   T.J. Burns "A Mechanism for High Strain Rate Shear Band Formation," *Shock Compression of Condensed Matter - 1989*, S.C. Schmidt, J.N. Johnson, and L.W. Davison (eds.) North-Holland 345 (1990).

16.   Y.C. Chi and J. Duffy "On the Measurement of Local Strain and Temperature During the Formation of Adiabatic Shear Bands in Steels" *Report No. 4, ARO Contract No. DAAL 03-88-K-0015*, (September 1989)

17.   C.H. Li "A Pressure-Shear Experiment for Studying the Dynamic Plastic Response of Metals at Shear Strain Rates of $10^5 sec^{-1}$" *Ph. D. Thesis, Brown University*, Providence, RI (1981).

18.   U.S. Lindholm, A. Nagy, G.R. Johnson, and J.M. Hoegfeldt "Large Strain, High Strain Rate Testing of Copper," *Jl. of Eng. Materials and Technology* (October 1980).

19.   S. Nemat-Nasser, University of California at San Diego, personal communication (December 1989).

20.   X.M. Hu, S.J.P. Palmer, and J.E. Field, "The Application of High Speed Photography and White Light Speckle to the Study of Dynamic Fracture," *Optics and Laser Technology*, 303 (December 1984).

21.   J.M. Huntley and J.E. Field, "Application of Moire Photography to the Study of Dynamic Fracture," Proc. Seventh Intl. Conf. on Fracture (March 1989).

22.   G.E.Cort, Los Alamos National Laboratory, personal communication (June 1990).

23.   L.H. Leme Louro and M.A. Meyers, "Stress Induced Fragmentation in Alumina-Based Ceramics," *Shock Compression of Condensed Matter - 1989, S.C. Schmidt, J.N. Johnson, and L.W. Davison (eds.)*, North-Holland 465 (1990).

24.   D.R. Curran, L. Seaman, and D.A. Shockey, "Dynamic Failure of Solids," *Physics Reports 147 (5 and 6)* 253 (1987).

25.   L.J. Gibson and M.F. Ashby, *Cellular Solids-Structure and Properties*, Pergamon, 66-67 (1988).

26.   J. Lankford, C.E. Anderson, and H. Coque "Microstructure Dependence of High Strain Rate Deformation and Damage Development in Tungsten Heavy Alloys," *International Conference on Shock-Wave and High-Strain-Rate Phenomena in Materials*, (1990).

27.   B.O. Reinders, "Influence of Mechanical Twinning on the Deformation Behavior of Armco Iron," ibid.

28.   C.E. Anderson, Jr., "An Overview of the Theory of Hydrocodes," *Int. J. Impact Eng. 5*, 33 (1987).

29. J.A. Zukas, G.H. Jonas, K.D. Kimsey, J.J. Misey, and T.M. Sherrick, "Three-Dimensional Impact Simulations: Resources and Results," *Computer Analysis of Large-Scale Structures*, K.C. Parck and R.F. Jones, Jr. (eds.), AMD, V. 49, ASME (1981).

30. J.A. Zukas, T. Nicholas, H.F. Swift, L.B. Greszczuk, and D.R. Curran, *Impact Dynamics*, Wiley (1982).

31. M.L. Wilkins, "Calculations of Elastic-Plastic Flow," *Methods of Computational Physics, 3*, B. Adler, S. Fernback, and M. Rottenberg (eds.) Academic (1964).

32. K.D. Kimsey and J.A Zukas, "Contact Surface Erosion for Hypervelocity Problems" *BRL-MR-3495*, U.S. Army Ballistics Research Laboratory (1986).

33. G.R. Johnson and W.H. Cook, "A Constitutive Model and Data for Metals Subjected to Large Strains, High Strain Rates, and High Temperatures," *Seventh International Symposium on Ballistics*, The Hague, The Netherlands (April 1983).

34. F.J. Zerilli and R.W. Armstrong "Dislocation Mechanics - Based Constitutive Relations for Material Dynamics Calculations," *Jl. App. Phys.* 61, 1816, (1987).

35. P.S. Follansbee, "Dynamic Deformation and Fracture in Armor/Anti-Armor Materials," *Proceedings of the DoD/DARPA Coordination Meeting on Armor/Anti-Armor Materials and Advanced Computational Methods, IDA-D-576*, G. Mayer (ed.), Institute for Defense Analyses, (March 1989).

36. E.P. Chen, Sandia Albuquerque National Laboratory, private communication (April 1990).

37. J.D. Immele (Chairman) *Report of the Review Committee on Code Development and Material Modelling LA-UR-89-3416,* (Los Alamos National Laboratory Report) Review Committee on Code Development and Materials Modelling, 15 (October 1989).

38. G.R. Johnson, Honeywell, Inc., private communication (December 1989).

39. W.E. Johnson and C.E. Anderson, Jr. "History and Application of Hydrocodes to Hypervelocity Impact," *Int. J. Impact Eng. 5*, 423, (1987).

40. J.S. Wilbeck, C.E. Anderson, Jr., J.C. Hokanson, J.R. Asay, D.E. Grady, R.A. Graham, and M.E. Kipp, "The Sandia Computerized Shock Compression Bibliographical Database," *Metallurgical Applications of Shock-Wave and High-Strain-Rate Phenomena*, L.E.. Murr, K.P. Staudhammer, and M.E. Meyers, (eds.) Marcel Dekker, 357, (1986).

41. J.S. Rinehart and J. Pearson, *Explosive Working of Metals*, MacMillan (1963).

42. G.R. Cowan and A.H. Holtzman, "Flow Configuration in Colliding Plates: Explosive Bonding," J. Appl. Phys. 34(4), 928, (1963).

43. P.S. DeCarli and J.C. Jamieson, "Formation of Diamond by Explosion Shock," *Science*, 133, 821 (1961).

44. P.S. DeCarli, "Method of Making Diamond," U.S. Patent 3,238,019, March 1, 1966.

45.   O.R. Bergmann and N.F. Bailey, "Explosive Shock Synthesis of Diamond," in
       High Pressure Explosive Processing of Ceramics, R.A. Graham and A.B.
       Sawaoka (eds.) Trans Tech 65 (1987).

46.   P.S. DeCarli, *Bull. Amer. Phys. Soc., 12,* 1127 (1967).

47.   Committee on Dynamic Compaction of Metal and Ceramic Powders, *Dynamic
       Compaction of Metal and Ceramic Powders*, National Materials Advisory Board,
       National Research Council, Washington, DC, NMAB-394, (1983).

48.   Committee on Shock Compression Chemistry in Materials Synthesis and Processing,
       Shock Compression Chemistry in Materials Synthesis and Processing, National
       Materials Advisory Board, National Research Council, Washington, DC,
       NMAB-414, (1984).

49.   A. Niiler, T. Kottke, and L. Kecskes, "Shock Consolidation of Combustion
       Synthesized Ceramics," in *Proc. of First Workshop on Industrial Applications of
       Shock Processing of Powders*, CETR, New Mexico Institute of Mining and
       Technology (June 1988)

50.   M.A. Meyers, University of California, San Diego, private communication
       (December 1989).

51.   R.L. Williamson and R.A. Berry "Microlevel Numerical Modelling of the Shock
       Wave Induced Consolidation of Metal Powders" in *Shock Waves in Condensed
       Matter - 1985*, Y.M. Gupta (ed.), Plenum, 341, (1986).

52.   T. Vreeland, Jr., P. Kasiraj, A.H. Mutz, and N.N. Thadhani, "Shock Consolidation
       of a Glass-Forming Crystalline Powder," in *Metallurgical Applications of Shock
       Waves and High Strain Rate Phenomena, L.E. Murr*, K.P. Staudhammer, and M.A.
       Meyers, (eds.) Marcel Dekker, 231, (1986).

53.   J. Weertman, "Moving Dislocations in a Shock Front," *Shock Waves and High-
       Strain-Rate Phenomena in Metals*, M.A. Meyers, and L.E. Murr (eds.) Plenum, 469
       (1981).

54.   M.A. Meyers and L.E. Murr, "Defect Generation in Shock Wave Deformation,"
       ibid., 487, (1981).

55.   M.P. Mogilevsky, "Defect Nucleation under Shock Loading," *International
       Conference on Shock-Wave and High-Strain-Rate Phenomena* (August 1990).

56.   S.V. Zemsky, Y.A. Ryabehikov, and G.D. Epshteyn, "Mass Transfer of Carbon
       Under the Influence of a Shock Wave," *Phys. Met. Metal. 46(1)*, 171, (1979).

57.   J.T. Dickinson, L.C. Jensen, and A. Jahan-Latibari, "Fracto-emission: The Role of
       Charge Separation," *Jl. Vac. Sci. and Tech* (February 1984).

58.   H. Fair, "Hypervelocity Then and Now," *Inst. Jl. Impact Eng. 5* 13 (1987).

59.   L.A. Glenn, A.L. Latter, and E.A. Martinelli, "Multistage Gasdynamic Launchers,"
       Shock Compression of Condensed Matter - 1989, S.C. Schmidt, J.N. Johnson, and
       L.W. Davison (eds.) North-Holland 977 (1990).

60.   *The Impact of Supercomputing Capabilities on U.S. Materials
       Science and Technology - NMAB-451*, National Academy of Sciences (1988).

4

Constitutive Equations at High Strain Rates

L.W. MEYER

Fraunhofer-Institut für angewandte Materialforschung - IFAM
Lesumer Heerstr. 36, D-2820 Bremen 77, Germany

In this review is presented which forms of semiempirical or physically-based constitutive equations are useful for high strain rate applications, how they match the real behavior, and where are the limitations.

I. INTRODUCTION

"Constitutive Equations are the vehicle by which our knowledge of material behavior enters engineering design." This sentence from Kocks [1] illustrates very appropriately the goal: To solve mechanical problems with numerical methods where the quality of the results depends strongly on a sufficient description of the involved material behavior.

This restriction (to 'sufficient') is needed because the material behavior cannot be expressed in a general way. It is too complex to find complete solutions. Our understanding of the micromechanics certainly has made welcomed progress, but many questions are still open.

Therefore we have to concentrate on specific deformation or loading conditions, and on certain temperature- and strain-rate ranges. Hot working, creep, cyclic loading, unloading and history effects, i.e. are not included. However cold working, monotonic straining transients, as well as temperatures between 50 and 500 K and strain rates between 10^{-5} to 10^5 s^{-1} are considered in the following.

II. SEMI-EMPIRICAL EQUATIONS

One of the simplest forms of describing the material behavior is the Ludwik equation, [2]:

$$\sigma = A + B\,\epsilon^n \quad \text{or} \quad \tau = C + D\,\gamma^n \tag{2.1}$$

Using the constants A and B and the so called strain hardening parameter n it is often used to characterize the stress σ or τ as a function of the strain ϵ or γ at ambient temperatures and quasistatic loading. Ludwik claims that the strain hardening develops monotonically with a constant to the power n. Indeed, this is often true, at least beyond a transient for the very first plastic deformation, up to a saturation state of stress. To include the influence of temperature, T, and strain rate, $\dot{\epsilon}$, or $\dot{\gamma}$, the power law (2.1) was extended in several ways:

a) by Klopp, Clifton, Shawki [3] $\qquad \tau = \tau_0\,\gamma^n\,T^{-\nu}\,\dot{\gamma}_p^m \tag{2.2}$

with the strain rate hardening parameter m, and the temperature softening parameter ν, both as power laws,

b) by Litonski [4] $\qquad \tau = C(\gamma_o + \gamma_p)^n\,(1-aT)(1+b\dot{\gamma}_p)^m \tag{2.3}$

Here the strain is split up into a constant γ_o (which approximately represents the elastic part) and the plastic strain γ_p. The temperature influence is expressed linearly instead of as a power function. This assumes to a first approximation a linear-decrease of the stress with increasing temperature.

c) by Vinh et al. [5] $\qquad \tau = F(\gamma)^n\,(\dfrac{\dot{\gamma}}{\dot{\gamma}_o})^m\,\exp(\dfrac{W}{T}) \tag{2.4}$

They found that an exponential expression of temperature provided a better fit to their results for aluminum, copper and mild steel.

d) Johnson/Cook [6,7] $\tau = [A+B\gamma^n][1+C\ln(\frac{\dot{\gamma}}{\dot{\gamma}_o})][1-(T^*)^m]$ (2.5)

where $T^* = (T-T_r)/(T_m-T_r)$

Johnson/Cook also used the Ludwik equation as a base and included strain rate and temperature influences. They chosed a logarithmic rate sensitivity with a constant C and normalized the strain rate with $\dot{\gamma}_o = 1\ s^{-1}$. Therefore, their constants of the Ludwik equation are related to a strain rate of $1\ s^{-1}$ and not to quasistatic loading. The temperature is included by a power expression of the homologous temperature T^*. Based on quasistatic and dynamic torsion and tension tests up to $\dot{\gamma} = 4 \times 10^2\ s^{-1}$, Johnson/Cook provided constants for 6 ductile and 6 less ductile materials, including Cu, Fe, brass, Ni, C-steel, tool steel, Al alloys, and DU. By this broad source of data, the Johnson/Cook equations are often used as constitutive equations for calculations.

e) Armstrong/Zerilli [8-10]

bcc: $\sigma = \Delta\sigma_G + B_0\exp[(-\beta_0+\beta_1\ln\dot{\epsilon})T]+K_0\epsilon^n+k_\epsilon\ell^{-\frac{1}{2}}$ (2.6)

fcc: $\sigma = \Delta\sigma_G + B_1\epsilon^{\frac{1}{2}}\exp[(-\beta_0+\beta_1\ln\dot{\epsilon})T]+k_\epsilon\ell^{-\frac{1}{2}}$ (2.7)

where $\Delta\sigma_0$, B_0, B_1, β_0, β_1, K_0, n, and k_ϵ are constants and ℓ is the average grain size. Armstrong and Zerilli distinguish between bcc and fcc materials. For fcc materials, as an important improvement, they coupled the strain hardening term with a temperature and strain rate dependence. The bcc materials are modeled similar to Clifton et al. and Johnson/Cook, where the strain or work hardening is independent from a temperature or strain rate influence. In a refinement [10], they included the effect of twinning at high strain rates and high strains, respectively, above a certain von Mises stress level, which leads to an increase of the Hall-Petch term of eq.(2.6 and 2.7):

$$\Delta(k\ell^{-\frac{1}{2}}) = k\ell^{-\frac{1}{2}}[(N+1)^{\frac{1}{2}}-1]$$ (2.8)

where N equals the average number of twins within the grains. With this improvement, they successfully modeled the Taylor-Impact of Johnson/Cook [7] for Armco iron without taking into account viscous drag. For copper, indeed, Armstrong and Zerilli [9] included a viscous drag influence in the thermal stress

$$\sigma_e = \frac{1}{2} \sigma' \ [1+\sqrt{1+ \frac{4C_0 \dot\epsilon T}{\sigma'}}]$$ (2.9)

with σ' the T- and $\dot\epsilon$-dependent component of eq. (2.7):

$$\sigma' = B_1 \ \epsilon^{\frac{1}{2}} \ \exp[(-\beta_0 + \beta_1 \ln\dot\epsilon)T]$$ (2.10)

They found good agreement with the results of Follansbee et al. [11] and Gourdin [12] on copper under compression and tensile loading.

Equations (2.1) to (2.5) were expressed mathematically as simply as possible or extended with some additional constants to meet the experimentally measured behavior as closly as possible. The advantage of such a point of view is that, with some effort, the mathematical curve fitting can meet the materials stress-strain behavior as close as the description allows. Sometimes the approximation gets close, as Vinh [5] or Campbell [13] demonstrated for fcc materials. In other cases, the curve fitting cannot approach the real behavior, because the strain hardening is not constant versus strain rate. Then a compromise has to be made (and it is on the mind of the user) in which strain rate region the empirical description has to fit best.

Another inherent restriction often is not noted: The description of the material behavior, the modelling and the named coefficients are based on test results determined up to a certain limited rate of strain. Therefore the users have to keep in mind, that only up to that strain rate the modelling is based on experimental results and that calculations of faster events are done as an extrapolation or as a guess, no more, no less.

It would be desirable if the semi-empirical equations could be established on a physical basis related to micromechanical behavior. But this is mostly not the case because the real behavior of a material is not a function of one state parameter, i.e. the strain. The reality is much more complicated. We have to consider that the material behavior is dependent on the evolution of the microstructure. This evolution depends on the initial structure and on the strain rate, respectively, the strain rate history and on the loading conditions such as compression or shear.

III. CONSTITUTIVE EQUATIONS BASED ON MICROSTRUCTURAL PROCESSES

The (external) stress σ leading to plastic deformation has to overcome the internal stresses from both far and short range obstacles. Written in a sum, one can assume

$$\sigma = \sigma_a + \sigma^* \, (T, \dot{\epsilon}) \tag{3.0}$$

Far range stress fields are determined mainly by the structure of the materials and less by the strain rate. Therefore, this part is called the athermal stress component, σ_a. The thermally influenced stress component, σ^* is related to the interaction between dislocations and short range obstacles which can be overcome with the help of thermal fluctuations. This includes the strong influence of the temperature, T and strain rate, $\dot{\epsilon}$.

With the Orowan equation for the dislocation velocity, the Arrhenius equation for the dislocation waiting time before an obstacle, and the equation for the free activation energy, the thermal activated stress σ^* is obtained:

$$\sigma^* = \hat{\sigma}\{1-[\frac{kT}{\Delta G_o} \ln \, (\frac{\dot{\epsilon}_o}{\dot{\epsilon}_p})]^n\}^m \tag{3.1}$$

Now σ^* can be expressed as a function of the external variables T and $\dot{\epsilon}_p$ and the constants $\hat{\sigma}$, ΔG_o, $\dot{\epsilon}_o$, n and m [14]. The importance of this expression is that it is based on micromechanics and that all the parameters have a physical background: the short range obstacles which can be overcome by the assistance of the thermal component σ^* are characterized by $\hat{\sigma}=$ stress amplitude at T = 0 K, named mechanical threshold stress, ΔG_o = the activation energy at T = T_o, n or 1/q and m or 1/p determine the form of the force-distance-curves of mobile dislocations in the vicinity of specific short-range obstacles and $\dot{\epsilon}_o$ is specified by the microstructure. Known values of m, n and $\dot{\epsilon}_o$ for some metals are listed in Table 1. These equations are valid for one type or one dominant type of dislocation interaction with the microstructure, and the microstructure is assumed to be constant.

The concept of "threshold stress $\hat{\sigma}$", developed by Kocks [15], Follansbee [16-18] and Follansbee and Gray [19], uses the same fundamental description, but extends the application of the thermal activation theory

Table 1   Constants for eq. 3.1 to 3.4 for different materials

Materials	Ref.	Constants			$\hat{\sigma}$
		$m=1/p$	$n=1/q$	$\dot{\epsilon}_0$	
Aluminum (Pure)	[14]	1	1		
Some Hexagonal Metals	[14]	1	1		
Titanium Alloys	[14]	1	1/2	10^7	
	[45]	1	1/2	10^{10}	
Copper, Homogeneous Alloy	[14]	2	2/3		
	[42]	2	2/3		
	[16]	3/2	1		
Iron, Pure	[34]	1.5	1	10^8	
	[14]	2	1		
	[38]	2	1		
Steel Carbon	[14]	4	1		
0.46C, 1.5Mn	[36]	2	2/3	10^7	1690
0.45C	[38]	2	2/3	$10^{6.4}$	
0.45C	[38]	2	1/2	$10^{6.4}$	
0.45C	[38]	3	2	$10^{6.4}$	
Steel: 1Si and NiCrMo	[38]			10^7	
9Cr-3Si	[39]			10^7	
0.3C-3Ni-1Cr-0.5Mo	[36]	2	1/2	10^8	2120
Austenitic Steel	[37]			$2 \cdot 10^9$	
23Cr-17Ni-3Mo	[36]			10^{10}	
18Cr-Stainless Steel	[38]			10^8	
21Cr-16Ni-5Mn-3Mo	[35]	2	1/2	$8 \cdot 10^9$	2750
Nitronic 40	[34]	1.5		10^9	

to higher strains taking the evolution of the microstructure into account. With splitting the threshold stress $\hat{\sigma}$ into an athermal and thermal activated component (subscript "a" and "t", resp.), Follansbee and Kocks [16] use basically the same expression as eq. (3.1) to describe the kinetics s_i [in square brackets] or the flow stress σ, normalized with the temperature dependent Young's modulus $\mu(T)$:

$$\frac{\sigma}{\mu} = \frac{\hat{\sigma}_a + (\hat{\sigma} - \hat{\sigma}_a)}{\mu} \{1 - [\frac{kT}{g_o \mu b^3} \ln(\frac{\dot{\epsilon}_o}{\dot{\epsilon}})]^{1/q}\}^{1/p} \qquad (3.2)$$

The evolution of the structure which determines $\hat{\sigma}$ is considered as the balance between dislocation accumulation (θ_0) and dynamic recovery (θ_r). To express that the strain is not assumed as a state variable, the strain hardening θ is described differentially

$$\frac{d\hat{\sigma}}{d\epsilon} = \theta = \theta_0 - \theta_r \quad \text{or} \quad \theta = \theta_0[1 - F(\frac{\hat{\sigma} - \hat{\sigma}_a}{\hat{\sigma}_s - \hat{\sigma}_a})] \qquad (3.3, \ 3.4)$$

where the temperature and strain rate dependence is referred to the saturation threshold stress $\hat{\sigma}_s$. For fcc pure metals and alloys over wide ranges of temperatures but narrow ranges of strain rates, first it was assumed [20] that θ_0 is roughly constant with strain rate and is found to represent the strain hardening in stage II deformation. Follansbee and Kocks [16] indeed concluded (from tests at 76K and up to high true strains of ~1) that θ_0 must increase with strain rate in the following form

$$\theta_0 = C_1 + C_2 \ln(\dot{\epsilon}) + C_3 \dot{\epsilon} \qquad (3.5)$$

in order to fit the experimental data to eq. (3.4) for accumulation and dynamic recovery. The determined strong increase of θ_0 above $\dot{\epsilon} = 10^3 \ s^{-1}$ indicates that the dislocation accumulation θ_0 must increase dramatically above that range of strain rate. This observation might explain why frequently above $\dot{\epsilon} = 10^3 \ s^{-1}$ a sharp increase in flow stress is reported.

On the other hand, the consideration of drag effects on the dislocation velocity is assumed as well to explain a sharp increase of stress-strain rate sensitivity [21]. At very high rates of strain for pure drag

and an ideal crystal, it is assumed that the dislocations move with a constant velocity and that the flow stress σ (which exceeds $\hat{\sigma}$) is directly
proportional to the strain rate $\dot{\epsilon}$. For the transition range between thermally activated and drag controlled deformation ($\sigma < \hat{\sigma}$), both mechanisms may
be operative. The drag influence can be included here by an influence on
the running time t_r of the dislocations between the obstacles. Under this
assumption, Hoge and Mukherjee developed a combined equation for the material behavior of tantalum which fits well with the measured behavior [22].
Burgahn et al. [23] showed that for a carbon steel C45 under the same assumption a similar relationship leads to a close agreement between predicted and measured behavior up to $\dot{\epsilon} = 10^3$ s^{-1} and that the expected higher
rate sensitivity should reasonably influence the flow stress at $\dot{\epsilon} >$
10^5 s^{-1}.

In all equations listed above, the temperature influence was not
included in the work hardening or even when regarded by a separate linear,
power or exponential term. Klepaczko [24,25] proposed not to neglect the
temperature influence and included it in the basic double power expression

$$\tau = C(\theta)(\Gamma_o+\Gamma)^{n(\theta)} \, Z^{m(\theta)} + \langle\eta(\dot{\Gamma}-\dot{\Gamma}^*)\rangle \tag{3.6}$$

Furthermore, he combined a specific form of the Arrhenius relation with
the strain rate sensitivity m

$$\tau = \hat{\tau}\left[\frac{\dot{\Gamma}}{\dot{\Gamma}_o} \, \exp\frac{\Delta H}{kT_m\theta}\right]^{m(\theta)} \tag{3.7}$$

with $\theta = T/T_m$ for the homologous temperature, T_m the melting temperature, Γ
the true shear strain, $\dot{\Gamma}$ the true shear strain rate, $\dot{\Gamma}_o$ the frequency factor, ΔH the apparent activation energy, k the Boltzmann constant, and $\hat{\tau}$
the threshold stress. The expression in the brackets is the well-known
Zener-Hollomon parameter, named Z. The functions $n(\theta)$ and $m(\theta)$ have to be
chosen appropriate to the material behavior.

The complete set of constants reaches fifteen for bcc materials or
thirteen for fcc materials. For three materials, polycrystalline pure
aluminum, copper and a cold rolled 0.18 % carbon steel (1018), Klepaczko
identified the material parameters from published results and found
reasonable agreement. Of course, the determination of this large number of

constants requires some effort, but an important step is done: the inclusion of physical-based expressions.

Contrary to all models mentioned before, Steinberg and Colleagues [26] proposed a model to predict the yielding stress Y and the shear modulus G under uniaxial shock conditions. This model includes temperature and pressure influence, but neglects the strain rate. Based on experimental observations, it predicts reasonably well the material behavior at conditions of strong shocks. The constants used are given by Steinberg for more than 14 materials.

To cover also weaker shock conditions, where the strain rate influence cannot be neglected, Steinberg extended his model [27] with a thermal activated and a viscous drag term similar to Hoge and Mukherjee [22]. Consequently, the calculated and experimental curves, i.e. for Ta at 5 GPa, got much closer than with the rate independent model.

Under the assumption of small strains, Bodner/Partom [28] established a set of constitutive equations which may consider isotropic and directional hardening, thermal recovery of hardening, temperature dependence and an isotropic and directional damage development. Including all effects, the number of material constants is large. Nevertheless, they can be obtained from standard uniaxial tests. For dynamic monotonic loading, only the basic rate dependence of plastic flow and isotropic hardening without thermal recovery is required.

The main equation governing the inelastic deformations is the kinetic equation D_2^P which is F (deviatoric stress invariants J_2) chosen essentially empirically, but mathematical flexible and with a physical basis to model different material behavior:

$$D_2^P = D_o^2 \exp[-(\frac{Z^2(1-\omega)^2}{3J_2})^n] \qquad (3.7)$$

with D_0 the limiting strain rate in shear, $n = f(T,p)$, a material constant that controls rate sensitivity by temperature and pressure influence and the overall level of flow stress. Z is interpreted as a load history dependent parameter, corresponding in a general way to the yield stress, and ω is a history dependent damage parameter.

For the one dimensional stress case, eq. (3.7) can be developed to

$$\sigma = \frac{Z}{(2 \ln[\frac{2D_o}{\sqrt{3}\dot{\epsilon}_{11}^P}])^{\frac{1}{2n}}} \qquad (3.8)$$

which is plotted in a normalized form in Fig. 1 for various values of n, [29].

Depending on the choice of n and D_o, from a nearly rate insensitive behavior (n > 5, $\dot{\epsilon}_p/D_o$ < 10^{-5}) to a strong rate sensitive behavior (n < 5, $\dot{\epsilon}_p/D_o$ > 10^{-3}), the flow stress can be modeled in close accordance to experimental results.

The dependence of the parameter Z on the plastic work W_p, generated from the initial state, is assumed to have the exponential form

$$Z(W_p) = Z_1 + (Z_o - Z_1) \exp(\frac{-m\,W_p}{Z_o}) \qquad (3.29)$$

where Z_1 is a limiting, saturation value of Z to ensure that resistance to plastic flow is limited. $Z_o = Z_1 - \Delta Z$ is the initial value of Z_1, m is a constant related to the rate of hardening. Bodner showed that isotropic as

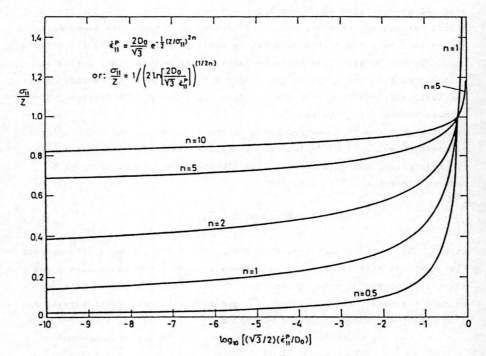

FIG. 1   *Dependence of the uniaxial flow-stress parameter on the strain rate parameter for different strain rate sensitivities n of the Bodner equation, after Bodner 1987.*

well as directional hardening and recovery at higher temperatures can be modeled rather well [29]. Anisotropic behavior is not included. For high strain rate application, only the basic dependence of plastic flow stress and isotropic hardening is required without thermal recovery. Then the following constants are needed: D_o, n, Z_o, Z_1, m and E. Rajendran, Bless and Dawicke [30] evaluated the material constants form Hopkinson-bar and flyer plate experiments for three initially isotropic materials. They showed that the Bodner-Partom modeling [28] successfully describes rate independent, rate sensitive, work hardening and non-work hardening material behavior.

IV. COMPARISON OF CONSTITUTIVE EQUATIONS WITH EXPERIMENTAL RESULTS

Clifton and co-workers [3] published shear stress data on pure aluminium from low to high rates of strain. Up to $\dot{\epsilon} = 10^4$ s^{-1} they found the rate sensitivity to be very low, but above $\dot{\epsilon} = 5 \cdot 10^4$ s^{-1}, the rate sensitivity increases rapidly, up to 190 MPa per order of magnitude of strain rate, as shown in Fig. 2. This rapid stress increase at high rates of strain is also predicted from the Bodner-Partom equations. Bodner [31] compared his equation with these experimental results, Fig. 2. To satisfy the physical background of the equations, his comparison is done with using the yield stress at 0 K of 230 MPa as a value for the hardening parameter Z in its saturated state Z_s. This yields a very good agreement in the strain rate range below $\dot{\epsilon} = 10^4$ s^{-1}. To match the data from experiments with extra thin, evaporated Al-film which leads to strain rates of about 10^7 s^{-1}, a limiting strain rate $2D_0$ of $5 \cdot 10^6$ or 10^7 s^{-1} seems to be the best choice. With these assumptions, however the increased rate sensitivity of the bulk material at $\dot{\epsilon} = 10^5$ s^{-1} is not well predicted because the stronger stress increase by the equations starts later between 10^6 and 10^7 s^{-1}.

For the precipitation hardened Al alloy 6061-T6, the results of Hoggart and Recht [32] and Li [33,3] indicate that the rate sensitivity (with 60 MPa per order of magnitude of strain rate) is not as high as for the weaker pure aluminium, Fig. 3. To model these results, Bodner [31] named parameter of Z_s = 550 MPa, n = 5 and $2D_0 = 10^7$ s^{-1} for a reasonable agreement. Indeed, regarding the full range of strain rates, a choice of Z_s = 550 MPa, n = 3 and $2D_0 = 5 \cdot 10^5$ brings the behavior from low strain rates up to $\dot{\epsilon} = 10^5$ s^{-1} closer together, Fig. 3.

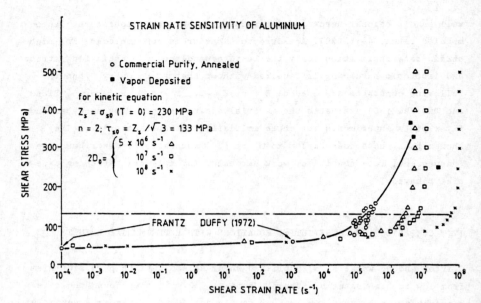

FIG. 2   Comparison of the strain rate dependence of the flow stress of
pure 1100-0 Aluminium, after Klopp, Clifton and Shawki 1985 and Bodner
1988.

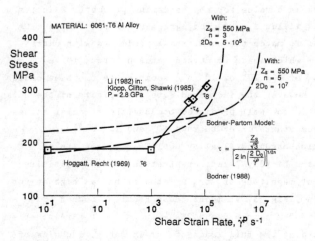

FIG. 3   Comparison of the strain rate dependence of the flow stress of
hardened 6061-T6 Aluminium-Alloy, Bodner 1988.

It should be noted ~~that~~ the "limiting strain rate $2D_0$" and the frequency factor $\dot{\epsilon}_0$ of the Arrhenius equation should not be interchanged. Of course, in both cases, a material-dependent saturation stress level is reached (so the physical background could be similar). Furthermore, for some metals like titanium, copper, carbon steels, the found frequency factors, Table 1, coincide with the value of 10^7 s^{-1} given by Bodner [31]. But for austenitic steels, the frequency factor ranges between 10^8 to 10^{10} s^{-1} [34-38,17-19]. Even in cases where the saturation stress at T=0 K and the frequency factor $\dot{\epsilon}_0$ are not estimated, but evaluated from results of temperature- and strain rate-varied tests [35,36], the use of $2D_0 = 10^7$ s^{-1} leads to a better agreement than the use of $\dot{\epsilon}_0 = 8 \cdot 10^9$ s^{-1} for $2 D_0$ of a high strength austenitic steel [34], or the use of $\dot{\epsilon}_0 = 5 \cdot 10^8$ s^{-1} for $2 D_0$ of a low alloyed CNiCr-steel [36], Fig. 4.

The same CNiCr-steel with bcc structure was used here to apply the empirical equations and to compare their ability to describe the measured material behavior. From the engineering stress-strain curves, true stress data were developed and from ln/ln plots, the constants of the Ludwik, the

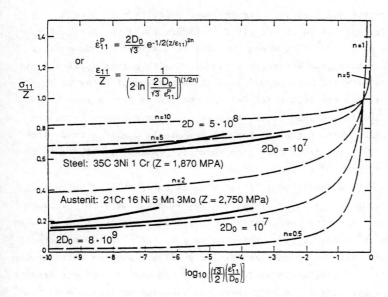

FIG. 4   *Comparison of the strain rate dependence of flow stress of a bcc- and fcc-steel (after Meyer/Staskewitsch, 1988 and Stiebler, et.al., 1989) with calculated values based on the Bodner-Partom Model (Bodner, 1988).*

Clifton, or Johnson/Cook equations are evaluated in order to get the best fit. The comparison at quasistatic loading between the measured and Ludwik-modeled behavior, Fig. 5, gives an excellent agreement from first yielding to 5 % strain and slightly under estimates, the stresses between 5 and 10 % strain. With increasing strain rates, the true flow stresses are initially insensitive up to $\dot{\epsilon} = 10^{-2} - 10^{-1}$ s^{-1}, and then display a mode-rate rate sensitivity [36]. Assuming that the effect of rate sensitivity can be included by a simple power law theorem, the ln/ln plot should be a straight line. Indeed, this is the case, however they have different slopes, Fig. 6. Therefore, the rate sensitivity $m = d\ln(\sigma - A)/d\ln\dot{\epsilon}$ is not the same for different flow stresses, for m varies between 0.09 and 0.04. Dependent on which flow strength is be modeled, different constants must be used. In this comparison, an m of 0.064 was chosen for Clifton's model-ing, marked with circles in Fig. 5. At low strains, the predicted stresses are too low; above 2 % strain, the predicted values are too high. A better correlation is reached with the Johnson/Cook eq. (2.5), but in the strict sense this holds only for this selected strain rate. The evaluation of the rate sensitivity parameter C shows that the constant C is a function of the yield strength and that it increases monotonically with strain rate $\dot{\epsilon}_p$, Fig. 7. In contrast to the original procedure [6,7] the constant $\dot{\gamma}_0$ is taken here to be 10^{-2} s^{-1}, where the rate sensitivity for this steel appears. Fortunately, the stress variation of C with higher strains dimin-ishes, but the rate influence remains. A certain value of C has to be selected to provide the best model for the particular strain rate region of interest. The same holds for the rate sensitivity constant β_1 of the Armstrong/Zerilli model (eq. 2.6) for the bcc steel: β_1 seems to be inde-pendent of strain, but drops down from about 10^{-3} K^{-1} at $\dot{\epsilon} = 10^0$ s^{-1}, to $3,5 \cdot 10^{-4}$ K^{-1} at $\dot{\epsilon} = 10^4$ s^{-1}, Fig. 8.

Regarding the material behavior at high rates of strain, there are no doubts that up to $\dot{\epsilon} = 10^3$ s^{-1}, besides the athermal behavior, in many cases, thermally activated deformation mechanisms govern the flow stresses. For the range $10^3 < \dot{\epsilon} < 10^5$ s^{-1}, the results are less clear, but there is strong evidence that the thermal activation still governs the main influ-ence to $\dot{\epsilon} = 10^4$ s^{-1} [36,38-40], or even higher. In addition to our results with high strength steels [36,39], where the behavior up to $\dot{\epsilon} = 8 \cdot 10^3$ s^{-1} was found to be fully thermally activated, several other investigations support this opinion: The often cited example of Campbell/Ferguson [41], who found a remarkable change in strain rate sensitivity of the lower

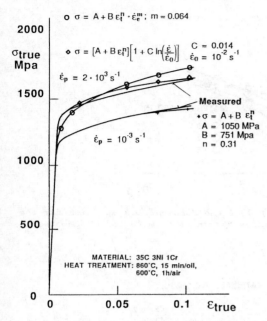

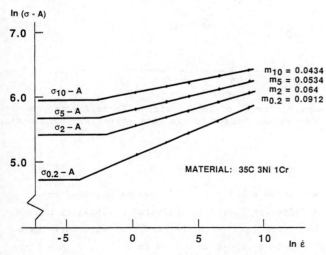

FIG. 5   *Comparison of measured material behaviour at low and high rate of strain with the prediction by Ludwik, double power and Johnson/Cook equations (material: HSLA-steel 35C 3Ni 1Cr).*

FIG. 6   *Strain rate sensitivity m of the double power equation for the .2, 2, 5 and 10 % flow-stress versus strain rate (Material: HSLA steel 35C 3Ni 1Cr).*

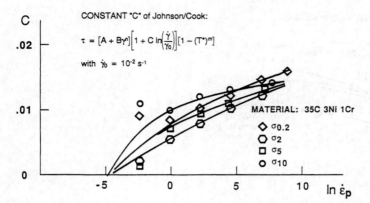

FIG. 7   *Strain rate sensitivity C of the Johnson/Cook equation for the*
.2, 2, 5 and 10 % flow stress versus strain rate (material: HSLA-steel
35C 3Ni 1Cr).

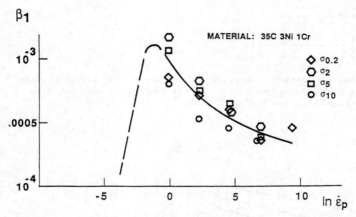

FIG. 8   *Strain rate sensitivity* β_1 *of the Armstrong/Zerilli equation for*
the .2, 2, 5 and 10 % flow stress versus strain rate (material: HSLA-steel
35 C 3Ni 1Cr).

yield stress of mild steel above $\dot{\epsilon} = 5 \cdot 10^3 \ s^{-1}$, was re-examined by
Nojima [38]. He analyzed different thermally activated dislocation mechanisms for carbon steel and found that, with the choice of structural parameters which are typical for these types of steels (m = 1/p = 2, n = 1/q =
2/3, H_0 = 0,62 eV), the yield stress of the carbon steels can be described
very well with the thermal activation equation (3.1). In addition, and

that is important here, he showed that even up to the highest strain rate of $\dot{\epsilon} = 4 \cdot 10^4$ s^{-1}, the flow stress of the 0.12 C-steel used by Campbell/ Ferguson is properly modeled with an equation of pure thermal activation without the inclusion of drag effects [38]. The only difference is seen at T=195 K above $\dot{\epsilon} = 10^0$ s^{-1}, where creations of twinning might have occurred.

Other examples are given by Follansbee [42] with the non-increasing strength at $\dot{\epsilon} = 10^4$ s^{-1}, when the stress is measured under the condition of a constant threshold stress or a constant microstructure, or by Armstrong et al. [43] with the aforementioned alpha-uranium, where the measured rate sensitivity at $\dot{\epsilon} = 10^4$ s^{-1} is in accordance with a model description which is based solely on the thermal activation theory.

VI. SUMMARY

The semiempirical equations (eqs. 2.2 to 2.5) on the basis of the Ludwik expression separate the influence of strain rate and temperature from the strain hardening. In reality, the influences are coupled, especially in bcc materials. To make the equations agree with experimental results, the constants often have to be adjusted for the desired area of validity. The advantage of these simple equations is their easy implementation in computer codes.

Armstrong/Zerilli and Klepaczko improved the description by distinguishing between fcc and bcc materials and by introducing temperature-dependent coefficients. In both cases, the number of constants increased, with Klepaczko up to 12 or 15, but an important step towards a meaningful physical description has been accomplished.

The Bodner/Partom model uses less coefficients and has the possibility of being applicable to different loading conditions, including high rates of strain. Follansbee introduced the concept of the threshold stress, and by a model for the evaluation of state of the material structure, he overcame the old limits of the thermal activation theory. Of course this can only be achieved through an intensive test program. But from the viewpoint of material characterization, this is acceptable. In order to document the material behavior under dynamic conditions, no one may believe that with a few tests the complexity of real materials can be evaluated satisfactorily. From the viewpoint of modeling and computation, the opinion of Alexander [44], with respect to metal forming processes can be adopted with the same intention for the high strain rate application: "Since even the classical theory requires sophisticated numerical computational techniques for the solution of all but the simplest of geometrical configurations, it seems worthwhile examining how [the procedure] can be modified to include the important effects [of real material behavior]."

Today this sounds more like an understatement. Let us assume, that everywhere the need for an accurate material description is accepted. The improvements in understanding will be worthwhile.

Acknowledgment

The assistance of the Fraunhofer-Gesellschaft München, the German Ministry of Defense, Bonn, both Germany, the Center of Excellence for Advanced Materials, and the Department of Applied Mechanics and Engineering, both at the University of California, San Diego, California, USA is greatly appreciated.

REFERENCES

1. U.F. Kocks, in Unified Equations for Creep and Plasticity, A.K. Miller (ed.), Elsevier Appl. Sci., N.Y., 1 (1987).

2. P. Ludwik, *Phys. Z.*, 10: 411 (1909).

3. R.W. Klopp, R.J. Clifton, T.G. Shawki, *Mech. Mat.*, 4: 375 (1985).

4. J. Litonski, *Bull. Acad. Pol. Sci.*, 25: 7 (1977).

5. T. Vinh, M. Afzali, A. Roche, in Mechanical Behavior of Material, Proc. ICM, K.J. Miller and R.F. Smith (eds.) Pergamon Press, Oxford & New York, 633 (1979).

6. G.R. Johnson et al., *J. Eng. Mat. Techn.*, 105: 42,48 (1983).

7. G.R. Johnson, W.H. Cook, in Proc. 7th Intl. Symp. on Ballistics, The Hague, 541 (1983).

8. F.J. Zerilli, R.W. Armstrong, *J. Appl. Phys.*, 61, No.5: 1816 (1987).

9. R.W. Armstrong, F.J. Zerilli, *J. d. Physique*, 49: C3-529 (1988).

10. F.J. Zerilli, R.W. Armstrong, in Shock Waves in Condensed Matter, S.C. Schmidt and N.C. Holmes (eds.), Elsevier Publ., Amsterdam, 273 (1988).

11. P.W. Follansbee, G. Regazzoni and U.F. Kocks, in 3rd Conf. on Mechanical Properties at High Rates of Strain, J. Harding (ed.), Inst. Physics, London, Conf.Ser.No.70: 71 (1984).

12. W.H. Gourdin, in Shock Waves in Condensed Matter - 1987, S.C. Schmidt, N.C. Holmes (eds.), Elsevier, Amsterdam, 351 (1988).

13. J.D. Campbell, A.M. Eleiche, and M.C.C. Tsao, in Fundamental Aspects of Structural Alloy Design, Plenum Publ. Corp., New York, 545 (1977).

14. O. Vöhringer, "Deformation Behaviour of Materials", Proc. Intl. Summer School, Univ. Nantes (France), Sept. 1989.

15. U.F. Kocks, *ASME J. Engrg. Mater. Tech.*, 98: 76 (1976).

16. P.S. Follansbee, U.F. Kocks, *Acta Metall.*, 36: 81 (1988).

17. P.S. Follansbee, in Proc. IMPACT'87 - Impact Loading and Dynamic Behavior of Materials, C.Y. Chiem, H.-D. Kunze, and L.W. Meyer (eds.), DGM Informationsges., Oberursel/Frankfurt, Vol. 1: 315 (1988).

18. P.S. Follansbee, in Shock Waves and Condensed Matter, S.C. Schmidt and N.C. Holmes (eds.), Elsevier Publ., New York, 249 (1988).

19. P.S. Follansbee, G.T. Gray III, *Met. Trans. A*, 20A: 863 (1989).

20. H. Mecking, U.F. Kocks, *Acta Metall.*, 29: 1865 (1981).

21. A. Kumar, F.E. Hauser, J.E. Dorn, *Acta Metall.*, 16: 1189 (1968).

22. K.G. Hoge and A.K. Mukherjee, *J. Mat. Sci.*, 12: 1966 (1977).

23. F. Burgahn, O. Vöhringer, E. Macherauch, Proc. Explomet 1990, this volume

24. J.R. Klepaczko, *J. Mechanical Working Technology*, 15: 143 (1987).

25. J.R. Klepaczko, P. Lipinski, and A. Molinari, in Proc. IMPACT-87 - Impact Loading and Dynamic Behavior of Materials, C.Y. Chiem, H.-D. Kunze, and L.W. Meyer (eds.), DGM Informationsgesellschaft, Oberursel/Frankfurt, Vol. 2: 695 (1988).

26. D.J. Steinberg, S. Cochran, and M. Guinan., *J. Appl. Phys.*, 51: 1498 (1980)

27. D.J. Steinberg and C.M. Lund, *J. de Physique*, 3, Sup.No.9, T49: C3-433 (1988).

28. S.R. Bodner and Y. Partom, *J. Appl. Mech.*, 42: 385 (1975).

29. S.R. Bodner, in Unified Equations for Creep and Plasticity, A.K. Miller (ed.), Elsevier, N.Y., 273 (1987).

30. A.M. Rajendran, S.J. Bless, and D.S. Dawicke, *J. Eng. Mat. Techn.*, 198: 75 (1986)

31. S.R. Bodner, in Proc. IMPACT'87 - Impact Loading and Dynamic Behaviour of Materials, C.Y. Chiem, H.-D. Kunze, and L.W. Meyer (eds.), DGM Informationsges., Oberursel, Vol. 1: 77 (1988).

32. C.R. Hoggatt and R.F. Recht, *Exp. Mech.*, 9: 441 (1969).

33. C.H. Li, Ph.D. Thesis, Brown University, Providence, RI (1982).

34. P.S. Follansbee, in 4th Conf. on Mechanical Properties at High Rates of Strain, J. Harding (ed.), Inst. Physics, London, Conf.Ser.No. 102: 213 (1989).

35. K. Stiebler, 4th Conf. on Mechanical Properties at High Rates of Strain, J. Harding (ed.), Inst. Physics, London, Conf.Ser.No.102: 181 (1989).

36. L.W. Meyer, E. Staskewitsch, in Proc. IMPACT'87 - Impact Loading and Dynamic Behaviour of Materials, C.Y. Chiem, H.-D. Kunze, and L.W. Meyer (eds.), DGM Informationsges., Oberursel/Frankfurt, Vol.1: 331 (1988).

37. R. Lagneborg, B.-H. Forsen, *Acta Met.*, 21: 781 (1973).

38. T. Nojima, in Proc. IMPACT'87 - Impact Loading and Dynamic Behaviour of Materials, C.Y. Chiem, H.-D. Kunze, and L.W. Meyer (eds.), DGM Informationsges., Oberursel/Frankfurt, Vol. 1: 357 (1988)

39. L.W. Meyer, 3rd Conf. on Mechanical Properties at High Rates of Strain, J. Harding (ed.), Inst. Physics, London, Conf.Ser.No. 70: 81 (1984)

40. T. Nicholas, *Exp. Mech.*, 138: 177 (1981).

41. J.D. Campbell and W.G. Ferguson, *Phil. Mag.*, 21: 63 (1970).

42. P.S. Follansbee, in Metallurgical Applications of Shock-Wave and High-Strain-Rate Phenomena. L.E. Murr, K.P. Staudhammer, and M.A. Meyers (eds.), M. Dekker, Inc., N.Y. 451 (1986).

43. R.W. Armstrong, P. Follansbee, and T. Zocco, in 4th Conf. on Mechanical Properties of Materials at High Rates of Strain, J. Harding (ed.), Inst. Phys. London, Conf.Ser.No. 102: 237 (1989).

44. J.M. Alexander, in Plasticity Today, A. Sawczuk, G. Bianchi (eds.), Elsevier Appl. Sci, Publ., London, 683 (1985).

5

Material Deformation at High Strain Rates

C. Y. CHIEM

Plasticité Dynamique, Laboratoire Matériaux
Ecole Nationale Supérieure de Mécanique
44072 Nantes Cedex 03, France

A brief review is presented of response of materials to various types of high strain rate loading.
Following collection of data from a large number of investigations, behavior of materials under the effect of strain rate and temperature will be discussed. This will include the correlation of mechanical properties to microstructures. Introducing micro-mechanisms of deformation will help to elucidate certain physical phenomena in this topic. Some models taking account of microstructural parameters will be discussed in order to clarify the range of applicability of each set of parameters. This will be discussed with various states of damage of materials.

I. INTRODUCTION

A wide range of engineering employ structural materials which are subjected to high or very high strain rate of loading such as impact of debris of aircraft composite panels or bird-strike on aero-engine, plastic flow close to fast propagating cracks, high speed metal-forming or machining processes and various military applications, etc. The way of approach in which this paper is developed is focalized in the illustration of the present state of research work in this domain.

Progress in technology is continuously challenged by requirements of new technical performance. These requirements offer motivations for the development of mechanical testing of materials and structures under severe environment including fast development of numerical tools. In the past, many keynote lectures or review papers[1-5] are involved with structural response to high strain-rate loadings; and recently, more and more review papers comment on the material behavior at high strain rate[6-12]. More recently, many non-metallic and composite materials including several reviews on these works have been published[13-15].

It is clear that most materials show a significant variation in the mechanical response underincreasing strain-rate loading or under temperature effects. Development of constitutive equations or equations of state which relate stress and strain to the usual conditions of strain rate and temperature allows techniques of numerical analysis to be applied in the modelling of structural design in function of known material properties. Recent progress shows the need to consider microstructural aspects or evolution in the material. Attempts have been made to introduce internal structure sensitive parameters and effects of initial structure and texture of the materials on the deformation mechanisms.

Extending those effects to high strain-rate processes becomes much more complex and the understanding of the dynamic response of materials is rendered much more difficult.

It is not possible for the time allotted to mention all the work realized by various scientists in this field of study. The choice is then made to describe this topic by selection of examples of important points and to refer to other sources cited in references to fill in the gaps.

II. HIGH STRAIN-RATE DEFORMATION OF METALLIC MATERIALS

A. *GENERAL CONSIDERATION*

Phenomenological theories including yield- and stability-criteria are not discussed in this review. We will only emphasize on deformation behavior of the materials and their metallurgical and physical effects. These latters are characterized mainly by the relationships between microstructures and residual mechanical properties and are resulted from the very complex correlations between stress, strain, stress and strain states, strain-rate and temperature. So, the major aim of doing experimental tests is to determine the stress required for a deformation to occur at strain-rate ε of the material of a specimen. This material has indeed a given structure s under fixed physical conditions such as temperature T and pressure P. By doing one of these parameters as a variable and the others as constant parameters, we set up a certain number of terms which help to describe as properly as possible a rather complex function which is:

$$\sigma = \sigma \ (\varepsilon, T, P, s) \tag{1}$$

This expression orientates mechanical modelling which has to describe changes in plastic stress. Mostly, mechanical tests are arranged to be under uni-axial stress or uni-axial strainconditions in order to simplify the expression and to facilitate the identification of the micromechanisms of deformation. Tendencies of the stress-strain behavior of materials versus strain-rate and temperature were discussed in [16,17].

This yields some constitutive equations which are developed by some authors referenced from [18] to [25].

It is clear that s from equation (1) is dependent on the initial structures of the materials (i.e., thermomechanical history prior to the test). The evolution of s during the test depends on the conditions of the test. This includes a range of microstructures such as planar dislocation arrays at lower strains which evolve into more dense arrays of dislocations at higher strains.

These different arrays can be composed of twin faults and α' martensite which occurs at the intersections of twin-fault bundles. The dislocation density changes are simply related to changes in stress (or strain) through the following kind of expressions:

$$\sigma = \sigma_o + K\sqrt{\rho} \tag{2}$$

$$\rho = \rho_o + M.\varepsilon \tag{3}$$

where σ_o, K and M are constants and ρ is the dislocation density (ρ_o is the initial dislocation density), $\sqrt{\rho}$ being the average length of dislocation-segments. Under these considerations, constitutive equations using microstructural approach must take into account microscopic parameters such as: dislocation densities, twin densities, the mean value of dislocation cell size, the mean value of grain diameter, the mean distance between twins, the volume fraction of dislocation cells and twin-faults respectively.

B. STRAIN-RATE SENSITIVITY

One of the major problems which was noticed in 1961 is the rapid increase in strain-rate sensitivity of aluminum tested at strain-rate higher than 10^3 s^{-1} [26]. Similar phenomenon has been shown with other materials [27-30]. In solid mechanics point of view, some authors attribute this increased rate sensitivity to an inertia effect due to the rapid increase of strain-rate, while in the physical point of view, this rate sensitivity is attributed to a sudden transition from a thermally activated mechanism (low velocity) to a linear viscous mechanism (very high dislocation velocities with phonon and/or electron damping effects). This difference of points of view is well discussed in [9]. Nevertheless, the existing data seem inconclusive to define the transition behavior of the stress from low and medium strain-rates to high strain-rates around 10^3 s^{-1}.

Figure 1 shows an example of this strain-rate sensitivity through results obtained from investigations on a 3% Si-Fe single-crystal [31] which is b.c.c. This sudden transition as we can notice in Fig. 2 should be predominantly due to rapid increase of twin density versus strain-rate. Dislocation multiplication contributes also to rate sensitivity but much more smoothly in the case of b.c.c. materials. For f.c.c. materials, most have a higher stacking fault energy and have less

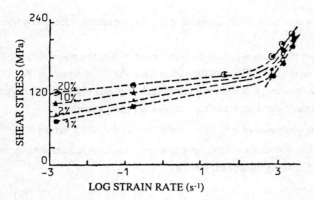

FIG. 1 *Strain-rate sensitivity of 3% Si-Fe single-crystals under high strain-rate shear loading [31].*

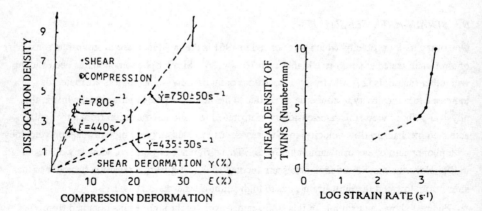

FIG. 2 *Results on uniaxial stress experiments of a 3% Si-Fe single-crystal subjected to rapid solicitations: dislocation density versus shear and compression strains at various strain-rates, linear density of twins versus strain rate [31] versus strain-rate.*

tendency to twin unless the temperature is decreased to very low value or the strain-rate is increased. At lower and medium strain-rates, f.c.c. materials as well as b.c.c. materials have more tendency to have dislocation cells as shown in Figs. 3 and 4.

C. *TWINNING AND DISLOCATION GLIDE*

Twinning is a rapid process which can occur only if pressure, shear stresses and temperature can reach together the required threshold conditions. In addition, crystallographic orientation, stacking fault energy, pulse duration, existing substructure and grain-size can also alter those conditions. The threshold stress of twinning [27] was proposed as:

$$\tau_c = \gamma_{fe} / nb \tag{4}$$

where γ_{fe} is the stacking fault energy of the materials, b, the Burgers vector of dislocations which generate the micro-twin and n, a stress concentration factor (1<n<3) as twinning is considered here as resulting from localized arrangement of dislocations. This localized phenomena can be observed macroscopically on the stress-strain curves of high strain-rate behavior of tungsten single-crystals shear-loaded in (112) planes as shown in Fig. 5. At high strain-rate, if all required conditions are satisfied, the stress concentration which is sufficient to generate twins is induced by microslip of dislocations in the early stage of deformation. Afterward, twinning occurs giving the serrated portion of the stress-strain curves [28].

Indeed, within the whole material, dislocation slip mechanism and twinning mechanism co-exist and are undergoing compromising arrangements in response to mechanical loading. Recent work [27] demonstrates this tendency as shown in Fig. 6 where the deformability of tungsten single-crystals is plotted versus ultimate shear stresses under three different strain-rates γ_1, γ_2, and γ_3 for the two tungsten single-crystals. The crystals are deformed at the initial predominance of either by slip of dislocations or by twinning. In the first case, the deformation augments with increasing strain-rate while in the case of twinning, it varies inversely with increasing strain-rate. This consequence shows the importance to take into account the rate of multiplication of dislocations and twins. By adding these two microstructural parameters to constitutive equations, the dynamic behavior of materials will be much more conveniently described.

To be consistent with this overall point of view, previous work in 3% Si-Fe single-crystals [25] proposed a derivation of the critical stress for twinning (Eq. 4) as follows:

$$\tau_{tw} = (1/n). \{ (\gamma_{fe}/b_s) + (G.b_s/2r) \} \tag{5}$$

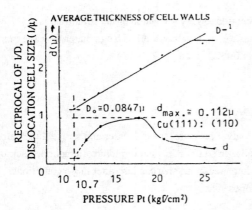

FIG. 3 *Reciprocal of dislocation cell size. 1/D and thickness of cell walls, d vs. pressure P_t induced by electromagnetic field in Cu single-crystal. The dislocation cell in the figure shows (---: mobile dislocations) and (___: dislocations at the cell walls) [31].*

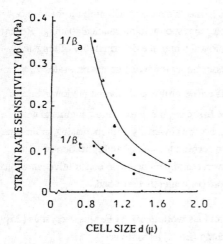

FIG. 4 *Apparent and true strain rate sensitivities as functions of dislocation cell size in Al single crystals [38,39].*

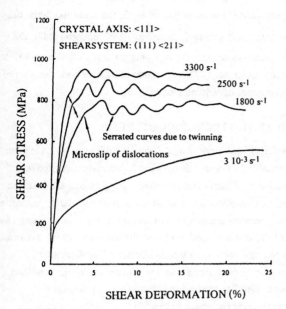

FIG. 5 *Stress-strain curves of a tungsten single-crystal of axis <111> subjected to various strain-rates of loading in shear. Serrated portion of the curves show twinning initiated by microslip of dislocations.*

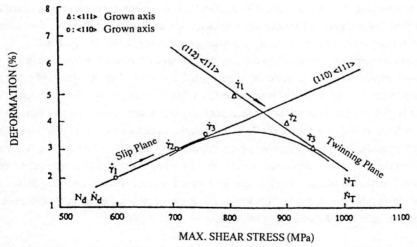

FIG. 6 *Deformability of the tungsten single-crystals with grown-axes <110> and <111> as a function of ultimate shear stress under three different strain-rates [34].*

with n as the coefficient of stress concentration at the tip of the twins, G, the shear modulus, b_s, the Burgers vector of screw-dislocations, γ_{fe}, the stacking-fault energy of the material and r, the radius of dislocation arc. This formula takes into account the twinning process as first components and the dislocation glide motion as second component. This approach allows for consideration of twinning as well as dislocation glide to be in permanent interactions.

III. STRAIN-RATE HISTORY EFFECTS AND MICROSTRUCTURES

History effects and temperature effects during high strain-rate deformation of materials have been studied in the aim of understanding material behavior and developing appropriate constitutive equations. It has been shown that two additional factors determining history effects are first, the lattice structure and then the development of the microstructure which is evolutive during straining. This supposes that two types of micromechanisms are involved in rate and temperature effects; one is the currently operating mechanisms of thermal activated motion of dislocations and the dynamic recovery of dislocations furnishing the physical "Storage-Flux-Annihilation" equilibrium of dislocation dynamics [35]. Experiments can be performed with a torsional Kolsky bar modified for simple shear loading of the specimens. Details of the test technique are mentioned in [38,39,55]. A recent review of history effects is presented in [56,57].

In this section only insight on the very few microstructural results are presented. The strain rate jump tests are generally viewed as particularly important in determining constitutive relations for dynamic plasticity because configurations are expected not to change significantly during the rise time of the jump in strain rate. Satisfactory correlation of experimental results with models [36,37] of the form suggested for thermally activated micromechanisms was shown. Results of incremental tests on aluminum single-crystals were presented in [38,39]. In Fig. 4 strain sensitivities are correlated to dislocation cell size; the later are important substructure related to dislocation densities and velocities, though effecting dislocation mobility. Saturation of cells formation induces a state of stability where all cells are having equal strength Burgers vectors. As reported in [40], the density of cell wall dislocations is approximately 50 times higher than the mobile dislocation density at the core of the cell. Evolution of dislocation cell wall thickness and cell size versus shock pressure generated by electromagnetic field [41] is shown in Fig. 3. One can notice the existence of a critical pressure for the cells to be nucleated and the cell wall thickness evolves to a maximum with pressure increase denoting a high level of work-hardening of the material. It was also shown that for high stacking fault energy metals, formation of dislocation cells and cross-slip of dislocations are easier.

IV. SHOCK INDUCED MICROSTRUCTURES AND MECHANICAL PROPERTIES

Microstructural observations of the recovered specimen of 3% Si-Fe single-crystals are done with TEM. In contrast to the normal microstructure after high strain rate testing, no twins are observed in the recovered crystal. Impact testing with explosives on 3% Si-Fe single-crystals shows a very large density of twins. It is shown previously that twinning occurs only if a critical stress is reached. It is clear that conditions for the nucleation of twins depend on five parameters: the incubation time of twinning, the pressure and duration of the shock, the stacking-fault energy of the material which induces the interactions between twinning and dislocation glide motions and the temperature. Also all these are dependent on dislocation velocities; this later can increase or decrease the threshold stress of twin nucleation.

For this material, twins and stacking faults are the main micromechanisms responsible for the rapid strain rate sensitivity transition at 10^3 s^{-1}. This seems to be in good agreement with the transition zone mentioned in [42]. The mention of "rapid evolution" at the rate sensitivity transition point of this figure is also consistent with twinning because this latter is a very fast and sudden process (Propagation occurs at sound velocity).

A. *DISLOCATION MOBILITY*

Relationship between the resolved shear stress on a slip plane and the velocity of mobile dislocations on that slip plane is fundamental for the development of constitutive equations in the field of dynamic plasticity. The uniaxial strain experiments performed by plate impact recovery tests are aimed to contribute a little in this direction. Details on these experiments are already described in [43] with work done on 3% Si-Fe single-crystals. Results of these investigations on dislocation velocities are plotted together with results of many investigators in Fig. 7 [44-52]. The curves exhibit a general tendency of stress dependence that implies various mechanisms. Two regimes are existing: (a) a low velocity regime which is strongly stress dependent and where thermally activated mechanisms are predominant; (b) a high stress regime where high velocity is linearly dependent on stress and viscous drag is the rate controlling mechanism.

The treatment of thermally activated motion of dislocations past barriers such as impurities and intersecting dislocations seems quite well established [51,52]. At very low stresses, it is much more appropriate to use a stochastic approach as proposed in [56]:

$$V = (L\tau/B).(U_o/kT)^{1/2} \exp(-U_o/kT) \qquad (6)$$

for the case $\tau \leq 0.65$ kT/br$_o$1 $\simeq \tau$. In this relation B is the damping constant, Uo, the maximum of the bonding energy, L, the distance travelled for each activation, r_o, distance between defect and the glide plane, 1, the dislocation segment length. At higher stresses where the dislocation velocity increases with increase in the resolved stresses. In this case the usual formula of the rate theory is only an approach:

$$V = Lv_o(\tau).\exp\{-\Delta G_{(\tau)}/kT\} \qquad (7)$$

case where $\tau \leq 0.65$ Uo/br$_o$1 and with a Debye frequency slightly dependent on stress, $v_{o(\tau)}$.

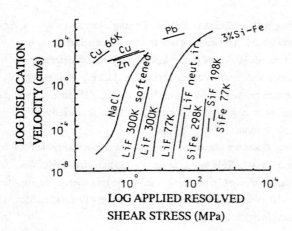

FIG. 7 *Dislocation mobility of various materials [38].*

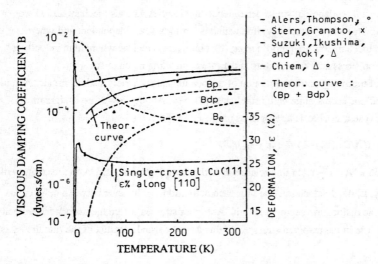

FIG. 8 *Temperature dependence of the viscous damping coefficient in copper single-crystal deformed by pulsed eectromagnetic field [41,53].*

At stress levels that are high enough to overcome the Peierls barrier associated with the periodic structure of the lattice, and low enough to cause dislocation velocities which are below elastic wave speeds for relativistic effects to be minimize, the resolved shear stress and the dislocation velocity then satisfy a linear viscous drag equation:

$$\tau b = BV_d \tag{8}$$

with the Burgers vector b as a lattice parameter that is equivalent of the displacement discontinuity of the dislocation and B is a drag coefficient. B is temperature dependent as shown in Fig. 8 [38,41,53]. At room or high temperature, interaction of thermal phonons and dislocations is responsible for viscous drag forces. When temperature is approaching the superconductivity transition point, mechanisms of electron drag become predominant. Phonon scattering is estimated to contribute to the drag coefficient. When materials are involved with very strong shock wave phenomena, evolution of the micromechanisms is also dependent on the type of waves as shown in Fig. 10[56]. We can summarize into four types of wave effects the metallurgical effects of the materials: *Behind the elastic waves, no imperfection; *Behind the plastic waves, generation of dislocations, stacking faults, twin boundaries and antiphaseboundaries; *Behind transformation waves (or plastic wave II), if the transformation $\alpha \Rightarrow \beta$ occurs, a new crystalline structure is formed containing the imperfections that were created by the preceding plastic wave and the transformation shear. If there is a reverse transformation, we have then ($\alpha \Rightarrow \beta \Rightarrow \alpha$); the imperfections are therefore due to double transformation by shear. It is difficult at the present to get better understanding in this field. More experimental work should be done at macroscopic level as well as at microscopic approach. Figure 9 shows the relationship between critical twinning pressure and stacking fault free energy for a number of f.c.c. metals and alloys as estimated and given from [57]. This again demonstrates the predominant effect of critical pressure (or stress) on the formation of twins or dislocation cells.

Many questions still exist in the region of very high dislocation velocities. Most of the work done has involved theoretical approaches regarding such phenomena as supersonic dislocations[41], the Smith shock interface [54] loading history dependent dislocation velocities[55] and dynamic nucleation of dislocations running at speeds higher than the Raleigh wave speed [62], etc. In contrast, very little work has been done in experiments of the dynamic of high velocity dislocation. This is mainly due to a lack of satisfactory means.

During the production of dislocations or other defects in the shock front, there is occurrence of heating. On the basis of work done on a solid during rarefaction, authors of ref.[58] obtained the following formula for the residual yield or flow stress of metals and alloys subjected to a "planar shock":

$$\sigma = \sigma_o + 2\alpha G\, b.\{\ C_R.\Delta T - \{(P_i + P_o).(V_o - V_i)\}/2 + \int_{V_i}^{V_f} P_{s(V)}\, dV\ \}^{1/2} \tag{9}$$

where σ_o, α, C_R are constants for any particular material, G is the shear modulus, b is the Burgers vector, ΔT is the temperature change in the material, V_o, V_i and V_f are the initial, intermediate and final volumes, and P_o and P_i are the initial and intermediate pressures. Actually P_i and V_i are intermediate values associated with the peak shock pressure (P), and $P_{s(V)}$ is the isentropic relief path (considering the hydrodynamic theory [63]).

V. SOME WORDS ON HIGH POLYMERS

In the past two decades the mechanical behavior of high polymers subjected to high strain-rate loading have been extensively studied in experimental as well as fundamental research field. The viscoelastic behavior of the polymers is well described in [64, 65]; the high strain-rate behavior os some polymers has been studied experimentally by the use of torsional impact machine [66] and the torsional Hopkinson-bar apparatus [67] in which a modified Norton-Hoff constitutive relation was proposed. In [68] and [69], the authors tend to relate the macroscopic behaviors of polymers to the rupture of molecular chains and the influence of the distribution of the chemical compositions has been studied in [70].

The polymers consist of long molecular chains of covalently bonded atoms. The arrangement of the molecular chains presents two states: - The amorphous state which is considered to be a random tangle of molecules and the crystalline state named by the high orientation of molecules. The observation by means of some microscopic equipments shows that the semi-crystalline polymers have the structure of spherulites or in modelisation, the paracrystalline structure (Fig. 10(a) and (b)). An attempt of constitutive equation modelling was done in [71]. In this model, the crystalline part and the amorphous part are assumed to be cubes assembled in parallel and according to the amount of deformation a generalized modulus is defined which takes into account the shear modulii corresponding respectively to the crystalline and the amorphous oarts of the polymer, also, the coefficient of crystallinity. Thermally activated mechanisms of dislocations are supposed to be actived in the crystalline part and a thermodynamic visco-plastic model is applied to the amorphous part. After the use of identification technique, a simplified form of the constitutive equation which is written as:

$$\tau = A\ \ln(1 + \alpha\gamma).\gamma^\beta.\{1 + (\Delta T^\theta/T_g)\} \tag{10}$$

where A is a constant of the material, α, the dislocation multiplication parameter, β, the strain-rate sensitivity, ΔT, the temperature increment during high strain-rate plastic straining due to localized

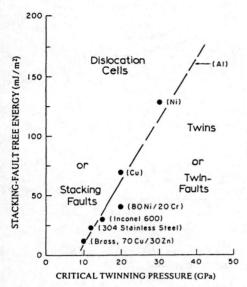

FIG. 9 *Critical twinning pressure versus stacking-fault free energy for a number of fcc metals and alloys. (Critical pressure values are estimated from shok-loading data of Ref. [57] stacking-fault free energy values are from [58]).*

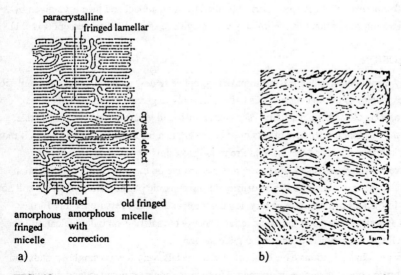

a)

b)

FIG. 10 *Microstructural model of high polymers: a) the model of paracrystalline structure; (b) an element of spherulite.*

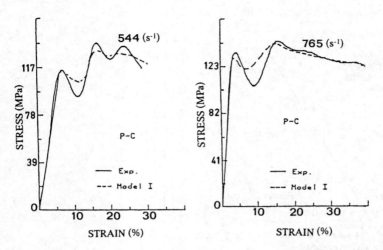

FIG. 11 *Attempt of modelisation on the stress-strain behavior of a polycarbonate PC deformed by dynamic torsion [71].*

adiabatic behavior of the polymers, T_g, the glassy transition temperature and θ, the stress sensitivity to temperature effects. All these parameters except A can be experimentally determined. An example of the model is applied to a polycarbonate [71] which is deformed at high strain-rate by torsion. Comparison of stress-strain curve obtained from the model and the curve given by experimental testing shows that this preliminary attempt gives a quite good agreement (Fig. 11).

VI. CONCLUSION

As previously mentioned, it is not possible to make a complete review under this topic and omission of results is expected; so the objective of this paper aims only to give a certain insight on the microscopic aspects of dynamic plasticity. The author wishes to give a certain warning in this occasion for the fact that tremendous efforts must be undertaken for microstructural studies if more powerful constitutive equations are wished. Progress is too slow in this field.

As the level of individual dislocations progress is sensitive on measurements of dislocation mobility but changes in mobile dislocation density during dynamic plastic deformation are still not understood. Influences of strain rate history, temperature, and shock pressure require further investigation as well as the role of the stacking fault energy of materials on twin nucleation, dislocation cell size formation and dislocation glide motion.

Until now most investigations have concentrated on metals, very few on single-crystals and almost none on composite materials or ceramics. Emphasis should be put to extend dynamic testing on the new trend of materials. Finally special care should be taken on the development of physically based models of a wide range of applicability and the achievement of this requires good understanding of the micromechanisms of deformation that enables further progress in this field of research.

ACKNOWLEDGMENT

The author is indebted to his colleagues and friends for efficient help in the realization of all the results necessary for this review. My thanks are addressed to the French Ministry of Defence for their continual support to the studies performed in our Laboratory.

The author is also indebted to Dr. K. P. Staudhammer for discussions on this paper and Ms. D. Vigil for her help in realizing this manuscript. Both are members of the Los Alamos National Laboratory. Ms. Faye Ekberg at the University of Texas at El Paso typed the final, camera-ready manuscript.

REFERENCES

1. N. Jones and T. Wierzbicki, Editors, Structural Crashworthiness, Butterworths Press, London and Boston (1983).
2. W. Johnson and A. G. Mamalis, Crashworthiness of Vehicules, 1st Ed., Mechanical Engineering Publications, London (1978).
3. S. R. Bodner, Proc. Oxford Conference on "Mechanical Properties at High Rates of Strain," 70:451 (1984).
4. P. S. Symonds, H. Kolsky, and J. M. Mosquera, Proc. Oxford Conference on "Mechanical Properties at High Rates of Strain," 70:479, 1984).
5. C. Y. Chiem, "Crash et Collision de Structures", Proc. "Entretiens Sciences et Défense", Ministère de la Défense, Délégation Générale pour l'Armenent, Paris Mai (1990).
6. J. Mescall and V. Weiss, "Material Behaviour under High Stress and Ultrahigh Loading Rates," Proc. Sagamore Army Materials Research", Conf. 29 (1982).
7. J. Duffy, Proc. Oxford Conference on "Mechanical Properties at High Rates of Strain," 47:1 (1979).
8. L. E. Malvern, Proc. Oxford Conference on "Mechanical Properties at High Rates of Deformation," 70:1 (1984).
9. C. Y. Chiem, "Proc. International Conference "IMPACT 87" I:57, Bremen (1987).
10. K. P. Staudhammer, Proc. International Conference "IMPACT 87", 1:93, Bremen (1987).
11. R. J. Clifton, A. Gilat, C. H. Li, "Metallurgical Behaviour under High Stress and Ultra-High Loading Rates," Plenum Publishing Corp. (1983).
12. L. E. Murr, Proc. "Materials at High Stran Rates", Ed. T. Z. Blazinski, Elsevier Appl. Sci., London and New York (1987).
13. J. Harding, "Materials at High Strain Rates," Elsevier Applied Science Publishers, London, 133 (1987).
14. C. Y. Chiem and Z. G. Liu, Proc. Euromech Colloquium 214 on "the Mechanical Behaviour of Composites and Laminates," Kupari, Yugoslavia, (1986).

15. B. Hwaija, C. Y. Chiem, and J. G. Sieffert, "Strain-Rate History Effects on the Mechanical Behaviour of Microconcrete, "European Conf. on Structural Dynamics, Bochum (FRG), (1990).

16. U. S. Lindholm, "Review of Dynamic Testing Techniques and Material Behaviour," Institute of Physics, Serie No. 21, 3, (1974).

17. I. N. Ward, "Mechanical Properties of Solid Polymers," J. Wiley & Sons, Ltd. (1979).

18. P. Ludwik, "Elemente der Technologischen Mechanik," Berlin (1909).

19. G. R. Johnson and W. H. Cook, 7th International Symposium on Baltistics, the Hague; April (1983).

20. U. S. Lindholm, J. Neck. *Phys. Solids,* 12:317 (1964).

21. L. E. Malvern, *J. Appl. Mech., 18*:203 (1951).

22. P. Perzyna, *Quant. Appl. Math., 20*:321 (1963).

23. D. J. Steinberg, S. G. Cochran, and N. W. Guinan, *J. Appl. Phys., 51*:3 (1980).

24. F. J. Erzrilli and R. W. Armstrong, *J. Appl. Phys., 30*:129 (1959).

25. L. W. Meyer, "Constitutive Equations at High Strain-Rate," this Proceeding, (1990).

26. F. E. Hauser, J. A. Simmons, and J. E. Dorn, "Response of Metals to High Velocity Deformation," Ed., P. G. Shewman et V. F. Zackay, 93 (1961).

27. J. D. Campbell, *Mater. Sci. and Eng.* 12-9 (1973).

28. J. F. Adler and V. A. Philips, *J. Inst. Metals* 83-80 (1954).

29. J. W. Edington, *Phil. Mag.,* 1741 (1969).

30. A. R. Rasenfield et G. T. Hahn, Trans. ASM, 963 (1966).

31. J. P. Dubois, P. Blinot, and C. Y. Chiem, *J. de Physique,* C5-13: 46 (1985).

32. J. Friedel, "Dislocations," Pergamon Press, London (1964).

33. W. S. Lee, "Corrélation entre la microstructure et les propriétés mécaniques de matériaux à base de tungstene soumis au cisaillement à grande vitesse," Thèse de Doctorat, E.N.S.M./Université de Nantes (France), June (1990).

34. C. Y. Chiem, and W. S. Lee, *Mater. Sci. and Engr.,* (to be published).

35. J. Klepaczko, Proc. Oxford Conf., 283 (1989).

36. J. R. Klepaczko and C. Y. Chiem: *J. Mech. Phys. Solids 34* (1986) 29.

37. J. R. Klepaczko, A. Rouxel, and C. Y. Chiem: *J. de Physique, 46* (1985) C5-29.

38. C. Y. Chiem and J. Duffy, *Matls. Sci. and Eng.* 57 (1983) 233.

39. C. Y. Chiem, Dr.-es-Sci. d'Etata, ENSM, Univ. de Nantes, Juin (1980).

40. J. Shioiri, K. Satoh, and K. Nishimura: Proc. "High Velocity Deformation of Solids," K. Kawata and J. Shioiri, (Eds.), Springerverlag, N.Y. (1978) 50.

41. C. Y. Chiem, Dr.-Ing. Thesis, ENSM, Nantes, Janvier (1976).

42. P. S. Follansbee, "Impact 87" (1987).

43. C. Y. Chiem, R. J. Cliffon, P. Blinot, and P. Kumar: Technical Report ENSM/ETCA NoP M/D G V/83/X-1 (1983).

44. K. M. Jassby and T. Jr. Vreeland: *Phil. Mag. 21* (1970) 1147.

45. V. R. Parameswaran and J. Weertman: *Met. Trans. 2* (1971)1233.

46. D. P. Pope and T. Jr. Vreeland: *Phil. Mag. 20* (1969) 1163.

47. E. Y. Gutmanas, E. M. Hadgornyi, and A. V. Stepanov: *Soviet Phys. Sol. State 5* (1963) 743.

48. D. W. Moon, and T. Jr. Vreeland: *J. Appl. Phys. 39* (168)1766.

49. W. G. Johnston, and J. J. Gilman: *J. Appl. Phys. 30* (1959)129.

50. J. E. Flinn, and R. F. Tinder: *Scripta Met. 8* (1974) 689.

51. U. F. Kocks, A. S. Argon, and M. F. Ashby: *Progr. Mat. Sci. 19* (1975), 1-288.

52. V. L. Indenbohm, Yu, Z. Estrin, In M. Balarin (Eds.): Reinststoffprobleme 5 (1977). Akademie-Verlag, Berlin, 683.

53. C. Y. Chiem, and M. Leroy, C. R. Acad. Sce., Paris 282 C(1976) 323.

54. C. Y. Chiem, and J. Duffy: Brown University Report NSF ENG75-18532/9 (1979).

55. J. Duffy: "Material Behaviour Under High Stress and Ultrahigh Loading Rates Sagamore Army Materials Research Conference Proceeddings, J. Mescall and V. Weiss (1982) 21.

56. E. Hornborgen, "High Energy Rate Working of Metals," Sandsfjord and Lillehammer (Eds.), NATO, 14-25 Sept. (1964) 345.

57. L. E. Murr, "Shock Wave and High Strain Phenomena in Metals," ed. L. E. Murr and M. A. Meyers Plenum Press, NY, 1981, pp. 607.

58. L. E. Murr, and M. A. Meyers, "In Explosive Welding, Forming and Compaction," ed T. Z. Blazynski, Applied Science Pub., London 1983, p. 83.

59. J. Weertman: "Metallurgical Effects at High Strain-Rate," N. Y., Plenum (1973) 319.

60. C. S. Smith: *Trans. Met. Soc. AIME 212* (1958) 574.

61. Y. Y. Earmme, and J. H. Weiner: *J. Appl. Phys. 48* (1977)3317.

62. J. H. Weiner, and M. Pearl: *Phil. Mag. 31* (1975) 679.

63. J. M. Walsh, and R. H. Christian, Phys. Rev. 97 (1955) p. 1544.

64. J. D. Ferry "Viscoplastic properties of polymers" New York, London, Sydney, Toronto, 1970.

65. I. M. Ward "Mechanical properties of solid polymers" - John Willey & Son, LTD 1971.

66. T. Vinh and T. Kha "Adiabatic and viscoplastic properties of some polymers at high strain and high strain rate" - Conference Series n°70. Institute of Physics Bristol and London 1984.

67. C. Y. Chiem, J. P. Ramousse and Z. G. Liu. "Le comportement d'adhésifs et de matières plastiques soumis aux grandes vitesses de déformation". Rapport technique de l'E.N.S.M. n°CRB-85-AP/F 1985.

68. E. H. Andres "Developments in polymer fracture-1" Applied Science Publisher 1979.

69. H. H. Kausch "Polymer-fracture" Springer-Verlag Berlin Heidelberg, New York 1978.

70. D. W. Von Krevelen "Properties of polymers" - Elsevier 1976.

71. C. Y. Chiem and M. Y. Liu, Proc. Of the "International Symposium on Intense Dynamic Loading and its effects 574-583, Ed. Science Press, Beijing, China (1986).

6

Microstructure and Fracture During High-Rate Forming of Iron and Tantalum

M. J. WORSWICK,* N. QIANG,† P. NIESSEN,† and R. J. PICK†

*Defense Research Establishment Suffield
Medicine Hat, Alberta, Canada, T1A 8K6

†Department of Mechanical Engineering
University of Waterloo
Waterloo, Ontario, Canada, N2L 3G1

The evolved microstructure and damage resulting from high strain rate deformation of high purity iron and tantalum have been studied in Taylor cylinder specimens and Explosively Formed Projectiles (EFPs). In the iron, deformation occurred through slip and mechanical twinning while fracture was through adiabatic shear localization and ductile fracture along these shear bands. The tantalum exhibited remarkable ductility and very low hardening rates. Deformation was entirely by slip leading to necking and ultimate chisel-type rupture. TEM work showed the deformed specimens to have a very complex cell structure with numerous dislocation loops.

* Current address: Department of Mechanical and Aerospace Engineering, Carleton University, Colonel By Drive, Ottawa, Ontario, Canada, K1S 5B6

I. INTRODUCTION

The design of optimal high-rate metal deformation processes depends on the determination of both accurate material constitutive relations and failure criteria. As a first step in the determination of failure criteria, a study of the deformation and fracture mechanisms in high purity iron and tantalum was undertaken. Two high rate deformation processes were considered: the Taylor cylinder and the Explosively Formed Projectile (EFP). In the Taylor cylinder experiments, specimens of Armco Iron and tantalum were impacted at velocities sufficient to cause damage and then sectioned for metallographic examination. EFPs fabricated from Armco Iron were studied in order to compare the deformation mechanisms under explosive loading to those experienced in the Taylor test.

II. EXPERIMENT

The Taylor specimens were 9 mm in diameter and 33 mm in length. For the iron specimens, a symmetric configuration was used in which one cylinder was fired into an identical, stationary target cylinder. The impact velocities of 420 and 525 m/s were sufficient to cause damage. The tantalum cylinders did not remain symmetrical after impact since flow instabilities developed. Consequently, they were fired against a hardened steel anvil at velocities in the range 182-316 m/s.

Explosively Formed Projectiles (EFPs) were fabricated from Armco Iron. These devices consist of a dish shaped metal liner backed by high explosive. Upon

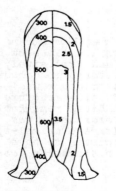   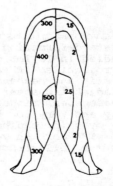

(a)   8 mm Wave Shaper   *(b)   16 mm Wave Shaper*

FIG. 1 *Predicted final shapes of iron EFPs. Contours of temperature(°C) and equivalent plastic strain are plotted on left and right, respectively.*

detonation of the explosive, the liner is formed into an elongated projectile. A tubular PVC "wave shaper" was incorporated into the explosive fill to tailor the propagation of the detonation wave and the liner deformation. Shorter wave shapers, 8 mm in length, promoted the formation of solid cross-section projectiles (Fig. 1a), while longer wave shapers, 16 mm in length, caused the liner to fold backwards during deformation to form a thin-walled, hollow projectile (Fig. 1b).

The deformed Taylor cylinders and EFPs were caught in a soft recovery pack, sectioned along their longitudinal axes, metallographically prepared and examined under optical microscopes. Scanning electron microscopy (SEM) and transmission electron microscopy (TEM) were also used to reveal submicron damage.

III. RESULTS

Prior to discussing the metallographic observations, it is worth examining results from hydrocode simulations of the cylinder impacts and EFP formations. The EFP calculations were performed using DYNA3D while the cylinders were modelled with EPIC-2. Figure 1 plots the predicted final shapes of two EFP designs and shows contours of equivalent plastic strain and temperature. Plastic strains in the range 100-300% are predicted causing a temperature rise of roughly 500-600°C. Deformation was more severe in the solid EFP as indicated by the higher strains and temperatures. The largest strains occurred along the axis within the neck region of both projectiles. The simulations also predicted that strain rates were initially in excess of 10^5 s^{-1} and decayed to levels of 10^4 s^{-1} within 50 μsec.

Strains and temperatures equal to those experienced by the EFPs were predicted for the iron Taylor cylinders; however, the regions of high strain were very small and were concentrated at the impact face on the cylinder axis. The strain rates also reached 10^5 s^{-1} but these levels were sustained for only 10 μsec.

A. ARMCO IRON CYLINDERS

Deformation occurred in the iron cylinders through dislocation slip and twinning. Voids were nucleated around the cylinder axis, close to the impact face. A tensile hydrostatic stress occurs in this region due to the radial expansion associated with the cylinder mushrooming.

Voids nucleated at grain boundaries and joined to form cracks as seen in Fig. 2. There was no indication that the nucleation of the voids required the presence of precipitates. It is likely that a viable void forms by accumulation of vacancies which are created by the temperature rise associated with the sudden highly localized deformation. That void formation near the impact face occurs by a

FIG. 2   *Cracks in Armco Iron (v = 525 m/s).*

vacancy mechanism rather than microcracking is indicated by the fact twin
intersections were never found to be associated with voids and cracks.

B.   TANTALUM CYLINDERS

The tantalum displayed excellent ductility, very high toughness and low strain
hardening. Plastic flow during impact was localized in the immediate vicinity of the
impact face (Fig. 3). This behaviour is indicative of an extremely low strain
hardening coefficient. Indeed, in room temperature tensile testing of this material
[2], the strain hardening exponent was only 0.08 at a strain rate of 0.042 s^{-1}. The
grains at the impact face were heavily deformed (Fig. 4); a rough estimation
suggested that the strain close to the impact face was approximately equal to 4.

FIG. 3 *Tantalum Cylinder
Impacted at 260 m/s.*

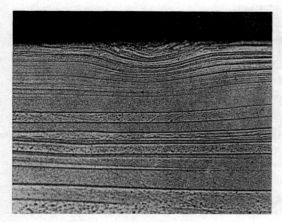

FIG. 4 *Heavily deformed*
tantalum grains at the impact
face (v = 260 m/s).

Surprisingly, even at this strain, no cavitation or cracking was detected using the optical microscope. Occasional grain boundary cavities were observed using the SEM. Through observation of the temper colours on the impact face, the temperature was estimated to have exceeded 450°C.

The extraordinary resistance to cavitation in tantalum is believed to be due to the material's high purity and high melting point. Due to the suppression of void nucleation, the impact specimens deformed without fracture into a "mushroom" shape. Only in the specimen that had been impacted at 316 m/s did the rim of the "mushroom" thin out sufficiently for fracture to occur. This fracture of heavily deformed tantalum was a chisel-type rupture. SEM examination of the area ahead of one crack tip revealed that massive slip bands in the necked region give rise to surface cracks; however, these cracks are quite shallow and open up to become part of the deep neck that develops into the chisel type failure.

To further examine the unusual deformation mechanism of tantalum, preliminary TEM studies were undertaken. In foils taken 0.5 mm from the impact face, very pronounced cell formation was observed. As shown in Fig. 5, the cell centres are almost free of dislocations that have now arranged themselves tightly in cell walls. Figure 6, taken from another grain, shows a cell structure with predominantly straight cell walls. A striking feature of the deformation structure is the presence of what appear to be dislocation loops (Fig. 6). Attempts are presently being made to verify these to be loops. Since the material is free of precipitates one must assume that these loops are generated by point defects such as collapsed vacancy disks. No twins were detected in any of the foils.

FIG. 5 *Cell structures in tantalum (0.5 mm from impact face,
197 m/s), 39000x.*

FIG. 6 *Cell structures in tantalum (0.5 mm from impact face,
v = 197 m/s), 39000x.*

C. *ARMCO İRON PROJECTILES*

Figure 7 shows cross-sections of the soft recovered Armco Iron EFPs. Examination
of the figure reveals a trend for the projectiles to become hollow and to elongate as
the wave shaper size increases, in agreement with the hydrocode predictions in
Fig. 1. Fracture occurred in the 16 mm wave shaper design and a large crack was

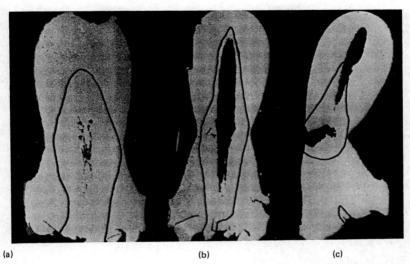

(a)   (b)   (c)

FIG. 7 *Profiles of Armco Iron EFPs. (a)   8 mm WS.*
(b)   12 mm WS.   (c)   16 mm WS.

observed in the 12 mm wave shaper design. These fractures were not predicted by
the model which neglects damage.

The deformation modes in the EFPs were the same as in the Armco Iron
cylinders, namely dislocation slip and twinning. Deformation generally increased
with increasing distance from the outermost surface of the EFPs with the most
severe deformation occurring in the central region, in qualitative agreement with the
predicted strains in Fig. 1. Recrystallization occurred in the centres of the EFPs,
due to the large amount of heat generated during the forming process. The
approximate boundaries of the 100% recrystallized zones are shown in Fig. 7. These
zones correlate well with the isotherms plotted in Fig. 1, with the recrystallization
boundaries lying near the 500°C isotherm. Data by Goldthorpe et al [2] shows that
recrystallization in cold worked iron occurs in approximately 10 minutes at 600°C.
In the EFPs the time for recrystallization would be very short and the temperature
lower than 600°C, yet, complete recrystallization occurred. This indicates that the
driving energy for recrystallization in deforming EFPs is very high.

Damage in the form of cavitation and adiabatic shear localization was
observed within all of the projectiles. Cavitation was predominantly intergranular
and was observed throughout the projectiles; the highest porosities occurred in the
neck region and reached 1-2%. The voids were spheroidal in the projectile nose and
were considerably elongated in the neck region. Fracture occurred in the 16 mm
wave shaper design and a large crack was observed in the 12 mm design, running at
34° to the longitudinal axis of the projectile. Fracture occurred through shear

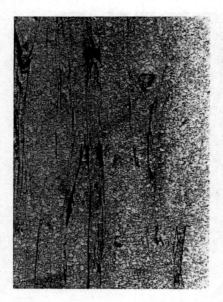

FIG. 8 *Shear bands developed in the centre of the solid core
projectile, 75x.   The vertical direction is aligned with the
projectile longitudinal axis.*

localization across the projectile wall causing coalescence of individual voids to form
a crack.

    Adiabatic shear bands developed in all three projectiles but were most
prevalent near the central region of the solid projectile (Fig. 8). As seen in Fig. 1,
this region experienced the highest strains and temperatures. Cracking through
adiabatic shear was not observed in the Taylor specimens, presumably due to the
lower strains and temperatures throughout most of the specimens.

IV. DISCUSSION

The hydrocode predictions have revealed that the maximum strains and
temperatures within the Taylor test and EFPs are similar; however, the EFPs see
larger gross deformation with all of the material deformed to strains exceeding
100%. The substantially higher plastic work and temperature rise in the EFPs lead
to larger recrystallized zones than in the iron cylinders. The larger high temperature

zones also lead to adiabatic shear bands within the EFPs. The most extensive network of adiabatic shear bands occurred in the solid EFP which experienced higher temperatures and strains than the hollow projectile.

Apart from the adiabatic shear bands, no other significant differences were observed between the deformation modes within the iron cylinders and EFPs. Both specimens deformed through dislocation slip and twinning and cavitation was intergranular.

Very few voids nucleated within the tantalum even as impact speeds reached 316 m/s. It was found that slip was the only deformation mechanism and the microtwins claimed by Moser et al [3] were not observed in this study. The present results agree with Bassett and Baskin [4] who reported that in impact deformation of 99.9% pure tantalum, twins were produced at -196°C but not at -77°C or 25°C. Because twinning requires high initiation stresses, the easy slip and low working hardening characteristics of tantalum prevent the attainment of a critical twinning stress.

V. REFERENCES

1. Pick, R.J. and Niessen, P., *Study of Material Behaviour at High Rates of Strain,* Final Report, SSC W7702-8-R009/01-SG, University of Waterloo, Canada, April, 1990.

2. Goldthorpe, B.D., Andrews, T.D. and Hogwood, M.G., The Structure, Deformation, and Thermal History of Explosively Formed Projectiles, NATO Defence Research Group, *28th Seminar on Novel Materials for Impact Loading,* Vol. 1, Bremen, Germany, ed. H.D. Kunze, Sept, 1988.

3. Moser, K.D., Chatterjee, T.K. and Kumar, P., The Effects of Silicon on the Properties of Tantalum, *JOM,* Oct, 1989, p50.

4. Miller, G. L., *Tantalum and Niobium,* Butterworths, London, 1959, p373.

7

High-Velocity Tensile Properties of Ti-15V-3Cr-3Al-3Sn Alloys

N. TAKEDA and A. KOBAYASHI

Research Center for Advanced Science and Technology
The University of Tokyo
4-6-1 Komaba, Meguro-ku, Tokyo 153, Japan

Department of Materials Science
The University of Tokyo
7-3-1 Hongo, Bunkyo-ku, Tokyo 113, Japan

Beta titanium (β-Ti) alloys, Ti-15V-3Cr-3Al-3Sn, were tested in high velocity tension up to the strain rate of 10^3 /s using "one bar method". It was found that the ultimate strength σ_{max}, the total elongation ε_t, and the absorbed energy E_{ab} increased significantly with increasing strain rate. The detailed microscopic observations revealed the microfracture mechanisms to explain the impact-energy absorbing capability of this material. Quasi-cleavage fractures in quasi-static loadings was mainly due to fine α-phase acicular structures near grain boundaries. Ductile fractures in impact loadings, on the other hand, was due to plastic deformation inside the prior β-phase grains.

I. INTRODUCTION

Titanium alloys have been used for many aerospace structures. While α-β alloys, such as Ti-6Al-4V, are the most popular at present, the desire to reduce total manufacturing costs has led to R & D projects aimed at developing a superior titanium sheet alloy [1]. β-alloys, which have BCC (body-centered cubic) structure, have been extensively studied for such purposes. Ti-15V-3Cr-3Al-3Sn (Ti-15-3, in short) is a cold-strip producible, cold forma-ble β-alloy designed to reduce both processing and fabrication costs [2]. It is also design-ed to offer high performance through improved mechanical properties.

Charpy or Izod tests are the most common impact tests, but the difficulty in inter-preting test data prevents us from quantitative evaluation of materials. This is due to the stress complexity in a short-beam specimen. Impact tests in simple stress or strain state (tension, compression or shear) are desirable because those data are applicable in designing real impact-resistant structures [3]. Such tests have been applied to pure-Ti [4], Ti-6Al-4V [5-9] and Ti-6Al-2Sn-4Zr-2Mo [10]. However, no data are available for Ti-15-3 alloys.

In the present study, high-velocity tension was achieved using the "one bar method" developed at the University of Tokyo [11]. Dynamic stress-strain curves up to final fracture were obtained for Ti-15-3 alloys up to the strain rate of 10^3/s. Microfracture mechanisms were then investigated using detailed scanning electron micrographs (SEM).

II.   EXPERIMENTAL PROCEDURE

A.   *TESTED MATERIALS*

A wide variety of microstructures due to phase transformations and attendant property variations can be achieved in Ti-15-3 alloys simply by changing processing conditions. The Ti-15-3 alloy used in the present study was hot-rolled, solution-treated and then aged to obtain balanced strength and toughness. Table 1 summarizes the chemical composition, processing conditions and resultant static mechanical properties at room temperature. An all-β structure just after the solution treatment is metastable and precipitates a fine α-phase acicular (HCP, hexagonal closed packed) structure upon subsequent aging. The aging was

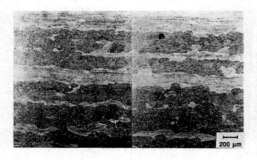
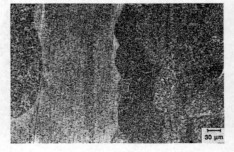

(a)   (b)

FIG. 1   *Specimen microstructure.*

almost completed in this material. The 0.2 % proof limit and the critical COD changed from 960 to 1150 MPa, and from 0.05 to 0.027 mm, respectively, after the aging.

The specimen microstructure after surface polishing and etching is shown in Fig. 1. The average grain size is approximately 80 μm in diameter. Some grains appear to have aged quickly than others. This effect is produced by some grains being incompletely re-crystallized during the anneal cycle, thus retaining some stored energy to accelerate the aging process resulting in a more rapidly aged grain [1]. In particular, fine α acicular microstructures are precipitated near prior β grain boundaries or sub-boundaries (Fig.1(b)).

B. EXPERIMENTAL APPARATUS − "ONE BAR METHOD"

The testing system of "one bar method" is shown in Fig. 2 and consists of a hammer, an impact block, a specimen and an output bar. This loading system is free from spurious oscillation effects in dynamic stress-strain curves such as seen in a short load-cell system. Since the details are given in Refs. 3 and 11, a brief summary is given here. The impact block is impacted by a hammer attached to a rotating wheel when the wheel reaches the desired speed (50 m/s maximum). A stress wave generated in the specimen propagates into the long output bar. The strain $\varepsilon_g(t)$ of the output bar is measured using semiconductor strain gages (4 in circumference) at a distance a from the impact end of the output bar, before the reflected wave reaches the strain gages. The impact-block velocity $V(t)$ is also measured electro-optically. The stress and strain of the specimen are calculated using the one-dimensional stress wave theory as follows:

$$\sigma(t) = (S_0/S)\, E_0\, \varepsilon_g(t+a/c), \qquad \varepsilon(t) = (1/l) \int_0^t [V(\tau) - c\, \varepsilon_g(\tau)]\, d\tau \qquad (1)$$

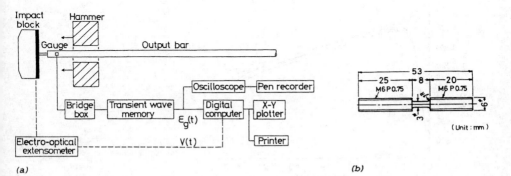

(a)   (b)

FIG. 2 *Testing system of "one bar method". (a) Schematic of measuring system. (b) Specimen dimensions.*

where l and S are length and cross-sectional area of the specimen, and S_0, E_0 and c are cross-sectional area, Young's modulus and elastic wave velocity of the output bar. A personal computer system was established to obtain stress-strain curves up to the strain rate of 10^3 /s. This testing system was found to give precise dynamic stress-strain curves up to final fracture for a variety of materials [10].

III. EXPERIMENTAL RESULTS

Quasi-static (strain rate: 10^{-3} /s) and dynamic (10^3 /s) stress-strain curves are shown in Figs. 3 and 4, respectively. The characteristic values such as the maximum stress or ultimate strength σ_{max}, the total elongation ε_t , and the absorbed energy E_{ab} are summarized in Fig. 5. All the three values significantly increase as the strain rate increases. The

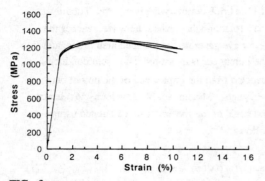

FIG. 3 *Typical static stress-strain curves.*

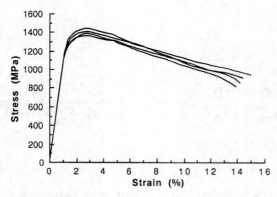

FIG. 4 *Typical dynamic stress-strain curves.*

ratio of each characteristic value at 10^3 /s to that at 10^{-3} /s is 1.11 in σ_{max}, 1.40 in ε_t, and 1.38 in E_{ab}. Thus the Ti-15-3 alloy gives highly desirable properties in high-velocity tension. The specimen photos after final fracture are shown in Fig. 6. The specimen in high-velocity tension shows higher elongation and larger reduction of area than that in quasi-static tension.

There are only a few data available for the strain-rate effect on mechanical properties of β-Ti alloys. Rosenberg [2] found that Ti-15-3 in the anneal condition was quite insensitive to the strain rate ranging from 10^{-4} to 2×10^{-3} /s, which was much lower than that in the present study. Several researchers noticed a high strain-rate sensitivity of the yield stress and flow stress of pure-Ti [4] and Ti-6Al-4V [5-9] in the strain rate from 10^{-3} /s to 10^3 /s, but the total elongation decreased or remained constant. Kawata *et al.* [10] tested Ti-6Al-2Sn-4Zn-2Mo (α+β type) using a similar testing apparatus, and found a noticeable increase in σ_{max}, but not in ε_t.

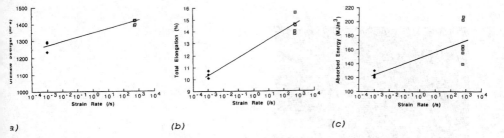

a)   (b)   (c)

FIG. 5 *Strain-rate dependence of characteristic values. (a) σ_{max}.*
(b) ε_t.   (c) E_{ab}.

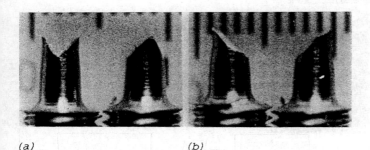

(a)   *(b)*

FIG. 6 *Photos after final fracture.   (a) Static.   (b) Dynamic.*

IV.   MICROSCOPIC OBSERVATIONS AND MICROFRACTURE MECHANISMS

The detailed SEM observations revealed some of the fundamental microfracture mechanisms governing the macroscopic fracture behavior. Effects of the strain rate on fracture
surface appearances can be observed in Figs. 7 to 10.

    In quasi-static fracture surfaces, the specimen center portion (Fig. 7(a)) exhibits
highly irregular quasi-cleavage fracture appearances, which mainly consist of microfractures at or near grain boundaries or sub-boundaries (Fig.7(b)). The fracture surface
becomes much smoother near the periphery (Fig. 8(a)). The reason for the low ductility or
occurence of quasi-cleavage fracture in this high-strength condition is the result of fine
grain boundary α precipitate. In the present specimen, grain boundary α seems much
weaker than the prior β matrix. As a result, essentially all of the plastic strain is concen-
rated in the grain boundary α which fails by ductile rupture [12, 13]. In intergranular
fracture surfaces, the grain boundary facets contain small dimples (≤ 3 μm in diameter)
consistent with ductile rupture of the grain boundary α, as shown in Fig. 8(b).

    In high-velocity tension, fracture surfaces exhibit relatively smooth surfaces,
which contain many dimples characteristic of the ductile fracture (Figs. 9 and 10). The

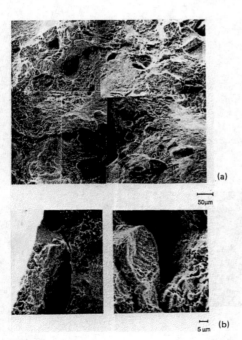

(a)

50μm

(b)
5 μm

FIG. 7 *SEM photos of fracture surfaces (Static, Center Region).*

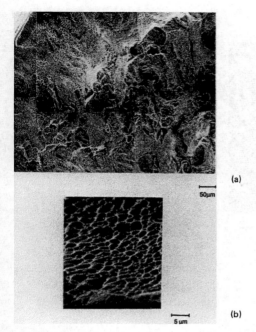

FIG. 8 *SEM photos of fracture surfaces (Static, Periphery Region).*

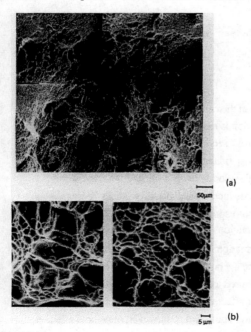

FIG. 9 *SEM photos of fracture surfaces (Dynamic, Center Region).*

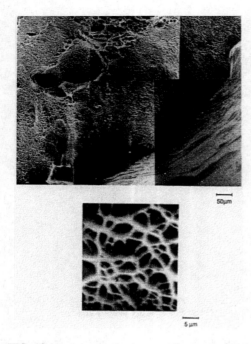

FIG. 10 *SEM photos of fracture surfaces (Dynamic, Periphery Region).*

average dimple diameter is more than 10 μm at the center, while approximately 5 μm near the periphery. The dimple diameter distribution is much wider in high-velocity tension than in quasi-static one. Etched specimen cross-sections are shown in Fig. 11. Irregular intergranular crack paths are evident in quasi-static tension, although ductile fracture is predominant in high-velocity tension. In high-velocity tension, different microfracture mechanisms appear to occur. As the strain rate increases, dislocation glides become much difficult at or near grain boundaries, which corresponds to the increase in the yield stress and σ_{max}. This induces much plastic deformation or dislocation glides inside the prior β grains. Thus, the characteristic ductile deformation leaves typical dimple fracture surfaces.

According to the energy calculation, the average temperature rise in the specimen is estimated to be 5-6 °C. Since the actual measured temperature rise is less than 3 °C, the obtained dynamic stress-strain curves are not affected by the overall temperature rise. Local or microscopic temperature rise near grain boundaries, however, may have affected the deformation and fracture behavior.

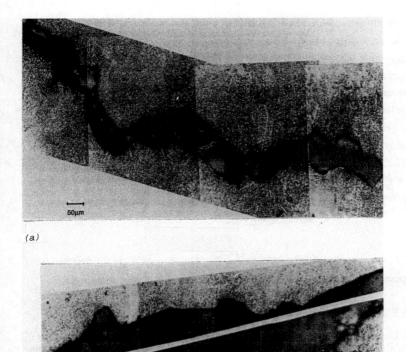

(a)

(b)

FIG. 11 Etched specimen cross-sections. (a) Static. (b) Dynamic.

V. CONCLUSIONS

Dynamic tensile stress-strain curves up to final fracture were obtained for Ti-15-3 alloys up to the strain rate of 10^3/s using the "one bar method". Microfracture mechanisms were then investigated using detailed scanning electron micrographs (SEM) to explain the strain-rate dependent fracture behavior. The characteristic values such as the ultimate strength σ_{max}, the total elongation ε_t, and the absorbed energy E_{ab} increased as the strain rate increased from 10^{-3} /s to 10^3 /s. Thus it should be noted that the present Ti-15-3 alloy gave highly desirable properties in high-velocity tension.

In quasi-static tension, irregular quasi-cleavage fractures prevailed, which mainly consisted of microfractures at or near grain boundaries or sub-boundaries. In the present high-strength specimen, grain boundary α seems much weaker than the prior β matrix. As a result, essentially all of the plastic strain was concentrated in the grain boundary α which failed by ductile rupture. In high-velocity tension, on the other hand, fracture surfaces contained many dimples. As the strain rate increases, it seems that dislocation glides become much difficult at or near grain boundaries, which corresponds to the increase in the yield stress and σ_{max}. This induced much plastic deformation or dislocation glides inside the prior β grains, leaving typical dimple fracture surfaces in high-velocity tension.

ACKNOWLEDGMENT

The authors wish to express their hearty thanks to Dr. Abo, Nippon Steel Corporation for providing specimens.

REFERENCES

1. P. J. Bania, G.A. Lenning and J.A. Hall, in *Beta Titanium Alloys in the 80's*, R. R. Boyer and H.W. Rosenberg (eds.), AIME, New York, 1983, p. 219.

2. H.W. Rosenberg, *ibid.*, p. 409.

3. K. Kawata, S. Hashimoto, and N. Takeda, in *Proc. 13th Cong. Int. Counc. Aero. Sci.*, AIAA, New York, 1982, p. 827.

4. A.-S. M. Eleiche, in *Titanium '80: Science and Technology*, H. Kimura and O. Izumi (eds.), AIME, New York, 1980, p. 831.

5. C. J. Maiden and S. J. Green, *J. Appl. Mech.*, 33: 496 (1966).

6. T. Nocholas, *Exper. Mech.*, 38: 177 (1981).

7. L. M. Meyer, in *Titanium Science and Technology*, G. Lutjering *et al.* (eds.) Deutsche Gesellschaft Metallkunde fur E. V., Adenauerallee, 1984, p. 1851.

8. L. M. Meyer and C. Y. Chiem, *ibid.*, p. 1907.

9. G. T. Gray and P.S. Follansbee, in *Proc. World Conf. Titanium*, P. Lacombe *et al.* (eds.), Societe Fracaise de Metallurgie, Cedex, 1988, p. 117.

10. K. Kawata, S. Hashimoto, S. Sekino, and N. Takeda, in *Macro- and Micro-Mechanics of High Velocity Deformation and Fracture*, Springer-Verlag, Tokyo, 1985, p. 2.

11. K. Kawata, S. Hashimoto, K. Kurokawa, and N. Kanayama, in *Mechanical Properties at High Rates of Strain 1979*, Inst. Phys., London and Bristol, 1979, p. 71.

12. J. C. Williams, F. H. Froes, J.C. Chesnutt, C. G. Rhodes, and R. G. Berryman, in *ASTM STP 651*, ASTM, Philadelphia, 1978, p. 64.

13. T. W. Duerig and J. C. Williams, in *Beta Titanium Alloys in the 80's*, R. R. Boyer and H.W. Rosenberg (eds.), AIME, New York, 1983, p. 19.

8

Mechanical Behavior of a High-Strength Austenitic Steel Under Dynamical Biaxial Loading

E. STASKEWITSCH and K. STIEBLER

Fraunhofer-Institut für Angewandte Materialforschung - IFAM - Lesumer Heerstraße 36, 2820 Bremen 77, West Germany

Uniaxial and biaxial experiments were carried out on the high strength austenitic steel X 2 CrNiMnMoNNb 21 16 5 3 at room temperature in the strain rate range from 10^{-5} up to 10^2 s^{-1}. With the use of a modified split-Hopkinson-bar tubular specimen were loaded by tension and torsion simultaneously. The dependence of flow stress from strain rate and stress state was investigated. The experimental data are described by Perzyna's constitutive equation, including a yield criterion as a function of strain and strain rate and a relationship based on thermal activation.

I. INTRODUCTION

For economical design and numerical simulations an exact knowledge of the material behaviour under service conditions is necessary. Particularly in the field of multiaxial loading with high strain rates only limited experimental data are available. This fact becomes currently more important, because the effiency of computer simulation is often limited by a lack of essential material properties.

Therefore the present paper is concerned with the uniaxial and biaxial quasistatic and dynamic strength properties of a high strength austenitic steel, and with the mathematical description of the results by means of a constitutive equation based on the structure-mechanical model of thermally activated deformation processes.

For biaxial material testing under high loading rates a modified split-Hopkinson-bar was built. By this testing apparatus thin-walled tubular specimens were loaded through combined tension and torsion simultaneously. Strain rates up to $\dot{\epsilon} \approx 2\cdot10^2$ s^{-1} were achieved. A detailed description of operation, measurement and data acquisition is given elsewhere [1,2,3].

II. MATERIAL

Our investigation deals with the elastic-plastic behaviour of the austenitic steel X2 CrNiMnMoNNb 21 16 5 3 (composition in wt%: C: 0.03, Cr: 21, Ni: 16, Mn: 5, Mo: 3, N: 0.3). This austenitic steel is characterized by high strength, non-magnetizability and good chemical resistance, and is therefore used for special ship and apparatus engineering. The material was delivered as a rod in a quenched condition.

The microstructure of this steel is dominated by a single phase solid solution, Fig. 1. The average grain size was found to be ASTM 9 - 10. Therefore about 40 grains are present across the wall thickness of the tubular specimen. This number of grains can be considered to be large enough for a quasi isotropic stress state.

III. EXPERIMENTAL RESULTS

A representation of the experimental results of the quasistatic ($\dot{\epsilon}_v^p = 10^{-3}$ s^{-1}) and dynamic ($\dot{\epsilon}_v^p = 10^2$ s^{-1}) combined tension-torsion tests is given in Fig. 2. Different symbols were choosen for the flow stresses at different equivalent plastic strains ϵ_v^p. The ellipses were fitted by the following equation

$$\sigma_v = \sqrt{\sigma^2 + \alpha\tau^2} \tag{1}$$

where σ_v is the equivalent stress and σ resp. τ are the axial resp. torsional flow stresses. The yield loci of the quasistatic tests (dashed lines) expand with increasing plastic strain due to strain hardening. An additional expansion of the ellipses is caused by the increase of the strain rate of about five orders of magnitude (full lines). There a good agreement between measured yield points and fitted ellipses when a variable shape-parameter α is used.

An exact analysis of the results established not only a difference in the size, but also a change in the shape of the yield loci. Fig. 3 shows

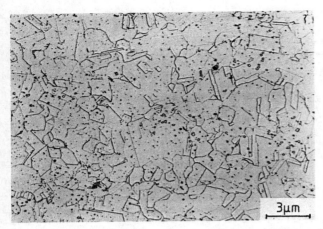

FIG. 1 *Microstructure of the austenitic steel X2 CrNiMnMoNNb 21 16 5 3.*

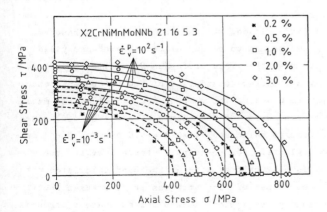

FIG. 2 *Yield loci of the quasistatic and dynamic combined tension-torsion tests corresponding to different proof strain definitions of yielding from* ε_v^p = 0.2 % *to 3.0 %.*

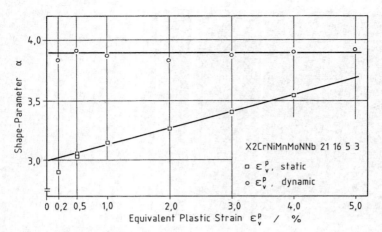

FIG. 3   *Shape parameter* α, *as a function of equivalent plastic strain.*
α *was found by fitting Eq.(1) to the measured flow stresses in Fig.2.*

the dependence of the shape-parameter α, see Eq. (1), on the equivalent
plastic strain ϵ_v^p. For the quasistatic tests the shape-parameter starts at
α = 3.0, which is the characteristic value for the von Mises yield criterion, and increases linearly with equivalent plastic strain. For the dynamic tests the value α is constant over the covered strain regime and
amounts 3.9. Therefore it is quite near the Tresca yield criterion which
predicts α = 4.0. Consequently, the yield criterion for the austenitic
steel X2 CrNiMnMoNNb 21 16 5 3 in the investigated condition is a function
of strain and strain rate.

The design of a multiaxial loaded structural component is based on
the equivalent stress σ_v, which has to be calculated by a proper yield
criterion. Fig. 4 verifies that the yield criterion Eq. (1) with a variable shape-parameter α gives a better description of the experiments than
the von Mises criterion with α = 3.0. The hatched areas include all measured equivalent stress-strain-curves of the biaxial tests. The difference
between the quasistatic results calculated with α = 3.0 and α = α(ϵ_v^p) is
small but should not be neglected. For the dynamic tests however, the
scatter band for α = 3.9 is only half as wide as the band calculated with
α = 3.0. Moreover, the mean values of flow stresses are increased about
50 MPa. Over the whole range of equivalent strain the equivalent stresses
of the dynamic tests exceed the quasistatic stresses about 220 MPa.

Additionally a number of uniaxial tests were performed to get more
information about the dependence of the proof stresses on strain rate.

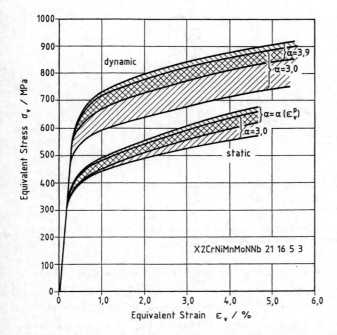

FIG. 4 *Equivalent stress-strain-curves for all combined tension-torsion tests calculated by the von Mises criterion and with the variable shape-parameter α from Fig. 3.*

These tests were performed by a universal servohydraulic testing machine in a strain rate regime from $\dot{\epsilon}^P = 10^{-5}$ to 10^0 s^{-1} and by a flywheel apparatus at strain rates of $\dot{\epsilon}^P \approx 10^2$ s^{-1}. For all tests special cylindrical specimen were used. The forces were measured on the specimen itself.

Fig. 5 represents the plastic strain rate dependences of proof stresses from 0.2 to 3.0 % equivalent plastic strain. The measured proof stresses figured by different symbols for different proof strains, are fitted by the following equation [4,5]

$$\sigma = \sigma_G + \sigma_o^* \left[1 - \left(\frac{kT}{\Delta G_o} \ln \frac{\dot{\epsilon}_o}{\dot{\epsilon}^P} \right)^q \right]^P \tag{2}$$

This equation is based on thermally activated flow processes and relates the flow stress σ to the current value of plastic strain rate $\dot{\epsilon}^P$ and

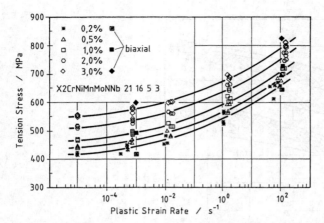

FIG. 5 *Dependence of different proof stresses on strain rate in uniaxial tension and biaxial tension-torsion loading.*

temperature T. The athermal stress σ_G was taken from the low strain rate tests. The values used for the free activation energy ΔG_o = 0.94 eV and the frequency factor $\dot{\epsilon}_o$ = 8.0 · 10^9 s^{-1} were obtained by jump tests at different temperatures as suggested by O. Vöhringer [6]. The other parameters were found by fitting procedure. They are nearly the same for all proof stresses: σ_o^* = 2750 MPa, p = 2.0 and q = 0.5. Therefore the lines in Fig. 5 are parallel to each other. Additionally, the mean flow stress values for 0.2, 1.0 and 3.0 % plastic strain determined by biaxial tests were drawn into Fig. 5. There is a good agreement with the uniaxial tests.

IV. CALCULATIONS

For the description of the quasistatic and dynamic flow behaviour of the austenitic steel under tension-torsion loading, Perzyna's constitutive equation [7] may yield the following set of coupled differential equations

$$\dot{\epsilon} = \frac{\dot{\sigma}}{E} + \eta \langle \phi(F) \rangle \frac{\partial \sigma_v}{\partial \sigma} \tag{3a}$$

$$\dot{\gamma} = \frac{\dot{\tau}}{G} + \eta \langle \phi(F) \rangle \frac{\partial \sigma_v}{\partial \tau} \tag{3b}$$

The total strain rates $\dot{\epsilon}$ and $\dot{\gamma}$ are split into an elastic and a viscoplastic part. The first term corresponds to the Hooke's law, where E resp. G stand for the Young's resp. shear modulus. The latter term contains the viscosity constant η, the function σ_v, which is the yield criterion in form of equivalent stress, as well as the function $\langle\phi(F)\rangle$. $\phi(F)$ describes the dependence of the overstress F (difference between dynamic and static equivalent stress) on plastic strain rate. Using the relations $\eta = \dot{\epsilon}_o$, $\dot{\epsilon}^P = \langle\phi(F)\rangle$ and $F = \sigma_v - \sigma_G$, the constitutive Eqs. (3) can be combined with the thermally activated deformation mechanism formulated in Eq. (2).

If the deformations ϵ and γ of the tubular specimen are given as functions of time, the stresses $\sigma(t)$ and $\tau(t)$ can be calculated by stepwise integration of Eqs. (3). The reason for this approach is the fact that only the strains of a plastically deformed component can be measured or estimated from boundary conditions, but generally not the stresses.

First calculations of uniaxial tension tests at different strain rates were carried out. In this special case only Eq. (3a) has to be solved. The results are represented as full lines in Fig. 6a. For comparison

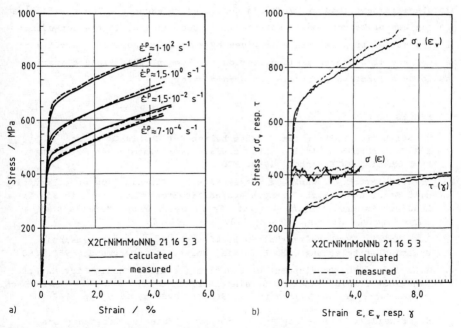

FIG. 6 *Comparison of calculated and measured stress-strain-relations of*
a) some uniaxial tests at different strain rates and
b) a dynamic combined tension-torsion test at a strain rate of 10^2 s^{-1}.

the experimentally measured stress-strain-curves are shown as dashed lines.
There is an excellent agreement for all four strain rates.

As it can be seen from the example of a biaxial test, Fig. 6b, the
calculated and the measured tensile stress-strain-curve $\sigma(\epsilon)$, the shear
stress-strain-curve $\tau(\gamma)$, and the equivalent stress-strain-curve $\sigma_v(\epsilon_v)$
agree within the experimental scatter.

V. SUMMARY

The yield loci of the quasistatic and dynamic combined tension-torsion
tests are described by ellipses, which correspond to different proof
strain definitions of yielding. The size and the shape of the yield
loci are functions of strain and strain rate.

The high strength austenitic steel X2 CrNiMnMoNNb 21 16 5 3, exhibits a strong increase of the flow stresses with increasing strain rate.
This dependence is described excellently by the theoretically founded
structure mechanical model of thermally activated flow processes.

The quasistatic and dynamic tension-torsion behaviour of the austenitic steel is described by Perzyna's constitutive equation. There will be a
good agreement between measured and calculated results, if the dependence
of flow stress on strain rate is given by a formula based on thermal activation, and if a proper yield criterion is choosen. This yield criterion
has to be a function of strain and strain rate.

REFERENCES

1. K. Stiebler, "Beitrag zum Fließverhalten der Stähle Ck 35 und
   X 2 CrNiMnMoNNb 21 16 5 3 unter zweiachsiger dynamischer Belastung",
   Ph.D.Thesis, RWTH Aachen, 1989, FRG

2. K. Stiebler, H.-D. Kunze, E. Staskewitsch, "Plastic Flow of a Ferritic
   Mild Steel and a High Strength Austenitic Steel under Dynamic Biaxial
   Loading" in *Proc. of the 4th Int. Conf. Mech. Prop. Materials at
   High Rates of Strain*, Oxford, 1989, pp. 181-188

3. L.W. Meyer, "Werkstoffverhalten hochfester Stähle unter einsinnig
   dynamischer Belastung", Ph.D.Thesis, Universität Dortmund, 1982, FRG

4. J.Harding, "The Development of Constitutive Relationships for Material
   Behaviour at High Rates of Strain" in *Proc. of the 4th Int. Conf. Mech.
   Prop. Materials at High Rates of Strain*, Oxford, 1989, pp. 189-203

5. E. El-Magd, W. Abdel-Gany and M. Homayun, "Material Constants and
   Mechanical Behaviour during Plastic Wave Propagation" in *Proc. of the
   NUMETA '85 Conf.*, Swansea, 1985, pp. 437-445

6. O. Völringer, *Z. Metallkunde* 65, 1974, pp. 32-36

7. P. Perzyna, "Fundamental Problems in Viscoplasticity" in *Advances in Applied Mechanics*, Vol. 9, Academic Press, 1966, pp. 243-377

9

Mechanical and Microstructural Response of Ni₃Al at High Strain Rates and Elevated Temperatures

H. W. SIZEK and G. T. GRAY III

MST-5 MS E546
Los Alamos National Laboratory
Los Alamos, NM 87545

In this paper, the effect of strain rate and temperature on the substructure evolution and mechanical response of Ni₃Al will be presented. The strain rate response of Ni₃Al was studied at strain rates from 10^3 s⁻¹ (quasi-static) to 10^4 s⁻¹ using a Split Hopkinson Pressure Bar. The Hopkinson Bar tests were conducted at temperatures ranging from 77K to 1273K. At high strain rates the flow strength increased significantly with increasing temperature, similar to the behavior observed at quasi-static rates. The work hardening rates increased with strain rate and varied with temperature. The work hardening rates, appeared to be significantly higher than those found for Ni270. The substructure evolution was characterized utilizing TEM. The defect generation and rate sensitivity of Ni₃AL are also discussed as a function of strain rate and temperature.

I. INTRODUCTION

While the influence of strain rate on the structure/property response of a variety of metals and alloys has been extensively studied, the effect of strain rate on ordered alloys remains largely unknown. Extrapolating trends in mechanical properties which are well known in conventional alloys to ordered systems is complicated by the unusual temperature dependence of the flow strength in some intermetallics such as Ni_3Al. In most materials the flow stress decreases with increases in temperature over all strain rate regimes. In Ni_3Al and other ordered intermetallics, the anomalous yield strength behavior has been recognized for many years [1-6]. The strain rates at which previous investigations have been conducted on Ni_3Al have been limited to rates less than $0.1 \ s^{-1}$ [1-6]. In these strain rate regimes the flow stress increased with temperature to about 1000K and then decreased with further increases in temperature [3-6]. The cause of the anamolous flow stress behavior in Ni_3Al is currently the topic of wide discussion. Most agree that this behavior is due to a change in planarity of the mobile dislocations and the locking of these dislocations by the kinks that form during cross-slip.

At high strain rates, $1 < \dot{\epsilon} < 10^4 \ s^{-1}$, the mechanism controlling deformation in FCC metals is thought to be thermally activated glide [7-10]. Although Ni_3Al is an ordered FCC metal, the anamolous flow stress behavior should still be observed at high strain rates if the rate controlling mechanism is thermally activated. The impetus behind this research was to examine whether or not this anamolous behavior was observed at high strain rates.

II. EXPERIMENTAL

Ni_3Al obtained from Idaho National Engineering Laboratory, with a composition of 75.9 at.% Ni, 24.1 at.% Al and 0.095 at% B, was used in this study [11]. This alloy is a powder metallurgy product fabricated by Homogeneous Metals of Clayville, NY. The powders were produced by vacuum gas atomization, sealed in evacuated steel cans and extruded at 1100 °C. Compression samples were cut with a wire EDM. The samples were annealed at 1100°C in sealed quartz tubes or in an inert atmosphere and were quenched in water. This heat treatment yielded an equiaxed microstructure with a 40 μm grain size. Tests were also conducted on samples that were furnace cooled, but no discernable differences in the mechanical behavior were found. Quasi-static compression tests were conducted on a screw-driven load frame at strain rates of 0.001 and $0.1 \ s^{-1}$. Dynamic tests, strain rates above $10^3 \ s^{-1}$, were conducted in a split-Hopkinson pressure bar [12]. High temperature tests were performed in a vacuum furnace mounted on the split-Hopkinson bar [13]. TEM foils were cut from the samples after testing using a low speed diamond saw.

The foils were thinned mechanically with 600 grit paper and then jet polished in a Fischione dual jetpolisher. The polishing conditions were -40°C, 60 volts and 10 mA in an ethanol-4% perchloric acid solution. The foils were examined in either a Phillips CM30 AEM or a JEOL 2000EX.

III. RESULTS AND DISCUSSION

At a strain rate of 3000 s⁻¹, the flow stress at six percent strain increases with temperature, between 300K and 1083K. (Fig. 1). The flow stress at 77K is slightly higher than that found at room temperature. This behavior is similar to that found in the low strain rate regimes [2-6]. Above 1083K the flow stress at six percent strain saturates and only decreases slightly at higher strains. This saturation is significantly different than that observed at this temperature in the quasi-static strain rate regimes for Ni₃Al, but similar to that seen in pure metals at high strain rates. At larger strains, in the high temperature range, the flow stress reaches saturation, i.e. Ni₃Al shows no work hardening, and the flow stress falls to levels less than those found at the same strains at lower temperatures.

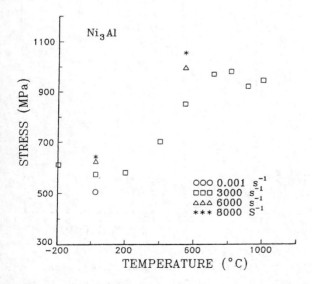

FIG. 1 *Flow stress at 6% strain versus temperature for Ni₃Al with 0.095 at % B.*

The strain rate sensitivity of Ni$_3$Al varies considerably with temperature. At 77K, Ni$_3$Al shows no strain rate sensitivity over strain rates ranging from 2000 to 7000 s^{-1}. Both at room temperature and 545°C, Ni$_3$Al shows significant increases in flow stress with strain rates ranging from 2000 s^{-1} to 8000 s^{-1}. The strain rate sensitivity at room temperature, however, is lower than that found at 545°C over an equivalent increase in strain rate. This increase in strain rate sensitivity with temperature is commonly observed in other metals [14].

The variations in work hardening rate with strain rate and temperature, perhaps because of the anamolous flow behavior, are in some respects different than those seen in Ni. For strain rates greater than 2000 s^{-1} at 77K the strain hardening rate (dσ/dϵ) increases with increasing flow stress (Fig. 2a). This increase in strain hardening rate is also observed in the lower strain rates at room temperature, although the rate of increase with flow stress is not as high as that

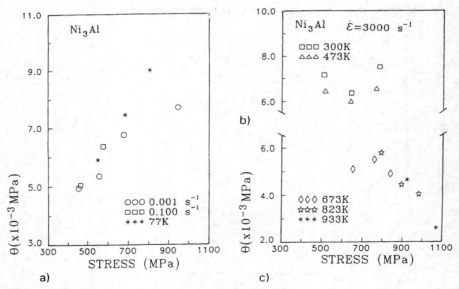

FIG. 2 *Work hardening rate versus flow stress for Ni$_3$Al with 0.095 at % B deformed at a) 77K ϵ > 2000s^{-1} and 300K low ϵ b) 300K, 473K c) 677K, 823K, 983K.*

seen at 77K (Fig. 2a). At temperatures between room temperature and 473K and a strain rate of 3000 s^{-1}, the strain hardening rate remains relatively constant with increasing flow stress, but decreases slightly with temperature (Fig. 2b). At temperatures greater that 823K the strain hardening rate decreases with increasing flow stress, which is similar to the behavior of most metals [7-8].

At room temperature the strain hardening rate increases with increasing strain rate but is insensitive to flow stress (Fig. 3). The value of the strain hardening rate for Ni₃Al at room temperature and quasi-static strain rates, is approximately 10% higher than that found for Ni270 [9]. The strain hardening rate for high strain rates is found to be considerably higher by about 40%, (μ/100 @ $\dot\epsilon$ =3000 s^{-1} for Ni₃Al versus μ/145 for Ni270) than that found for Ni270 [9-10]. Some caution must be taken in making a direct comparisons of work hardening rate values between Ni₃Al and Ni because of possible differences in internal structure (dislocation density, grain size, etc.), the dependence of strain hardening on strain rate [6-9] and the observation that ordered materials strain harden at a

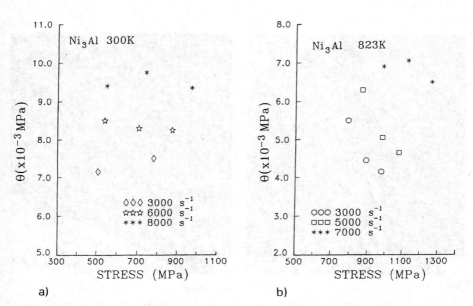

a)   b)

FIG. 3 *Work hardening rate versus flow stress for Ni₃Al deformed at various strain rates for a) 300K b) 823K.*

higher rate than non-ordered materials [4-6]. At 545°C the strain hardening rate also increases with strain rate. At strain rates less than 5000 s⁻¹ the strain hardening rate decreases with increasing flow stress and falls in the region of decreasing work hardening rates shown in Figure 2c. At strain rates higher than 5000 s⁻¹ the strain hardening rate remains relatively constant with increasing flow stress.

A. TEM Observations

After annealing the Ni$_3$Al was relatively dislocation free. Samples deformed at a strain rate of 0.001 s⁻¹ had an even distribution of dislocations which were predominately of the <110>{111} type (Fig. 4a). Stacking faults were also present in the material deformed at this strain rate. The sample deformed at 3000 s⁻¹ showed a similar dislocation structure but a higher dislocation and stacking fault density. At a strain rate of 8000 s⁻¹ the dislocation density was quite high and coarse planar slip bands lying on {111} planes were observed in many grains (Fig. 4b).

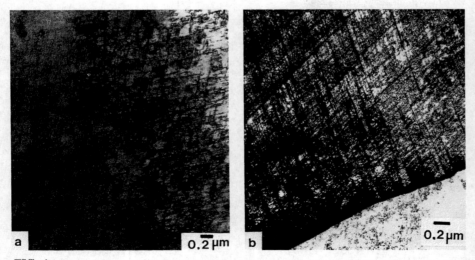

FIG. 4 *TEM bright field of Ni$_3$Al deformed at a) 300K and ε = 0.001 s⁻¹ b) 300K, ε = 8000 s⁻¹ c) (Dark field) 77K, ε = 3000 s⁻¹ d) 1273K, ε = 3000 s⁻¹.*

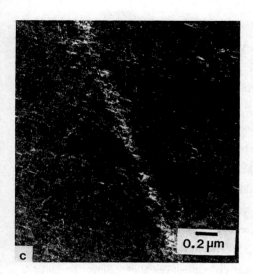

The variation in dislocation structure with temperature at a strain rate of 3000 s^{-1} is similar to that found at quasi-static rates. At liquid nitrogen temperature the deformation is limited to <110>{111} type dislocations with a very high density of stacking faults. This high density of stacking faults most likely contributes to the lack of strain rate sensitivity. At 673K, <110>{111} type dislocations are predominate but <110>{100} type dislocations and stacking faults are also observed. The stacking fault density decreases with increasing temperature, with only a few stacking faults observed in the material deformed at 1273K. In the samples deformed at 973K and 1273K the number of <110>{001} dislocations increases, however the predominate dislocations are still <110>{111} type (Fig. 4d). Also observed in these samples were <110>{110} dislocations; a configuration reported by Caron et.al. [15]. The materials deformed at higher temperatures also have considerably lower dislocation densities. Since the material is recovering faster than the dislocations can accumulate, it is difficult to say which dislocations are controlling the deformation behavior at the high strain rates and high temperatures. One might speculate that if the <110>{001} dislocations were controlling deformation then recovery in this system is much more rapid than in the <110>{111} system. A study is currently under way to examine this possibility.

IV. SUMMARY

The anomalous flow strength reported for Ni$_3$Al at quasi-static strain rates was also observed at strain rates of 3000 s^{-1}. The work hardening rate found in Ni$_3$Al was higher than Ni270 at room temperature. The temperature dependence of the work hardening rate versus flow stress was qualitatively similar to that found in most metals; with temperature increases the work hardening rate decreases. Ni$_3$Al had a strain rate sensitivity that was higher than that observed in Ni270.

Acknowledgments - The authors would like to acknowledge the assistance of W. Wright and C. Trujillo for performing the Hopkinson bar tests and and M. Lopez for conducting the quasi-static tests. This research was performed under the auspices of the United States Department of Energy.

REFERENCES

1. D.P. Pope and S.S. Ezz, Int. Met. Rev., 29: 136 (1984).

2. P.A. Flinn, Trans AIME, 218: 145 (1960).

3. P.H. Thornton, R.G. Davies, and T.L. Johnston, Met. Trans. A, 1: 207 (1970).

4. A.E. Vidoz and L.M. Brown, Phil. Mag., 7: 1167 (1962).

5. R.A. Mulford and D.P. Pope, Acta Met., 21: 1375 (1973).

6. P.K. Sagar, G. Sundararajan, and M.L. Bhatia, Scr. Met., 24: 257 (1990).

7. P.S. Follansbee, *"High-Strain-Rate Deformation of FCC Metals and Alloys," Metallurgical Applications of Shock-Wave and High-Strain-Rate Phenomena*, L.E. Murr, K.P. Staudhammer, M.A. Meyers, eds., Marcel Dekker Inc., New York, 1986.

8. P.S. Follansbee, U.F. Kocks, Acta Met., 36: 81 (1988).

9. P.S. Follansbee and G.T. Gray III, J. Plasticity, in press, (1990).

10. P.S. Follansbee, J.C. Huang, and G.T. Gray III, Acta Met., in press, (1990).

11. R.N. Wright, V.K. Sikka, J. Mater. Sci., 23: 4314 (1988).

12. P.S. Follansbee, "The Hopkinson Bar," *The Metals Handbook 9th edn.*, vol. 8, 198-203, ASM, Metals Park, OH, (1985).

13. C.E. Frantz, P.S. Follansbee and W.J. Wright, "New Experimental Techniques with the Kolsky Bar," *High Energy Rate Fabrication*, I. Berman and J.W. Schroeder, eds., Am. Soc. Mech. Engr., New York, 1984, 229.

14. *Metal Forming*, W.F. Hosford and R.M. Caddell, Prentice Hall, Englewood Cliffs, NJ, (1983).

15. P. Caron, T. Khan, and P. Vessiére, Phil. Mag. A, 60: 267 (1989).

10

Influence of Mechanical Twinning on the Deformation Behavior of Armco Iron

B.-O. REINDERS and H.-D. KUNZE

Fraunhofer-Institut für Angewandte Materialforschung - IFAM -
Lesumer Heerstr. 36, 2820 Bremen 77, Germany

A correlation between the deformation behaviour and the microstructure evolution of Armco iron, in particular the occurrence of mechanical twinning and its influence on the deformation mechanisms is presented. The mechanical behaviour of Armco iron in a special clean quality with three different grain sizes was investigated under quasistatic and dynamic tensile loading in a strain rate regime from $5 \cdot 10^{-4}$ to $1 \cdot 10^{4}$ s^{-1}. The developement of the microstructure as a function of plastic strain and strain rate was characterized by optical, scanning and transmission electron microscopy.

I. INTRODUCTION

There is considerable interest in the material behaviour under dynamic loading, in particular in the effect of strain rate and temperature on the deformation mechanisms in materials. Especially bcc metals, which show two different deformation mechanisms under dynamic loading and low temperatures, are of fundamental concern. Under these conditions bcc materials deform in addition to gliding through mechanical twinning. Several investigations concerning this phenomena has been carried out, but the influence

of mechanical twinning on the strength and deformation behaviour of bcc
metals has still not been clarified [1-4].

The purpose of this paper is to describe the material behaviour of
bcc metal, particularly Armco-iron, in a wide strain rate regime under
tensile loading with varying grain sizes. The results of mechanical testing are correlated with the evolution of the microstructure, to contribute to the clarification of the influence of mechanical twinning on the
deformation behaviour of materials. The results are part of an extensive
investigation program into Armco iron, as yet not finished, concerning a
correlation between the mechanical behaviour and the microstructure.

To determine the grain size dependence of the material behaviour, the
grain size of the same charge of Armco iron in a special clean quality was
set at about 15, 80 and 300 μm. The grain size at the state of delivery
was about 80 μm. 300 μm was adjusted by coarse-grain annealing for 5 hours
at 1200 °C, 15 μm by two-stage cold rolling with strain values of about 80
resp. 60 % and following recrystallization for one hour at 550 °C.

II. EXPERIMENTAL

The apparatus and set-up used for the tensile experiments within a strain
rate regime from $5 \cdot 10^{-4}$ to $1 \cdot 10^4$ s^{-1} were described earlier [5,6].

The twin density was determined by optical microscopy, counting the
amount of mechanical twins and calculating the average amount of twins per
grain.

The transmission observations were carried out with a 300 kV Philips
electron microscope at an operating voltage of 250 kV. The dislocation
density was measured by grid lines in two directions, counting the intersections of dislocations with these grid lines using an expression derived
by Heimendahl [7].

III. RESULTS

The material behaviour of the Armco iron investigated is characterized by
a continuous increase of the stresses with increasing strain rate. Fig. 1
shows an example for the coarse grained iron. Under quasistatic strain rates the grain size dependence follows approximately a Hall-Petch relation,
under dynamic rates this relation is not valid, Fig. 2. Under these conditions the grain size dependence shows a minimum, the material with the
largest grain size is nearly as strong as the smallest one.

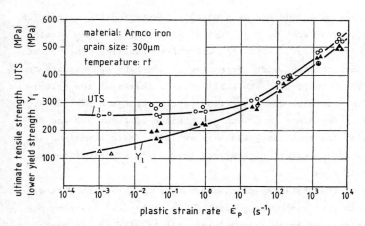

FIG. 1 *Lower yield strength and ultimate tensile strength of Armco iron with a grain size of about 300 μm versus strain rate.*

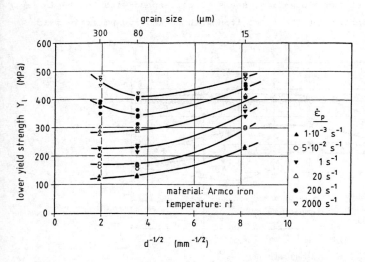

FIG. 2 *Lower yield strength of Armco iron as function of grain size and strain rate*

The deformation behaviour was determined by measuring the elongation e_{f3} and the reduction of area q. Fig. 3 gives an overview of the strain rate dependence of the deformability for all three qualities. The reduction of area is almost constant with increasing strain rate and amounts to 90 to 95 % with the exception of the small grained charge at low velocities. In opposition to this, the elongation passes through a maximum at medium strain rates for Armco iron with grain sizes of about 300 and 80 μm. The charge with the smallest grain size exhibits also a maximum in elongation at low strain rates, although at high rates a further increase in deformability is shown.

Micrographs of the tested specimens were made to determine whether and at which strain rate resp. elongation, first mechanical twinning occurs. At low strain rates Armco iron deforms only by gliding, at higher rates mechanical twinning also occurs. Investigations as function of plastic strain, demonstrate that whenever mechanical twinning occurs, it happens just after exceeding the elastic limit. The limit velocity at which first twinning takes place increases with decreasing grain size, Fig. 4. The occurence of mechanical twinning correlates with the strength and deformation behaviour. With the appearance of mechanical twinning the slope of the strain rate hardening, changes to a higher value for all three qualities, Fig. 1 and 4. Moreover the maximum of the elongation, in particular the decrease of the elongation, seems to correspond with the beginning of mechanical twinning for the Armco iron with 80 and 300 μm grain size,

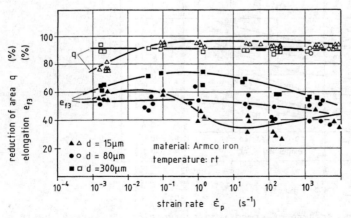

FIG. 3 *Deformability of Armco iron versus strain rate*

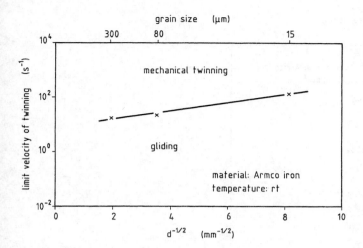

FIG. 4 *Limit velocity of mechanical twinning of Armco iron versus strain rate*

Fig. 3 and 4. For the third quality the occurence of mechanical twinning probably causes a renewed increase of the elongation.

The quantitative evaluation of twin density as function of plastic strain and grain size, demonstrates a distinct rise with increasing grain size, even at low plastic strains, Fig. 5. Provided that twin boundaries have the same effect as grain boundaries with respect to dislocation motion, a higher density of mechanical twins leads to a smaller effective grain size, in particular to a smaller mean free path of dislocations. Because the strength is proportional to the mean free path according to the Hall-Petch relation, a higher twin density might lead to higher stresses. So the higher twin density of the Armco-iron with a grain size of about 300 µm, might be the cause of the higher stresses relativ to the smaller grained charge under dynamic loading, see Fig. 2.

To verify whether mechanical twinning has really a strengthening effect, and to decide whether mechanical twinning improves the deformability of the material or not, further investigations have to be carried out, to separate the influence of mechanical twinning and dislocation motion on the material behaviour. One possibility of suppressing mechanical twinning is quasistatic pre-work [8]. First experiments have shown that only about 1 % quasistatic pre-work can suppress mechanical twinning in this Armco

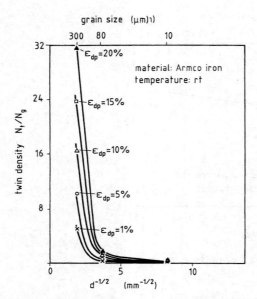

FIG. 5 *Twin density of Armco iron as function of plastic strain and grain size*

iron. First results of these investigations demonstrate that a suppression of mechanical twinning in the fine grained iron leads to a decrease of the elongation of about 20 %. These results correspond with the renewed increase of the elongation with the occurence of mechanical twinning, see Fig. 3.

In accord with these results, is the influence of mechanical twinning on the failure behaviour of the Armco iron. All three qualities failed through ductile fracture under dynamic loading. Mechanical twinning did not initiate any microcracks, the stress concentrations in the neighbourhood of the incoherent twin boundaries could be demolished by glide mechanisms or further twinning. Opposite to which, mechanical twinning leads causatively to failure under low temperatures, in this case the deformability of the matrix is not sufficient to prevent cracking in the twin surroundings. Metallographical investigations into the ground section of a specimen near the fracture surface, demonstrated twins directly in the fracture surface, Fig. 6.

To discover whether there is an effect of mechanical twinning on the dislocation structure, transmission electron microscopical examinations of

FIG. 6 *Micrograph of the ground section near the fracture surface of a tested Armco iron specimen demonstrating twins in the fracture surface*

the tested specimens were performed as function of plastic strain and strain rate. These investigations established evident differences in the evolution of the dislocation structure between quasistatic and dynamic strain rates. For example the evolution of the dislocation structure of the Armco iron with a grain size of about 300 μm is shown in Fig. 7. Under quasistatic strain rates even a small plastic deformation leads to a sub-grain structure. At 7 % plastic strain a deformed cell structure has al-ready developed. With further deformation the anisotropy of the elongated cell structure increases. Opposite to which, under dynamic loading the ma-terial shows a homogeneous planar dislocation structure of strain rates up to 10 % and more. The structure is determined by long screw dislocations. Only at 20 % deformation an elongated cell structure was observed similar to that under quasistatic strain rates. The reason for the differences is, thermally activated processes like cross slip for instants can not contri-bute to the deformation at small elongations under dynamic loading.

The quantitative evaluation of the dislocation density established an increase with decreasing grain size under quasistatic as well as under dy-namic loading. The density rise between quasistatic and dynamic loading also increases with decreasing grain size, Fig. 8. In the coarse grained material, first strains proceed predominantly by mechanical twinning under dynamic loading. In fine grained material the deformation is more control-

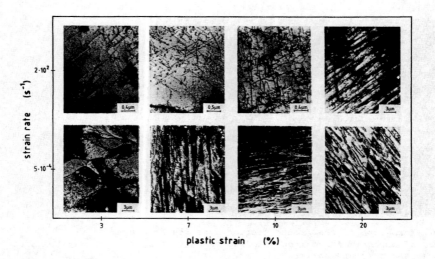

FIG. 7 *Evolution of the dislocation structure of Armco iron with a grain size of about 300 μm as function of plastic strain and strain rate*

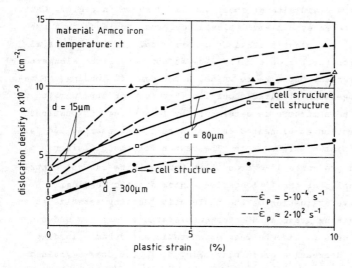

FIG. 8 *Dislocation density of Armco iron as function of plastic strain, strain rate and grain size*

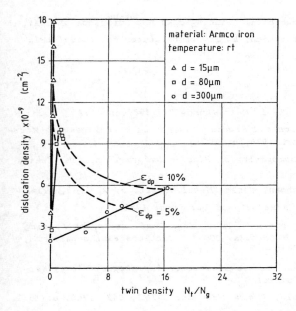

FIG. 9 *Dislocation density versus twin density of Armco iron with different grain sizes*

led by dislocation motion. There may be a correlation between mechanical twinning and the dislocation density as shown in Fig. 9. For constant plastic strains the dislocation density decreases with increasing twin density.

IV. CONCLUSIONS

The mechanical behaviour of Armco iron was characterized under tensile loading as a function of plastic strain, strain rate and grain size at room temperature. Optical, scanning and transmission electon microcopical investigations were used to clarify the influence of mechanical twinning on the strength and deformation behaviour. First results can be summerized as follows:

- Armco iron deforms under dynamic loading and low temperatures in addition to gliding through mechanical twinning
- mechanical twinning seems to have a strengthening effect
- mechanical twinning seems to improve deformability
- the dislocation density decreases with increasing twin density

- under dynamic loading, mechanical twinning does not influence the failure mechanism, opposite to which, at low temperatures mechanical twinning leads causatively to failure

REFERENCES

1. D. Löhe and O. Vöhringer, *Z. f. Metallkunde 77*, 1986, p. 557

2. R.W. Hertzberg, *Deformation and Fracture Mechanics of Engineering Materials*, John Wiley, New York 1976, p. 101

3. R.E. Reed-Hill in *The Inhomogenity of Plastic Deformation*, ASM, Metals
   Park, Ohio, 1973, p. 285

4. G.F. Bolling and R.H. Richman, *Canad. Metal. Quart. 5, 2*, 1966, p.143

5. L.W. Meyer, *PhD Thesis, University of Dortmund*, 1982

6. F.-J. Behler and B.-O. Reinders in *Proc. Int. Conf. on Impact Loading
   and Dynamic Behaviour of Materials*, DGM Informationsgesellschaft mbH,
   Oberursel, FRG, 1987, p. 677

7. M.v. Heimendahl, *Einführung in die Elektronenmikroskopie*, Vieweg Verlag, Braunschweig, 1970, p. 161

8. C.E. Richards and C.N. Reid, *Trans. Met. Soc. AIME, 242*, 1968, p. 1831

11

Microstructure Dependence of High-Strain-Rate Deformation and Damage Development in Tungsten Heavy Alloys

J. LANKFORD, H. COUQUE, A. BOSE, and C. E. ANDERSON

Southwest Research Institute
P. O. Drawer 28510
San Antonio, Texas 78228-0510

The effects of microstructure and strain rate upon the tensile and compressive deformation and failure of tungsten heavy alloys are characterized. It is found that matrix micro- structure and residual damage induced by swaging beyond a certain point can significantly alter strength, ductility, and deformation stability.

I. INTRODUCTION

Tungsten heavy alloys are two phase metal matrix composites having unique combinations of strength, ductility, and density. They are produced by liquid phase sintering of powder compacts comprising mixed elemental powders of 90-98 wt.% tungsten with nickel and other elements such as iron, cobalt, or copper; presently the majority of commercial heavy alloys are fabricated from tungsten, nickel, and iron. The final microstructure consists of a contiguous network of nearly pure bcc tungsten grains embedded in a ductile fcc matrix, which contains about 24 wt% tungsten. The tungsten grain size normally varies between 20 to 50 μm depending on the alloy composition and processing parameters.

The major application of tungsten heavy alloys is in the core of kinetic energy penetrators used for the purpose of defeating modern armor materials. Surprisingly, however, heavy alloy research to date has focussed predominantly on the improvement of quasi-static mechanical properties [1,2]; moreover, this research has been concentrated within a narrow range of classic heavy alloy compositions and grain sizes. Thus, relatively little research effort has gone into the evaluation of the dynamic mechanical behavior of either "classic" or

new heavy alloys. Most of the dynamic work to date has involved compression [3,4] or torsional [5] testing, which generally provides information regarding flow behavior only. Tensile failure is also important in the practical application of these alloys, but is even less
explored for rapid loading conditions. Thus, the objective of this paper is to present some of
the results of a study aimed at establishing the degree to which it is possible to influence
resistance to high strain rate tensile and compression loading by means of thermomechanical
microstructural alteration.

II.   MATERIALS

Four commercially available liquid phase sintered tungsten heavy alloys were selected:
their designations, compositions, treatments, and mean grain sizes are given in Table I, and
representative microstructures are shown in Figure 1. Since each alloy contains W and Ni,
its designation reflects the remaining elemental constituent(s), and the percent of swaging
for the Co alloys. It should be noticed that small precipitates are present in the matrices of
Co(7) and Co(25), but are absent in the Fe-Co and Fe alloys.

III.   EXPERIMENTAL PROCEDURES

Quasistatic testing was performed at a strain rate of 10^{-4} s^{-1} in both tension and compression
using a servo-controlled hydraulic test machine, while high strain rate experiments were performed by means of a split Hopkinson pressure bar (or Kolsky) apparatus. Tensile tests in
the latter corresponded to a strain rate (ε) of 1000 s^{-1}, while in compression
$\dot{\varepsilon} = 2000\text{-}5000$ s^{-1}. Most of the tensile testing involved specimens of gage length (GL) 8.9
mm, and a gage diameter (D) of 3.18 mm. Exceptions were the quasistatic experiments on
the Fe and Co(7) materials, for which GL=50.8 mm, and D=6.4 mm. Cylindrical compression specimens measured 12.7 mm in gage length, and 6.35 mm in diameter.

Table I. Material Conditions

Alloy Designation	Composition (wt. %)	Treatment/Condition	Average Grain Size (μm)
Fe	90W-8Ni-2Fe	As sintered	23·5
Co (7)	91W-6Ni-3Co	Swaged 7%* and aged	24·8
Co (25)	91W-6Ni-3Co	Swaged 25%* and aged	22·9
Fe-Co	91W-4·4Ni-1·9Fe-2·7Co	Swaged 20%*	17·1

* reduction in area

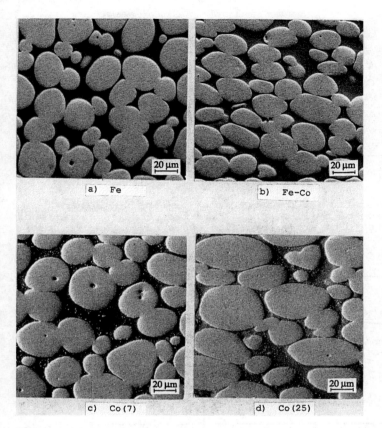

FIG. 1 *Typical Microstructures: a) Fe, b) Fe-Co, c) Co(7), d) Co(25); Swaging Direction is Horizontal.*

Following testing, tensile fracture surfaces were characterized by scanning electron microscopy (SEM). Compression specimens were sectioned parallel to the load axis, polished, and likewise subjected to SEM.

IV. RESULTS

The tensile data, in terms of true stress (σ_y^t) and natural strain (ε), are summarized in Figures 2 and 3. Results of the study by Matic, et al. [6] on the effect of tensile specimen size on the flow stress of a ductile steel were used to correct the failure strains of the larger quasi-static tensile specimens (Co(7) and Fe). In Figure 2, it is evident that dynamic loading suppresses the ductility of Co(7) and Fe to a significant degree. For Co(25) and Fe-Co, on

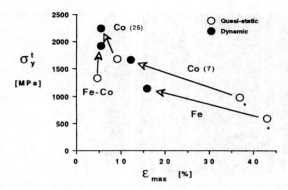

FIG. 2 *Quasi-static and dynamic values of tensile yield stress as a function of failure strain. Marked data (*) corresponds to normalized data with respect to the dynamic tensile specimen.*

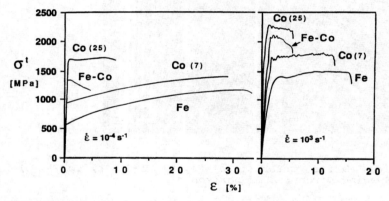

FIG. 3 *Quasi-static and dynamic tensile stress-strain curves.*

the other hand, the principal effect is to increase the yield strength. Several other interesting features are shown in Figure 3. In particular, two of the materials (Co(7) and Fe) strain harden; Fe-Co strain softens; and Co(25) is perfectly plastic under quasi-static conditions, while strain softening at $\dot{\varepsilon}=10^3 \mathrm{s}^{-1}$. This strain softening is occurring quite uniformly along the gage sections, since none or little necking is observed (radial strain to axial strain ratio at failure of 0.25 for the dynamically loaded Fe-Co microstructure.)

To quantify the hardening and softening occurring under quasi-static and dynamic tensile loading conditions, the data were fit using a simple power law relation of the form

$$\sigma = K(\varepsilon^P)^N$$

where σ and ε^P represent the flow stress and plastic strain, respectively, and K and N the power law parameters. The fit was performed for strains ranging from the yield point to $\varepsilon^P=0.10$; softening was characterized with the same power law. Thus, the tensile yield stress for a strain offset of 0.002 versus the exponent N is plotted in Figure 4. Two facts are immediately evident, i.e., regardless of microstructure, the effect of higher strain rate is to 1) raise the yield strength, and 2) decrease N. A hardening material therefore becomes less so, and a softening alloy more so.

Experiments involving compression are reported in greater detail elsewhere [7]. However, an example of compression versus tensile deformation is shown in Figure 5. It is clear that for $\varepsilon=10^{-4}s^{-1}$, the compressive flow stress exceeds that measured in tension, and does so by an amount which increases with strain. At high strain rates, however, σ-ε curves for tension and compression overlay one another (the (dotted) oscillations in the experimental compression curve are caused by dispersion of the longitudinal waves in the Hopkinson pressure bars, a phenomenon increasing in amplitude with applied stress, i.e., strain rate, in the pressure bars, while the dashed curve represents the peak-to-peak best fit). Generally, compressive specimens do not fail, although for certain alloys, adiabatic shear bands develop at high strains (on the order of 0.5) and subsequently fracture. To date the latter phenomenon has been documented for the Fe-type alloys.

Fractographic evidence shows (Figure 6) that regardless of microstructure, the predominant mode of failure at quasi-static rates of loading is cleavage of tungsten grains (although the cleavage facets, for the highly swaged Fe-Co and Co(25) (Figure 6b and 6d, respectively) appear much more brittle than do those for the other two less-worked alloys). Under dynamic conditions, however, failure proceeds along very different lines. For Fe and

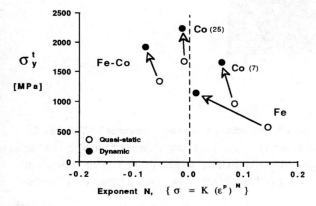

FIG. 4 *Tensile yield stress versus power law exponent.*

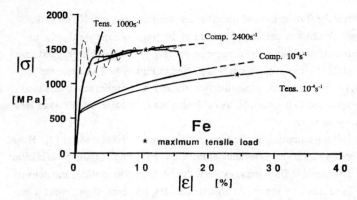

FIG. 5 *Tensile and compression stress-strain response as a function of strain rate for the Fe microstructure.*

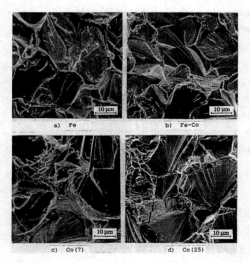

FIG. 6 *Fracture Surfaces,* $\dot{\varepsilon} = 10^{-4}s^{-1}$.

Fe-Co alloys, damage occurs initially by microfracture along contiguous W-W facets, followed by debonding along the W-matrix interface and flow of the matrix to form continuous ligaments. These ligaments constitute large dimples centered on the W-grain facets, final failure corresponding to coalescence of these dimples via microvoid formation within the ligament [8].

The dynamic failure of Co(7) and Co(25) represents a variation on this theme. In particular, there is less tendency to form macrovoids about the initial W-W facets. Instead, sheets of microvoids are quickly nucleated by precipitates within the matrix elements separating the facets; failure occurs when these microvoids coalesce. Energy dispersive spectroscopy indicates that the precipitates are pure tungsten.

V. DISCUSSION

Two principal issues will be addressed in the following discussion: 1) the role of micro-structure in the strain rate dependence of tensile deformation/failure for the four alloys, and 2) compressive versus tensile deformation for the standard sintered (Fe) alloy. It should be noted that the strain hardening of bcc metals like tungsten is usually much more strain rate sensitive than for fcc metal and alloys, which remain fairly ductile over a wide range in strain rate [9]. Body-centered-cubic metals, on the other hand, tend to lose their ductility with increasing strain rate, and generally undergo a ductile-to-brittle transition.

Considering first the as-sintered Fe alloy, its long region of stable hardening at $\dot{\varepsilon}=10^{-4}s^{-1}$ probably reflects the dominant influence of tungsten deformation. This is supported by the fracture surface (Figure 6a), which is ductile in appearance. At high strain rates, W deformation is depressed, hence the lower rate of hardening, and W-W contiguous boundaries open up, isolating subsequent deformation within the matrix, which fails via void nucleation. The behavior of the lightly-swaged Co(7) alloy is qualitatively similar, but differs in detail. Specifically, ultimate failure is caused by precipitation-induced voids formed in matrix ligaments separating failed W-W facets.

The remaining two highly-swaged alloys also behave similarly to one another, but differently from the Fe and Fe-Co materials; in particular they soften, rather than harden (except for Co(25) at $\dot{\varepsilon}=10^{-4}s^{-1}$, which is elastic-plastic). This behavior can be rationalized as follows.

As the applied stress is increased, the Fe-Co alloy yields, and immediately begins to soften. Since the material is already worked to a significant degree (20% RA), some of the W-grains fail via shear cleavage (one recalls the brittle appearance of the fractured W-grains shown in Figure 6b) once a critical stress is reached. This raises the local stress on surrounding grains that are not yet critically-stressed, enabling them to flow at a lower applied stress. As these individual grains exhaust their residual (from swaging) ductility, and fail in sequence, the load drops further, until the ensemble of microfractured grains coalesces at failure. As the strain rate increases to $10^{3}s^{-1}$, tungsten flow is prohibited, and swaging-induced dislocation damage produces rapid failure of matrix ligaments. The brittle quasi-static microfracture facets characteristic of the highly worked Co(25) alloy (Figure 6d) suggest the same basic scenario. In this case, however, the matrix precipitates block slip, and thereby provide some hardening capacity versus the single-phase Fe-Co matrix.

The following conclusion derives from the foregoing. Based on the ductile appearance of the dynamic fractures for all four alloys, it is apparent that the matrix never actually "ex-

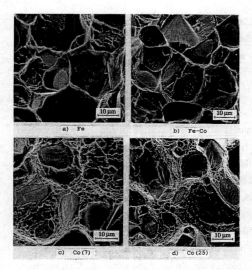

FIG. 7 *Fracture Surfaces,* $\dot{\varepsilon} = 10^3 s^{-1}$.

hausts" it ability to flow. That is, failure of the matrix phase is caused by cumulative, deformation-induced damage in the form of voids. This suggest that post-swaging "residual ductility" in the present case is more a reflection of residual dislocation damage following the swaging operation.

   As noted earlier, the quasi-static compressive flow stress curve for the as-sintered Fe alloy lies above the corresponding tensile curve. Based on the relatively high rate of compressive hardening, and the observation of uniformly deformed tungsten grains in sectioned specimens, it seems likely that matrix deformation dominates the flow stress. Now, since it is well known that fcc metals are anisotropic in their flow behavior versus bcc metals, it is likely that the texture of compressed matrix will be quite different from that of matrix deformed in tension. If so, and if the tensile stress-strain curve also is dominated by matrix (as seems to be the case), a similar but offset curve would be expected; this is indeed observed. At high strain rates, however, these textural developments surely would be suppressed, and flow would tend to reflect dislocation generation rather than the long range motion required to produce a texture. Dislocation nucleation would be relatively insensitive to crystallographic anisotropy, and the resulting tensile and compressive stress-strain response should be similar.

VI.   ACKNOWLEDGMENTS

The authors gratefully acknowledge the support of the Defense Advanced Research Projects Agency through Army Research Office Contract DAACO3-88-K-0204.

VII. REFERENCES

1. K. S. Churn and R. M. German, *Met. Trans.*, 15A:331 (1984).

2. W. J. Bruchey, Jr. and D. M. Montiel, "Mechanical Property Evaluation of a Series of Commercial Tungsten Alloys," Memorandum Report BRL-MR-3606, U. S. Army Ballistic Research Laboratory (1987).

3. R. L. Woodward, N. J. Baldwin, I. Burch, and B. J. Baxter, *Met. Trans.*, 16A:2031 (1985).

4. R. Tham and A. J. Stilp, *J. de Physique*, 49:C3-85 (1988).

5. G. R. Johnson, J. M. Hoegfeldt, U. S. Lindholm, and A. Nagy, *J. Eng. Mats. Tech.*, 105:48 (1983).

6. P. Matic, G. C. Kirby III, and M. L. Jolles, "The Relationship of Tensile Specimen Size and Geometry Effects to Unique Constitutive Parameters for Ductile Materials," Naval Research Laboratory Report No. 5936 (1987).

7. J. Lankford, H. Couque, A. Bose, and R. German, *Met. Trans.* (in preparation).

8. H. Couque and J. Lankford, *Inst. Phys. Conf. Ser.*, 102:89 (1989).

9. U. S. Lindholm, L. M. Yeakley, and R. L. Bessey, "An Investigation of the Behavior of Materials Under High Rates of Deformation," AFML-TR-68-194 (1968).

12

Short and Long Transients in Dynamic Plasticity of Metals, Modeling and Experimental Facts

J. R. KLEPACZKO

Laboratory of Physics and Mechanics of Materials
The University of Metz
57045 Metz, France

Consistent approach to the kinetics of plastic deformation of metals has been applied to describe evolution of the mobile and immobile dislocation densities. The modeling of short and long transients of flow stress, instantaneous rate sensitivity, rate sensitivity of strain hardening and temperature effects can be rigorously performed in terms of the constitutive formalism discussed in this paper.

The formalism in the proposed form can be applied in variety of loading conditions including dynamic plasticity near the threshold stress, plastic wave propagation including HEL, etc. Generally it can be used to take into account all transient phenomena in dynamic plasticity.

Results of some numerical calculations are shown for an aluminum in the form of figures.

I. INTRODUCTION

Plasticity in pure metals and also alloys is determined by generation, motion and storage of dislocations. Those dislocation processes constitute microstructural aspects of modeling in rate dependent plasticity. Since metals deformed plastically at different strain rates and temperatures undergo a complicated microstructural evolution, more exact modeling is now the priority task in dynamic plasticity, [1].

On the other hand, during last decades vast experimental evidence has been collected that plastic strain is accumulated in slightly different manner at different strain rates and temperatures. This is the source of so-called strain rate and temperature history effects, for example [2-8]. Those effects can be decomposed into two cases. The first one is the instantaneous, or short-time, reaction of a material to changes in strain rate or temperature, and is manifested, for example, by instantaneous rate sensitivity, [1]. Usually an abrupt change of strain rate and temperature is accompanied by so-called short transient of flow stress. The second one is due to different rates of defect accumulation, for example forest dislocations, at different strain rates and temperatures. Those effects are manifested by different rates of strain hardening at different strain rates or temperatures and can be determined, for example, via the rate sensitivity of strain hardening, [1,9]. The different rates of strain hardening at different strain rates and temperatures are manifested as so-called long transients of flow stress, [10]. Definitions of "short" and "long" transients are related to increments of strain within which the transients are detected experimentally.

An ample experimental evidence of short as well as long transients in FCC metals has been demonstrated previously in the review paper, [8]. Additional evidence are reported in [11] and [12] for aluminum. Representative low strain rate and medium strain rate stress-strain curves obtained in tension are shown for copper in Fig. 1. Three curves are shown corresponding to three types of loading : low strain rate loading at constant strain rate $\dot{\varepsilon}_i = 2.23 \times 10^{-5}$ s^{-1}, medium strain rate loading at constant strain rate $\dot{\varepsilon}_r = 6.28 \times 10^{-1}$ s^{-1}, and low strain rate loading to a strain ε_i, followed by medium rate loading at constant strain rate $\dot{\varepsilon}_r = 6.3 \times 10^{-1}$ s^{-1}. The latter strain rate jump experiment provides evidence of both short and long tran-

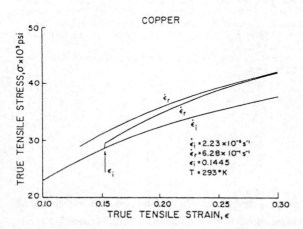

FIG. 1 *Experimental evidence of short and long transient of flow stress for polycrystalline Cu, [14].*

sient of flow stress. The first jump in stress is expected to be a direct measure of the change of flow stress with change in strain rate at fixed dislocation structure. Then the result of change in strain rate (or temperature) can be interpreted as a test of materials response for dislocation microstructures that existed prior to the increment of strain rate (or temperature). The instantaneous stress increment $\Delta\sigma_s$ is a measure of the instantaneous rate sensitivity β_s, [1,7]. Whereas the rate sensitivity of strain hardening determined at constant strain ε_i can be found by splitting the entire stress difference $\Delta\sigma$ at two constant strain rates as follows $\Delta\sigma = \Delta\sigma_s + \Delta\sigma_h$. The entire stress difference $\Delta\sigma_s + \Delta\sigma_h$ refers to the same strain ε_i but to two slightly different microstructures, and consequently $\Delta\sigma_h$ is the result of deformation history, [1]. It is obvious that $\Delta\sigma_h$ can be associated with the rate sensitivity of strain hardening β_h, [1]. The dependence of strain hardening on strain rate or temperature is the source of long transients in plasticity of metals, sometimes called the fading memory, [13].

Another example of the short transient for copper is shown in Fig. 2. This time the torsion Kolsky apparatus has been applied and the incremental part of the flow curve is obtained at $\dot{\Gamma}_r \approx 3.4 \times 10^2$ s^{-1}, in addition $\Gamma_i = 0.15$ and $\dot{\Gamma}_i \approx 2 \times 10^{-3}$ s^{-1}. Further demonstration of short and long transients for polycrystalline Al can be found in [12], where short transients are well pronounced. In the case of BCC metals the short transients are typically observed in the form of the higher and lower yield point. Direct analogy of the yield point developed by the mobile dislocation flux $\rho_m v$ are the wave profiles in uniaxial strain, for example for pure polycrystalline Fe, [15]. The velocity-time profile shows the relaxation behind the precursor (a decrease of acceleration) which indicates for the existence of the short transient.

In conclusion, after more careful examination of experimental data with large increments of strain rate, almost everytime the short transients are observable. The long transients,

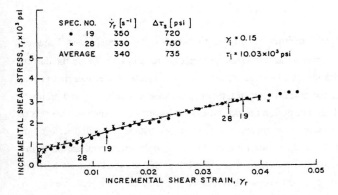

FIG. 2 *Experimental evidence of short transient for polycrystalline Cu at large increment of strain rate, [14].*

i.e. strain-rate and temperature history effects, are specially visible for FCC structures, albeit for BCC and HCP they are also present.

II. THE FORMALISM IN CONSTITUTIVE MODELING

Recently, microstructural aspects in constitutive modeling of plastic deformation are more frequently considered. Typical approach is represented by one internal state variable assumed as the total *mean* dislocation density ρ, [1,15,16,17]. Such approach to constitutive modeling was proposed, for example, by Klepaczko in 1975, [15]. It can be shown, however, that only the long transients can be described by the one state-variable models. Since some reviews of on-parameter models are at present available, [1,16], no attempt is offered to discuss this subject.

In order to take into account both the short and long transients two-state-variable approach must be introduced. Such approach to constitutive modeling, with the mobile and immobile dislocation densities, ρ_m and ρ_i, as the state variables, has been recently introduced in [10]. A unified theoretical concept is employed in which rate-sensitive strain hardening, temperature and instantaneous rate sensitivity is rigorously treated in terms of kinetics of dislocation multiplication, glide and annihilation.

The notion is adopted that plastic deformation in shear is the fundamental mode in metal plasticity and appropriate Taylor factors should be employed to find macroscopic quantities, [16]. However, in this study the macroscopic quantities are used as discussed in [1,15,16]. The flow stress in shear τ at constant structure is given to a good approximation by

$$\tau(\dot{\Gamma},T,s_j)_{STR} = \tau_\mu [s_j (\dot{\Gamma},T)]_{STR} + \tau^* [\dot{\Gamma},T,s_j (\dot{\Gamma},T)]_{STR} \tag{1}$$

where τ_μ and τ^* are respectively the internal and effective stress components, $\dot{\Gamma}$ and T are respectively the current values of strain rate and absolute temperature, and $s_j (\dot{\Gamma},T)$ are the current values of j internal state variables s_j. The structure, or the current state of the material, is characterized by s_j internal state variables. In general, the internal state variables depend on history of plastic deformation. The assumption of additivness (1) implies existence of two sets of obstacles opposing the dislocation movements. The first set of obstacles coupled with τ_μ are supposed to be strong obstacles to dislocation motion like cell walls, grain boundaries, twins, etc. The secondary defects, while more numerous, like forest dislocations, Peierls barriers, second phase particles, are supposed to be weak obstacles and they can be overcome by moving dislocations with assistance of thermal vibration of crystalline lattice and the effective stress τ^*. The kinetics of dislocation movements interrelates at constant structure, characterized by s_j state variables, the instantaneous values of effective stress τ^*, plastic strain rate $\dot{\Gamma} = d\Gamma/dt$ and temperature T, by generalised Arrhenius relation, [1,7,16],

$$\dot{\Gamma} = v_i \, (T,s_j) \exp \left[- \frac{\Delta G_i \, (T, \tau^*, s_j)}{kT} \right] \qquad (2)$$

or after inversion

$$\tau^* = f^* \, [\dot{\Gamma}, T, s_j \, (\dot{\Gamma}, T)] \qquad (3)$$

where v_i is the frequency factor, ΔG_i is the free energy of activation, kT has its standard meaning. The subscript i indicates the i-th, so far unspecified, thermally activated micromecha-
nism of plastic deformation.

Since the microstructure undergoes an evolution, and the state of microstructure is defined by s_j variables, the structural evolution is assumed in the form of a set of j differential
equations of the first order

$$\frac{ds_j}{d\Gamma} = f_j \, [s_k, \dot{\Gamma} \, (\Gamma), T \, (\Gamma)], \qquad k = 1 \ldots j \qquad (4)$$

Solution of the set (4) provides current values of s_j to be introduced into eq. (1). Thus, the
flow stress τ can be calculated for any deformation history.

III. IDENTIFICATION OF INTERNAL STATE VARIABLES

Flow stress in polycrystalline metals can be related to characteristic spacing of obstacles to dislocation motion, [18]. The class of athermal obstacles which increases the internal stress τ_μ is
identified by eq. (5), [10,16],

$$\tau_\mu = \alpha_1 \mu \, b \, \rho^{1/2} + \alpha_2 \left(\frac{b}{d(\rho)} \right)^\delta + \alpha_3 \, \mu \left(\frac{b}{D} \right)^{1/2} + \alpha_4 \, \mu \left(\frac{b}{\Delta(\rho)} \right)^{1/2} \qquad (5)$$

where μ is shear modulus, α_j (j = 1... 4) are constants which characterize dislocation / obstacle strength, b is magnitude of Burgers vector. The mean distance between immobile dislocations is L_i and related immobile dislocation density is $\rho_i = 1/L^2_i$, whereas the mean distance

between mobile dislocations is L_m, and related density of mobile dislocations is $\rho_m = 1/L^2_m$. It is assumed here that

$$\frac{1}{L^2} = \frac{1}{L_i^2} + \frac{1}{L_m^2} \quad \text{or} \quad \bar{\rho} = \bar{\rho}_i + \bar{\rho}_m \tag{6}$$

The first term in eq. (6), i.e.

$$\tau_{\mu 1} = \alpha_1 \mu b (\rho_i + \rho_m)^{1/2} \tag{7}$$

is the main contribution to the internal stress τ_μ due to dislocation/dislocation interaction. The next two terms in eq. (5) are related respectively to evolution of subgrain $d(\rho)$ as discussed in [1] and [16], and to the effect of grain diameter D - so called Hall-Petch term. The exponent δ is equal 1 for cells and small subgrains with large misorientations, and $\delta = 1/2$ for "ideal" subgrains in thermally recovered metals, [19]. It is well known that at low temperatures and at high strain rates some metals produce deformation twins, the mean spacing Δ between twins obviously will increase the internal stress τ_μ as it is predicted by the fourth term in eq. (5).

Thus, five quantities : ρ_i, ρ_m, d, D and Δ characterize microstructure and canditate as internal state variables. In the present study *two* internal state variable model will be analysed, [10], and the variables are the mean immobile dislocation density ρ_i and the mean mobile dis-location density ρ_m. The last two components of the internal stress τ_μ are neglected in this ana-lysis, i.e. $\alpha_3 = 0$ and $\alpha_4 = 0$.

IV. EVOLUTION OF MOBILE AND IMMOBILE DISLOCATION DENSITIES

Since plastic deformation is the result of dislocation generation, movement and storage the rates of those quantities must be balanced. Equation (6) leads to the balance equations (8)

$$\dot{\rho} = \dot{\rho}_m + \dot{\rho}_i \quad \text{and} \quad \frac{d\rho}{d\Gamma} = \frac{d\rho_m}{d\Gamma} + \frac{d\rho_i}{d\Gamma} \tag{8}$$

If ρ_m does not change, the frequent assumption in one-parameter models, then the accumula-tion rate of ρ is equal to the storage rate of ρ_i. For example, when evolution equations for ρ_i and ρ_m are assumed in the form

$$\frac{d\rho_m}{d\Gamma} = M_m - f_{ma} (\rho_m, \dot{\Gamma}, T) \tag{9}$$

$$\frac{d\rho_i}{d\Gamma} = M_i - f_{ia} (\rho_i, \dot{\Gamma}, T) \tag{10}$$

where M_m and M_i are multiplication factors for ρ_m and ρ_i and f_{ma} and f_{ia} are respectively anni-hilation functions, the condition (8) leads to the relations

$$M_i = M_m - f_m (\rho_m, \dot{\Gamma}, T) \quad \text{or} \quad M_i = \frac{d\rho_m}{d\Gamma} \tag{11}$$

and the total multiplication rate is

$$\frac{d\rho}{d\Gamma} = M_m - [f_{ma} (\rho_m, \dot{\Gamma}, T) + f_{ia} (\rho_i, \dot{\Gamma}, T)] \tag{12}$$

The simplest practical form of eq. (12) has been derived in [16] and discussed in [1],

$$\frac{d\rho}{d\Gamma} = M_{II} (\dot{\Gamma}) - k_a (\dot{\Gamma}, T) (\rho - \rho_o) \tag{13}$$

with $M_{II}(\dot{\Gamma}) = 1/b\,\lambda\,(\dot{\Gamma})$, where $k_a\,(\dot{\Gamma},T)$ is the annihilation factor. The generation rate of dislo-cations $M\,(\dot{\Gamma})$ can be related to the mean free path $\lambda\,(\dot{\Gamma})$. The mean free path is a diminishing function of $\dot{\Gamma}$, [16,20]. A more advanced model with two internal state variables, ρ_m and ρ_i, has been derived in [10]. Evolution of the mobile dislocation density must reflect behavior of short transients during mechanical tests. Evolution equation for ρ_m, as derived in [10], can be written in the form †

$$\frac{d\rho_m}{d\Gamma} = M_{II}(\dot{\Gamma}) + (\rho_m - \rho_{mo}) \left(\frac{\eta_1}{\Gamma} - \eta_2\right) \tag{14}$$

where η_1 is constant characterizing activity of dislocation sources after change in strain rate, η_2 is probability of slowing down of dislocation production or immobilization, ρ_{mo} is the ini-tial mobile dislocation density. General solution of eq. (14) is postulated in the form

---

† In [10] M_{II} was omitted in the typing.

$$\rho_m - \rho_{mo} = M_{II}(\dot{\Gamma})f(\Gamma) + g(\dot{\Gamma},\Gamma)\,\Gamma^{\eta_1} \exp(-\eta_2\Gamma)\,\text{sign}\,\ddot{\epsilon} \tag{15}$$

where $f(\Gamma)$ is unspecified function of Γ. Introducing (15) to (14) the explicit form of differential evolution equation for ρ_m is obtained.

$$\frac{d\rho_m}{d\Gamma} = M_{II}(\dot{\Gamma})\,h(\Gamma) + g(\dot{\Gamma},\Gamma)\left(\frac{\eta_1}{\Gamma} - \eta_2\right)\Gamma^{\eta_1} \exp(-\eta_2\Gamma)\,\text{sign}\,\ddot{\epsilon} \tag{16}$$

with

$$h(\Gamma) = \left[\frac{f(\Gamma)}{\Gamma}(\eta_1 - \eta_2\,\Gamma) + 1\right] \tag{17}$$

In eq. (15) the second term with $g(\Gamma,T)$ describes "bursts" of the mobile dislocations when strain rate is increased, or reduction in production when strain rate is reduced. The g-function is scaling the effect of Γ and T on multiplication activity. In general, g-function can be specified as, [10]

$$g(\dot{\Gamma}, T) = A(T)\,M_{II}\left(\frac{\dot{\Gamma}}{\dot{\Gamma}_o}\right) \quad [\text{cm}^{-2}] \tag{18}$$

where M_{II} is the multiplication factor at $\dot{\Gamma} = 1 \times 10^{-4}\ \text{s}^{-1}$ at RT, and Γ_o is normalisation constant. The constant A, which indicates for the "rate sensitivity" of mobile dislocation bursts, can be determined, for example, by ultrasonic methods [10]. The second term in eq. (14) shows transient character, i.e. at $\Gamma = 0$, ρ_m has its current value, next, a maximum occurs at $\Gamma_m = \eta_1/\eta_2$, and finally at larger strains, again the current value of ρ_m is reached. Mechanical tests indicate the following limits for Γ_m, $0.01 \leq \Gamma_m \leq 0.08$. Analysis of experimental data for FCC metals show that $\eta_1 \approx 2$ and $\eta_2 \approx 40$, for $\Gamma_m \approx 0.05$, [10]. The second part of the transient character of the ρ_m-production, i.e. diminishing rate of production, stems from the fact that created pile-ups develop back stresses on the dislocation sources and production of ρ_m slows down.

The evolution and rate of storage of immobile dislocations produce strain hardening. Once the evolution of immobile dislocations is defined one can predict the current level of the long range stress reached by accumulation of ρ_m. At present there is no agreement which form of evolution equation provides the best results. Relatively general relations for evolution of ρ

(total density, one-parameter model) have been discussed in [1] and [20]. However, in the present case the simplest relation will be employed, [10]

$$\frac{d\rho_i}{d\Gamma} = M_i\,(\dot{\Gamma},\,T,\,\Gamma) - k_o\left(\frac{\dot{\Gamma}}{\dot{\Gamma}_o}\right)^{-2m_oT}(\rho_i - \rho_{io}) \tag{19}$$

where k_o is the non-dimensional annihilation factor at $T = 0K$, m_o is the absolute rate sensitivi - ty due to annihilation. The condition (11) leads to the relation

$$\frac{d\rho_i}{d\Gamma} = \frac{d\rho_m}{d\Gamma} - k_o\left(\frac{\dot{\Gamma}}{\dot{\Gamma}_o}\right)^{-2m_oT}(\rho_i - \rho_{io}) \tag{20}$$

where $d\rho_m/d\Gamma$ is defined by eq. (16). Since the annihilation term in the form of eq. (20) is well established in the one-parameter models, it accounts here for the total rate of annihilation, i.e. $\rho^-_m + \rho^-_i$.

It is to be noted that in the case of eq. (16) and eq. (20) the mean free path λ is a com - plicated function of $\dot{\Gamma}$, T and Γ. It is also clear that the two-state-variable formulation gives more flexibility in constitutive modeling. In the first place, both short and long transients can be modelled at the same time.

V. APPLICATION OF THE MODEL

In order to apply the complete model for numerical analyses it is necessary to formulate Ar - rhenius relation in the explicit form, [1,20]. In the inversion form of the Arrhenius relation, effective stress τ^* can be written, [1,10,16,18,20], as follows

$$\tau^* = \tau_o^*\left[1 - \left(\frac{kT}{G_{io}}\ln\left(\frac{b^2v_D\rho_m}{2\dot{\Gamma}}\right)\right)^{1/q}\right]^{1/p} \tag{21}$$

all symbols have their standard meaning, but τ^*_o is the thermally activated part of the total threshold stress τ_o ; $\tau_o = \tau_{\mu o} + \tau^*_o$. The internal stress τ_μ is defined by eq (5), but only two first terms in eq (5) were employed. Finally, the total stress τ can be calculated in the for - malism by integration of the evolution equations for ρ_m and ρ_i for desired deformation history and introduction of ρ_m into eq. (20), and $\rho_m + \rho_i$ into eq. (5).

Numerical simulations have been performed for polycrystalline aluminum. Simula - tions of long transients for aluminum with one state variable as the *total* dislocation density ρ

were reported earlier, [1,16,20]. Some preliminary results for two state variable model with evolution of ρ_m and ρ_i were reported in [10]. All constants and parameters which enter into equations (5), (16), (17), (20) and (21) have been given in the papers cited above. The set of differential equations for structural evolution, eq (16) and eq (20), have been integrated unumerically by 4-th order Runge-Kutta procedure with $\rho_{io} = 2.5 \times 10^8$ 1/cm² and the initial fraction $f_o = \rho_{mo}/\rho_{io}$, $f_o = 0.02$. Evolution of the increments of $\Delta\rho_m$ (the second term in eq. (15)) is shown in Fig. 3 as a function of $\Delta\Gamma$ for six strain rates. Whereas Fig. 4 shows evolution of the increment of $\Delta\rho_m$ $(\Delta\Gamma)_\Gamma^{\cdot}$ where the initial strain rate $\dot{\Gamma}_i = 50$ s⁻¹ is increased to 5×10^2 s⁻¹ at different values of strain Γ_i.

Since the forest dislocation density ρ_i is much higher in comparison to ρ_m relatively small variation of ρ_m does not change much variation in ρ_i. Consequently, evolution of ρ_i (Γ) is similar to the evolution of the total ρ (Γ) in one-state-variable models, for example eq. (13), [1,16,20], and ρ_i (Γ) will not be here discussed. Introduction of ρ_m $(\Gamma)_\Gamma^{\cdot}$ into eq. (21) permits for calculation of τ^* $(\Gamma)_\Gamma^{\cdot}$. The result of such calculation is shown in Fig. 5, where the evolution of effective stress τ^* is shown as a function Γ when the initial strain rate $\dot{\Gamma}_i = 1$ s⁻¹ is changed abruptly to $\dot{\Gamma}_r$ at $\Gamma_i = 0.1$. It is evident that the nature of the short transients is well reflected by the model.

The final results of calculations are shown in Fig. 5 where the shear stress τ is plotted as a function of strain. In addition, the strain rate jumps at different temperatures were also simulated. The parameters in numerical simulations are the same as of experiments for poly-

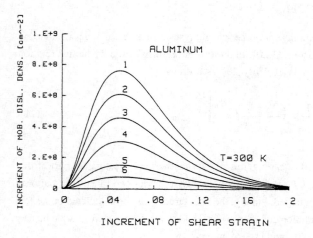

FIG. 3 *Numerical simulation of incremental evolution for* $\Delta\rho_m$ *(Γ) at different strain rates,* $\dot{\Gamma}_1 = 1 \times 10^{-3}$ *1/s,* $\dot{\Gamma}_6 = 1 \times 10^2$ *1/s, increments of* $\dot{\Gamma}$: $\Delta\dot{\Gamma} = 2 \times 10^2$ *1/s.*

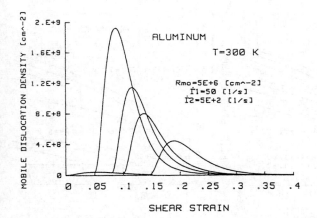

FIG. 4 *Numerical simulation of incremental evolution for* $\Delta\rho_m$ (Γ) *with increment of* $\dot{\Gamma}$ *from* $\dot{\Gamma}_i = 50$ *1/s to* $\dot{\Gamma}_r = 5 \times 10^2$ *1/s.*

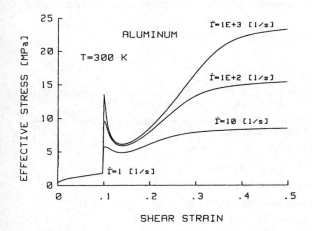

FIG. 5 *Evolution of effective stress* τ $(\Gamma)\dot{\Gamma}$ *before and after jumps in strain rate,* $\dot{\Gamma}_i = 10$ *1/s at* $\Gamma_i = 0.1$ *to values indicated in the figure.*

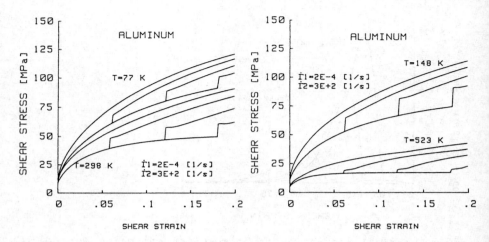

FIG. 6 *Results of calculations of* τ *(Γ)* $\dot{\gamma}$ *with two-parameter forma-lism ; constant strain rates and incremental simulations.*

crystalline Al reported in [21]. Also similar calculations for the same parameters were perfor-med earlier for one-parameter model, [1,16]. Results presented in Fig. 6 indicate that the for-malism presented here permits for more or less exact description of short and long transients when strain rate or temperature is changed in programmed manner. The formalism enables, in general, for an exact modeling of transient phenomena during plastic deformation. For exam-ple, the HEL and the short and long transients in plastic wave propagation can be accounted for.

REFERENCES

1. J.R. Klepaczko in *Impact : Effects of Fast Transients Loadings,* W.J. Amman, W.K. Lin, J.A. Studer and T. Zimmermann (eds.), Balkema, Rotterdam, 1988, p. 3.

2. J.R. Klepaczko, *Effects of Strain Rate Changes on Strain Hardening in Aluminum,* Ph. D. Thesis, Polish Academy of Sciences, Warsaw, 1964.

3. U.S. Lindholm, *J. Mech. Phys. Solids,* 12 : 317 (1964).

4. J.R. Klepaczko, *Arch. of Mech.,* 19 : 211 (1967).

5. J.R. Klepaczko, *J. Mech. Phys. Solids,* 16 : 255 (1968).

6. J.D. Campbell, *Dynamic Plasticity of Metals,* Springer-Verlag, Vienna, 1970.

7. J.R. Klepaczko, *Mater. Sci. Engng.,* 18 : 121 (1975).

8. J.R. Klepaczko, R.A. Frantz and J. Duffy, *Engng. Trans.,* 25 : 3 (1977).

9. U.F. Kocks, *in Mechanical Testing for Deformation Model Development,* R.W. Rohde and J.C. Swearengen (eds.), ASTM-STP 765, ASTM, Philadelphia, 1982, p. 121.

10. J.R. Klepaczko, in *Constitutive Laws of Plastic Deformation and Fracture*, Proc. 19-th Canadian Fracture Conference, A.S. Krausz et al. (eds.), Kluwer academic press, Rotterdam, 1990, in print.

11. H. Neuhauser, in *Dislocation in Solids*, Vol. 6, North Holland, Amsterdam 1983, p. 408.

12. R.A. Frantz and J. Duffy, *J. Appl. Mech.*, 39 : 939 (1972).

13. J.R. Klepaczko and J. Duffy, in *Mechanical Properties at High Rates of Strain*, J. Harding (ed.), The Institute of Physics, London and Bristol, 1974, p. 91.

14. J.R. Klepaczko, *Strain Rate Incremental Tests on Copper*, Brown Univ. Tech. Report, No GK-40213/6, Brown University, 1974.

15. J.R. Klepaczko, *Mater. Sci. Engng.*, 18 : 121 (1975).

16. J.R. Klepaczko, in *Constitutive Relations and Their Physical Basis*, Proc. 8-th Risø International Symp. on Metallurgy and Materials Science, Risø Matl. Lab., Roskilde, 1987, p. 387.

17. Y. Estrin and H. Mecking, *Acta Metall.*, 32 : 57 (1984).

18. U.F. Kocks, A.S. Argon and M.F. Ashby, *Thermodynamics and Kinetics of Slip*, Pergamon Press, Oxford, 1975.

19. F.R.N. Nabarro, in *Strength of Metals and Alloys*, Proc. ICSMA-7, Vol. 3, Pergamon Press, London, 1986, p. 1667.

20. J.R. Klepaczko, "Constitutive Modeling in Dynamic Plasticity Based on Physical State Variables", in *Proc. Int. Conf. DYMAT*, Les éditions de Physique, Les Ulis, 1988, p. C3-553.

21. P.E. Senseny, J. Duffy and R.H. Hawley, *J. Appl. Mech.*, 45 : 60 (1978).

13

High-Strain-Rate Titanium Compression: Experimental Results and Modelization

S. GABELOTAUD, C. NGUY, P. BENSUSSAN, M. BERVEILLER*, and P. LIPINSKI*

DGA/Centre de Recherches et d'Etudes d'Arcueil, 16 bis avenue Prieur de la Côte d'Or, 94114 ARCUEIL CEDEX, FRANCE. *Laboratoire de Physique et de Mécanique des Matériaux (U.A.CNRS), ISGMPENIM Ile du Saulcy, 57045 METZ Cedex 01 FRANCE

Both macroscopic and microscopic response of titanium to static and dynamic loading have been assessed. Compression tests show that flow stresses are higher under dynamic loading than under static loading. Observed active deformation systems are the same under both static and dynamic loading but the number of twinned grains is higher in the latter case. Numerical simulations based on a self-consistent model have been performed. Calculated stress-strain curves are found to be in good agreement with experimental ones. Nevertheless, additional microscopic observations are needed in order to establish fully satisfactory correlations with predicted numerical results.

I. INTRODUCTION

While many studies are available for materials of cubic structures, it is not the case for hexagonal close-packed metals such as titanium. Especially, few models exist to describe both mechanical behavior and active deformation mechanisms at high strain rate.

The purpose of this paper is to present first experimental results for high strain rate titanium compression and secondly results of a model which allow to correlate microscopic and macroscopic phenomena.

II.   EXPERIMENTS

A.   *EXPERIMENTAL METHODS*

Compression tests have been performed on pure commercially α-titanium which chemical composition in weight ppm is: 78 C, 257 Fe, 1176 O, 3.86 H, 73.4 N.   After a one hour annealing treatment at 700°C under high vacuum the metal is found to be dislocation-free.   The grain size is about 60 μm. Material texture shows that c-axes lie in the short transverse - long transverse plane; a-axes are distributed randomly around c-axes.   Samples have been machined such as the compression axis is parallel to the transverse direction of the plate.   A split Hopkinson pressure bar [1] has been used for dynamic compression tests.

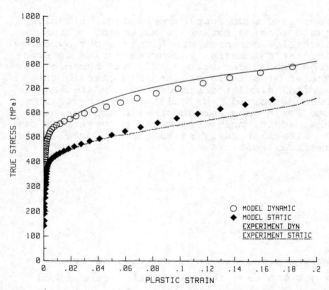

FIG. 1 *Static and dynamic stress-strain curves, solid lines and symbols indicate experimental and numerical results respectively.*

B. EXPERIMENTAL RESULTS

Results already described elsewhere [2] are summarized here stress-strain curves for static ($\overset{\circ}{\varepsilon} = 10^{-4}\text{s}^{-1}$) and dynamic ($\overset{\circ}{\varepsilon} = 10^{+3}\text{s}^{-1}$) compression loading are presented on Fig. 1. Flow stress appears to be higher under dynamic loading than under static one, but strain hardening is observed to be the same in both cases.

Microstructural characterizations using optical microscopy show that twinning is an important deformation mode. The fraction of twinned grains increases with strain (Fig. 2). The curve for static loading is found to be in good agreement with Mullin's results on a titanium of similar purity than that in this study [3]. Slip and twinning modes have been identified by transmission electron microscopy.

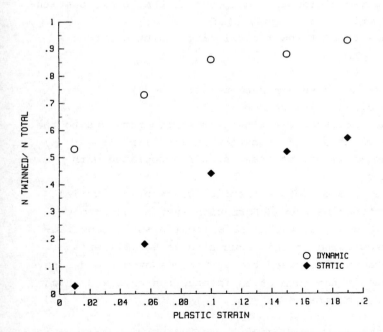

FIG. 2 *Fraction of twinned grains versus plastic strain for static and dynamic loading (experimental results).*

III. MODEL

A. MODEL DESCRIPTION

A self-consistent model first established for cubic structures [4] has been adapted to hexagonal structures. The
main assumption of this model is that material is macro-
homogeneous and micro-inhomogeneous.

First of all, local behavior is determined and is then
homogeneized in terms of global behavior using scale transition laws. The full description of this model has been
given in reference [4]. The adaptation to hexagonal structures is described here.

The main difference between cubic and hexagonal structure lies in the role of twinning in the latter structure.
It is then needed to identify evolution laws for deformation
by twinning and for interaction between twinning and glide
deformation modes.

For a given dislocation slip system the local constitu-
tive law is written as: $\underline{\dot{\varepsilon}}^P = \underline{R}\dot{\gamma}$ (1)

The same type of law for twinning can be written:
$\underline{\dot{\varepsilon}}^P = \underline{R}g\dot{f}$ (2)

where

$\underline{\dot{\varepsilon}}\pi$=the plastic strain rate tensor

$\underline{R}$ = an orientation tensor

$\dot{\gamma}$ = the amplitude plastic shear strain rate due to the
displacement of a dislocation in its slip plane

g = the amplitude of shear strain associated with
twinning

$\dot{f}$ = the rate of volume fraction of twinned material

Interactions between deformation systems have to be
considered in order to establish a local plastic constitutive law. According to the assumption of Berveiller's
model, interactions between two systems denoted r and s have
been included in the form of an interaction matrix H which
coefficients are defined as [5]:

$H_{rs} = B_1 + B_2.\det(\epsilon^{Pr} - \epsilon^{Ps})$ (3)

where B_1 and B_2 are material-dependent constants. It can be
seen that weak interaction coefficient are reached when det

$(\epsilon^{pr} - \epsilon^{ps}) = 0$, this condition being a necessary condition for the existence of common habit plane and shear direction for both systems. On the other hand, strong interaction coefficients may be obtained when det $(\epsilon^{pr} - \epsilon^{ps}) \neq 0$.

The interaction between two deformation systems is then taken into account in the relation between the critical resolved shear stress rate for a system $r : \overset{\circ}{\tau}{}^{r}_{crss}$ and the amplitude of plastic shear strain rate for a deformation system s $(\overset{\circ}{\gamma}{}^{s})$ as:

$$\overset{\circ}{\tau}{}^{r}_{crss} = \Sigma H_{rs}\overset{\circ}{\gamma}{}^{s} \quad (4)$$

On the other hand a deformation system r is active if, according to Schmid's law:

$$\tau^{r}_{crss} = \sigma_{ij}R^{r}_{ij} \quad (5) \quad (\sigma_{ij} = \text{the applied stress tensor compo-}$$
nents

Consequently from the expressions (1) (2) (4) and (5) the following law can be obtained:

$$\epsilon_{ij} = \Sigma_{rs}R^{r}_{ij}K^{rs}R^{s}_{kl}\sigma_{kl} \quad (6) \qquad (\underline{K} = \underline{H}^{-1})$$

B. NUMERICAL RESULTS

In order to characterize the single crystal microscopic behavior, active deformation systems have been identified. Using experimental results thirty six systems have been found to be possibly active.

6 {1012} <1120>glide systems
12 {1011} <1120>glide systems
6 {1012} <1011>twinning systems
6 {1122} <1123>twinning systems and
6 {1121} <1126>twinning systems

The polycrystal has been described by 100 spherical grains and an isotropic texture is described by 30 orientations. For static loading, critical shear stresses are as follow [3]: 40 MPa for prismatic glide, 140 MPa for pyramidal glide and {1012} twinning and at last 186 MPa for {1122} and {1121} twinning.

Macroscopic Behavior Considering the numerical results obtained with different values of the strain hardening matrix and of the critical shear stresses for dynamic loading, those which are the closest to the experimental ones

have been obtained with the following values and are the
ones which are presented here.

$$H_{mn} = \frac{\mu}{240} + \frac{\mu}{24} \times A_{mn}$$ (μ is the shear modulus = 40 000 MPa)

τ_c values for prismatic glide, pyramidal glide and 1012
twinning under dynamic loading = 102 MPa, 214 MPa and 186
MPa respectively.

Calculated stress-strain curves obtained after static
compression are presented on Fig. 3. Polycrystals were
loaded along three axes 1, 2 and 3 which are at 90° one from
another. Although the initial texture should be isotropic,
the response is not exactly the same along the three compression axes, the difference between two curves being of
about ten per cent.

This phenomenon may be due to the fact that the number
of grains and orientations chosen are not sufficient.
Nevertheless, the stress-strain curves are similar for the
three orientations although the yield stress is lower for
axis 3. The response along axis 1 is the closest to the
experimental results. This axis has thus been chosen for
microscopic phenomena analysis and to study the influence of
strain rate (Fig. 1).

Material data given above lead to a good agreement
between calculated and experimental curves. Considering
that dislocation motion is more difficult during dynamic
than during static loading, increase critical resolved shear
stresses with strain rate can be justified. The use of the
same strain hardening matrix for static and dynamic loading
seems to lead to a very good agreement between experimental
and numerical results. To confirm this good agreement it is
necessary to verify microscopic active deformation mechanisms predictions.

Microscopic Behavior Predicted active deformation systems
for static and dynamic loading are shown in Fig. 4. From
the beginning of the deformation up to 0.35% strain, only
prismatic glide is predicted to be active. The four other
systems begin only for strain above 0.35%. As the strain

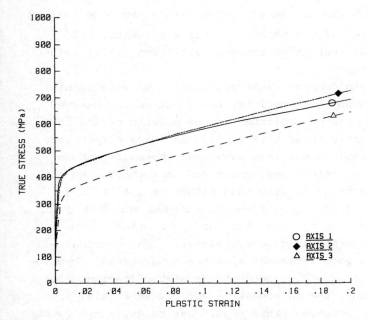

FIG. 3 *Numerical stress-strain curves for static loading.*

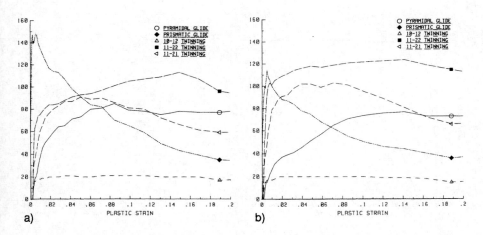

FIG. 4 *Numerical prediction for active deformation systems: a - static loading; b - dynamic loading.*

rate increases, the number of active {1012} twinning systems does not change but the number of active prismatic and pyramidal glide systems decreases, {1122} and {1121} twinning ones increase.

In order to compare these predictions to experimental optical microscopy observations, the fraction of observed twinned grains has been plotted versus plastic strain for static and dynamic loading (Fig. 5). For each strain rate, three conventional conditions have been chosen for these measurements, a grain being considered as twinned when the fraction of twins is greater than either 0%, 5% or 7.5% respectively. Considering $F_T > 0\%$, the curves for static and dynamic loading are similar, but for $F_T > 5\%$ and 7.5% the curves under dynamic loading are higher. The numerical results are in good agreement with the experimental observations for dynamic loading but not for static loading. For this latter case, the agreement is good if an initial F_T value of 5% is assumed. This result may be explained by the appearance of a large number of micro-twins during static loading which may not be observable by optical microscopy.

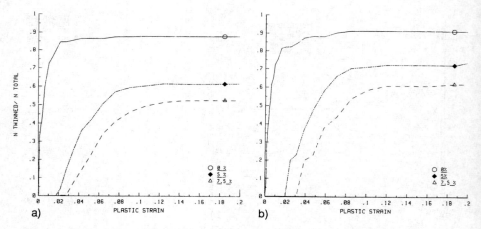

FIG. 5 *Numerical prediction for the fraction of twinned grains: a - static loading; b - dynamic loading.*

IV. DISCUSSION

This model allows a good prediction of the macroscopic behavior (σ - ϵ curve) of pure titanium. As far as the microscopic behavior is concerned, satisfactory predictions of activated deformation mechanisms under static as well as dynamic loading are obtained.

The predicted number of twinned grains is the same for static and dynamic loading contrary to experimental results. This can be explained by the fact that assumed critical resolved shear stresses (τ_{crss}) chosen for dynamic loading may not be realistic. Additional computing should be performed using greater τ_{crss} for glide system but smaller ones for twinning systems.

Finally, microstructural observations show that in the majority of grains, only one or two twinning systems have been activated. When a first twinning system is activated, it may then be difficult to activate other ones at later deformation stages. This has not been thoroughly taken into account in the H matrix.

Complementary computing have to be performed: first to estimate the model sensibility to the τ_{crss} parameter, and secondly to take into account the texture of such anisotropic structure. Some experimental tests should also be devised to determine the dynamic critical resolved shear stresses.

V. CONCLUSIONS

As well as in the cubic structure materials where only one kind of deformation mechanism (glide) is active, flow stress in hexagonal α-titanium increases with strain rate. In the case of titanium, there are two types of mechanisms which are activated (glide and twinning). Observed active deformation systems are the same under both static and dynamic loading, but the number of twinned grains is more important after dynamic one.

Calculated stress-strain curves obtained by numerical simulations based on a self-consistent model have shown that

it is possible to describe the macroscopic behavior, but
parameters (τ_{crss}, H matrix) still need to be fitted or
measured in order to correlate experimental and numerical
microscopic phenomena. The texture effect has to be introduced in the model using a non isotropic initial orientation
distribution of grains and a larger number of grains.

REFERENCES

1.   H. Kolsky, *Proc. Phys. Soc. B62,* 676(1949).

2.   S. Leclercq, C. Nguy and P. Bensussan, Int. Phys. Conf.
     Ser. N° 102: session 6b, 299(1989).

3.   S. Mullins, B.M. Patchett, *Met. Trans. 12A,* 853(1981).

4.   P. Lipinski, M. Berveiller, *Int. J. of Plasticity, 5,*
     149(1989).

5.   S. Gabelotaud, Thèse Université d'Orsay - Paris XI,
     1990, (to be published).

14

Modeling of Flow Stress in Steels as a Function of Strain Rate and Temperature

F. BURGAHN, O. VÖHRINGER and E. MACHERAUCH

Institut für Werkstoffkunde I
Universität Karlsruhe (TH)
D-7500 Karlsruhe, FRG

A constitutive equation between flow stress, temperature and strain rate will be presented which is based on the description of thermally activated glide of disloca-tions over short-range obstacles. A new method to determine the properties of this equation will be developed and discussed.
   Deformation tests at conventional strain rates and temperatures T $\leq$ 0.3 T_m (T_m = melt. temperature) are carried out on steel Ck 45 (SAE 1045) and X 45 Cr Si 9 3 (SAE HNV 3). The modelling of the flow stress to high strain rates using the con-stitutive equation shows a good agreement with experimental data.

I. INTRODUCTION

The relation between flow stress σ, temperature T and strain rate $\dot{\varepsilon}$ of metallic ma-terials is controlled for T $\leq$ 0.3 T_m (T_m = melting temperature in K) by the thermally activated slip of dislocations over short-range obstacles. The quantitative descrip-tion of the influence of T and $\dot{\varepsilon}$ on the flow stress is given by a constitutive equation [1,2,3] based on the results of Seeger's flow stress theory [4].

   In the present work a method will be described which allows the estimation of the properties of this constitutive equation at conventional $\dot{\varepsilon}$-values. Corresponding investigations were carried out using the normalized steel Ck 45 (SAE 1045) and the quenched and tempered valve steel X 45 CrSi 9 3 (SAE HNV 3).

II. SOME FUNDAMENTAL ASPECTS

If the thermally activated slip of mobile dislocations over short-range obstacles dominates the deformation behaviour, the flow stress can be separated below a specific, strain rate $\dot\varepsilon$ dependent temperature T_0 into athermal flow stress σ_G and thermal flow stress σ^*. This is schematically shown in Fig. 1. Derived from dislocation theory and statistic mechanics, the following relationship between the local occurrence of the activation enthalpy ΔG, which is responsible for the thermally activated overcoming of short-range obstacles, and the present strain rate [5,6] is valid:

$$\dot\varepsilon = \dot\varepsilon_0 \cdot \exp\left[-\frac{\Delta G}{k \cdot T} \right].$$

(1)

$\dot\varepsilon_0$ is $(\rho_m \cdot b \cdot L \cdot \nu_0)/M_T$. ρ_m is the density of mobile dislocations, b the magnitude of the Burgers vector, L the average distance between the short range obstacles, ν_0 the Debye frequency, M_T the Taylor-factor, and k the Boltzmann constant. In case of a short-range obstacle a description of ΔG is given by the force-distance curve as schematically illustrated in Fig. 2a showing a dislocation in interaction with an obstacle. The force F*, which is proportional to the thermal flow stress σ^*, is plotted as a function of the distance x. At $0\,K < T < T_0$ the activation enthalpy is given by

$$\Delta G = \Delta G_0 - f(T, F^*).$$

(2)

After carrying out a coordinate transformation, a relation between the short-range force F*(d) and the obstacle breadth d results, which is plotted in Fig. 2b. Fig. 2c illustrates that the relation between the thermal flow stress $\tau^* = \sigma^*/M_T$ and the activation volume $v = b \cdot L \cdot d$ is derived from Fig. 2b by multiplication of F* with $b \cdot L/M_T$ and of d with $b \cdot L$. It can be seen that at $T = 0\,K$ the thermally available activation enthalpy ΔG is zero. At $T > 0\,K$, the free thermal enthalpy ΔG is available for the dislocation and therefore only the short range force $F^* < F_0^*$ or $\sigma^* < \sigma_0^*$ is necessary to overcome the obstacle. Thus, Eq. 3 is valid:

$$\Delta G_0 - \Delta G = \frac{1}{M_T} \int_V^{V_0} \sigma^*(v)\ dv + \frac{1}{M_T}\,\sigma^* \cdot v .$$

(3)

At

$$T = T_0 = \Delta G_0/(k \cdot \ln(\dot\varepsilon_0/\dot\varepsilon))$$

(4)

the thermally available activation enthalpy $\Delta G(T_0) = \Delta G_0$ and allows the overcome of the obstacle without external force F*; therefore σ^* is zero. T_0 increases to higher values with increasing strain rate $\dot\varepsilon$. If ΔG is fitted with the potential law [1,2,3,5]

$$\Delta G = \Delta G_0 \cdot \left[1 - \left[\frac{\sigma^*(T)}{\sigma_0^*} \right]^{1/m} \right]^{1/n},$$

(5)

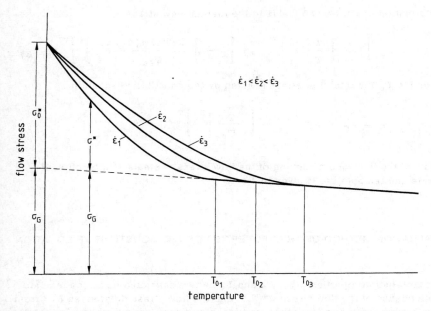

FIG. 1 *Flow stress as a function of the temperature at different strain rates*

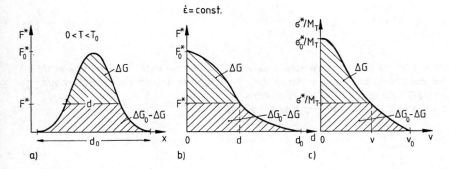

FIG. 2 *Force-distance curve of a short-range obstacle and thermal activated overcome by a dislocation*

a transformation of Eq. 1 and 5 yields to the thermal flow stress

$$\sigma^* = \sigma_0^* \cdot \left[1 - \left(\frac{T}{T_0} \right)^n \right]^m = \sigma_0^* \cdot \left[1 - \left(\frac{k \cdot T \cdot \ln(\dot{\varepsilon}_0/\dot{\varepsilon})}{\Delta G_0} \right)^n \right]^m . \qquad (6)$$

Thus, for $T \leqslant T_0$ the total flow stress is given by the constitutive equation

$$\sigma = \sigma_G + \sigma_0^* \cdot \left[1 - \left(\frac{T}{T_0} \right)^n \right]^m . \qquad (7)$$

The basis of the following modelling of the flow stress of steels as a function of strain rate and temperature is given by Eq. 7.

III. EVALUATION METHOD OF THE PROPERTIES OF THE CONSTITUTIVE EQUATION

The characteristic properties ΔG_0, $\dot{\varepsilon}_0$ and T_0 can be estimated using the temperature dependence of the flow stress $\sigma^*(T, \dot{\varepsilon})$ and the flow stress differences $\Delta \sigma^*$ from strain rate jump tests. The algorithm used for their determination is given in the following:

1) Determination of T_0 and $\sigma_G(T_0)$ from the flow stress-temperature curves at $\dot{\varepsilon} =$ const. and $\varepsilon_p =$ const. (see Fig. 1).

2) Determination of the thermal flow stress component σ^* from the temperature dependence of the flow stress at $\dot{\varepsilon} =$ const. and $\varepsilon_p =$ const. with

$$\sigma^*(T, \dot{\varepsilon}) = \sigma(T, \dot{\varepsilon}) - \sigma_G . \qquad (8)$$

3) Determination of the activation volumina from strain rate jump tests using the estimated flow stress jumps $\Delta \sigma^*$ at constant temperature and plastic strain with

$$v = M_T \cdot k \cdot T \cdot \frac{\Delta \ln \dot{\varepsilon}}{\Delta \sigma^*} . \qquad (9)$$

4) Plotting of the thermal flow stress as a function of the activation volume at different temperatures and integration of the σ^*-v-graph to determine the difference $\Delta G_0 - \Delta G(T)$ as a function of the temperature.

5) Establishing Eq. 10 from Eq. 1 and 4 yields

$$\frac{\Delta G(T)}{\Delta G_0} = \frac{T}{T_0} . \qquad (10)$$

This allows the extrapolation of the integrated values of $\Delta G_0 - \Delta G(T)$ for $T = 0$ K and therefore the estimation of ΔG_0.

A computer program which enables the application of numerous sets of data from deformation experiments at different strain rates was developed [7]. First, the values ΔG_0, $\dot{\varepsilon}_0$ und T_0 are estimated. Then, a complete iteration procedure to calculate m, n and σ_0^* using the constitutive equation in modified forms, also allows the approximation of measured values from strain rate jump tests. The approximation of the thermal flow stress as a function of temperature at $\dot{\varepsilon}$ = const. and as a function of strain rate at T = const., respectively, is accomplished using Eq. 6. The flow stress jumps as a function of temperature are approximated at constant origin strain rate with the relation

$$\Delta\sigma^*(T,\dot{\varepsilon}) = \sigma_0^* \cdot m \cdot n \frac{\Delta\ln\dot{\varepsilon}}{\ln\dot{\varepsilon}_0/\dot{\varepsilon}} \left[1 - \left[\frac{T}{T_0}\right]^n \right]^{m-1} \cdot \left[\frac{T}{T_0}\right]^n \quad (11)$$

resulting from the transformation of Eq. 6 and 9. Thus, a nonlinear equation system with the unknown parameters σ_0^*, m and n approximates the experimental data, which can be solved after linearizing at the investigated parameter positions.

IV. EXPERIMENTAL RESULTS AND DISCUSSION

In the following the single steps to determine the characteristic properties of Eq. 7 will be presented exemplarily for the steel Ck 45 (SAE 1045). Round specimens with a diameter of 3.8 mm and a gauge length of 60 mm were normalized (850 °C/1h). The mean ferrite grain size, determined by a linear intercept method, was 8 μm. Quasistatic tensile tests were carried out on a 500 KN universal testing machine (Zwick) which is equiped with a low temperature cooling box (Cyroson). Stress-strain curves obtained from tensile tests carried out at $\dot{\varepsilon} = 10^{-4}$ s^{-1} and 10^{-3} s^{-1} in the temperature range 81 K $\leq$ T $\leq$ 423 K were used to determine the 1.5-proof stress, illustrated in Fig. 3, as a function of temperature. To calculate the activation volumina with Eq. 9, strain rate jump tests were carried out. The observed flow stress jumps were evaluated as shown by [8]. Fig. 4 shows the estimated flow stress jumps for two different origin strain rates as a function of temperature. The amount of the strain rate changes was always a factor of ten. Fig. 5 shows the relationship between the thermal flow stress σ^*/M_T and the activation volume v which is necessary to determine $\Delta G_0 - \Delta G(T)$ using Eq. 3. An integration at defined temperatures leads to the curve of the activation enthalphy $\Delta G_0 - \Delta G(T)$ as a function of temperature at the strain rates $\dot{\varepsilon}_1$ and $\dot{\varepsilon}_2$. Thus, at T = 0 K, ΔG_0 = 0.63 eV and at $\Delta G_0 - \Delta G$ = 0, $T_0(\dot{\varepsilon}_1)$ = 276 K and $T_0(\dot{\varepsilon}_2)$ = 302 K. Also Eq. 4 yields $\dot{\varepsilon}_0 = 4.05 \cdot 10^7$ s^{-1}. From the solution of the nonlinear equation system one obtains m = 1.78, n = 0.5 and σ_0^* = 2000 N/mm^2. By modelling with the constitutive Eq. 6 and 7, it is now possible to extrapolate flow stress data to lower temperatures and higher $\dot{\varepsilon}$-values. Fig. 7 exemplarily shows the extrapolation to highest strain rates. A comparison with the results obtained from high strain rate deformation tests carried out at room temperature [9] reveals a good agreement between the experimentally measured and the calculated flow stress data.

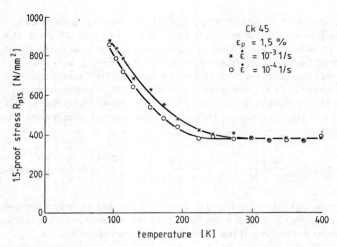

FIG. 3 *1.5-proof stress as a function of the temperature*

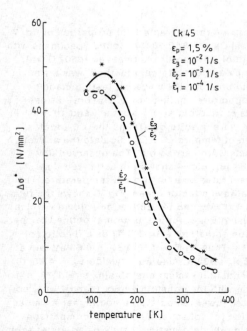

FIG. 4 *Flow stress jump $\Delta\sigma^*$ from strain rate jump tests as a function of the temperature*

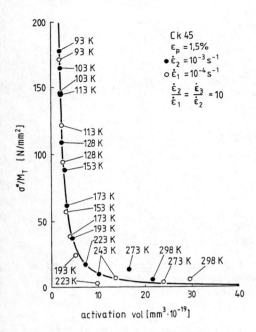

FIG. 5 *Relationship between the thermal flow stress* σ^*/M_T *and the activation volume v*

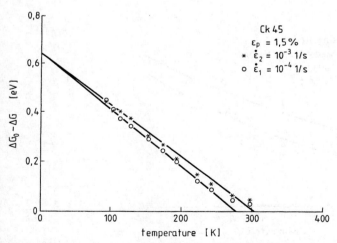

FIG. 6 $\Delta G_o - \Delta G$ *as a function of temperature at* $\dot{\varepsilon} = 10^{-3}$ s^{-1} *and* 10^{-4} s^{-1} *at* $\varepsilon_p = 1.5\%$

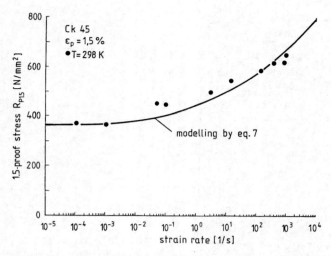

FIG. 7 *Measured and calculated 1.5-proof stress as a function of the strain rate at Ck 45 (SAE 1045)*

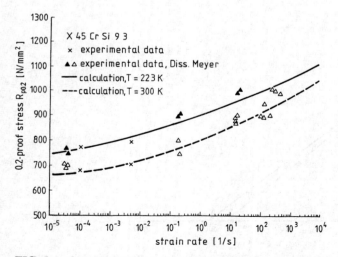

FIG. 8 *Measured [11] and calculated 0.2-proof stress as a function of temperature and strain rate at X 45 Cr Si 9 3 (SAE HNV 3)*

Analogous experiments were carried out at conventional strain rate values $\dot{\varepsilon}$ and temperatures T $\leq$ 423 K using the quenched and tempered steel X 45 Cr Si 9 3 (SAE HNV 3). The analysis of the temperature dependence of $\sigma*$ at ε_p = 0.2% (0.2-proof stress) and the $\Delta\sigma*$-values evaluated from strain rate jump tests gives the parameters of the constitutive eq. 6 with $\sigma_0^* = 552$ N/mm^2, $\dot{\varepsilon}_0 = 1.04 \cdot 10^6$ s^{-1}, $\Delta G_0 = 0.66$ eV just as m = 1,75 and n = 1. The parameters, with the exception of σ_0^*, are similar to the values of Ck 45. σ_0* is essentially smaller than in the case of normalized Ck 45. This finding is explained by the so called "alloy softening", where a high amount of solute atoms causes a reduced temperature dependence of the thermal flow stress [10]. Extrapolations to higher strain rates with Eq. 7 are also possible. Fig. 8 shows a comparison between the measured flow stress with identical material state [11] and the calculated curves, using Eq. 7, at T = 223 K and T = 300 K. The agreement between the modelling and the experimental findings is also satisfying.

V. CONCLUSION

A new method to estimate the properties of a constitutive equation which describes the flow stress as a function of temperature and strain rate is given. The exact knowledge of the relationship between the thermal flow stress component and the activation volume allows, by means of integration, to determine the activation enthalpy necessary for the thermally activated overcoming of short range obstacles as a function of the temperature.

An exemplary investigaton of the two steels SAE 1045 und SAE HNV 3 at conventional $\dot{\varepsilon}$-values and low temperatures shows that the constitutive equation enables the extrapolation of flow stress data to highest strain rates, which shows a good agreement with the measured flow stress values.

REFERENCES

1. Ono, K., *J. Appl. Phys.* 39 (1968), pp. 1803-1806
2. Vöhringer, O., *Z. Metallkde.* 65 (1974), pp. 32-36
3. Kocks, U.F., Argon, A.S., Ashby, M.F., *Progr. Mat. Sci.* 19 (1975)
4. Seeger, A., *Z. Naturforschung* 9a (1954), pp. 758-775, 865-869, 870-881
5. Macherauch, E., Vöhringer, O., *Z. Werkstofftechnik* 9 (1978), pp. 370 - 391
6. Gerthsen, C., Kneser, H., Vogel, H., *Springer- Verlag,* Berlin (1986)
7. Schulze, V., *Studienarbeit,* Universität Karlsruhe (1988)
8. Munz, D., Macherauch, E., *Z. Metallkunde,* 57 (1966), pp. 442-451
9. Reinders, B.-O. , *unpublished results,* IFAM , Bremen (FRG)
10. Pink, E., *Z. Metallkde.* 67 (1976), pp. 564-567
11. Meyer, L.W., *Dissertation,* Universität Dortmund (1982)

15

Ductile Fracture of Cu-1% Pb at High Strain Rates

C. DUMONT and C. LEVAILLANT

Centre de Mise en Forme des Matériaux
Ecole Nationale Supérieure des Mines de Paris
Sophia Antipolis, 06560 VALBONNE, FRANCE

In order to study ductile damage at high strain rates, dynamic tensile tests are performed on notched round bars. A model material is elaborated to present homogeneous soft inclusions (lead) in a high purity matrix (copper). Two complementary approaches have been undertaken from these experimental results. The tests are analysed with a finite element code taking into account inertia effects and quantitative damage analysis has been performed using ionic abrasion on longitudinal section of deformed specimen.

I. INTRODUCTION

Fracture mechanisms for ductile materials, especially in the quasistatic strain rate range, have been widely studied. Three steps were identified: nucleation, growth and coalescence of microvoids around inclusions. Several laws are proposed in the literature for each stage:

- for quasistatic strain rates, two parameters are involved: the equivalent strain and stress triaxiality (ratio of the mean stress on the equivalent stress). The more they increase, the more damage develops. This general result concerns as well nucleation phenomenon (Argon [1], Beremin [2]), as growth (Rice and Tracey [3]) and coalescence (Latham and Cockroft [4], Oyane [5]) stages. With the aim of showing stress triaxiality influence, Hancock and Mackenzie [6] performed tensile tests on notched round bars with different notch radii.

Meanwhile, few works deal with high ductile materials such as copper [7,8].

-for dynamic strain rates, damage problems have been studied by Seaman et al. [9] using the plate impact test. A cumulative function represents material damage for a given state of strain:

$$N = N_0 \exp(-\frac{R}{R_1}) \tag{1}$$

where N is the number of microvoids with radius greater than R, N_0 the total number of microvoids and R_1 the characteristic size of the distribution. Such experimental device leads to very high strain rates above 10^6 s^{-1}. For that strain rate range, we find in the literature, peculiar damage laws arising parameters such as viscosity. Other authors use damage models, partially deduced from quasistatic data, for numerical simulation of very high strain rates experimental tests such as cylinder explosion (Nash and Cullis, [7]) or cylinder impact (Johnson and Cook, [8]) with rather good results. Nevertheless, Zureck et al. [10] show discrepancies between quasistatic data [7] and expanding ring tests performed on OFHC copper.

All these considerations lead us to characterize with more accuracy damage evolutions at high strain rates both from a global point of view (ductility evolution) and from a local one (growth of microvoids). Then we will show the links between these two approaches. Moreover, we propose to explore the strain rate range between 10^2 s^{-1} and 10^4 s^{-1} where few experimental data exist.

II. EXPERIMENTAL DEVICE AND METHODOLOGY

It is well known that the ductility of OFHC copper increases between quasistatic strain rates and dynamic strain rates [11].

In order to study damage mechanism involved at high strain rates, we have even chosen high purity copper (CuC1 type) in which soft inclusions have been intentionally added, that is one percent volumic of lead. The material has been elaborated by casting, then extruded to 22 mm diameter, and drawn to 8 mm diameter before annealing to 675 °C during two hours. Finally, we obtain a mean grain size of 30 μm. The lead inclusions are generally located at the triple points and are characterized by a mean diameter of 3 μm.

As in Hancock and Mackenzie works [6] we perform tensile tests with different state of stress using notched round bars with three different notch radii (Fig. 1). But in order to study dynamic damage we use an original tensile device available at CEA : a crossbow. A projectile, thrown at a velocity varying between 5 m/s and 30 m/s, impacts the specimen mobile head while the load is measured at the fixed head. An optical extensometer enables us to measure sample elongation. Finally, we obtain the load vs time and the displacement vs time curves. In

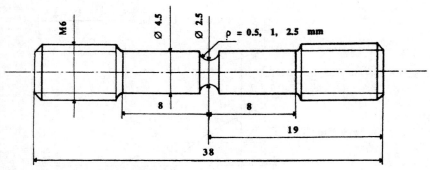

FIG. 1 *Specimens geometry*

order to characterize the material rheology, dynamic tensile tests are also performed on smooth round bars. Moreover, quasistatic tensile tests are also performed on the same samples in order to establish comparison between the two strain rates ranges.

III. DYNAMIC TENSILE TESTS ANALYSIS AND RESULTS

A DYNAMIC TENSILE TEST ANALYSIS FOR SMOOTH ROUND BARS

For a better understanding of the dynamic tensile test development, especially with regard to inertia effects or plastic strain waves, numerical simulation was done using the finite element method [12,13]. A one dimensional code and a modified version of the two dimensional code FORGE2® are used, both computations taking into account inertia effects.

After a transient period, strain rate is homogeneous along the specimen and equal to the value obtained if we analyse the test as a quasistatic one .The same conclusion can be given on the load or the strain evolution.

B DYNAMIC TENSILE TEST ANALYSIS FOR NOTCHED ROUND BARS

We are able to achieve local values of parameters which may influence damage evolution such as strain, stress triaxiality or strain rate.

Strain rate amplification Owing to strain localization in the notched region, higher strain rates are reached than in the case of smooth round bars for a same mobile head velocity. The multiplicative factor is about ten and strain rates exceed 10^4 s^{-1} in our configuration.

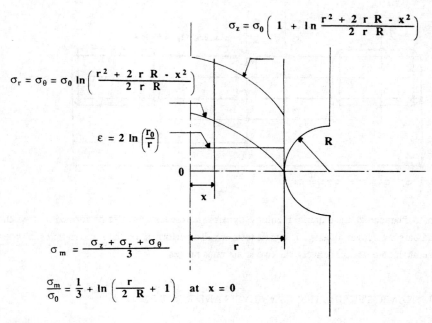

FIG. 2 *Bridgman analysis*

Strain values In the literature, the authors use the analysis of Bridgman (Fig. 2) to describe tensile tests performed on notch round bars [6,7]. The analysis involves a constant strain in the minimum section of the specimen and one of the assumptions is a low radius of the curvature of the notch. Clausing [14] has criticized this analysis.

Although the mean strain in the minimum section equals the value calculated by Bridgman (Fig. 2), our computation shows that the local strain increases from the specimen axis to the notch edge (Fig. 3).Whatever the strain rate, for a given geometry, a same relative variation of the minimum diameter leads to equivalent values of strain. Then we confirm that no difference exists during the quasistatic or the dynamic test development after a transient period. For this reason, a classical computation code (COMPACT2®), which does not take into account inertia effects, gives the same local strain evolution.

Stress triaxiality values In the same way, FORGE2® does not show any difference between quasistatic and dynamic strain rates (Fig. 4). However, as for Bridgman analysis, this computation code underestimates stress triaxiality values. As a conclusion, we will better prefer to use results obtained with COMPACT2®.

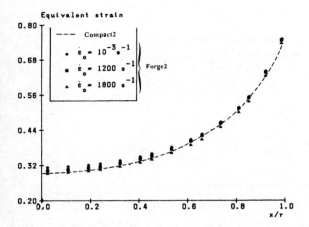

FIG. 3 *Strain evolution along the minimum radius of the specimen for a mean strain equal to 0.49 - ρ = 0.5 mm*

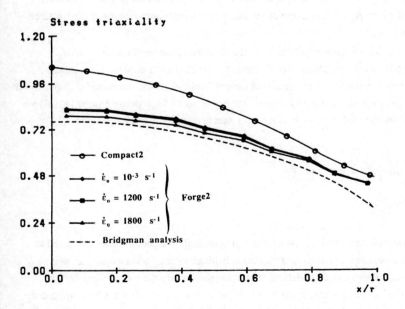

FIG. 4 *Stress triaxiality evolution along the minimum radius of the specimen for a mean strain equal to 0.49 - ρ = 0.5 mm*

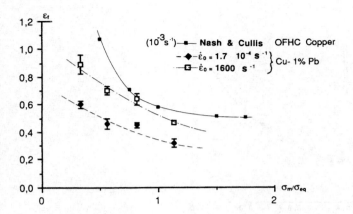

FIG. 5 *Influence of stress triaxiality on the fracture strain for various strain rates*

IV. DUCTILITY

Ductility is characterized by the strain ε_f to fracture in the minimum section of the specimens $\varepsilon_f = \ln(A_0/A)$, where A_0 is the initial minimum area and A the minimum area of the fractured specimens (Fig. 5).

In order to compare our results to those found in the literature [7], stress triaxiality values are obtained with the analysis of Bridgman (Fig. 2). So, we observe the classical influence of stress triaxiality on ductility. Because of lead inclusions, our material shows lower strain to fracture than for high purity copper. Moreover, as for high purity copper [11,12] we obtain higher ductility values at high strain rates than at quasistatic ones.

V. LOCAL DAMAGE APPROACH

A METHODOLOGY

Owing to numerical simulations, local values of mechanical parameters can be compared to local damage evolutions. The fractured specimen, is longitudinally polished till the median plane is reached. After an ionic polishing, in order not to modify the morphology and the dimensions of microvoids, they are mapped in adjacent regions of 230 μm x 310 μm near the fractured zone, using the Noesis image analysis system. As they correspond to inclusions that do not induce cavity growth until fracture, particles whose area are less than ten μm² are eliminated. Finally, as in Seaman's works, we use a cumulative function $N = f(S)$ to characterize local damage, where N is the number of microvoids with area greater than S.

Moreover, the mean grain shape aspect ratio leads us to assess the local strain which can be compared to the numerical results. Assuming the initial grains to be parallelepipedic, there is no change in the shape of the grain section and using the incompressibility, the equivalent strain is $\varepsilon = 2/3 \ln (F/F_0)$, where F_0 and F are the initial and the current grains aspect ratios.

B. RESULTS AND DISCUSSION

The good agreement between the computed and the measured strain along the minimum radius validates our numerical simulation(Fig. 6). In particular, we confirm that the dynamic tensile test can be considered as a quasistatic one after a transient period.

It appears that the damage is maximum at the notch edge (Fig. 7) i.e. at the maximum strain (Fig. 3) and at the minimum stress triaxiality locations (Fig. 4).

The major influence of strain on the microvoids growth is confirmed by increase of the mean area of microvoids in each region with strain (Fig. 8). On the other hand, there is a

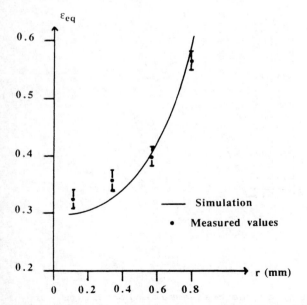

FIG. 6 *Comparison between the computed and the measured strain along the minimum radius for a fractured specimen at a tensile velocity equal to 25.1 m/s - ρ = 0.5 mm*

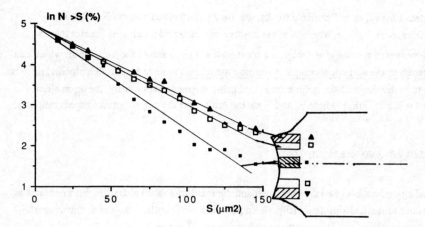

FIG. 7 *Evolution of the cumulative damage near the fractured region for a tensile velocity equal to 25.1 m/s − ρ = 0.5 mm*

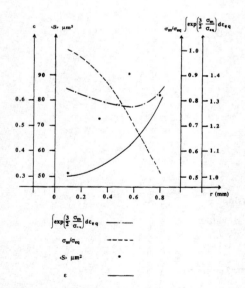

FIG. 8 *Evolution of strain, of the mean area of porosities, of the stress triaxiality and of the critical cavity growth according to Rice et Tracey [3] along the minimum radius for a tensile velocity equal to 25.1 m/s − ρ = 0.5 mm*

complete disagreement with the classical growth laws of microvoids such as Rice and Tracey relation [3] which emphasizes stress triaxiality influence:

$$\frac{\dot{V}}{V} = 0.287 \exp\left(\frac{3}{2}\frac{\sigma_m}{\sigma_{eq}}\right) \Delta\varepsilon \tag{2}$$

The same conclusion can be deduced for quasistatic tensile tests. We see no influence of strain rates on microvoids morphology or on damage evolution, as the major parameter may be the local equivalent strain.

However, this result is inconsistent with the global approach (see §IV). In fact, the influence of stress triaxiality is only minor. The higher the stress triaxiality, the lower the notch radius and the less homogeneous is the strain along the minimum radius (Fig.9). So, we obtain higher local strain values for the same mean strain when the notch radius decreases. Finally, if we compare strain evolution along the minimum radius for different notch radii, it appears that fracture occurs when the local strain overtakes a limit value for a given nominal strain rate (Fig.10).

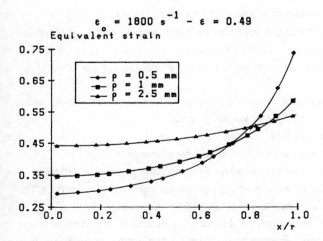

FIG. 9 *Strain evolution along the minimum radius for the different specimens geometries and for a same mean strain equal to 0.49 at* $\dot{\varepsilon}_0 = 1800$ *s⁻¹*

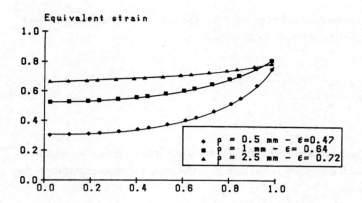

FIG. 10 *Strain evolution along the minimum radius when fracture occurs for the different specimens geometries at $\dot{\varepsilon}_0$ = 1600 s⁻¹*

VI. CONCLUSION

The use of notched specimen enables us to achieve strain rates exceeding 10^4 s⁻¹ with a dynamic tensile device usually leading to strain rates ten times lower. This mechanical test can be exploited as a quasistatic one after a transient period due to inertia effects and to plastic or elastic waves propagation. This result arises from numerical simulation, that are validated by the local strain measurement.

Moreover, it appears that the Bridgman analysis does not give satisfactory results with regard to tensile notched bars, especially for low notch radius:

-it underestimates stress triaxiality

-strain cannot be considered as a constant in the minimum section.

If ductility increases with strain rate, whatever the stress triaxiality, the local damage analysis does not show peculiar differences between quasistatic and dynamic strain rates. The increase of the fracture strain leads only to higher mean area of microvoids.

The stress triaxiality influence is indirect. The local approach shows that damage is only influenced by the local strain value. Fracture occurs when a critical strain, depending on strain rate, is achieved. High stress triaxiality or low notch radius leading to a less homogeneous strain in the minimum section, the critical strain will be reached for a lower mean strain.

REFERENCES

1.   A.S. Argon, J. IM and R. Safoglu, Met. Trans., 6A: 825 (1975)

2.   F.M. Beremin, Met. Trans., 12A: 723 (1981)

3.   J.R. Rice and D.M. Tracey, J. Mech. Phys. Solids, 17: 201 (1969)

4.   M.G. Cockroft D.J. LATHAM, J. Inst. Metals, 96: 33 (1968)

5.   M. Oyane, Bull J.S.M.E. 15, 90: 1507 (1972)

6.   J.W. Hancock and A.C. Mackenzie, J. Mech. Phys. Solids, 24: 147 (1976)

7.   M.A. Nash and I.G. Cullis, Proc. 3rd Conf. Mech. Prop. High Rates of Strain, Eds. J. Harding, Inst. Phys. Conf. Ser. No. 70: 307 (1984)

8.   G.R. Johnson and W.H. Cook, Eng. Fract. Mech., 21: 31 (1985)

9.   L. Seaman, D.R. Curran and D.A. Shockey, J. Appl. Phys., 47: 4814 (1976)

10.   A.K. Zurek, J.N. Johnson and C.E. Frantz, Proc. Int. Conf. on Mech. and Phys. Behavior of Mater. under Dynamic Loading, Les editions de Physique, 47-C3: 269 (1988)

11.   G. Regazzoni and F. Montheillet, Proc. Int. Conf. on Mech. and Phys. Behavior of Mater. under Dynamic Loading, Les editions de Physique, 46-C5: 435 (1985)

12.   C. Dumont, C. Levaillant, M. Arminjon and J.L. Chenot, Proc. Int. Conf. on Mech. and Phys. Behavior of Mater. under Dynamic Loading, Les editions de Physique, 47-C3: 505 (1988)

13.   C. Dumont, C. Levaillant, M Arminjon, J.P. Ansart and R. Dormeval, Proc. 4th Conf. Mech. Prop. High Rates of Strain, Eds. J. Harding, Inst. Phys. Conf. Ser. No. 102: 65 (1989)

14.   D.P. Clausing, J. Mater. JMLSA, 4:566 (1969)

16

Plastic Flow Localization at High Strain Rates

D. DUDZINSKI, M. EL MAJDOUBI, and A. MOLINARI

Laboratoire de Physique et Mécanique des Matériaux U.A CNRS n°1215
Faculté des Sciences, Ile du Saulcy, 57045 METZ cedex 01, FRANCE

*The influence of inertia and heat conduction on the deve-
lopment of plastic flow instability is studied for a plate submitted to a high strain rate biaxial loading.*

I - INTRODUCTION

The localized necking of a plate under biaxial loading has been a quite active field of research in the past two decades, however mainly quasi-static loadings were considered. In that case, the localization of plastic flow has been investigated with the bifurcation approach [1,2], the initial defect theory of Marciniak and Kuczynski,[3], and more recently with a linear perturbation analysis by Dudzinski and Molinari, [4]. With this last approach, general material behaviour can be accounted for, including initial anisotropy, strain rate sensitivity, thermal softening and heat conduction.

In this paper, we use the linear perturbation analysis to predict the plastic instability of an unbounded plate submitted to a dynamical biaxial loading. The sheet is assumed initially homogeneous. At any stage of the postulated homogeneous deformation process a perturbation is superimposed. The Lagrangian formulation appears to be the more suitable one and the perturbation is expressed in terms of the material coordinates. The choosen form of the spatial part of the perturbation allows to analyse plastic instabilities in form of bands. At high strain rates, we observe that the effects of heat generation and of inertia have a very important influence on the plastic flow localization. The critical wavelength corresponding to the highest rate of growth of the perturbation is conditioned by the inertia and the conduction effects.

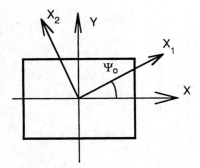

FIG. 1 *Geometry of the plate.*

For the deformation process of the sheet, the deformation gradient is assumed given by :

$$\mathbf{F} = \begin{bmatrix} F_{11} & F_{12} & 0 \\ F_{21} & F_{22} & 0 \\ F_{31} & F_{32} & F_{33} \end{bmatrix} \qquad (5)$$

with

$$F_{33} = \frac{\partial x_3}{\partial X_3} = \frac{h}{h(0)} = e^{\varepsilon_3} \qquad (6)$$

where ε_3 is the principal logarithmic deformation through the thickness h, h(0) is the initial thickness.

Using the relation (4), the flow law (1) takes finally the following form :

$$D_{11} = \dot{F}_{11}\, F_{11}^{-1} + \dot{F}_{12}\, F_{21}^{-1} = \dot{\varepsilon}\, T_{11}$$

$$D_{22} = \dot{F}_{21}\, F_{12}^{-1} + \dot{F}_{22}\, F_{22}^{-1} = \dot{\varepsilon}\, T_{22} \qquad (7)$$

$$D_{12} = \frac{1}{2}\left[\dot{F}_{11}F_{12}^{-1} + \dot{F}_{12}F_{22}^{-1} + \dot{F}_{21}F_{11}^{-1} + \dot{F}_{22}F_{21}^{-1}\right] = \dot{\varepsilon}\,T_{12}$$

II - GOVERNING EQUATIONS

The process of localized sheet necking in biaxial loading is carried out within the context of plane stress. Due to this assumption the non vanishing components of the Cauchy stress tensor in any frame (x_1,x_2) defined in the plane of the sheet are: σ_{11},σ_{22} and σ_{12}.

The strain rates, as well as the stress components are considered as uniform through the thickness of the plate. We restrict attention to incompressible rigid plastic and isotropic materials. The J2-flow theory used in this analysis is taken to be:

$$D_{\alpha\beta} = \lambda \frac{\partial \bar{\sigma}}{\partial \sigma_{\alpha\beta}} \qquad \alpha,\beta=1,2 \qquad (1)$$

where $\bar{\sigma}$ is the effective Von Mises stress used as the flow potential :

$$\bar{\sigma} = \left[\sigma_{11}^2 + \sigma_{22}^2 - \sigma_{11}\sigma_{22} + \frac{3}{2}\sigma_{12}^2 + \frac{3}{2}\sigma_{21}^2 \right]^{1/2} \qquad (2)$$

From the condition that the plastic deformation rate of work satisfies $\sigma_{ij} D_{ij} = \bar{\sigma}\,\dot{\bar{\varepsilon}}$, where $\dot{\bar{\varepsilon}}$ is the effective strain rate, it follows that:

$$\lambda = \dot{\bar{\varepsilon}} \qquad (3)$$

The strain rate tensor is defined by:

$$D = \frac{1}{2}(\dot{F}F^{-1} + F^{-1T}\dot{F}^{T})$$

or

$$D_{km} = \frac{1}{2}\left[\frac{d}{dt}(x_{k,K})\, X_{K,m} + \frac{d}{dt}(x_{m,K})\, X_{K,k} \right] \qquad (4)$$

$x_{k,K} = \partial x_k / \partial X_K$ are the components of the deformation gradient tensor $\mathbf{F}$, X_K are the material coordinates. The plate is submitted to an overall strain rate tensor $\mathbf{D}^o$ with principal axes X and Y. The reference frame (X_1,X_2) has a fixed angle Ψ_0 with respect to the axes (X,Y) (figure 1). By (X_1, X_2) and (x_1, x_2) are denoted the initial and present position of a particle in the reference frame.

with :

$$T_{11} = \frac{\partial \overline{\sigma}}{\partial \sigma_{11}} = \frac{2\,\sigma_{11} - \sigma_{22}}{2\overline{\sigma}} \quad ; \quad T_{22} = \frac{\partial \overline{\sigma}}{\partial \sigma_{22}} = \frac{2\sigma_{22} - \sigma_{11}}{2\overline{\sigma}}$$

$$T_{12} = \frac{\partial \overline{\sigma}}{\partial \sigma_{12}} = \frac{3}{2}\,\frac{\sigma_{12}}{\overline{\sigma}} \tag{8}$$

In a rigid viscoplastic solid, a part β of the plastic rate of work $\overline{\sigma}\,\dot{\overline{\varepsilon}}$ is converted into heat. The conservation of energy leads to the equation :

$$c\,\dot{\theta} = k\left(\frac{\partial^2 \theta}{\partial x_1^2} + \frac{\partial^2 \theta}{\partial x_2^2}\right) + \frac{k}{h}\left(\frac{\partial \theta}{\partial x_1}\frac{\partial h}{\partial x_1} + \frac{\partial \theta}{\partial x_2}\frac{\partial h}{\partial x_2}\right) + \beta\,\overline{\sigma}\,\dot{\overline{\varepsilon}} \tag{9}$$

where θ is the absolute temperature and k the thermal conductivity.

$\dot{\theta}$ and $\dot{\overline{\varepsilon}}$ are material derivatives of the temperature and the effective strain respectively. The Taylor-Quinney coefficient β is taken equal to 0.9 .

The thermoviscoplastic response of the material is described through a power law relating the effective stress $\overline{\sigma}$, the effective strain $\overline{\varepsilon}$, the effective strain rate $\dot{\overline{\varepsilon}}$ and the temperature θ :

$$\overline{\sigma} = \mu\,\overline{\varepsilon}^{\,n}\,\dot{\overline{\varepsilon}}^{\,m}\,\theta^{v} \quad ; \quad \overline{\varepsilon} = \int_0^t \dot{\overline{\varepsilon}}\,(\tau)\,d\tau \tag{10}$$

The exponents n, m, and v denote the strain hardening, the strain rate sensitivity and the thermal softening ($v < 0$) coefficients .

Under the assumption of uniform stresses and strain rates through the thickness, the equations of motion are :

$$\frac{\partial\left(h\,\sigma_{11}\right)}{\partial x_1} + \frac{\partial\left(h\,\sigma_{12}\right)}{\partial x_2} = \rho_0\,h\,\frac{dv_1}{dt}$$

$$\frac{\partial\left(h\,\sigma_{12}\right)}{\partial x_1} + \frac{\partial\left(h\,\sigma_{22}\right)}{\partial x_2} = \rho_0\,h\,\frac{dv_2}{dt} \tag{11}$$

With the incompressibility condition :

$$D_{11} + D_{22} + D_{33} = 0 \qquad (12)$$

a complete set of 9 equations is obtained: (2), (7), (9), (10), (11), (12). The boundaries of the plate are adiabatic. It is assumed that a basic homogeneous or quasi-homogeneous solution $\boldsymbol{\sigma}^{\,0}, \bar{\sigma}^0, \dot{\bar{\varepsilon}}^0, \theta^0$ corresponds to an applied overall strain rate $\mathbf{D}^{0\cdot}$ This occurs when the acceleration of particles resulting from the applied $\mathbf{D}^0$ are negligible (but the acceleration can be very important when a flow instability occurs). The unknowns are x_1, x_2, the stresses $\sigma_{11}, \sigma_{12}, \sigma_{22}, \bar{\sigma}$, the strain rate $\dot{\bar{\varepsilon}}$, the logarithmic deformation ε_3 and the absolute temperature θ.

The straining path is assumed linear and defined by the ratio ρ:

$$\rho = \frac{D_{YY}^0}{D_{XX}^0} = \frac{\dot{\varepsilon}_Y^{\,0}}{\dot{\varepsilon}_X^{\,0}} = \frac{\varepsilon_Y^0}{\varepsilon_X^0} = \text{Const.} \qquad (13)$$

where $\varepsilon^0{}_X$ and $\varepsilon^0{}_Y$ are the principal logarithmic deformations. For the homogeneous solution, the following deformation gradient expressed in the frame (X,Y,Z) is given by :

$$\mathbf{F} = \begin{bmatrix} e^{\varepsilon_X^{\,0}} & 0 & 0 \\ 0 & e^{\varepsilon_Y^{\,0}} & 0 \\ 0 & 0 & e^{-\left(\varepsilon_X^{\,0} + \varepsilon_Y^{\,0}\right)} \end{bmatrix} \qquad (14)$$

The basic solution is denoted by the superscript ^{0} :

$$\mathbf{S}^0 = \left(x_1^0, x_2^0, \sigma_{11}^0, \sigma_{12}^0, \sigma_{22}^0, \bar{\sigma}^0, \dot{\bar{\varepsilon}}^0, \varepsilon_3^0, \theta^0 \right)$$

In particular, the homogeneous principal stresses satisfy the following relations :

$$\sigma_{11}^0 = \sigma_X^0 \cos^2 \Psi_o + \sigma_Y^0 \sin^2 \Psi_o$$

$$\sigma_{22}^0 = \sigma_X^0 \sin^2 \Psi_o + \sigma_Y^0 \cos^2 \Psi_o \qquad (15)$$

$$\sigma_{12}^0 = -\left(\sigma_X^0 - \sigma_Y^0\right) \sin \Psi_o \cos \Psi_o$$

with :

$$\frac{\overset{o}{\sigma_X}}{\overline{\sigma}} = \frac{2 + \rho}{\sqrt{3}\left(1 + \rho + \rho^2\right)^{1/2}} \quad ; \quad \frac{\overset{o}{\sigma_Y}}{\overline{\sigma}} = \frac{1 + 2\rho}{\sqrt{3}\left(1 + \rho + \rho^2\right)^{1/2}} \tag{16}$$

The homogeneous temperature is obtained by integration of the energy equation (9), where all terms in the second member are disappearing, except $\beta \, \overline{\sigma} \, \dot{\overline{\varepsilon}}$:

$$\overset{o}{\theta} = \left[\frac{\beta \mu \overset{\star o}{\varepsilon}{}^m}{c} \frac{1 - \nu}{1 + n} \overset{-o}{\varepsilon}{}^{(1+n)} + \overset{o}{\theta}{}^{(1-\nu)}(0) \right]^{1/(1-\nu)} \tag{17}$$

$\theta^o(0)$ is the initial homogeneous temperature.

III - EFFECTIVE INSTABILITY ANALYSIS

The stability of the basic solution S^o is tested by superposition at any time t_0 of a small perturbation $\delta S = (\delta x_1, \delta x_2, ..., \delta \theta)$. The following form is stated for the components of δS :

$$\delta S_j = \delta S_j^i \, f_i \, \exp\left[\eta \left(t - t_o \right) \right] \tag{18}$$

δS_j is expressed in terms of f_i (i=1,2), which are unit base vectors defined by the relations :

$$\begin{aligned} f_1 &= \cos \xi \, X_1 \\ f_2 &= \sin \xi \, X_1 \end{aligned} \tag{19}$$

for example, the perturbation superimposed to the homogeneous temperature is written as :

$$\delta \theta = \overset{1}{\delta \theta} \cos \xi X_1 + \overset{2}{\delta \theta} \sin \xi X_1 \tag{20}$$

The spatial modulation of the perturbation is periodic and defined by the wave number ξ. The coefficient η in the relation (18) characterizes the growth of the perturbation if Re $(\eta) > 0$ (instability) or the decay if Re $(\eta) < 0$ (stability). When instability occurs the growth of the perturbation is viewed to represent the early stages of a localization process in form of bands.

The solution is :

$$S_j = S_j^o + \delta \, S_j^i f_i \exp\left[\eta \left(t - t_o \right) \right] \tag{21}$$

After substitution of (21) into the previous set of 9 equations presented, linearization and projection on the basis f_i, 18 linearized equations are obtained where the unknowns are the components δS^i_j ($i=1,2$; $j=1,9$) of the perturbation.

A non trivial solution exists if and only if the determinant of this system vanishes. A polynomial equation in η is then obtained. The instability is effective when one of the roots of the polynomial equation in η satisfies the relation:

$$\text{Re}\,(\hat{\eta}) = e \quad ; \quad \hat{\eta} = \frac{\eta}{\overset{\cdot\,o}{\varepsilon}} \qquad (22)$$

where e is a large enough positive number [4].

For a given orientation Ψ_0, a given straining path ρ, a given wave number ξ, the value of the homogeneous effective strain $\bar{\varepsilon}^c$ for which the determinant vanishes, is calculated. The corresponding principal homogeneous strains ε^c_X and ε^c_Y are :

$$\varepsilon^c_X = \frac{\sqrt{3}\,\bar{\varepsilon}^c}{2\left(1+\rho+\rho^2\right)^{1/2}} \quad ; \quad \varepsilon^c_Y = \rho\,\varepsilon^c_X \qquad (23)$$

The first mode to attain an effective instability of intensity e is characterized by the initial orientation Ψ^e_0 minimizing ε^c_X :

$$\varepsilon^e_X(e,\rho,\xi) = \text{Inf}_{\Psi_0}\,\varepsilon^c_X\left(e,\Psi_0,\rho,\xi\right) = \varepsilon^c_X\left(e,\Psi^e_0,\rho,\xi\right) \qquad (24)$$

The calculations are carried out for the 1020 HRS steel, which behaviour can be represented by a law of the form (10), where the parameters are defined by :

$$v = -0.51 \quad n = 0.12 \quad m = 0.0133 \quad \mu = 1.41 \times 10^{10}\ \text{SI}$$
$$\rho_0 = 7800.\ \text{kg/m}^3 \quad c = 3.6 \times 10^6\ \text{J/m}^3.\text{K} \quad k = 15\ \text{W/m.K}$$

In figure 2, the overall strain rate is fixed to $\overset{\cdot}{\bar{\varepsilon}}^0 = 10^4\ \text{s}^{-1}$, the effective instability parameter is choosen equal to 10, the straining paths are $\rho = 0$ (plane strain) and $\rho = -0.5$ (uniaxial tension). In order to access the influence of heat conduction and inertia on the stability of plastic flow, the effective critical strain ε^e_X is plotted against the wave number ξ for different conditions of deformation:

1- isothermal ($\beta=0$ and $\delta\theta^o=0$) and quasi static ($\rho_o=0$) process, $\varepsilon^e{}_X$ does not depend on the wave number ξ.

2–non conducting material (k=0) and quasi-static process, $\varepsilon^e{}_X$ keeps a constant value smaller than the previous one. This corresponds to the wellknown destabilizing effect of thermal softening material ($v < 0$).

3- heat conducting material and quasi-static process. Heat conduction refrains the temperature localization and stabilizes the plastic flow for large wave number ξ.

4- heat conducting material and dynamic process. Perturbations with a small wave number are stabilized by the inertia forces while those with large wave numbers are stabilized by heat conduction. This results to critical wavelength ξ^c for which $\varepsilon^e{}_X$ is minimum. The perturbation associated to ξ^c is likely to develop in the material.

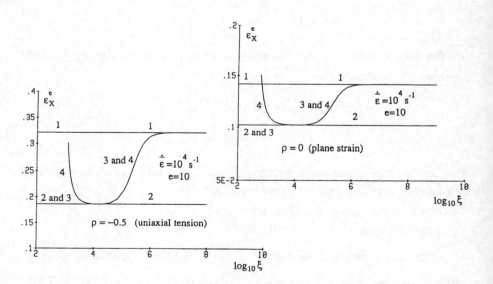

FIG. 2 *Effective critical strain $\varepsilon^e x$ as a function of wave number ξ. Influence of inertia and thermal effects.* $\dot{\varepsilon}^o = 10^4 s^{-1}$

For increasing strain rates, the inertia effects are more important and the heat conduction effects are transfered to higher wavelengths as shown in figure 3.

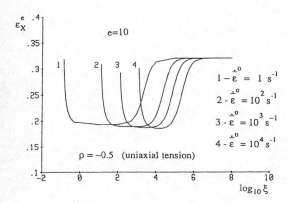

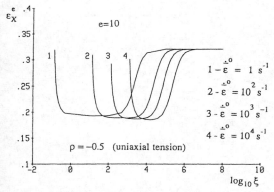

FIG. 3 *Influence of the strain rate.*

REFERENCES

1. S. Stören and J.R. Rice J. Mech. Phys. Solids 23: 421 (1975).
2. J.W. Hutchinson and K.W. Neale, Mechanics of sheet metal forming. Plenum Press, New York (1978).
3. Z. Marciniak and K. Kuczynski, Int. J. Mech. Sci. 9: 609 (1967).
4. D. Dudzinski and A. Molinari, Int J. Solids Struct. (in print).

17

The Deformation of Tungsten Alloys at High Strain Rates

R. S. COATES and K. T. RAMESH

Department of Mechanical Engineering
The Johns Hopkins University
Baltimore, MD 21218, U.S.A.

Compression and torsional Kolsky bars were used to subject several tungsten-nickel-iron (W-Ni-Fe) heavy alloys to high rates of deformation. The results indicate a distinct rate sensitivity over a range of strain rates from 10^{-4} sec^{-1} to 7×10^3 sec^{-1}, which is influenced by tungsten content and degree of prior swaging but is almost independent of tungsten grain size. Adiabatic shear localization has been observed in high-rate shearing tests; narrow shear bands are formed, followed immediately by catastrophic fracture. Metallographic analyses and microhardness measurements were performed to study the microstructural evolution with increasing strain at high rates.

I. INTRODUCTION

A typical tungsten heavy alloy (made by liquid phase sintering mixtures of tungsten and other elemental metal powders) is *W-Ni-Fe,* with tungsten contents in the range of 80-97% (by weight). These materials have a two phase microstructure, with hard tungsten particles in a tungsten-nickel-iron matrix. A relatively high tungsten content yields high strength and density while the ductile matrix provides reasonable deformability; this combination of properties makes these materials well suited to application in armor piercing kinetic energy penetrators. Since penetrator materials are subjected to high strain rates and large deformations, it is necessary to develop an understanding of the behavior of these alloys under such conditions.

203

II. BACKGROUND

Krock and Shepard [1], in one of the earliest studies of the mechanical properties of *W-Ni-Fe* heavy alloys, obtained quasistatic stress-strain curves in tension at room temperature for a range of tungsten contents (80-92%). Ekbom [2] performed quasistatic tensile tests over a range of temperatures on an as-sintered 90%W-6%Ni-4%Fe tungsten heavy alloy. Woodward *et al* [3], using drop tower compression tests, observed that at higher strains and intermediate strain rates the flow stress appeared to decrease with strain. Meyer *et al* [4] measured the dynamic properties in tension and compression of a tungsten heavy alloy in the as-sintered and swaged conditions. In tension, both process conditions showed an increase in flow stress with strain rate, with the as-sintered material showing a much higher rate sensitivity. Bose, *et al* [5] found that the rate sensitivity and flow stress in tension of an as-sintered 90W-7Ni-3Fe heavy alloy were inversely proportional to temperature. Nicholas [6] observed that the flow stress in tension of a 90W-7Ni-3Fe alloy increased with strain rate. Johnson *et al* [7] showed that the flow stress in shear of a 90%W alloy increased significantly with strain rate over a range from 10^{-3} to 10^{2} *sec*$^{-1}$; they also observed a strong softening response, attributed to localization.

The microstructure of a typical liquid phase sintered W-Ni-Fe alloy (Fig. 1) consists of nearly spherical BCC tungsten grains in an FCC W-Ni-Fe matrix. Woodward *et al* [8], in a comparative study of a 95%W alloy and an alloy of the matrix composition, observed that the matrix material had a greater ductility and work-hardened more rapidly. German *et al* [9, 10]

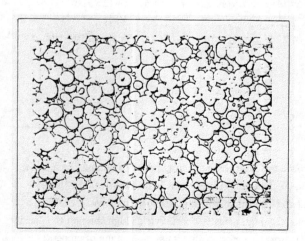

FIG. 1 *Microstructure of typical tungsten-nickel-iron heavy alloy (90%W-7%Ni-3%Fe, 15% swaged, supplied by GTE)*

studied the influence of W content and grain size on the mechanical properties in tension; in general, a coarser microstructure leads to a lower quasistatic strength. Clearly the microstructure of the material has a significant effect on the quasistatic properties; it is to be expected that this microstructural sensitivity will also extend into the high-rate regime.

The W-Ni-Fe heavy alloys evaluated in this work had a nickel to iron ratio of 7:3 and are commercially available materials manufactured by the Teledyne Firth Stirling Corp. and the GTE Corp. (GTE). The alloys studied had tungsten contents of 90, 91, 93, and 97%; samples of the 91%W alloy were swaged 0, 10, 15, and 20% and had two distinct grain sizes.

III. DESCRIPTION OF EXPERIMENTS

Kolsky [11] in 1949 developed a technique for the measurement of the properties of materials undergoing homogeneous compressive deformations at fairly high strain rates, from 10^2 to 10^4 sec^{-1}. The technique has also been adapted to the study of shearing deformations (the torsional Kolsky bar is described, for instance, by Hartley *et al* [12]). Several reviews of the theory and application of this technique have been published, *e.g.* by Lindholm [13].

The Kolsky compression bar consists of two long metal bars that are designed to remain elastic throughout the test. These bars sandwich a small cylindrical specimen. One end of one of the bars (the input bar) is impacted by a projectile fired from a gas gun. The impact generates a compressive stress wave that propagates down the input bar into the specimen. This wave reverberates within the specimen, sending a transmitted wave into the second bar (the output bar) and a reflected wave back into the input bar. Both reflected and transmitted pulses are measured using strain gages placed on the bars. If certain conditions are satisfied (discussed, *e.g.*, by Davies & Hunter [14] and Bertholf & Karnes [15]), the reflected pulse provides a measure of the strain rate imposed on the specimen, and the transmitted pulse provides a measure of the stress state within the sample. The strain gage signals thus provide the history of the strain rate and of the stress within the specimen during a test; by combining the two signals it is possible to obtain the stress-strain relation for the specimen material under moderately high strain rates. The adaptation to shearing deformations uses a torsional pulse rather than a compressive pulse, and the specimen is a thin tube. Detailed descriptions of the techniques and analyses used in this work may be found in Coates & Ramesh [16, 17].

In an effort to understand the relative contributions of each phase to the strength and deformation of these alloys, microhardness measurements were made in the tungsten grains and matrix for several specimens of a single alloy that had been taken to different strains at a nominal strain rate of 2×10^3 sec^{-1}. Measurements were only made in the largest grains and in the largest regions of the matrix in order to reduce the influence of nearest neighbors.

IV. RESULTS AND DISCUSSION

A true stress - true strain curve obtained for the GTE material during compression at a strain
rate of 2×10^3 *sec*[-1] is shown in Fig. 2. Very little strain hardening is observed for this
(swaged) material. Fig. 3 is a rate - sensitivity (at fixed strains) diagram for this alloy, showing
the results of the Kolsky bar tests (1.8×10^3 to 7×10^3 *sec*[-1]) as well as of quasistatic
compression tests performed for us by Dudder [18]. This figure also includes data from
Nicholas [19] and Follansbee [20] for similar swaged tungsten heavy alloys. It is seen that the
rate sensitivity of this tungsten heavy alloy is quite significant over this range of strain rates.

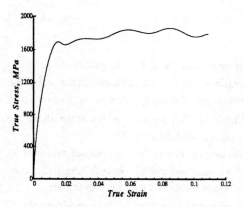

FIG. 2 *True stress – true strain curve obtained during
compression at a strain rate of 2010 per second (GTE alloy)*

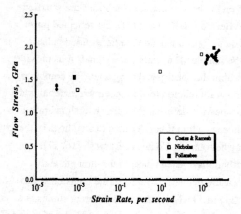

FIG. 3 *Rate sensitivity diagram for GTE alloy; comparison with
results of Nicholas [19] and Follansbee [20]*

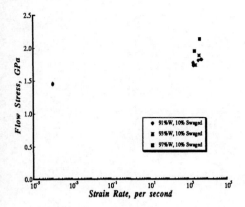

FIG. 4 *Rate sensitivity diagram for TFS alloys; note the effect of tungsten content*

The effects of tungsten content, swaging, and grain size on the high-rate behavior were studied through tests on the alloys provided to us by Teledyne Firth Stirling. Compressive stress-strain curves for alloys of varying tungsten content, obtained at a nominal strain rate of $4 \times 10^3 \ sec^{-1}$, are shown in Fig. 4. It is apparent that the higher tungsten contents lead to much stronger materials that are still fairly ductile (in compression). The influence of tungsten content on the rate sensitivity is shown in Fig. 5; the limited data available indicates an increasing rate sensitivity with increasing volume fractions of tungsten. Considering the high tungsten volume fractions involved, this result is consistent with the results for tungsten single crystals presented by Meyer and Chiem [21]. However, since the response of the matrix phase

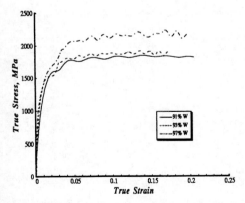

FIG. 5 *Effect of increasing tungsten content on the dynamic stress-strain behavior in compression (TFS alloys)*

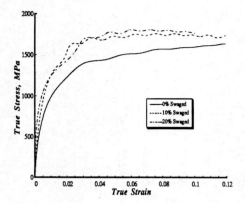

FIG. 6 *Effect of increasing degrees of prior swaging on the dynamic stress-strain behavior in compression*

to high rates of deformation has not yet been investigated, its contribution to the rate sensitivity is not known. Fig. 6 presents true stress - true strain curves at a strain rate of $2 \times 10^3 \ sec^{-1}$ in compression for the 91% W alloy in the as-sintered, 10% swaged, and 20% swaged conditions. The as-sintered material shows relatively strong work-hardening in comparison with the swaged materials; in addition, while a 10% swaging provides a dramatic increase in the flow stress, further swaging does not appear to provide as much incremental benefit, at least at these high strain rates. A secondary effect of increased swaging appears to be a reduction in the rate sensitivity. Finally, Fig. 7 shows that an increase in the tungsten grain size from approximately 40 to 65 microns appears to have negligible effect on the compressive

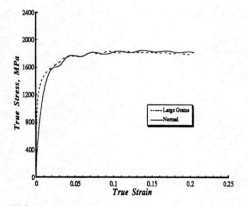

FIG. 7 *Effect of variation in tungsten grain size on the dynamic stress-strain behavior in compression (TFS alloys)*

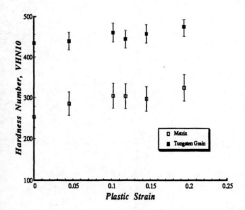

FIG. 8 *Microhardness data for tungsten grains and matrix for specimens taken to various strains at a constant nominal strain rate of 2000 per second (GTE alloy)*

stress-strain behavior at these strain rates. While these result contrast with the strong effect the tungsten grain size has on the tensile behavior with respect to the fracture strain [9], this is to be expected, since for these materials in compression the micromechanisms of large inelastic deformation are very different from those in tension.

Microhardness measurements have been used by several workers [2, 22, 10, 8] to study the contribution of the two phases to the strength and deformation of these alloys. The results of microhardness measurements made in the tungsten grains and the matrix for several specimens of the GTE alloy that had been taken to different strains at a nominal strain rate of $2 \times 10^3 \ sec^{-1}$ are shown in Fig. 8. The results indicate a greater degree of hardening of the matrix (than of the tungsten grains) with increasing deformation at this strain rate. However, there is too much scatter in the measurements to draw any firm conclusions regarding the relative contributions of the phases to the observed macroscopic behavior. Part of the scatter arises from inaccuracies in the measurements, part from nearest neighbor constraints, and part from the random orientations of the tungsten grains; Stephens [23] has reported different degrees of work hardening depending on the crystallographic orientation of tungsten.

Some preliminary tests have been conducted with the torsional Kolsky bar on a 91%W, 25% swaged W-Ni-Fe alloy. A shear stress - shear strain curve obtained on this material at a nominal shear rate of $260 \ sec^{-1}$ is shown in Fig. 9. No catastrophic failure was observed in this particular test; the specimen remained intact. However, adiabatic shear localization was observed in tests conducted at higher shear rates. As an example, Fig. 10 shows the surface of a torsion specimen that was subjected to a shearing deformation at a shear rate of $1650 \ sec^{-1}$.

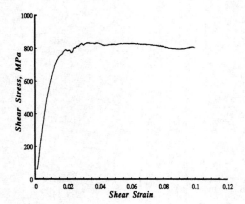

FIG. 9 *Shear stress – shear strain curve obtained during high rate shearing of a 91%W, 25% swaged tungsten-nickel-iron alloy. The nominal shear rate is 260 per second*

FIG. 10 *Micrograph showing region adjacent to a shear band developed in tungsten heavy alloy during shearing at a nominal shearing rate of 1600 per second*

Most of the material is relatively undeformed; however, over a length scale of 200 microns the tungsten grains begin to show evidence of shearing, and this localizes into a very narrow band of extremely intense shearing deformation. The band appears to involve the tungsten grains as well as the matrix. An interesting feature of adiabatic shear localization in this tungsten alloy (as opposed to that in 1018 CR steel, *e.g.*) that we have observed is that the initiation of localization appears to be followed by an immediate complete loss of load-bearing capacity.

V. CONCLUSIONS

Compression and torsional Kolsky bars were used to subject several tungsten-nickel-iron (W-Ni-Fe) heavy alloys to high rates of deformation. The results indicate a distinct rate sensitivity over a range of strain rates from 10^{-4} sec^{-1} to about 7×10^3 sec^{-1}. The rate sensitivity appears to increase with increasing tungsten content and decrease with increasing amounts of swaging; however, changes in the tungsten grain size had no effect on either the compressive flow stress or the rate - sensitivity. Stress - strain curves at these high rates show very little work hardening of the swaged alloys after small strains. High-rate shearing tests resulted in the formation of relatively narrow shear bands, followed immediately by catastrophic fracture.

VI. ACKNOWLEDGMENTS

Thanks are due to T. Penrice of Teledyne Firth Stirling and J. Spencer of GTE for the materials; G. Dudder of the Battelle Pacific NW Lab. for the quasistatic data; and the U.S. Army Ballistic Research Laboratory for providing RSC the opportunity to participate in this research.

REFERENCES

1. Krock, R.H., & Shepard, L.A., *Trans. Met. Soc. AIME*, 227: 1127 (1963).

2. Ekbom, L., *Modern Dev. Powder Met.,* MPIF, Princeton, 14: p. 177, 1981.

3. Woodward, R.L., *et al, Met. Trans. A*, 16A: 2031 (1985).

4. Meyer, L.W., *et al, Proc. 7th Int. Symp. Ballistics*, Hague, p. 289, 1983.

5. Bose, A., Sims, D., & German, R.M., *Met. Trans. A*, 19A: 487 (1988).

7. Nicholas, T., *AFWAL-TR-80-4053*, 1980.

8. Johnson, G.R., *et al, Jnl. Eng. Matl. Tech.*, 105: 48 (1983).

9. Woodward, R.L., *et al, Jnl. Matl. Sci. Lett.*, 5: 413 (1986).

10. Rabin, B.H., & German, R.M., *Met. Trans. A*, 19A: 1523 (1988).

11.   Bourguignon, L.L., & German, R.M., *Int. Jnl. Powd. Met.*, 24: 115 (1988).

12.   Kolsky, H., *Proc. Phys. Soc. Lond.*, 62B: 676 (1949).

13.   Hartley, K.A., *et al*, *ASM Metals Hdbk*, Vol. 8, p. 218, ASM International, 1985.

14.   Lindholm, U.S., *Jnl. Mech. Phys. Sol.*, 12: 317 (1964).

15.   Davies H., & Hunter S.C., *Jnl. Mech. Phys. Sol.*, 11: 155 (1963).

16.   Bertholf, L.D., & Karnes, C.H., *Jnl. Mech. Phys. Sol.*, 23: 1 (1975).

17.   Coates, R.S., & Ramesh, K.T., *Johns Hopkins Tech. Rep. DIL-9001*, 1990.

18.   Coates, R.S., & Ramesh, K.T., *Johns Hopkins Tech. Rep. DIL-9002*, 1990.

19.   Dudder, G., Battelle Pacific Northwest Lab., *Personal Communication*, 1990.

20.   Nicholas, T., *AFWAL-TR-80-4053*, 1980.

21.   Follansbee, P., Los Alamos National Laboratory, *Personal Communication*, 1989.

22.   Meyer, L.W., & Chiem, C.-Y., *Mech. Prop. High Rates*, Oxford, p. 227, 1989.

23.   Baosheng, Z., *et al*, *Horizons of Powder Metallurgy*, Part II, p. 1131, 1986.

24.   Stephens, J.R., *Met. Trans.*, 1: 1293 (1970).

18

Texture-Induced Anisotropy and High-Strain-Rate Deformation in Metals

S. K. SCHIFERL and P. J. MAUDLIN

Los Alamos National Laboratory
Los Alamos, New Mexico 87545

We have used crystallographic texture calculations to model anisotropic yielding behavior for polycrystalline materials with strong preferred orientations and strong plastic anisotropy. Fitted yield surfaces were incorporated into an explicit Lagrangian finite-element code. We consider different anisotropic orientations, as well as different yield-surface forms, for Taylor cylinder impacts of hcp metals such as titanium and zirconium. Some deformed shapes are intrinsic to anisotropic response. Also, yield surface curvature, as distinct from strength anisotropy, has a strong influence on plastic flow.

I. INTRODUCTION

Crystallographic texture is a property of increasing interest for a number of high strain-rate processes, including anti-armor designs. The peculiar behavior of textured materials has been exploited in a number of ways, from spin compensation in shaped-charge liners to desirable yield anisotropy in new materials.

"Texture" refers to the preferred orientation of single-crystal grains in a polycrystalline solid. For hexagonal-close-packed metals, such as titanium and zirconium, plastic anisotropies due to texture can be very large. Yield anisotropies can be > 2:1 [1], and plastic strain ratios (R-values) typically range from 3 to 7 [2]. For cubic metals, yield anisotropies tend to be small, but strain anisotropies can be significant (R ~ 0.6 for rolled copper; R > 2.5 for some steels) [2]. A strong preferred orientation is typically the result of large deformation (> 50%); the patterns of deformation textures depend on both crystal structure and deformation path.

We have previously modeled changes in texture and corresponding changes in yield anisotropy during liner collapse in titanium shaped-charge jets [3], a system where inertial effects dominate the flow. In the present study, we investigate the role of texture in a different kind of system – a high-strain-rate regime where inertial effects no longer dominate. Examples of this kind of system include the Taylor anvil test (end-on impact of a cylinder on a flat target) and EFPs (explosively-formed penetrators: collapse of a metal liner to a stable shape). Both systems are known to be sensitive to material properties, particularly details of the yield strength. The simulation of anisotropic effects in this class of problem requires a coupled approach: the deviatoric stress states encountered by the material have a significant effect on the kinematics of the deformation, and these stress states need to be part of the yield function.

We will concentrate on appropriate constitutive relations for Taylor impacts of anisotropic cylinders. For such a moderate strain deformation, we assume constant anisotropy (constant yield-surface shape) as a first approximation. The scale of the yield surface will not, in general, remain constant; strain hardening, strain-rate hardening, and thermal softening change the overall yield within much smaller strains than texture evolution changes the anisotropy. We also assume that our material has inhibited twinning; this subject is discussed in more detail in Sec. IV.

Determining the shape of the yield function is more complex. An anisotropic yield function is, by its nature, directional. Ideally, our data would include probes of the material with many different stresses. In practice, we have at most a few yield measurements and/or R-values, plus some constitutive rules for the isotropic case. In this situation, texture calculations, which model the underlying cause of single-phase anisotropy, can be useful. A texture code [3] "samples" an oriented collection of grains by applying a strain tensor; deformation of each grain is via the single-crystal mechanisms of slip and twinning. We consider the activation of any deformation system to require a certain critical-resolved-shear-stress (CRSS); the stress on each grain is the sum of the CRSSs for the active systems. To calculate yield surface points, critical stresses are averaged over the grains. A fitting function for the stress points and the modifications required for the continuum code are described in Sec. II, along with an orthogonal representation introduced to help visualize and fit the yield surface.

In the present work, we investigate the effect of large plastic anisotropy for Taylor cylinder tests. The material properties are typical of α-titanium (hcp) and some of its alloys, if twinning is inhibited. We consider features of the deformed shapes, for anisotropic orientations $\beta = 0°, \pm 45°, 90°$ (where β is the initial angle between the material's weak axis and the cylinder axis), and for different curvatures of the yield surface. The anisotropic simulations exhibit significant differences from isotropy, particularly for $\beta = \pm 45°$, where shear-strengthening (or shear-weakening) produce shapes specific to anisotropic response. The results also indicate that anisotropic response is strongly influenced by yield-surface curvature, rather than merely by the strength anisotropy that determines yield-surface eccentricity.

II. MODELING PLASTIC ANISOTROPY

To use texture information for a constitutive model, we need to generate a yield surface. As described above, a textured yield surface is a collection of stress points, not an analytic function. We need to find a function that reproduces the main features of the point distribution, but does not incorporate detail that a continuum simulation cannot resolve. We use the texture code only for the shape of the yield surface, not the scale; we normalize the yield surface by applying the texture code to a randomly oriented collection of grains. We also note that a general texture code requires no special symmetry, and is therefore adaptable to 3D simulations. For simplicity we consider axisymmetric cases, which can be addressed with 2D codes.

To visualize and fit the yield surface, we introduce a 3D stress space. For a 2D simulation, there are three independent components of the deviatoric stress tensor: two diagonal components and one off-diagonal component. We refer to the orthogonal set $\{Sx, Sy, Sz\}$, with Sx and Sy in the pi-plane, the plane in which the diagonal deviatoric stresses are at 120° to each other. One possible linear combination of tensor components is:

$$Sx = -\sqrt{3}\left(\frac{1}{2}S_{11} + S_{22}\right); \quad Sy = \frac{3}{2}S_{11} \; ; \quad Sz = \sqrt{3}\,S_{12} \; . \tag{1}$$

This representation has a number of advantages. It allows for easy fitting of a yield function, and easy visualization of stress paths on the yield surface. It also allows for translation, to accommodate unequal yield strengths in tension and compression (a significant feature of many materials when twinning is involved). A spherical yield surface corresponds to isotropy, and an ellipsoidal yield surface, without translation, corresponds to Hill's [4] quadratic surface.

The characteristic features of the yield surface for a particular material depend heavily on the crystal symmetry, as well as the grain orientation distribution. Figures 1-2 show projections of the polycrystalline yield surface on the pi-plane for two cases: a compression texture for hcp crystals (α-titanium) [2] and a rolling texture for fcc (copper) [5]. Neither form suggests a simple shape: the former resembles a flattened ellipse, the latter a faceted border.

A useful form for an hcp yield surface is the superquadric function [6]:

$$F = \left[\left\{\left(\frac{Sx - b_1}{a_1}\right)^{2/\varepsilon_2} + \left(\frac{Sy - b_2}{a_2}\right)^{2/\varepsilon_2}\right\}^{\varepsilon_2/\varepsilon_1} + \left(\frac{Sz - b_3}{a_3}\right)^{2/\varepsilon_1}\right]^{\varepsilon_1} - 1 \quad , \tag{2}$$

where a_1, a_2, a_3 define the eccentricity; b_1, b_2, b_3 are offsets in the x, y, z directions; and ε_1 and ε_2 are curvature parameters in the polar and azimuthal directions. This form gives a smooth function, and can fit 3D shapes ranging from a needle to an ellipsoid to a box. As Fig.1 indicates, a superquadric fits the hcp yield surface much better than a simple ellipsoid.

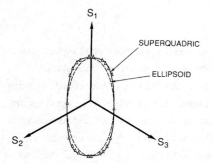

FIG. 1   *Deviatoric pi-plane yield locus for a (twinning) titanium compression texture. Note unequal yields in tension and compression. Points are from texture calculations. Lines are analytic forms.*

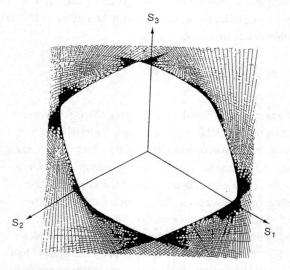

FIG. 2   *Deviatoric pi-plane yield locus for rolled copper, 85 % reduction (from Ref. 4). Texture calculation.*

The fcc yield surface presents a problem. A superquadric function has the wrong symmetry, and a fit will also tend to smooth out the vertices. However, we can reproduce some salient features of an fcc yield surface – the flat spots and vertices – by choosing superquadric parameters to fit a box-like shape. The box (see Fig.3d) is globally incorrect, but can give us some insight into the effect of a faceted yield surface on plastic deformation.

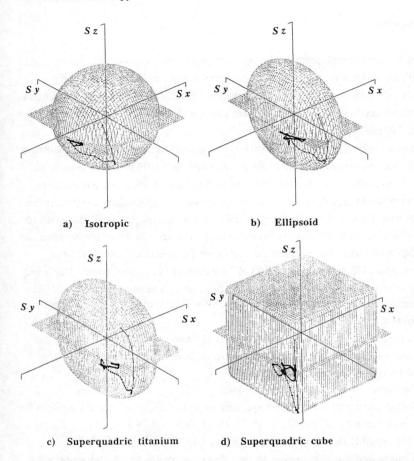

a) Isotropic   b) Ellipsoid

c) Superquadric titanium   d) Superquadric cube

FIG. 3 *Yield surfaces and deviatoric stress paths for Taylor cylinder impacts, β = 0 °, in a 3D representation (see Sec. II). β is the initial angle between the material weak axis and the cylinder axis.*

Finally, to use these yield surfaces it is necessary to make several major modifications to a continuum code. First, we evaluate the constitutive relations in a reference material frame, and use explicit rigid-body rotations to connect laboratory and material frames. The necessary rotation matrix is obtained from the polar decomposition of the deformation gradient tensor [7]. Also, we use an associative flow algorithm [8] to update the stresses. If the stress state is plastic, associative flow guarantees a plastic strain normal to the yield surface.

III. PROCEDURE

Our method, to calculate high-strain-rate response for anisotropic materials, is based on a yield function with a constant anisotropic shape, incorporated into a continuum code. Taylor cylinder tests were simulated with a modified version of EPIC2-88 [9], an explicit Lagrangian 2D finite-element code. The cylinders were 1-inch long, with an L/D of 10/3. The initial velocity was 190 m/s.

Our anisotropic material was modeled from the general features of heavily cross-rolled (transversely isotropic) nontwinning titanium or zirconium sheet [10]; this sheet gives an R value of ~ 6.5, which corresponds to an anisotropy ratio of z/x ~ 1.95, where z is the compressive yield in the through-thickness direction and x is the average compressive yield in the plane of the sheet. Details of the yield surface, including shear strengths, were obtained with a texture simulation, using single-crystal properties of hcp titanium at moderate temperatures and relatively high strain-rates. The details of the nontwinning texture calculation have been described elsewhere [3]; here we use a CRSS of 3 for the secondary slip systems. Hardening and thermal softening were approximated by a rate-dependent constitutive model [11]. Since our main interests in these simulations concern systems with large plastic strains, we treat elastic response simply as isotropic.

Several different yield surfaces were constructed for these calculations; their 3D representations are shown in Figs. 3a-d. The parameters, in terms of a normalized superquadric function (compare Eqn. 2) are as follows:

1) Isotropic: $a_1 = a_2 = a_3 = 1$, $\varepsilon_1 = \varepsilon_2 = 1$
2) Ellipsoidal model, using the strength anisotropies as major and minor axes (a_3 is the shear strength): $a_1 = 0.55$, $a_2 = 1.20$, $a_3 = 1.09$, $\varepsilon_1 = \varepsilon_2 = 1$
3) Superquadric fit: $a_1 = 0.55$, $a_2 = 1.20$, $a_3 = 1.09$, $\varepsilon_1 = 0.75$, $\varepsilon_2 = 0.76$
4) Cubical superquadric, to evaluate the effect of corners on an otherwise isotropic yield function: $a_1 = a_2 = a_3 = 1$, $\varepsilon_1 = \varepsilon_2 = 0.1$

The simulations were performed for the orientations $ß = 0°$, $\pm 45°$, and $90°$, where $ß$ is the angle between the material weak axis and the cylinder axis.

IV. RESULTS AND CONCLUSIONS

The deformed shapes from simulations of anisotropic and isotropic Taylor cylinder impacts are shown in Figs.4–5. For Fig.4, we have used an anisotropic yield surface corresponding to the standard ellipsoidal form. For $ß = 0°$, the weak direction is along the cylinder axis; this orientation realized the most compressive strain. For $ß = \pm 45°$, the profiles in both cases are

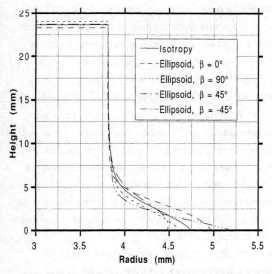

FIG. 4 *Taylor cylinder profiles, using an ellipsoidal yield surface (β is defined in Fig. 3). The abscissa is expanded to show detail.*

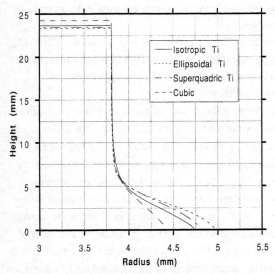

FIG. 5 *Taylor cylinder profiles, using a superquadric yield surface, β = 0° (β is defined in Fig. 3). The abscissa is expanded to show detail.*

qualitatively different than the isotropic results, and the differences are large; the special nature of anisotropic response can maximize or minimize resistance to shear. Practically, orientation ß = 90° could be cut from a thick cross-rolled billet and ß = 0° from an extrusion. Diagonal cuts from the billet could give ß = ± 45°, but only in one plane; complete profiles of such cylinders are part of a 3D problem, and beyond the scope of the present study.

The cylinder profiles from a superquadric function, including a fit to the textured yield surface, are shown in Fig.5. The differences from isotropy are actually smaller than for the ellipsoidal surface, even though the yield anisotropies are the same in both cases. Some insight into these differences can be obtained by constructing deviatoric stress paths on the yield surfaces for particular computational cells. Figures 3a-c show these paths for the isotropic, ellipsoidal, and superquadric models, for ß = 0°, and for a cell near the edge of the cylinder toe. It is clear that the anisotropic paths differ from the isotropic path. What is more striking is the amount of time spent near parts of the surfaces with high curvature. It is the lower curvature in the superquadric that appears to modify the effects of anisotropy.

This effect can be described as an "attraction" of stress states to regions of high curvature [12]. That the "attraction" is due to curvature, not eccentricity, can be seen in Fig.3d, which shows the stress path for a cubical yield surface with slightly rounded corners. Figure 5 includes the final cylinder profile for this yield surface; the differences from isotropy are large, and qualitatively different than for the other anisotropic functions. This behavior has important implications for the behavior of highly-textured cubic materials, particularly for fcc metals like copper and aluminum, where yielding behavior exhibits minimal yield strength anisotropy but suggests flat spots and vertices.

It should be noted that these curvature effects are seen only with associative flow [12], where the plastic strain is normal to the yield surface. Nonassociative flow laws (e.g., radial return on a nonspherical surface) and anisotropic models with spherical yield surfaces (kinematic hardening) will not produce these effects.

The major discrepancy in this modeling is the absence of any texture (and yield surface shape) change. Gradual evolution, from unequal grain rotations due to slip, is probably not the major difficulty here; the Taylor cylinder strains, except perhaps at the edge of the toe, are typically < 30%. Also, while texture does change with deformation, the yield anisotropy changes more slowly. A more serious problem is the possibility of deformation twinning.

For many hcp materials, and for some deformation modes (particularly tension in the strong direction), twinning can effectively "flip" a texture within a strain of only a few percent. For simplicity, we have assumed that twinning is inhibited. This is possible for a number of alloys, for very small grain size, for certain initial dislocation structures, and for materials with certain impurities [13]. For an hcp material where twinning does operate, the ß = 0° orientation would be affected: compression in the weak direction would effectively rotate the anisotropy, making the deformation along the axis and in the toe less extreme. For such materials, and for large deformations, the effects of texture evolution need to be addressed.

We have concentrated here on the effects of plastic anisotropy on high-strain-rate deformation. For moderate strains such effects can be large – and are not amenable to isotropic treatments. Our simulations show that the major effects stem from differences in the stress paths, including attraction to regions of high curvature on anisotropic yield surfaces. To model such behavior, it is necessary to have detailed information about yielding behavior. Texture studies, as a prescription for yield surface models, can help provide insight into this difficult area of constitutive modeling.

ACKNOWLEDGMENTS

This work was supported by the Defense Advanced Research Projects Agency, and through the Joint DoD/DOE Munitions Technology Development Program.

REFERENCES

1. R. G. Ballinger and R. M. Pelloux, *J. Nucl. Mater.*, 97:231 (1981).

2. W. F. Hosford and R. M. Caddell, *Metal Forming: Mechanics and Metallurgy*, Prentice-Hall, Englewood Cliffs, New Jersey, 1983.

3. S. K. Schiferl, *J. Appl. Phys.*, 66:2637 (1989); S. K. Schiferl, in *Shock Compression of Condensed Matter - 1989* (1990).

4. G. R. Canova, V. F. Kocks and C. N. Tome, *J. Mech. Phys. Solids*, 33:371 (1985).

5. R. Hill, *The Mathematical Theory of Plasticity*, Oxford University Press, London, 1950.

6. F. Solina and R. Bajcsy, *IEEE Trans. Pat. Anal. Mach. Intell.*, 12:131 (1990).

7. J. K. Dienes, *Acta Mech.*, 32:217 (1979). For implementation, see: D. P. Flanagan and L. M. Taylor, *Comp. Meth. Appl. Mech. Eng.*, 62:305 (1987).

8. R. Hill, *Plasticity*, Clarendon Press, Oxford, 1956.

9. G. R. Johnson and R. A. Stryk, Air Force Armament Lab. Report TR-86-51 (1986).

10. F. Larson and A. Zarkades, Metals and Ceramics Inf. Center Report 74-20 (1974).

11. G. R. Johnson and T. J. Holmquist, Los Alamos Report LA-11463-MS (1989).

12. T. J. R. Hughes, in *Proc. of the Workshop on the Theoretical Foundations for Large-Scale Computations of Nonlinear Material Behavior* (1984).

13. H. Conrad, *Prog. Mater. Sci.*, 26:123 (1981).

19

Shock-Wave Deformation of W-Ni-Fe Heavy Alloys at Elevated Temperatures

A. PEIKRISHVILI, L. JAPARIDZE, G. GOTSIRIDZE, and N. CHIKHRADZE

Institute of Mining Mechanics
Academy of Sciences of Georgian SSR
Tbilisi, Georgia, USSR, 380086

W-Ni-Fe heavy alloys were investigated after shock wave treatment in the range of parameters: pressure 7-20 GPa, temperature 20-1000°C. The investigation showed that application of heating before loading prevents formation of cracking which is eliminated at temperatures above 300°C. It has been established that with an increase of pre-heating temperature, the alloys W-Ni-Fe are strengthened more, which results in hardness increase and broadening of line 321 of tungsten. It has been found that for the given conditions the maximum increase of hardness is 55%. Investigation of bonding materials of Ni-Fe phase shows that the shock wave treatment leads to an increase of microhardness and the amount of tungsten content by 22-34% within it.

I. INTRODUCTION

In the works on shock wave strengthening of metallic materials [1-5] there are observed mainly one-phased materials. Heterogeneous alloys are investigated to a lesser extent and the works on strengthening of such materials on the basis of tungsten are practically absent.

In the present work the influence of shock wave treatment was carried out on alloys of the W-Ni-Fe system representing composite material, consisting of tungsten grains in a Ni-Fe-W matrix.

Preliminary experiments showed, that even weak shock waves influence on W-Ni-Fe alloy and there are formed a lot of cracks, at stronger influence the alloy is changed into powder. Taking into consideration this and also the experience of hard alloys strengthening of WC-Co type [6-7], the samples heated in the range of 300-1000°C were caused by shock wave treatment and explosive material was used of ammonite and RDX kinds (detonation velocity 3.6-6.1 km/sec respectively).

For the strengthening of W-Ni-Fe alloys it has been selected the scheme of axis-symmetrical loading, which is brought in [7].

II. RESULTS OF EXPERIMENTS

The samples investigation loading at different temperatures of preliminary heating shows that after high temperature (at heat temperature of 300°C and below the samples are destroyed) shock wave treatment the alloys W-Ni-Fe are strengthened, that is expressed in increasing of the hardness and broadening of halfwidth of line (321) of W phase, Table 1.

As it is seen from Table 1 with the increase of preliminary heating temperature, the hardness and halfwidth of line (321) of W phase occur, that testifies about plastic deformation of tungsten grains and increase of a number of defects in the crystalline structure. With this the maximum hardness (HRC 43) is close to the hardness of alloy strengthened by static deformation (HRC 44.5). From the table it is also seen that the increasing of preheating temperature promotes tungsten content increase in the bonding after shock wave loading. The dissolved W content in Ni-Fe bonding is increased from 21 to 32%.

'Figure 1 represents the microstructure of W-Ni-Fe alloy treated by shock waves at 600°C. It is seen that in tungsten grains slip lines are observed. This testifies about the process of plastic deformation and as a result the hardness increase and halfwidth increase of the line (321) take place.

TABLE 1

Hardness and halfwidth of line (321) of W phase changes for W-Ni-Fe alloy at shock wave treatment by ammonite.

Magnitudes	Initial state	Preheating temperature, °C			
		400	600	800	900
Hardness HRC	26	33	35	41	43
Halfwidth of line (321) of W phase, mm	0.51	0.68	0.70	0.83	0.86
W content in bonding, %	21	30	-	32	32

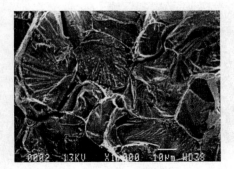

FIG. 1 *Microstructure of W–Ni–Fe alloy after shock wave treatment at 600°C (P=10 GPa). The lines of sliding are observed.*

Investigation of the diffraction patterns of the W-Ni-Fe alloy showed that in their initial state there are observed dotted electron diffraction patterns. It should be mentioned that the crystals in the initial state are quite large, as it is possible to obtain the picture of more perfect diffraction (Fig. 2a). Figure 2 illustrates the diffraction patterns from W-Ni-Fe alloy correspondingly initial (Fig. 2a) and treated by shock wave at elevated temperature (Fig. 2b,c).

The analysis shows that in the treated alloys it is practically impossible to obtain dotted electron diffraction patterns. With increase of deformation amount of fragments per unit of square increases. This indicates to the number of spots appeared at the diffracted rings (Fig. 2b). It should be mentioned that this process due to the volume occurs obviously inhomogeneously, apparently the degree of deformation inside the sample greatly varies by value. The dimensions of the fragments also differ from each other. The described changes in the structure of alloy in equal extent concern all investigated samples deformed by shock wave, however, to establish the quantitative difference between them is practically difficult.

In Fig. 2c is shown a diffraction pattern of W-Ni-Fe alloy at T=800°C, P=10 GPa. Interpretation of the electron diffraction pattern shows that in the given case we have ßW in stable condition. Apparently, the conditions brought above are advisable for its formation in stable type, as in other conditions there are no signs of ßW formation.

Further investigations pursued the aim of discovering the possibility of hardness of alloy increase by means of shock wave treatment, preliminary strengthened by static deformation. The results are shown in Tables 2 and 3.

As it is seen from the given Table 2 shock wave treatment leads to the some increase of hardness and halfwidth of line (321) of W phase at the temperatures of preliminary heating

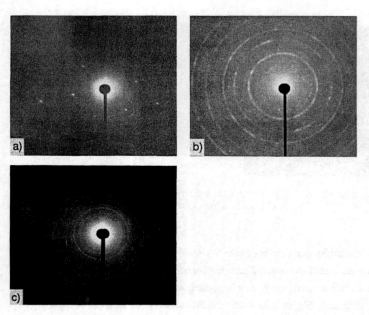

FIG. 2 *Diffractional pictures of W-Ni-Fe alloy: a) dotted electron diffraction pattern of the initial alloy. b) ring kind electron diffraction pattern of W-Ni-Fe alloy worked by shock wave at high temperature. T=800°C, P=10 GPa. c) electron diffraction pattern got from W after shock wave treatment at T=800°C, P=10 GPa.*

TABLE 2

Hardness and halfwidth of line (321) of W phase changes in static conditions of strengthened alloy after shock wave treatment by ammonite.

Magnitudes	Initial state	Preheating temperature, °C			
		400	600	800	1000
Hardness, HRC	44.5	47	46	48	42
Microhardness, HV 0.05 of W_2 phase, kg/mm	502	662	664	712	591
Microhardness of bonding, HV 0.05, kg/mm²	320	424	387	437	351
Halfwidth of line (321) of W phase, mm	0.85	1.00	1.05	0.88	0.68

TABLE 3

Hardness and halfwidth of line (321) of W phase changes after shock wave treatment of different intensity (preheating temperature 600°C)

Magnitudes	Initial state	Pressure, GPa		
		5	10	20
Hardness, HRC	44.5	41	46	44
Microhardness, HV 0.05 of W	502	665	664	637
Microhardness of bonding, HV 0.05, kg/mm²	320	351	387	283
Halfwidth of line (321) of W phase, mm	0.85	0.80	1.05	0.95

from 400 to 800°C, however, the degree of this increase is not large. At the temperature 1000°C a strength loss is observed.

As it is seen from Table 3 the shock wave treatment leads to the alloy strengthening only at P=10 GPa.

The discovered character of hardness change of strengthened alloy (existence of maximum, Tables 2 and 3) is a result of two processes taking place: deformational strengthening by shock wave and losing the strength as a result of defects annealing at elevated temperatures. At high heat temperatures (1000°C) or low pressures (5 GPa) the process of losing the strength predominates.

The investigation of microstructure of preliminary strengthened alloy and then at high temperatures (800°C) additionally worked by shock wave it was discovered that at some parts of the alloy it's occurred the fragmentation of tungsten grains (Fig. 3). It had been observed in those cases when container was produced not from steel, as in the rest experiments, but from W-Ni-Fe alloy, that increases essentially the stresses in the work rod.

As it is seen from microstructures the tungsten grains are crushing into some subgrains, which are divided by obvious bounds. Microanalysis showed that in the cracks between the separate grains there are present Ni and Fe, i.e. the cracks "are healed" and it is practically more fine granular alloy in this part. However, the hardness of the section is reduced (402 HV 5) at the same time the initial hardness was 460 HV 5, that testifies about incomplete "healing".

FIG. 3 *Microstructure of fragmentational section of preliminary strengthened W-Ni-Fe alloy in static conditions, deformed by shock wave at T=800°C.*

Formation of such fragmentational sections it is not characteristic for the W-Ni-Fe alloys in the whole and appear only in the exceptional cases. However, it requires additional investigations and explanations of the reasons of its formation.

Regarding ßW it should be mentioned, that the preliminary strengthened alloys treated by shock wave in stable conditions its presence was also fixed at T=600°C, P=10 GPa. At other parameters of treatment ßW was not observed.

Taking into consideration the mentioned above and the data of preliminary worked W-Ni-Fe alloys in static conditions it should be supposed that the conditions of the formation of ßW are the following: T=600-800°C, P=10 GPa.

III. CONCLUSIONS

1. Usage of preheating temperature of the samples before shock wave loading enables to eliminate the cracking and to increase hardness of W-Ni-Fe alloys. The temperature should be not lower than 400°C.

2. High temperature shock wave treatment of sintered alloys W-Ni-Fe increases hardness of these materials from HRC 26 to HRC 43, practically 60%.

3. Hardness of preliminary strengthened W-Ni-Fe alloys in static conditions treated by shock wave at high temperature increases insignificantly on 3-4 units of HRC.

4. Formation of stable ßW after shock wave treatment at high temperatures can't cause

apparently the essential influence on the properties of the alloy. However, from the point of view of phase changes it is interesting phenomena and it needs additional investigations for more precise determination of the parameters and conditions of its formation.

REFERENCES

1. A. A. Deribas, Physica uprochnenia i svarka vzriza. Nauka, 1980.

2. M. Meyers, L. Murr Udarnie volni i javlenia visokotemperaturnoi deformatsii metallov. M., Metallurgia.

3. I. N. Gavriliev, A. A. Deribas i dr. Struktura i mekhanicheskie svoistva austenitnoi stali posle nagruzhenia udarnimi volnami. Proceedings of the Xth International Conference on High Energy Rate Fabrication. Lujbljiana, Yugoslavia, 1989.

4. M. A. Mogilevski, Vlijanie parametrov udarno-volnovogo nagruzhenia na uprochnenie i strukturnie izmenenia. Proceedings of the IXth International Conference on High Energy Rate Fabrication. Novosibirsk, USSR, 1986.

5. A. A. Deribas, Ispolzovanie vzrivnoi obrabotki materiallov v promishlennosti USSR. Proceedings of the IXth International Conference on High Energy Rate Fabrication. Novosibirsk, USSR, 1986.

6. E. O. Mindeli, G. P. Licheli i dr. Uprochnenie metallokeramicheskikh tverdikh splavov vzrivov v nagretom sostojanii. Physiks gorenija i vzriva (FGV), N2, 1971.

7. A. Peikrishvili and N. Chikhradze Explosive working of some metals and alloys at high temperatures. Proceedings of the International Conference "Explomet 85", p. 915, 1985.

Section II
Shock and Combustion Synthesis

20

Diamond Formation in Aluminum-Nickel/Graphite Under Shock Loading

I. SIMONSEN*, Y. HORIE†, T. AKASHI‡, and A.B. SAWAOKA**

*Department of Materials Science and Engineering, and

† Department of Civil Engineering, North Carolina State University, Raleigh North Carolina, USA

‡Sumitomo Coal Mining Co., Akabira, Japan

**Research Laboratory of Engineering Materials, Tokyo Institute of Technology Yokohama, Japan

Characterization of aluminum-nickel/graphite composite powders subjected to high pressure shock wave compression was performed using optical and transmission electron microscopy. Formation of a polycrystalline diamond phase was observed which was found as an apparent continuation of graphite fibers. A lattice transformation process is suspected. Among various resulting Ni-Al alloy phases Ni_3Al is predominant and has the characteristics of rapidly solidified Ni_3Al.

I. INTRODUCTION

The shock compaction of nickel-aluminum and nickel/graphite powders has been studied previously, and the formation of resulting nickel-aluminum phases and a diamond phase, respectively, has been established (Ref. 1). The present work examines the results of shock loading experiments with aluminum-nickel/graphite powders.

The motive for adding aluminum to the nickel/graphite mixture which had previously formed a diamond phase was threefold: one, to gain higher temperature during the otherwise identical shock conditions by inducing the exothermic reaction between nickel and aluminum; two, to see whether the precipitation of carbon in metal is influenced by the strong affinity of aluminum to carbon; three, to gain insight into the time requirement for possible diamond formation from the formation speeds of nickel aluminides.

II. EXPERIMENTAL PROCEDURE

The experimental conditions have been described elsewhere (Ref. 2) but can be briefly summarized as a mousetrap-type plane wave generator used to explosively compact the prepressed powders with an iron flyer plate (Fig. 1). While a diamond phase was found in only one of the samples shock loaded under various conditions in the nickel/graphite study (50% initial density, 1.6k/s flyer plate velocity), this one resulted in two specimens with diamond: #6, which was compacted under identical conditions as the nickel/graphite sample, and #2, which had the same initial density but was compacted at lower flyer plate speed (1.1k/s).

III. RESULTS

Metallographic study preceded the examination of the specimens by transmission electron microscopy (TEM). By this method none of the spherulites which had been the site of the diamond phase in the nickel/graphite sample were found. Alloying between nickel and aluminum was observed: mostly Ni_3Al in sample #6, and various Al-rich phases besides NiAl and Ni_3Al in sample #2.

TEM investigation of the two samples did confirm the presence of spherulites, however, though they were less frequent, much smaller and did not display any of the graphite-ring-with-diamond-center characteristics (Figs. 2,3). Mostly, they appeared in the form of clusters of embryonic spherulites which were nestled at the metal phase boundaries, mainly between nickel and Ni_3Al (Figs. 4-7). The lower frequency and different appearance may be due to the influence of the carbon affinity to aluminum.

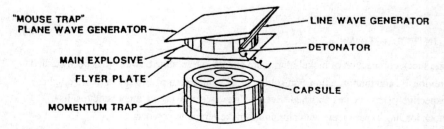

FIG. 1 *Sketch of shock loading apparatus*

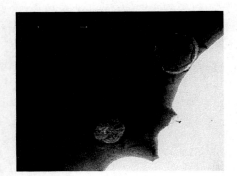

FIG. 2 *Small graphite spherulites in nickel matrix.*
Bar = 1μm

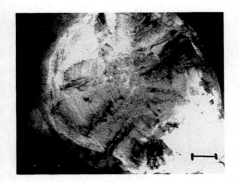

FIG. 3 *Enlargement of spherulite showing mini spheres at center. Bar = 100nm*

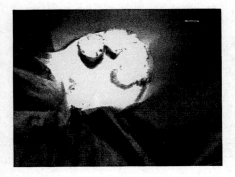

FIG. 4 *Nest of spherulites between Ni₃Al and nickel phases.*
Bar = 1μm

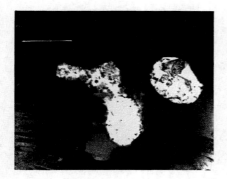

FIG. 5 *Same as Fig. 3.   Bar = 1µm*

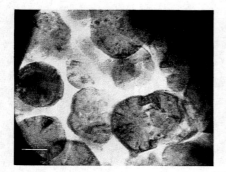

FIG. 6 *Enlargement of spherulite cluster.   Bar = 100nm*

FIG. 7 *Unidentified cluster in Ni₃Al-martensite, maybe diamond.   Bar = 100nm*

Evidence of a diamond phase was not found in connection with the spherulites. If it does exist there, the areas were too small for us to identify it. Both specimens, we found, displayed a change in the appearance of the graphite flakes, however. The normally parallel fibers of the flakes were disturbed in many places as if they had experienced violent turbulence (Fig. 8). A diffraction pattern of such an area (Fig. 9) shows mostly the full rings of random orientation and only a hint of the arcs of the preferred orientation that is typical of the graphite flakes.

Closer scrutiny of these regions resulted in the observation that the graphite fibers seem to "dissolve" into another morphology at the edge of the flakes (Figs. 10,11). This new phase did not match any of the metallic phases. Electron diffraction proved it to be polycrystalline diamond which could be distinguished from polycrystalline nickel by the

FIG. 8 *Graphite turbulence. Bar = 100nm*

FIG. 9 *Diffraction pattern of perturbed graphite*

FIG. 10 *Enlargement of graphite turbulence.* *Bar = 100nm*

FIG. 11 *Transition from graphite to diamond*

missing (200) reflection and from graphite by the missing (0002) reflection. Many such regions were found, and diffraction patterns showed that the crystallites were often large enough to cause spotty rings. Some spots even display a symmetry as to suggest a twinned crystal (Figs. 12,13). One pattern (Figs. 14,15) is curiously oval; this is presumed to result from magnetic distortion of the electron beam by a nearby nickel phase area.

Additional findings include the alloying phases of nickel and aluminum, mostly in the form of the well known Ni_3Al-martensite. This phase grows dendritically into the nickel matrix (Fig. 20) and has the typical herringbone features of the martensite (Fig. 21), and in some areas the so-called tweed structure (Fig. 22). These appearances of Ni_3Al have been found and studied extensively in rapidly solidified nickel aluminides (Ref. 3).

FIG. 12 *Diamond twinned crystal. Bar = 100nm*

FIG. 13 *Diffraction pattern to Fig. 12*

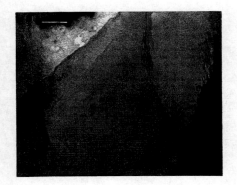

FIG. 14 *Diamond phase. Bar = 100nm*

FIG. 15 *Diffraction pattern to Fig. 14*

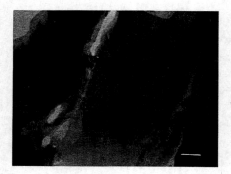

FIG. 16 *Diamond phase.   Bar = 100nm*

FIG. 17 *Diffraction pattern to Fig. 16*

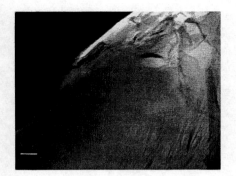

FIG. 18 *Graphite-diamond transition. Bar = 100nm*

FIG. 19 *Lower magnification of Fig. 16, showing surrounding phases: Ni at top, Ni₃Al at left and graphite fibers at right. Bar = 1μm*

FIG. 20 *Dendritic growth of Ni₃Al-martensite in nickel matrix. Bar = 1μm*

FIG. 21 *Enlargement of Ni₃Al—martensite phase.   Bar = 100nm*

FIG. 22 *Tweed structure of Ni₃Al—phase.   Bar = 100nm*

The phenomenon of the perturbed graphite reminded the authors of an earlier observation in the aluminum-nickel compaction study. In that investigation, swirls of the NiAl phase had been noted. Since this phase has a wide compositional range its color under light optical observation changes with varying Al-content. These color differences could be seen in swirls in small, isolated areas of that phase. Such perturbations attest to the great mechanical stresses the explosively compacted materials undergo. Also, the addition of aluminum and the exothermic reaction with nickel cause temperatures to rise to at least the melting point of the NiAl-phase (1638°C).

High resolution electron microscopy could answer the question whether there is an amorphous transitional stage between the graphite and the diamond. The possibility of a direct transformation through the high shear forces in the manner proposed by Kleiman et al (Ref. 4) exists, where a carbon atom is presumed to change position within the molecule to cause the transition from graphite to diamond ("puckering").

The results of this experiment indicate that the formation of diamond is not by way of diffusion or precipitation. In the nickel/graphite study we found carbon diffusion through the molten nickel and its precipitation as graphite nodules. A part of these had turned into diamond. It seems that the formation process is not the same in these two studies.

IV. CONCLUSION

These experiments make clear that the formation of diamond can take place under conditions much less severe than previously thought. The maximum pressures attained in sample #2 under shock loading with 1.1k/s flyer plate velocity has been calculated to be approximately 10 GPa. Temperatures reached probably range around the NiAl melting point. The formation of nickel aluminide morphologies similar to those found in rapidly solidified samples suggests reaction times in line with this process.

REFERENCES

1. Y. Horie, R. A. Graham and I. K. Simonsen, *Mater. Lett.* 3(1985) 354-359.

2. I. Simonsen, S. Chevacharoenkul, Y. Horie, T. Akashi and H. Sawaoka, *J. Mater. Sci.* 24(1989) 1486-1490.

3. I. Baker, J. A. Horton and E. M. Schulson, *Metallography* 19(1986) 63-74.

4. J. Kleiman, R. B. Heimann, D. Hawken and H. M. Salansky, *J. Appl. Phys.* 56(5) (1984) 1440-1454.

21

Shock-Initiated Chemical Reactions in 1:1 Atomic Percent Nickel-Silicon Powder Mixtures

B. R. KRUEGER and T. VREELAND, JR.

W. M. Keck Laboratory of Engineering Materials
California Institute of Technology, 138-78
Pasadena, CA 91125

A series of shock initiated chemical reaction experiments have been performed on 1:1 atomic percent mixtures of nickel and silicon powders. It has been observed that only minor surface reactions occur between the constituents until a thermal energy threshold is reached above which the reaction goes to completion as evidenced by large voids, bulk melting, and scanning electron microscopy and x-ray diffraction results. The experiments show the energy difference between virtually no and full reaction is on the order of 5 percent. A sharp energy threshold indicates that with the particular morphology used, the bulk temperature determines whether or not the reaction occurs rather than local, particle level, conditions.

I. INTRODUCTION

A series of experiments have been performed on two mixtures of 1:1 atomic percent elemental nickel and silicon powders in a well characterized propellant driven plate system. The experiments reveal the existence of a thermal energy threshold below which little or no reactions occur and above which full reaction occurs as evidence by bulk melting and x—ray diffraction and scanning election microscopy (SEM) results. The experiments also show the width of the threshold to be on the order of 5 percent. From the existence of a sharp thermal energy threshold, certain conclusions can be made about the parameters which determine whether or not bulk reactions occur.

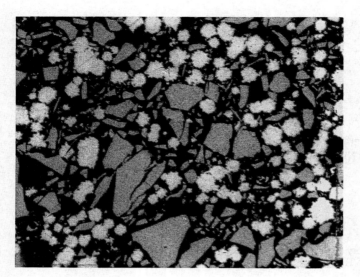

FIG. 1a *Optical image of Mix A.*

FIG. 1b *Optical image of Mix B.*

II. EXPERIMENT

Two mixtures of 1:1 atomic percent (67.6 wt % Ni) powders of Ni and Si were used. The first, Mix A, consisted of 15 μm spherical Ni from Inco Metals and −325 mesh irregular Si of unknown purity. The second, Mix B, was 20 μm − 45 μm spherical nickel (Aesar Stock # 10581) and −325 mesh irregular silicon (Cerac Stock # S−1052). The elemental powders were mechanically mixed in petroleum ether to avoid particle agglomeration and then dried. No special care was taken to remove or prevent formation of oxides on the particle surfaces. Optical images of the two mixtures are shown in Figures 1a and 1b.

The shock facility used is the Keck Dynamic Compactor, a 35 mm smooth bore cannon. Experiments with a metallic glass have shown that the gun and full cavity target assemblies result in highly one dimensional shock conditions.[1] A 5 mm 303 stainless steel plate was used as the flyer. The geometry of the target assembly limits the shock duration by the reflection from the back of the flyer plate, and therefore the duration is governed by the flyer thickness. The effects of duration have not been explored. The shock facilities are discussed in further detail in Ref. 1.

Preliminary experiments were conducted with porous bronze inserts, pressed into a target fixture, which contain 4 smaller cavities. These experiments have the advantage of identical impact conditions for each sample and four samples per shot. A disadvantage is that the impedance mismatch between the inserts and samples may give rise to two dimensional effects. To insure that two dimensional effects were not the determining factor, critical results were confirmed using a full cavity.

III. RESULTS AND ANALYSIS

The shock conditions were determined using an averaging method which assumes that the mixture's bulk modulus is linear with pressure and the mixture's Grüneisen parameter to specific volume ratio is constant. With these assumptions, the Rankine−Hugoniot relationships and the known Hugoniot of the flyer, the mass averaged shock conditions can be determined.[2] The homogeneous shock temperature was determined by matching the thermal energy to the integration of the mixture's heat capacity of the form $C = a + 10^{-3}bT + 10^5c/T^2$, where C is the heat capacity and T is temperature. No attempt was made to correct the heat capacity for pressure effects. The thermodynamic parameters used are listed in Table I, and the calculated shock results of the experiments discussed below are listed in Table II.

Table I

Density (g/cm^3)	β_S (GPa)	$\left.\dfrac{\partial\beta_s}{\partial P}\right|_s$	γ_0	a	b	c
Ni(α) 8.90	192.50	3.940	1.91	17.00	29.48	0
Ni(β)				25.12	7.540	0
Si 2.33	97.88	4.186	0.74	23.95	2.470	−4.14

Table 1 *Thermodynamic parameters used to determine the Hugoniots of the Ni/Si mixtures. The isentropic bulk modulus of Ni and the Grüneisen paramaters for Ni and Si were taken or calculated from V.N. Zharkov and V.A. Kalinin, Equations of State for Solids at High Pressures and Temperatures, Consultants Bureau, New York, 1971. The pressure derivative of the isentropic bulk modulus of Ni was determined by fitting solid Ni Hugoniot data. Silicon's isentropic bulk modulus and its pressure derivative were taken from O.L. Anderson, J. Phys. Chem. Solids, Vol 27, p. 547. The heat capacity coefficients were taken from, E.A. Brandes, (Ed.), Smithells Metal Reference Book, Sixth Edition, Butterworth & Co., (1983), p. 8-42. The units of a, b and c are J/(mole-K), J/(mole-K²) and J-K/mole, respectively.*

Table II

Mix	Por. (%)	Vel. (m/sec)	P (GPa)	E (kJ/Kg)	E_T (kJ/kg)	T_H (°C)	React. (Y/N)
A	37.5	1000	5.37	364	352	581	N
A	42.8	1040	4.95	413	404	660	Y
A*	35.9	1020	5.87	374	359	594	N
A*	39.7	1020	5.21	389	378	622	Y
B	37.5	1020	5.60	380	367	605	N
B	41.2	1060	5.38	421	410	670	Y
B*	37.5	1050	5.86	398	384	631	N
B*	39.9	1050	5.46	407	396	648	Y

Table 2 *The calculated shock conditions of the experiments discussed in the text. The column headings correspond to the mixture, porosity, flyer velocity, pressure, total energy, thermal energy, homogeneous temperature, and whether or not the reaction occurred, respectively. An asterix indicates a four cavity experiment.*

The first set of experiments were conducted with Mix A. The green's porosity was typically 40%, and flyer velocities were varied from 700 to 1600 m/sec. Optical microscopy of the compacts recovered from low energy shots showed no evidence of chemical reactions, and the compacts were poorly bonded. In higher energy shots, the reaction apparently went to completion as evidenced by large voids, bulk melting and a homogeneous appearance under the optical microscope.

Further experiments showed the energy difference between no and full reaction to be small. In a four cavity experiment at a flyer velocity of 1.02 km/sec, no reactions occurred in a sample pressed to a porosity of 35.9 ± 0.8 percent while the reaction went to completion in the sample pressed to a porosity of 41.1 ± 0.8 percent. These porosities and impact conditions correspond to calculated homogeneous temperatures of 594° C and 622° C, respectively, or a total energy difference of less than 4 percent. Two full cavity experiments confirmed the energy difference to be less than 12 percent. No attempt was made using full cavities to narrow down the threshold width further.

An analogous procedure was conducted with Mix B. Since the morphologies of the two mixtures are similar, it is not surprising that the results for Mix B are nearly identical. A four cavity experiment showed the energy threshold to lie between thermal energies corresponding to homogeneous temperatures of 631° C and 648° C, or a total energy difference of less than 3 percent. Full cavity experiments confirmed that the separation in total energy between no and full reaction was less than 10 percent. No attempt was made using full cavities to narrow down the threshold width further. It is interesting to note that, except for the very small amount of surface reactions discussed below, no compacts of either mixture were recovered in an intermediate condition between no and full reaction.

Scanning electron microscopy of compacts shocked to just below the reaction threshold revealed that minor surface reactions had occurred which are not detectable optically or with x—ray diffraction. As can be seen in the back scattered electron image shown in Figure 3, the extent of the reaction was very limited. The nature of the reacted region differed slightly between Mix A and Mix B. The reactions in Mix A were very uniform along the interfaces where they occurred with a typical thickness of 0.5 μm. The Mix B compact also had a uniform reaction zone on a small portion of the interfaces, but there also existed pools of reacted material. A back scattered electron image of a typical reacted interface for mix B is shown in Figure 4, which also contains the far more typical interface where no reaction can be seen.

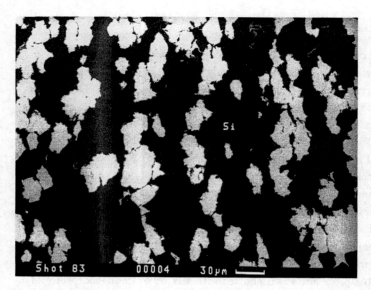

FIG. 2 *SEM back scattered image of Mix B shocked in a full cavity experiment to an energy just below where bulk reaction occurs. At this magnification, there is little evidence of any interaction between the Ni and Si. The shock propagated from right to left.*

FIG. 3 *SEM back scattered image of Mix B shocked in a full cavity experiment to an energy just below where bulk reaction occurs. At this magnification, small interfacial reactions can be seen as well as pools of reacted material. Also shown are interfaces where no reactions are detectable, by far the majority.*

IV. DISCUSSION

From the existence of a sharp threshold, it can be reasoned that the shock parameter governing whether or not bulk reactions occur in a 1:1 mixture of this particular morphology is the homogeneous temperature. The experiments rule out the possibility that the threshold is a pressure or elastic energy effect since the more porous compacts react but are shocked to a lower pressure and elastic energy.

One may argue that local particle level conditions change significantly across the threshold, however any such explanation is unlikely since it must exclude the possibility that the same local conditions exist anywhere at lower shock energies. For example, the existence/kinetics of shock initiated chemical reactions has been explained by local mass mixing and other similar terms describing local differences in the particle velocities of the constituents.[3–6] However, as can be seen in Figure 2, there is no evidence of mass mixing, and it is unlikely that increasing the energy by as little as 3 percent will greatly enhance mass mixing, especially with a lower shock pressure since one would expect greater constituent mixing at greater pressures.

Another local condition which may arguably change as the threshold is crossed is that a critical melt pool size is attained, however this is also not likely. Since the energy difference between practically no and full reaction is small, there is a high probability that there exists some local pre–shock particle configuration in the less porous green which will result in a "critical" melt pool size upon compaction. Furthermore, if the reaction is determined by some local pre–shock particle configuration, reactions would occur sporadically, depending on local particle placement during the pressing of the greens. Another argument may be that a critical density of melt pools is attained above the threshold, however this can not be true since, in the time it takes to "communicate" between melt pools through heat conduction, the melt pools no longer exist.[7] It is possible to argue other particle level explanations, however, as mentioned above, a necessary feature of such an approach would be that the same local conditions can not exist in lower energy samples.

We therefore conclude that the homogeneous temperature determines whether or not reactions occur in 1:1 atomic percent Ni/Si mixtures of the particular morphologies used here. This is the only parameter which undoubtedly varies across the threshold and is a reasonable explanation as to why lower energy compacts do not react while slightly higher energy compacts react fully. Since the homogeneous temperature determines whether or not bulk reactions occur, one can also conclude that the reaction kinetics are slower than the time required for temperature equilibration. Assuming a linear heat conduction time constant of, $\tau = r^2/\kappa$, where r is the Ni particle radius and κ is nickel's thermal diffusivity, implies that the reaction occurs on a time scale greater than several microseconds.[7]

V. CONCLUSION

Experiments on two mixtures of similar morphology of 1:1 atomic percent Ni and Si powders reveal the existence of a sharp energy threshold below which no significant reactions occur and above which the reaction goes to completion. From the existence of a sharp energy threshold, it can be reasoned that the homogeneous temperature determines whether or not the bulk reaction occurs rather than particle level conditions. One can also conclude that the reaction occurs on a time scale greater than several microseconds in Ni and Si particles larger than 20μm.

ACKNOWLEDGMENTS

This work was supported under the National Science Foundation's Materials Processing Initiative Program, Grant No. DMR 8713258. We would like to thank Phil Dixon, formally at the New Mexico Institute of Technology for preparing Mix B.

REFERENCES

1. A.H. Mutz and T. Vreeland, Jr., This Proceedings.

2. The averaging method used was similar to M.B. Boslough, J. Chem. Phys. 92, (3), (1990), p. 1839.

3. S. Batsanov, *et.al.*, Combustion, Explosions and Shock Waves, Vol 22,1986, p. 765.

4. Graham, R.A, *et.al.* in Shock Waves in Condensed Matter, Gupta, Y.M. (Ed.), Plenum Press, New York, (1986) p. 693.

5. Thadhani, N.N., Costello, M.J., Song, I., Work, S., and Graham ,R.A., in Proceedings of the TMS Symposia on Solid State Powder Processing, Indianapolis, October 1–4, 1989. To be published.

6. M.B. Boslough and R.A. Graham, Chem. Phys. Lett., Vol. 121, (1985), p. 446.

7. R.B. Schwarz, P. Kasiraj and T. Vreeland, Jr., in Metallurgical Applications of Shock Waves and High–Strain–Rate Phenomena , Murr, Meyers and Staudhammer (Eds.) , Marcel Dekker, Inc., New York, (1986), p. 313.

22

Thermochemistry of Shock-Induced Exothermic Reactions in Selected Porous Mixtures

M. B. BOSLOUGH

Structural Physics and Shock Chemistry Division
Sandia National Laboratories
Albuquerque, New Mexico 87185, U. S. A.

Steady detonation theory has been used to describe shock-induced chemical reactions in a variety of powder mixtures. By introducing a number of simplifications, pressure and temperature increases caused by a reaction can be estimated by considering only the initial powder density, porosity, heat of reaction and thermophysical properties of the product phases. For a given mixture, the pressure increase is greater for higher initial powder packing density. Materials with large volume increases associated with the reaction will demonstrate the largest pressure increases, as expected. Shock temperatures can be approximated by a linearly increasing function of pressure, where the zero-pressure intercept is the adiabatic reaction temperature and the slope depends on initial porosity, volume change and product specific heat. The approximation provides a means of characterizing shock-induced chemical reactions in terms of zero-pressure material properties.

I. INTRODUCTION

Many methods of shock processing of materials involve rapid, exothermic, shock-induced chemical reactions [1-3]. Full understanding of such reactions requires time-resolved measurements of their progress during a shock experiment. For reactions that lead to products in the condensed state, the measurable variable that is most sensitive to the extent of reaction is the temperature [4]. Unfortunately, radiation pyrometer measurements are dominated by the highest temperatures present in a sample [5]. Thus, application to powders will always be complicated by the fact that the temperature distribution is heterogeneous for a period of time after the shock wave arrives, until thermal equilibrium is achieved.

Pressure equilibrium is reached much more rapidly, so its measurement avoids complications associated with heterogeneities. However, its value tends to be a weak function of the extent of the reaction for shock events leading to condensed product phases. Moreover, the presence of volatile impurities, and uncertainties in the initial porosity of the powder compact, have disproportionately large effects on the resulting pressure [4]. Nevertheless, time-resolved pressure measurements can potentially be used to observe the progress of shock-induced reactions, such as Sn + S → SnS, for which the difference between reactant and product Hugoniots has been resolved experimentally [6].

This paper presents a simple approximation that can be used to determine shock pressures and temperatures associated with exothermic reactions without knowing the details of the equation of state of the reactant or product mixtures. The approximation is based on the fact that, for initially porous powders, the change in specific volume associated with the reaction is usually a small fraction of the total volume change due to compression by the shock wave. It assumes that, to first order, the volume decrease due to shock compression is offset by thermal expansion. In all cases considered here, the product phases are in the liquid state, and material strength is ignored.

II. BACKGROUND: THE HEAT DETONATION DESCRIPTION

If the reaction takes place behind a steady wave, it can be described as a type of detonation in which the chemical energy is converted mostly to heat, rather than to work as it would be in a conventional detonation. This steady-wave picture of shock-induced exothermic reactions has been referred to as the "heat detonation" model and is described by Boslough [7], where it is shown that such a reaction is fully equivalent, thermochemically speaking, to an overdriven detonation in an explosive. As is the case for any detonation, the thermodynamic states reached by the initiating shock wave can be determined from the Hugoniots of the reactant and product mixtures, which in turn can be calculated by making use of the Rankine-Hugoniot equation, the heat of reaction, and the equations of state of the reactants and products.

This calculation was performed for the standard thermite reaction (Fe_2O_3 + 2Al → Al_2O_3 + 2Fe) by considering the known equations of state of the reactant and product components, which were added according to a simple mixing model to give a bulk equation of state. Since shock pressures of less than 10 GPa were considered, Bridgman's quadratic equation was used for simplicity to avoid complications associated with a higher-order finite strain theory that would be required at higher pressures.

At moderate pressures, initial porosities on the order of 50% are required to initiate reactions [8]. For 50% porous thermite, the specific volume of states on the calculated product Hugoniot varies by less than 0.5% between zero pressure and 10 GPa, and the

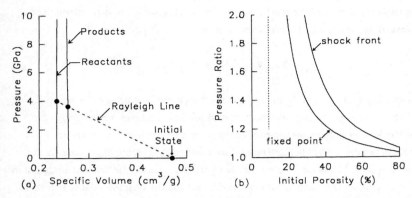

FIG. 1   *(a) Reactant and product Hugoniots for stoichiometric Al/Fe$_2$O$_3$ thermite with initial porosity of 50%, calculated using Bridgman and Mie-Grüneisen equations [7], (b) maximum pressure increases for stoichiometric Al/Fe$_2$O$_3$ thermite at a fixed Lagrangian point and at the shock front, calculated using the constant-volume Hugoniot approximation.*

product Hugoniot varies by less than 1.5% over the same pressure range. This variation is much smaller than the approximately 10% difference between specific volumes of reactants and products, and the approximately 40% decrease in specific volume associated with shock compression. When these calculated Hugoniots are plotted in the pressure-volume plane at a scale that includes the initial state (Fig. 1a), it is clear that the curves can be approximated with vertical lines, and calculated values for the waste heat and for the shock pressure difference between reactant and product Hugoniot will not be significantly affected.

III. CONSTANT-VOLUME APPROXIMATION

For highly porous reactive solids shocked to moderate pressures, the errors introduced by replacing reactant and product Hugoniots with isochores are small. There are other sources of uncertainty (e.g. strength, porosity and volatile impurities) that are likely to have larger effects. The assumption of constant-volume Hugoniots provides a greatly simplified method of calculating shock-induced pressures and temperatures. High pressure equation of state parameters that were previously used in thermite calculations [4] are unnecessary. Instead, only properties of reactant and product phases at atmospheric pressure are required.

The specific volumes of the reactant and product Hugoniots are determined from a straightforward application of stoichiometric mixing:

$$V_r = \Sigma n_{ri} V_{ri}, \quad V_p = \Sigma n_{pi} V_{pi}, \tag{1}$$

where n_{ri} and V_{ri}, and n_{pi} and V_{ri} are the stoichiometric coefficients and the specific volumes of the ith components of the reactant and product mixtures, respectively. To determine the specific volumes of the product components, the adiabatic reaction temperature is required. This is the temperature reached by a complete reaction in the absence of heat flow, and is identical to the shock temperature on the product Hugoniot in the weak-shock limit. The temperature is found by setting the negative heat of reaction (ΔH_r) equal to the enthalpy required to heat the mixture to the adiabatic reaction temperature T_a:

$$-\Delta H_r = \sum_i n_{pi} \left[\int_{T_o}^{T_a} C_{pi}(T)\,dT + \sum_j \Delta H_{trji} \right], \tag{2}$$

where ΔH_{trji} is the heat of transition to the jth phase of the ith product component, and $C_{pi}(T)$ is the constant-pressure specific heat of the ith product component at temperature T. Once the adiabatic reaction temperature is known, the specific volumes of the product components can be found in published tables of high temperature properties, and they can be added according to equation (1) to give the specific volume of the product Hugoniot. For many reactions, the adiabatic temperature exceeds the ambient-pressure vaporization temperature of one or both of the products. In these cases the volume of the product phases are determined by extrapolating the liquid state properties into the vapor field; the products remain condensed as long as the pressure is sustained.

IV. SHOCK PRESSURES

The heat detonation model assumes that the shock wave is steady, so all states reached by it must lie along the same Rayleigh line. For shock pressures much greater than the initial pressure, a simple geometric construction in the pressure-volume plane gives the equation:

$$P_r/P_p = (V_{oo}-V_r) / (V_{oo}-V_p), \tag{3}$$

where P_r and P_p are the pressures on the reactant and product Hugoniots, respectively, and V_{oo}, V_r and V_p are the initial (porous) volume, and the specific volumes of the reactant and product Hugoniots, respectively. The pressure ratio given by equation (3) is the maximum that can be measured by an *in situ* stress gage experiencing a steady reaction shock.

Application of the jump conditions to the isochoric Hugoniots allows an equation to be written for the maximum possible growth in pressure at a given point due to a shock-induced reaction. For a fixed particle velocity (equivalent to impact against an infinite-impedance boundary) the ratio of final to input pressure is:

$$P_f/P_i = (V_{oo} - V_r)/(V_{oo} - V_p) . \tag{4}$$

Since the maximum final pressure P_f given in equation (4) lies on the product Hugoniot, it can be set equal to P_p in equation (3). Combining the two equations gives an expression for the maximum possible growth in pressure at the shock front:

$$P_r/P_i = (V_{oo} - V_r)^2/(V_{oo} - V_p)^2 . \tag{5}$$

Fig. 1b shows calculated maximum pressure increases for the standard thermite reaction at a fixed point and at the shock front. The vertical dotted line denotes the porosity at which $V_{oo} = V_p$, where the approximation breaks down. The parameters for a number of selected exothermic reactions have been calculated, and appear in table 1.

V. SHOCK TEMPERATURES

Shock temperatures have been calculated using the Bridgman equation and a Mie-Grüneisen thermal equation of state in the context of the heat detonation model [7]. For 50% porous thermite, the shock temperature was found to be an approximately linearly increasing function of Hugoniot pressure up to 10 GPa. The post-shock temperature (after release of pressure) differed from the shock temperature by only a few percent.

The constant-volume Hugoniot approximation provides a convenient method of calculating shock temperatures in exothermically reacting porous solids. An adaptation of the Rankine-Hugoniot equation to this situation gives the difference in specific internal energy ΔE between a state at pressure P on the product Hugoniot and that of the product phases at the adiabatic reaction temperature and zero pressure:

$$\Delta E = \frac{1}{2} P (V_{oo} - V_p) . \tag{6}$$

This internal energy difference can also be written in terms of temperature by the approximation:

$$\Delta E = \int_{T_a}^{T_H} C_p(T) \, dT . \tag{7}$$

TABLE 1 *Properties of Selected Reactive Mixtures*

Reactants	Products	$-\Delta H_r$ (kJ/g)	T_a (K)	V_r (cm³/g)	V_p (cm³/g)	ΔV (%)	C_p (J/g K)	Fig. 2 key
$Cr_2O_3 + 2Al$	$Al_2O_3 + 2Cr$	2.63	2330	0.239	0.248	4	1.05	(a)
$3ZnO + 2Al$	$Al_2O_3 + 3Zn$	2.11	2550	0.213	0.246	16	0.78	(b)
$3SnO + 2Al$	$Al_2O_3 + 3Sn$	1.23	3430	0.181	0.227	26	0.50	(c)
$3Fe_3O_4 + 8Al$	$4Al_2O_3 + 9Fe$	3.73	3440	0.235	0.255	9	1.02	(d)
$Fe_2O_3 + 2Al$	$Al_2O_3 + Fe$	4.00	3630	0.235	0.264	12	1.23	(e)
$3Cu_2O + 2Al$	$Al_2O_3 + 6Cu$	2.45	3470	0.190	0.199	5	0.67	(f)
$3SnO_2 + 4Al$	$2Al_2O_3 + 3Sn$	2.86	4100	0.188	0.262	40	0.66	(g)
$3MnO_2 + 4Al$	$2Al_2O_3 + 3Mn$	4.87	4130	0.249	0.311	25	1.12	(h)
$3Ag_2O + 2Al$	$Al_2O_3 + 6Ag$	2.13	4820	0.156	0.191	22	0.43	(i)
$3CuO + 2Al$	$Al_2O_3 + 3Cu$	4.09	4940	0.194	0.253	31	0.79	(j)
$3Ag_2O_2 + 4Al$	$2Al_2O_3 + 6Ag$	3.88	6980	0.164	0.275	68	0.54	(k)

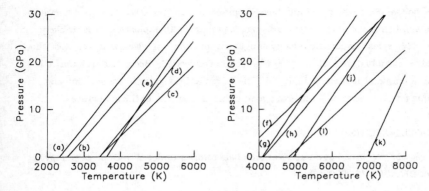

FIG. 2   *Calculated shock temperatures for selected reactive solids with initial porosities of 50%, with key to reactions given in table 1.*

Combining equations (6) and (7), and taking the specific heat to be independent of temperature gives a linear equation for the shock temperature as a function of pressure for the product Hugoniot:

$$T_H = T_a + cP, \tag{8}$$

where the zero pressure intercept is the adiabatic reaction temperature and the slope is given by the constant

$$c = (V_{oo} - V_p)/2C_p. \tag{9}$$

Shock temperatures calculated using equations (2), (8) and (9) for several selected reactions in powders with 50% initial porosity are displayed in Fig. 2.

Since the Hugoniots are approximated by isochores, there is no reversible work associated with the shock compression, and the entire internal energy increase given by equation (6) is retained upon pressure release. Thus, the shock and postshock temperatures in this approximation are identical.

VI. SUMMARY

A method is outlined for characterizing non-gas-forming, shock-induced chemical reactions in porous solids in terms of zero-pressure thermophysical and thermochemical properties of reactant and product components. The method provides a convenient way of determining shock pressures and temperatures along the product Hugoniot of such reactive solids. Since the maximum possible pressure changes associated with chemical reactions in highly porous powders are small (compared with explosives), it is unlikely that *in situ* stress or particle velocity gauges can provide a precise picture of most shock induced reaction histories whose products remain condensed. Because shock and postshock temperatures are so high, their measurement remains the most promising technique for time-resolved studies of shock-induced reactions in these mixtures.

ACKNOWLEDGMENTS

This work was performed at Sandia National Laboratories under U.S. Dept. of Energy contract DE-AC04-76DP00789.

REFERENCES

1.   N. N. Thadhani, *Adv. Materials & Manufacturing Processes* 3: 493 (1988).

2.   H. Kunishige, Y. Oya, Y. Fukuyama, S. Watanabe, H. Tamura, A. B. Sawaoka, T. Taniguchi and Y Horie, *Res. Lab. of Eng. Matl. Rept. 15*, Tokyo Institute of Technology (1990).

3.   L. H. Yu and M. A. Meyers, *J. Matl. Sci.*, in press (1990).

4.   M. B. Boslough, *Proceedings of the Ninth Symposium (International) on Detonation*, in press (1990).

5.   M. B. Boslough and T. J. Ahrens, *Rev. Sci. Instrum.* 60: 3711 (1989).

6.   S. S. Batsanov, G. S. Doronin, S. V. Klochkov and A. I. Teut, *Combustion, Explosion and Shock Waves* 22: 765 (1986).

7.   M. B. Boslough, *J. Chem. Phys.* 92: 1839 (1990).

8.   R. A. Graham, *Sandia Report SAND88-1055*, Sandia National Laboratories (1988).

23

Reaction Synthesis/Dynamic Compaction of Titanium Carbide and Titanium Diboride

J. LASALVIA, L. W. MEYER*, D. HOKE, A. NIILER**, and M. A. MEYERS

Department of Applied Mechanics and Engineering Sciences
University of California, San Diego
La Jolla, California 92093, U.S.A.

*IFAM, Bremen, Germany

**Ballistic Research Laboratory, Aberdeen Proving Ground, Maryland, U.S.A.

A novel approach for producing dense ceramics and inter-metallics is described. This approach combines reaction syn-thesis with a high velocity forging step (10-15 m/s impact velocity) to achieve densification and near-net shape. Com-paction is necessary because reaction synthesized titanium carbide and diboride are extremely porous. By combining these two processes, titanium carbide and diboride disks with densities greater than 96% were produced.

Elemental powders of titanium and carbon or boron are dry mixed in a jar with a grinding medium. The powder is then pressed into a cylindrical green body which is placed into an insulated cavity within the forging die of the forging machine (Dynapak). Because the reaction is strongly exothermic (44.1 kcal/mol for TiC and 66.8 kcal/mol for TiB$_2$), the final product is raised to a temperature above its ductile-to-brittle transition. The forging step accomplishes two objectives: (1) densification, and (2) shaping. After the forging step, the titanium carbide (or boride) is removed from the forging die and placed within a furnace for slow cooling.

The major problem encountered in this study has been thermal shock. Insulating the titanium carbide during and after the hot forging step and raising the temperature of the sur-roundings by use of a furnace were used to decrease cracking.

SEM observation of titanium carbide reveals the grains are equiaxed with an average size of 44 μm. This is indicative of crystallization or recrystallization after the forging step.

Low strain-rate fracture surfaces are primarily intergranular, while at higher strain rates, the fracture mode is primarily transgranular. Vickers microhardness values are comparable with those obtained from a commercially hot-pressed material. Preliminary dynamic compressive strength measurements indicated a value greater than 1.7 GPa at a strain-rate of 10^2 per second for TiC.

I.   INTRODUCTION

Reaction synthesis (or combustion synthesis, or SHS) is a materials processing technique by which ceramics, ceramic composites, and intermetallic compounds may be produced. The reaction synthesis process results from an exothermic, self-propagating reaction among elemental powders or solid reactants immersed in a reacting gaseous atmosphere. A review of the subject is provided by Munir and Anselmi-Tamburini [1]. Extensive research efforts in the Soviet Union by Merzhanov and co-workers [2-5] have led to the industrialization of this method and to the synthesis of hundreds of materials. In the U.S., the principal efforts in SHS are by Holt and co-workers [6,7] at Lawrence Livermore National Laboratory, Niiler and co-workers [8,9] at BRL, Munir and co-workers [10] at the University of California, Davis, and K. Logan [11] at Georgia Institute of Technology. Of particular relevance to the research described herein is the work of Niiler and co-workers [8,9] in which the reaction synthesis is followed by explosive compaction. Recent calculations by Wilkins [12] have shown that the pressures generated in titanium carbide and boride by explosive detonation process in the fixture used by Niiler and al. are low: 3-5 GPa. This was the motivation for the substitution of dynamic compaction in a high-speed forging machine for explosive compaction.

II.   EXPERIMENTAL MATERIALS AND TECHNIQUES

The powders used in this research consisted of elemental titanium, boron, and carbon. The titanium powder size is 325 mesh (maximum diameter of 44 μm); the average graphite particulate size in 2 μm; the boron particle size is ~325 mesh (<44 μm). The powders were mixed in a grinding jar for two hours in order to produce a homogeneous powder mixture. The powders were then compacted into disks (3- and 4-in diameter) at a pressure of approximately 50 MPa, yielding a green compact with a density of approximately 60% of the theoretical. The titanium carbide was produced from the reaction Ti + 0.9 C, while titanium diboride was produced from the mixture Ti + B, to which 20 wt pct of pre-synthesized TiB_2 was added. The addition of inert TiB_2 reduced the velocity of the reaction front and the heat output, therefore ensuring the integrity of the compact after reaction. A typical microstructure

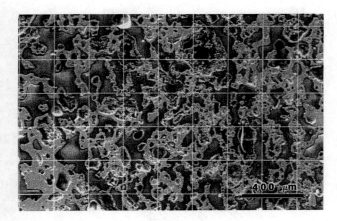

FIG. 1 *Porous microstructure of reaction synthesized TiC.*

after reaction synthesis is shown in Figure 1. The clear regions are TiC and the dark regions are the mounting epoxy. The body is approximately 50 pct porous.

Compaction was achieved in a suitably modified DYNAPAK unit. This is a quick release high-energy forging machine developed for metal-working applications. The hammer derives its kinetic energy from compressed nitrogen gas. The machine was modified for the specific objectives of this investigation. It is schematically shown in Figure 2. The hammer is propelled down and impacts the workpiece at velocities ranging from 10 to 15 m/s. Thus, the dynamic pressures generated are very low. The maximum energy output of the DYNAPAK in this investigation was 25 kJ. Figure 2 shows a schematic of the machine in the "ready" and "fired" position.

Special dies were designed for the forging of the compact. They contained refractory insulation to minimize heat losses from compact during and after forging. The details of the assembly are described by LaSalvia [12]. After the reaction was completed, the forging hammer was accelerated against the hot, porous ceramic. The forging hammer was then raised and the extractor activated. The specimen was then removed, by means of special tongs, and inserted in a furnace, under protective argon atmosphere, at the temperature of 1,100 °C. The specimens were allowed to cool in the furnace, over a period of 12 hours.

From the compacts, specimens for optical and scanning electron microscopy were obtained by sectioning with a diamond saw. Mechanical property measurements consisted of

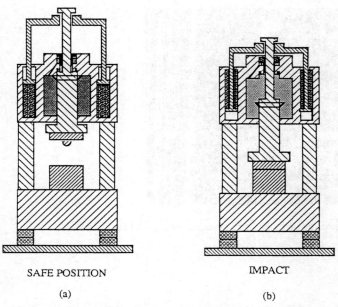

SAFE POSITION

IMPACT

(a)

(b)

FIG. 2 *High-speed forging machine (impact velocity: 10-15 m/s); (a) ready and (b) fired positions.*

the determination of the Vickers microhardness and dynamic compressive failure strength. Dynamic compressive strength was determined using the split Hopkinson bar technique. Cubes with 10 mm sides were sectioned from the compacts and placed between the transmitter and incident bars. The pulse rise was shaped by means of a copper disk placed between the striker and the incident bars. This ensured a sufficiently large rise time, not necessary for metallic specimens.

III.   RESULTS AND DISCUSSION

It was possible, after process development, to eliminate a great fraction of the thermal cracks, very prevalent in titanium carbide. The titanium diboride compacts did not exhibit the same propensity for cracking. Figure 3 shows titanium carbide and diboride compacts; the cracks observed for titanium carbide are only surface features and are due to hot tearing (breaking of the thin surface layer that is rendered brittle very early in the forging process). The circumferential cracks observed for the titanium diboride specimen are due to insufficient lateral confinement.

FIG. 3 *Appearance of (a) titanium carbide (4 in.) and (b) titanium diboride (3 in.) compacts.*

Figure 4 shows both the bulk and center density as a function of specific energy of the DYNAPAK (energy divided by mass of compact) for titanium carbide. Because of heat losses, the surface material is cooler and less ductile than the center material. The difference between the two curves shows this. A specific energy of 70 J/g is required to produce compacts that have a density of 96% throughout. This specific energy can be used to estimate the flow stress of the ceramic at the imposed strain rate ($\sim 2 \times 10^2$ s^{-1}). Carroll and Holt [13,14] developed an expression for the energy required to consolidate a porous material. Equating this energy to the kinetic energy of the machine, one has:

$$E_K = \frac{2}{3} \sigma_y V_s \{\alpha_0 \ln \alpha_0 - (\alpha_0 - 1) \ln (\alpha_0 - 1)\} \tag{1}$$

where σ_y is the flow stress of the material, α_0 is the initial (before compaction) distention (ratio between specific volumes of porous and densified material), and V_s is the specific volume of the titanium carbide. From Figure 1, one can estimate α_0 (=2). Hence,

$$E_K = 0.942 \, \sigma_y \, V_s \tag{2}$$

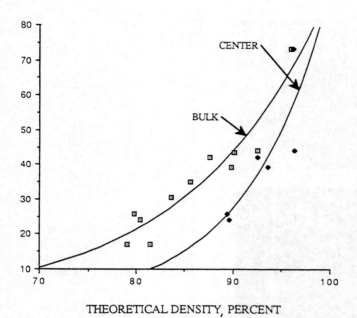

THEORETICAL DENSITY, PERCENT

FIG. 4   *Compact (bulk) and center density as a function of specific energy for TiC.*

Equation 2 yields:

$$\sigma_y = 425 \text{ MPa.}$$

This is the flow stress of the titanium carbide immediately upon completion of the reaction. The temperature is estimated to be in the 2000-3000 ^{0}C. range. Further work is required to determine the dynamic mechanical response of a hot, porous, plastic ceramic.

The microstructure of titanium carbide is shown in Figure 5(a). The average grain size, as measured by the linear intercept method, is 44 μm. The grains are equiaxed, indicating that crystallization, or recrystallization occurred after plastic deformation. This supports the synchrotron radiation experiments performed by Wong and Holt [15], that showed that the final TiC structure formed only one minute after the reaction was completed. Two intermediate structures were observed. Porosity is also observed in Figure 5(a); it is more prevalent at the grain boundaries. Because of heat losses, the grain size close to the surface was smaller than in the center of the compacts. The titanium diboride microstructure is shown in Figure 5(b). The voids that can be seen are due to the particle pullout during cutting and polishing operations. The grain size is approximately 8 μm. This indicates that either bonding between

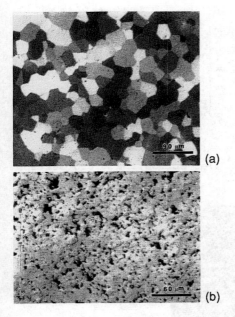

FIG. 5 *Optical micrographs showing the grain structure of the reaction synthesized dynamically compacted (a) TiC and (b) TiB₂.*

the pre-synthesized and combustion synthesized material is not very good, or that grain-boundary cohesion is poor.

Vickers microhardness measurements across the thickness are shown in Figure 6. The indentor weight was 300 g. These values are quite high: HVN 2250 for TiC; HVN 2425 for TiB$_2$. These values compare favorably with hot pressed TiC and TiB$_2$ (HVN 2235 for hot pressed TiC and HVN 2780 for hot pressed TiB$_2$). The microhardness profiles across the cross-section show that the values are consistently high.

The dynamic strengths of the titanium carbide specimens was established at strain rates that fluctuated between 50 and 100 s^{-1}. The compressive strengths varied considerably, and the highest one was equal to 1.7 GPa. The variation is due to pre-existing cracks in the specimens. The fracture morphology differed considerably from the one obtained from the thermal cracks. While dynamic fracture tended to be of a mixed transgranular-intergranular mode, the slow fracture, induced by thermal stresses, was totally intergranular. These morphological differences are due to the higher energy available under dynamic crack propagation. Figure 7 shows the two morphologies. It can also be noted that the thermal

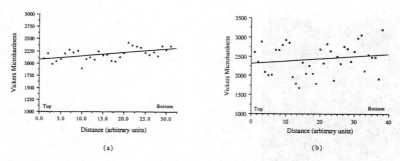

(a)

(b)

FIG. 6   *Vickers microhardness profiles across the cross-section*
         for (a) TiC and (b) TiB₂.

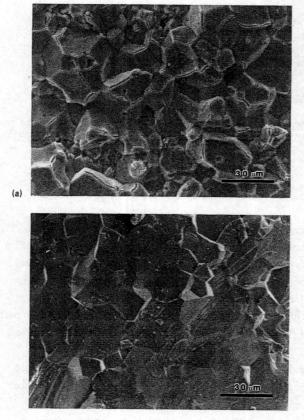

(a)

FIG. 7   *(a) Fracture surface generated by thermally-induced*
         stresses.
         (b) Fracture surface generated in Hopkinson bar
         experiment.

fracture surface seems to be covered by a surface layer, possibly an oxide or a nitride. This layer is cracked along the grain-boundary edges.

V. CONCLUSIONS

It is demonstrated that dynamic forging is a feasible process to consolidate ceramics produced by combustion synthesis. Dynamic compaction in a rapid forging machine has considerable advantages over hot pressing (lower turn-around time) and over explosive compaction (it is readily adaptable to automated industrial production).

VI. ACKNOWLEDGMENTS

This research was supported by the U. S. Army Research Office under Contract Number DAAL03-88-K-0194. The guidance provided by Dr. G. Mayer (IDA) and of Dr. A. Crowson (ARO) is gratefully acknowledged. The use of the facilities of the Center of Excellence for Advanced Materials is greatly appreciated. Mr. Klaus Blueggel was an important element of this team, and it is through his ingenuity and decisiveness that the DYNAPAK was rendered operational and modified. Mr. Jon Isaacs was greatly helpful, performing the dynamic testing experiments.

REFERENCES

1. Z. A. Munir and U. Anselmi-Tamburini, *Mater. Sci. Rep.* 3: 281 (1989).

2. E. I. Maksimov, A. G. Merzhanov, and V. M. Shkirko, *Comb. Explo. Shock Waves*, 1: 15 (1965).

3. N. P. Novikov, I. P. Borovinskaya, and A. G. Merzhanov, "Thermodynamic Analysis of Self-Propagating High-Temperature Synthesis Reactions," in *Combustion Processes in Chemical Technology and Metallurgy,* Ed. A. G. Merzhanov, Chernogolovka, 1975 (English Translation).

4. V. M. Maslov, I. P. Borovinskaya, and A. G. Merzhanov, *Comb. Explo. Shock Waves*, 12: 631 (1976).

5. Y. M. Maksimov, A. T. Pak, G. B. Lavranchuk, Y. S. Naviborodenko, and A. G. Merzhanov, *Comb. Explo. Shock Waves*, 15: 415 (1979).

6. J. B. Holt and Z. A. Munir, *J. Mat. Sci.* 21: 256 (1986).

7. A. Niiler, T. Kottke, and L. Kecskes, "Shock Consolidation of Combustion Synthesized Ceramics," in *Proc. First Workshop on Industrial Applications of Shock Processing of Powders*, New Mexico Tech., Socorro, NM, p. 901 (1988).

8. A. Niiler, L. J. Kecskes, T. Kottke, P. H. Netherwood, Jr., and R. F. Benck, "Explosive Consolidation of Combustion Synthesized Ceramics: TiC and TiB$_2$," *Ballistics Research Laboratory Report* BRL-2951, Aberdeen Proving Ground, MD, December 1988.

9. A. Niiler, L. J. Kecskes and T. Kottke, this volume.

10. Z. A Munir, *Am. Ceram. Soc, Bull.* 67 [2] 343 (1988).

11. M. L. Wilkins, Lawrence Livermore National Laboratory, private communication, (1990).

12. J. C. LaSalvia. "Production of Dense Titanium Carbide by Combining Reaction Synthesis with Dynamic Compaction," M.Sc. Dissertation, University of California, San Diego, 1990.

13. M. M. Carroll and A. C. Holt, *J. Appl. Phys.* 43: 759 (1972).

14. M. M. Carroll and A. C. Holt, *J. Appl. Phys.* 43: 1626 (1972).

15. I. Wong and J. B. Holt, "Time-Resolved Diffraction Studies using Synchroton Radiation," First Intl. Ceramic Science and Technology Congress, Am. Cer. Soc., Oct. 31-Nov. 3, 1989, Anaheim, California.

24

Shock-Induced Chemical Reactions and Synthesis of Binary Compounds

N.N. THADHANI,[1] A. ADVANI,[1] I. SONG,[1] E. DUNBAR,[1] A. GREBE[1] and R.A. GRAHAM[2]

[1]CETR, New Mexico Tech, Socorro, New Mexico 87801, USA.

[2]Sandia National Laboratories, Albuquerque, New Mexico 87185, USA.

The results of an experimental program on shock-induced chemical reactions and synthesis of binary compounds are presented. Binary powder mixture systems that are investigated include: (i) intermetallic forming compounds (e.g., Ni-Al, Ni-Si, Nb-Si, etc.) associated with large negative heats of reaction; and (ii) isomorphous (e.g., Ni-Cu) and fully immiscible (e.g., Nb-Cu) systems associated with zero (or positive) heat of reaction. The extent of shock induced chemical reactions and the type of shock synthesized compounds formed in these systems are observed to be dependent on (i) shock-loading conditions, (ii) the relative volumetric distribution of the mixture constituents, and (iii) differences in respective material properties which affect relative particle flow.

I. INTRODUCTION AND BACKGROUND

Shock compression of powder mixtures can induce extensive plastic deformation and fluid-like flow of individual particles, leading to enhancement of reactivity and sufficiently intimate contact between clean surfaces to permit combustion-type chemical reactions within the duration of shock loading. These features in fact form the basis of a conceptual framework characterizing shock-induced chemical synthesis, as identified by Graham [1]. Shock compression conditions control the degree of plastic flow and mechanical mixing between reactant powders and the extent of enhancement of reactivity. Constituent powder properties such as morphology, density, melting temperature, strength, sound speed, etc. and the energetics of the reaction (e.g., heat of reaction) also strongly affect the synthesis process [2,3]. Thus, the objectives of our overall

program are: (a) to study shock-induced chemical reactions in various
binary alloys; (b) establish the mechanisms of chemical reactions; and (c)
investigate shock synthesis of novel compounds unattainable by other
conventional techniques.

II.   EXPERIMENTAL APPROACH AND PROCEDURE

The selection of materials was based on the objective of understanding the
mechanisms of shock-induced chemical reactions in:

(a)   intermetallic forming systems associated with a large negative heat
      of reaction ($H_R \ll 0$), e.g., Ni-Al, Ni-Si, and Nb-Si; and

(b)   isomorphous and immiscible systems with zero or positive heats of
      reaction ($H_R \geq 0$), e.g., Cu-Ni and Cu-Nb.

Thus, the Ni-Al system was used for establishing the effects of shock
conditions, powder particle morphology, and the volumetric distribution of
starting reactant powders. The Ni-Si and Nb-Si systems were used to
investigate the effects of physical (melt temperature) and mechanical
property (hardness) differences between constituents in systems having
otherwise similar reaction initiation characteristics at ambient pressure.
Cu-Ni was selected as a typical isomorphous system, while Cu-Nb was
selected as a typical immiscible system, both associated with zero or
positive heats of reaction. The shock synthesis experiments were conducted
using the CETR/Sawaoka 12-capsule plate impact system as well as the Sandia
Momma Bear A (with Comp B explosive) fixtures [2-4]. Two dimensional
radial effects (due to impedance mismatch between the steel holder and the
porous compact) dominate the shock loading process in these and similar
recovery fixtures. Numerical simulations conducted on the CETR/Sawaoka
fixtures [4] show pressure and temperature conditions far in excess of
those in the Sandia Momma Bear fixture.

III. RESULTS AND DISCUSSION

A.   *SHOCK SYNTHESIS OF INTERMETALLIC TYPE SYSTEMS*
Ni-Al system: It was shown in our earlier work [3] that the extent of
shock-induced chemical reactions and type of the shock synthesized products
formed are influenced by the morphology of starting powders and the
intensity of shock loading conditions. These results show that higher
(more in tense) shock conditions and a flaky (irregular) powder morphology,
promote plastic deformation and flow, and mixing of dissimilar particles.
In such an intimately mixed and activated configuration, and with rapid

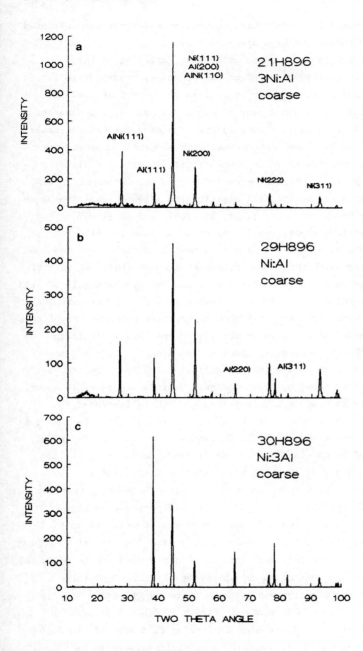

FIG. 1 *XRD results of shock synthesized Ni–Al compacts of coarse powders mixed in (a) 3Ni:Al, (b) Ni:Al, and (c) Ni:3Al volumetric distribution.*

shock-induced increases in temperature, combustion reactions are initiated and self-sustained during the high pressure state.

In experiments which build upon prior work, mixtures of three different volumetric distributions of coarse Ni and Al powders (similar to those used in earlier work) [3] were shock treated under identical conditions using the Sandia Momma Bear A Comp B set-up. It was observed (as shown in the XRD results in Figure 1) that the mixtures with volumetric ratios which yield an Ni_3Al and NiAl atomic stoichiometry, undergo bulk chemical reaction. Powders mixed in a volumetric distribution yielding Ni_3Al [Figure 1(a)] show a greater degree of bulk chemical reaction, in contrast to powders mixed in the NiAl distribution [Figure 1(b)] [based on comparison of peak heights of NiAl (110), Ni (200), and Al (110)]. Practically no reaction is observed in the powders mixed in $NiAl_3$ distribution [Figure 1(c)]. Due to the vast density difference between Ni and Al, the starting Ni_3Al stoichiometry powder mix has almost 67 vol% Ni and balance Al, the NiAl stoichiometry powder mix has approximately 40 vol% Ni and balance Al, while the $NiAl_3$-stoichiometry powder mix has only 18% Ni and 87% Al. Thus, with the $NiAl_3$ stoichiometry mixture there is very little Ni available which can participate in the mixing and form an intimate mixture with Al, thereby undergoing chemical reaction.

Ni-Si and Nb-Si: Similar to the intermetallic-forming Ni-Al system, are the Ni-Si and Nb-Si systems, however, the melt temperature of Si is known to be substantially reduced at higher pressures. The differences in melting temperature and hardness of the elemental constituents in the Nb-Si system are larger than those in the Ni-Si system. However, differential thermal analysis of unshocked Ni-Si and Nb-Si powder mixtures revealed identical reaction behavior for both with reaction initiating at approximately 1250°C, which corresponds to the eutectic melting temperature of a few percent Ni (or Nb) in Si, typical of self-propagating high-temperature combustion reactions. Shock synthesis experiments on Ni-Si and Nb-Si conducted using the 12 capsule CETR/Sawaoka fixture, at an impact velocity of 0.9 km/s, revealed that complete chemical reaction to $NiSi_2$ and Ni_2Si compounds was observed in the Ni-Si powder mixture [Figure 2(a,b,c)], while no bulk chemical reaction was evident in the Nb-Si powder mixture [Figure 3(a,b,c)]. In the Nb-Si compact [Figure 3 (a,b,c)] Si powder particle boundaries were totally absent indicating that most of the Si had melted and resolidified. The individual Nb particles appear to be contained in the matrix of molten and resolidified Si, although localized 1 μm thick reaction layer around the Nb particles is observed. These

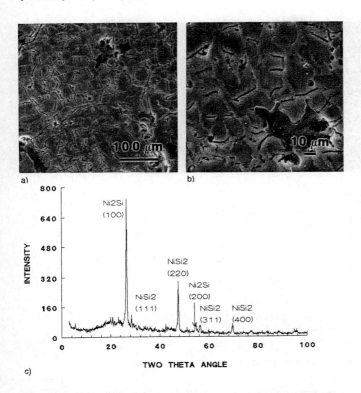

FIG. 2 *(a,b) SEM micrographs and (c) XRD results showing complete reaction in shock synthesized Ni-Si compacts.*

observations suggest that the presence of a melt phase **may not** be necessary for shock induced chemical reactions. In fact, it is also possible that melting of Si may limit particle flow and intimate mixing between Si and the harder Nb particles which can otherwise be attained while both Nb and Si are in the solid state.

B. SHOCK SYNTHESIS OF ISOMORPHOUS AND IMMISCIBLE SYSTEMS

Cu-Ni: The Cu-Ni system, which has a lens shaped phase diagram that runs from the low melting (1083°C) Cu end to high melting (1453°C) Ni end, is a typical isomorphous-forming system [5]. Shock-compression processing of fine (0.3-0.7μm), flaky (0.37 x -44μm) and coarse (-44μm) Ni powders with 50 - 100μm rounded Cu powders was performed for 1:1 packing stoichiometry using 65% packed compacts in the CETR/Sawaoka capsules at impact velocities

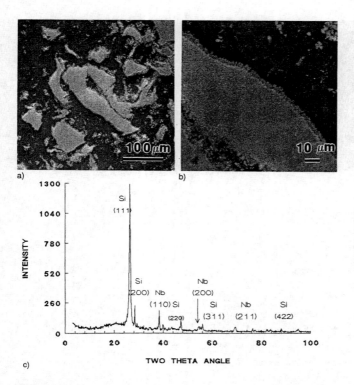

FIG. 3 *(a,b) SEM micrographs and (c) XRD results showing no bulk reaction in Nb-Si compacts.*

of 0.9 and 1.6 km/s. Compacts prepared from fine, flaky and coarse Ni-containing powder mixtures at 0.9 km/s showed etching contrast, as created by presence of reaction zones, for all morphologies examined. Increasing the impact velocity to 1.6 km/s resulted in presence of Cu-Ni chemical reactions in regions close to the center of the non-impact surface (high peak-pressure regions). The reaction zone in the fine and flaky morphologies appeared to have a solidification structure, and no Cu or Ni powder particles were visible in this region [Figure 4 (a,b)]. The coarse specimen also showed a solidification structure but, in addition, had small (variable) amounts of unmelted Ni particles at isolated locations in the microstructure (Figure 4c).

        SEM analysis of the reaction zone verified presence of a dendritic solidification structure in the reaction zone (Figure 5). Microchemical analysis using SEM-EDS showed that the Cu:Ni ratio in reaction

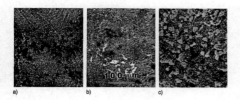

FIG. 4 *Optical micrographs of shock processed Cu-Ni compacts of (a) fine, (b) flaky, and (c) coarse powder morphology.*

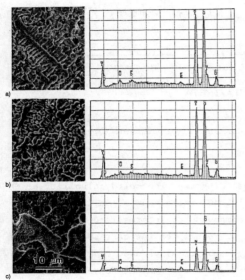

FIG. 5 *SEM photographs and corresponding EDS analysis of shock processed (a) fine, (b) flaky, and (c) coarse Cu-Ni powder compacts.*

(solidification) zones in fine and flaky samples was approximately 1:1, and in the coarse powder morphology sample, it varied from 3:7 to 1:1, as shown in Figure 5. XRD analysis confirmed the presence of an additional phase in the reaction region. As shown in Figure 6, fcc peaks from Cu and Ni powders can be seen along with additional peaks present between the original Cu and Ni peaks. This indicates that the phase formed in the reaction region has a lattice parameter in between that of Cu and Ni. *Cu-Nb:* The Cu-Nb immiscible system [5] has a typical S-shaped solidus curve, which runs from the low-melting Cu end (at 1080°C) to the high-

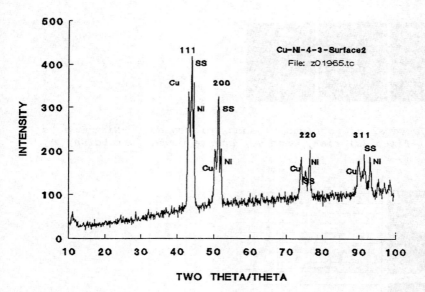

FIG. 6 *XRD analysis showing presence of an additional phase (marked `SS´ - solid solution in shock processed Cu-Ni compact.*

melting Nb end (2469°C), in the phase diagram. The solid solubility of Nb in Cu is less than 0.1 atomic % at 1080°C, while that of Cu in Nb is a maximum of 1.2 atomic percent. No compounds or intermediate phases occur in this system. Shock synthesis of Cu-Nb samples was carried out in a 1:1 stoichiometry at 65% packing density. Powder sizes used in mixtures were less than 44 μm, while the impact velocity (for the CETR/Sawaoka experiments) was approximately 1.9 km/s.

Optical examination of the Cu-Nb compact showed small particles of Cu and Nb in some locations, while other regions showed a splitting of Cu into finer dendrite-like fragments, as well as a solidification microstructure with no particles of Cu or Nb at locations particularly near the center of the non-impact surface [Figure 7(a),(b)]. SEM-EDS and electron microprobe analysis (EMPA) techniques further substantiated reaction/solid-solution formation in Cu-Nb compacts. A solidification microstructure containing dendrites with variable amounts of Cu and Nb was observed on the non-impact surface as shown in Figure 8(a). Dendrites analyzed were noted to be primarily Nb-rich containing as high as 96% Nb versus 4% Cu, while other dendrites containing significantly higher amounts of copper (up to 60%) were, however, also present [Figure 8(c),(d)]. XRD analysis [Figure

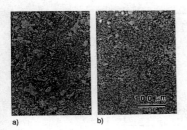

a)   b)

FIG. 7 *(a), (b) Optical micrographs of shock processed Cu-Nb compacts.*

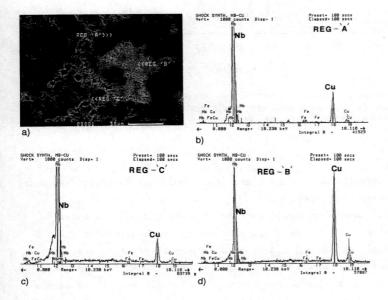

a)   b)

c)   d)

FIG. 8 *(a) SEM microstructure showing solidification microstructure of a reaction product in Cu-Nb compact, (b) and (c) EDX traces revealing regions containing 96% Nb and 4% Cu (Regions `A´ and `C´), and (d) EDX trace revealing region containing 60% Cu and balance Nb (Region `B´).*

9(a),(b)] on the non-impact surface also showed presence of peaks (between 35° and 45°) in addition to those of elemental Nb and Cu, confirming the presence of phases other than that of Cu and Nb. The intensity of these reaction product peaks was low, but could be reproduced on subsequent runs, especially for slow scans between 35° and 45°. The nature of product

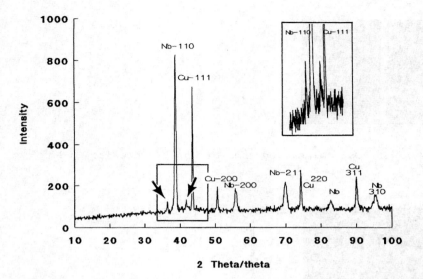

FIG. 9 *XRD analysis showing presence of additional peaks corresponding to Cu-Nb shock-induced reaction product. (Insert shows enlarged slower scan from 30º to 50º).*

phases (solid-solution or intermetallic-type) could not be identified using the techniques employed above.

The shock-induced chemical reaction in Cu-Ni and Cu-Nb samples shocked at 1.6 km/s and 1.9 km/s are produced by melting, mixing and solidification of the constituent powders. Since there is no heat of reaction evolved in isomorphous Cu-Ni and immiscible Cu-Nb system, it requires both Cu and Ni (or Cu and Nb) powders to melt and mix to form a uniform solidification region. Impetus for melting and mixing is provided by shock-compression of powders.

IV.   SUMMARY OF SHOCK-INDUCED REACTION MECHANISMS

Shock-induced chemical reactions in two types of binary powder systems are investigated. Intermetallic forming systems such as Ni-Al, Ni-Si, and Nb-Si, (associated with large negative heats of reaction), undergo solid-state combustion reactions in highly activated and intimately mixed powders, a configuration produced due to shock-induced plastic deformation and particle flow, mixing of reactants, and temperature increases associated with void collapse during shock compression. Critical in such reactions is the degree of flow and mixing between dissimilar particles permitted by prevailing shock conditions (shock pressure and temperature), powder properties (morphology, hardness, etc.) and the volumetric distribution of reactants. For isomorphous (Cu-Ni) and fully immiscible (Cu-Nb) systems (associated with zero or positive heats of reaction) chemical reactions require complete melting and mixing of powder particles to form a uniform solidification volume. The rapid quench rates then make it possible to retain compounds formed in melt state, thus providing a technique for forming compounds such as CuNb alloys which are otherwise immiscible.

ACKNOWLEDGMENTS

Funded by NSF Award No. DMR-8713258 and Sandia National Laboratories Grant No. 41-5737.

REFERENCES

1.   R.A. Graham, "Issues in Shock-induced Solid State Chemistry," in *Proc. of 3rd Int. Symposium on High Dynamic Pressures*, LaGrande Motte, France, June 5-9, 175, (1989).

2.   R.A. Graham, "Shock Compression of Solids as a Physical-Chemical-Mechanical Process, in S.C. Schmidt and N.C. Holmes, Editors, *Shock Waves in Condensed Matter - 1987*, North Holland, 11: 11 (1988).

3.   N.N. Thadhani, "Shock-induced Chemical Synthesis of Intermetallic Compounds," in *Shock Compression of Condensed Matter*, eds. S.C. Schmidt, J.N. Johnson, L.W. Davison, North-Holland, 1990, p. 503.

4.   F.R. Norwood and R.A. Graham, "Numerical Simulation of a sample Recovery Fixture for High Velocity Impact," (this volume).

5.   T.B. Massalski, "Binary Alloy Phase Diagrams: Volumes I and II," American Society for Metals, Metals Park, Ohio, 1986.

25

Dynamic Compaction of Combustion Synthesized TiC-Al$_2$O$_3$ Composite

G. E. KORTH, R. L. WILLIAMSON, and B. H. RABIN

Materials Technology Group
Idaho National Engineering Laboratory, EG&G Idaho, Inc.
Idaho Falls, Idaho 83415-2218, U.S.A.

Physics and Mathematics Group
Idaho National Engineering Laboratory, EG&G Idaho, Inc.
Idaho Falls, Idaho 83415-2211, U.S.A.

Materials Technology Group
Idaho National Engineering Laboratory, EG&G Idaho, Inc.
Idaho Falls, Idaho 83415-2218, U.S.A.

A dispersed phase TiC-Al$_2$O$_3$ composite was densified and consolidated by dynamic compaction using explosives immediately after it was formed by a combustion synthesis reaction from TiO$_2$, Al, and C powders. By taking advantage of the self-generated heat from the exothermic reaction, this one-step processing technique has the potential for considerable cost savings over conventional processing methods. Near full densification was achieved with samples from 25 to 76 mm in diameter x 6 mm thick. The high-rate densification is necessary to consolidate the ceramic material before heat loss to the containment fixture reduces the temperature to a level below that necessary for bonding. Numerical modeling was used to assist in the design of the fixture and determine the dynamic compression conditions. Good density and strength are achieved by selecting the correct compaction parameters, addressing heat loss, and allowing for evolution of impurity gases.

I. INTRODUCTION

Ceramic composites offer the potential for considerable improvement over monolithic ceramics in properties such as hardness, strength, and fracture toughness. A dispersed phase composite of alumina

(Al_2O_3) reinforced with approximately 30 wt.% titanium carbide (TiC) exhibits these property improvements and is an excellent candidate for ceramic cutting tool applications [1,2]. To fabricate a TiC-Al_2O_3 composite by conventional methods, either hot pressing powders above 1600°C [3,4] or pressureless sintering in excess of 1800°C [5-7] followed by hot isostatic pressing is necessary. These approaches are relatively costly. Another possible approach is that used by Adachi [8], high-pressure self-combustion sintering. This method took advantage of low cost reactants and combustion synthesis, but still required the use of a specialized hot press capable of 3 GPa.

This paper describes a novel processing method for producing a Al_2O_3-TiC composite material through a combination of combustion synthesis and dynamic consolidation. The composite was produced from TiO_2, Al, and C according to the reaction:

$$3TiO_2 + 4Al + 3C \rightarrow 3TiC + 2Al_2O_3 \ . \tag{1}$$

This reaction produces a high quality composite material containing approximately 41 vol.% TiC, but if not densified, it contains over 50% porosity and is useful only as a starting material for further processing. Alternatively, dynamic compaction can be used to densify the material while it is still hot and able to deform plastically. In this process, the heat generated from the exothermic reaction is also used as the process heat to consolidate/densify the material during a high-rate compaction using small charges (few grams) of explosives. Other investigators have previously used a similar approach to produce monolithic TiC, TiB_2 [9-11] and HfC [12] from elemental powders, but in the current work only inexpensive reactants have been used to make the process more cost effective. A previous paper [13] reported the initial investigation and feasibility of this processing method for the TiC-Al_2O_3 composite. The work reported in this paper includes improvements in the containment fixture to eliminate thermal quench cracks, scaling up the sample size from 25 to 76 mm in diameter, and numerical modeling of the dynamic process.

II.   EXPERIMENTAL PROCEDURE

Stoichiometric mixtures of commercially available TiO_2, Al, and C were prepared by wet milling in ethanol. The mixture was then dried, re-milled, screened, and pressed into cylindrical pellets either 25 or 76 mm in diameter by 22 to 25 mm in height. The nominal particle sizes and sources of the starting powders were: TiO_2, 0.2 μm (reagent grade rutile or anatase, Fisher Scientific); Al, 10 μm (commercially pure, inert gas atomized aluminum powder, Valimet, Inc.); and C, 0.02 μm (acetylene black, Chevron Chemical Co.). The green density of the pellets pressed without binder at 20 MPa was approximately 1.8 g/cm³. Some of the pellets were outgassed by vacuum baking at 280°C prior to reaction and densification.

The experimental assembly is illustrated in Fig. 1 for the 76 mm sample size. The 25 mm sample fixture was similar, but used a slightly different configuration of explosive. The experiments were conducted by first igniting a small (~2 g) charge of loose Ti + B powder with an electric match (Estes, Inc. model rocket igniter) which in turn initiated the combustion of the sample. After thermocouples located at the opposite end from the ignition indicated the reaction was complete (5-7 s), the sample was allowed to

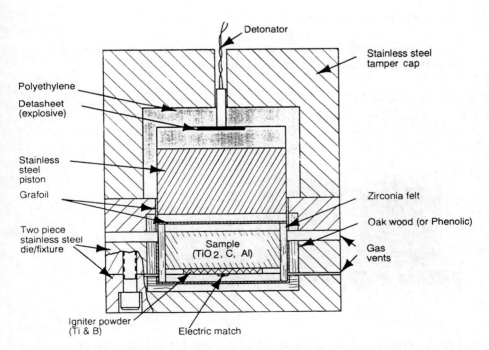

FIG. 1 *Experimental assembly for 76 mm sample. All steel parts are reusable.*

thermally equilibrate and outgas for 5-15 s, and then the explosive
charge was detonated. The apparatus provided two levels of thermal
insulation, rigid confinement from the steel fixture, and an
explosively driven piston to perform the densification. The 96%
porous zirconia felt provided high temperature insulation during the
combustion and outgassing step, and the wood (or phenolic) provided
some short-time insulation that retains some strength during the
high-rate densification step. The layered graphite (Grafoil, Union
Carbide Corp.) provided short term thermal protection for the wood
(or phenolic) during the compression phase. The polyethylene in the
explosive driver area provided an energy transfer medium that
protected the steel fixture so that it could be reused.

    To assist in fixture design and scale-up, numerical simulations
were performed to compute the thermal response of the reacted powder
prior to densification and to predict the dynamic behavior of the
driver assembly during compaction.

III. RESULTS AND DISCUSSION

Fig. 2 shows the 25 mm sample before reaction, after reaction without
densification pressure, and after reaction with dynamic

FIG. 2 *Pressed 25 mm pellet of TiO_2, Al, and C reactants (left),
$TiC-Al_2O_3$ sample reacted without external densification (center), $TiC-Al_2O_3$ sample reacted and densified using explosive consolidation (right).*

FIG. 3 *Densified TiC-Al₂O₃ 76 mm sample as removed from the fixture.*

consolidation. Fig. 3 shows a 76 mm reacted sample after removal
from the dynamic compression fixture. X-ray diffraction analysis of
a consolidated sample showed that the reaction had proceeded to
completion since only TiC and Al_2O_3 peaks were present.

The fixture used in the earlier study [13] contained only the
zirconia insulation, and quench cracks were present in the
consolidated samples. The addition of the secondary insulation of
wood or phenolic largely eliminated the quench cracks, but layered
porosity remained a problem. It is believed that the bands of
porosity correspond to the layering created by the combustion
reaction as it propagates through the pressed pellet (note the
layering produced on the sample reacted without densification shown
in Fig. 2). The reaction front did not simply progress from the
ignition end to the opposite end, but was observed to propagate
through the sample in a spiraling mode a layer at a time. The layers
are believed to be formed from thermal cracks due to the volume
change between pressed reactants and synthesized products. Fig. 4
shows a compacted sample at three levels of magnification. In the
macrophoto (left) the layering is evident. The center photo shows
bands of porosity and the right shows the regions between the bands
where a fine dispersed phase microstructure with nil porosity was

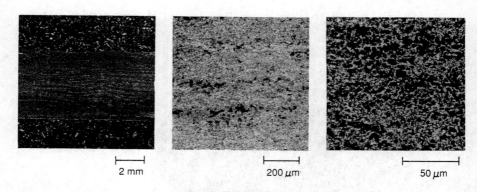

FIG. 4 *TiC-Al₂O₃ composite at various magnifications.*
Macrophotograph showing layered structure (left), microstructure of
same sample showing bands of porosity (center), and higher
magnification between porosity bands illustrating fine dispersed
phase composite (right). Dark phase in right micrograph is Al₂O₃ and
light phase is TiC.

obtained. Hardness of the high density area was found to range

between 22 and 25 GPa with an average of 23 GPa.

In scaling up from the 25 mm to the 76 mm sample size, it was

found that the general quality of the sample and the microstructure

of the two sizes correlated well with piston velocity. Numerical

simulations, using the CSQII computer code [14], were performed to

determine the piston velocity at impact with the reacted powder.

Fig. 5 shows the computed relationship between the mass charge of

Detasheet C (DuPont flexible sheet explosive) and the kinetic energy

($\frac{1}{2}mv^2$) of the piston for the 76 mm sample size fixture shown in

Fig. 1. Piston velocities of 15 to 20 m/s were found to be the best

compromise for a good specimen. Higher velocities extruded the near-

fluid sample material into the compressible secondary insulation

(wood or phenolic) and also resulted in through-the-thickness

delamination separations from rebound. Lower velocities did not

satisfactorily densify the sample and residual porosity was

significant.

A containment fixture without the secondary thermal insulation

provided higher densities in the core of the sample, but thermal

quench cracks and unbonded material near the surface were much more

prevalent. If the sample surface cools rapidly due to intimate

contact with the metallic fixture, the temperature drops below the

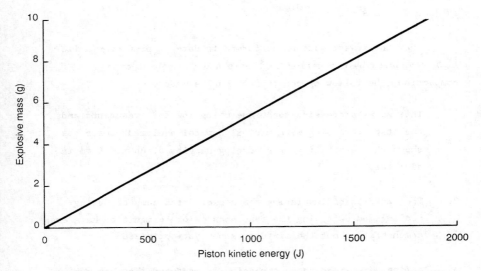

FIG. 5 *Relationship between mass of explosive and kinetic energy of piston for 76 mm fixture. Piston mass is 1.94 kg.*

level where bonding can occur and the local integrity of the material is significantly reduced. Further improvements in the sample quality could undoubtedly be attained if another secondary insulation material could be identified that has a low thermal conductivity, similar to phenolic or oak wood, but has a higher bulk modulus. Also, this material would have to withstand very high temperatures, at least momentarily; the calculated adiabatic temperature of the reaction is approximately 2100°C.

        Another factor that appears to be very important in attaining a high density composite is the elimination of gases during processing. Although there are no gaseous products identified in the combustion synthesis reaction, impurity gases are liberated. The sources of these gases include the residual ethanol used in wet milling the mixture, adsorbed water, and oxides/hydroxides on the original powder reactants that thermally decompose at the high reaction temperatures. Vacuum degassing the pressed pellet of reactants at 280°C for several hours prior to combustion and densification decreased the porosity of the final product compared to a sample given the same densification treatment without degassing.

IV. CONCLUSIONS

From the investigations performed to date on producing a TiC-Al_2O_3 composite by combustion synthesis and subsequent dynamic compaction, the following conclusions can be made.

1. This novel processing technique using low cost reactants and one step processing with minimal capital equipment costs has shown the feasibility of producing samples of up to 76 mm in diameter.

2. The benefits of insulating the consolidated compact during cooling and degassing the reactants prior to ignition have been demonstrated, but the fixture is not yet optimized.

3. A fine, dispersed phase composite is produced that has a hardness equivalent to or greater than that of material produced by the more expensive processing methods.

4. Delamination in the through-thickness direction appears to be the major problem still unresolved at this time. Further development of techniques to increase the retention time of the compression pulse during cooling appears to be the most promising solution.

V. ACKNOWLEDGMENTS

The authors gratefully acknowledge the contributions of S.T. Schuetz, G.L. Fletcher, V.L. Smith-Wackerle, M.D. Harper, and R.C. Green for their assistance in the experimental and sample preparation work. This work was supported by the U.S. Department of Energy under DOE Contract No. DE-AC07-76ID01570.

VI. REFERENCES

1. A.G. King, *Bull. Am. Ceram. Soc.*, 43[5], 395-401, 1965.
2. D. Bordui, *Bull. Am. Ceram. Soc.*, 67[6], 998-1001, 1988.

3. R.P. Wahi and B. Ilschner, *J. Mater. Sci.*, 875-885, 1980.

4. S.J. Burden, *Bull. Am. Ceram. Soc.*, 67[6], 1003-1005, 1988.

5. M. Lee and M.P. Borom, *Adv. Ceram. Mater.*, 3[1], 38-44, 1986.

6. R.A. Cutler, et al., *Mat. Sci. and Eng.*, A105/106, 183-192, 1988.

7. T. Isigaki, et al., *J. Mat. Sci. Lett.*, 8, 678-680, 1989.

8. S. Adachi, et al., *J. Am. Ceram. Soc.*, 73[5], 1451-52. 1990.

9. A. Niiler, et al., *Proc. First Workshop on Industrial Applications of Shock Processing of Powders*, CETR, Socorro, NM, June 1988.

10. A. Niiler, et al., *U.S. Army Ballistic Research Lab BRL-TR-2951*, Aberdeen Proving Ground, MD, December 1988.

11. L.J. Kecskes, et al., *J. Am. Ceram. Soc.*, 73[5], 1274-82, 1990.

12. L.J. Kecskes, et al., *J. Am. Ceram. Soc.*, 73[2], 383-87, 1990.

13. B.H. Rabin, et al., *J. Am. Ceram. Soc.*, 73[7], 2156-58, 1990.

14. S.L. Thompson, *Sandia National Laboratory, SC-RR-710714*, 1972.

26

**Explosive Compaction Processing of
Combustion Synthesized Ceramics and Cermets**

A. NIILER, L. J. KECSKES, AND T. KOTTKE

Ballistic Research Laboratory
Terminal Ballistics Division
Aberdeen Proving Ground, MD 21005, USA

*A method which combines explosive compaction with combustion
synthesis (CS) to fabricate fully consolidated, mostly crack
free TiC, TiB$_2$ and their cermets is described. The method
takes advantage of the approximately 2500°C temperatures that
these CS products reach during the reaction to eliminate the
~50% sample porosity with shock pressures. A 2-D calculation
by Wilkins has shown that local pressures can exceed 3.5 GPa
for more than 100 μsec in some samples when 75mm of Amatol is
used. The fabrication procedure is described. Results on the
systems Ti+C and Ti+2B, along with the effects of stoichiometry and the addition of components like Cu, Fe, Mo, W, Al$_2$O$_3$,
and ZrO$_2$ on the product characteristics are presented.*

I. INTRODUCTION

High-technology, structural ceramics are being utilized in a widening range of applications which require light-weight, high-temperature, high-performance materials.
Commercial processing of these ceramics, typically the borides, carbides, nitrides and
oxides of a variety of metals is generally done by sintering, hot pressing or hot
isostatic pressing. These ceramics can also be produced by Combustion Synthesis
(CS), a process by which component elemental powders are chemically reacted to
form the product. The reactions are characterized by high exothermicities, reaction
temperatures in excess of 2500°C, and reaction front velocities on the order of a few
centimeters per second. Frankhouser[1] has compiled a comprehensive listing of the

early work in the Soviet Union in this area and Munir[2] has recently written an informative review article which contains a extensive bibliography.

Although CS has proved to be a convenient, and in the Soviet Union a preferred, way to make non-oxide ceramic powders, the fabrication of full density monolithic ceramics has been difficult. Typically, when a compacted green body has reacted, it retains about 50% porosity. The major causes for this porosity are precursor impurity evolution and the difference in the specific volumes of precursor compact and reacted ceramic product. In the absence of a continuous liquid phase in the product, full density material can be achieved only by application of external pressure after the synthesis reaction is complete.

In this paper we describe a processing method which utilizes the CS reaction to form porous ceramic bodies which, while still hot, are Dynamically Compacted to high density by explosives (hence the term CS/DC for the technique). There are many advantages to this method. The heat released by the exothermic reactions can raise the ceramic sample's temperature to well above its ductile-brittle transition point and in some cases, to above its melting point, thus making the consolidation by external pressure relatively easy. The heat is also useful in liberating volatile impurities from the sample prior to the application of the pressure. The fixture in which both the reaction and compaction take place is made of common, easily machined materials having a good impedance match to the hot ceramic. A detailed description of this experimental procedure is given by Niiler et al,[3] while a brief summary follows here. Some results from a calculation by Wilkins[4] are discussed and physical as well as microstructural features of a number of samples fabricated by this process are presented.

II. EXPERIMENTAL DESCRIPTION

The bulk of BRL's work on the CS/DC method was performed on the binary systems Ti+2B and Ti+C. A boron to titanium ratio of 2.0 was used for the TiB_2 samples and a C to Ti ratio of 0.8 was used for most of the TiC samples. Experiments on TiC were also carried out with C:Ti ratios ranging from 0.7 to 1.0 to study the effect of carbon lattice vacancies on the TiC microhardness and on both TiC and TiB_2 with 5 at% of the additives Cu, Fe, Mo, W, Al_2O_3, and ZrO_2 to study possible improvements to compactability and/or microstructures.

The powders were stored and mixed under argon, then uniaxially pressed into 5cm dia. by 2.5cm thick disc shaped green compacts. The green compact was placed in a gypsum block made of layers of wallboard epoxied together with the center cored to a diameter slightly larger than that of the compact. A mild steel ring with 1mm thick wall, lined with graphite sheet, was placed in the air space between the compact and the gypsum to serve as lateral containment during the reaction and compaction. The graphite sheet prevented impurities from diffusing into the hot sample. Both the steel ring and the gypsum block had matching holes to the outside of the block to vent the gases evolved during the reaction. The top and bottom of the containment vessel were made of high hardness steel plates. Figure 1 shows a diagram of this assembly.

The green compact was ignited by initiating a loose Ti-B powder mixture at the top of the sample with an electric match. The heat released by the reaction not only propagated the combustion wave through the whole sample, but also heated the sample to temperatures in excess of 2500°C. Experiments performed on samples reacted in containment fixtures as described above show temperatures in excess of 1200°C for at least one minute, allowing an extended time window during which the compaction pulse can be applied. The detonation of the explosive was generally initiated immediately after the reaction was complete, usually about 3 to 10 seconds following the ignition. The explosive used in this experiment was Amatol, an 80/20 mixture of ammonium nitrate and TNT whose detonation velocity is 3850 m/sec and

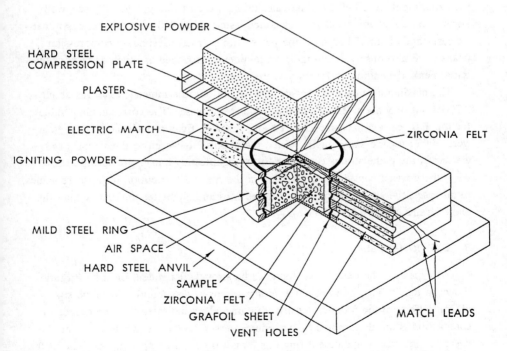

FIG. 1 *Schematic diagram of the fixture used in the Combustion Synthesis, Dynamic Compaction (CS/DC) of ceramics.*

whose density is 0.86 g/cc. A range of explosive to driver plate areal density ratios (c/m) from 0.1 to 0.5 was investigated to determine the effect of shock duration on product characteristics. Detonation was initiated either by a Detasheet linewave generator at one side of the square explosive container to realize a sweeping wave compression, or by a Detasheet booster at the top center of the container to obtain a symmetric compression approximating a plane wave.

III. RESULTS AND DISCUSSION

A. 2D CALCULATION

Mark Wilkins has performed a series of calculations on these BRL materials and geometry using the general explicit finite element code for transient analysis, EFHYD 2D. The Reaugh[5] powder equation of state was used for TiC and TiB$_2$ and appropriate EOS models were used for the other materials in the fixture. Calculations were done for both the sweeping wave (plane strains) and center initiated wave (axisymmetric) cases with 25.4mm of Amatol for TiB$_2$ and 75.35mm of Amatol for the TiC. These Amatol thicknesses are close to the optimum density points from the experiments (see below). The calculations were carried out to 200 μsec, well beyond the initial compression peak which lasted about 100 μsec for the axisymmetric cases and about 150 μsec for the plane strains cases. Secondary compression peaks were also observed and their magnitudes approached as much as 50% of the initial peak, depending on location in the sample.

The maximum calculated pressure for TiC in the axisymmetric case was about 3.7GPa but only about 2.4GPa in the plane strains case. The corresponding values for the TiB$_2$ calculations were 2.5GPa and 1.5GPa. Even though the peak pressures were higher for the axisymmetric cases, the impulse delivered to the samples was greater in the plane strains cases due to the significantly longer pressure pulses. Also, there was a considerable variation in the spacial distribution of pressure values. In all cases, the axisymmetric cases showed substantially better uniformity than the plane strains cases.

B. EXPERIMENTAL

As expected, the density was found to be strongly dependent on the c/m ratio shown in fig. 2. The mode of initiating the explosive was found to have no significant effect on the product density, a finding which indicates that the degree of compaction is not determined by maximum pressure alone. The thickness and the design of the steel containment ring was found to be quite important as some of the work done by the explosive goes into compressing that ring. The densities are

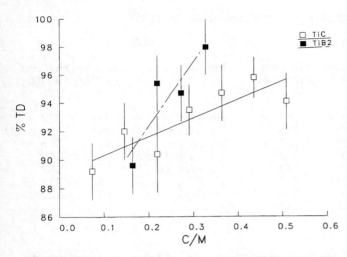

FIG. 2 *The dependence of TiC and TiB₂ densities on c/m. The error bars indicate sample to sample variation at a given c/m and the lines are least squares fits to the points to indicate the trends.*

determined from a core taken from the center of the sample and in most cases, the reported values are averages from a number of different samples. The best sample structures (compromising between density and cracking) were obtained for TiB$_2$ at c/m=0.22 and for TiC at c/m=0.44. When c/m values beyond the 0.22 and 0.44 values were used, the product density was reduced while increased fracturing and delamination indicated greater edge effects and stronger rarefaction waves from the shock. In order to avoid distorting the compression plate, it was necessary to use at least 1.91 cm thick high hardness steel for TiC while 1.27 cm was sufficient for TiB$_2$.

The two requirements, higher c/m and thicker compression plate, indicate that the TiC is harder to compact than the TiB$_2$. At 1800°C, the compressive yield strength of TiC is 50 MPa[6] while that of TiB$_2$ is 441 MPa[7]. Thus if yield strength, or plasticity, were the only factor involved in compactability, TiC should compact more easily that TiB$_2$. Since it does not, another mechanism must be invoked to explain the difference. It is speculated that the TiB$_2$ melting point of 2800°C was exceeded but that the TiC melting point, 3150°C, was not reached even with the additional energy provided by the shock. Thus, TiB$_2$ compacted more easily than TiC because it contained at least some liquid phase while the TiC did not.

The highest density, as a percentage of full theoretical density (TD) and microhardness results for some of the samples that have been made by the CS/DC process at BRL are summarized in Table 1. The samples' densities are given to only

TABLE 1.   TiC and TiB₂ RESULTS

TiC			*TiB₂*		
Material	Density (% TD)	Hardness HK(100g) GPa	Material	Density (% TD)	Hardness HK(100g) GPa
Hot Pressed	98+	25.5	Hot Pressed	99+	33.6
C/Ti Ratio 0.8	98+	20.4	No Additives	99+	32.6
C/Ti Ratio 1.0	95	26.7	5% Copper	90	24.6
5% Copper	93	24.4	5% Molybdenum	96	27.2
5% Molybdenum	94	25.1	5% Tungsten	95	28.2
5% Tungsten	94	26.0	5% ZrO₂	88	23.9
5% ZrO₂	91	23.8	5% Al₂O₃	88	12.7
5% Al₂O₃	92	16.0			

two significant figures. This is so because in the case of TiC, the uncertainty in the theoretical density is affected by the C/Ti ratio and in both by the actual amount of additive remaining in the product (an undetermined quantity). In this table, commercial hot pressed materials are also included for comparison.

Recent work at BRL has concentrated on improving the properties of TiC and TiB₂ based ceramic materials by controlling microstructural variables such as residual porosity, grain size, and intergrain bonding by the addition of small amounts of metal or metal oxide or by varying the compositions. The effect of small amounts of Cu, Mo, W, Al₂O₃ and ZrO₂ on the TiC and TiB₂ product characteristics has been investigated. Figure 3 shows SEM micrographs of the polished surfaces of TiC with the five additives and figure 4 shows the same for TiB₂. As can be seen, the porosity and intergrain bonding depend mainly on the melting temperature of the additive relative to the combustion reaction temperature (2500-3000°C). A considerable amount of data (see ref 3) has been accumulated with 2-3% iron showing that it remains in the product at the same level as it is in the precursors. These data also show that the presence of iron tends to arrest the grain growth and that it accumulates only in the grain boundaries. Mo and W also remain in the product and seem to improve the intergrain bonding while samples with Al₂O₃ and ZrO₂ exhibit a significantly reduced grain size but relatively poor intergrain bonding. Ultimately, a system using both metals for improved intergrain bonding and metal oxides for grain size control may provide the most desirable combination of material properties.

A very exciting possibility with CS produced ceramics is the ability to easily tailor the product composition. The phase diagram of TiC shows that it can be formed over a wide range of C to Ti compositions. The CS/DC method allows investigation of the properties of TiC with various C:Ti ratios simply by varying the composition of

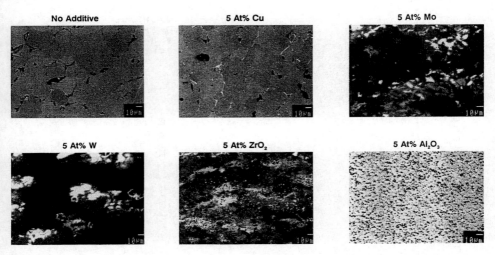

FIG. 3   *SEM micrographs of polished surfaces of TiC and TiC with 5at% of Cu, Mo, W, ZrO₂, and Al₂O₃.   Cu is mostly vaporized from the sample, Mo and W remain in triple junction points and improve bonding, ZrO₂ decreases density without affecting hardness very much while Alumina is detrimental to both density and hardness while reducing grain size considerably.   See Table 1.*

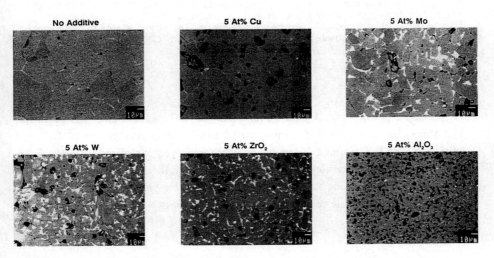

FIG. 4   *SEM micrographs of polished surfaces of TiB₂ and TiB₂ with 5at% of Cu, Mo, W, ZrO₂, and Al₂O₃.   With the possible exception of W, all the additives are detrimental to hardness and to a lesser extent to density.   The oxides are especially deleterious to sample density.   See Table 1.*

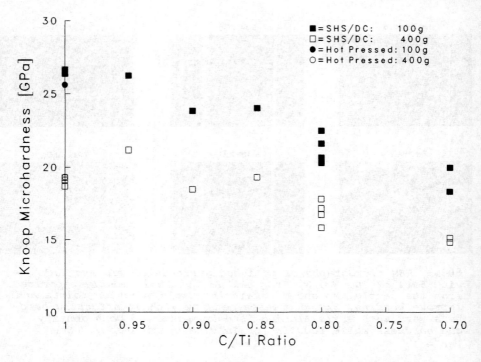

FIG. 5 *Dependence of the TiC microhardness on the C:Ti ratio.*
Microhardness measurement results with both 100g and 400g
loads are shown to illustrate the similarity between intrinsic
and extrinsic behavior of the hardness in these TiC samples.

the precursor powder mix. Figure 5 shows that the TiC microhardness can be varied 33% simply by varying the C:Ti ratio from 1.0 to 0.7. SEM micrographs of TiC fabricated with the various C:Ti ratios show that improved consolidation seems to be correlated to reduced carbon content, thus softer material. These results also point to the desirability of using a third component, a 'sintering aid' as it were, to help consolidate the TiC while retaining its maximum hardness.

REFERENCES

1. W. L. Frankhouser, *System Planning Corp. Report* 81-4082: 1982.
2. Z. A. Munir and U. Anselmi-Tamburini, *Mater. Sci. Reports* 3: 277, 1989.

3. A. Niiler, L. J. Kecskes, T. Kottke, P. H. Netherwood, Jr., and R. F. Benck, *Ballistic Research Laboratory Report* BRL-TR-2951: Aberdeen Proving Ground, MD, 1988.

4. M. L. Wilkins, *private communication.*

5. J. E. Reaugh, *J. Appl. Phys.* 61: 962, 1987.

6. L. E. Toth, *Transition Metal Carbides and Nitrides:* Academic Press, New York, NY, 1971.

7. J. R. Ramberg and W. S. Williams, *J. Mat. Sci.* 22: 1815, 1987.

27

Shock Synthesis of Silicides

L. H. YU and M. A. MEYERS*

Department of Materials and Metallurgical Engineering
New Mexico Institute of Mining and Technology
Socorro, New Mexico 87801, U. S. A.

Department of Applied Mechanics and Engineering Science
University of California, San Diego
La Jolla, California 92093, U. S. A.

Shock-induced chemical synthesis of high temperature materials (silicides) were investigated. Niobium, molybdenum, titanium powders mixed with silicon powders were chosen as reactant materials for shock-induced synthesis of silicides. Shock processing was carried out using a modification of the experimental set-up developed by Sawaoka and Akashi [U.S. Patent 4,655,830]. The shock waves were generated in the materials by the impact of a flyer plate at velocities of 1.2 and 2 km/s. Both room temperature and high temperature (500°C) experiments were conducted. The passage of shock waves of sufficient pressure and temperature induced a highly exothermic and self-sustaining reaction between reactant materials. The profuse formation of voids indicates that melting of the material occurred; in contrast, unreacted regions did not exhibit porosity. The distribution of the reacted regions follows contours of temperature, does not correlate very well with pressure contours and seems to indicate that there is a threshold energy required to initiate shock-induced reactions.

* This paper is dedicated to B. Krueger, a bright young student of Cal Tech who precociously lost his life; he proposed the concept of the threshold energy for shock synthesis.

I.   INTRODUCTION

Shock compression processing has been applied to synthesize materials since 1961, when DeCarli and Jamieson [1,2] synthesized diamond from rhombohedral graphite. Since then, there has been considerable research activity in materials development using this process to synthesize compounds [3-9]. This unique process can not only induce the chemical reactions between the elemental materials, but also the high pressure can produce fully dense compacts in time durations of the order of microseconds. Three experimental conditions were used : high impact velocity (2 km/s) at room temperature, low impact velocity (1.2 km/s) at room temperature, and low impact velocity at high temperature (500°C). It was demonstrated that the shock waves can induce chemical reactions between reactant materials.

II.   EXPERIMENTAL PROCEDURES

Shock-induced chemical synthesis experiments were conducted on mixtures of Nb-Si, Mo-Si and Ti-Si to determine the extent of shock induced reactions and the nature of reaction products. The composition of reactant materials was chosen based on the stoichiometric composition of the silicides: $NbSi_2$, $MoSi_2$, and Ti_5Si_3. The powders were blended and loaded into the capsules in argon atmosphere. A modified version of the Sawaoka set-up was used in the shock experiments. The detonation is initiated from the detonator at the top of an explosive lens; a conical lens consisting of explosives with two detonation velocities

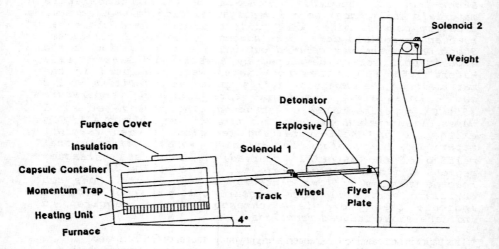

FIG. 1 *Experimental set-up for high temperature shot; "Sawaoka" set-up with twelve capsules used.*

was used to generate a planar wave in the main explosive charge. The flyer plate is accelerated by the detonation of the explosive charge and impacts the capsules embedded in a steel fixture. The shock waves are transmitted through the capsules into the powders. A momentum trap is used to trap the reflected tensile waves in order to recover the specimens. The experimental set-up for high temperature test is schematically shown in Fig. 1. The set-up containing the capsules is heated inside a discardable furnace. The explosive charge, on top of the flyer plate, is lowered down a ramp, when the temperature in the capsules reaches the desired level. It is detonated immediately afterward. Solenoid 1 initiates the movement of the charge, while solenoid 2 is used as a safety. The recovered specimens were characterized by optical and scanning electron microscopy and x-ray diffraction. Only the Nb-Si and Mo-Si systems are discussed in detail in this paper. The reader is referred to recent papers by Yu et al. [10] and Yu and Meyers [11] for greater details.

III. RESULTS AND DISCUSSION

The packing density of Nb-Si and Mo-Si powders were 60% of the theoretical density. The Nb, Mo, and Si powders all have irregular shapes prior to compaction. The powder size is below 325 mesh ($\sim$45 μm) for all materials. Three regions could clearly be identified in the capsules : unreacted, partially reacted, and fully reacted regions. These regions are shown in Figs. 2(a), 2(b), and 2(c), respectively. In the unreacted regions, one can clearly distinguish the Nb and Si particles. The Si particles underwent substantial deformation and totally envelop the Nb particles. In the partially reacted regions, shown in Fig. 2(b), one sees some very interesting phenomena, that allow us to start to understand the sequence of events occurring in shock synthesis. Silicon plastically deforms and most probably melts, surrounding the the niobium particles. $NbSi_2$ nucleates at Nb-Si interface (small particles indicated by arrow A); when these particles reach a critical size, they are dragged away by the flow of silicon (arrow B). Concomitantly, diffusion of Si into Nb creates the region marked by arrow C. The fully reacted region exhibited profuse porosity, indicating that it had totally melted. These pores are due to three possible causes : (a) entrapped gases; (b) shrinkage during solidification; (c) tensile stresses. Figure 3 shows the distribution of the three regions for the different experimental conditions. In the low impact velocity/high temperature shot, the fully reacted region took the most of specimen; only small fractions of unreacted and partially reacted regions were found. In the low impact velocity/room temperature shot, the partially reacted region took the most of specimen; the unreacted and fully reacted regions were decreased. In the high impact velocity/room temperature shot, the fully reacted region was larger than in the low impact velocity/room temperature shot, showing that an increase in pressure enhances the reaction. The shape of the fully reacted region seems to parallel maximum temperature contours (see

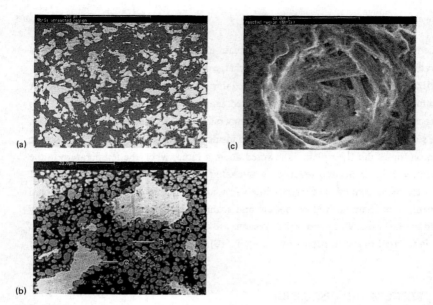

FIG. 2 *Scanning electron micrographs of (a) unreacted, (b) partially reacted, and (c) fully reacted regions for Nb-Si capsules.*

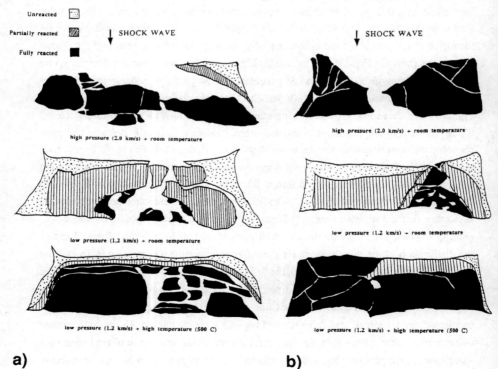

FIG. 3 *The profiles of three different reaction distributions for three shock conditions; (a) Nb-Si; (b) Mo-Si.*

Yu et al.[10] and Yu and Meyers [11]). This is indicative of the importance of temperature (energy) in initiating the reaction. There seems to be a threshold energy level required for the shock-induced reaction as suggested by Krueger and Vreeland [14].

A preliminary analysis of shock-induced chemical reactions was conducted for the present investigation. A complete analysis has been recently developed by Horie and Kipp [12], and Boslough [13], but the simple analysis suffices for the purposes of this work. The basic equation of energy conservation was modified by adding an energy of reaction term. The assumptions of heat of reaction independent of pressure, zero volume change (with reaction), and no melting are implicit in this equation:

$$E_2 - E_{00} = \frac{1}{2} P (V_{00} - V) + E_R \tag{1}$$

E_R is the energy of reaction at constant volume. The thermodynamic state after complete reaction can be determined by assuming steady-state. Figure 4 shows Hugoniot pressure-volume curves for the solid material, the powder, and the powder with reaction. The curve for the reacted powder is transposed to the right from the unreacted powder. The internal energy for the reacting material, at that volume, is given by:

$$E_2 - E_{00} = \frac{1}{2} P_2 (V_{00} - V) = \frac{1}{2} P_1 (V_{00} - V) + Q \tag{2}$$

Figure 4 shows the three pressures P_1, P'_1, and P_2. If no reaction occurs, the shock pressure is given by P_1. The additional reaction energy, E_R, increases the pressure in the

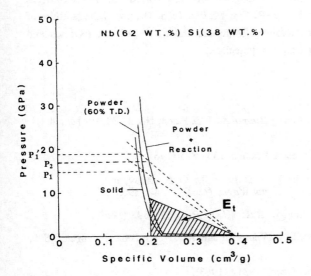

FIG. 4 *Hugoniots for solid, powder, and reacting Nb-Si powder (62 wt.%-Si wt.% mixture with 60 % T. D.).*

powder to P'_1. If one draws a Rayleigh line through P'_1, one obtains a pressure P_2 after reaction (both shock and reaction front move at the same velocity in steady-state). One concludes that shock synthesis leads to an acceleration of reaction velocity with an increase in pressure. With the equation of state for the porous reacting material one can compute shock propagation velocities by means of Rayleigh lines. Figure 4 also shows that E_t is required to induce reaction; if $E < E_t$, no reaction would occur.

IV.   CONCLUSIONS

It has been demonstrated that niobium, molybdenum, and titanium silicides can be synthesized by shock compression. The detailed analysis of the Nb-Si system showed that the reacted regions have a large fraction of porosity that can be due to several reasons. The influences of shock pressure, shock temperature, and pre-heating temperature on the extent of reaction were established and it is suggested that a threshold energy is required for the initiation, consistent with the model of Krueger and Vreeland [14]. A preliminary calculation was conducted that shows that shock-induced reaction may cause an increase in shock pressure and velocity.

ACKNOWLEDGMENTS

This research was supported by the National Science Foundation under Grant CBT 87-13258. The help of N. N. Thadhani is gratefully acknowledged. The explosive experiments were conducted at the CETR, New Mexico Tech. The use of the facilities of the Center of Excellence for Advanced Materials and of the Electron Optics and Microanalysis Facility at UCSD is greatly appreciated.

REFERENCES

1.   P. S. DeCarli, *Method of Making Diamond*, U. S. Patent No. 3,238,019 March 1, (1966).

2.   P. S. DeCarli and J. C. Jamieson, *Science*, 133 : 821 (1961).

3.   S. S. Batsanov, A. A. Deribas, E. V. Dulepov, M. G. Ermakov, and V. M. Kudinov, *Comb. Expl. Shock Waves USSR*, 1 : 47 (1965).

4.   A. N. Dremin and O. N. Breusov, *Russ. Chem. Rev.*, 37 : 392 (1968).

5.   R. A. Graham, B. Morosin, E. L. Venturini, and M. J. Carr, *Ann. Rev. Mater. Sci.*, 16 : 315 (1986).

6.   Y. Kimura, *Japan. J. Appl. Phys.*, 2 : 312 (1963).

7.   Y. Horie, R. A. Graham, and I. K. Simonsen, *Materials Letters*, 3 : 354 (1985).

8. I. K. Simonsen, Y. Horie, R. A. Graham, and M. J. Carr, *Materials Letters*, 5 : 75 (1987).

9. A. B. Sawaoka, and T. Akashi, *High Density Compacts*, U. S. Patent No. 4,655,830 (1987).

10. L. H. Yu, M. A. Meyers, and N. N. Thadhani, *J. Matls. Res.*, 5 : 302 (1990).

11. L. H. Yu and M. A. Meyers, *J. Mat. Sci.*, 26 : 601 (1991).

12. Y. Horie and M. J. Kipp, *J. Appl. Phys.*, 63 : 5718 (1988).

13. M. B. Boslough, *J. Chem. Phys.*, 92 : 1839 (1989).

14. B. Krueger and T. Vreeland, California Insititute of Technology, 1990, private communication.

28

Shock Recovery Experiment of Carbon

T. SEKINE

National Institute for Research in Inorganic Material
Tsukuba, Ibaraki 305, JAPAN

(Presently at the Seismological Laboratory, California Institute of Technology, Pasadena, CA 91125, USA)

Shock recovery experiments on artificial graphite and vitreous carbon were conducted in a pressure range of 25 to 85 GPa, utilizing multiple reflections of shock wave to reduce the shock temperature and post-shock temperature. The results on vitreous carbon, combined with the Hugoniot data, may suggest two transformations of the first graphitization and a second diamondization.

I. INTRODUCTION

Modern experimental techniques cast a great interest on the behavior of carbon at extreme conditions, as well as theoretical considerations [1], such as at high temperature and high pressure. Carbon is most refractory as graphite in low pressure region up to about 12 GPa and as diamond at higher pressures than 12 GPa. Diamond also is the hardest material known to us.

Reliable Hugoniot measurements have been accumulated for various kinds of carbons with different densities and structures up to 120 GPa [2; 3]. If a discontinuity in the shock velocity U_s vs. particle velocity u_p plot for the various carbons is related to initiation of a phase transformation, a mixed phase region, or complete transformation to the denser phase, there appears to be a general trend that the initiation pressure of transformation to the denser form becomes lower with decreasing the initial density of the starting carbon. The initiation pressure also is affected by the structural factor of the initial carbon form, e.g. the pyrolitic graphite is less compressible and has a higher initiation pressure than the other carbon forms with a similar density.

Previous shock recovery experiments [4; 5; 6; 7; 8] on graphite or graphite-metal mixtures indicated that the shock-induced high pressure carbon was cubic diamond, lonsdaleite (hexagonal diamond), or both. Calculated Hugoniot temperatures of 2000 K on pyrolitic and Ceylon graphites require a pressure of 90 and 68 GPa, respectively [3]. The residual post-shock temperature is significantly lower than the shock temperature and is below the rapid atmospheric regraphitization temperature (RAGT) where diamond formed by shock compression could survive in the process of pressure relaxation because of sufficiently rapid cooling [6; 9]. Therefore, the effect of residual temperature may be negligible for such shock conditions. Hugoniot temperature calculations, however, indicate that 2000 K could be given at shock pressure of 10 GPa for the graphite with a density of 1.0 g/cm^3 and 20 GPa for the graphite with 1.5 g/cm^3 [3]. If we employ the multiple shock reflection method to increase pressure within these samples, we may expect to reduce considerably, the Hugoniot temperature which depends mainly upon the first shock state (pressure) of the sample. This prevents regraphitization of the shock-induced dense phase by the residual temperature, even if we employ less dense graphite or carbon as the starting material.

We report here the shock recovery results on graphite and glassy carbon in order to characterize the phase transformation from graphite and amorphous state to denser form(s). The effect of temperature on shock-induced transformation of carbon is considered separately by carrying out a recovery experiment on preheated carbon. There is another problem, whether the glassy carbon converts into the denser form directly or through an intermediate phase such as graphite. This also may be related to the effect of crystal perfection on the phase transformation during shock compression. These effects can be understood by the mechanism of phase transformation during shock compressions and the adiabatic release state.

II. EXPERIMENTAL

Artificial graphite and glassy carbon were employed as starting materials in the present study. Artificial graphite (T-5 Ibiden) is a rod with a density of 1.8 g/cm^3 and cut to disks of 1.2 mm in diameter and 1 mm thick. It consists of an average grain size of 10 μm. Glassy carbon (GC-10, Tokai Carbons) is a solid amorphous plate of 1 mm thick and with a density of 1.49 g/cm^3, and cut to disks of the same size as artificial graphite. These disks were sandwiched between stainless steel 304, molybdenum (Mo), or tungsten (W) disks of 12 mm in diameter and 3 mm thick, and then the whole assemblies were set into stainless containers.

Each container was embedded in a central hole of a large metal holder which was used as a target. In a preheated experiment, the container was surrounded by a heating element and a thermal insulator [10]. Once the thermocouple inserted near the sample within the container indicated a given temperature, the projectile collided with the hot container target.

After impact by metal flyers of stainless steel 304, Mo, and W, the stainless container was machined open to remove the sample carefully. Recovered samples were investigated by X-ray powder diffraction techniques and electron microscopy observations.

Pressure estimation was carried out by the impedance match method using the measured impact velocity of flyer and known Hugoniot data for stainless steel 304, Mo, or W, and sample [2]. Pressure within a sample disk increased by multiple reflections. Temperature was estimated by correcting the calculated temperature based upon the pressure-volume relation during the multiple shock compression.

III. EXPERIMENTAL RESULTS AND DISCUSSION

Table 1 summarizes experimental conditions investigated for carbons in this study. In Figure 1, the pressure-temperature diagram plots boundaries of stability fields for graphite,

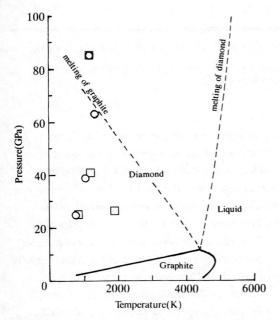

FIG. 1 *A proposed phase diagram of carbon [1] and plot of estimated P-T conditions for artificial graphite (circles) and glassy carbon (squares).*

Table 1. Experimental conditions of shock recovery on carbons

Run #	impact velocity km/sec	flyer thickness mm	peak pressure GPa	shock duration μsec	temperature K	
					sample	capsule
Artificial graphite (T-5)						
215	1.18	SUS 304 4	25.5	1.6	750	430
200	1.72	SUS 304 4	39.5	1.3	1100	550
201[a]	1.94	Mo 4	62	1.3	1350	700
202[b]	1.73	W 4	85	1.3	1200	720
216[b]	1.78	W 4	85	1.3	1200	720
Glassy carbon (GC-10)						
197	1.20	SUS 304 8	26.0	2.5	800	430
196	1.80	SUS 304 8	40	2.5	1200	570
198[b]	1.73	W 4	85	1.2	1200	720
217[b]	1.77	W 4	85	1.2	1200	720
203[c]	1.23	SUS 304 8	26.7	2.5	1900	1480

a; sample was sandwiched between molybdenum (Mo) plates in stainless steel (SUS 304) container,
b; samples were sandwiched between tungsten (W) plates in SUS 304 containers,
c; sample was preheated to 1050 K just before impact.

diamond and liquid carbon, and metastable extension of melting curve of graphite in the diamond stability field, as well as our estimations of shock-recovery conditions. Samples subjected to relatively low shock pressures were successfully recovered, but most of the samples sandwiched between W disks were lost because of reaction of carbon with tungsten to form tungsten carbides [11]. Only several discrete carbon grains were recovered. In run #217, the central portion of sample disappeared to react out completely while the edge of the initial sample remained between W disks. This indicates that the central portion was at higher temperature and pressure than the edge. The temperature and pressure gradient in the sample is in good agreement with a two-dimensional analysis by Norwood et al. [12].

Figure 2 gives a comparison of powder X-ray diffraction patterns for post-shock artificial graphite at various pressures. In the starting graphite, the diffraction peaks (002), (100), (101), (004), and (110) are observed between 20° and 80° in 2θ by radiation of Cu Kα. The diffractions (002) and (004) of these peaks were always observed in all run products from

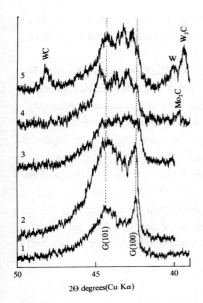

FIG. 2 *Powder x-ray diffraction patterns for the starting artificial graphite (1) and post-shock samples #215 (2), #200 (3), #201 (4) and #216 (5). Note that the strongest diffraction peak for diamond is between G (100) and G (101) which is hard to be identified in the post-shock samples. Samples 4 and 5 contains metal and carbides.*

the artificial graphite. The width at the half height of the (002) diffraction increased with increasing shock pressure, but its position changed very little. As shown in Fig. 2, the diffraction of graphite (100) and (101) were not separated from run #200, and the identification of diamond or lonsdaleite by X-ray diffraction method was difficult. The diffraction patterns of run products #201 and 216 revealed contamination of carbide or metal. These X-ray diffraction studies did not indicate that the run products from the artificial graphite contains, positively, the denser form(s). The starting glassy carbon was powdered and measured by X-ray diffraction to give very broad diffractions corresponding to graphite (002) and (004). The diffraction of graphite (002) in run product #197 was about 23.5 in 2θ and about 8.5° wide at the half height, while those of preheated run product #203 was about 25.5 in 2θ and about 3° wide at the half height. This indicates that preheating to 1050 K enhanced graphitization but gives no information of presence of denser phase(s). Then we investigated powdered post-shock samples through the electron microscopy method. Transmission electron microscopy (TEM) observations, as well as electron diffraction method for the recovered artificial graphite subjected to the lowest shock pressure (#215), did not indicate the presence of the other phase(s) than graphite. The other run products from the graphite contained diamond or lonsdaleite. The amount of diamond in run products appeared to be quite small. There was not observed any spherical particles to show melting in the recovered samples from the artificial graphite.

The starting glassy carbon gave very weak diffused diffraction rings as shown in Fig. 3b and looked homogeneous in terms of electron transmission (Fig. 3a). The run product #197 was very similar to the starting material and no significant change was detected in the electron diffraction patterns and transmission images. Some grains (Figs. 3c and d) from run product #196 gave sharp ring patterns (Fig. 3d), of which interplanar spacings correspond to diamond. Run products #198 and 217, subjected to most intensive shocks, contained rounded, hollow particles (Fig. 3e) and graphite structure (Fig. 3f) in the TEM image. The number of grains giving the sharp diffraction rings (Fig. 3d) appears to be rather rare. On the other hand, preheated run product #203 contained a number of hollow particles (Fig. 4d) and a large area of graphite structure (Figs. 4a and b). In addition, some grains (Fig. 4e) gave a spotty diffraction pattern (Fig. 4c) of which interplanar spacings agree to those of chaoite [13], as listed in Table 2.

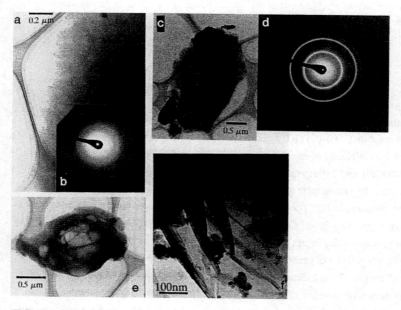

FIG. 3 *Transmission electron microphotos and their electron diffraction patterns for vitreous carbon. Starting material (a and b), post-shock sample #196 (c and d), rounded and hollow grain (3) showing rapid cooling of melted portion, and local crystallization of graphite (f).*

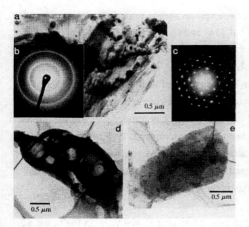

FIG. 4 *Transmission electron microphotos and their electron diffraction patterns for post-shock sample of preheated vitreous carbon to 1050 K. Local crystallization of graphite (a and b), hollow structure (d) and a chaoite (c and e). Measured interplanar spacings for the chaoite are given in Table 2.*

Table 2. Electron diffraction data for chaoite

h k l	d_{obs} (nm)	d_{cal} (nm)
1 1 0 2 $\bar{1}$ 0	4.46	4.46
3 0 0	2.56	2.57
2 2 0 4 $\bar{2}$ 0	2.23	2.23
4 1 0 5 $\bar{1}$ 0	1.71	1.68
3 3 0 6 $\bar{3}$ 0	1.49	1.50
6 0 0	1.28	1.28

The diamond formation from glassy or amorphous carbon under static high pressure conditions has been investigated by Hirano et al. [14] and Onodera et al. [15]. Their results indicate that the crystallization temperature of diamond decreases with increasing pressure, e.g. 2300°C at 9 GPa, 1800°C at 10 GPa, and 1640°C at 15 GPa. The process of diamond crystallization from glassy carbon may follow the formation of well-crystallized graphite [14], even in the stability field of diamond. This would be due to a kinetic reason; the metastable crystallization of graphite would be activated more easily at a low-temperature and the initiation of diamond crystallization occurs when it reaches a higher temperature. Further, Onodera et al. [15] have investigated the effect of heat treatment temperature of carbon precursor on crystallization of graphite and diamond, suggesting that the nature of local bonding such as graphitizable and non-graphitizable species may control the crystallization behavior and process.

Gust [3] interpreted the discontinuity at $u_p = 2.4$ km/sec (P=24 GPa) in the u_s vs. u_p plot for the vitreous carbon as a probable relation of the phase transition to the hexagonal diamond and the slope for $u_p > 3.25$ km/sec (P>33 GPa) indicative of probable phase transformation to cubic diamond. Our post-shock vitreous carbon at the peak pressure of 26 GPa (#197) did not indicate any significant difference from the starting carbon, and instead preheated sample (#203) subjected to a similar peak pressure contained graphite structure. Taking into account the shock temperature difference, these shock data on the vitreous carbon suggest that it converts first into the graphite structure and then into diamond.

The metastable melting curve of graphite has been investigated by experiments [16], as well as theoretical consideration [1]. Our results on artificial graphite did not indicate any melting up to 85 GPa for discrete post-shock graphite. However, it is not certain whether the metastable melting curve is located above 85 GPa at a temperature of 1200 K or not, because it was found that graphite had reacted with tungsten.

REFERENCES

1.   M. van Thiel and F.H. Ree, *Intern. J. Thermophys.*, 10: 227 (1989).

2.   S.P. Marsh (Ed.), *LASL Shock Hugoniot Data*, pp. 1-327, University of California Press, Berkeley, CA, (1980).

3.   W.H. Gust, *Phys. Rev.*, B 22: 4744 (1980).

4.   P.S. DeCarli and J.C. Jamieson, *Science*, 133: 1821 (1961).

5.   L.F. Trueb, *J. Appl. Phys.*, 42: 503 (1971).

6.   D.G. Morris, *J. Appl. Phys.*, 51: 2059 (1980).

7.   J. Kleiman, R.B. Heimann, D. Hawken and N.M. Salansky, *J. Appl. Phys.*, 56:1440 (1984).

8.   T. Sekine, M. Akaishi, N. Setaka and K. Kondo, *J. Mat. Sci.*, 22: 3615 (1987).

9. G.R. Cowan, B.W. Dunnington and A.H. Holtzman, *U. S. Patent*, 3,401,091: (1968).

10. T. Sekine, *J. Mat. Sci. Lett.*, 8: 872 (1989).

11. Y. Horiguchi and Y. Nomura, *J. Less-Comm. Metals*, 11: 378 (1966).

12. F.R. Norwood, R.A. Graham and A. Sawaoka, in *Shock Waves in Condensed Matter*, Y. M. Gupta (ed.), Plenum Press, New York, 1986, p. 837-842.

13. A.E. Goresy and G. Donnay, *Science*, 161: 363 (1968).

14. S. Hirano, K. Shimono and S. Naka, *J. Mat. Sci.*, 17: 1856 (1982).

15. A. Onodera, K. Higashi and Y. Irie, *J. Mat. Sci.*, 23: 422 (1988).

16. F.P. Bundy, *J. Geophys. Res.*, 85: 6930 (1980).

Section III
Dynamic Consolidation

29

Shock-Compression Processing in Japan

A. B. SAWAOKA AND Y. HORIE

Center for Ceramics Research
Tokyo Institute of Technology
Box 7908, North Carolina State University
Raleigh, NC 27695, U.S.A

Current research activities and an early history of the shock processing of condensed matter in Japan have been reviewed. The first Japanese shock processing experiment was conducted in 1959 at the National Defense Academy of Japan. In the 1970's, the number of institutions engaged in shock recovery experiments had increased to nine. National projects sponsored by the Science and Technology Agency played a major role in the expansion of shock wave research in the 70's as well as in the 80's. In 1990, a new university program for shock wave research was initiated by the Ministry of Education, Science and Culture. At the same time, a major academic society, the Japan High Pressure Science and Technology, was created to bring together a large group of people in universities and industrial and government institutes. For the last several years, about 200 papers were presented at the annual meetings of high pressure scientists and engineers. About 15% of these papers were in the field of shock wave science and technology.

I. INTRODUCTION

After World War II, Japan had a long dormant time in the field of shock wave research, especially in condensed matter. Early attempts to explore the applications of explosively generated shock waves began in the late 50's at the

National Defense Academy with special focus on powder compaction and synthesis. This work will be discussed in some
detail in section 5. It needs to be emphasized that in
Japan shock wave research has no connection to military R &
D during and after WWII and has been pursued by physicists
and chemists as part of their academic studies on materials
response under the conditions of high-dynamic pressures.

A rapid increase in the number of organizations engaged
in shock wave research was seen in the 70's and may be
attributed partially to projects sponsored by the Science
and Technology Agency. Through those projects the first
two-stage gas gun driven by gun powder was constructed at
the Tokyo Institute of Technology in 1972. At the same
time, explosive facilities for recovery experiments were
established at the Tohoku University and the National Chemical Laboratory for Industry. The latter focused on the
development of a liquid explosive lens. Other institutions
that began the study of shock wave and its applications in
condensed matter in the 70's are Kumamoto University, Nippon
Oil & Fats, and Asahi Chemical Industry. Historical tracks
of the institutions engaged in shock wave research in Japan
is shown in Fig. 1. It may be concluded from the figure
that a new era of modern shock wave science and recovery
technology in Japan has truly begun in the early 70's.

During the first half of the 80's, the second series of
research projects sponsored again by the Science and Technology Agency had been undertaken at the Tokyo Institute of
Technology and the National Chemical Laboratory for Industry. The noteworthy results of the projects were the developments of a rail gun and explosive generators. The former
was developed at TIT for achieving very high speed impact of
small projectiles in the range of several to tens of km/sec.
The goal of the latter at NCLI was the development of reliable plane wave generators.

In 1990 a new three year university research program on
shock wave was initiated under the sponsorship of the Ministry of Education, Science and Culture. The program is
headed by Prof. K. Takayama of Tohoku University and in-

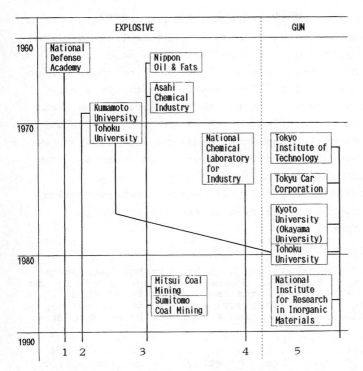

FIG. 1 *Historical tracks of the institutions engaged in shock wave research in Japan.*

cludes research topics in not only condensed matter, but also gases. The condensed matter program consists of the teams headed by A. B. Sawaoka (TIT) and Y. Shono (Tohoku), K. Nagayama (Kyushu), and T. Mashimo (Kumamoto). The topical areas of research are residual effects (Tohoku), shock wave behavior in composite structures (TIT), shock wave in polymers (Kyushu), and shock wave in inorganic materials (Kumamoto).

II. CURRENT RESEARCH ACTIVITIES AND EXPERIMENTAL FACILITIES

Presently, there are eleven major institutions that are engaged in shock wave research in condensed matter in Japan. Research activities at these institutions can be quickly surveyed by looking at the proceedings of self-organized

annual meetings of high pressure scientists and engineers in
Japan. For example, Table 1, shows a partial listing of the
program of the 29th High Pressure Conference held in
Fujisawa, November 16-18, 1988 as it relates to dynamic high
pressure research. For the last several years there were
about 200 papers presented at the annual meetings and about
15% of them were in the field of shock waves in condensed
matter. Beginning 1991 the annual meeting will be sponsored
by a newly established academic society, the Japan Society
of High Pressure Science and Technology.

The following is a summary of current research activities and experimental facilities at the above described
eleven institutions.

A. Universities

1. Kumamoto Universities, High Rate Energy Laboratory
    Facility: Explosive chamber (max. 200 gr of HE) and
    a 40mm keyed powder gun (max. velocity of about
    2km/sec). Areas of study: EOS of stabilized
    zirconia and shock compaction of oxide superconduc-
    tors.
2. Okayama University
    Facility: A 30mm powder gun and a 20mm two-stage
    light gas gun. Areas of study: Powder compaction
    and shock wave structure.
3. Tohoku University, Institute for Material Science
    Facility: A 30mm powder gun and a 20mm two-stage
    light gas gun. Areas of study: Microstructures of
    shock compressed materials, especially oxide super-
    conductors.
4. Ibid, Institute for Fluid Science
    Facility: A 25mm two-stage light-gas gun. Areas of
    study: Precision measurements of shock wave.
5. Tokyo Institute of Technology
    Facility: 20mm and 40mm powder guns, 80mm and 20mm
    two-stage light-gas guns, an 8mm rail gun and a
    plasma gun. Areas of study: New materials synthe-
    sis, dynamic powder compaction with exothermic
    chemical reactions, and shock wave in composite
    structures.

B. National Institutes

1. National Chemical Laboratory for Industry
    Facility: Explosive chamber (max. 5kg of HE) and a
    rail gun. Areas of study: Diamond synthesis and
    shock compaction of non-oxide ceramic powders.
2. National Institute for Research in Inorganic
    Materials
    Facility: A 30mm powder gun. Areas of study:
    Material synthesis.

TABLE 1

Partial Listing of the Program of the 29th High Pressure
Conference of Japan, November 16-18, 1988, Fujisawa. Numbers in parentheses correspond to the institutions listed at
the end of the table.

1. "Shock Compression Experiment for the Formation of
   Ureilite." M. ARAKAWA, M. KATO, Y. TAKAGI (1).

2. "Shock Compression of Zirconia." T. MASHIMO, S.
   MATSUZAKI, M. KODAMA (2), K. KUSABA (3), K. FUKUOKA (4),
   Y. SHONO (4).

3. "Shock-Induced Changes in La_2Ni_4 Type Compounds." H.
   TAKEI, H. TAKEYA, K. KUSABA (5), Y. SHONO (4).

4. "Reitveld Analysis of Shock Induced Phase by X-Ray
   Powder Diffraction." K. KUSABA, T. KAJITANI, Y. SHONO
   (4).

5. "Shock Induced Phase Transition of $TaFeO_4$." K. KUSABA,
   M. KIKUCHI, K. FUKUOKA, Y. SHONO (4).

6. "Shock Induced Phase Transition of M_2O_3 Type Compounds."
   T. ATOU, K. KUSABA, K. FUKUOKA, M. KIKUCHI, Y. SHONO
   (4).

7. "Shock Deformation of Alpha-Quartz." H. TATTEVIN, Y.
   SHONO, M. KIKUCHI, K. KUSABA, K. FUKUOKA, E. AOYAGI (4).

8. "Behaviors of Projectile and Armature in the Electromagnetic Accelerator." S. USUBA, Y. KAKUDATE, M. YOSIDA,
   K. AOKI, K. TANAKA, S. FUJIWARE (6).

9. Hypervelocity Acceleration of Microprojectiles by Two-
   Stage Plasma Accelerator (II). H. TAMURA, K. UEDA, A.
   B. SAWAOKA (7).

10. "Separation Method of Projectile and Sabot (II)." T.
    USUI, H. TAMURA, H. KUNISHIGE, A. B. SAWAOKA (7).

11. "Numerical Studies for a Dual Element Meteoroid Shield
    Structure." T. USUI, H. KUNISHIGE, H. TAMURA, A. B.
    SAWAOKA (7).

12. "Heterogeneous Deformation and Radiation Spectrum by
    Shock Compression." T. HIRAMATSU (8), K. KONDO (7).

13. "First Principle Calculation of Shock Wave Equation of
    State of Solid LiH." J. HAMA, K. SUITO (9).

14. "Measurement & Recovery Experiments of Compression-Shear
    Shock Waves." T. MASHIMO (2).

15. "Two-Stage Shock Loading Path of Sodium Chloride." K.
    KANE, H. TAMURA, A. B. SAWAOKA, K. KUSABA (7), K.
    FUKUOKA, K. OH-ISHI, Y. SHONO (4).

16. "Multi-Stage Shock Loading Path of Graphic-like BN." K.
    KANE, T. TANIGUCHI, H. TAMURA, A. B. SAWAOKA (7).

17. "Morphology of Carbon Particles Formed by the Conically
    Converging Shock Wave Technique." K. YAMADA, S.
    TOBISAWA (10).

Table 1 (Con't.)

18. "Ceramic Superconducting Film Synthesized by Electric Discharge Explosion." R. MATSUDA, A. B. SAWAOKA (7).

19. "Dynamic Loading Synthesis of Chevrel Phase Compound." Y. GOTOH, Y. OOSAWA, S. FUJIWARA (6).

20. "Cylindrical Implosion Fixtures for High Pressure Generation and Recovery." M. YOSHIDA, S. USUBA, Y. KAKUDATE, K. AOKI, S. FUJIWARE (6).

21. "Shock Synthesis of Diamond." M. YOSHIDA, K. AOKI, S. USUBA, Y. KAKUDATE, S. FUJISAWA (6).

22. "Carbon from Shocked Adamantane." T. SEKINE (11).

23. "Diamond Synthesis in Graphite/Nickel Composite Powders Subjected to Shock Compression." Y. HORIE, I. SIMONSEN, S. CHEVACHAROENKUL (12), T. AKASHI, A. B. SAWAOKA (7).

24. "Shock Reaction Between Magnesite and Metallic Iron." T. SEKINE (11).

25. "Effect of Explosive Compression of Metal and Oxide Powder at High Temperature." Y. KUMURA, H. YAMANOTO (10).

26. "Effect of Dynamic Compaction on $\alpha-Al_2O_3$ Powder." Y. FUKUYAMA (17), T. AKASHI (13), A. B. SAWAOKA (7).

27. "Shock Compression Behavior of $\alpha-Al_2O_3$ Powders and its Numerical Simulation." H. KUNISHIGE, S. WATANABE, H. TAMURA, A. B. SAWAOKA (7).

28. "Shock Consolidation of Difficult-to-Combine Ceramics." K. KONDO, S. SAWAI, K. KOSIKAWA (7).

29. "Effect of Initial Diamond Powders on Shock Compaction." S. SAWAI, K. KONDO (7).

30. "Diamond Ceramics of Nanometer-Size Particles." K. KONDO, S. SAWAI (7).

31. "Shock Compaction of Diamond/Ceramic/Metal Composite." S. SAWAI, K. KONDO (7).

Organizations:

(1)   Nagoya University
(2)   Kumamoto University
(3)   Kumamoto Institute of Technology
(4)   Tohoku University, Institute of Materials Research
(5)   University of Tokyo, Institute of Solid State Physics
(6)   National Chemical Laboratory of Industry
(7)   Tokyo Institute of Technology, Res. Lab. of Engr. Materials
(8)   Asahi Glass Co., Ltd.
(9)   Osaka University
(10) National Defense Academy of Japan
(11) National Institute for Research in Inorganic Materials
(12) North Carolina State University
(13) Toshiba Tungroy Co.

C. Industries

1. Asahi Chemical Industry Co.
    Facility: Explosive chamber (max. 900kg of HE).
    Areas of study: Compaction of amorphous powders.
2. Mitsui Coal Mining Co.
    Facility: Explosive chamber (max. 10kg of HE).
    Areas of study: Diamond synthesis and powder compaction.
3. Nippon Oil & Fats Co.
    Facility: Explosive chamber (max. 10kg of HE).
    Areas of study: Synthesis of w-BN and cladding of amorphous foil.
4. Sumitomo Coal Mining Co.
    Facility: Explosive chamber (max. 10kg of HE).
    Areas of study: Diamond synthesis and powder compaction.

III. DEVELOPMENT OF RECOVERY TECHNOLOGY IN JAPAN

The technique of sample recovery plays a key role in materials research under shock wave loading. Both solids and liquids can be subjected to shock compression and recovered for analysis by using appropriate recovery systems.

The systems developed at NDA over a period of thirty years beginning in the late 50's are of cylindrical type where a tube containing sample powder is held in a concentric explosive charge. Ignition of the explosive from one end results in the passage of shock wave axially along the tube of powder. A current system is designed to explore crystal growth under the conditions of very high pressures and temperatures by exploiting conically converging shock wave.

In 1974 Shono et al [1] developed a mouse trap type recovery fixture in combination with an explosive lens system where a flyer impacts on a die that contains sample powder. However, this system was abandoned shortly thereafter because of a tragic accident.

Also in 1974, Sawaoka et al developed a recovery fixture for use with a two-stage light gas gun. In this system a projectile impacts on a target assembly containing sample material [2]. The container is protected by a spall ring and a momentum trap. This was used extensively for the investigation of w-BN synthesis. The amount of mass that be produced on a gun is extremely limited, but is sufficient for research purpose.

For industrial applications, one must resort to explosive systems that can handle volumes of material. The system developed by Nippon Fats & Oil for production of w-BN [3] is a double-tube cylindrical fixture where the outer tube is accelerated to a high speed by an explosive charge. This configuration is known to be capable of generating higher pressures than those by direct contact of an explosive charge.

Another advantage of explosive systems is that they are capable of loading multi-capsules simultaneously for rapid canvassing of a variety of material combinations very efficiently. For example, TIT and Nippon Oil & Fats used in their early experiments a recovery system where three to four capsules embedded in a cylindrical container were impacted simultaneously by a flyer plate [4]. Presently, this system is extended to twelve capsules in a single shot.

Presently, the limit of impact velocity for recovery is known to be about 3km/sec for explosive systems involving a flyer plate. For guns, the velocities are 4km/s and 2.5km/s for metallic and ceramic materials respectively. There are other limitations that apply to most recovery systems. For instance, the loading conditions in sample powder, especially in low impedance materials, are not homogenous because of multidimensional wave interactions [5]. But, sometimes this can be exploited to obtain a great deal of information from a single sample over a wide range of loading conditions.

IV.   POWDER COMPACTION RESEARCH IN JAPAN

The first explosive compaction of powder was conducted at the NDA in 1959 by Kimura and Nomura. Initially, they used water as a pressure transmitting medium for the compaction of iron powder. The subsequent development at the NDA in the 60's is reviewed in the next section. The present authors had a thrilling experience of opening a "time capsule" when they visited the NDA in 1989; we had an opportunity to open and examine one of the capsules that were fired in 1960 and preserved in the laboratory. The consolidated iron bar of about 1m long and 3cm in diameter was found in perfect conditions.

Shock wave compaction is a unique process for densifying powders, especially amorphous and strongly covalent materials which are difficult to sinter using conventional methods. However, the cost of shock compaction is estimated to be about ten times that of hot isostatic method on a test plant basis. Therefore, shock compaction is not a competitive process for oxide ceramics such as alumina and zirconia which can be sintered by other methods such as hot pressing to obtain compacts with excellent mechanical properties. However, judgement on the processing of non-oxide ceramics such as silicon carbide is still out, because as yet the conventional sintering of such materials has not been successful without additives that weaken mechanical properties.

Theoretically, silicon carbide is said to possess the highest mechanical properties at elevated temperatures. Therefore, in recent years it has been studied extensively in Japan, USA, and Europe for possible applications in gas turbine engines to increase fuel efficiency operating at 1300 to 1400 °C. But, so far the production of dense sintered compacts has not been successful without deleterious additives. Therefore, much attention is given in Japan toward the possibility of dynamically compacting non-oxide ceramics and diamond [6-10]. However, there still remain two vexing problems that were already identified by Kimura and Nomura in the 60's. These problems are cracking in and heterogeneous properties of recovered compacts.

Powder materials are inherently heterogeneous because of particle shape, void and size distributions, etc. Therefore, loading under shock compression is rarely homogeneous even in mono-sized powder. Many studies have been made to deal with the quantification of process conditions in powder under shock wave loading, but there is not yet a definitive answer to the problem [8-10].

Diamond has many potential applications, but it is the most difficult material to sinter. It is likely that dia-

mond will be the ultimate dream of shock wave compaction
technology. Several Japanese groups, including the present
authors, are working toward this challenging goal [6-10].

Recently, there have been two workshops on the feasibility of industrial applications of shock processing technology in Japan. Both were held at TIT in 1986 and 1988. The
third one will be held in the Mt. Zao immediately after the
18th Int'l Symp. on Shock Waves in Sendai, July 26-27, 1991.

Shortly after the discovery of high T_c superconductors
in 1986, people in Japan as those in the U.S.A. and the
U.S.S.R., began experimenting the shock consolidation of
these new materials. However, the loss of superconductivity, cracking and material heterogeneity have not yet been
overcome.

V. AN EARLY HISTORY OF SHOCK PROCESSING AT THE NATIONAL
    DEFENSE ACADEMY OF JAPAN

It is well recognized in both the U.S.A. and the U.S.S.R.
that the first shock synthesis of chemical compounds from
elemental powders was carried out in Japan at a little-known
institution called the National Defense Academy. However,
the background of this pioneering work is little known even
in Japan and may soon fall into oblivion because of the
retirement of the major participants: Profs. Horiguchi,
Nomura, Tobisawa and Kimura. Presently, only Kimura is
still active at the Academy. Therefore, we thought it is
important to take a look at the early history of NDA and to
highlight the background of this singular achievement. This
article is a synthesis of three source materials: inter-
views with Horiguchi, Kimura, and Tobisawa, a bound copy of
Nomura's publications on the applications of explosive shock
effects, and selected papers of the other three. Explosive-
related work at NDA began shortly after Nomura joined the
newly established Academy in 1956 from Tohoku University
[11-24]. It is important to note that in contrast to those
in the U.S.A. and U.S.S.R., the work at NDA were not extension of military R & D during World War II. They were low
budget, academic research as any others in Japan with spe-

cial focus on exploration of material modifications by
explosive charges. Nomura recalls in a recent article [25]
that at that time he knew little about explosives and was
induced, in spite of his protest, to take charge of such
studies. At Tohoku University he was studying the kinetics
of orth to para transition in hydrogen and his only connection to explosives was a study of discharge reaction in
liquid air.

Nomura describes in a 1963 article [26] that "the development of explosive metal working abroad in the late 50's
taught him the possibilities of novel explosive applications," and that "their experimentations began with the idea
that explosive charges might be used as a part of a process
for producing useful materials." They were interested in
"material modifications and promotion of chemical reactions
as those by radiation and ultrasonics." Table I provides a
summary of early explosive processing of materials at NDA in
the 60's.

Judging from the references cited in the papers, the
group seems to have been influenced by the work of people
such as Pearson, La Rocca, and Bridgman, and also a Du Pont
patent (in Japan) on explosive compaction of powders such as
uranium oxide, Be, Zr, etc. But, they immediately recognized almost insurmountable technical difficulties that
result from cracking and the heterogeneity of material
properties in recovered compacts. While admitting the
immaturity of their techniques, Nomura already cast doubt on
the practicality of explosive powder compaction in 1963
[26]. Therefore, the studies at NDA thereafter were focused
on what he calls, "frozen" phenomena (or effects) in explosively loaded materials and their exploitation for useful
purposes. From the earlier work on metallurgical effects
and powder compaction, they recognized that "frozen phenomena" such as lattice distortion, defects, and surface reactivity might be useful for material processing.

They knew the synthesis of minerals by Michel-Levy and
also of diamond by De Carli and Jamieson. Interestingly,
Nomura dismissed their own results on diamond synthesis by

TABLE 2:

Explosive Processing of Materials at National Defense Academy of Japan.

Year	Materials	Note	Reference
1961	Zn and Cd ferrites	Compaction	(11)
1962	Fe, Ti	Compaction	(12)
1963	Ti+Carbon Black	Synthesis	(13)
	$ZnO+Fe_2O_3$	Synthesis	(14)
	Graphite	Compaction	(15)
	Co, Ti, Fe,	Compaction	
	ZnS (Cu, Cl)	Use of temp. indicating compounds	(16)
1964	(Ti, W, Al)+C	Synthesis	
	$ZnO+SiO_2$		
	$MnCl_2+ZnO+SiO_2$		(17)
	Acetylene Black	Activation	(18)
	Ferrites, Glass	Crushing	
	$BaTiO_3$		(19)
1965	ZnS (CU, Cl)	Magnetic Susceptibility	(20)
1966	Acetylene Black	Magnetic Susceptibility	(21)
1967	Quartz	Magnetic Susceptibility	(22)
1968	Ti+Acetylene Black	Synthesis	(23)
1969	(Ti, Si, Zr)+Coll. Carbon,	Synthesis,	
	ZrO_2 + C	Reaction,	
	Zr, Si, ZrO_2	Compaction,	(24)

stating that they were able to observe only the strongest diffraction peak and that there is no time for diamond growth under explosive loading. He thought that explosive technique alone would not produce useful diamonds.

According to Nomura [26], the group pooled their instruments and shared the task of probing "frozen" phenomena in as many materials as possible. Kimura conducted electron microscopy and electron diffraction analysis in the Department of Mechanical Engineering, Tobisawa carried out magnetic susceptibility measurements to high temperatures in the

Chemistry Department where Nomura was a professor.
Horiguchi, a senior scientist at the Institute of Physical
and Chemical Research in Tokyo, undertook X-ray analysis and
matters related to chemical experiments. He was a class
mate of Nomura at Tohoku University. Their selection of
materials was based upon common interest and familiarity.
For instance, both Kimura and Tobisawa had some experience
dealing with ferrite sintering. Carbon, having unpaired
electrons on its surface was the catalyst in Nomura's experiments on the orth to para transition in hydrogen.
Tobisawa was studying the role of unpaired electrons in
solid state reactions using magnetic properties. Horiguchi,
stimulated by Bridgman's work, was looking for a way to
conduct experiments on high pressure chemical reactions
without developing expensive facilities.

The work at NDA were mostly of exploration type, but
they were aware of needs to better characterize experimental
conditions under explosive loading. The use of ZnS was part
of an attempt to estimate pressures using the static result
that its luminescence intensity decreases drastically when
subjected to pressures above 4 GPA. They also used hexa-
methylenetetramine cobalt and nickel salts for residual
temperature measurements. The kinds of instrumentation
developed in the U.S.A. at the time were beyond their financial means. Nevertheless, they attained qualitative understanding of the shock compression processing that is very
similar to those we have today. For instance, they discussed shock induced lattice distortions, appearance of free
radicals, crushing of powders, removal of oxide layers,
increase of surface areas, etc. Regarding shock induced
reactions, they suggested that once the exothermic reactions
were initiated under shock loading by, say, the creation of
dispersed hot spots resulting from (1) the collapse of
voids, (2) the temperature rise due to shock compression,
and (3) the high speed flow of powders, then the reactions
could be sustained by the heat of reaction alone.

There had been few shock processing activities at NDA in
the 70's. However, recently there has been a resurgence of

interest in the area of crystal synthesis and shock consolidation of superconductors. Noteworthy is a series of work
by Tobisawa and Yamada on crystal synthesis using a conically converging shock wave in a cylindrical recovery fixture
[27-28].

REFERENCES

1. Y. Syono, T. Goto, J. Nakai, Y. Nakagawa and H. Iwasaki,
    J. *Phys. Soc. Japan, 37,* 442(1974).

2. T. Soma, A. Sawaoka and S. Saito, *Mat. Res. Bull., 9,*
    755 (1974).

3. A. Sawaoka, T. Soma and S. Saito, *Japan J. Appl. Phys.,*
    13, 891(1974).

4. F. R. Norwood, R. A. Graham and A. Sawaoka, "in Proc. of
    the 4th APS Topical Conference on Shock Waves in Con-
    densed Matter." Plenum Press, New York, p. 837 (1986).

5. M. Araki and Y. Kuroyama, Industrial Explosives Society,
    Japan (in Japanese) 49, 250(1988).

6. T. Akashi and A. B. Sawaoka, *J. Mat. Science, 22,* 3276
    (1987).

7. A. B. Sawaoka and T. Akashi, U. S. Patent 4,655,830 and
    4,695,321, (1987).

8. K. Kondo and S. Sawai, *J. Am. Ceram. Soc., 72,* 837
    (1989).

9. K. Kondo and S. Sawai, *J. Am. Ceram. Soc., 73,* 1983
    (1990).

10. Y. Horie, Shock Compression of Condensed Matter-1989,
    North-Holland, Amsterdam, 479(1990).

11. Y. Kimura, *Powder and Powder Metall. (in Japanese)* 8, 58
    (1961).

12. Y. Nomura, *J. Ind. Explo. Soc. of Japan, 22,* 321(1962).

13. Y. Horiguchi and Y. Nomura, *Bull. Chem. Soc. Japn., 36,*
    486(1963).

14. Y. Kimura, *Japn. J. Appl. Phys., 2,* 312(1963).

15. Y. Nomura and Y. Horiguchi, *Bull. Chem. Soc. Japn., 39,*
    26(1963).

16. Y. Horiguchi and Y. Nomura, *Rept. Inst. Phys. Chem.*
    Res., 39, 211(1963).

17. Y. Horiguchi and Y. Nomura, Preprint for the Workshop
    "Applications of High Dynamic Pressures in Chemistry,"
    held at R.I.P.C., April 23, 1964.

18. Y. Horiguchi and Y. Nomura and T. Funayama, *Kogyo Kagaku*
    Zasshi, 68, 1966.

19. Y. Kimura, Preprints for the Annual Mtg. of Japn. Soc. of Precision Engr., 1964.

20. Y. Nomura and S. Tobisawa, *Appl. Phys. Lett.*, *7*, 126 (1965).

21. S. Tobisawa, *Memoirs of the Defense Academy (Japan)*, *6*, 493(1967).

22. S. Tobisawa, ibid, 6, 485(1967).

23. Y. Horiguchi and Y. Nomura, *Kogyo Kagaku Zasshi*, *71*, 1419(1968).

24. H. Suzuki, H. Yoshida, and Y. Kimura, *Yogyo Kyokai Shi*, *77*, 278(1969).

25. Y. Nomura, *J. Ind. Explos. Soc. of Japan*, *49*, 291(1988).

26. Y. Nomura, *Kagaku to Kogyo*, *16*, 123(1963).

27. S. Tobisawa and K. Yamada, Memoirs of the Defense Academy (Japan) 25, 17(1985).

28. K. Yamada and S. Tobisawa, *Carbon*, *26*, 867(1988).

30

Shock Consolidation of Diamond

THOMAS J. AHRENS[1], G. M. BOND[2], W. YANG[1], and G. LIU[2]

[1]Lindhurst Laboratory of Experimental Geophysics, Seismological Laboratory, California
    Institute of Technology, Pasadena, CA 91125, USA
[2]Department of Materials and Metallurgical Engineering, New Mexico Institute of Mining and
    Technology, Socorro, NM 87801, USA

*Shock consolidation of powders with 50 to 60% of crystal
density of diamond and boron nitride, both pure and admixed with
SiC and Si_3N_4 whiskers and graphite, occurs upon application
of <1 μsec duration shock pulses of 15 to 20 GPa. For powders,
10 to 100 μm in diameter, frictional sliding during shock
consolidation gives rise to surface heating to temperatures of
~4000K and results in a molten layer several μm thick. This
molten layer freezes via heat conduction into the interior of
the grains on a time scale short compared to the shock pulse
duration. Using the geometric model of compaction of Kaker and
Chaklader [1] for spherical or cubic grains, we estimate that the
irreversible work carried out against material strength in the
sample is ~0.4% of the total shock energy. Therefore, most
shock energy goes into surface heating. The recently reported
consolidation of < 1 μm sized diamond powders is attributed to
shock induced nearly uniform heating and resulting plastic
deformation above 10 GPa and 1300K. Consolidation of diamond
admixed with graphite is assisted by the transformation of
graphite to amorphous carbon which occurs above ~20 GPa as a
result of graphite in the diamond field crossing the metastable
extension of its melting line.*

I. INTRODUCTION

The consolidation of diamond and diamond-structured cubic boron nitride (C-BN) from
powders in the initial size range of 5 to 50 μm [2; 3; 4; 5; 6] and more recently of diamond
powder in the submicrometer size range [7] shows considerable promise in producing

technologically useful material. Diamond is the hardest material presently known with a micro Vickers hardness of ~ 120 GPa. Sapphire (corundum) has a hardness of ~ 50 GPa and single-crystal C-BN has a hardness of ~75 GPa. Diamond powders are abundant, both from natural and man-made sources. As a technological material, polycrystalline pore-free consolidated diamond has many potential uses, in addition to those stemming from its great hardness on account of its high longitudinal elastic velocity (18 km/sec) and thermal conductivity (10 W/(cm °K.). Similarly, isostructural C-BN which is synthesized via static or dynamic high pressure, also, has a high longitudinal wave velocity (16 km/sec) and thermal conductivity. C-BN is also more resistant to chemical reaction in an oxidizing environment at high temperature than diamond. In addition, both cubic and hexagonal diamond powder [8], and the dense wurtzite-phase of BN may be synthesized (in large quantities) by dynamic compression from the graphite phase [9]. Nanometer-sized single-crystal domains result from such shock synthesis. This material may be shock consolidated [7]. Single-crystal powders of diamond and C-BN with initial densities of 60 to 70% of crystal density can be consolidated to crystal density by driving shocks of ~15 to 30 GPa amplitude into samples within steel recovery containers.

Experiments in which SiC, Si_3N_4 whiskers and graphite phase material are admixed with both diamond and C-BN are also reported [10; 11].

We review the different physical mechanisms which act to give rise to shock consolidation processes in these materials. We present a new analysis evaluating the energy taken up during shock consolidation by deformation processes, describe the thermal state upon consolidation of sub-micron diamond particles and suggest an additional mechanism by which the admixture of graphite with diamond facilitates consolidation.

II. SURFACE HEATING MODEL

Previous modeling of the dynamic consolidation processes of porous metal powders [12; 13; 14] suggested that surface heating, by a particle-particle sliding produces a molten layer of sample material on each grain. This model is motivated both by microphotographic observations of shock consolidated spherical, usually metallic, particles (e.g. [15; 16]) (Fig. 1) and by a large number of high-speed frictional sliding experiments on metals and non-metallic materials at speeds approaching a km/sec which demonstrate that interface melting occurs in virtually every medium [17].

The simplest model for energy partitioning upon shock compaction assumes: 1) The energy of the shock wave all goes into melting a surface layer of the compacted particles. 2) The Hugoniot curve for the porous media is represented by the two straight lines 0'0 and 0C as an approximation to the porous Hugoniot 0'0B (Fig. 2). 3) The (constant pressure) specific heat (c_p), melting temperature (T_m) and enthalpy of melting (H_m) are independent of pressure

←— Initial shock propagation

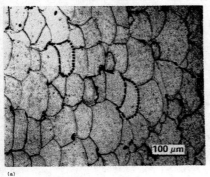

(a)

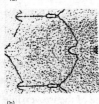 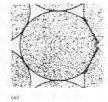

(b)   (c)

FIG. 1 *Characteristic shape of shock consolidated, initially spherical powder. (a) Microphotograph of 37 μm diameter copper spheres dynamically compressed with a 5.6 GPa stress wave which propagated from right to left (from Gourdin [13]). (b) and (c) computer simulation, 48 ns and 30 ns after 1 km/sec stainless steel impact onto 10% porous steel rod. Array shows shock consolidation morphology (from Williamson and Berry [27]).*

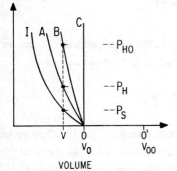

FIG. 2 *Pressure-Volume plane sketch of principal isentrope, OI, and principal Hugoniot, OA, porous Hugoniot O'OB, and pore-collapse Hugoniot O'OC. Isentropic, principal Hugoniot and porous Hugoniot pressures, P_S, P_H, P_{HO} are indicated.*

and temperature. Within these assumptions an upper bound to the mass fraction of melted materials is

$$L = \frac{P_H V_0 \, (V_{oo}/V_0 - V_0)/2}{c_p \, (T_m - T_0) + H_m}$$

(1)

where T_0 is the initial temperature and V_{oo} and V_0 are the specific volumes of the porous powder and crystal, respectively.

Some typical values, appropriate for the shock consolidation of diamond [2] are $P_H = 10$ GPa, $V_{oo}/V_0 = 1.85$, $c_p = 2 \times 10^7$ erg/(g°K), $T_m - T_0 = 4000K$, and $H_m = 9.2 \times 10^{10}$ erg/g yields values of L of 0.07.

Because calculations such as those shown in Figs. 1b and c, in which 75 μm diameter 304 stainless steel rods are treated as elastoplastic solids, assume no frictional sliding and heating, surface temperatures of only 900K, considerably below the 1650K melting point of stainless steel, are calculated. Temperatures within the interior of the rods are raised only to 400K. This is below the continuum shock temperature of ~650K calculated for 10% porosity.

For spherical or near spherical geometry, we calculate the magnitude of the irreversible work done against the material strength. Kakar and Chaklader [1] modeled the quasistatic compaction of solid spheres packed in different geometries to infer the densification induced by mutual indentation. Although the isotropic geometry of their model only approximates the characteristic anisotropy of particle deformation in shock compaction, the initial packing density they examined for the hexagonal prismatic case is close to the initial density of many dynamic consolidations. For this packing geometry, each initially spherical particle deforms as in Fig. 3. The unit volume of each particle of radius R is given by

$$V_p = 4\pi \, R^3/3 - Z\pi \, h^2 \, (3R - h)/3$$

(2)

where Z is the coordination number, and h is the height of the flattened spherical segment. In terms of the radius of the flattened faces of the sphere of radius, a, this becomes

$$V_p = \alpha \, (2R^2 + a^2) \, (R^2 - a^2)^{1/2} - \beta \, R^3$$

(3)

For orthorhombic hexagonal prismatic deformation, $Z = 8$, $\alpha = 8\pi/3$ and $\beta = 4\pi$. Deformation of spheres occurs until the flat circular faces on the prismatic surfaces touch. At this point the radius of the spherical particle has increased from $R_0 = 0.62035$ (which corresponds to $h = 0$ and $V_p = 1$) to $R_c = 0.64321$. At this point, solution of Eq. 2 yields $h_c = 0.086165$.

At $R = R_c$ and $h = h_c$, the volume of a sector is $V_c = 0.01433261$. Referenced to an initial radius, R_0, the initial sector height, h_0 is 0.087835. Use of Eq. (3) to calculate the

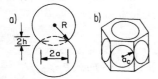

SPHERES IN CONTACT

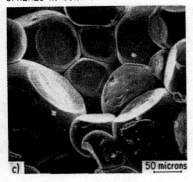

FIG. 3 *Compaction geometry of spherical particles: a) Geometric definition of parameters. (b) hexagonal prismatic packing; critical contact surface radius, a_c, is shown (after Kakar and Chaklader[1]. (c) explosively compacted Ti alloy powder from Linse [28].*

initial and final flat radii, a_o and a_c corresponding to R_o and R_c, yields $a_o = 0.318217$ and $a_c = 0.32159$. The bulk density increases from 60.46 to 83.51% of crystal density as R increases from R_o to R_c. Use of these values to calculate the principal strains in cylindrical coordinates (see e.g. [18], p. 114) yields:

$$\varepsilon_{\theta\theta} \cong 2(a_c - a_o)/\bar{a} = 0.02108; \quad \varepsilon_{rr} \cong (a_c - a_o)/\bar{a} = 0.10545; \quad \text{and} \quad \varepsilon_{zz} \cong h_c/\bar{R} = 0.13638$$

The bar indicates average values between initial and final configurations. For an ideally plastic material, the stresses acting with these strains are equal within a multiplicative constant, of order unity, to the dynamic yield stress, Y. Thus the plastic strain energy per unit volume of a spherical particle is

$$E_{plastic} \cong \left[\varepsilon_{\theta\theta} + \varepsilon_{rr} + \varepsilon_{zz}\right] Z \, V_c \, V_0 \, Y/2 \tag{4}$$

For $Z = 8$ and $V_c = 0.01433$, this yields: $E_{plastic} = 0.0151 \, V_0 \, Y$.

We will now compare $E_{plastic}$ to the shock energy, $E_H = P_H (V_{oo}-V)/2$, for the diamond consolidation example given above. The diamond powder is, of course, actually

344 *Ahrens et al.*

small cubes, not spheres. Therefore, we expect that more deformational energy will actually be expended than we calculate for a spherical case. In the present example, we use: $V_o = V_o'/0.8351$.

Here V_o' is specific volume of diamond, 0.2849 cm^3/g, and 0.8351 is the density fraction corresponding to $R = R_c$. We also assume that $Y = 1$ or 10 GPa. Thus, $E_{plastic} = 0.5 \times 10^7$ or 5×10^7 ergs/g. This compares to 1.2×10^{10} ergs/g for E_H. Even if Y is as large as 10 GPa, we infer that the shock energy is some 240 times greater than the deformation energy. We conclude that most of the energy of the shock is, in fact, dissipated in sliding as in Eq. 1 and does not reside in deformational energy. However, for highly angular powders [19] strain energy can play a more substantial role.

Given that a molten layer forms on the surface of shocked particles of initially spherical or cubic shape, if dynamic consolidation is to be successful, the molten layer must freeze before stress unloading takes place. What we envision is that the exterior surface of the crystal powder follows the path indicated as "surface" in Fig. 4, whereas the interior achieves cool states on the solid diamond Hugoniot labeled "interior" in Fig. 4. The time for cooling provides a vital constraint on the shock consolidation process. This time scale for cooling or freezing, t_f, of an initially molten layer of mass fraction, L, is given by [14]

$$t_f = \pi D \{LdH_m/Dc_p (T_m-T_o)\}^{-2}/16 \qquad (5)$$

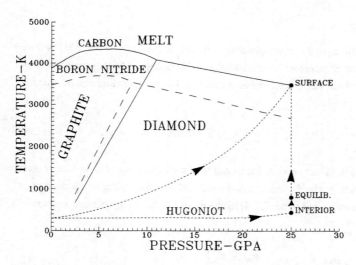

FIG. 4 *Phase relations for diamond and boron nitride. Shock consolidation to 25 GPa indicates thermodynamic path of "surface" point achieves a temperature of 3600 K whereas "interior" point remains cold on principal Hugoniot. Path for thermal equilibrium is indicated.*

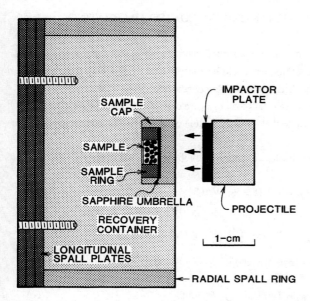

FIG. 5 *Diamond shock consolidation experimental assembly. Sapphire umbrella prevents contamination of sample with metal.*

where D is the thermal diffusivity and d is crystal diameter. For $D = 1 \times 10^{-4}$ m^2/sec, d = 10μm and $L = 0.05$, we find $t_f = 6.5 \times 10^{-10}$ sec. This time is short compared to the duration of shock produced by the impact of 1-2 mm thick steel flyer plates used by the present authors and others (Fig. 5) to conduct shock experiments. Moreover, the thermal conduction time: $t_t \sim d^2/D$, is on the order of 10^{-6} sec for 10μm diameter crystals. As discussed by Potter and Ahrens [2], if the thermal conduction time equals the shock propagation time through the crystal: $d_{min} = D/U$, cracking of the sample should occur and samples are predicted not to consolidate via the mechanism of Fig. 4. Here U is shock velocity. For diamond, with U = 15 km/sec, $d_{min} = 7$ nm. However, samples with these grain sizes [7] <u>do</u> consolidate as discussed in Sect. III.

If the freezing time, t_f, becomes comparable with the shock duration time, t_d, in the sample, consolidation will not occur. For a shock duration of $t_d = 0.8$ μsec $= t_f$, solution of Eq. 5 yields a value of $d = 250$ μm. Thus a larger crystal than this will consolidate only for longer shock durations.

III. SHOCK CONSOLIDATION OF DIAMOND VIA CONTINUUM HEATING

Recently, Kondo and Sawai [7], demonstrated that submicron sized diamond powders, made by either static or dynamic techniques, could be consolidated by shocking them to 20 to 25 GPa pressures. These powders had initial packing densities of 52 to 58% of crystal density. Since powders with values of d less than d_{min} are expected to achieve nearly continuum temperatures, is it possible that partially molten carbon is produced and gives rise to the shock consolidation?

We calculate a continuum shock temperature with the usual equations.

$$T_H = T_s + (P_H - P_S) \, V/(C_V \, \gamma) \tag{6a}$$

$$P_S = (P_H \, V/\gamma - E_H - \int_{V_o}^{V} P_S \, dV) \, \gamma / V \tag{6b}$$

$$T_S = T_o \, \exp \int_{V}^{V_o} (\gamma \, dV / V) \tag{6c}$$

$$P_{HO} = (E_H - V \, P_H/\gamma)/((V_{oo}-V)/2 - V/\gamma) \tag{6d}$$

and

$$T_{HO} = T_H + V \, (P_{HO} - P_H)/(\gamma C_v) \tag{6e}$$

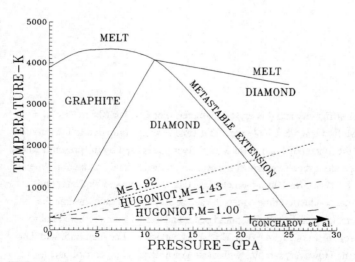

FIG. 6 *Phase diagram of carbon. Intersection of principal graphite Hugoniot (m = 1.00) with metastable extension of graphite melting line is predicted to induce formation of carbon glass. Graphite forms carbon glass upon static compression above 20 GPa as demonstrated by Goncharov et al. [23]. Hugoniots for porous diamond, with distensions of m 1.43 and m = 1.92 are shown.*

where P_H, V, V_0, and T_H are the pressure, compressed and initial volume, and temperature along the solid Hugoniot, and P_{HO}, V, V_{oo}, and T_{HO} are the same quantities along the porous Hugoniot (Fig. 2). Also, P_S and T_S are the isentropic pressure and temperature at volume, V. From the shock wave equation of state for porous diamond [20], the shock temperature for various values of distentions, $m = V_{oo}/V_0$, for diamond were calculated and are shown in Fig. 6. As can be seen, continuum temperatures of only ~1500K are achieved in porous diamond even for m = 1.92. Plastic deformation can be achieved at such low temperatures as demonstrated by the shock recovery experiments of Kondo and Sawai [7]. Recent observations of plasticity in diamond at 10 GPa and 1300K [21] are consistent with the plastic deformation consolidation experiments. We conclude that consolidation occurs as a result of a combination of small quantities of melt, possibly the result of interparticle jetting (e.g. [15]), followed by plastic deformation. The temperature and stress conditions required to induce plastic flow in diamond (and C-BN) are, however, not as yet well defined (e.g. [22]).

IV. SHOCK CONSOLIDATION OF DIAMOND VIA ADMIXTURE WITH GRAPHITE

Previous experiments [4] demonstrated that 13-16% graphite when admixed with diamond powder in the < 5, 4-8 and 100-150μm diameter range all resulted in enhanced compaction and conversion of graphite to diamond. Although at least 2 mechanisms for the enhanced compaction produced by the addition of graphite were previously suggested [4], recently Goncharov et al. [23], reported carbon glass formation at room temperature in the 20 to 30 GPa range. In analogy with the formation of amorphous H_2O and SiO_2 from the low pressure phases as a result of crossing the metastable melting curve of H_2O ice (I_h) [24] under static conditions and SiO_2-coesite under static and dynamic conditions [25; 26], we believe graphite, when shocked, also forms carbon glass, metastably, in the equilibrium field of diamond. The Hugoniot temperature pressure relation of crystal graphite as it crosses the metastable extension of the graphite melting curve at ~25 GPa is shown in Fig. 6. Subsequent to conversion of graphite to glass material, partial crystallization of diamond from the hot glass is expected to occur and we believe this assists shock consolidation.

V. SHOCK CONSOLIDATION OF DIAMOND AND C-BN via ADMIXTURE WITH SiC and Si_3N_4 WHISKERS.

Recently, Tan and Ahrens [10] and Yang et al. [11], have demonstrated that high quality dynamically consolidated material can be obtained when 10 to 30% SiC whiskers (SCW) and/or Si_3N_4 whiskers (SNW) are admixed with diamond and C-BN prior to consolidation in the 15 to 20 GPa shock pressure range. Scanning (SEM) and transmission electron microscopy (TEM) demonstrates that the SCW or SNW are melted and recrystallized

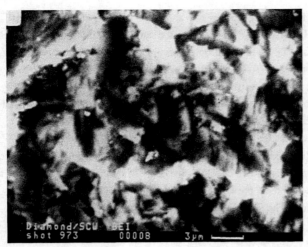

FIG. 7 *SEM image of diamond (75%) plus SCW (25%) shocked to 21 GPa.*

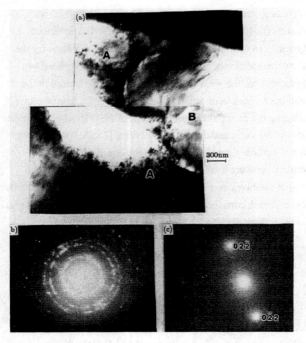

FIG. 8 *TEM observatios on same sample as Fig. 7 (a) bright-field image of sub-micron SiC crystals (A) surrounding undeformed diamond crystal (B). (b) and (c) diffraction pattern from regions (A) (lower) and (B).*

FIG. 9 *TEM bright-field image of dislocations in diamond shocked to 21 GPa in admixture with 30% SCW.*

in these composite materials. SCW and SNW have zero pressure melting points of 2173 and 3100 K, substantially below the 4000 and 3600 K values for diamond and C-BN. Typical textures are shown in Figs. 7 - 9 for material which was 70-75% diamond and 25 -30% SCW pressed initially to 60% of crystal density. This mixture was shocked to ~20 GPa. From the results of rod-sphere mixing experiments, the SCW is assigned a lower bulk density, 46% of crystal density, versus 71% for the diamond [11]. We calculate from Eq. 1 that some 7% of the diamond melted whereas 46% of the SCW is melted upon shock loading. The diamond demonstrates both brittle and ductile deformation features. In general, the SCW is either in the form of submicron-sized crystallites or is amorphous. The effects of both pressure and temperature on the shock response of diamond, C-BN, SCW and SNW require further study.

VI. CONCLUSIONS

Experimental evidence, from shock-consolidated spherical particles, indicates that the deformational energy expended during shock consolidation is on the order of 0.4% of the shock energy. Most of the energy dissipated upon shock consolidation of spherical and (probably) cubic powders is dissipated as a result of grain boundary sliding giving rise to 5 to 10% melt fractions of diamond and C-BN. Submicron powders such as the 52% crystal density samples of Kondo and Sawai [7] achieve temperatures of ~1500K and consolidate upon thermally induced plastic deformation. Diamond admixed with graphite consolidates via the mechanism of graphite transformation to carbon glass upon compression to pressures greater than ~20 GPa. Subsequent conversion of hot amorphous material to diamond within its stability field facilitates consolidation. Consolidations of diamond and C-BN with admixtures of SCW and BNW are bonded with melted and recrystallized whisker material. The extensive melting of the whisker material results from the initial enhanced microporosity which occurs around the whisker material and the lower melting point of SCW and SNW, relative to diamond or C-BN. The deformation mechanisms under dynamic compression require more study.

ACKNOWLEDGMENTS

Research supported by California Institute of Technology, New Mexico Institute of Mining and Technology, and U.S. Army Research Office, Contract DAAL-03-88K0199, Contribution #4921, Division of Geological and Planetary Sciences, California Institute of Technology, Pasadena, CA 91125.

REFERENCES

1. A. K. Kakar and A. C. D. Chaklader, *J. Appl. Phys.*, 38: 3223-3230 (1967).
2. D. K. Potter and T. J. Ahrens, *App. Phys. Lett.*, 51: 317-319 (1987).
3. T. Akashi, V. Lotrich, A. Sawaoka and E. K. Beauchamp, *J. Amer. Ceram. Soc.*, 68: 322-324 (1985).
4. D. K. Potter and T. J. Ahrens, *J. Appl. Phys.*, 63: 910-914 (1988).
5. S. Sawai and K. Kondo, *J. Am. Ceram. Soc.*, 71: 185-188 (1988).
6. T. Akashi and A. B. Sawaoka, *J. Mater. Sci.*, 22: 1127-1134 (1987).
7. K. Kondo and S. Sawai, *J. Am. Ceram. Soc.*, 73: 1983-1991 (1990).
8. P. S. DeCarli and J. C. Jamieson, *Science*, 133: 1821-1822 (1961).
9. G. A. Adadurov, Z. G. Aliev and L. O. Atovmyan, *Soviet Phys.-Doklady*, 12: 173 (1967).
10. H. Tan and T. J. Ahrens, *J. Mater. Res.*, 3: 1010-1020 (1988).
11. W. Yang, G. M. Bond, H. Tan, T. J. Ahrens and G. Liu, *J. Mat. Res.*, submitted (1990).
12. T. J. Ahresn, D. Kostka, P. Kasiraj, T. Vreeland Jr., A. W. Hare, F. D. Lemkey and E. R. Thompson, in *Rapid Solidification Processing of Principles and Technologies, III,* R. Mehrabian (eds.), National Bureau of Standards, 1983, p. 672-677.
13. W. H. Gourdin, *J. Appl. Phys.*, 55: 172-181 (1984).
14. R. B. Schwarz, P. Kasiraj, T. Vreeland Jr. and T. J. Ahrens, *Acta Metall.*, 32: 1243-1252 (1984).
15. T. Taniguchi, K. Kondo and A. Sawaoka, in *Metallurgical Applications of Shock-Wave and High-Strain-Rate Phenomena,* L. E. Murr, K. P. Staudhammer and M. A. Meyers (eds.), M. Dekker, New York, 1986, p. 293-311.
16. W. H. Gourdin, *Mat. Sci. and Engineering,* 67: 179-184 (1984).

17. F. P. Bowden and D. Tabor, *Friction and Lubrication of Solids*, Oxford Press, 1968.

18. Y. C. Fung, *Foundations of Solid Mechanics*, Prentice Hall, Englewood Cliffs, N. J., 1965.

19. T. Vreeland Jr., P. Kasiraj, A. A. Mutz and N. N. Thadhani, in *Metallurgical Applications of Shock-Wave and High-Strain-Rate Phenomenon*, L. E. Murr, K. P. Staudhammer and M. A. Meyers (eds.), M. Dekker, New York, 1986, p. 231-246.

20. M. N. Pavlovskii, *Soviet Physics - Solid State*, 13: 741-742 (1971).

21. C. A. Brookes, V. R. Howes and A. R. Parry, *Nature*, 332: 139-141 (1988).

22. T. Evans and R. K. Wild, *Phil. Mag.*, 12: 479-486 (1965).

23. A. F. Goncharov, I. N. Makarenko and S. M. Stishov, *JETP*, 96: 670-673 (1989).

24. O. Mishima, L. D. Calvert and E. Whalley, *Nature*, 310: 393-395 (1984).

25. D. R. Schmitt and T. J. Ahrens, *J. Geophys. Res.*, 94: 5851-5871 (1989).

26. R. J. Hemley, A. P. Jephcoat, H. K. Mao, M. C. Ming and M. H. Manghnani, *Nature*, 334: 52 (1988).

27. R. L. Williamson and R. A. Berry, in *Shock Waves in Condensed Matter*, Y. Gupta (eds.), Plenum Press, New York, 1986, p. 341-346.

28. V. D. Linse, in *Metallurgical Applications of Shock-Wave and High-Strain Rate Phenomena*, L. E. Murr, K. P. Staudhammer and M. A. Meyers (eds.), M. Dekker, New York, 1986, p. 29-55.

31

Shock Compaction of Diamond Powder in Reactive Mixtures

H. KUNISHIGE, Y. HORIE* and A. B. SAWAOKA

Center for Ceramics Research, Tokyo Institute of Technology, Yokohama 227, JAPAN
*Department of Civil Engineering, North Carolina State University, Raleigh, N.C. 27695 USA

Shock recovery experiments of diamond powder in reactive mixtures were carried out using a "mouse trap" device, metallographic observations were performed to investigate microstructures of recovered samples. Vicker's hardness measurements were performed on the sample containing 40 vol. % diamond to investigate the extent of chemical reaction. Numerical simulations of the recovery fixture were also conducted using the two dimensional hydrocode named PISCES and a reactive powder mixture model.

I. INTRODUCTION

There have been two serious problems in dynamic compaction of hard-to-sinter ceramics. One is macro- and micro-cracking and the other is the lack of uniformity in recovered samples.

The use of exothermic chemical reaction is supposed to be one of hopeful methods for alleviating these problems. For example, two recent U.S. patents showed the improvement on Vicker's hardness in diamond compacts bonded with SiC which was synthesized from elemental silicon(Si) and graphite(C) under shock loading[1]. But, the mechanisms were not yet understood.

This study was an attempt to gain better understanding of such mechanisms. Shock recovery experiments were performed using a "mouse trap" device on powder mixtures consisting of silicon, graphite and diamond. This diamond composite can be expected to show a high hardness value because shock synthesized SiC was known to be a hard material. Special attention was focused on the effect of diamond content on reactive powder compaction. Scanning electron microscopy (SEM) observations were performed to investigate microstructures. Vicker's hardness measurements were also performed on the sample containing 40% diamond in volume. Vicker's hardness value was expected to be influenced by the amount of reacted components.

   Numerical simulation were also conducted by the two dimensional hydrocode named PISCES. A reactive mixture model was developed based on the mechanical and thermodynamical equilibrium[2]. Calculations were also focused on the effect of diamond content.

II. EXPERIMENTAL PROCEDURE

A schematic of the recovery assembly fixture is shown in Fig. 1. Starting mixed powders are summarized in Table 1.

   The reactive components silicon and graphite were first milled in a vibration mill for two hours. Then diamond powder and the reactive components were hand-mixed in a

Table 1:

Composition of the starting powders.

	Inert species (% Diamond)	Reactive species (% Silicon+Graphite)	Packing Density (% Theoretical)
1	40	60	60
2	50	50	60
3	60	40	60
4	70	30	60

Average particle size	diamond	10 μm
	silicon	0.023 μm
	graphite	0.1 μm

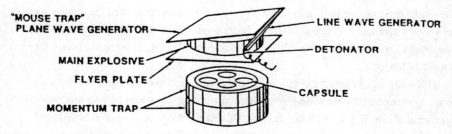

FIG. 1 Schematic of a mouse trap fixture.

FIG. 2 SEM microphotograph of starting powder, diamond content is 40% in volume. Mean particle sizes are diamond 10 micrometer. Silicon 0.023 micrometer and graphite 0.1 micrometer.

mortar with ethyl alcohol in helium atmosphere. Special attention was paid not to scratch the mortar with the hard diamond powder. After mixing, they were heat treated at 400°C for an hour to remove volatile components.

The SEM photographs of starting powder are shown in Fig. 2. Sample powders were pressed into a cylindrical steel capsule in helium atmosphere to prevent the oxidation of fine silicon powder.

Packing density was fixed at 60% in volume. Impact velocity was also fixed at 2.1 km/s.

After shock loading, recovered sample were carefully removed from the capsule using a lathe. Samples were first fixed in regin and halved along the longitudinal plane of symmetry, then polished for hardness measurements. Vicker's microhardness was measured at 500 g weight.

III. EXPERIMENTAL RESULTS

SEM photographs of fractured surface of recovered samples are shown in Figs. 3(a), 3(b) and 3(c). It may be seen that diamond particles of about 10 micrometer diameter were surrounded by a mixture of synthesized SiC and unreacted components. Diamond particles seemed not to be fractured under shock loading. There seems little difference in appearance among these figures in spite of different diamond contents.

Vicker's hardness measurements were performed with the sample containing 40% diamond. The results are shown in Fig. 4.

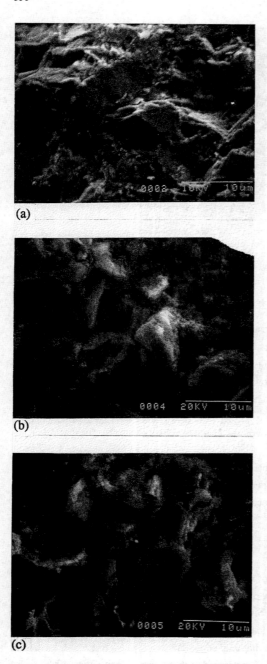

FIG. 3 SEM microphotographs of fractured surfaces of the diamond+Si+graphite compact. Each diamond contents are (a) 40 vol. %, (b) 60 vol. % and (c) 70 vol. %, respectively.

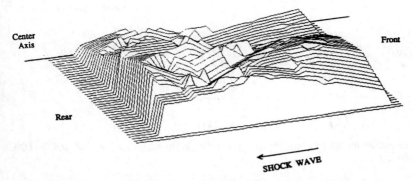

Center
Axis

Front

Rear

SHOCK WAVE

FIG. 4 Vicker's hardness profile in the sample containing of 40 vol. % diamond.

The maximum Vicker's hardness of 2,352 kg/mm^2 was observed in a bottom peripheral region. Low hardness values were observed in a middle center region.

IV. NUMERICAL SIMULATION

To gain better understanding of the shock compaction process involving exothermic chemical reaction, numerical simulations were conducted by the two dimensional hydrocode named PISCES. The powder mixture model was that discussed in the previous paper [2].

Figs. 5(a) and 5(b) show the calculated maximum pressure contours in the samples of 40% diamond content and 80%, respectively. The high pressure building due to the focusing of radial waves discussed in reference [2] is seen in a rear central region for both cases. Also, the higher the diamond content, the higher the maximum pressure.

Figs. 6(a) and 6(b) show the maximum temperature contours in the samples of 40% and 80%. A low temperature region is observed in the middle center of the sample for both cases. These low temperature regions were first compacted by a weak shock wave from the front surface, which result in a low temperature increase [2].

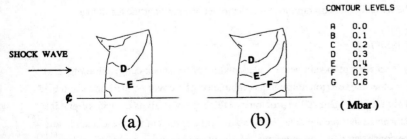

CONTOUR LEVELS

A	0.0
B	0.1
C	0.2
D	0.3
E	0.4
F	0.5
G	0.6

(Mbar)

SHOCK WAVE

(a)   (b)

FIG. 5 The calculated maximum pressure contours are shown in the samples of (a) 40% and (b) 80% diamond content.

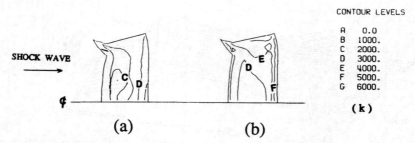

CONTOUR LEVELS

A	0.0
B	1000.
C	2000.
D	3000.
E	4000.
F	5000.
G	6000.

(k)

(a)   (b)

FIG. 6 The maximum temperature contours are shown in the samples of (a) 40% and (b) 80% diamond content.

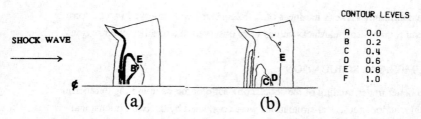

CONTOUR LEVELS

A	0.0
B	0.2
C	0.4
D	0.6
E	0.8
F	1.0

(a)   (b)

FIG. 7 The amount of reacted SiC are shown in the samples of (a) 40% and (b) 80% diamond content.

Figs. 7(a) and 7(b) show the amount of reacted SiC in the sample of 40% and 80% diamond. In the present calculation the chemical reaction is assumed to depend on temperature only. So, a low reacted region is observed in the middle center of the sample, similar to temperature contour.

According to the Vicker's hardness measurements, a low hardness region is observed in the middle center of the sample. This result is consistent with the calculated results, because the less silicon and graphite react, the lower Vicker's hardness indicates.

V. DISCUSSIONS

This study is an attempt to gain better understanding of the shock compaction process involving exothermic reaction, through a combination of recovery experiments and their numerical simulations. One of the problems of this type of study is the lack of precise material data and experimental data. But, experimental results in the present study are found to be consistent with computational results in spite of many assumptions in the model calculation and too few material data.

To make model calculations more accurate, it is necessary to measure and to examine these samples in more detail. But cutting and polishing of samples have proved to be a very difficult task because of their hardness. So, it is necessary to establish a way to treat such hard samples in situ.

REFERENCES

1. A. B. Sawaoka and T. Akashi, "High Density Compacts of High Hardness Materials", U.S. Patent 4,655,830 and 4,695,321 (1987).

2. H. Kunishige et al. Report of the Research Laboratory of Engineering Materials, Tokyo Institute of Technology, No. 15 (1990) pp. 235-263.

32

Method for Determining Pressure Required for Shock Compaction of Powders

A. FERREIRA* and M. A. MEYERS†

*Departamento de Engenharia Mecânica e de Materiais
Instituto Militar de Engenharia
Praça General Tibúrcio no. 80, Rio de Janeiro, RJ, 22290,
Brasil

†Center of Excellence for Advanced Materials
University of California, San Diego
La Jolla, CA 92093, U.S.A.

A model for prediction of the pressure required to shock consolidate a general porous material is presented. The energy required to shock consolidate the material is computed by calculating the various contributions: collapse of voids to densify the material (plastic deformation work), melting of interparticle regions, and generation of defects (dislocations) in shock hardened regions. Based on the total energy involved and by applying the Mie-Grüneisen equation of state to the porous material one can predict the pressure required to consolidate the powder (at a given distension). The general implications of the analysis are presented and discussed. These calculations complement the method developed by Schwarz et al. [1] that predicts the minimum pulse duration for shock consolidation.

I. INTRODUCTION

It is well established that interparticle melting plays an important role in the bonding that occurs between the powder particles in shock compaction of metals. In ceramics, on the other hand, the intense pressure and surface heating might be sufficient to ensure bonding between particles.

Gourdin [2] calculated the energy deposition during shock compaction and concluded that most of it is expended in the melting of interparticle layers.   Both Schwarz et al.   [1] and Gourdin [2,3] developed energy deposition models at the particle surfaces, which predicted melting fractions in shock consolidated materials.   While the models proposed by Schwarz et al. [1] and Gourdin [2,3] are predictions of shock consolidation parameters for soft materials, they do not incorporate the strength effect, that plays an increasingly important role as the strength increases.   Figure 1 shows that there exist a linear relationship between the experimental pressure required for consolidation and the yield strength (Y) of the starting material.   One can note that pressure necessary to consolidate the material increases as yield strength increases, indicating that the strength of the material has significant influence in the consolidation process.

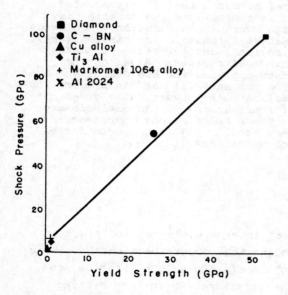

FIG. 1 *Correlation between yield stress and experimental pressure required for shock consolidation.*

II. APPROACH

The energy developed during the shock consolidation process is dissipated under several mechanisms. Although all mechanisms occur simultaneously, Figure 2 schematically separates the different processes taking place during shock consolidation process. As the shock wave progresses, void collapse process occurs and maximum densification is obtained. Bonding among the particles is provided by melting of a thickness (t) in the interparticle region and its subsequent resolidification. At the same time, as the shock wave progresses, defects are being generated in the particle interior by the passage of the shock wave, whose amplitude significantly exceeds the dynamic yield strength. Shock-induced defects include point defects, dislocations, twins, and phase transformations. At the final stage the particle material acquires a residual velocity that is related to the kinetic energy transferred to the material. Figure 3 sche-

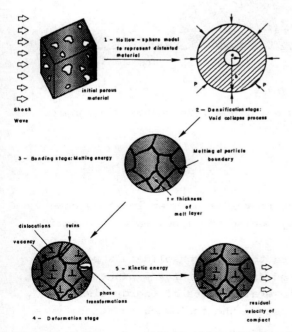

FIG. 2 *Scheme describing the different stages taking place during shock consolidation process.*

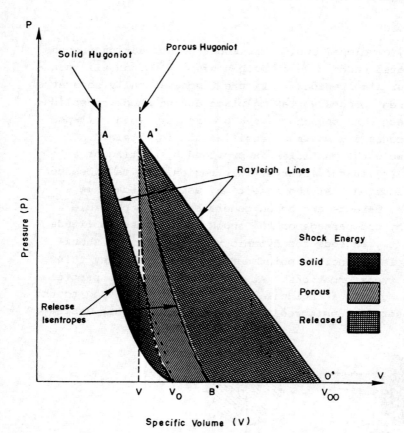

FIG. 3 *Schematic Pressure vs. specific volume diagram for solid and porous materials.*

matically shows the pressure-volume diagram in which compares the specific energy behind the shock wave in a solid and porous materials.

The pressure required to shock consolidate a porous material will be estimated from the energy expended in this process. Based on the several sources of energy dissipation, one can therefore set up the following equation:

$$E_T = E_{v.c} + E_m + E_d$$

(1)

where E_T is the total energy involved in the shock consolidation process, $E_{v.c}$ is the energy necessary to collapse the voids, E_m is the energy due to melting process, E_d is the energy of deformation and is related to shock hardening process.

III. FORMULATION

A. *VOID COLLAPSE ENERGY*

The description of the mechanical response of the porous material will be based on Carroll and Holt's theory [4]. The relative values of the internal radius (a) and of the external radius (b) (Figure 2) define the average distention of the material ($\alpha=b^3/b^3-a^3$). The distention ratio (α) is thus defined [4] as the relation between the total volume (V) and the volume of solid material (V_s). Based on the hollow-sphere model the energy involved in the void collapse process has following expression [4]:

$$E_{v.c} = \frac{2}{3} Y V_s \{ [\alpha_0 \ln\alpha_0 - (\alpha_0-1) \ln(\alpha_0-1)] - [\alpha\ln\alpha - (\alpha-1)\ln(\alpha-1)] \} \quad (2)$$

where α is the final distention, α_0 is the initial distention, Y is the yield strength, V_s is matrix volume, $E_{v.c}$ is the total energy or the change in internal energy that is involved in the densification process. In this calculation it was assumed that the final density of the material after passage of the shock wave is 98% of the theoretical density. This value is in agreement with values described by Gourdin [2].

B. *MELTING ENERGY*

Melting requires the addition of an extra energy. The expression of the required energy necessary to produce a given melting fraction can be written in the following form [1]:

$$E_m = [\overline{C}_p(T_m - T_0) + H_m] L \quad (3)$$

where $\overline{C}_p$ is the average value of specific heat, H_m is the latent heat of fusion, L is the mass fraction melted, T_m is the melting temperature of the solid material, and T_0 is the initial temperature. This previous expression does not take into consideration the raise of the initial temperature due to plastic deformation work that is done when the material is densified during the void collapse process ($\Delta T = E_{v.c}/c_p$). One can write the final expression for melting energy as follows:

$$E_m = [\overline{c}_p(T_m - T_0 - \frac{E_{v.c}}{\overline{c}_p}) + H_m] L \qquad (4)$$

The energy for melting is treated separately from the void collapse energy because a great deal of redundant plastic deformation (jetting, friction) takes place in shock compaction, leading to additional energy deposition. According to Gourdin [2] the thickness of melt layer ranges from 0 to 2.5µm. If one considers that for obtaining a good compaction and mechanical properties the particles surface should have a layer of melting of constant thickness, the melting fraction for monosized spherical powders can be expressed as:

$$L = \frac{V_m}{V_T} = 1 - [\frac{D_p - 2t}{D_p}]^3 \qquad (5)$$

where V_m is the melted volume, V_T is the total volume of the particle, D_p is the particle diameter and t is the thickness of the melt layer. The thickness of melting layer obtained by Gourdin [2] provides a guideline for the fraction of melting needed. In the computations that follow, it will be assumed that metals require an average melt layer of 1.5µm, while ceramics do not require interparticle melting.

C. DEFORMATION ENERGY

The energy associated with dislocations generated by shock wave passage can be estimated from the energy of a dislocation line [5]:

$$E_d = (\frac{Gb^2}{10} + \frac{Gb^2}{4} \ln \frac{\rho_d^{-1/2}}{5}) \frac{\rho_d}{\rho_0} \qquad (6)$$

where G is the shear modulus, b is the Burgers vector, ρ_d is the dislocation density, and ρ_0 is the density of the con-

solidated material. The density of the consolidated material takes part in order to obtain the specific energy.
According to the literature [6,7] one finds that the dislocation density for shock loaded materials in the pressure
range where shock consolidation occurs is approximately
equal to $5 \times 10^{10} cm^{-2}$. This average value will be used in the
calculations.

IV.   RESULTS AND DISCUSSION

The pressure calculated using the previous formulation will
be compared with experimental results. Figure 4 shows the
total energy required for shock consolidation as a function
of the distention. From this plot one can note that for
high strength materials, such as Ti_3Al, the void collapse
energy dominated the process, while for low strength materials, such as Al, In-718, and Markomet 1064 the melting
energy controls the process. The other materials present an
equilibrium between both types of energy.

    With help of the Mie-Grüneisen equation of state for
porous material and the release isentrope curve, the relationship between shock pressure and energy as a function of
distention can be established. The details of these calculations are described elsewhere [8]. The required pressure
for shock compaction of different powder materials was
calculated at several distentions and same particle diameter
(40μm), in which the hardness values range from 1.2 GPa to
98 GPa from these results one can obtain a master plot. The
pressure vs hardness curves were fitted to a best strength
line, as shown in Figure 5. The results show that as distention decreases, higher pressure is required to shock
consolidate the material. It is also shown that at same
melting fraction (same particle diameter) and at same distention, the greater is the strength the greater is the
pressure required for shock compaction. Thus, as a first
approximation, these calculations can be used as a starting
point for prediction of the shock consolidation pressures.
However, it is worth commenting that is a very first initial
idea for prediction of the pressure values. Some simplifi-

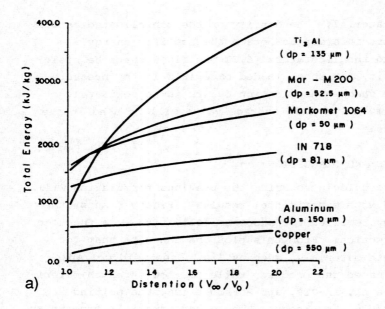

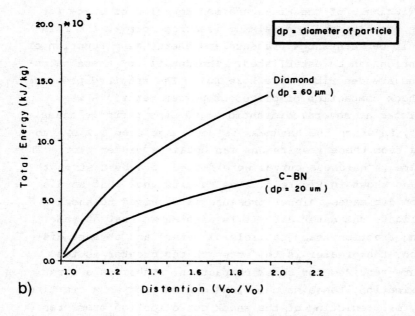

FIG. 4 *Total energy involved during the shock consolidation vs distention for: a) several materials, and b) diamond and C-BN.*

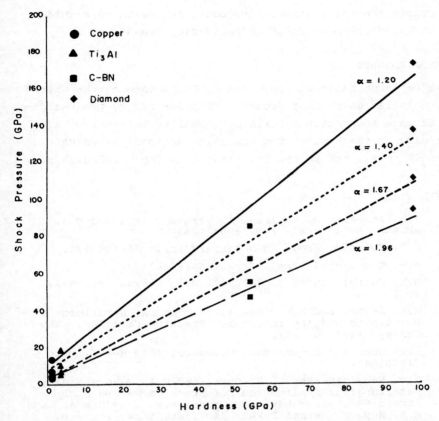

FIG. 5 *Pressure required for shock consolidation vs hardness at several distentions.*

cations were introduced in the calculations in order to become easier to solve a complex phenomena.

V. CONCLUSIONS

The model based on energy deposition predicts shock consolidation pressures that are a function of particle size, powder strength, and distention. The predictions of the model are in fair agreement with experimental results and therefore the calculational procedure can serve as a guide in the prediction of shock consolidation pressures.

ACKNOWLEDGMENT

This research has been supported by DARPA through the United Technologies Government Products Division and by the National Science Foundation Materials Processing Initiative, A. Ferreira is thankful to the Brazilian National Research Council (CNPq) and to the Brazilian Army for a fellowship.

REFERENCES

1.   R.B. Schwarz, P. Kasiraj, T. Vreeland Jr., and T.J. Ahrens, *Acta Metall.: 32*(1984)1243.

2.   W.H. Gourdin, *Prog. in Materials Sci.: 30*(1986)39.

3.   W.H. Gourdin, *J. Appl. Phys.: 55*(1984)172.

4.   M.M. Carroll and A.C. Holt, *J. Appl. Phys.: 43*(1972) 1626.

5.   M.A. Meyers and K.K. Chawla, "Mechanical Metallurgy Principles and Applications," Prentice-Hall Inc., New Jersey, 1984, p. 241.

6.   L.E. Murr and D. Kuhlmann-Wilsdorf, *Acta Metall.: 26* (1978)847.

7.   C.Y. Hsu, K.C. Hsu, L.E. Murr, and M.A. Meyers, in *Metallurgical Applications of Shock Waves and High-Strain Phenomena;* Eds. L.E. Murr, K.P. Staudhammer, and M.A. Meyers; Marcel Dekker Inc., New York, 1986, p. 433.

8.   M.A. Meyers and S.L. Wang, *Acta Metall.: 4*(1988)925.

33

Effect of Internal Gas Pressure on the Shock Consolidation of 304 Stainless Steel Powders

N. E. ELLIOTT and K. P. STAUDHAMMER

Los Alamos National Laboratory
Materials Science and Technology Division
Los Alamos, New Mexico 87545 U.S.A.

Capsules of 304 SS powders having a pre-compacted density of 67% were shock consolidated at peak pressures of 100 GPa. Initial internal N_2 gas pressures from 7×10^{-4} Pa to 0.1 GPa were employed. However, as the internal N_2 gas pressure in the powders was increased, the quality of the compacted density decreased. While it is intuitive that high internal gas pressures in porous materials do not enhance their consolidation, a greater understanding of the consolidation process and the part entrapped (intentional or unintentional) gas plays is elucidated.

I. INTRODUCTION

When powder in a container is consolidated, the initial bulk density dictates the volume change that is required to produce a solid compact. The arrangement of the particles and the consequent distribution of voids between the particles has a major influence on the subsequent behavior of the powder mass. In the consolidation process powder must undergo rapid deformation and flow to eliminate the void space and form strong interparticle bonds. This process of densification and interparticle bonding is very complex and is strongly dependent on any residual (intentional or unintentional) gas in the void pockets. Gas presence in a pre-compacted sample can have a significant effect on the quality of the resulting compacted monoliths. In fact, earlier attempts [1] to shock consolidate metal powders using optimum shock pressures under atmospheric gas pressures in the pre-compacted powder resulted in poor consolidation and/or fractional melting of the powders. Initial solutions to this employed vacuum outgassing of consolidated monoliths [5], and to evacuate the initial compact prior to shock consolidation. To a large extent both methods appeared to work. However, vacuum outgassing after consolidation did not prove to be convenient or practical. On the other hand, pre-evacuating, did produce better results and was thought to be the best obtainable. The vacuum that many investigators were achieving, however, was not a "hard" vacuum, but rather just a reduced internal pressure. The residual pressure resulting from gas trapped in the pores of the metal powders produced variations in the results and interpretation of vacuum prepared specimens. In fact, pre-evacuation of capsules containing samples of metal powders in the size range of 50 to 250 μm at approximately 65% packing density act as "virtual leaks" to a pumping system [7,8]. It is extremely difficult to remove the entrapped gases by pumping on them, even to the extent of a few days. To truly obtain very low internal pressures ($<10^{-3}$ Pa) other methods such as moderate temperature bake out under vacuum must be employed.

The effect of residual internal gas pressure on the shock consolidation process can be demonstrated by the use of fig.1. In fig.1 the Hugoniot compressibility of a porous material is shown generically. This Hugoniot curve is for a porous system having an "ideal vacuum"

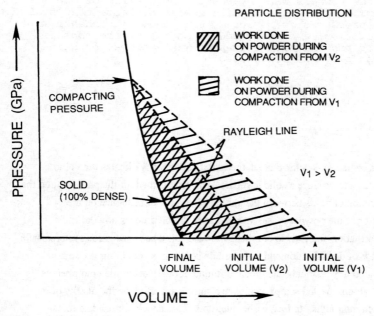

FIG. 1 *Schematic of shock pressure versus volume Hugoniot.*

environment in the pores. Any residual gas environment even at 10^{-1} Pa would shift the release isotrope towards the right (lower compaction) due to the compressibility of the entrapped gas that cannot leave the system. Thus the Hugoniot as shown in fig.1 is only an approximate representation as one approaches very low residual gas pressures. Practically, as we will show with the experiments performed here, residual pressure as low as 10^{-1} Pa has a profound effect on the consolidated monolith, preventing full consolidation. In fact, an understanding of internal gas pressure and its accumulative effect on the consolidation process is essential in the production of optimum monoliths. However, such an understanding is not only applicable to the consolidation process but also among other things, to shock synthesis.

II. EXPERIMENTAL

Samples of Valimet 304SS 170/320 mesh powder were exposed to similar shock conditions. The pre-shock internal gas pressure varied over the range of best obtainable vacuum to roughly 0.1 GPa of N_2 gas. The conditions are summarized in the following table:

Pressure N_2 gas (Pa)	Molar ratio $N_2/304SS$	Starting density	range of shock pressure
7×10^{-4}	10^{-12}	~67% theoretical	10^8 Pa
10^5	10^{-4}	~67% theoretical	to
10^7	10^{-2}	~67% theoretical	4×10^8 Pa
10^8	10^{-1}	~67% theoretical	

The internal gas environments of the samples were produced as follows:

∗7×10^{-4} Pa- The 304SS powder was placed in a 304SS tube and evacuated to roughly 10^{-3} Pa. The system was then backfilled to atmosphere with dry nitrogen gas. This process was repeated 3 times. The sample was then left on the vacuum system for 4 days with periodic external vibration to increase pumping speed in the fine pores of the powder. At this point the sample was valved off from the pumping system and the pressure rise with time noted. After four days of pumping, "virtual leaks" from the porous powder were still being observed. The pressure recorded was not the lowest obtained but an estimate of the equilibrium resulting from the trapped gas. A pinch weld in a brazed copper end piece was made to preserve the vacuum. The sample was explosively compacted roughly one hour after being removed from the pumping system.

∗10^5 Pa- Were prepared similarly to the vacuum sample by backfilling with dry nitrogen from roughly 10^{-3} Pa and pinch welded.

∗10^7 and 10^8 Pa- The samples were placed in commercial 316 stainless steel high pressure tubing and pressurized with dry N_2. Commercial high pressure valves were used to maintain internal pressure until explosive compaction. The design of the assembly allowed the passing shock wave to pinch off the tubing after passage through the sample, separating the valve assembly and leaving the sample for post shock analysis and is shown schematically in fig.2.

Various versions of the shock design have been described previously [8-10]. Post shock analysis was carried out by a variety of techniques. The primary method of analysis was Scanning Electron Microscopy (SEM) supported by optical microscopy. Energy Dispersive Analysis (EDX) was used to confirm chemical analysis. X-ray diffraction (XRD) was used to determine texturing of the powders produced by the shock event. The post shocked samples were cut perpendicular to the central axis at selected distances correlating to specific shock pressures. Metallographic samples were chemically etched by standard techniques prior to analysis.

III. RESULTS AND DISCUSSION

Shown in fig.3 are the overview cross section at equivalent shock pressures of 0.4 GPA for the range of internal gas pressures form $7x10^{-4}$ Pa (fig.4a) to 10^8 Pa (fig.3d). The best consolidation was achieved with the "best vacuum" condition of $7x10^{-4}$ Pa. At 10^5 Pa and 10^7 Pa (fig.3b,c) the resultant structure shows less consolidation, however, reasonably well

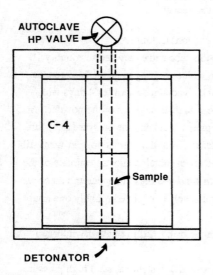

FIG. 2 *Schematic of high internal gas pressure shock system.*

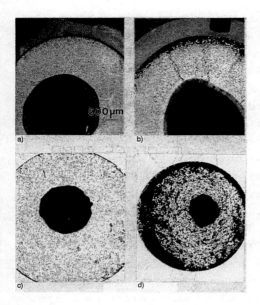

FIG. 3 *Optical micrographs of 0.4 GPa shocked samples as a function of a) 7 X 10^{-4} Pa, b) 10^5 Pa, c) 10^7 Pa, and d) 10^8 Pa internal N_2 gas pressure.*

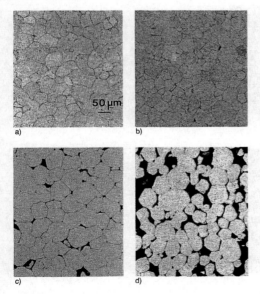

FIG. 4 *Optical micrographs of 0.1 GPa shocked samples as a function of a) 7 X 10^{-4} Pa, b) 10^5 Pa, c) 10^7 Pa, and d) 10^8 Pa internal N_2 gas pressure.*

compacted regions exist within each sample. At 10^8 Pa the structure is very loosely compacted with minimal particle bonding. In fact many of the powder particles were lost, some due to metallographic preparation, but in general most were lost as a result of shock venting of the sample holder. It should be noted that the sample holder for the 10^8 Pa, N_2 pressure (fig.3d) had a smaller internal diameter than those shown in figs.3a-c. This was necessary to accommodate the high internal gas pressure. In comparison to the shock pressure, an internal pressure of 10^8 Pa has a significant cooling effect upon release just after shock wave passage. Thus, melting at the inner core nearest the Mach stem that was present in fig.3b with an initial pressure of 10^5 Pa is not present here. In addition, the vented gas exits with a far greater velocity at the higher pressure and has the capability to carry with it those particles which are not sufficiently anchored in the matrix. Samples consolidated at a lower shock pressure of 0.1 GPa are shown in fig.4. The quality of consolidation is evident for these series of micrographs, as well as, the degree of porosity. Clearly, the effect of internal N_2 gas pressure is particularly evident in 4c and d, where consolidation was not achieved. A more detailed SEM microstructure is shown in fig.5 taken from fig.4d. The regions marked A are backfilled epoxy material used for the metallographic preparation of the samples. The structure of the particles shows little deformation and no melting of the particle surfaces or near interstices of the pore collapse as was evident in portions of fig.4a. The

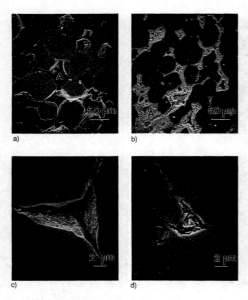

FIG. 5   *SEM backscatter electron images of shocked 10^8 Pa internal N_2 gas pressure as a function of a,c) 0.1 GPa shock pressure and b,d) 0.2 GPa shock pressure.*

obvious lack of melting particularly at the pore sites is shown in fig.5c-d. Even at the higher pressure (fig.5d) melting is not observed. Figure 6 is a schematic representation of the increase in temperature resulting from the shock event. There are actually three major contributions; the adiabatic, residual and strain which are discussed in an earlier paper [12]. The combined effect of these factors is shown as the line labelled "Solid". However, for a porous sample, the temperature rise is much greater due to the collapse in volume and the resulting large strain temperature contribution as depicted in fig.6. Additionally, the temperature increases with shock compression of an enclosed or entrapped gas in the pores. The entrapped gas has different compressibility than the matrix metal powder and upon passage of the shock wave the gas remains behind in a smaller volume, at greater pressure and at a higher energy state. This results in a net increase in the overall temperature. The internal gas pressure decreases the "quality" of consolidation via excess temperature resulting in localized melting and/or cracking of the consolidated compact. At very high internal gas pressures, the gas can have a very detrimental effect and prevent particle to particle bonding even though the adiabatic temperature may exceed melting. This results from the high internal gas pressure preventing particle contact.

An attempt to explain the increase in temperature resulting from shock compression of the higher internal gas pressure can be illustrated in fig.7. Temperature profiles as a function of time, for shocked samples involves a complex combination of events. The initial adiabatic compression causes the temperature of all samples to rise along a similar path. However, after passage of the shock wave the amount of residual strain, non-adiabatic gas heating and heat of reaction cause the sample to cool along different paths as indicated in fig.3. The time

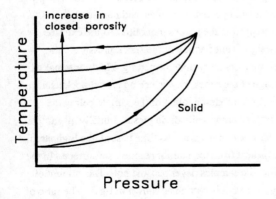

FIG. 6 *Schematic representation of temperature increase with pressure for solid and porous materials. For porous materials the final temperature is strongly dependent on the closed porosity.*

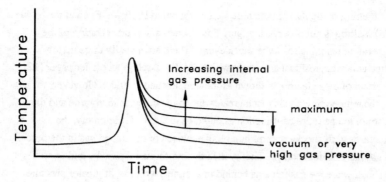

FIG. 7 *Schematic of temperature versus time as a function of internal gas pressure. Illustrating a maximum temperature effect as a function of increasing internal gas pressure.*

scale for these events is much longer than that of the shock wave itself and the specimen temperature is lowered. None the less, it is in this regime where many of the observed events of chemical and metallurgical significance occur.

The relationship of strain and non-adiabatic heating is illustrative of this point. In the case of a hard vacuum, a monolith with no porosity can be obtained. The post shock temperature is a result of heating caused by the deformation of individual particles to fill the interparticle voids. As the pressure of internal trapped gas in collapsing voids increases, the situation changes. At lower starting internal pressures, the individual particles still deform although not to complete void collapse. As the pressure in the collapsing void reaches the shock pressure the internal gas becomes trapped. This results in heating from a combination of local strain in the particles filling collapsing voids, and non-adiabatic heating resulting from compressed gas trapped in the porous monolith. At the extreme where the starting internal pressure is a significant fraction of the shock pressure, little strain is observed. Individual particles are essentially undeformed fig.4d,5d. Internal gas pressure vented immediately after passage of the shock wave since no closed pockets have formed. The rapid drop in pressure results in modest cooling of the sample below the starting conditions. This effect was observed empirically during tests. Initially all samples could be handled immediately after explosive compaction. As time passed, the high strain, low internal pressure samples warmed as heat from the interior reached the surface. In the case of low strain, high internal pressure, the samples never became hot. The amount of this strain was measured more directly by x-ray diffraction as shown in fig.8. The ratio of the intensity of the (220)/(200) reflections are indicative of internal particle strain and are plotted in fig.9. The overall strain in all samples was about 3% as measured by the elongation of the sample holders. This amount would normally not be detected by x-ray

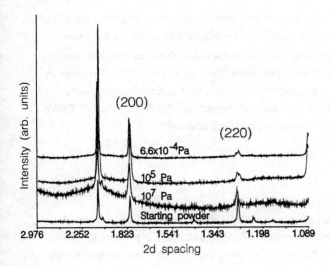

FIG. 8 *X-ray diffraction pattern of the starting powder and shocked (0.4 GPa) samples containing the indicated internal gas pressure.*

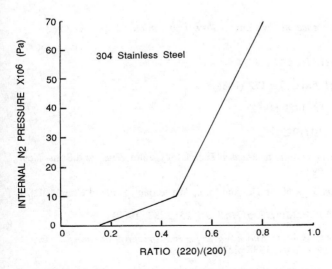

FIG. 9 *Ratio of the (220)/(200) peaks as a function of the internal gas pressure from the data shown in fig.8. The ratio is indicative of the amount of strain in the powder.*

diffraction. The amount of deformation in individual particles is much greater due to void collapse and can be detected. The sample with the highest internal pressure showed essentially no difference from the original starting powder. The sample with the lowest internal pressure exhibited significant preferential texturing as a result of deformation as measured by XRD. The type of texturing produced in the samples with low internal pressure, high strain is similar to that observed for mechanically drawn material. This is consistent with the cylindrical geometry of the shock assembly.

IV. CONCLUSIONS

Consolidation is directly proportional to the initial internal gas environment of powder specimens. We have observed measurable effects with starting gas pressure less than atmospheric. Because of the difficulty in obtaining vacuum in powder specimens, care must be exercised in interpreting results from experiments conducted under nominal but uncertain vacuum conditions. The use of *controlled* gas environments can significantly enhance a variety of shock consolidation processes, in particular, shock synthesis can successfully employ these results.

REFERENCES

1. *Fifth International Conference on High Energy Rate Fabrication*, Chap. 5, Denver, CO, June 1975

2. D. Raybould, *J. Mat. Sci., 16:* 589 (1981)

3. W. H. Gourdin, *J. Appl. Phys., 55:* 172 (1984)

4. D. Morris, *J. Mat. Sci., 17:* 1789 (1982)

5. R. Prümmer, *Vortrag 4:* 10 (1972)

6. John F. O'Hanlon, *A User's Guide to Vacuum Technology,* John Wiley and Sons, New York, NY, Chap.2/3, 1980

7. A. Roth, *Vacuum Technology*, North Holland Pub., Amsterdam, Holland Chap.4, 1979

8. K. P. Staudhammer and L. E. Murr, *J. of Matl. Sci. 25:* 2287 (1990)

9. K. P. Staudhammer and L. E. Murr, *Shock Waves for Industrial Applications*, L. E. Murr ed., Noyes Pub., New York, 1988 p.237.

10. K. P. Staudhammer and K. A. Johnson,*Metallurgical Applications of Shock-Wave and High Strain Rate Phenomena*,L. E. Murr, K. P. Staudhammer and M. A. Meyers eds., Chap.7, Marcel Dekker, New York, 1986 p. 149.

11. K. P. Staudhammer and K. A. Johnson, *Proceedings of the International Symposium on the Intense Dynamic Loading and its Effects*, Z. Zhemin and D. Jing eds., Science Press, Beijing China, 1986

12. K. P Staudhammer, *Shock Compression of Condensed Matter -1989*, S. C. Schmidt, J. N. Johnson and L. W. Davidson eds., Elsevier Science Pub. B.V.,1990, p.519.

ACKNOWLEDGMENTS

The authors wish to thank A. D. Bonner and R. S.Medina for their helpful support and logistics at the firing site, and to A. J. Gray for the metallographic preparation. This work was supported by the DOE under contract W-7405-ENG-36.

34

Dynamic Consolidation of Intermetallics

M. S. VASSILIOU, C.G. RHODES, and M.R. MITCHELL

Rockwell International Science Center
1049 Camino Dos Rios
Thousand Oaks, California 91360

Dynamic Consolidation is a potentially useful process for consolidating advanced intermetallics, because of its ability to retain nonequilibrium microstructures, and to fabricate composites with thin reaction zones. Dynamic consolidation of rapidly-solidified gamma-composition titanium aluminide produced a fully dense and well-consolidated compact which retained the nonequilibrium alpha-2 microstructure of the starting powder. In addition, dynamic consolidation of a titanium-aluminide/silicon carbide powder blend produced a composite material with a reaction zone on the order of 50 nm thick between the titanium-aluminide matrix and the silicon carbide particles. Finally, dynamic consolidation of mechanically-alloyed chromium and silicon powders produced a consolidated material containing equilibrium chromium-silicide, and hence did not retain the nonequilibrium properties of the mechanically-alloyed starting powder.

I. INTRODUCTION

Intermetallic compounds are attractive for application as structural materials. Their use, however, is limited because of the difficulty of fabricating components. The characteristic that makes these materials attractive for many structural applications, i.e., high-temperature strength, is counterbalanced by the fact that both primary and secondary processing temperatures are raised inordinately high. In addition to making fabrication more difficult, high processing temperatures applied for long times can lead to the loss of nonequilibrium properties which might be desirable for some applications. They can also lead to the formation of detrimental reaction products between the matrix and the reinforcing material in composites. One approach to reducing these problems is to employ dynamic (shockwave) consolidation. This is a very rapid process, which does not expose the bulk of the powder to high temperatures [1]. Recently, the pace of research on dynamic consolidation of intermetallics has quickened (e.g. [2], [3]).

II. DYNAMIC CONSOLIDATION OF γ TITANIUM ALUMINIDE WITH A NONEQUILIBRIUM MICROSTRUCTURE

The starting powder of γ Ti-48 Al was prepared by Nuclear Metals, Inc., using the Plasma Rotating Electrode Process (PREP). This rapid solidification process produced a nonequilibrium α-2 (DO_{19}) microstructure. Fig. 1a shows the nonequilibrium fine dendritic single-phase α-2 structure in the starting powder. Fig. 1b shows what happens to this structure after hot-pressing for 4 hours at 1100C and 68.95 MPa. The equilibrium lamellar mixture of α-2 (10-15 vol %) and γ (85-90 vol %) phases is produced, and the nonequilibrium microstructure is lost.

In the dynamic consolidation experiment, the starting powder (14.19 g) was placed in an evacuated steel container, which was covered with a steel driver plate 2.54 mm thick and placed in a larger sample assembly wherein it was surrounded on the sides by lateral momentum traps, and underlain by several steel spall plates. The experimental design was guided by the discussion in [4]. The porosity of the starting powder was 41.3%. The sample assembly was impacted by a stainless steel flyer plate 6.4 mm. thick, launched by a C-4 explosive charge. The impact velocity was estimated to be 1.0 km/sec, and the initial impact pressure was calculated to be 5.2 GPa. As can be seen in fig. 1c, the resulting dynamically consolidated powder has a dendritic microstructure, similar to that of the nonequilibrium starting powder (the nonequilibrium α-2 structure was confirmed by electron diffraction). These results have been described more fully in a previous paper [3].

III. DYNAMIC CONSOLIDATION OF TITANIUM ALUMINIDE/SiC COMPOSITE

One theory regarding the mechanism by which a brittle reaction zone can reduce tensile strength in a continuous fiber-reinforced metal-matrix composite has been put forward by Ochiai and Murakami [5]. Their treatment assumes a strong bond between fiber and reaction zone and early fracture of the reaction zone on applying a tensile stress. The crack in the

FIG. 1 *(a) Starting powder of Ti-48Al, showing nonequilibrium fine dendritic single-phase α-2 structure. (b) Starting powder after hot-pressing. Nonequilibrium structure has been transformed to equilibrium lamellar structure of 10-15 vol % α-2, and 85-90 vol % γ. (c) Starting powder after dynamic consolidation. Nonequilibrium microstructure, similar to that in (a), has been retained.*

reaction zone then acts as a notch in the fiber. From this, they developed an expression for the maximum reaction zone thickness, below which a crack in the reaction zone will not act as a notch:

$$C = (G^* / \pi\epsilon_{fu}^2 E_f) / \{F(a/b)^2\} \tag{1}$$

where, for the fiber, G^* = critical strain energy release rate, ϵ_{fu} = fracture strain, and E_f = Young's modulus. $F(a/b)$ is a function of a, the fiber radius, and b, the total thickness of fiber plus reaction zone. It can be seen that the critical reaction zone thickness is a function of the actual reaction zone thickness. The critical reaction zone thickness vs. actual reaction zone thickness for SiC fibers is plotted in Fig. 2. The important features are (1) that the critical reaction zone thickness is less than one micron for an actual reaction zone thickness up to 5 microns, and (2) that the "safe" thickness, i.e. where the actual reaction zone thickness is less than the critical thickness, is less than one micron.

Fig. 3a shows a typical SiC fiber-reinforced titanium aluminide (Ti-25Al-11Nb) consolidated by hot pressing. The reaction zone, which consists largely of titanium carbide close to the fiber, titanium silicide close to the matrix, and titanium aluminum carbide in between, is one to two microns thick. It is not feasible, then, to use conventional fabrication techniques to consolidate a SiC fiber-reinforced titanium aluminide composite having a "safe" reaction zone thickness. Conventional fabrication techniques would probably not be effective for consolidating SiC particle-reinforced titanium aluminide composites either, since the reaction could literally consume SiC particles.

A blended mixture of Ti-25Al-5Nb powder with 35 vol % SiC particulates and an initial porosity of 22.5% was consolidated by means of a shock wave generated by the impact from a gun-launched projectile. The powders were encapsulated in a stainless steel container. The sample assembly was impacted at a velocity of 1.27 km/s by a projectile carrying a stainless steel flyer plate. The resulting composite consisted of a uniform dispersion of SiC in the

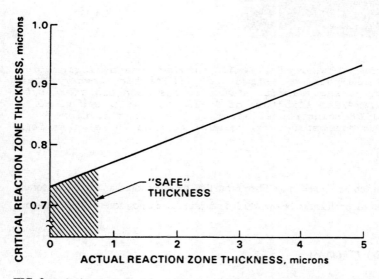

FIG. 2 *Critical Reaction Zone thickness for SiC-reinforced titanium aluminides. In "safe" region, a crack in the reaction zone will not act as a notch in the fiber.*

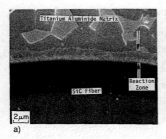
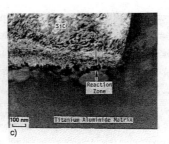
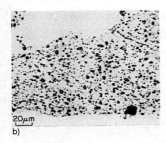

FIG. 3 *(a) SiC-reinforced Ti-24Al-11Nb, consolidated by hot pressing. Reaction zone between titanium-aluminide matrix and SiC fiber is 1 micron or greater. (b) Dynamically consolidated SiC-reinforced Ti-24Al-5Nb (c) Dynamically consolidated SiC-reinforced Ti-24Al-5Nb. Reaction zone is about 50 nm thick.*

titanium aluminide matrix (fig. 3b). Fig. 3c shows that the reaction zone in the dynamically consolidated composite is on the order of 50 nm thick, compared to 1000-2000 nm in the conventionally consolidated composite of fig. 3a. We have not identified the exact composition of the reaction zone, but it probably consists of titanium silicide(s). In any case, 50 nm is well below the critical thickness predicted by Eq. (1). Dynamic consolidation clearly offers the potential for fabricating titanium aluminide composites without tensile property degradation.

IV. DYNAMIC CONSOLIDATION OF MECHANICALLY-ALLOYED CHROMIUM-SILICON

One of the intermetallic compounds being studied for potential high-temperature structural applications is Cr_3Si. To avoid the need for high-temperature melting and processing equipment, mechanical alloying is being studied as a technique for producing Cr_3Si powder [6]. The mechanically alloyed powder can subsequently be vacuum-hot-pressed by conventional methods to produce a fully dense solid.

The process of mechanical alloying consists of ball-milling elemental powders in an inert atmosphere. During the process the dissimilar powders cold weld together. After sufficient milling time, individual powder particles have the composition of the powder blend (in this case, Cr_3Si). The process can produce amorphous metallic alloy powders [7-9], although in the case described here, X-ray diffraction patterns indicate elemental Cr and Si after long-time milling.

Examination of the mechanically alloyed powders reveals large particles that are clearly two-phase, and smaller particles that appear to be single-phase yet are composed of both Si and Cr (Fig. 4a). Transmission electron microscopy (TEM) of these powders confirmed that the larger particles had large zones of elemental Cr and large zones of elemental Si. The smaller particles contained Cr and Si in roughly the Cr_3Si composition, but X-Ray diffraction showed no equilibrium Cr_3Si to be present. Not surprisingly, the as-mechanically-alloyed powders were heavily deformed (fig. 4b). The presence of the larger two-phase particles indicates that the mechanical alloying process was not perfect (The larger particles were present even after 300 hr of milling [6]).

The conventional consolidation procedure for mechanically-alloyed powders is vacuum hot pressing at 1100C and 48.27 MPa for 2 hours. This conventional processing resulted in a two-phased structure consisting of Cr particles in a Cr_3Si matrix (fig. 5a). The high temperature promoted formation of the equilibrium silicide phase that did not exist in the starting powders.

Dynamic consolidation is an alternative fabrication technique that requires no high-temperature processing. It thus has the potential to retain the non-equilibrium starting

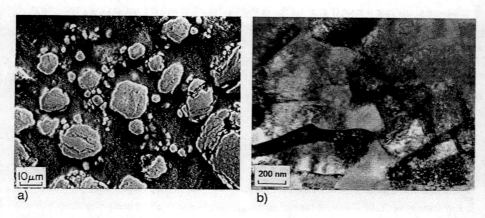

a)   b)

FIG. 4 *(a) Ball-milled Cr and Si powders. Larger particles are two-phased, smaller particles are single-phased. (b) TEM photograph of ball-milled Cr particle showing heavy deformation.*

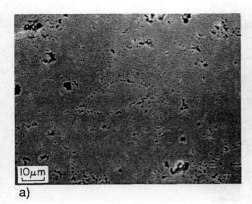 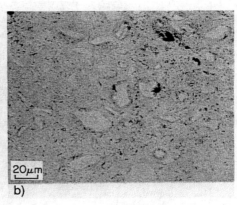

a)   b)

FIG. 5 *(a) Vacuum-hot-pressed mechanically-alloyed* Cr_3Si. *Large grains are pure Cr, matrix is* Cr_3Si. *(b) Mechanically-alloyed* Cr_3Si *after dynamic consolidation. Large grains are pure Cr, matrix is* Cr_3Si.

powder structure, whether that structure be one of mixed elemental powders or amorphous single-phase powders. We have used dynamic consolidation to test the possibility of retaining the starting structure in Cr_3Si. The starting powder (21.71 g) was placed in an evacuated steel container, which was covered with a steel driver plate 2.54 mm thick and placed in a larger sample assembly, wherein it was surrounded on the sides by lateral momentum traps and underlain by several steel spall plates. The porosity of the starting powder was 40%. The assembly was impacted by a stainless steel flyer plate 1.93 mm. thick, launched by a C-4 explosive charge. The impact velocity was calculated to be 1.5 km/sec.

The densified solid had a microstructure similar to that observed in the vacuum hot-pressed material: large particles contained in a fine grained matrix (fig. 5b). Thin-foil TEM reveals that the matrix, which has a grain size on the order of 500 nm, is Cr_3Si that is slightly Si-rich (fig. 6). The large particles are Cr and contain no Si. There are numerous small voids, or bubbles, in both the Cr particles and the Cr_3Si grains. These may be the result of entrapped Ar (from the ball milling operation) or incomplete densification.

There is no evidence of deformation in the fine-grained matrix or in the larger Cr grains, which is somewhat unusual for shockwave consolidation of powders [10]. The starting powders had a high dislocation density (fig. 4b). Thus, not only did the consolidation process fail to introduce deformation-- it actually removed existing deformation. The most logical explanation of this phenomenon is that there has been excessive heating during shock-wave passage and/or very slow cooling following the shock. The presence of large amounts of equilibrium Cr_3Si supports this concept.

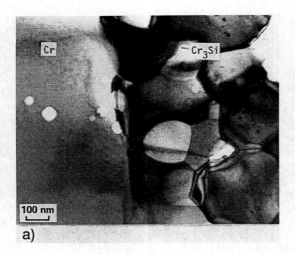

a)

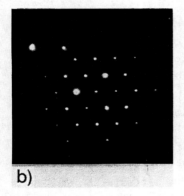

b)

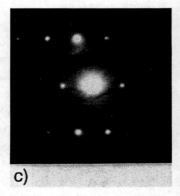

c)

FIG. 6 *TEM of dynamically consolidated Cr_3Si. (a) Boundary region between Cr (large grain) and Cr_3Si (small grains). (b) Convergent beam diffraction pattern confirming Cr_3Si structure, [111] zone. (c) selected area diffraction pattern confirming Cr structure, [311] zone.*

The experiments reported here were unsuccessful in retaining the as-ball-milled structure. This does not, however, preclude the use of dynamic consolidation for retaining non-equilibrium Cr_3Si structures, such as amorphous phases. There are several experimental variables that can be adjusted to optimize the shock-wave impact on the powders. Studies should continue towards that optimization.

CONCLUSIONS

Overall, dynamic consolidation may be a viable method for consolidating advanced intermetallics with a good range of properties. Research must continue into improving the basic process, so that products are uniform and crack free.

ACKNOWLEDGMENTS

We are grateful to Dr. Thomas J. Ahrens for the use of his facility to conduct the projectile impact experiment of Section III, and for helpful discussions. We are also grateful to Drs. J.A. Graves of Howmet Corp. and C.C. Bampton of the Science Center who supplied the mechanically alloyed powder of Section IV. Mr. E. Lockwood and Mr. John Gray of Rocketdyne provided invaluable technical assistance in performing the explosive experiments of Sections II and IV.

REFERENCES

1. R.R. Schwarz, P. Kasiraj, T. Vreeland, and T.J. Ahrens, *Acta Met.*, 32: 1243 (1984).

2. L.H. Yu, M. A. Meyers, and N. N. Thadhani, *J. Mat. Res.* 5: 302 (1990).

3. M.S. Vassiliou, C.G. Rhodes, M.R. Mitchell, and J. Graves *Scripta Met.* 23: 1791 (1989).

4. P. S. DeCarli and M. A. Meyers, "Design of Uniaxial Strain Shock Recovery Experiments," in *Shock Waves and High-Strain-Rate Phenomena in Metals, Concepts and Applications*, M. A. Meyers and L. E. Murr (eds.), Plenum Press, NY, 1981, Chap. 22, p. 341.

5. S. Ochiai and Y. Murakami, *J. Mat. Sci.*, 14: 831 (1979).

6. C.C. Bampton, C.G. Rhodes, J. A. Graves, M.R. Mitchell, and M. S. Vassiliou, "Synthesis of Chromium Silicide by Mechanical Alloying," in *Proc. ASM Conference on Structural Applications of Mechanical Alloying*, ASM, 1990.

7. R.L. White and W.D. Nix, *New Developments and Applications in Composites*, D. Kuhlman-Wilsdorf and W.C. Harrigan, eds., TMS, Warrendale, PA, 1979, p 78.

8. R.B. Schwarz, R.R. Petrich, and C.K. Saw, *J. Non-Cryst. Solids*, 76: 281 (1985).

9. C.C. Koch, O.B. Cavin, C.G. McKamey, and J.O. Scarborough, *Appl. Phys. Let.*, 43: 1017 (1983).

10. L.E. Murr, A.W. Hare, and N.G. Eror, *Adv. Mat. and Proc.*, 132: 36 (1987).

35

Shock Densification/Hot Isostatic Pressing of Titanium Aluminide

SHI-SHYAN SHANG and MARC A. MEYERS

Materials Science Program
University of California, San Diego
La Jolla , CA. 92093 , U.S.A.

Consolidation of rapidly solidified titanium aluminide powders (Ti₃Al) employing explosive shock pressure followed by hipping was carried out successfully. Shock densification was achieved by using a double tube design in which the flyer tube was explosively accelerated, impacting the powder container. HIP-activated reactions were used to chemically induce bonding between Ti₃Al particles. Elemental mixtures of Ti(15wt%) and Al(15wt%) powders were added to intermetallic compound powders (Ti₃Al). The highly exothermic reactions were activated by hipping at 1000 °C and enhanced the bonding between the inert intermetallic powders. Successful consolidation was obtained. Compression test indicated strong bonding between Ti₃Al particles. Well bonded Ti₃Al compacts having average ultimate compressive strength of 2.0 GPa and compressive fracture strain 20.38% were produced by this technique.

I. INTRODUCTION

Titanium aluminide (Ti_3Al) intermetallics have properties which make them desirable candidates for applications in aircraft turbine engines [1-3]. These ordered intermetallic compounds are less dense and stiffer than conventional titanium alloys. However, a main problem with titanium aluminides is their low ductility at room temperature, making them difficult to fabricate and damage-intolerant. Shock consolidation of rapidly solidified Ti_3Al (RSP) powder is a process that has considerable potential. Shock-densification processing is an attractive alternative for RSP materials since long term thermal exposures are not essential for processing. However, it has been widely reported that there remain two unsolved problems in the shock compaction technique [4]. One is cracking of the compacts at both the microscopic and macroscopic levels. The other is a lack of uniformity in microstructure and mechanical properties within resulting compacts. At three recent workshops held in the US [5], the Soviet Union [6] , and Japan [7], cracking was identified as a major unresolved problem. These two problems tend to increase with an increase in the shock pressure used. In order to alleviate these two problems, low shock pressures are desirable. The objectives of this investigation were (1) to eliminate or minimize cracks in resulting compacts; and (2) to use the heat generated from exothermic reaction to help consolidation. Thus, RSP Ti_3Al powders are compacted to pressures as low as possible to just ensure densification of the powder. Then, the highly exothermic reactions of elemental powders (Ti and Al) added to the intermetallics were activated during the hipping stage. The heat produced is dissipated on the surfaces of inert powders and the reacted elemental powders act as a cement to enhance bonding between the powders.

II. EXPERIMENTAL PROCEDURES

A schematic of the processes of shock densification and of HIP-induced chemical reaction to help consolidation is illustrated in Fig. 1. For shock densification, the cylindrical axisymmetric double-tube system was used in this study. A detailed description of the system is presented elsewhere [8]. Commercially available ANFO (Ammonium Nitrate-Fuel Oil), modified to produce a lower detonation velocity, was used as the explosive. The detonation

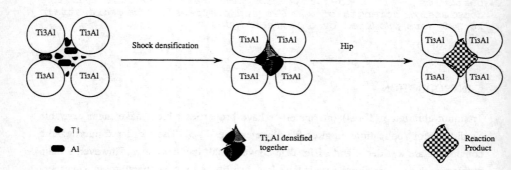

FIG. 1 *Schematic illustration of shock-densification and hip-induced reaction consolidation.*

velocity was approximately equal to 2200 m/s. The shock pressure for 65% packing density was calculated to be 9.5 GPa. The explosive was initiated at the top and caused the implosion of the tube containing the powder, as the detonation front ran downward. Well densified $Ti_3Al+Ti+Al$ compacts were machined for subsequent hipping and annealing. Hot isostatic pressing (HIP) was conducted at Degussa Electronics Inc. at $600^\circ C$ and $1000^\circ C$. The pressure, 200 MPa, was used for one hour. Specimens were encapsulated in quartz under vacuum and annealed to different temperatures, ranging from $600^\circ C$ to $1000^\circ C$. Both original and consolidated compacts were characterized by optical and scanning electron microscopy, X-ray diffraction, and microhardness and compression tests.

III. RESULTS AND DISCUSSION

A. MATERIAL CHARACTERIZATION

The morphology of the Ti_3Al powder is almost perfectly spherical. The composition of these powders is actually quite different from stoichiometric Ti_3Al: Ti- 21wt%Nb- 14%wtAl- 1%wtEr. The Ti_3Al powders were produced through rapid solidification processing (RSP) and were supplied by United Technologies Government Products Division. In rapidly solidified Ti-Al-Nb alloys, the high temperature phase can be quenched to room temperature

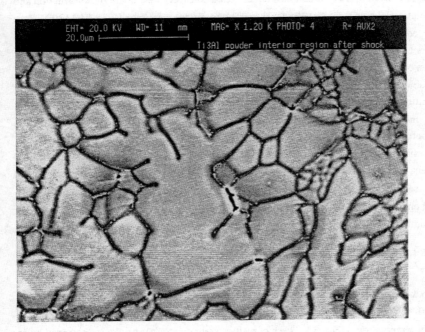

FIG. 2 *Scanning electron micrograph of Ti_3Al powder cross-section which possessed of microcellular structure.*

under sufficient cooling rate [9]. However, the β phase transforms to martensite (α') which can eventually transform to the ordered α_2 phase [10]. Addition of β-stabilizing elements, such as niobium, deppresses the M_s temperature. The etched cross-section of the original Ti_3Al powder reveals a microcellular structure (Fig. 2). Fine round white precipitates uniformly decorating the cell are observed. These precipitates are believed to be Er_2O_3. Minor additions of Er_2O_3 particles were made to obtain a homogeneous of fine particles in the matrix. The Ti-Al-Nb alloys strengthened by erbia (Er_2O_3) have been the object of considerable study.

B. *SHOCK DENSIFICATION EXPERIMENTS WITH CYLINDRICAL GEOMETRY*

Earlier work by Ferreira et al. [11,12] showed that excessive detonation pressure produces overcompaction and results in flaws and cracks. Thus, a low detonation velocity (2200 m/s) for $Ti_3Al+Ti+Al$ was chosen in order to minimize cracking. Successful densification was obtained using this explosive with low detonation velocity, as compared to the compacts produced using explosive with higher detonation velocity (3500 m/s). The explosive Chapman-Jouguet pressure is directly proportional to the square of the detonation velocity. The crack density was reduced by approximately 40%. SEM analysis (Fig. 3) of recovered compacts revealed that Ti_3Al particles underwent a considerable amount of deformation and that the material was fully densified. Ti and Al X-ray dot mappings [Figs. 3(a), 3(b)] were performed in order to identify unreacted regions. The white irregular particulates correspond to unreacted Ti and the dark irregular particulates correspond to unreacted Al, while the small amounts of grey particulates are the reaction products of Ti and Al. These show particles that were subjected to shock densification with weak bonding. (insufficient pressure for jetting and melting between particles and for extensive reaction). The interior of Ti_3Al particle still retains microcellular structure. Transmission electron microscopy (TEM) revealed the substructural features of the shock-densified alloys. Fig. 4. shows fine grain size (approximately 500 nm) and dislocations in high densities that are typical of shocked materials. This region is characteristic of the interiors of the powders. Er_2O_3 dispersoids are clearly seen. Their size and spacing is such that they are effective high temperature strengtheners. There seems to be a tendency for them to arrange themselves in rows (along cell boundaries).

C. *CHEMICALLY-INDUCED ANNEALING AND HOT ISOSTATIC CONSOLIDATION OF Ti_3Al*

The shock densified Ti_3Al compacts were annealed to different temperatures ranging from 600 °C-1000 °C. The unreacted regions (Ti-Al) and porosities were observed in specimens at different temperatures (Fig. 5.). These porosities were due to (a) shrinkages occurred when Ti and Al reacted (b) trapped gases causing expansion of voids. Consequently, hot isostatic

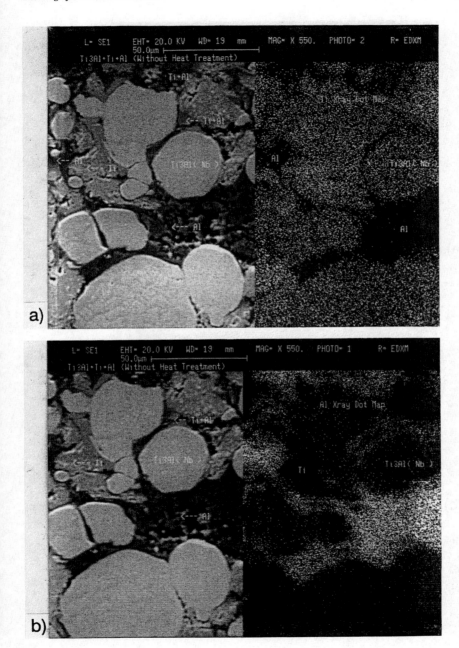

FIG. 3 *(a) Scanning electron micrograph of shock-densified Ti₃Al + Ti + Al alloy and Ti X-ray dot mapping, and (b) Al X-ray dot mapping.*

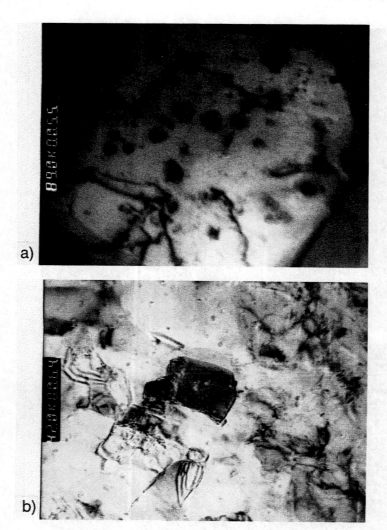

FIG. 4 *Bright field transmission electron micrograph from shock-densified Ti₃Al + Ti + Al alloy showing: (a) grain size approximately 500 nm, and (b) presence of dislocations and Er₂O₃ precipitates.*

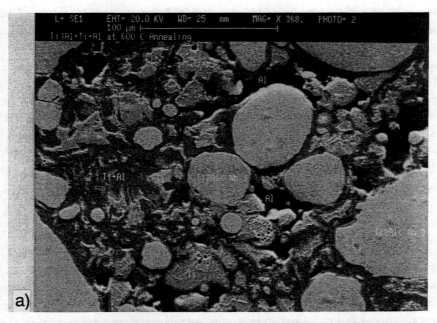

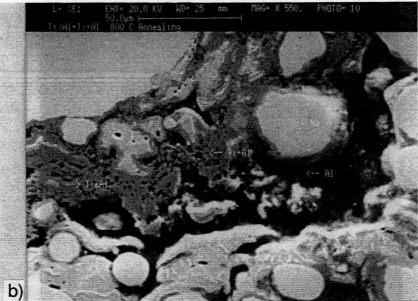

FIG. 5 *Scanning electron micrograph from shock-densified Ti₃Al + Ti + Al alloy (a) annealed at 600˚C (b) annealed at 800˚C in which the different regions were identified by EDS analysis.*

pressing (HIP) was conducted at 600 °C and 1000 °C for one hour. SEM analysis of the material recovered from hipping at 600 °C revealed that Ti and Al powders reacted with each other [Fig. 6(a)]. The dark spheres correspond to the voids and the greyish particulates correspond to the reaction products of Ti and Al. Analysis of the material recovered after hipping at 1000 °C [Fig. 6(b)] revealed that the reaction products bonded the Ti_3Al powders very well, as compared to the compacts produced using hipping at 600 °C. Fig. 7(a) and 7(b) show the microstructures of the Ti_3Al compacts which were hipped at 600 °C and 1000 °C respectively. EDS analysis [Fig. 7(b)] revealed that white regions in the particles are Nb rich , and the black regions are Nb lean. These black regions are believed to be α_2 phase,while white regions are believed to be β phase (Nb rich regions). One can also see stringers of Er_2O_3 dispersions that mark the boundaries of the previous cells.

D. COMPRESSIVE PROPERTIES OF Ti_3Al ALLOYS

Compression tests were carried out in the MTS 810 testing machine at a strain rate of 3.5 × 10^{-4} s^{-1}. The specimens were cylindrical, with a diameter of 6mm and a length of 12mm.The compressive data are plotted in Fig. 8. The data are from shock-densified specimens which were hipped or annealed at various temperature. The data demonstrate that Ti_3Al exhibits an apparent increase in ductility with increasing the temperature of annealing or hipping. The lower ultimate compressive stress and ductility of annealed specimens, compared to the 1000 °C hipped specimen, could be due to three reasons: (1) Ti and Al were not totally reacted in annealed specimens (2) shrinkage occurred when Ti and Al reacted ,with the production of voids (3) phase transformation. Hipping at 600 °C resulted in the α_2 structure and this structure exhibited a high yield stress but low ductility.

E. MICROHARDNESS OF Ti_3Al ALLOYS

The microhardness of the Ti_3Al powders before and after annealing or hipping was measured to establish the effects of subsequent processing. The microhardness of interparticle melted and resolidified regions was also measured in order to determine whether the interparticle regions had a similar rapidly solidified structure as that of the original powder. A load of 200 gf for 15 seconds was applied to determine the microhardness. The microhardness values of the Ti_3Al particles in as-received condition were determined as 347 ± 17 VHN. Fig. 9 shows the variation of microhardness as a function of processing and temperature. The Ti_3Al particles , after shock densification , have greater microhardness value than the original powders (340VHN); so , the shock hardening effect is confirmed. The microhardness values of Ti_3Al particles increased after annealing at 500 °C due to cell refinement of β phase. As the temperature of processing is raised above 500 °C , the microhardness of particle interiors decreases in value, due to β phase transformation to α_2 phase and annealing of the shock-

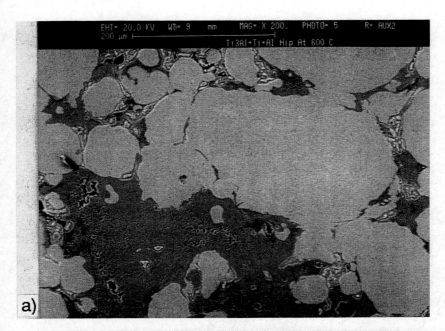

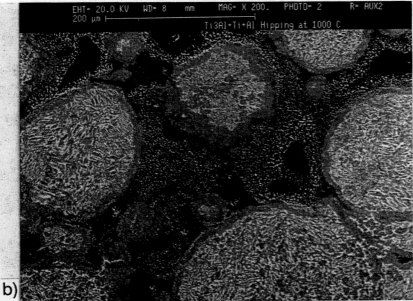

FIG. 6 *Scanning electron micrograph from shock-densified Ti₃Al + Ti + Al alloy (a) hipped at 600°C (b) hipped at 1000°C in which the different regions were identified by EDS analysis.*

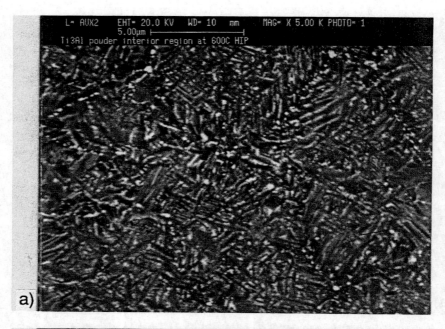

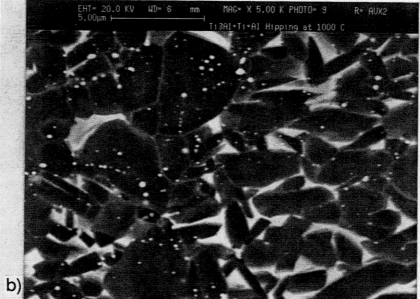

FIG. 7 *Scanning electron micrograph showing the microstructures of the Ti₃Al powder (a) hipped at 600°C (b) hipped at 1000°C.*

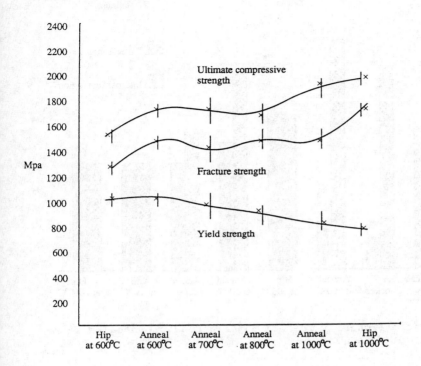

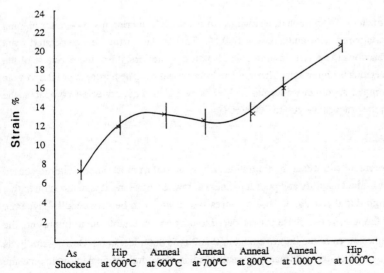

FIG. 8 *Compressive properties of Ti₃Al hipped or annealed from 600°C to 1000°C (a) Yield strength, ultimate compressive strength, and fracture strength (b) Maximum compressive strain.*

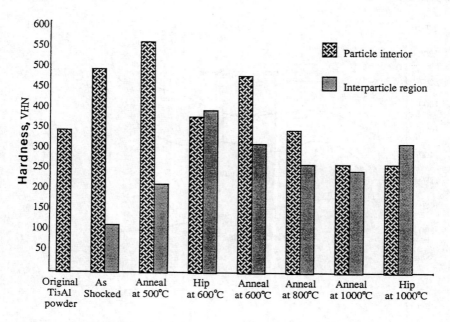

FIG. 9 *Microhardness values of Ti₃Al hipped or annealed from 600˚C to 1000˚C.*

induced substructure. The microhardness values of the particle interior are lower after hipping at 600 ^{0}C , compared with annealing at 600 ^{0}C. That is due to the heat generated from exothermic reactions (Ti + Al) flowing towards particle interiors. Also, the hardness of the interparticle regions is higher after hipping than after annealing. The hardness of interparticle regions for hipped specimens exceeds 300 VHN while in most annealed compacts, the hardness is in the order range 200 to 300 VHN.

IV. CONCLUSION

The generic problem with titanium aluminide alloys, as well as most other aluminide based alloys, is their limited ductility and poor toughness at low temperature. It is demonstrated that Ti₃Al with high ductility (about 20% compressive strain) can be produced by dynamic compaction / hipping. The β phase is considered to play an important role in improving the ductility, but the source of the improvement of ductility is not understood. One possibility is that the improved ductility is primarily a result of microstructural refinement which has the effect of reducing slip length. This would reduce dislocation pile-up length caused by the planarity of slip in the α_2 phase and subsequently require more deformation to increase localized stresses to a level at which crack initiation can occur, thus prolonging the onset of fracture [13]. This is just one microstructural feature of many which can affect flow and fracture behavior. The morphology, volume fraction, and chemistries of the α_2 and β phase, all of which influence flow and fracture behavior, are also affected by processing and heat treatment in multiphase titanium aluminide alloys.

ACKNOWLEDGMENTS

This research was sponsored by the National Science Foundation Materials Processing Initiative Award No. DMR 8713258. The authors would like to thank Dr. Arnaldo Ferreira, Dr. Kenneth Vecchio,Dr. N.N. Thadhani, Mr. Li-Hsing Yu, Dr. Soon-Nam Chang and Mr. Jerry Lasalvia for their assistance during the work. The shock densification experiments were conducted by Expolsive Fabricators Inc. , in Louisville , Colorado. The help provided by Mr. W. Sharpe, President, is gratefully acknowledged. The use of the facilities of the Center of Excellence for Advanced Materials is gratefully acknowledged.

REFERENCES

1. D. Shechtman, M. J. Blackburn, and H. A. Lipsitt; *Met. Trans.* 5: 1373 (1974).

2. H. A. Lipsitt, D. Shechtman, and R. E. Schafrik; *Met. Trans.* 11A: 1369 (1980)

3. I. I. Kornilov and T. T. Nartova; *Dokl. Adad, Nauk SSSR.* 114: 829 (1961)

4. T. Akashi and A. B. Sawaoka; *Adv. Ceram. Mater.* 3: 288 (1988)

5. Proc. First Workshop on Industrial Applications of Shock Processing of Powders, CETR , New Mexico Institute of Mining and Technology, Socorro, NM, June 1-3 , 1988.

6. Proc. Seminar on High Energy Rate Working of Rapidly Solidified Materials, Novosibirsk, USSR , 10-14 October 1988.

7. A. Sawaoka, Proc. Second Workshop on Industrial Applications of Shock Processing of Materials , Tokyo Institute of Technology, Japan, December,1988

8. M. A. Meyers and S. L. Wang; *Acta Metall.* 4: 925 (1988)

9. E. M. Schulson; *Int. J. Powder Metall.* 23(1): 25 (1987)

10. M. J. Blackburn; *Trans. Metall. Soc.* 239: 1200 (1967)

11. A. Ferreira; Ph.D. Thesis, New Mexico Institute of Mining and Technology, Socorro , New Mexico , 1989.

12. A. Ferreira, M. Meyers and N. N. Thadhani; *Met. Trans.* , in press (1990)

13. D. A. Lukasak; M.Sci Thesis , Penn. State University,(1988)

36

Computer Simulations of Laser Shock Compaction of Powders

J. P. ROMAIN and D. ZAGOURI

Laboratoire d'Energétique et de Détonique - URA - CNRS 193
ENSMA, Rue Guillaume VII
86034 POITIERS Cedex , France.

Computer simulations are used to study the propagation of laser-induced shock-waves in a porous medium, and compared with experimental results. The simulations reveal a complex wave structure and a rapid decrease of shock amplitude, due to the short duration of the laser pulse. Experiments show that laser shock loaded aluminum powders are compacted on about 200 μm thickness under the irradiated surface. This thickness is related with the distance of shock propagation without attenuation of pressure. Experimental and numerical results confirm that laser-induced shocks provide an efficient technique for surface densification of porous materials.

I. INTRODUCTION

Dynamic consolidation of powders is achieved by shock-waves, usually generated by explosives or by projectile impact. In these processes, the shock duration is long enough for compacting samples of large dimensions. Recently, the possibility of compacting powders with the use of laser-generated shock waves was demonstrated [1]. In such a method, the pressure is applied on the target surface during a very short time, typically 10 to 100 ns, depending on the laser pulse length. A consequence of the extreme briefness of applied pressure is a rapid decrease of shock pressure. The powder is then compacted on a small depth under the irradiated surface : laser shock-waves may be used as a technique for surface densification of porous materials.

407

The purpose of this paper is to present results of computer simulations, describing the behavior of a porous medium under the action of a laser-generated shock-wave, particularly for predicting the consolidation depth as a function of irradiation conditions, nature and porosity of materials. Experimental results are also compared with calculations.

II. POROUS MODEL

The behavior of a porous medium under shock loading is described through a simple and classical model, considering the material as homogeneous and assuming that the pressure needed for closing the voids is negligible in comparison with involved shock pressures. The transition from initial state at density $\rho_0^* = \frac{1}{V_0^*}$ to the corresponding crystal density $\rho_0 = \frac{1}{V_0}$ occurs at zero pressure without increase of energy. The Hugoniot curve $P_H^*(V)$, is derived from the relations of conservation for mass, momentum and energy through the shock. Using the Mie-Grüneisen equation of state, with reference to the solid state Hugoniot :

$$P(V) - P_H(V) = \frac{\Gamma}{V} [E(V) - E_H(V)]$$

where P_H and E_H denote respectively pressure and energy in the shocked state, then P_H^* expresses as :

$$P_H^*(V) = P_H(V) \ \frac{1 - \frac{\Gamma_0}{V_0}\left(1 - \frac{V}{V_0}\right)}{1 - \frac{\Gamma_0}{2}\left(k - \frac{V}{V_0}\right)}$$

where Γ is the Grüneisen coefficient, dependent on volume through the classical relation :

$$\frac{\Gamma}{V} = \frac{\Gamma_0}{V_0} = cste$$

and k is a coefficient related to the porosity : $k = V_0^*/V_0$.

This model is introduced in a 1-D Lagrangian finite difference code (SHYLAC) and is validated by comparison of computed results with available experimental data [2] on the pressure dependence of volume, material velocity (u) and shock velocity (D) of various materials.

III. PROPAGATION OF A SHORT PRESSURE PULSE IN A POROUS MEDIUM

A computational study of shock propagation at short pulse duration has been made, using aluminum of porosity k = 1.5 (66 % crystal density) as test material. Initial conditions for applied pressure on the target surface are P = 30 GPa , t = 10 ns. Successive computed shock profiles are represented on Fig.1 and compared wilth computed profiles in compact aluminum in same conditions of applied pressure. The results illustrate the much faster shock decay in the porous medium. The reason for this is the high velocity of the leading edge of the release wave. Namely c* + u* > c + u, whereas D* < D, where c and u are respectively the sound velocity and the material velocity behind the shock front of velocity D in the solid material, and the asteriks denote the corresponding quantities in the porous medium.

The onset of pressure decay occurs when the release wave overtakes the shock front, at a distance X*$_c$ defined by :

$$X^*_c = \frac{\tau \ c^* \ D^*}{c^* + u^* - D^*}$$

Then, for t = 10 ns, X*$_c$ = 130 μm, in exact agreement with the computed result.

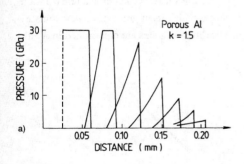

a)

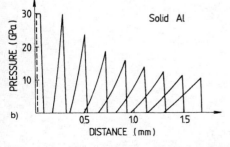

b)

FIG. 1 *Computed evolution of shock profile in porous aluminum (k = 1.5). Comparison with solid aluminum at fixed conditions of applied pressure : 30 GPa during 10 ns.*

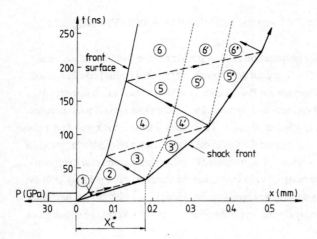

FIG. 2 *Computed space-time diagram describing the multiple wave system : shocks (full lines) and release waves (dashed lines) in porous aluminum. Applied pressure conditions are those of Fig. 1. Numbers refer to states defined in Fig. 3.*

The computed space-time diagram of shock propagation, shown in Fig. 2, reveals successive decelerations of the shock front, due to a complex system of multiple release waves and secondary shock waves. The basic mechanism is that, at each time a release wave overtakes the shock front, it generates a reverse shock, which itself generates a new release wave as reflecting at the front face of the target. The different states are determined in (P, u)

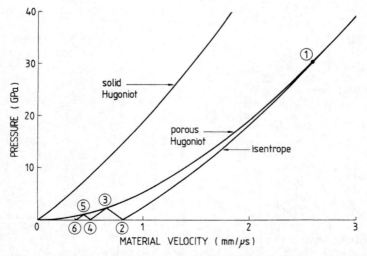

FIG. 3 *Representation of states in pressure - material velocity coordinates. Numerotation is in correspondance with Fig. 2.*

coordinates (Fig. 3). For a given state number, with or without superscripts' or" the corresponding zones are characterized by the same pressure and same material velocity, but by different specific volume and temperature. From this analysis, the main feature is the existence of two principal zones, in correspondance with the P(X) profiles of Fig.1: at a distance $X < X^*_c$, the porous medium is submitted to a high shock pressure (e.g. 30 GPa) generating suitable conditions for compaction and consolidation, whereas at a distance $X > X^*_c$, the shock pressure is strongly reduced (e.g. $P_3 = 2.2$ GPa, $P_5 = 0.9$ GPa, $P_7 = 0.5$ GPa) and becomes unsufficient for consolidating the material.

IV - LASER SHOCK COMPACTION OF AN ALUMINUM POWDER. EXPERIMENT AND ANALYSIS

Laser-shock compaction experiments have been performed using a device schematically described on Fig. 4a. The powder samples are covered with a transparent glass window acting as a confinement for the plasma generated by absorption of incident laser radiation on the metallic surface. The shock is generated by the rapid pressure increase of the plasma at the glass-target interface. In comparison with irradiation of bare targets, the confinement increases both the pressure and the duration of applied pressure [3,4,5], providing better conditions for compaction. The code describes the variation of pressure in the plasma treated as an ideal gas. This depends on the target nature and irradiation conditions. A computed profile is represented on Fig. 4b. The coupling efficiency was assumed to be about 10 %, as determined from previous measurements [5].

A sample of aluminum powder at 80 % crystal density ($k = 1.25$) was submitted to a

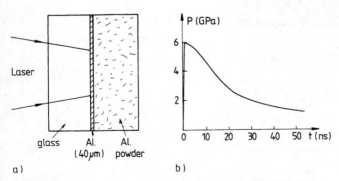

a)   b)

FIG. 4 *Device for laser shock compaction of aluminum powders (a) ,and computed pressure history at the glass-aluminum interface (b). Irradiation conditions are : 22 J energy in a semi-gaussian laser pulse of 8.8 ns width at half maximum, focused on a 6 mm diameter spot. Coupling efficiency is assumed to be 10 %.*

laser shock in the conditions indicated on Fig. 4. A foil of 40 μm thickness was introduced between the powder and the glass window. After shock loading, the cross-section of the target, Fig.5, shows that the foil is welded on the powder surface and clearly reveals three successive zones of compaction :

    - at a distance less than about 70 μm under the powder surface, the compaction is complete. No voids subsist and the grain boundaries almost disappear.

    - at a distance between 70 μm and 300 μm, partial compaction is observed

    - at a distance larger than 300 μm, very reduced or no compaction occurs.

The computed shock pressure evolution (Fig. 5) is well correlated with these observations. the induced pressure (5 GPa) is maintained without attenuation up to about 50 μm distance of propagation. Then it decreases rapidly and becomes lower than 1 GPa at 250 μm depth.

So, experimental results confirm the strong rate of shock decay in the porous medium, and validate the code calculations for predicting the thickness of compaction as related with the zone of the target submitted to a non attenuated shock. However, the simple model used in the calculations neglects the strength of the material and should be improved for predicting partial or non compaction of the porous medium in the very low pressure regime.

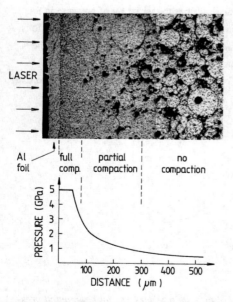

FIG. 5 *Cross-section of an aluminum powder (k = 1.25) after shock loading in conditions described on Fig. 4, and computed shock pressure evolution ,illustrating the correlation between the degree of compaction and the shock amplitude.*

V. CONCLUSION

Laser generated shocks are characterized by short pressure pulses capable of compacting a small thickness of porous material. From experiments performed on aluminum powders, the typical scale should be a few hundreds of microns, depending on irradiation conditions. Experimental results are correlated with the extremely rapid shock decay, as evidenced and quantified by computer simulations. The code may be used to predict the compaction depth of porous materials in a variety of situations related with the nature of the materials, their porosity, and irradiation conditions.

ACKNOWLEDGMENTS

The authors are grateful to B. Dubrugeaud and M. Jeandin for their cooperation in performing the metallurgical observations. Laser shock experiments were performed at the LULI. The work has been supported by DRET contract number 89/1189.

REFERENCES

1 - P. Darquey, J.P. Romain, M. Hallouin et F. Cottet, *Journal de Physique, Supp. n° 9, 49*, C3-425 (1988).

2 - R.G. Mc Queen, S.P. Marsh, J.W. Taylor, J.N. Fritz and W.J. Carter, in *"High Velocity Impact Phenomena"*, ed. by R. Kinslow, Academic Press (1970).

3 - L.C. Yang, *J. Appl. Phys.* 45, 2601 (1974).

4 - R.D. Griffin, B.L. Justus, A.J. Campillo and L.S. Goldberg, *J. App. Phys. 59*, 1968 (1986).

5 - J.P. Romain and P. Darquey, *J. Appl. Phys.* to be published.

37

Underwater-Shock Consolidation of Difficult-to-Consolidate Powders

A. CHIBA, M. FUJITA, M. NISHIDA, K. IMAMURA and R. TOMOSHIGE

Department of Materials Science & Resource Engineering
Kumamoto University
Kumamoto 860, Japan

Department of Mechanical Engineering
Kumamoto University
Kumamoto 860, Japan

A powder compaction method using underwater-shock wave was developed for difficult-to-consolidated powders. Titanium aluminide compound and Si_3N_4 powders were chosen as model materials. The underwater-shock pressure at the top of powder container was estimated to be in the range of 10-20 GPa depending on the mass and/or types of explosive used. The relative density of an as-compacted titanium aluminide specimen reached 95% and more, and it showed substantial interparticle bonding. The relative density of an as-compacted Si_3N_4 specimen was about 95% or more without any densification aids but the interparticle bonding was not sufficient. By post-sintering at 1923K, the relative density and the average hardness of Si_3N_4 compact reached 100% and 1700 Hv, respectively, which are comparable with those of commercially hot-pressed and pressureless sintered Si_3N_4 pellets with sintering aids.

I. INTRODUCTION

Shock consolidation is a process by which the particle surfaces are highly deformed and often molten and resolidified producing interparticle bonding in a one-step-process.

This technique allows powders to be consolidated at room temperature, so avoiding exposure to the high temperatures which are characteristic of other consolidation techniques.

This is one of advantageous over other techniques and can endow compact with unique microstructures and mechanical properties.

We have previously reported that a cylindrical axisymmetric explosive consolidation with a pressure medium is a suitable technique for producing a high density compact of metallic powders such as Al, Ti and TiNi without any cracks or a central melt hole [1,2,3]. Nevertheless, there was the difficulty in producing difficult-to-consolidate powders compact without crack and central hole due to the lack of shock pressure level and pressure duration.

In the present study, a new consolidation technique which can be applied to difficult-to-consolidate powders has been developed. Rapidly solidified titanium aluminide powder and Si_3N_4 powder were selected as the model materials of the difficult-to-consolidate materials.

Titanium aluminide is known as a candidate for advanced aircraft engines and light-weight, high-temperature structures. Although room temperature toughness of this intermetallic compound is limited, its static strength and stiffness degrade only slowly with increasing temperature. Also, it is lighter and has improved oxidation resistance superior to conventional titanium alloys. Since the titanium aluminide powder is difficult-to-consolidate material mentioned above, the establishment of newly developed consolidation technique is required.

So far, Si_3N_4 ceramic material which is of great technological interest has the difficulty in consolidation, so that in most cases the high fabrication temperature as well as the use of densification aids are required. The densification aids are considered to be the origin of the decrease of high temperature strength. Therefore, the fabrication of Si_3N_4 compact was aimed without using any densification aids.

II. EXPERIMENTAL PROCEDURE

A. *Materials*

Rotation-electrode processed titanium aluminide compound powders (Ti-48at%Al) produced by Daido Steel Co.,Ltd. Nagoya Japan were used in this study. The powder was about 150μm in diameter and nearly spherical with a dendritic structure as shown in Fig.1-a.

Si_3N_4 powders used consist of 95% alpha and 5% beta-types, which were produced by Onoda Cement Co.,Ltd. Onoda City Japan. The powders were observed to be a mixture of spherical and angular shapes and the average particle size was about 0.9μm in diameter as shown in Fig.1-b.

B. *Underwater-shock loading*

Figure 2 shows the experimental assembly for underwater-shock consolidation. The experimental configuration consists of three parts. They are explosive container, water tank and powder container from top to bottom.

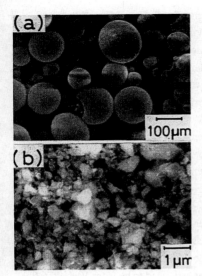

FIG.1 *Scanning electron micrographs showing (a) the original titanium aluminide powder and (b) the original Si₃N₄ powder.*

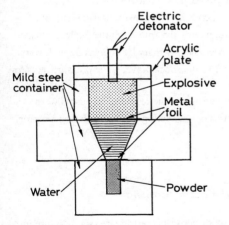

FIG.2 *Experimental assembly for underwater-shock consolidation.*

The shock wave produced by the detonation of explosive travels from the top to bottom of the water tank. The energy of shock wave increases by convergence due to the reduction of cross section area. Increased shock pressure and shock duration facilitates the consolidation of the powder.

The pressure level and shock duration of the consolidating powder can be easily controlled by adjusting the mass and detonation velocity of explosive and/or the conical angle of the water tank. These are the prominent features of this newly developed technique.

Titanium aluminide powders were poured into the powder container and the pouring density was about 40%. Si_3N_4 powders were poured into the powder container up to the depth of about two fifths after pouring metallic alloy powders at the bottom of the container as a momentum trap. Subsequently, metallic alloy powders were poured again as the upper layer to prevent the spatter of Si_3N_4 during compaction. Two kinds of the powder containers were used. These are 25mm and 50mm in depth. The pouring density of Si_3N_4 powders was about 50%.

Plastic explosive (SEP, provided by Asahi Chemical Industry Co., Ltd. Chiyoda, Tokyo Japan) was chiefly used, which consists mainly of nitric ester. The detonation velocity of the explosive was about 6900 m/s. In addition, Composition B (provided by Chugoku Kayaku Co., Ltd. ,Kure City Japan) was used for the consolidation of Si_3N_4 powders. The detonation velocity was about 7800 m/s.

The shock velocity in water at the top of powder container was determined to be 5000-6000 m/s by pin-contact method. Considering the pressure vs. particle velocity relationship for water and explosives used [4,5] and pressure vs. shock velocity relationship for water [6,7], the pressure level obtained in this experiment was estimated to be from 10 to 20 GPa.

With the present consolidation method, the underwater-shock wave can be applied directly to the powder and characteristics of this method are as follows:
(1) The pressure in water increases abruptly with increasing the density of water [6] and therefore the high pressure is easily obtained by regulating the amount of the explosive used. The duration of holding high temperature in compact is momentary by depressing with water.
(2) This method induces the homogeneous transfer of the shock wave. The pressure acts for a longer duration.

The microstructural and mechanical characteristics of the resulting compacts were determined by density measurement, optical microscopy, scanning and transmission electron microscopies, and hardness testing.

III. RESULTS AND DISCUSSION

A. Characteristics of titanium aluminide compact

Figure 3 shows a microstructure of titanium aluminide compact consolidated with 45 mm in thickness of the explosive. The dark and bright grains mainly consist of TiAl and Ti_3Al phases, respectively. In the top side, many melt zones were recognized, which lie in the interstices of powder particle. Each titanium aluminide particle was deformed into the structure of Japanese folding fan-like shape stacked each other to produce the well-consolidated microstructure indicated by arrows in Fig.3. This structure is the same as that found by Taniguchi, Kondo and Sawaoka [8] and by Williamson and Berry [9].

It is clearly seen that substantial interparticle bonding was achieved by melting of powder surfaces and of powder itself. The relative density of an as-compacted titanium

FIG.3 *Optical micrograph of as-compacted titanium aluminide powder.*

aluminide specimen reached 95% or more by using more than 25 mm thickness of explosive (SEP) layer.

Figure 4 shows the change of hardness with the thickness of explosive layer for TiAl and Ti₃Al grains and melt zone at the middle part of compact. The hardness of all three constituent phases decreases with increasing the thickness of explosive because of temperature increment during consolidation. At the thinnest thickness of the explosive layer i.e. 20 mm, the hardness shows the maximum value more than 1.5 times of that in original powder.

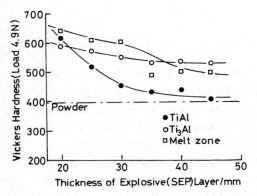

FIG.4 *Hardness change of constituent phases in titanium aluminide with the thickness of explosive (SEP) layer.*

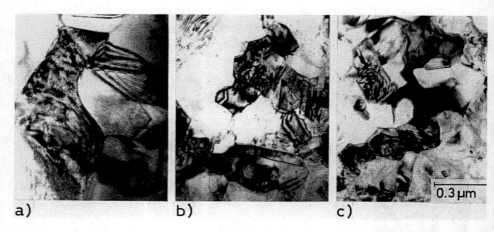

FIG.5 *Transmission electron micrographs of melting zone of as-compacted titanium aluminide in the (a) top, (b) middle and (c) bottom regions of a compact.*

Figure 5 shows transmission electron micrographs of melting zone along the original powder surface in the thin foil prepared from the same as-compacted titanium aluminide in Fig.3. The average grain size is 500, 300 and 100 nm in the top, middle and bottom regions, respectively. There are no internal defect features in crystal grains in the top region, while the internal defect features such as micro twins are seen in the middle and bottom regions. The change of microstructure of melt zone is due to the difference of thermomechanical processes and the extent of heat generation during the explosive consolidation. A high density of dislocations and deformation twins were always seen in the grain interior, especially in middle and bottom regions. It was also observed that the rapidly solidified structure was retained in the grain interior.

B. Characteristics of Si_3N_4 compact and effect of sintering

Figure 6 shows the change of hardness for as-compacted Si_3N_4 specimen with varying the thickness of explosive layer and the pouring depth of powder container. In case of using a deeper pouring depth, i.e. 50 mm, sound compacts without cracks were obtained but their hardness was generally low about 200 Hv, regardless of thickness of the explosive layer. The specimen size obtained soundly by using composition B with a deeper depth of powders container was 30 mm in diameter and 10 mm in height. On the other hand, using a shallower pouring depth (25 mm) of powder, hardness of compacts increased with increasing the thickness of explosive but the formation of profuse cracking in the compact was inevitable. Therefore, following experiments were carried out by using a 50 mm pouring depth.

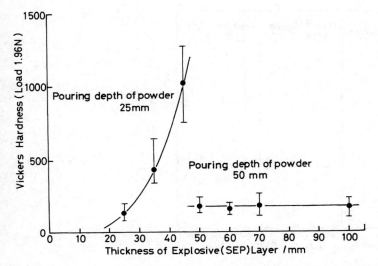

FIG.6 *Hardness change of as-compacted Si₃N₄ with the thickness of explosive layer and with the pouring depth of powder.*

Relative density and hardness of an as-compacted and post-sintered Si₃N₄ specimen were shown in Fig.7 and Fig.8, respectively. Relative density of an as-compacted one is about 95%. However, it was recognized that the interparticle bonding was not perfect as a whole but partially bonded by SEM observation of a fractured surface.

Therefore, to improve the strength of Si₃N₄ compact, the post-sintering is necessary. Post-sintering of Si₃N₄ compacts was carried out with packing in Si₃N₄ powders at 1773K or

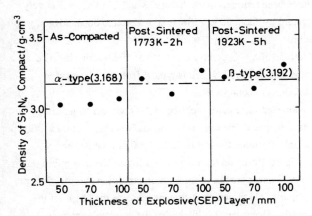

FIG.7 *Density change of Si₃N₄ compact with the thickness of explosive layer in various conditions.*

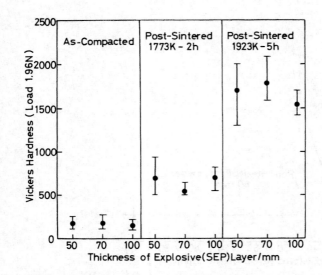

FIG.8 *Hardness change of Si₃N₄ compact with the thickness of explosive layer in various conditions.*

1923K. By post-sintering at 1773K for 2h, relative density reaches about 100% but the hardness is about 600 Hv since the crystal structure was still alpha state. By post-sintering at 1923K for 5h, the Si_3N_4 powder compact fully transformed to beta-type. It was confirmed by X-ray and electron diffractions. As a result, the relative density and the hardness of the Si_3N_4 compact increased about 100% and 1700 Hv, respectively, which are comparable with those of commercial pellet. Sintering temperatures for high density Si_3N_4 compact by hot press and gas-pressure sintering methods have been required more than 1973 and 2173K, respectively. The reduction of sintering temperature by about 50 to 250K is considered to be an advantage of the underwater-shock consolidation.

Figure 9 shows fracture surface of Si_3N_4 compact sintered at 1923K for 5h. The mixture of plate-like and fine grains was observed, which is typical features of densified structure with transformation from alpha to beta-Si_3N_4.

Figure 10 shows transmission electron micrographs of Si_3N_4 compact sintered at 1923K for 5h. The dominant microstructural features are hexagonal in cross section as shown in Fig.10(a). The grain size is largely scattered from 100 nm to 1000 nm, while the densification is fully proceeded. Figure 10(b) and (c) are high resolution electron micrographs of low and high angle tilt boundaries. These are not exact edge-on condition and insufficient to recognize the grain boundary impurity phase, which is considered to be the origin of the decrease of high-temperature strength. However, it is likely that the good coherency is achieved in the both boundaries as same as in reaction bonded Si_3N_4 without any densification

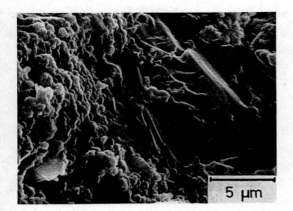

FIG.9 *Scanning electron micrograph of fractured surface of Si₃N₄ compact post-sintered at 1923K for 5h.*

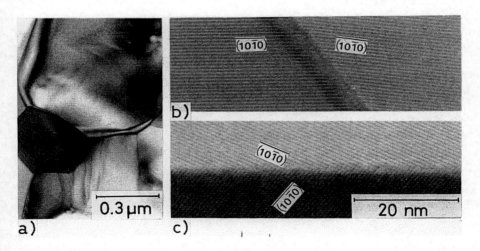

FIG.10 *Transmission electron micrographs showing grain-boundary structures, (a) low magnification image, (b) and (c) high resolution images for a low and high angle boundaries in Si₃N₄ compact post-sintered at 1923K for 5h.*

aids [10]. Although further careful examination will be needed on the grain boundary structures of explosively consolidated Si_3N_4, it is concluded that the underwater-shock consolidation technique is suitable for producing the high density compact of Si_3N_4 and other ceramics powders without any densification aids.

ACKNOWLEDGMENTS

The authors would like to express their appreciation to Dr. T. Watanabe of Tohoku University for his critical reading of the manuscript and valuable comments. Explosive experiments were performed at High-Energy-Rate Laboratory of Kumamoto University and they also express their thanks to Mr. N. Tsukimata of Kumamoto University for kind assistance in the explosive experiment. The titanium aluminide powders were kindly provided by Mr. K. Kusaka of Daido Steel Co.,Ltd..

REFERENCES

1. A. Chiba, M. Nishida, Y. Yamaguchi and J. Tosaka, *Scripta Metall.*, 22: 213(1988).

2. M. Nishida, A. Chiba, K. Imamura, T. Yamaguchi and H.Minato, *Metall. Trans.A*, 20: 2831 (1989).

3. M.Nishida, A. Chiba, R. Tomoshige and M. Uchida, in *Proc. MRS Int'l. Mtg. on Adv. Mats.*, K. Otsuka and K. Shimizu (eds.), Vol.9, Materials Research Society, Pittsburgh, 1989, p.617.

4. B. Crossland, in *Explosive Welding of Metals and its Application,* Clarendon Press, Oxford, 1982, p.70.

5. M. H. Rice and J. M. Walsh, *J. Chem. Phys.*, 26: 824, (1957).

6. I. I. Grass and L. E. Heuckroth, *Phys. Fluids,* 6: 543, (1963).

7. R. H. Cole, in *Underwater Explosion,* Princeton University Press, Princeton, 1948, p. 40.

8. T. Taniguchi, K. Kondo and A.Sawaoka, in *Metallurgical Applications of Shock-Wave and High-Strain-Rate Phenomena,* L. E. Murr, K. P. Staudhammer and M. A. Meyers (eds), Marcell Dekker Inc., New York, 1986, p. 293.

9. R. L. Williamson and R. A. Berry, in *Shock Waves in Condensed Matter,* Y. M. Gupta (ed.) Plenum, New York, 1986, p. 341.

10. H. Yoshinaga, *Materials Transaction, JIM,* 31: 233, (1990).

38

Several Techniques for One-Dimensional Strain Shock Consolidation of Multiple Samples

A. H. MUTZ and T. VREELAND, Jr.

W. M. Keck Laboratory of Engineering Materials
California Institute of Technology
Pasadena, California 91125, U. S. A.

We explored three methods of shock wave powder consolidation which retain the one−dimensional nature of a plane shock wave and allow multiple samples to be consolidated. The first technique uses a porous sintered metal cylinder as a shock fixture. The sintered material is chosen to match as closely as possible the solid density and compressibility of the powder to be investigated. The second method does not require a compatible porous material. A cylindrical target cavity is separated into multiple regions by thin sheet metal dividers. The dividers are of the same scale thickness as the powder size to retain a one dimensional condition in most of the compact. The third method may be the most interesting technologically. A powder media of near impedance match to the material under study is selected which resists bonding under the shock conditions to be used. Pressed greens of the material to be consolidated are then embedded in this pressure−transmitting media. The greens are then shocked along with the 'non−stick' media. In our experiment, a discontinuously reinforced metal matrix composite (MMC) was shock consolidated to near net shape by this method. Ti powder was mixed with SiC powder, pressed into a green with corners and radii, and embedded in fine zirconia powder. The shock wave generated by a 304 stainless steel flyer plate accelerated to 1.0 km/s fully consolidated the MMC without bonding the zirconia. The compact was recovered with well defined corners and flat surfaces.

I. INTRODUCTION

Multiple cavity experiments are usually done by encapsulating samples of powder in a solid, rigid fixture, and subjecting the entire assembly to a flyer plate impact or an explosive pressure pulse. The samples undergo equivalent pressure and energy conditions (if the samples are themselves similar) but the shock history of each is fairly complex; waves traveling through the target wrap around the capsules and subject different portions of a cavity to widely varying pressures [1]. Extensive numerical simulations have facilitated interpretation of resulting compacts but are not easily performed for every compact and geometry [2]. Simplifying the shock conditions eases analysis of the resulting compacts [3].

Striking only porous materials of similar impedance to the encapsulated samples creates samples subject to nearly one–dimensional shock conditions; the waves in the porous media don't 'wrap around', but proceed at approximately the same velocity as the shock wave in the sample. A similar line of reasoning led to the embedded off–center tube technique used in a cylindrical geometry [4].

Using this design philosophy, three types of plane wave experiments have been conducted: the effect of morphology on the consolidation of carbonyl Ni powders, the effect of particle size distribution on energy deposition in a maraging steel, and the near net shape consolidation of a discontinuously reinforced metal matrix composite material. The specific geometries and procedures are somewhat different, but the idea is consistently the same. The advantages in analytic experimentation are paralleled by possible technological advantages.

II. EXPERIMENT

A. *POROUS SINTERED BRONZE FIXTURE*

Commercially made porous bronze filter material (Pacific Sintered Metals F–100) was machined into 32mm dia. x 9.5mm thick disks with four 10 mm holes. The sintered bronze has a mean density of 4.7 g/cm^3 corresponding to a porosity of 46%. (See Fig. 1.) The cavities were separately loaded with carbonyl Ni powders of differing morphologies and pressed to 46% porosity. The carbonyl process nickel powders, obtained from Inco Metals, were 7–10μm spheres, 5.5–6.5 μm spiked spheres, 1 μm thick x 30 μm flakes, and 2μm thick x 10 μm long filaments. Shot 25 was performed

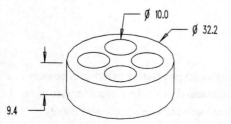

FIG. 1. *Drawing of a porous bronze four cavity target insert.*

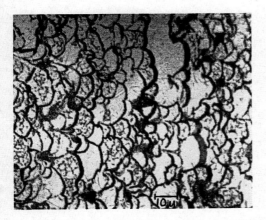

FIG. 2. *Micrograph of polished and etched shock consolidated spherical Ni.*

using the Keck Dynamic Compactor, a propellant–charged gas gun accelerating a 5mm
303 SS flyer plate to 1.17 km/s. This impact shocked the Ni powders to a calculated
pressure of 9.1 GPa and deposited approximately 430 J/g of thermal energy into the
powder. Three well bonded pills were recovered. A micrograph of the compacted
spherical powder is shown in Fig. 2. The flaky powder did not bond well, and was
apparently contaminated by the roller lubricant used in manufacturing.

B. STAINLESS STEEL DIVIDERS

We surmised that cavity partitions with thicknesses on the order of a particle diameter
or less would have minimal effect on the shock loading conditions of the material in the
fixture. If the materials themselves were nearly the same, the shock conditions in each
would not interfere with the others, thus each would be subject to the same uniaxial
shock conditions. The 304 stainless steel (SS) shim stock used was 0.13 mm thick.
The C350 maraging steel powder was sieved into various fractions from 44–300 μm dia.
The first cavity was loaded with 9.5g 125–150μm plus 9.5g 53–74μm powder. The
second cavity had 9.5g 180–300μm and 9.5g 44–53μm powder. The third cavity was
filled with 19g of 180–300 μm powder. The powders were consolidated by a 5mm thick
303 SS flyer plate accelerated to 1.28 km/s. The effect of particle size on energy
distribution and bond quality is being explored. The shot resulted in three sectors of
well bonded material (See Fig. 3). Uniform consolidation is demonstrated by the lack
of flyer plate deformation, divider buckling, and retention of compact shape.

FIG. 3. *Photograph of recovered M350 steel sectors of Shot 63.*

C. ZIRCONIA POWDER SHOCK TRANSMITTING MEDIA

A mixture of 10% Vol. 90 μm irregular SiC powder (Electro Abrasives 180 grit) and balance 10 μm Ti 6Al 4V alloy (Powder Metals Inc.) was pressed to a green shape shown in Fig. 4a under a static load of 150 MPa. The green compact was sufficiently strong to retain shape during careful handling. It was placed flat side up on a bed of −325 mesh zirconia powder (Cerac Z−1041), and more ZrO_2 powder was added to completely embed the green. The loaded target was covered with a 0.13mm thick 304 SS cover plate to retain the powder and then evacuated and impacted by a 5mm thick 303 SS plate at 1.0 km/s (Shot #59). The impact was sufficient to densify and bond the compact but not the zirconia. The resulting compact (Fig. 4b) is strained nearly uniaxially with 6% of uniform radial expansion. The radial expansion is caused by an insufficient packing density of the zirconia of only 47% (53% porosity), relative to the Ti + SiC, which is pressed to 59% of solid density.

III. DISCUSSION

The three experiments described above each rely on a combination of rigid radial confinement and uniform strain in the shock direction to retain a one−dimensional shock front. The effects of edge wave interactions at the rigid target and flyer plate edges are therefore of critical interest. Determining the extent of inter−cavity interactions is also important.

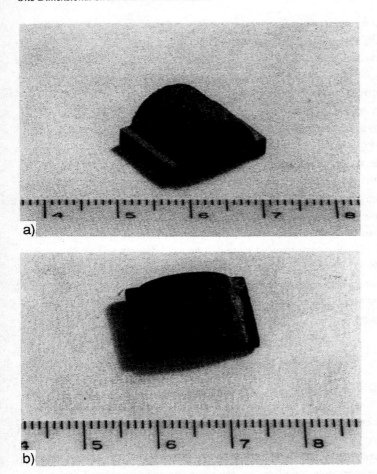

FIG. 4. *Photographs of (a) the pressed green compact of Ti + SiC and (b) the shock consolidated compact.*

When a flyer plate strikes a porous media contained in a rigid fixture (and not the fixture itself), a plane shock wave is generated in the media, as well as a radial compression wave in the fixture at the powder–fixture interface. The pressure of the wave in the fixture at the interface is determined by the shock impedance match between the shocked powder and the fixture material and the geometric boundary conditions. It may be a considerable fraction of the shock pressure in the powder. Note that the pressure in the shocked powder is now being released to an intermediate level. The wave in the cylinder attenuates in both radial and longitudinal directions, and results in minimal (1%) expansion of the fixture in our experiments. The three—element interactions of the flyer plate, powder, and rigid cylinder have not been simulated, however flyer plate experiments conducted using a metallic glass powder have confirmed the retention of a well—defined, nearly planar consolidation front across the entire compact surface [5]. We surmise from this evidence that the edge wave dissipation is relatively minor in the geometry used.

The cavity to cavity interactions are more easily understood. Edge wave interactions between neighboring shocked regions proceed at an angle determined by Al'tshuler, et. al [6]. Given a comparable shock speed and pressure, the angle is quite small and the interaction minor. In the case of rigid dividers, the shock 'ringing' is more intense, but it will dissipate over a length of several divider thicknesses as the the shock waves in the divider ring down to the powder shock pressure.

The ceramic—composite shock impedance match affected the condition of the MMC part fabricated. Matching the shock properties of a ceramic and metal precisely proved quite difficult. We compromised by matching solid density and shock speed as closely as possible. The resulting compact has not been mechanically tested, but shows no signs of macroscopic cracking, buckling or bending. A careful porosity match would result in a uniform strain in the shock direction. The shock speed match prevents 'wrap—around' waves.

IV. CONCLUSIONS

Three techniques have been demonstrated which maintain the plane wave condition in the shock wave used to consolidate powders. They are useful for the simultaneous consolidation of multiple samples. Valuable benefits of one—dimensional consolidation are nearly uniform processing conditions for the compacts and a simplified design and recovery of complex, near net shape parts from a single cylindrical target design. These techniques are also useful in the study of shock initiated chemical reactions in powders mixtures [7] .

ACKNOWLEDGMENTS

This work was supported by the National Science Foundation under the Materials Processing Initiative Program, Grant No. DMR 8713258. We would like to thank Kent Heady for preparing the Ti + SiC powder, and Norm Kenyon, at Inco Metals, for providing Ni powder data.

REFERENCES

1. G. E. Korth, J. E. Flinn, and R. C. Green, *Metallurgical Applications of Shock Wave and High Strain Rate Phenomena*, Marcel Dekker, New York 1986, p 129.

2. R. A. Berry and R. L. Williamson, *Metallurgical Applications of Shock Wave and High Strain Rate Phenomena*, Marcel Dekker, New York 1986, p 167.

3. M. L. Wilkins and C. F. Cline, *Second Workshop on Industrial Application Feasibility of Dynamic Compaction Technology*, p. 235, (1988).

4. P. S. DeCarli and M. A. Meyers, *Shock Waves and High Strain Rate Phenomena in Metals*, Plenum Press, New York, 1981, p. 341.

5. J. Bach and B. Krueger, unpublished.

6. L. V. Al'tshuler, S. B. Kormer, M. I. Brazhnik, L. A. Vladimirov, M. P. Speranskaya, and A. I. Funtikov, *Sov. Phys. JETP*, 11: 766 (1960).

7. B. Krueger and T. Vreeland, Jr, this proceedings.

39

Dynamic Compaction of Copper Powder: Experimental Results and 2D Numerical Simulations

T. THOMAS*, P. BENSUSSAN*, P. CHARTAGNAC**, Y. BIENVENU***

*   DGA/Centre de Recherches et d'Etudes d'Arcueil
        94114 ARCUEIL CEDEX, FRANCE
**   DGA/Centre d'Etudes de Gramat
        46500 GRAMAT, FRANCE
***   Centre des Matériaux
        Ecole Nationale Supérieure des Mines de Paris
        BP 87, 91003 EVRY CEDEX, FRANCE

A method for plate-impact dynamic compaction of copper powder has been developped. The optimization of the experimental set-up (impedance adjustments, tensile wave traps, relative thickness of impactor and target,...) is presented.

2D axisymetrical numerical simulations have been performed with a Lagrangian finite element code. These simulations show that, due to the difference in shock velocities in the container and in the powder, the powder is submitted to 2D loading waves. 1D simulations are shown not to evaluate properly the stress history and the energy deposition in the powder sample.

Metallographic observations as well as X-ray tomography experiments and tensile tests have been performed on consolidated samples. A very good agreement has been found between results of 2D numerical simulations and the observed final shape and density maps of the samples .

I.   INTRODUCTION

Dynamic consolidation of rapidly solidified powders can lead to very interesting structures [1-4]. As a matter of fact, in such a process the energy needed for powder consolidation may be deposited very rapidly (in a few s) so that very high heating and cooling rates (10^6 to 10^{10} K/s) can be reached . This energy is essentially deposited at the interfaces [5] of particles and the mean bulk temperature remains low enough for the initial structure and initial chemical composition of the powder to be retained. Thermally induced metallurgical and microstructural changes can thus be minimized. Hence, advantages of rapid solidification processing can be fully exploited. This is not the case in more conventional processes such as Hot-Isostatic-Pressing because they involve long times at high temperature.

The plate impact technique has been chosen to generate the shock wave necessary for powder consolidation. It allows the control of many experimental parameters e.g. the amplitude and duration of the pressure pulse or the impact obliguity.

Numerical simulations of the set-up have been performed. It is shown that 2D loading waves must be taken into account in order to fully understand and modelize the compaction processes .

II.   2D NUMERICAL SIMULATIONS

The optimized experimental set-up used for compaction tests is shown in fig.1. This set-up is made of an anvil container filled with statically pre-compacted 80 m in mean diameter [6] powder under a 1Pa vacuum and of 3 autofrettaged cylinders used to trap lateral rarefaction waves.

2D numerical simulations have been performed with a Lagrangian finite element code (EFHYD2D of ESI). Geometrical characteristics of the experimental set-up as well as the dynamic response of the powder (Reaugh equation of state) and of the material of the set-up have been taken into account. These simulations show that due to the differences in shock velocities in the container and in the powder, the powder is submitted to a non planar shock wave propagating in the as expected direction as well as by a sweeping wave initiated at the bottom of the powder container (fig 2) and propagating obliquely from the bottom up. The second wave loads the bottom of the powder before the direct one . 1D simulations are thus not reliable as they cannot predict these phenomena.

In order to avoid or limit the effects of lateral loading, the influence of impact area and of the anvil material should be investigated. Given the required mechanical resistance of the anvil, only the former parameter has been considered. Numerical simulations have thus been performed with several impactor diameter over powder diameter ratios.

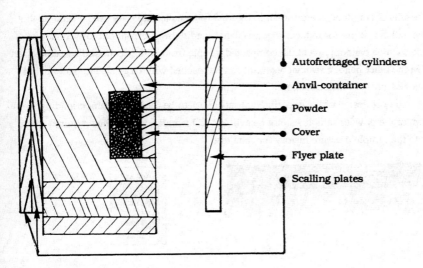

FIG 1 *Experimental set-up.*

- Autofrettaged cylinders
- Anvil-container
- Powder
- Cover
- Flyer plate
- Scalling plates

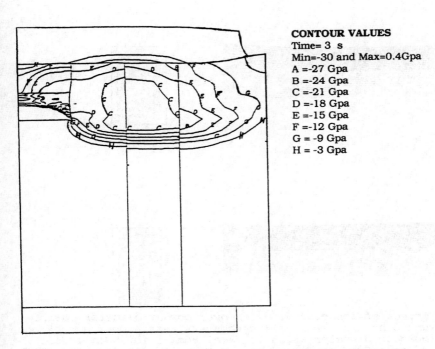

CONTOUR VALUES
Time= 3 s
Min=-30 and Max=0.4Gpa
A =-27 Gpa
B =-24 Gpa
C =-21 Gpa
D =-18 Gpa
E =-15 Gpa
F =-12 Gpa
G = -9 Gpa
H = -3 Gpa

FIG 2 *Contours of zz stress (z being the axisymmetry axis).*

Results of simulations are shown in figure 3 for 3 such ratios. As can be seen on this figure, a small size impactor induces a quasi-planar loading shock front but low constraint of the powder, which cannot prevent the compacted sample from being destroyed by rarefaction waves. On the other hand, a strong constraint can be reached with large size impactors, but curved shock fronts are then obtained.

A trade-off design of the experimental set-up has to be found, for which loading from the bottom of the powder sample cannot be avoided. 2D simulations are thus required in order to evaluate the sample pressure history for each test.

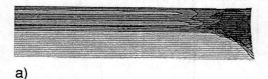

a)

b)

c)

FIG 3 *Effect of impactor diameter over powder diameter ratio. Pressure contours (MPa). All parameters remain constant except the impactor diameter which is : (a) 80mm ; (b) 50mm ; (c) 56,6mm for a 50mm powder diameter. Time is 3 μs in each case.*

III   EXPERIMENTAL RESULTS

The used powder is a gaz atomized copper powder of commercial purity . Its structure is mainly equiaxed with a grain size of about 10 μm. Some dendritic area have been observed before and after shock compaction (fig 4).

Metallographic observations have led to the indentification of the different consolidation mechanisms : mechanical hooking, interparticle welding and interparticle sintering (fig 5).

Particle's deformations can be clearly seen on figure 3 and 4. These deformations can be related to the predictions of the 2D simulations e.g. the existence of a sweeping loading wave.

Tensile specimens have been machined at different depths from the front face in as-consolidated samples free of cracks (fig 6). Tensile curves (example on fig 7) show that the material is highly hardened, but exhibits a very low ductility (A% < 1%). The apparent Young's modulus E is found to vary through the compacted sample (cf Table 1). U.T.S. is comparable to that of the dense material. The variation of E could be explained by inhomogeneous residual porosity as predicted by numerical simulations.

Fracture surface observations can lead to qualify the fracture mode as macro-brittle because of decohesions of particles and micro-ductile (fig 8) because of the ductile dimple fracture of welding zones (fig 9).

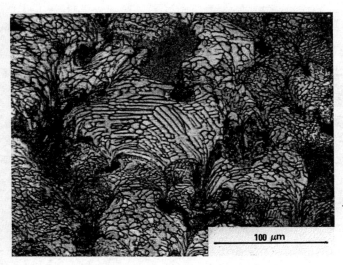

FIG 4 *Metallographic observation of the as-consolidated copper*

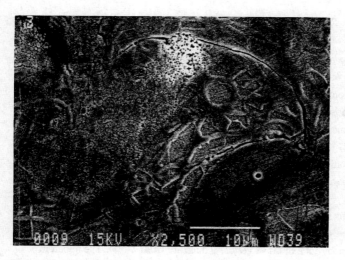

FIG 5 *Welding and sintering zones*

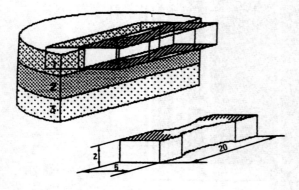

FIG 6 *Tensile specimens machined at different depths*

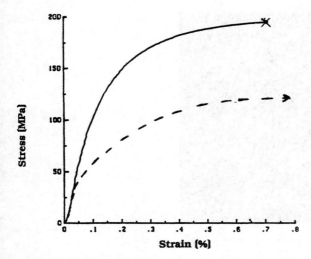

FIG 7 *Example of a conventional tensile curve of a specimen machined in the as-consolidated material*
___ *Dynamicaly consolidated copper*
-- *Annealed dense copper*

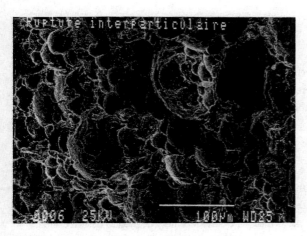

FIG 8 *Macro-brittle fracture mode*

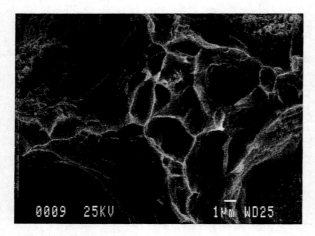

FIG 9 *Micro-ductile fracture mode*

TABLE 1 *Tensile tests results*

SPECIMEN	1	2	3	Annealed copper
E (MPa)	97000	63000	113000	125000
Rm (MPa)	-	210	195	230
$Rp_{0.1}$ (MPa)	-	185	155	$Rp_{0.5}=60$
UTS (MPa)	-	-	195	230

Micro hardness tests have also been performed. Values measured in the 60 to 300 Hv 20_g range show that the compacted samples are micro inhomogeneous. An annealed copper Hv hardness is in the 40 to 70 range.

IV.   DISCUSSION

2D numerical simulations show that, in order to obtain a quasi-planar shock wave, an optimization of the geometry of the set-up and particularly of the impactor diameter/powder container diameter ratio must be performed. They are definitely needed to describe the loading history of the powder.

Mechanical tests have been performed on the as-consolidated material. It appears that the ductility of the material should be improved by pre- and post-compaction thermomechanical treatments to be optimized in order to retain the refined initial microstructure.

Only pure copper has been studied here but such 2D simulations would obviously be essential for other applications such as reactive dynamic compaction.

ACKNOWLEDGMENTS *The authors gratefully acknowledge the participation of Engineering Systems International for some numerical simulations.*

REFERENCES

1. N.N. Thadhani , A. Mutz , T. Vreeland Acta Metall. VOL 37 pp 897-908 (1989)
2. L.H. Yu , N.N. Thadhani , M. A. Meyers DYMAT 88 Journal de Physique, VOL 49, colloque C3, pp 659-666 (1989)
3. R.N. Wright , J.E. Flinn , G.E. Korth MRP pp 281-286 (1986)
4. Y. Sano , K. Miyagi , K. Tokushima Journal of Engineering Materials and Technology VOL 111 pp 183-191 (1989)
5. W.H.W. Gourdin Journal of Applied Physics VOL 55 pp 172-181
6. J.E. Smuresky , T.J. Mc Cabe , R.A. Graham Shock waves in condensed matter 1987, S.C. SCHMIDT and N.C. HOLMES Eds, pp 411-414

40

Microstructure of Explosively Compacted Ceramic Materials

P. BOOGERD AND A. C. VAN DER STEEN

TNO Prins Maurits Laboratory
Rijswijk, The Netherlands

The relation between explosive compaction of ceramics and the resulting microstructure was investigated. Hydroxylapatite was explosively compacted to 99% TMD, and subsequently sintered to 100% TMD. The bulk ionic oxygen conductivity of a 93% TMD explosively compacted $YBa_2Cu_3O_{7-x}$ was studied with impedance spectroscopy. And the shock wave velocity during compaction of AlN powder to 86% TMD was measured and fitted to a first order model.

I. INTRODUCTION

Explosive compaction seem to develop as a tool to obtain ceramics with unique properties. The advantages of explosive compaction over conventional compaction techniques are the relatively low bulk temperature and the very short times high pressures are applied. However, these high pressures do affect the microstructure of the compact.

At the TNO Prins Maurits Laboratory research on explosive compaction is carried out, in collaboration with the Subfaculty of Anorganic Chemistry of the Technological University Delft. Special interest is focused on the relation between the applied explosive compaction technique, and the microstructure of the compacted ceramics.

In this paper three recently investigated examples will be presented; the 100% TMD compact of the pure biomedical ceramic hydroxylapatite, the conductivity at high temperature (> 300°C) of the HT$_c$S YBa$_2$Cu$_3$O$_{7-x}$, and results for the shape of the compaction wave in AlN.

A. Ca$_5$(PO$_4$)$_3$OH COMPACTION

Hydroxylapatite, Ca$_5$(PO$_4$)$_3$OH is a biomedical ceramic can be used for implants. It is an active biomedical ceramic, which means that it's fully compatible with our body.

A major drawback of this ceramic is that it decomposes at sintering temperatures above 1400°C to calciumoxide and calciumphosphate, that are rapidly dissolved in the body leaving a porous and brittle implant, and that it is difficult to compact to TMD by conventional means.

For these reasons explosive compaction was tried [1]. For it is known that during explosive compaction high temperatures only occur for a short time at the grain boundaries and that high densities can be obtained.

Experimental Several starting powders of hydroxylapatite with different grain sizes were used. We used the cylindrical, direct compaction method [2]. The hydroxylapatite was pressed in titanium tubes (D$_i$ = 10.7mm and D$_o$ = 12.7mm). Trimonite (AN/TNT/Al, 80/10/10) was used as explosive, with a detonation velocity of 3.6km/s. The density of the compact was determined by measuring the diameter reduction and Hepycnometry.

Results and Discussion As discussed by Prümmer [2] the results of the explosive compaction depends very strongly on the mass ratio of explosive and ceramic (E/M).

From preliminary experiments we learned that very fine powders (D$_{50}$ = 10-20µm), which could only be pre-pressed up to 40% TMD, would be overcompacted in our configuration. The best results were obtained using powders with a D$_{50}$ of 51 or 104µm, that could be more easily pre-pressed, see Table 1.

However, using only 1mm thick titanium tubes resulted in overcompaction. Instead of varying the E/M by changing the powder density, we increased the inert layer between the ceramic and the explosive. In this way we obtained homogeneously compacted samples. As an additional step sintering of the compact was carried out at 1000°C. The density increased and still no cracks could be observed.

This final, sintered material showed no signs of decomposition when examined with XRD or IR-spectrography. At this very moment some long term tests are carried out to investigate how these explosively compacted materials behave in a simulated biological environment.

B. $YBa_2Cu_3O_7$ COMPACTION

$YBa_2Cu_3O_{7-x}$ is a well known HT_cS material. Beside it's special properties at low temperatures, it is also an electronic and ionic conductor at temperatures above 300°C. In this example we will present [3] effects on it's oxygen conductivity, by explosive compaction, determined with impedance spectroscopy.

Experimental The starting $YBa_2Cu_3O_{7-x}$ powder was synthesized by calcinating three times a mixture of Y_2O_3, $BaCO_3$ and CuO. Subsequent heat treatment resulted in a material with x = 0.1, and an average grain size of $10\mu m$. The explosive used was Trimonite. Explosive compaction was carried out in steel tubes (D_i = 14mm and D_o = 17mm), with E/M = 3.5 and a starting density of 70% TMD. The same cylindrical configuration as in hydroxylapatite compaction was used.

The high temperature ionic conductivity was measured in a solid state cell: $Pt(O_2)/YSZ/YBa_2Cu_3O_{7-x}/YSZ/Pt(O_2)$ as a function of temperature (400-800°C), over the frequency range of 1-65000 Hz.

Results and Discussion After compaction, a tube of $YBa_2Cu_3O_{7-x}$ was obtained with only a very small hole (D < 0.5mm) and some cracks, indicating slight overcompaction, with a density of 93% TMD.

XRD showed slightly broadened peaks, but no substantial shift in the peak positions was observed indicating that the x value was still close 0.1. Scanning electron micrography of the compact showed no grain structure and hardly any cavities.

As expected, after compaction the $YBa_2Cu_3O_{7-x}$ was semi-conducting instead of superconducting down to 10 K.

The best fit of the impedance data was the equivalent circuit shown in Fig. 1. In this circuit the bulk resistance of $YBa_2Cu_3O_{7-x}$ is represented by R_2. Q and R_1 represent the capacitive element and the resistance associated with the $YSZ/YBa_2Cu_3O_{7-x}$ interfaces.

From R_2, the resistance corresponding to the ionic motion in $YBa_2Cu_3O_{7-x}$ is obtained. An Arrhenius plot of $\log(\sigma_i T)$ vs $1/T$ is shown in Fig. 2. For comparison the data obtained through similar measurements on a HIPped tablet of $YBa_2Cu_3O_{7-x}$ (density 83% TMD) are also shown. XRD results combined with these resistivity measurements make clear there is a difference between the sintered and the compacted material. Thus, a simple comparison cannot be made.

If the resistance would be dependant on the bulk conductivity a lower porosity would yield a lower resistance. Also ionic flow along the grain boundaries improves with decreasing porosity. This means that the difference in resistance cannot be explained by grain boundary effects.

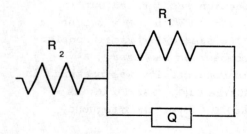

FIG. 1   A theoretical circuit fitting the impedance data of the compacted $YBa_2Cu_3O_{7-x}$. R_2 represents the resistance of the $YBa_2Cu_3O_{7-x}$ tablet. Q and R_1 correspond with the capacitive element and the resistance of the $YSZ/YBa_2Cu_3O_{7-x}$ surface.

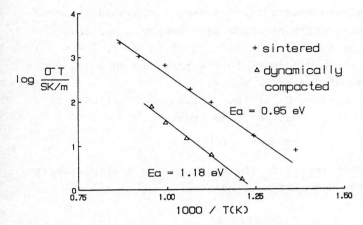

FIG. 2 *An Arrhenius plot of the resistance to ionic motion in sintered (83% TMD) and explosively compacted YBa₂Cu₃O₇₋ₓ (93% TMD).*

These results and the small change in the x-ray pattern might indicate that the lower ionic conductivity is caused by a disordered oxygen lattice, resulting in superconducting islands, separated by clusters of semiconducting $YBa_2Cu_3O_{7-x}$.

This kind of microstructure could occur if high temperatures at the grain boundaries during compaction caused local decomposition of $YBa_2Cu_3O_{7-x}$. Impedance spectroscopy seems to be a promising method to study these microstructural effects. Further investigations with different conditions have to be carried out to support these conclusions.

C. AlN COMPACTION

To understand more of the compaction process, we have started a series of experiments, with AlN as a model material. At the moment we are investigating the amount of energy absorbed in the ceramic during explosive compaction. Therefore we have developed a model, which describes the absorption of shock wave energy in the ceramic powder.

In addition we investigate the relation of the shock wave to the E/M. Scanning electron micrography is used to reveal the microstructure of the compact.

Model We have developed a first order, 1-dimensional model
based on the absorption of shock wave energy. For this
model we use the energy per area of a shock wave, e = P u τ,
in which P is the pressure, u is the particle velocity and τ
is the duration of the shock pulse.

The total energy, E, at the surface of a cylinder of
infinitesimal height, can be found from:

$$E = 2\pi re = 2\pi rPu\tau \tag{1}$$

in which r is the radius of the slab. Using a simple representation of the Hugoniot we find:

$$E = 2\pi rP\frac{U - C_o}{S}\tau \tag{2}$$

in which S and C_o are material dependent constants. For
this model we now propose that the amount of energy absorbed
is proportional to the amount of energy before absorption:

$$\frac{dE}{dr} = \frac{fE}{2\pi r} \tag{3}$$

in which f is the absorption factor depending on external
properties, e.g. microstructure, and internal properties,
e.g. thermal conductivity. We presume that the absorption
factor, f, can be considered constant in a pressure domain
in which compaction takes place by a specific combination of
mechanisms.

By solving (dU/dr) = (dU/dE) · (dE/dr), we find:

$$\frac{dU}{dr} = \frac{(f - 2\pi)}{2\pi r}\frac{U(U - C_o)}{3U - C_o} \tag{4}$$

And this yields after integration:

$$\frac{U(U - C_o)^2}{U_o(U_o - C_o)^2} = \left(\frac{r}{R}\right)^{\left(\frac{f - 2\pi}{2\pi}\right)} \tag{5}$$

Experimental The starting powder was a AlN powder containing 92% AlN, with an average grain size of 3-8μm. An aluminum container was used (D_i = 16mm and D_u = 20mm) in the same
cylindrical configuration as in hydroxylapatite compaction.

The radical velocity of the shock wave in the ceramic
was measured by placing sensors in line, in a horizontal

plane in the ceramic. Two experiments were conducted with
E/M = 4.5.

Results and Discussion The AlN was compacted to 86% TMD. A
scanning electron micrograph of the compact shows three
stages of compaction over the radius of the compact. The
outer region is fully compacted showing hardly any traces of
the original microstructure, see Fig. 3. The middle region
shows a distorted microstructure (Fig. 4) with a smaller
dimension than the starting powder. Particle breaking, and
plastic deformation could cause this kind of microstructure.
The inner region shows particles with the same sizes as the
starting powder, see Fig. 5. Apparently, the shock wave
energy was not sufficient to compact this region fully. The
compact appears undercompacted.

The results of the shock wave measurements are shown in
Fig. 6. In the fit we have used C_o = 2.5km/s and U_o = 2.0
km/s. With these values we find: f = 25 (= ca. 8π). The
lack of Hugoniot data for the AlN powder introduces a problem in interpreting the experimental results.

FIG. 3 *A scanning electron micrograph of explosively compacted AlN (86% TMD, sample 900628). The corresponding
shock wave velocity measurement can be seen in Fig. 6. The
micrograph shows the outer region of the sample. As can be
seen most of the original microstructure is lost.*

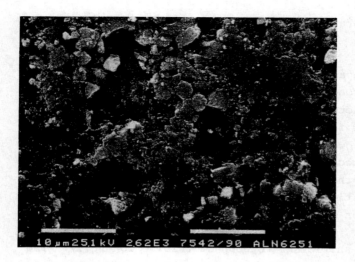

FIG. 4   A scanning electron micrograph of explosively compacted AlN (86% TMD, sample 900628).   The corresponding
shock wave velocity measurement can be seen in Fig. 6.   The
micrograph shows the middle region of the sample.   Notice
that part of the microstructure is an order of magnitude
smaller (0.1 - 1.0μm) than the original powder size (3-8μm).

FIG. 5   A scanning electron micrograph of explosively compacted AlN (86% TMD, sample 900628).   The corresponding
shock wave velocity measurement can be seen in Fig. 6.   The
micrograph shows the inner region of the sample.   Grains
with the original powder grain size (3-8μm) are clearly
visible.

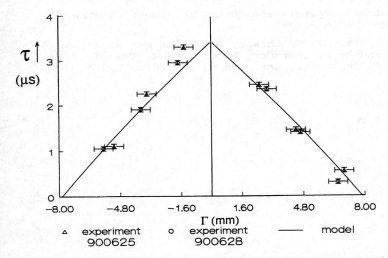

FIG. 6 *The results of the radial shock wave velocity mea-surements fitted to a first order model. Notice the conical shape of the profile, that might indicate optimum compaction [2].*

The conical shape of the velocity profile would indi-cate almost perfect compaction. The fact that SEM results indicate undercompaction might be explained by the fact that the width of the shock pulse and thus E/M have a relatively small effect on the shock wave velocity, compared to their effect on the compaction. If shock wave velocity measure-ments, can be connected with microstructural changes they might turn out to be a useful method for analysis in explo-sive compaction.

ACKNOWLEDGMENT The authors wish to express their gratitude to: H. H. Kodde, R. Oostdam, and W. A. H. van Willigen for their preliminary work, and to J. Schoonman, N. J. Kiwiet, H. J. Verbeek, M. A. Schrader, E. G. de Jong, W. C. Prinse and A. M. Maas for their advice and assistance.

REFERENCES

1.   H. H. Kodde and W. A. H. van Willigen, TNO Report PML 1990-IN4 (1990; unpublished, language; Dutch).

2.   R. Prümmer, Explosivverdichtung pulvriger Substanzen,
      Springer Verlag (1987).

3.   N. J. Kiwiet, J. Schoonman, A. C. van der Steen, H. H.
      Kodde and M. A. Schrader, Proc. 1st Int. Ceramic Sci-
      ence and Technology Conference, Annaheim, U.S.A.
      (1989).

41

Hot Shock Consolidation of Diamond and Cubic Boron Nitride Powders

K. HOKAMOTO, S. S. SHANG, L. H. YU and M. A. MEYERS

Department of Applied Mechanics and Engineering Sciences
University of California, San Diego
La Jolla, California 92093-0411, U.S.A.

Diamond and cubic boron nitride powders were explosively compacted at high temperature (873K) by using planar impact system at 1.2km/s. Silicon or graphite was added to the mixture to provide enhanced bonding through chemical reaction. Hot consolidated specimens exhibited decreased surface cracks as compared with the specimens consolidated at room temperature. Some of the materials consolidated showed the evidence of melting at the particle surfaces and hot consolidated diamond admixed with graphite powder improved the bonding between particles.

I. INTRODUCTION

Shock consolidation of extremely hard non-oxide ceramics powders has already been attempted by many researchers; diamond [1-5], cubic boron nitride (c-BN) [6] and some other powders [7, 8] have been consolidated, but there still are some problems to be improved. The main problems are remaining cracks, weak interparticle bonding and inhomogeneity of consolidated material.

In this investigation, we tried to shock consolidate diamond and c-BN powders at high temperature to eliminate the problems mentioned above. The technique of hot shock consolidation is expected to eliminate the cracks due to the increase of plastic flow and surface melting of powders and is also expected to improve the bonding between particles by the heating of particle surfaces [10]. The effect of particle size on the consolidated diamond and c-

BN powders is examined by comparing the results at room temperature [1-3, 6]. The effect of the addition of graphite or silicon powders is also examined as a means to enhance the bonding with the assistance of reaction [2, 4, 11-13].

II. EXPERIMENTAL PROCEDURES

Three sizes of natural diamond (4-8, 10-15 and 20-25μm) and two sizes of c-BN powders (10-20 and 40-50μm) were used as starting material. The composition of the specimens are shown in TABLE 1. The effect of addition of graphite or silicon powders is examined in these experiments. The powders were pressed to 60% of the theoretical density with the shape of 5mm thick and 12mm in diameter into a stainless steel capsule.

The planar impact system developed by Sawaoka and Akashi [14] was used for the consolidation. The schematic illustration of this set up is shown in FIG.1 (a). Twelve capsules can be compacted simultaneously. Figure 1 (b) shows the set up for hot consolidation developed by Yu and Meyers [15]. The capsules are preheated at 873K and compacted by the flyer plate at 1.2km/s.

The materials consolidated were characterized by optical microscopy, scanning electron microscopy (SEM), X-ray diffraction and Vickers micro-hardness.

III. RESULTS AND DISCUSSION

Figure 2 shows bottom quarter surfaces of diamond and c-BN powders after hot shock consolidation. Shock consolidated materials at room temperature showed many cracks at the bottom surface [5, 6], but some of the specimens hot consolidated showed only a few cracks

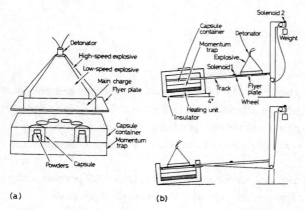

(a)   (b)

FIG.1 *Schematic illustration of planar impact system (a) and set up of hot consolidation (b).*

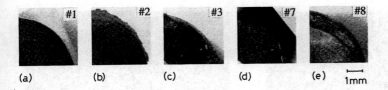

FIG.2 *Quarter bottom surfaces of consolidated diamond ((a)-(c))
and c-BN powders ((d),(e)).*

(FIG.2 (a), (b), (d), (e)). This result is due to heating. High temperature causes the melting
of particle surfaces easily and decreases the energy required for consolidation . In this case a
flyer plate velocity of 1.2km/s was chosen for the consolidation of diamond and c-BN
powders, while 1.8-3.0km/s is required for the consolidation of these powders at room
temperature [1-6]. The decrease of shock pressure induces the decrease of surface cracks.

A. *Characterization of hot consolidated diamond powders*

Figure 3 (a) shows the half of transverse cross section and (b) shows the Vickers microhardness distribution of consolidated 20-25μm diamond powders (specimen # 3). This
specimen shows many cracks in the bottom region, where the micro-hardness values show the
higher value. As reported earlier [5, 12] the micro-hardness distribution of shock consolidated
material shows the same tendency of the temperature distribution during the passage of shock
wave. Higher hardness values are also obtained in the high temperature area in our specimen.
Figure 4 shows the SEM of polished surface of 4-8μm diamond (# 1). No microcracks and voids can be seen in this photograph. Small round dark areas about 1-2μm in
diameter are the trace of the diamond particles pulled out. The enlarged SEM in such a dark
area generated in 20-25μm diamond (# 3) is shown in FIG.5. This area is composed of small
grains about 0.5 - 1μm in diameter. Since the original diamond powder is monocrystalline,
the small grains are thought to be generated during cooling of molten surface of the particles.
Such small polycrystalline regions in a diamond particle are already found in shock

Shock wave

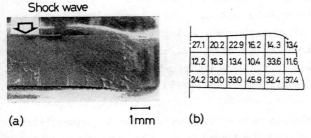

27.1	20.2	22.9	16.2	14.3	13.4
12.2	18.3	13.4	10.4	33.6	11.6
24.2	30.0	33.0	45.9	32.4	37.4

(a) 1mm (b)

FIG.3 *Cross section of consolidated 20-25μm diamond (#3) (a), and
its Vickers hardness distribution in GPa.*

FIG.4 *SEM of polished surface*   FIG.5 *Enlarged SEM of*
of consolidated 4-8μm   *consolidated 20-25μm*
diamond (#1).   *diamond (#3).*

consolidated [3] and shock synthesized diamond powders [16, 17]. The size of these small grains is decreased with the decrease of the size of the original diamond particle. This result depends on the cooling rate of the molten layer [18, 19]. Generation of small amount of graphite is detected by X-ray diffraction in 10-15μm diamond (# 2). Figure 6 shows the cracks generated in one particle consolidated 10-15μm diamond (# 2).

Akashi and Sawaoka [5] reported that 2μm-20μm diamond showed the optimum condition shock consolidated at room temperature. Potta and Ahrens [1] reported the same the effect of particle size. When the original particle size is too small, surface temperature is not sufficient to generate the molten layer to enhance the bonding by the dispersion of shock pressure. When the particle size is too large, many cracks are generated in the particles by the high pressure and a transformation from diamond to graphite is caused by the thick molten layer. In hot consolidated materials, cracks still are formed and the transformation to graphite is generated in 10-15 μm diamond specimens (# 2). The transformation to the graphite is usually generated in larger particle size specimens consolidated at room temperature [5] but this transformation is enhanced in hot consolidated material with smaller particle size by the help of

FIG.6 *Cracks generated in one particle of consolidated 20-25μm diamond (#3).*

heating. In case of the hot consolidation of pure diamond, smaller particle size is required for the good bonding compared with the consolidation at room temperature. Micro-hardness values in TABLE 1 assure this consideration.

TABLE 1 Experimental conditions of hot consolidation.
Average Vickers hardness values are also listed.

Specimen #	Composition	Average Vickers hardness* / GPa
1	4-8μm natural diamond	29.4
2	10-15μm natural diamond	26.0
3	20-25μm natural diamond	24.4
4	4-8μm natural diamond + graphite (15mass%)	>55
5	4-8μm natural diamond + silicon (7.5mass%)	15.8
6	10-20μm c-BN	37.5
7	40-50μm c-BN	53.0
8	40-50μm c-BN + graphite (15mass%)	24.4

*; Load 4.9 or 9.8N for 15sec.

Consolidated diamond with graphite shows the highest hardness value (TABLE 1). Due to the difficulty of the measurements, we only describe the Vickers hardness value more than 55GPa. The improvement of the bonding by the addition of graphite to diamond was already demonstrated for shock consolidation at room temperature by Potta and Ahrens [2]. The explanation provided by them was that the graphite, which has the lower thermal diffusivity, delays thermal equilibrium between molten surface and interior of diamond. This heat enhances the bonding between particles. Deformation of the softer component graphite around the diamond powder also decreases the pressure concentration and fills the pores between diamond particles. On the other hand, degradation of consolidated material appeared with the addition of silicon powders. We expected the heat generated during reaction between silicon and carbon to enhance the bonding between particles [2, 11-13], but in this case the excess heat generated induced the excess melting of the particle surfaces.

B.   Characterization of hot consolidated c-BN powders

Figure 7-9 show SEM of hot consolidated c-BN powders. Figure 7 is a fracture surface of 10-20μm c-BN (# 6) and this fracture surface suggests the melting of particle surfaces. In some areas of this specimen, small voids can be seen caused by the gas generated by the decomposition of c-BN (FIG.8). Consolidated 40-50μm c-BN (# 7) shows high micro-

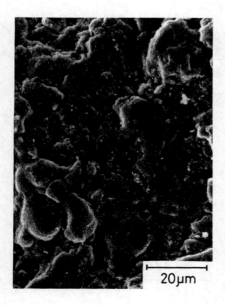

20μm

FIG.7 *SEM of fracture surface of consolidated 10-20μm c-BN (#6).*

 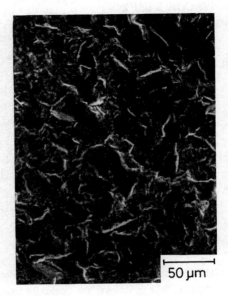

FIG.8 *Small voids generated in consolidated 10-20μm c-BN (#6).*

FIG.9 *SEM of fracture surface of 40-50μm c-BN (#7).*

hardness value (TABLE 1). SEM of fracture surface in FIG.9 shows the evidence of good bonding. Some of the grains in shock consolidated 40-50μm c-BN show the transgranular fracture. As Akashi and Sawaoka [6] reported on the shock consolidated c-BN powders at room temperature the specimen with larger particle size also showed higher hardness value in hot consolidation (TABLE 1). In contrast with the case of diamond, a larger melting layer is thought to be better for consolidation. It is reported that c-BN is more stable than diamond at high temperature in the atmosphere of oxygen [20]. This suggests that the transformation from cubic to hexagonal structure boron nitride is hard to occur even when the cooling rate is slow compared with the case of diamond.

The consolidated c-BN admixed with graphite shows low micro-hardness value compared with the c-BN without graphite (TABLE 1). It is considered that the rapid cooling rate could not be obtained for the generation of diamond because of the thick molten layer. The remaining graphite causes the degradation of consolidated material.

IV. CONCLUSIONS

Diamond and cubic boron nitride powders were explosively consolidated at high temperature by using a planar impact system. Hot shock consolidated materials exhibited a decrease of surface cracks as compared with specimens consolidated at room temperature. The consolidated diamond specimen showed evidence of surface melting of the particles by small crystallized grains generated during cooling of this molten layer. 4-8µm diamond consolidated showed good bonding. Excess cracks and generation of graphite were observed by the increase of particle size. Hot consolidated diamond admixed with graphite showed the highest hardness value, but the improvement of the bonding was not confirmed by the specimen admixed with silicon powders. Hot consolidated 40-50µm c-BN showed good bonding compared with the specimen which has small particles. The improvement of the bonding of c-BN powders by the addition of graphite could not be recognized in the present investigation.

ACKNOWLEDGMENTS

This research was supported by National Science Foundation under Grant CBT. 87 13258. The use of the facilities of the Center for Explosives Technology Research is gratefully acknowledged.

REFERENCES

1. D. K. Potter and T. J. Ahrens, *Appl. Phys. Lett.*, 51: 317 (1987).

2. D. K. Potter and T. J. Ahrens, *J. Appl. Phys.*, 63: 910 (1988).

3. S. Sawai and K. Kondo, *J. Am. Ceram. Soc.*, 71: C-185 (1988).

4. K. Kondo and S. Sawai, *J. Am. Ceram. Soc.*, 72: 837 (1989).

5. T. Akashi and A. B. Sawaoka, *J. Mater. Sci.*, 22: 3276 (1987).

6. T. Akashi and A. B. Sawaoka, *J. Mater. Sci.*, 22: 1127 (1987).

7. K. Kondo, *High Pressure Explosive Processing of Ceramics*, R. A. Graham and A. B. Sawaoka (eds.), Trans. Tech. Publications Ltd., Switzerland, 1987, p.227.

8. T. Akashi et al., *Shock Waves in Condensed Matter*, Y. M. Gupta (ed.), Plenum Press, New York, 1986, p.779.

9. S. L. Wang, M. A. Meyers and A. Szecket, *J. Mater. Sci.*, 23: 1786 (1988).

10. T. Taniguchi and K. Kondo, *Adv. Ceram. Mater.*, 3: 399 (1988).

11. A. B. Sawaoka and T. Akashi, *U.S. Patent* 4,655,830 (1987).

12. L. H. Yu, M. A. Meyers and N. N. Thadhani, *J. Mater. Res.*, 5: 302 (1990).

13. Y. Horie, *Shock Compression of Condensed Matter - 1989*, S.C.Schmidt, J.N.Johnson and L.W.Davision (eds.), Elisver Science Publishers, North-Holland, 1990, p.479.

14. T. Akashi and A. B. Sawaoka, *Mater. Lett.*, 3: 11 (1984).

15. L. H. Yu and M. A. Meyers, private communication.

16. L. F. Trueb, *J. Appl. Phys.*, 30: 4707 (1968).
17. L. F. Trueb, *J. Appl. Phys.*, 42: 503 (1971).
18. D. G. Morris, *Metal Science*, 16: 457 (1982).
19. D. Raybould, *Int. J. Powder Metall. and Powder Technol.*, 16: 9 (1980).
20. K. Ichinose et al., Proc. 4th Int. Conf. on High Pressure - 1974, Special Issue of the Review of Physical Chemistry of Japan, J. Osugi (ed.), Kawakita Printing Co. Ltd., Kyoto, 1975, p.436.

42

Importance of Preheating at Dynamic Consolidation of Some Hard Materials

L. JAPARIDZE, A. PEIKRISHVILI, N. CHIKHRADZE, and
G. GOTSIRIDZE

Institute of Mining Mechanics
Academy of Sciences of Georgian SSR
Tbilisi, Georgia, USSR, 380086

Adiabatic heat of self propagating high temperature synthesis (SHS) was used for shock wave consolidation of titanium carbides and borides, and electrical current was used for tungsten carbides based alloys in order to heat the consolidated materials. The temperature of preheating varied up to 2000 °C.
The investigation showed that the use of SHS simultaneously with the shock loading allows to obtain high density samples with high hardness values. In practice the Mach's zone is formed after the loading because of high preheating temperature. The study of tungsten carbides and its alloys showed that the formation of perfect structure and correct geometry essentially depends upon preheating temperature and prior density.

I. INTRODUCTION

In comparison of metallic materials the shock wave compaction of the ceramic materials is significantly difficult. This is apparently caused by brittleness of compacting particles as while loading they are deformed difficultly, clinging and fusing to each other and as a result there present cracks in the samples.

In the work [1] it is shown the use of preheating directly before the loading enables to obtain the high density materials from ceramic and refractory powders. It has been established that only the use of preheating enables one to obtain perfect structure in the compacting samples.

In existing work [2,3], the compaction of ceramic materials (on TiC and TiB basis) has been carried out at elevated temperatures, however, for preheating the process of preliminary synthesis and then the product loading occurred. Here it is shown the principal possibility of high dynamic stresses action on the process of structure formation after SHS.

However, for effective process control it is necessary to carry out further systematic structure formation investigations.

The aim of the present work was to investigate the role of preheating on the compaction process and structure formation of ceramic materials on TiC, TiB and WC basis at dynamic conditions.

Dynamic consolidation of these materials was carried out by an axissymmetrical scheme of loading at temperatures 600-2000°C. For consolidation of TiC and TiB, at first, the synthesis was carried out in the process of combustion and then after TiC and TiB formation - shock wave loading.

For compaction of powders on WC basis the temperature was changed in the range of 600-1200°C and heating occurred by electrical current as in [4]. In both cases the loading intensity on the wall of the container did not exceed 20 GPa.

II.   THE RESULTS OF EXPERIMENTS

COMPACTION OF THE POWDERS ON TiC AND TiB BASIS.

Investigations show that in all cases of compaction of TiC and TiB is caused by Mach's zone formation in the center of the sample, characterized by a number of pores and free spaces.

At Fig. 1 the structure of compacted TiC in the condition of SHS + shock wave loading is brought.

As it is seen from the microstructure (Fig. 1a) the I Mach's zone due to the structure sharply differ from the II adjoining layer which has obviously perfect view (Fig. 1a). The pores and free spaces are absent there. The formed grains of TiC have the perfect structure (Fig. 1b).

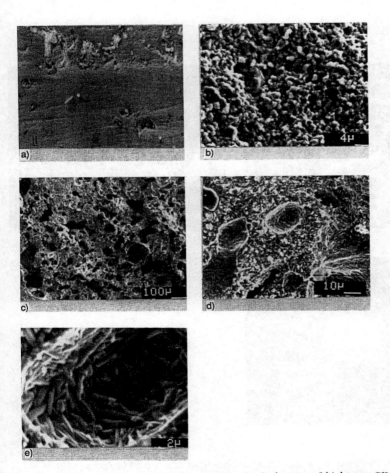

FIG. 1 *Structure of TiC compacted in conditions SHS + shock wave loading : a) general view, mag. x 100, b) microstructure of compacted sample (II zone), c), d), e) microstructure of central zone at different magnifications.*

   Thorough investigations of the compacted TiC in the
Mach's zone the forming pores in reality have the definite
structure (Fig. 1c,d,e). At large magnifications there are
observed the presence of needle-like crystals (Fig. 1d,e).
Analysis shows that in these needles it is titanium. This
fact enables one to conclude that they are the grains of TiC.

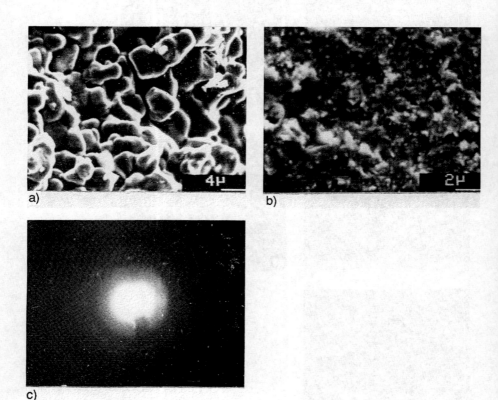

FIG. 2 *Microstructure of TiB compacted in conditions SHS +
shock wave loading: a) microstructure of TiB compacted by
10 GPa intensity, b) microstructure of TiB compacted by 20
GPa intensity, c) electron diffraction pattern of TiB at
loading 20 GPa.*

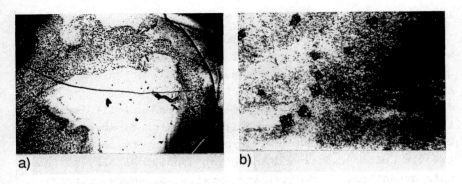

FIG. 3 *Structure of dynamically compacted WC-10%Co, preliminary density 60%, P=10 GPa, T-1100°C: a) geometry disturbance and cracking, mag. x 32, b) formation of (Co₂W₃C)
phases, mag. x 00.*

Appearance of such grains of TiC, apparently, is caused
by the processes of recondensation with stoichiometry disturbance.   Identical results were obtained in the case of
TiB.

The Mach's zone has not practically the hardness and
regarding the dense compacted II zone the hardness is equal
to 1800-2500 HV 0.05.

It is necessary to mention that the perfect structure
formation shown by the appearance of clear grain boundaries,
essentially depends on the stress. For instance, at the
loading of synthesizing TiC and TiB by the pressure of 10
GPa, the separate grains are characterized by the grain
boundaries and they may easily differ from each other (Fig.
2a). At a pressure of 20 GPa the boundaries of the grains
are difficult to observe. And in some cases they are even
practically absent (Fig. 2b), that is confirmed by studying
the diffraction patterns (Fig. 3b).

Putting passive additions and some metals (Ni,Fe) into
the compacting powders essentially decreases the Mach's
zone. That is apparently caused by the lower residual
temperature.

III.   COMPACTION OF POWDERS ON WC BASIS

Investigations show that when compaction of the powders on
WC basis at room temperature it is practically impossible to
get qualitative, highly dense samples of high values of
hardness.

The use of preheating temperature essentially improves
the situation and enables qualitative, high dense samples
with perfect structure and correct geometry to be obtained.
However, other factors in the given conditions play not an
unimportant role: preliminary density, pressure (stress)
and type (quality) of the powders.

As it was mentioned in [2], when compaction of alloys
on W basis, low predensity (60%) led to the disturbance of
the sample geometry. In these conditions the form of the
compacted sample becomes oval and it appears there new
undesirable structural components η (Co_3W_3C).

Figures 4 and 5 show several examples of WC-Co shock
compaction at different temperatures.

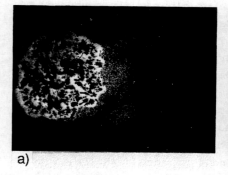 

a)   b)

FIG. 4 *Structure of surface of WC-15%Co compacted at 10
GPa, T=800°C: a) compaction by Co of WC plated powders,
predensity 80%, b) compaction of mechanical mixture of WC
and Co, predensity 60%.*

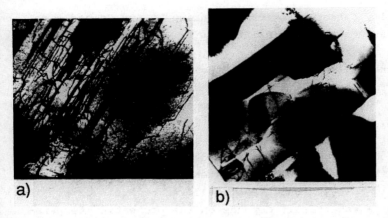

FIG. 5 *Fine structure of WC-Co alloy: a) sintered alloy loaded at P=20 GPa, T=1000°C, mag. x 35000, b) consolidated powder at P=10 GPa, T-1200°C, mag. x 13000.*

One of the principal problems during materials compaction by the axissymmetrical scheme is elimination of the Mach's zone formation. The problem concerns the powder from WC-Co, the samples without defects requires the very thorough experiments. It is thought that during compaction of the two-phased materials and it is the presence of phases of different densities in the process of loading produces a displacement of slight phase under the influence of shock wave in the direction of the center. Under the influence of high pressure melting occurs and vaporization produces free space and pores in the center of the sample.

With the purpose of testing of the given supposition there were used the powders of WC-Co of two types: mechanical mixture WC-Co and WC fully plated by Co. By experiments it has been established that in reality the type (quality) of the compacted powder influences the Mach's zone formation and thereby the quality of the compact obtained. The use of plated powders essentially makes easier the task of obtaining qualitative samples. At high temperatures of the compaction in identical conditions the loading in the case of plated powders, the dimensions of the Mach's zone are essentially less and in most cases it is practically absent. It is naturally caused by stronger bonding WC and Co and also more resistance to the displacement of Co from WC to the center of the sample.

Figure 6 visually confirms the mentioned above. As it is seen from the pictures in the case of powder WC plated by Co, after the loading, in receiving sample the Mach's zone is absent, while at dynamic compaction of the mechanical mixture of the same content and at the same parameters, there is obviously observed the Mach's zone [2].

Investigation of bonding content shows that it represents solid solution on Co basis alloyed by atoms of W. The content of W essentially depends on the preheating temperature. For instance, at 400°C the W is absent in bonding, at 600°C-2%, at 800°C-3%, at 1000°C-6% and at 1200°C-10%.

Figure 5 represents fine structure of WC-Co alloy at different conditions of loading. As it is seen from the fine structure in WC grains of dynamically compacted alloy WC-Co, there are observed the traces of plastic deformation rather in less extent (Fig. 5a) then in deformed sintered alloy (Fig. 5b). However, at compaction at high temperatures (Fig. 5a), in WC grains there are obviously observed the fragmentation and disorientation of the separate fragments due to each other, that should not be said about WC

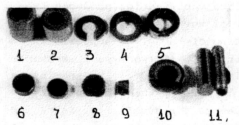

FIG. 6 *General view of samples on TiC, TiB and WC basis, produced by shock wave compaction at elevated temperatures: 1-7 - compaction of different WC-Co powders at different conditions of loading, 8 - onestaged compaction of pure WC at T-700°C, P=10 GPa, predensity 60%, 9 - samples from pure TiB produced in conditions of SHS + shock wave loading, 10 - sample TiC, produced in conditions of SHS + shock loading, 11 - compacted samples from WC-10%Co and WC-8%Co+50%W.*

grains of sintered WC-Co alloys dynamically deformed at high temperatures.

It is necessary to mention that the above mentioned parameters have important role when dynamic consolidation of the powders on WC basis. However, preheating temperature is the main parameter in combination of which all the rest have essential influence on the quality of the final product. The experiments in the range of pressure (stress) 5-50 GPa without the use of preheating of the powders on TiC, TiB and WC basis, do not lead to the positive results. In dependence of the alloy content, the optimal temperature of preheating is changed and has the different values of hardness and density. For instance, for WC-10%Co alloy at 1000°C the hardness is 68-70 HRC, while for WC-8%Co+50%W alloy at 1200°C the hardness is 59-61 HRC. The minimal temperature for TiC and TiB consolidation is 1400-1500°C, that is practically attainable only with SHS. At low temperatures for today's day more favorable results couldn't be obtained.

Figure 6 represents some types of the dynamically with preheating compacted samples on TiC, TiB and WC basis.

IV. CONCLUSION

1. Preheating temperature is the main parameter at dynamic compaction of hard materials on TiC, TiB and WC basis, providing the receiving of high dense samples with high values of hardness and perfect structure. At compaction of TiC, TiB and alloys on their basis, the minimal temperature is 1400-1500°C and in the case of WC basis alloys 900-1000°C.

   The process of SHS as a preheating method is the single method for producing high density samples by shock wave loading in the range of intensity 5-20 GPa for TiC and TiB based alloys.

2. Preliminary density and the quality of the compacting powders have not unimportant role at high temperature dynamic compaction of WC based powders and play an

essential role in the formation of correct geometry without defects of final products.

3. The typical peculiarity for WC grains during dynamic deformation at high temperature compacted by shock wave is strong fragmentation of separate particles and disorientation of individual fragments relative to each other.

REFERENCES

1. A. Peikrishvili, F. Tavadze i dr. Pressovanie nekotorikh karbidov udarnimi volnami pri visokikh temperaturakh. Proceedings of the International Conference of Using the Energy Explosure for Production of High Properties Materials, Pardubice, Czechoslovakia, 1985.

2. L. Japaridze, A. Peikrishvili et al. Dynamic consolidation of W based alloys at elevated temperatures. Proceedings of the International Conference on High Energy Rate Fabrication "HERF '89", Ljubljana, Yugoslavia, 1989.

3. Y. Gordopolov, V. Fedorov i dr. Vzrivnaja obrabotka SVS produktov. Proceedings of the International Conference on High Energy Rate Fabrication "HERF '89", Ljubljana, Yugoslavia, 1989.

4. F. Tavadze, A. Peikrishvili. Isledovanija vlijanija predvaritelnogo nagreva na vzrivnoe pressovanie karbidov volframa. Soobshenie AN GSSR, 109, 02, 1983.

43

Equation of State of Porous Metals in Explosive Compaction

SHAO BINGHUANG, WANG XIAOLIN, and LIU ZHIYUE

Institute of Mechanics
Chinese Academy of Sciences
Beijing, 100080, China

In this paper, the state equations of loading and unloading process have been discussed. According to the elasto-plasto-fluid model, we suggest that the relationship of Us ~ Up is a power function, $U_s = C_{op} + \lambda U_p^n$. A method to describe the coefficients is introduced by considering p-α model, Carroll-Holt sphere model and Hugoniot data of solid material. The key point is how to determine the coefficient Cop. In this paper, the Cop represents the plastic wave velocity of powder material, and it can be determined by using analytical and experimental methods.

In isentropic unloading process, it is assumed that the space of pores does not recover, and the expansion of explosive consolidated powder is caused only by matrix material volume deformation. According to this hypothesis, the unloading equation of consolidated powder can be deduced. The state equations of both the loading and unloading states of powder material have the same form as equations of solid materials, and so are simple and convenient for analysis.

1. INTRODUCTION

In order to investigate the phenomena of explosive consolidation of powder materials, state equation is a precondition. But up to now, we still need a simple and suitable equation to describe the process of explosive consolidation. The main reason is that the properties of powder material are more complicated than those of solid or fluid matter. Next, for most metal powders, the pressure of explosive consolidation is just in the transition region from elasto–plastic (E–P) state to fluid. So it is difficult to describe the consolidation process only by either E–P model or fluid model.

In the last twenty years, some kinds of important constitutive equations have been suggested to describe the behavior of porous materials under shock loading. Such as flow–dynamics model by Zeledovich [1], McQueen et al. [2], Herrmann's P~α model [3, 4, 5] and Carroll–Holt's hollow–sphere model (1972–1986) [6, 7, 8].

II. STATE EQUATION OF POWDER MATERIAL

A. BASIC IDEA

In this paper, the basic ideas of P~α model and hollow–sphere model are developed from a different viewpoint. The McQueen's formula [2] is still adopted, but the shock wave velocity U_s is taken as

$$U_s = C_{op} + \lambda_o U_p^n \tag{1}$$

where U_p is the particle velocity, λ_o, n are constants of the material, C_{op} is the disturbed wave velocity of powder material. The term $\lambda_o U_p^n$ denotes the effect of shock–loading. If the shock wave becomes weak enough, U_s would approach to the sound velocity C_{op}. Otherwise, the term $\lambda_o U_p^n$ would play an important role. So the formula (1) can reflect high or low pressure properties.

B. THE PROPERTIES OF C_{op}

There are two kinds of sound velocity in the powder materials, the elastic wave as a precursor–wave usually propagates before the shock wave, but deformation induced by it is small, and its effect can be neglected. The compacting of powder material is mainly a plastic deformation process propagating with a plastic wave, and its velocity is C_{op}. The shock wave is developed from the plastic wave. And the U_s can be denoted as C_{op} plus a revision term $\lambda_o U_p^n$. According to the shock wave equations, the U_s, U_p could be expressed as follows:

$$U_s^2 = -v_{oo}^2 \left(\frac{P - P_o}{v - v_{oo}} \right) H \tag{2a}$$

$$U_p^2 = (P - P_o)(v_{oo} - v) \tag{2b}$$

where V_{oo} is the initial specific volume of porous material. If velocity U_p approaches zero, the U_s would be approached C_{op}:

$$C_{oP}^2 = -v_{oo}^2 \left(\frac{dP}{dv}\right)_H \tag{3}$$

According to Eq. (3) a general expression was deduced by Herrmann[3]

$$C_{oP}^2 = -v_{oo}^2 \frac{\frac{1}{\alpha}\left(\frac{\partial P}{\partial v_s}\right) + P_e\left(\frac{\partial P}{\partial E}\right)}{1 + v \cdot \left(\frac{\partial \alpha}{\partial P}\right)\left(\frac{\partial P}{\partial v_s}\right)/\alpha^2} \tag{4}$$

where E is the specific internal energy. V_s is the specific volume of matrix material, P_e represents the initial pressure.

According to Carroll and Holt's sphere model, the expression of $p=p(\alpha)$ could be deduced. If velocity U_p approaches zero, i.e. the state is in quasi–static condition, the relationship of P~α is quite simple [6]:

$$P = \frac{1}{\alpha}\frac{2}{3}Y\left[\ln\left(\frac{\alpha}{\alpha-1}\right) - \ln\left(\frac{\alpha_0}{\alpha_0-1}\right)\right] \tag{5}$$

where α_o is natural bulked porosity of powder material. In Fig. 1, the P~V/V_o curves are shown. The solid line denotes rapidly solidified aluminum powder. It is compared with the results of Eq.(5), denoted by dashed line, with $Y_{Al}=0.15$GPa, $\alpha_{oAl}=2.12$. Substituting Eq.(5) into Eq.(4), and neglecting the high order term $P_e \cdot \partial P/\partial E$, the following relation is obtained

$$C_{oP}^2 = \frac{\alpha C_o^2}{1 + \frac{3}{2}\frac{C_o^2}{Y}\frac{\alpha}{v_o}\left[\ln\left(\frac{\alpha}{\alpha-1}\right) + \frac{1}{\alpha-1}\right]^{-1}} \tag{6}$$

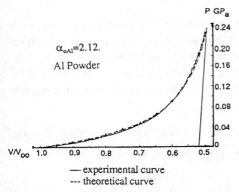

— experimental curve
--- theoretical curve

Fig. 1 The comparison of theoretical and experimental curves.

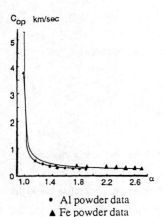

• Al powder data
▲ Fe powder data

Fig. 2 The curves of wave velocity C_{op} and porocity α, Al and Fe powders.

where C_o is a constant of solid material, representing the body wave velocity. When $\alpha \to 1$, then $C_{op} \to C_o$. V_o is the initial specific volume of solid material.

C. THE DETERMINATION OF COEFFICIENTS λ_o, n

If the coefficients C_o, λ of solid material are known, the coefficients λ_o, n can be determined by combining, Eq.(2) a,b , Eq.(6) and McQueen's formula [2]:

$$P = \frac{\rho_o C_o^2 (1-v/v_o)}{[1-\lambda(1-v/v_o)]^2} \frac{1-\gamma_o(1-v/v_o)/2}{1-\gamma_o(v_{oo}/v_o - v/v_o)/2} \tag{7}$$

In Figs. 3a, 3b, the aluminum and iron powders $U_s\sim U_p$ calculated curves are compared with experimental data taken from McQueen et al. [2]. In a wide pressure region even to the order of 100 GPa, these curves coincide well with experimental data. In the practice of explosive compaction the particle velocity U_p is usually less than 1 km/s, and pressure P<10 GPa. In this condition the $U_s\sim U_p$ relation can be reduced to a linear formula $U_s=C_{op}+\lambda_o U_p$ without causing obvious error.

According to Eqs (2) a, b , (6) and (7), the copper powder (ρ_{oo}=72% TD) and aluminum powder (ρ_{oo}=80% TD) $U_s\sim U_p$ relationships are given as follows:

Cu Powder: $U_s=0.24+2.76 U_p^{0.83}$ $(0 < U_p < 3 \text{ Km/s})$

$U_s=0.24+2.76 U_p$ $(0 < U_p < 1 \text{ Km/s})$

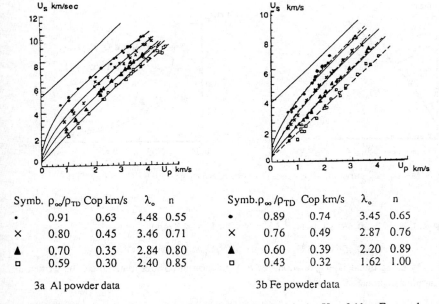

Symb.	ρ_{oo}/ρ_{TD}	Cop	km/s	λ_o	n
•	0.91	0.63		4.48	0.55
×	0.80	0.45		3.46	0.71
▲	0.70	0.35		2.84	0.80
□	0.59	0.30		2.40	0.85

Symb.	ρ_{oo}/ρ_{TD}	Cop	km/s	λ_o	n
•	0.89	0.74		3.45	0.65
×	0.76	0.49		2.87	0.76
▲	0.60	0.39		2.20	0.89
□	0.43	0.32		1.62	1.00

3a Al powder data   3b Fe powder data

Fig. 3 The curves of shock wave velocity U_s and particle velocity U_p of Al or Fe powder, experimental data taken from [2].

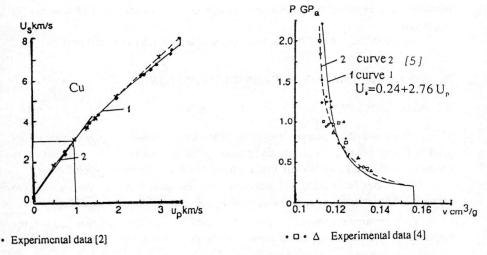

• Experimental data [2]   • □ • Δ Experimental data [4]

Fig. 4 (a) $U_s \sim U_p$ curve of Cu powder, (b) p~v curve of Cu powder (=72% TD).

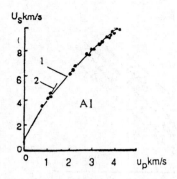

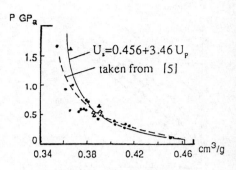

Fig. 5 (a) $U_s \sim U_p$ curve of Al powder, (b) $p \sim v$ curve of Al powder (=80% TD)

Al Powder: $U_s = 0.456 + 3.46 U_p^{0.717}$ $(0 < U_p < 4 \text{ Km/s})$

$U_s = 0.456 + 3.46 U_p$ $(0 < U_p < 1 \text{ Km/s})$

By using linear $U_s \sim U_p$ relation, the P~V relation of powder material can be written in the familiar manner.

$$P = \frac{C_o^2 P (1 - v/v_{oo})}{v_{oo}[1 - \lambda(1 - v/v_{oo})]^2} + \overline{P}$$

$$(8)$$

where initial pressure $\overline{P}$ is determined by Eq.(5).

In Fig. 4 and Fig. 5, the $U_s \sim U_p$ and P~V curves of Cu, Al powders are given and compared with experimental data taken from McQueen [2] Boade [4] and Butcher [7] respectively.

In Figs 4b, 5b, we find that the errors of the linear relation can be accepted.

III. ISENTROPIC UNLOADING OF CONSOLIDATED POWDER

A. *MECHANICS MODEL AND BASIC HYPOTHESES*

The isentropic unloading process of explosive consolidated powder is quite different from solid or liquid. Because the compacting process of powder material is irreversible. It is necessary to utilize the properties of matrix material to solve it. The basic hypotheses are:

(1) The powder material and matrix material are applied by the same shock wave pressure and have the same internal energy ($E_2 = E_1$), but their deformation are different.

(2) In isentropic unloading process, the space of void does not recover, and the expansion of consolidated powder is caused only by the deformation of matrix material, i.e. $\Delta V_{mat.} = \Delta V_{cons}$. In Fig. 6, the $\widehat{V_{oo} P}$, $\widehat{V_o M}$ and $\widehat{V_o H}$ curves represent Hugoniut curves for

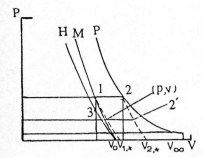

Fig. 6 Schematic diagram of isentropic unloading curve.

the powder, matrix and solid material, respectively. The dashed line $\widehat{1v_{1*}}$, $\widehat{2v_{2*}}$ represent the unloading curves of matrix and powder materials, respectively.

According to the hypothesis (2), if the unloading curve of matrix material is known, then the unloading curve of consolidated powder can be obtained.

B. HUGONIOT P~V RELATION OF MATRIX MATERIAL

The P_1~V_1 relation of matrix material was given by Boade [4]:

$$P_1 = \frac{P_3(v_o - v_1 - 2 \cdot v_o/\gamma_o) - \overline{P}(v_{oo} - v_2)}{(v_{oo} - v_2 - 2 \cdot v_o/\gamma_o)} \qquad (9)$$

where $P_1 = P_2$ $\qquad (10)$

$$P_2 = \frac{C_{oP}^2(1 - v_2/v_{oo})}{v_{oo}[1 - \lambda_o(1 - v_2/v_{oo})]^2} + \overline{P} \qquad (11)$$

and

$$P_3 = \frac{C_o^2(1 - v_1/v_o)}{v_o[(1 - \lambda(1 - v_1/v_o)]^2} \qquad (12)$$

the definitions of V_1, V_2, V_o, V_∞, P_1, P_2, $\overline{P}$ and P_3 are shown in Fig. 6.

C. ISENTROPIC UNLOADING EQUATION

If the matrix material starts unloading at state point 1 (see Fig. 6), the isentropic equation is shown as follows:

$$P = A\rho^K - B \qquad (13)$$

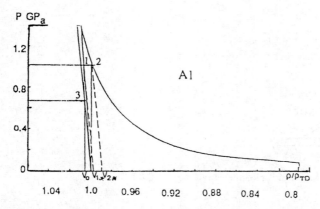

Fig. 7 Isentropic unloading curve of Al powder (=80% TD)

The unknown coefficients A, K and B are determined by the condition of state point 1 [9]. The coefficients can be expressed as the functions of C_1 / C_o and ρ_1 / ρ_o

By using Gruneisen equation and isentropic condition, $dE/dV=-P$, $\eta=(1- V_1/V_o)$ the sound velocity C_1 can be expressed in the form:

$$C_1^2 = C_o^2 (\frac{v_1}{v_o})^2 \{ \frac{(1-\gamma_o \eta/2)(1+\lambda \eta)}{(1-\lambda \eta)^3} - \frac{\gamma_o}{2 \rho_o C_o^2} P_3 - \frac{\gamma_o}{\rho_o C_o^2} P_1 \} \tag{14}$$

According to the hypothesis (2), if consolidated powder expands to state point 2′ (shown in Fig. 6), the specific volume V_2' would be

$$v_2' = v + (v_2 - v_1) \tag{15}$$

By combining with Eq (13), the unloading equation of consolidated powder is shown as following:

$$P = A (\frac{1}{v_2' + v_1 - v_2})^K - B \tag{16}$$

where V_2' is variable, and V_1, V_2 have been determined.

A typical unloading process is shown in Fig. 7.

IV. CONCLUSION

In this paper a suitable and simple equation of state for powder material is suggested. In a wide pressure region, it coincides with experimental data quite well. In lower pressure region, it can be reduced to a linear relationship. Thus the related equations of powder material have the same form as equations of solid material and give us great conventience, especially in analytical solution.

In isentripic unloading process, we suppose that the expansion of consolidated powder is caused only by the deformation of matrix material volume. According to this hypothesis, an unloading state equation of powders can be deduced with the similar form of solid material unloading equation. The results for unloading explain reasonably the experimental phenomena.

ACKNOWLEDGMENT

The authors gratefully acknowledge the support provided by Chinese National Natural Sciences Foundation.

REFERENCES

1.   Y.B. Zeledovich and Y. P. Raegel, *Physics of shock wave and high temperature fluid dynamics phenomena, Academic Press, Chapter 11,   10* (1966; Russian).

2.   R.G. McQueen, S.P. Marsh, J. W. Taylor, J.N. Fritz and W.J. Carter, *High Velocity Impact Phenomena, edited by R. Kinslow, Academic, Press N. Y.:* 297 (1970).

3.   W. Herrmann, *J. Appl. Phys., 40:* 2490 (1969).

4.   R. P. Boade, *Shock Waves and the Mechanical Properties of Solid, edited by John J. Burke, Volker Weiss, Syracuse University Press, NY:* 263–285 (1971).

5.   Ki–Hwan Oh, Per–Anders Persson, *J. Appl. Phys., 66(10):* 4736 (1989).

6.   M. M. Carroll, A. C. Holt, *J. Appl. Phys. 43(4) :* 1626–35   (1972).

7.   B. M. Butcher, M. M. Carroll and A. C. Holt, *J. A. P. 45 (9):* 3864–75 (1974).

44

Influence of Environment Condition on Conservation of Explosively Obtained Diamond

E. A. PETROV, G. V. SAKOVITCH AND P. M. BRYLYAKOV

Laboratory of Dynamic Synthesis of Superhard and Ceramic
Materials, NPO "Altai", Biysk, 659322, USSR

Director General, Professor
NPO "Altai", Biysk, 659322, USSR

Department of Synthesis of Superhard and Ceramic Materials
NPO "Altai", Biysk, 659322, USSR

The environment influence in explosion chamber on
diamond production from explosives was investigated. Depending on volume, composition and environment conditions in
the chamber the diamond losses because of graphitization and
gasification have been found to be significant. It has been
shown that for certain conditions there exists its own
explosive mass in the chamber, with which the diamond losses
will be minimal. Based on summation of experimental results
optimum conditions have been defined.

I. INTRODUCTION

Production of diamonds from explosives represents one of the
novel trends in explosive research and synthesis of materials with new properties. This trend is based on experimental production of diamond modification of carbon with detonation decomposition of explosives or their mixtures such as
C-H-N-O [1,2].

The main conditions of detonation production of ultra-
fine diamonds (UFD) from explosives are described [1]. The
UFD production in the process of detonation decomposition of
explosives is carried out when achieved necessary parameters
of pressure P and temperature T and with free carbon available in detonation wave. The subsequent conservation of
produced diamond is achieved in the result of scattering of
detonation products in the atmosphere being inert towards
carbon.

While thermodynamic parameters and explosive structure
define the formation of diamond phase nuclei and their
subsequent growth, environment conditions in explosion
chamber, i.e. volume, composition and properties of medium,
exert an effect on UFD conservation during the scattering of
detonation products.

II.   EXPERIMENTAL RESULTS

The results obtained when various gas media were used showed
that UFD and condensed carbon (CC) outputs towards starting
explosive mass have been changed depending on gas composition (see Table 1).

With density and explosive mass being equal, the
highest output is observed with carbon dioxide and falls off
in the following order: nitrogen, argon and helium, practically disappearing completely in vacuum. The similar
relationship is run down for UFD output with respect to CC.

With pressure rise in the chamber CC and UFD outputs
are increased in all examined gases. In this case UFD
output in nitrogen, argon and helium approximates to that
being achieved with carbon dioxide. But for similar output
to obtain they must be at substantially high pressure. So
long as detonation parameters and explosive composition have
not been essentially changed, the obtained results might be
explained by different cooling conditions of detonation
products, being realized in chamber depending on gas composition and properties. Cooling conditions of chamber, in
turn, determine CC losses in "secondary" chemical reactions

TABLE I

UFD and CC output from trotyl - hexogen in 1:4 ratio (TH 50/50) in controlled gas media

Medium	Density of the explosive charge $(kg/m^3 T)$	Explosive mass (kg)	CC/ Explosive (%)	UFD/ Explosive (%)	UFD/CC (%)
Carbon	1670	0.1	8.53	3.50	41.03
Dioxide	1660	0.05	9.89	5.32	53.79
Nitrogen	1670	0.1	6.12	0.79	12.91
	1550	0.1	7.47	1.10	14.73
	1670	0.05	12.30	2.75	22.36
Argon	1670	0.1	2.46	0.05	2.11
	1550	0.1	4.31	0.42	9.74
Helium	1670	0.1	2.42	0.19	7.85
Vacuum is no less than 0.89 KPa	1670	0.1	0.03		

Volume of spherical type explosion chamber - 0.175m³; explosive charge diameter - 40mm; primer - 0.005kg of tetryl.

and evaluate diamond losses in the result of its reverse going to graphite. This is evidenced by UFD output data towards CC.

The analysis showed that UFD and CC output was increased with the rise of thermal capacity of environment in explosion chamber. It may be expressed as a product of density p, specific thermal capacity c_v, pressure P and chamber volume V_{ch}. Fig. 1 shows UFD output from TH 50/50 composition against thermal capacity of used medium with equal values of explosive mass and density (1600-1670kg/m³). It is evident that UFD output with respect to CC also achieves constant value and makes up 62%. For experiments with less explosive mass UFD output reaches maximum with lower values of medium thermal capacity.

Based on SEM data and measurements of specific area S_{BET} particle size is remarkably reduced when UFD output falls off. So for instance, S_{BET} values of diamond powders, equal to 315 and 272m²/g, correspond to A and B on the curve I

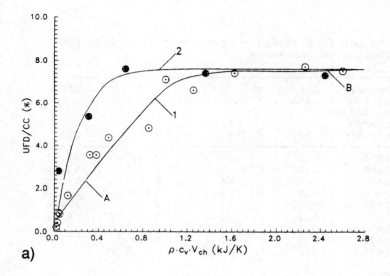

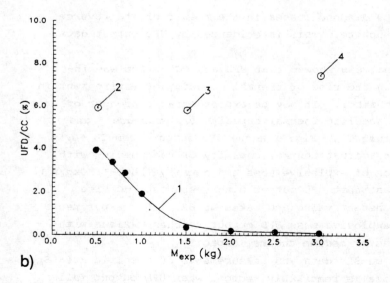

FIG. 1 *Dependence of UFD output on medium thermal capacity*
in chamber for M_{exp}=0.100kg: curve I; M_{exp}=0.05kg: curve II.

(Fig. 1). These results corroborate that part of UFD losses
is provided by its graphitization with detonation products
scattering in chamber.

When thermal capacity of the environment is increased,
the "secondary" chemical reactions in detonation products
are suppressed. This permitted to find conditions of UFD
and CC conservation when they scattered in reactive oxygen-
containing environment.

One can predict with certainty that increase of CC and
UFD output with rise of the environment thermal capacity is
reached with the help of more intense cooling of detonation
products when scattered in chamber. The environment facilitates the conservation of obtained UFD and doesn't effect on
the process of its formation. At the initial stage of
scatter cooling of the detonation products is concerned with
their adiabatic expansion and mainly depends on density and
compressibility of gases in chamber. At the following stage
thermophysical properties of environment exert essential
effect.

For each explosive formulation V_{ch}, parameters and
properties there exists its own mass of the explosive, with
which the UFD losses are minimal, i.e. they comply with B
conditions on the curve I. Based upon data (Fig. 1) this
explosive mass may be found from the following relation

$$M_{exp} = \frac{K \, p \, c_v \, V_{ch} \, P}{Q_{vexp} \, P_0},$$

where Q_{vexp} is the explosion thermal effect; K is coefficient
of proportionality.

According to proposed relation M_{exp} will grow when using
media with high thermophysical parameters and rising p in
chamber. This conclusion is confirmed by experiments in
chamber with volume equal to $3.0m^3$ (Fig. 2). With explosive
mass and chamber pressure being equal, UFD output goes
higher in medium with higher thermophysical parameters.
Rise of P in chamber increases significantly M_{exp}. As it
should be expected the significant rise of M_{exp} was achieved

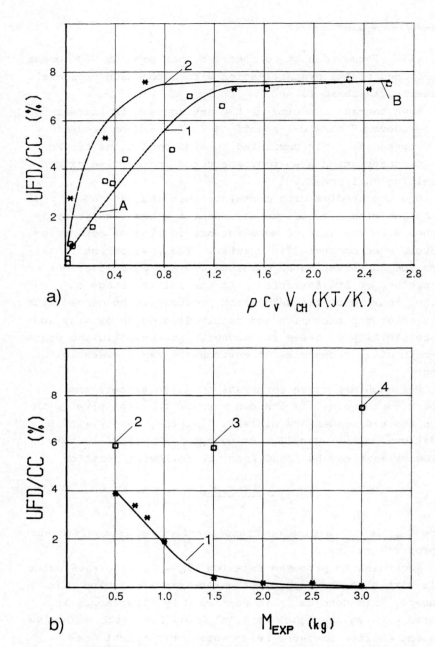

FIG. 2 Dependence of UFD output on explosive mass with various medium cooling conditions in chamber: 1- own explosion products at pressure of 0.1 MPa; 2- nitrogen at pressure of 0.3 MPa; 3- nitrogen at pressure of 0.4 MPa; 4- air-blown foam. Explosive composition- TH 60/40; charge density- 1640kg/m³; diameter- 90-110mm.

in the air-blown foam, having thermophysical properties on one order of magnitude or more higher than with gases. In this case M_{exp} may still grow.

III. CONCLUSION

Rise of M_{exp} and reduction of UFD during production is one of the most important problems in increasing the detonation synthesis efficiency. The described results may serve as foundation for optimum condition selection of UFD production and the ways of its further efficiency increase.

REFERENCES

1.   A. J. Lyamkin et al., *DAN USSR.* *3:* 302(1988).

2.   N. Roy Greiner, D. S. Phillips, J. D. Jonson and Fred Volk, *Nature,* *3:* 332(1988).

Section IV
Shaped Charge Phenomena

45

Shaped Charge Jetting of Metals at Very High Strain Rates

FRED I. GRACE

United States Army Ballistic Research Laboratory
Terminal Ballistics Division
Aberdeen Proving Ground, Maryland 21005-5066

Jets are formed from shaped charges when a metallic liner is subjected to extreme pressures of an explosive detonation. Liner deformation is examined with emphasis placed on metal flow velocities, strains, strain rates, temperature and role of metallurgical parameters in the jet formation process. Flash x-ray photographs are presented for copper lined shaped charges showing explosive detonation, liner collapse, jet elongation and jet break-up. Related studies such as shock-induced defect structures and efforts to examine recovered jet particles are discussed.

I. INTRODUCTION

In a shaped charge, the metal liner that ultimately forms the jet is subjected to severe deformation processes at extremely high strain rates and shock pressure. To some extent, these processes are sequential, beginning with 1) shock loading of the metal liner during explosive detonation, 2) high rate plastic flow as the liner moves towards the charge axis, 3) abrupt pressurization and release as liner metal flows into and out of the stagnation region, 4) rapid stretching of the formed jet, and finally 5) fracture of the elongated jet into various particles. These complex processes are not easily addressed from a metallurgical viewpoint since they exist for such short times (microseconds). Fig. 1 illustrates the five deformation stages described above.

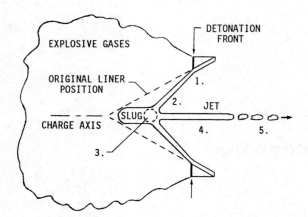

FIG. 1 *Shaped charge undergoing explosive detonation. Numerals designate deformation regions as indicated in text.*

In the past, physics of jet formation treated the metal as a fluid and that successfully predicted liner and jet motion. However, a wide range of observations clearly indicate that metallurgical properties control various aspects of the intervening processes and final jet behavior. For example, flash x-ray shadowgraphs have shown that jet break-up resembles classical failure modes in crystalline solids, i.e., brittle fracture and necking. Flash x-ray diffraction techniques of Jamet and Thomer [1] and Green [2] established that jets from metallic liners exhibit degrees of crystallinity. Examinations of recovered slugs (liner portion not going into jet) suggested that recrystallization occurred in some localized areas [3]. Studies on the starting liner metal indicate that jet break-up may be influenced by purity, grain size and texture [4].

Significant recent efforts include those of Zernow [5] for recovering jet particles intact. With this technique, Zernow [6] and Murr and co-workers [7] are investigating final jet metallurgical properties. Also, Chokshi and Meyers [8] are applying elements of dynamic recrystallization to explain high ductility of jets.

This paper reviews the physics of shaped charge jet formation and identifies some important associated macroscopic conditions (strains, temperatures, etc.) that can exist. Examples of liner collapse and jet break-up in the form of flash x-rays are provided. In addition, some current and previously reported high rate shock deformation studies and their possible relevance is discussed.

II. JET FORMATION AND PENETRATION

In the theory of jet formation, the metal liner is treated as a fluid. As such, liner collapse and flow resulting in a jet and slug are represented by an application of

Bernoulli's equation for streamline flow, i.e., $1/2 \rho v^2 + P = $ constant. Birkhoff et al [9] developed the theory for one-dimensional, incompressible, steady flow where the pressure was considered to be zero, i.e., $1/2 \rho v^2 = $ constant. Conserving mass and momentum in the collapse geometry of Fig. 2 gave jet and slug masses and jet and slug velocities as functions of liner mass, velocity and collapse angle β. Here β is related to the original cone half angle α and the Taylor "throw off" angle ϕ (2δ). Liner velocity is determined by the explosive-metal masses and is calculated using Gurney's formula, for example. To deal with real shaped charges where the explosive mass varies over the charge length, Pugh et al [10] developed a "non-steady" theory wherein gradients in collapse variables were introduced. This accounted for velocity gradients and jet stretching as observed in flash x-rays of experimental jets.

Grace et al [11] extended the theory to a more general set of solutions by allowing the pressure to vary along streamlines, i.e., $1/2 \rho v^2 + P = $ constant. Thus:

$$V_j = V_0 \left[\text{CSC}(\beta/2) \, \text{COS}(\beta/2 - \delta) + (\lambda - 1) \, \text{CSC}(\beta) \, \text{COS}(\beta - \delta) \right]$$

$$V_s = V_0 \left[\text{SEC}(\beta/2) \, \text{SIN}(\beta/2 - \delta) - (\lambda - 1) \, \text{CSC}(\beta) \, \text{COS}(\beta - \delta) \right]$$

$$dM_j = dM_1 \left[\text{COS}^2(\beta/2) + (1/2)(1/\lambda - 1) \, \text{COS}(\beta) \right]$$

$$dM_s = dM_1 \left[\text{SIN}^2(\beta/2) - (1/2)(1/\lambda - 1) \, \text{COS}(\beta) \right]$$

$$\lambda = \left[1 + 2P / \rho v^2 \right]^{1/2} \quad \text{and} \quad \beta = \alpha + 2\delta$$

The form of λ shown corresponds to the case where incoming flow is partially stagnated. When pressures in all streamlines are equal or zero, these equations reduce

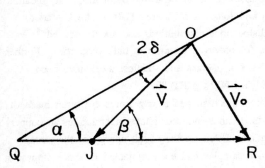

FIG. 2 *Geometry of collapse for a shaped charge, showing charge axis along QR, original liner inclined along QO, liner flow along OJ at velocity v, liner collapse angle β and the collision point J. Detonation front propagates from left to right and has advanced to the point O. Liner has aquired a velocity v_0 at point O as a result of the detonation pressure.*

to those of Birkhoff and Pugh. There have been any number of refinements. Godunov et al [12] treated the liner as visco-plastic where $1/2 \, \rho v^2 + \sigma =$ constant and $\sigma = \sigma_0 + \mu \dot{\varepsilon}$. Lampson [13] applied the Birkhoff approach to the case where the liner, during convergence, develops a velocity gradient through its thickness. These suggest that strain rate effects and shear strains can be important in jet formation.

The simplest penetration theory is based on a hydrodyamic view of the jet penetration process since jet velocities are large (5-10 mm/μs) and impact pressures far exceed material strengths. Birkhoff et al [9] applied Bernoulli's equation to this problem also. Results provided penetration depth of jets as a function of jet and target densities and jet length. Diperso et al [14] extended the theory to account for jet stretching and jet particulation. A simple representation is:

$$p = [\, \rho_j/\rho_t \,]^{1/2} \, (V_{jtip} - V_{jtail}) \times t_B$$

where t_B is the time lapse between the origin of the jet and break-up. It can be readily observed that jet penetration is increased with increased jet ductility.

III. EXPLOSIVE SHOCK LOADING OF LINER METAL

The passage of the detonation wave over the liner results in an explosive shock loading of the liner metal. Although the geometry is not planar, mechanisms involved in planar shock loading should apply. Smith [15] and Dieter [16] observed that mechanical hardening and microstructural defects were induced when metals were subjected to shock waves comparable to those of an explosive detonation. Application of TEM by Nolder and Thomas [17] and Johari and Thomas [18] to shock loaded samples of Ni and Cu showed that dislocations distributions and shock-induced mechanical twinning were responsible for observed changes in bulk properties. Further studies regarding shock response of Ni, Cu, α-brass and stainless steel have been summarized by Meyers and Murr [19] in EXPLOMET 80.

Copper and its alloys are of interest since shaped charge liners are often made of copper. Grace and Inman [20] examined shock-induced defect substructures in copper and a series of α-brasses. Results indicated that stacking fault energy of the metal controlled dislocation distributions when samples had been subjected to shock impact at 5.5 GPa (55 kb). Studies at higher peak shock pressure (20-40 GPa), as mentioned previously, showed that shock-induced mechanical twinning occurs. Data taken collectively suggests that metals with higher stacking fault energy require higher peak shock pressures to induce twinning.

Theoretical developments include those of Cowan [21], showing that shearing stresses in the shock front can exceed the critical shear stress required to activate twin-like displacements. Also, Grace [22] developed methods to calculate shock wave strengthening in metals, based on work done in the shock-relief compression cycle and a partitioning of that work into thermal and athermal components. Earlier, McQueen and Marsh [23] established theoretical arguments for temperature increases in shock compression and subsequent post-shock or residual temperatures.

These past studies should provide a basis for establishing effects of explosive detonation shock on metallic liners. Thus, liner metal ought to be hardened, have high defect density and be at 100-200 °C in this first stage of deformation.

IV. LINER COLLAPSE PROCESS

Metallurgical details of liner collapse have not been studied to any extent, however overall metal flow has been observed via flash x-ray and calculated by hydrocode [24]. These show that liner metal flows inward and speeds vary along the liner length. Thus, for the BRL 81mm shaped charge as shown in Fig. 3, inward velocities vary from 3.5 mm/μs near the apex to 1.0 mm/μs at the basal area. Liner elements near the apex undergo strains of 500 %, while basal elements can extend to 1500 %. Strain rates can reach 10^6/s. Strain and strain rates vary through the liner thickness. Generally, those on the liner inside portion are higher. Figures 4 & 5 show experimentally observed liner collapse [25] and computed collapse [24].

The collapse can produce significant pressurization of the liner [11]. Thus, metal

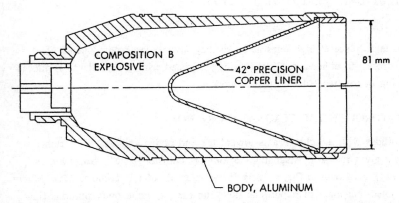

FIG. 3 *BRL 81mm precision shaped charge. Typical values include V_J(tip) at 7.8 km/s, V_J(tail) at 1.5 km/s and break-up time of 116 μs.*

FIG. 4 *Flash x-ray of the BRL 81mm shaped charge taken during liner collapse at a delay of 31 μs after detonator function [25].*

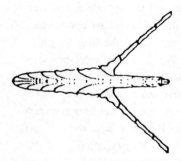

FIG. 5 *Hydrocode calculation of liner collapse and jet formation for the BRL 81mm shaped charge [24]. Run time is at 25 μs after main charge initiation.*

deformation takes place at high rates and also may involve large components of hydrostatic stress. In either case, large strains associated with liner collapse results in further heating of the metal due to plastic work.

V.   FLOW THROUGH THE STAGNATION REGION

The liner collides with the charge axis toward the end of its collapse. This rather abrupt impact gives rise to a "stagnation region" or "collision zone". Sedgewick's calculation [24], as shown in Fig. 5, provides an example of this complex flow pattern. It is to be noted that inner liner portions flow into the jet, while outer portions flow into the slug. Further, the flow is such that jet and slug receive different amounts of strain. Impact pressures can be estimated using Bernoulli's equation, $P = 1/2 \ \rho v^2$, where v is the incoming flow velocity. Typical values range from 70 GPa (700 kb) near the apex to 4.5 GPa (45 kb) at the base. Chapayak's hydrocode calculation [26] shows that abrupt pressurization during collision creates temperatures exceeding 700 °C. However, that temperature is reduced to 400-500 °C during liner flow into jet and slug under conditions of adiabatic expansion as the pressure drops.

It is not clear to what extent previous shock loading data relate to impact at collision. Hooker et al [27] examined copper after planar shock loading to 100 GPa (1.0 mb) and found no twin-type substructures. The metal appeared to be composed of refined subgrains with high dislocation densities in the subgrain walls. Generally, stagnation pressures are lower and they vary along the charge length. Thus, a gradient of substructures should exist in the metal that emerges from the stagnation region.

VI. JET STRETCHING AND BREAK-UP

Velocity gradients in shaped charges cause jets to stretch. Flash x-ray observations show details of this behavior as well as eventual break-up of the jet into a co-linear series of particles. Fig. 6 shows typical jet stretching and ductile break-up for the BRL 81mm shaped charge [25]. Before break-up a strain of 550 % is obtained. A break-up time for this jet was 116 μs, thus, an average strain rate would be close to 5 x 10^4/s. Of course, values of strain rate, strain and temperature vary along the jet length since each jet element has been formed under slightly different conditions. Von Holle and Trimble [28] measured the jet temperature to be 432 °C using IR techniques.

The jet stretching process is one that involves large strains and takes place at high strain rates and elevated temperature. Although, flash x-ray diffraction of aluminum jets indicate a degree of crystallinity [1,2], current efforts are examining whether local melting exists within the jet [6]. This might support the notion that dynamic recrystallization is responsible for large strains involved [8]. Recent observations of recovered jet particles by TEM are showing redefined grain boundaries [7]. These

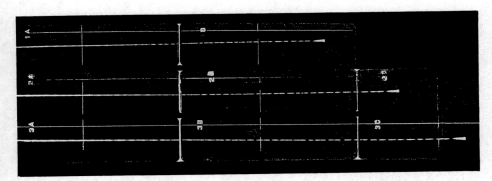

FIG. 6 *Flash x-rays of jets from BRL 81mm shaped charge showing jet elongation and jet break-up. Flash times are 168, 186 and 205 μs after detonator function.*

possibly reflect effects of either in-situ or post deformation thermal processes. A detailed examination of recovered jet particles [6] show fracture cones that exhibit both ductile and brittle features. However, since gradients exist, these results depend highly upon particle location along the jet length. The mechanisms responsible for jet elongation and failure are not yet established, but current efforts are addressing this difficult problem.

VII. SUMMARY

Jet formation involves a complex interplay between thermal-mechanical deformation cycles imposed on the liner and the metal response to those deformations. The physics and mechanics of the problem appear to be well understood, but improved penetration depends upon increasing jet ductilities beyond high amounts already obtained. Table I. summarizes the previous discussion by listing some important conditions that exist in each deformation stage. Conclusions are that each deformation stage subjects the metal to formidable loading and starting material for each stage depends on cumulative deformations of prior stages. Although values in the table are for one point, actually, they vary along the shaped charge due to gradients. Thus, metallurgical states are expected to vary along the jet length.

Probably, there are sufficient previous data to estimate metal response to the first stage of liner shock loading by the detonation wave. Much less is known about details of metal response in subsequent stages. Although jet stretching and break-up have been documented extensively using flash x-rays, the operative lattice mechanisms to explain those observations, namely jet ductility, have not been fully established. Finally, it is expected that more recent efforts, some examples of which follow in these proceedings, will provide insight into the role of metallic properties and the deformation mechanisms acting in shaped charge jet formation.

Table I. Deformation Conditions Related to BRL 81mm Shaped Charge
(Evaluated at Liner Midpoint: Radius 20mm)

DEFORMATION STAGE	TOTAL STRAIN (%)	STRAIN RATE (/sec)	PRESSURE GPa (kb)	TEMPERATURE (°C)
Shock Loading	<10	5.0×10^8	45 (450)	155
Liner Collapse	1267 inside	8.7×10^6	3 (30)	NC
	438 outside	2.6×10^6		
Stagnation Point	NC	NC	71 (712)	700+
Jet Elongation	556	4.6×10^4	tension	432
Jet Breakup	35	3.5×10^3	tension	(432)

NC - Not Calculated

REFERENCES

1. F. Jamet and G. Thomer, *C. R. Acad. Sci. (Paris) B279*, 501 (1974).

2. R. E. Green, Jr., *Rev. Sci. Instrum. 46(9)*, 1257 (1975).

3. F. Jamet, Proc. 8th Int. Symp. on Ballistics, (ADPA), Oct., 1984.

4. F. Witt and J. Pearson, ARSCD-TR 81022, ARDEC, Dover, NJ, Oct., 1981.

5. L. Zernow, Proc. 11th Int. Symp. on Ballistics, (ADPA), May, 1989.

6. L. Zernow and L. Lowry, These proceedings: Shock-Wave and High-Strain-Rate Phenomena in Materials, M. A. Meyers, L. E. Murr and K. P. Staudhammer (eds), Marcel Dekker Inc., New York, 1991.

7. A. Gurevitch, L. E. Murr, S. K. Varma, S. Thiagarajan and W. W. Fisher: These proceedings.

8. A. H. Chokshi and M. A. Meyers, *Scripta Metall. et Materia. 24*, 605 (1990).

9. G. Birkhoff, D. P. MacDougall, E. M. Pugh and G. Taylor, *Jour. Appl. Phys. 19*, 563 (1948).

10. E. M. Pugh, R. J. Eichelberger and N. Rostoker, *Jour. Appl. Phys. 23*, 532 (1952).

11. F. I. Grace, B. R. Scott and S. K. Golaski, Proc. 8th Int. Symp. on Ballistics, (ADPA), Oct., 1984.

12. S. K. Godunov, A. A. Deribas and V. I. Mali, *Fizika Goreniya i Vzryva*, 11 (1975).

13. M. Lampson, Proc. 10th Int. Symp. on Ballistics, (APDA), Oct., 1987.

14. R. Diperso, J. Simon and A. B. Merendino, Ballistic Research Laboratory MR 1296, 1965.

15. C. S. Smith, *Trans. Met. Soc. AIME 212*, 574 (1958).

16. G. E. Dieter, Response of Metals to High Velocity Deformation, 9, 409, Interscience Publishers Inc., New York, 1961.

17. R. L. Nolder and G. Thomas, *Acta Met. 12*, 227 (1964).

18. O. Johari and G. Thomas, *Acta Met. 12*, 1153 (1964).

19. M. A. Meyers and L. E. Murr, Shock-Wave and High-Strain-Rate Phenomena in Metals, M. A. Meyers and L. E. Murr (eds), Plenum Pub. Corp., New York, 1981.

20. F. I. Grace and M. C. Inman, *Metallography 3*, 89 (1970).

21. G. R. Cowan, *Trans. Met. Soc. AIME 233*, 1120 (1965).

22. F. I. Grace, *Jour. Appl. Phys. 40*, 2649 (1969).

23. R. G. McQueen and S. P. Marsh, *Jour. Appl. Phys. 31*, 1253 (1960).

24. R. G. Sedgewick, private communication, 1980.

25. S. K. Golaski, unpublished at BRL, 1979.

26. J. Chapayak, as reported by L. Zernow and L. Lowry: These proceedings.

27. S. V. Hooker, J. V. Foltz and F. I. Grace, *Metall. Trans. 2*, 2290 (1971).

28. W. G. Von Holle and J. J. Trimble, *Jour. Appl. Phys. 47*, 2391 (1976).

46

High-Strain-Rate Deformation of Copper in Shaped Charge Jets

L. ZERNOW and L. LOWRY

Zernow Technical Services Inc.
425 West Bonita Avenue
San Dimas, California, U.S.A. 91773

California Institute of Technology
NASA - Jet Propulsion Laboratory
4800 Oak Grove Drive
Pasadena, California, U.S.A. 91109

Optical, x-ray diffraction, SEM and preliminary TEM and SAD techniques
have been applied to "softly" recovered copper shaped charge jet particles.
Results of these examinations are noted. The hypothesis of momentary
melting of the jet core, as it passes through the collision zone, suggested
by the photomicrographs, is not presently supportable by the results of
computational analysis. Alternate explanations, like dynamic recrystallization may also have difficulty in accounting for the specific observations, including the presence of an apparent axial shrinkage cavity, and
the concentric rings of varying grain size.

I. INTRODUCTION

It has long been known that the lined shaped charge was a device capable of
penetrating deeply into steel armor [Ref.1-1948]. It has also long been
known, but not as widely recognized that, because of the velocity gradient
along the jet, the shaped charge device can also be inversely used as a research tool, to subject materials to high strain rate deformation [Ref.2-
1953]. This confluence of interests has, over the years, provided a
stimulus for the research aspects. It has more recently been recognized

and acknowledged that improved understanding of the finer details of the
physics of the jet formation process can pay substantial dividends in the
practical area of improving the penetration performance of shaped charges.

II. EARLY OBSERVATIONAL METHODS

A. *FLASH RADIOGRAPHY*

One of the most useful research tools for studying shaped charge jets, has
always been flash radiography. This technique permitted successive, precisely timed, multiple source observations of the jet formation and
stretching process with x-ray exposure times ranging downward from about
1 microsecond [in the 1940's] to durations, now available, of the order of
10 nanoseconds. These observations revealed that for copper jets,
stretching under the influence of the velocity gradient, the initial jet
was continuous and capable of undergoing extremely large [superplastic]
strains [> 20] before particulating into separate jet particles. The
particulation process could show either ductile or brittle fracture
characteristics. The amount of jet stretching prior to breakup is related
to the target penetration performance by virtue of the basic [velocity
independent] relationship for the approximate hydrodynamic penetration
depth at high jet velocities.

$$P = L \sqrt{\frac{\rho_j}{\rho_t}} \qquad (1)$$

where P = penetration depth ρ_j = density of the jet
L = effective length ρ_t = density of the
of the jet   target

Longer breakup times generally resulted in deeper penetrations, because of
the increase in effective jet length. The slug [which formed from the
outer portion of the liner during the liner collapse process] was easily
recoverable because of its relatively low velocity in a ground fixed coordinate system, and has been studied [e.g. Ref.3-1952] at considerable
length over the years. However, until fairly recently [Ref.4-1987], it
was not considered feasible to collect the jet particles themselves in a
relatively undamaged condition.

B. *FLASH X-RAY DIFFRACTION*

Although the concept of flash x-ray diffraction dates back to 1942
[Tsuckerman & Avdeenko, USSR], it was not until 1974 that Green [U.S.A] and
Jamet & Thomer [France/Germany-ISL] both demonstrated, essentially simul-

taneously, the ability to obtain dynamic flash x-ray diffraction data by transmission, through jets from small aluminum shaped charges. These results supported the earlier conclusion [Ref.2] that even for aluminum, the jet was essentially solid and crystalline, thus adding another piece of supporting information about the state of the actual jet. In addition, Jamet concluded, from an analysis of his diffraction patterns, that there was a strong [111] orientation along the jet axis. There were also continuing observations by Jamet & Schall [ISL], in other experiments using simple flash x-rays and film densitometry, that the density of the dynamically stretching jet could be as much as 15% lower than the theoretical density of the original liner. Later flash x-ray diffraction experiments by Jamet [Ref.5] concluded that within the limits of the precision of the measurements, estimated values of some of the lattice parameters in the stretching jet were consistent with the other observations that there was a density deficit in the dynamically stretching jet, that appeared to be too large to be attributed to temperature increases.

The intensity limitations of the existing flash x-ray diffraction sources did not permit transmission x-ray diffraction observations through higher density materials like copper. Therefore the interior structures of copper jets were not accessible for study by flash x-ray diffraction.

III. DEVELOPMENT OF THE SOFT RECOVERY TECHNIQUE FOR JET PARTICLES

It was the desire to study in more detail the interior structure and the properties of essentially undamaged copper jet particles, that provided the stimulus for the development of the soft jet recovery technique, which was first reported in 1987 [Ref.4].

There was full recognition of the differences between the start of a dynamically stretching continuous jet and the state of recovered jet particles. Clearly, the continuous jet cannot be recovered intact because of the high axial velocity gradient, which generates the axial strain. After the superplastic axial extension of the continuous jet, this velocity gradient ultimately leads to jet particulation. Flash x-ray and flash x-ray diffraction techniques therefore provide the only currently known methods for studying the structure of the continuous stretching jet. It was expected however, that if the individual jet particles could be recovered relatively undamaged, they could provide some clues regarding the condition of the interior of the copper jet and possibly new information regarding

the processes occurring during jet formation and particulation. The opportunity of being able to examine recovered jet particles with all of the
modern techniques available, represented an exciting prospect. The re-
mainder of this paper will be concerned with the observations made on
recovered copper jet particles.

IV.   OBSERVATIONS ON RECOVERED COPPER JET PARTICLES

A.   *EXTERIOR CHARACTERISTICS*

The exterior view of a recovered copper jet particle at low magnification
[13.75X] is shown in Fig. 1.* Note that the outside surface shows an
approximately orthogonal double network of what appears to be slip plane
traces, never seen before.   These traces appear to be symmetrically
located with reference to the jet particle axis.

Selected recovered particles were sectioned approximately "transversely" and "longitudinally".   The exposed cut surfaces were polished,
etched and examined optically as well as by x-ray diffraction, SEM and TEM.
Some of the results of these observations will be discussed under their
separate headings.

---

*Throughout, magnification is that of original photograph, not as reproduced.

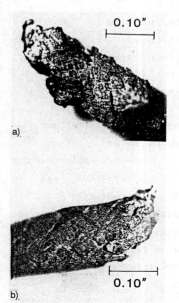

a)

b)

FIG. 1 *Exterior of two recovered jet
particles showing orthogonal surface markings.*

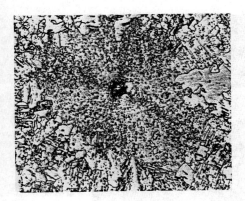

FIG. 2 *Center of jet particle #4 showing axial hole, micro crystalline region and first annular ring of large grains (200X). Transverse face.*

B. *SELECTED OPTICAL OBSERVATIONS*

Transverse cross sections of particle #2 at 100X magnification and particle #4 at 200X magnification, are shown in Figs. 2 and 3. Both show a central hole surrounded by a microcrystalline area, which is in turn circumscribed by distinct annular regions of larger grains and smaller grains. Higher magnification views [1000X and 2000X] of jet particle #2 are shown in Figs. 3a, 3b, 3c, 3d and 3e, which were taken successively across the transverse section, including the region around the central hole, the microcrystalline area around it and the annular rings of larger and smaller grains. Longitudinal cross sections of particles #2 and #5 are shown in Fig. 4 displaying the end locations at which the particles fractured.

C. *SELECTED SEM OBSERVATIONS*

The tip fracture region on particle #5 is shown at successively higher magnifications, on a longitudinal cross section, in Figs. 5 and 6. Clear evidence of elongated as well as spherical porosity appears to be associated with this fracture process, in the vicinity of the separation region.

The separate observation of apparent micron and submicron porosity on the interior of the recovered jet particles, could be explained by the mechanism of vacancy generation by rapidly moving intersecting dislocations, followed by coalescence into larger voids. However, these observations are

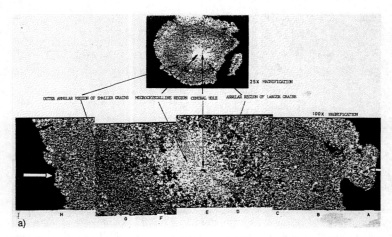

FIG. 3a *Complete cross section of jet particle #5 showing axial hole and three annular rings of varying grain size (100X). Arrow and letter coordinate axes are shown.   Transverse face.*

FIG. 3b *Regions A and B on Fig. 3 shown at 1000X.*

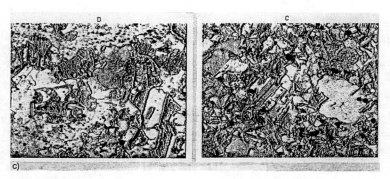

FIG. 3c *Regions C and D on Fig. 3 shown at 1000X.*

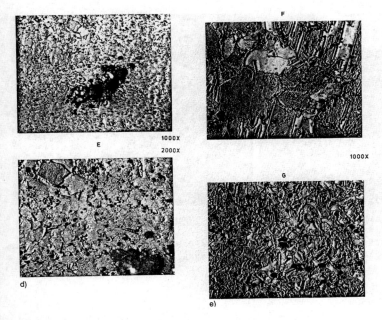

FIG. 3d *Region E on Fig. 3 shown at 1000X and 2000X, in the vicinity of the hole.*

FIG. 3e *Regions F and G on Fig. 3 shown at 1000X.*

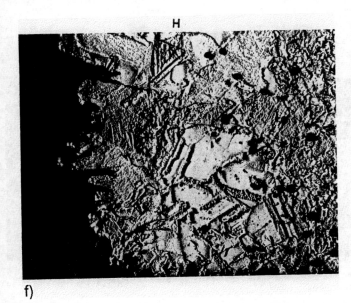

FIG. 3f *Region H on Fig. 3 shown at 1000X.*

 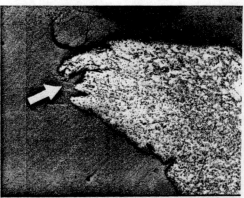

PARTICLE NO. 2   **PARTICLE NO. 5**

FIG. 4 *Jet particles #2 and #5, longitudinal sections, showing the fracture tips on each particle and potential fit matchup.*

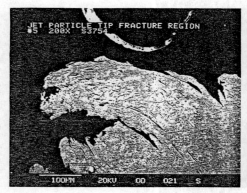 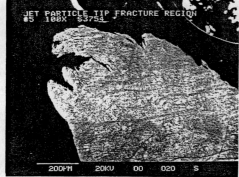

FIG. 5 *Jet particle #5, longitudinal section, SEM at 100X and 200X, showing porosity, at fractured end.*

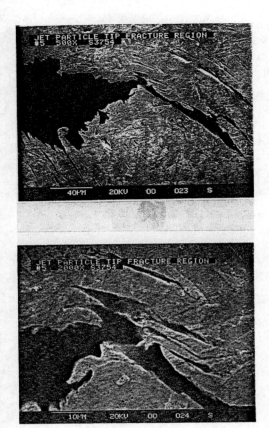

FIG. 6 *Jet particle #5, longitudinal section, SEM at 500X and 2000X, showing porosity, at fractured end.*

still equivocal, because at least some of the apparent porosity may be attributed to artifacts. Precision density measurement will be used to help clarify this problem. This question is of interest in connection with the previously noted density deficit observed in the stretching continuous jet. The density deficit could not be observed in the jet particles after particulation, within the sensitivity limits of the densitometric measurement techniques.

D. *SELECTED TEM AND SAD OBSERVATIONS*

Preliminary TEM and SAD observations were made on the longitudinal section of particle #2. The initial SAD observations shown in Figs. 7 and 8

FIG. 7 *TEM and SAD observations at center of longitudinal section, particle #2.*

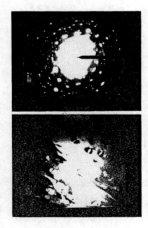

FIG. 8 *TEM and SAD observations off center longitudinal section, particle #2.*

indicate significant variation in structure across the particle diameter, that require further detailed study.

E.   *SELECTED X-RAY DIFFRACTION OBSERVATIONS*

Recovered particles #5 and #2 were examined at the Jet Propulsion Laboratory by x-ray diffraction techniques using a Siemens D500 x-ray diffractometer. The experimental geometry is shown in Fig. 9. Fe Kα radiation was used. The entire particle cross section was illuminated, so that the tabu-

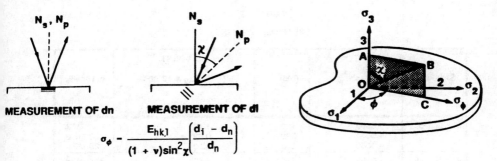

$$\sigma_\phi = \frac{E_{hkl}}{(1 + v)\sin^2\chi}\left(\frac{d_i - d_n}{d_n}\right)$$

where

E_{hkl} = Youngs Modulus for the crystallographic plane
v = Poisson's ratio
χ = angle of inclination from the sample surface
d_n = lattice spacing at $\chi = 0$
d_i = lattice spacing at $\chi \neq 0$
N_s = Sample Normal
N_p = Crystal Plane Normal

B.D. CULLITY
ELEMENTS OF X-RAY DIFFRACTION
ADDISON-WESLEY, 1978

FIG. 9 *Experimental setup for biaxial stress analysis by x-ray diffraction.*

lated values given represent averages over the entire particle cross section.

Selected outputs from these observations are shown in Tables I and II and Figs. 10,11 and 12. These indicate the presence of residual biaxial and possibly triaxial stresses as well as evidence of preferred orientation

TABLE I

RELATIVE PERCENTAGES OF CRYSTAL ORIENTATION ALONG
PARTICLE (WIRE) AXIS OF 3 PRINCIPAL CRYSTAL DIRECTIONS
IN RECOVERED COPPER JET PARTICLE SAMPLE NO. 5

RUN NO.	SAMPLE DESCRIPTION	% OF CRYSTALS WITH DIRECTIONS SHOWN PARALLEL TO PARTICLE (WIRE) AXIS			PARTICLE SCHEMATIC
		(111)	(110)	(100)	
1	TRANSVERSE FACE ROUGH CUT DISTURBED SURFACE	37.8%	28.7%	33.4%	ROUGH FACE
2	TRANSVERSE FACE MECHANICALLY POLISHED AND ETCHED	49.5%	11.5%	39%	POLISHED FACE
3	REORIENTED TO MAXIMIZE (111)	73.6%	15.7%	10.7%	
	LITERATURE DATA ON FIBER TEXTURE FOR DRAWN COPPER WIRE REF. 7	60%	0%	40%	POLISHED FACE WIRE

TABLE II

RESIDUAL STRESSES

SAMPLE	SLOPE OF $d(\text{Å})$ vs. $\sin^2\chi$	(hkl)	ELASTIC CONST. [+] $(\frac{E}{1+\nu})_{hkl} \times 10^5 MPa$	$\sigma(10^5 MPa)$
# 2 LONG.	-.0690	111	1.43	-460 MPa
	-.0040	220	0.62	- 20 MPa
	-.0060	200	0.50	- 20 MPa
# 2 TRANS.	-.0100	111	1.43	-690 MPa
	.0012	220	0.62	70 MPa
	.0030	200	0.50	80 MPa

*A negative stress value = compressive stress

[+] R. W. Hertzberg, Deformation and Fracture Mechanics of Engineering Materials, Wiley, New York (1976).

$$\text{Bulk } (\frac{E}{1+\nu}) = .804 \times 10^5 \text{ MPa}$$

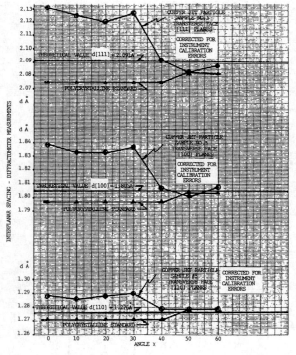

FIG.10   Interplanar spacing measurements for particle #5, from x-ray diffraction data.

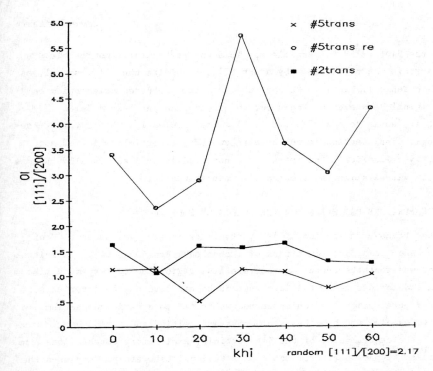

FIG. 11   *Orientation index for particles #5 and #2 showing effect of ∿ 5°*
reorientation of particle axis on particle #5.

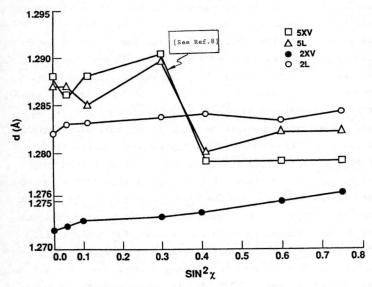

FIG. 12   *[220] lattice spacings as a function of orientation angle χ, for*
particles #5 and #2, showing two different types of correlation.

along the "jet axis" in particle #5. This was particularly noticeable after
the particle was reoriented [by about 5°] to maximize the (111) signal, as
noted in Table I. Particle #2 does not show the preferred orientation on
its original "transverse" face, but no attempt has yet been made to
orient it, so as to maximize the (111) signal. Clearly, when a transverse
or longitudinal section is cut on a recovered jet particle which is not
perfectly cylindrical, the jet axis is not readily defined visually. Errors
of 5° in visual alignment are readily understandable.

V.   TEMPERATURE ESTIMATES FOR THE COLLISION ZONE

When the transverse section of jet particle #2 [Fig.3] was shown to Prof.
T. Vreeland [California Institute of Technology, Pasadena, CA], his initial
comment was that the center microcrystalline region looked very much like
molten and rapidly recrystallized regions he had seen at the interfaces of
powdered metal particles, which he had subjected to dynamic compaction. In
order to examine the hypothesis of momentary melting and rapid resolidification of the central core of the jet particle, exploratory computations were
undertaken by personnel from Los Alamos National Laboratory. They used the
PINON code initially but found an error in the material model. They have
now gone to the MESA-2D code. Material models used have included the
Johnson-Cook model with thermal effects, as well as some simple elastic-
perfectly plastic models. As part of the analytical output, they have
generated instantaneous isothermal contour plots for the region of interest
in and around the collision zone, e.g. as shown in Fig.13 for a standard
BRL 81mm charge. Peak temperature plots along the axis of the jet have also
been generated, for specific collapse times during jet formation, as shown
in Fig.14 for the actual charge design used to generate particles #2 and
#5. As expected, the maximum temperatures occur in the collision region on
the axis of jet symmetry. This "bump" in the temperature distribution is
clearly seen in Fig.14. Since the particle in Fig.3 passed through the
collision region during its formation, it is precisely this region in the
jet formation process which is seen in the interior of Fig.3. The maximum
temperatures are found there, but the computed peak temperatures fall substantially short of those necessary to obtain melting. For the specific
shaped charge configuration that generated particle #2, the computed
pressure peak in the collision region is slightly under 0.3 megabar and the
computed temperature peak is about 700°C [Fig.14]. Upon emerging from the

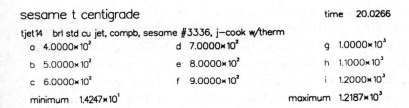

sesame t centigrade   time   20.0266

tjet14   brl std cu jet, compb, sesame #3336, j—cook w/therm

a	4.0000×10^2	d	7.0000×10^2	g	1.0000×10^3
b	5.0000×10^2	e	8.0000×10^2	h	1.1000×10^3
c	6.0000×10^2	f	9.0000×10^2	i	1.2000×10^3

minimum 1.4247×10^1 maximum 1.2187×10^3

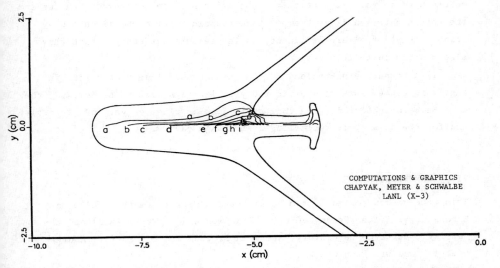

COMPUTATIONS & GRAPHICS
CHAPYAK, MEYER & SCHWALBE
LANL (X-3)

FIG. 13 *Isothermal contours after 20 microseconds for a BRL standard 81mm charge, using PINON code.*

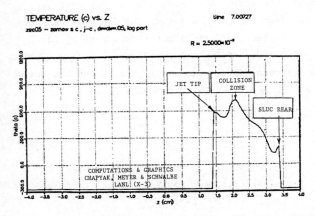

FIG. 14 *Axial temperature distribution after 7 microseconds for charge used to generate particles #2 and #5, using MESA-2D code.*

high pressure collision zone, the pressure drops quickly and the temperature quickly drops to the release temperature. The ambient pressure melting
point of copper is 1083°C. Therefore the momentary melting and recrystallization hypothesis is clearly not supported by these computations.

There are other thermophysical processes occurring at very high
pressures and strain rates, which have not been included in the material
models used so far. In particular, the effects of inhomogeneous heating and
the strain rate sensitive "temperature trapping" phenomena discussed by
Grady & Asay[Ref.6] are pertinent, as is pressure hardening, which they
also mention in their paper.

If alternate explanations for the microcrystalline center zone in
Fig.3 and Fig.4, such as dynamic recrystallization, are to be invoked, they
will have to account for the concentric annular rings of varying grain size
and for the axial hole resembling a shrinkage cavity.

VI. ACKNOWLEDGEMENTS

This work was supported by the Naval Surface Warfare Center [White Oak
Laboratory], Silver Spring, MD [Don Phillips and K. Wayne Reed] and the
Ballistic Research Laboratory, Aberdeen Proving Ground, MD (Dr. Andrew
Dietrich and Dr. Fred Grace].

The preparation of metallurgical samples and photomicrographs was done
by Paul Jacoy, NASA, JPL, Cal. Tech. and also by Sam DiGiallonardo of
Lawrence Livermore National Laboratory. The x-ray diffraction work was
carried out by Lynn Lowry, co-author of this paper. The SEM work was done
by Ron Ruiz of NASA, JPL, Cal. Tech. and the TEM work was done by Carol
Garland of the California Institute of Technology, Pasadena, CA. The computational study of jet temperature was carried out at Los Alamos National
Laboratory by Jay Chapyak, Ken Meyer and Larry Schwalbe [X-3].

REFERENCES

1. Birkhoff, MacDougall, Pugh & Taylor, *J. App. Phys.* _19_ (1948)

2. L. Zernow, *Doctoral Dissertation, Johns Hopkins University* (1953)

3. Eichelberger & Pugh, *J. App. Phys.* _23_ (1952)

4. L. Zernow, *Proc. Tenth International Symposium on Ballistics
    (Vol. II)* Oct. 1987 (ADPA)

5. F. Jamet, *Proc. Eighth International Symposium on Ballistics,* Oct. 1984 (ADPA)

6. D. E. Grady & J. R. Asay, *J. App. Phys.* <u>53</u> 1982

7. Barrett, *"Structure of Metals",* McGraw Hill (1952) pp. 442-443

8. I. Noyan & J. Cohen, *Residual Stresses, Springer and Verlag* (1988)

47

Comparative Studies of Shaped Charge Component Microstructures

ALAN GUREVITCH, L. E. MURR, S. K. VARMA, S. THIAGARAJAN, AND
W. W. FISHER

Department of Metallurgical and Materials Engineering
The University of Texas at El Paso
El Paso, Texas 79968

Techniques are being developed for preparing electron transparent thin sections from recovered, copper shaped charge jet fragments and other representative components of the shaped-charge regime - copper precursor rods and liner cones formed from these precursors - for observations by analytical transmission electron microscopy. Here we illustrate the methodologies and provide examples of TEM microstructures in variously drawn copper precursor bars, shaped charge liner cone locations, plane-wave shock loaded copper, shaped charge jet fragment, and native copper fragments. Microstructure comparisons are intended to form a basis or context from which to begin to assess the microstructural implications in the shaped charge regime.

I. INTRODUCTION

A shaped charge consists of a high explosive which surrounds a metal-lined cavity which can have various shapes. A conical liner is a very common configuration where the detonating explosive wave projects the liner elements along the cone axis where they collide

521

to form a slug and a high velocity jet accelerated axially as shown schematically in Fig. 1. As the jet accelerates outward from the slug, it becomes unstable and begins to break up. As first shown by Plateau [1], when the length of a cylinder exceeds a critical value, its surface energy is decreased by breaking into spheres. Rayleigh [2] expanded this into a theory of liquid jet instability which showed that a cylinderical liquid jet of radius, r, could continuously decrease its surface energy by breaking into droplets. Nichols and Mullins[3] extended this concept to solid rods within a solid body, and more recently Miller and Lange [4] observed these features experimentally in polycrystalline fibers.

A knowledge of the physical state of the jet is important in determining its ductility in relation to its breakup at the high rate ($>10^4$ s^{-1}) it is elongated. A few studies have been devoted to the recording of x-ray diffraction patterns of aluminum and copper shaped charge jets [5-8]. Jamet and Sharon [8] have shown that both aluminum and copper shaped charge jets are in the solid state, and aluminum exhibited a preferred orientation.

As pointed out in these proceedings by Grace [9], the shaped-charge jet involves both extreme deformation (strain, ε) at very high strain rates ($\dot{\varepsilon}$). In addition, the liner (Fig. 1) is initially subjected to shock waves (20-80 GPa) which help focus the liner material to form a slug from which the jet emerges. While there is evidence that the grain structure of the liner has a marked influence on the stability of the shaped charge jet [10], there has been neither a systematic study of fundamental microstructures in liner materials and jet fragments, nor of their relationship. In this study, we explore a range of microstructures in deformed

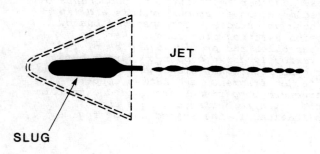

DETONATED
SHAPED CHARGE

FIG. 1 *Schematic view of detonated shaped charge. A high explosive is detonated around the metal liner (shown dotted) which collapses to form the slug and the jet accelerated along the liner axis opposite to the slug. The slug and jet are associated with the outside and inside liner surfaces respectively.*

copper over a range of strains and strain rates, and compare these observations with those in a copper shaped-charge liner, and a jet fragment-using transmission electron microscopy (TEM).

II. EXPERIMENTAL DETAILS

In an effort to survey and compare a range of microstructures in copper shaped charge liners, liner precursor bars, and the jet fragments, standard 3 mm discs were punched and electropolished to electron transparency from representative samples of variously drawn (and deformed) copper bars which can be used to form liner cones, as well as from various locations on the liner [11]; as illustrated in Fig. 2(a). Both in-plane and through-section (longitudinal or transverse) specimens were prepared for transmission electron microscopy (at 200kV in a Hitachi H-8000 TEM) and optical metallography for comparison. A 50/50 nitric acid/water etch was employed for optical metallography.

Sections were sliced from jet fragments which were built up with electrolytic copper [12] to allow for a sufficiently large mass from which to prepare the standard 3 mm discs for transmission electron microscopy. Native copper fragments served to develop this technique, and to allow for the observation of additional microstructures for comparison. The shaped charge jet fragments were recovered by Zernow [13] using a novel method. As shown schematically in Fig. 2(b), a particular interest involved comparing microstructures in the liner with those observed in the jet fragment.

III. RESULTS AND DISCUSSION

Since there are few if any published efforts to examine microstructures in shaped-charge precursor bars and within the shaped-charge regime represented by Fig. 2, we thought it would be instructive to develop a microstructural context in which to make comparisons, and begin to elucidate the physical nature of the shaped-charge regime, and to establish a methodology which would culminate in TEM observations of shaped-charge jet fragments [Fig. 2(a)]. Figure 3 begins this process by comparing a series of OFHC copper bars extruded at a rate of 100 inches/min. (4.2 cm/s) at strains indicated. As the strain is increased, the dislocation cell size (or cell spacing, d) decreases with increasing strain while the cell wall dislocation structure also changes. This feature is also demonstrated quantitatively in the insert in Fig. 3 which shows a similar relationship for cell diameter (d) versus true drawing strain for drawing rates of 40 to 330 in/min (1.7 to 138 cm/s).

Figure 4 shows for comparison with Fig. 3 that plane-wave shock loaded copper exhibits essentially the same trend in microstructural modification with peak shock pressure as the extruded bars in Fig. 3, where dislocation density and cell diameter increase and decrease respectively with increasing pressure over a range of 5 to 20 GPa [14]. The

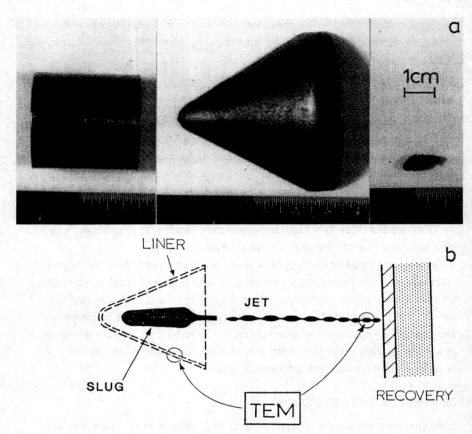

FIG. 2 *(a) Examples of (from left) copper bar stock, copper liner cone, and recovered Cu jet fragment (shown magnified). (b) Schematic representation of selective TEM of liner regions and jet fragment. The arrow suggests some connectedness.*

relationships between dislocation cell size, strain, and dislocation density implicit on comparing the insert in Fig. 3 and Fig. 4(b) illustrate the principle of dislocation similitude [15]. Comparing Fig. 4(a) and Fig. 3 for shock loaded and drawn copper illustrate this principle phenomenologically.

Figure 5 compares microstructures in the copper bar shown in Fig. 2(a) with a sample extracted from the tip of the liner cone in Fig. 2(a). The optical micrographs attest to differences in grain size (a factor of 5) while the TEM images attest to a significant difference in dislocation-related microstructures as a consequence of heat treatment of the

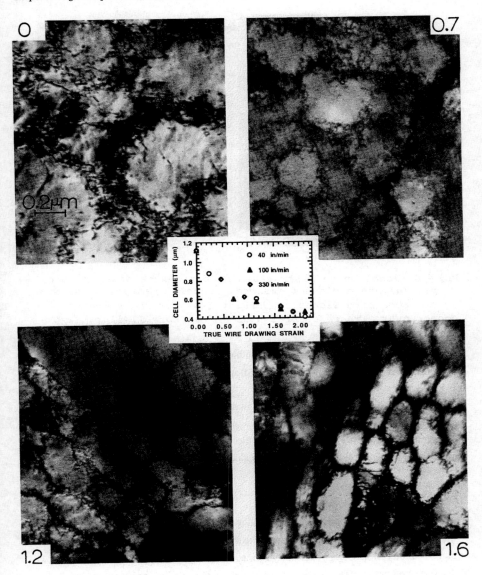

FIG. 3 *Examples of dislocation cell structures in variously drawn (extruded) OFHC copper bars at strains (ε) noted. The drawing rate was 4.2 cm/s. The insert shows dislocation cell size versus true drawing strain for several rates noted.*

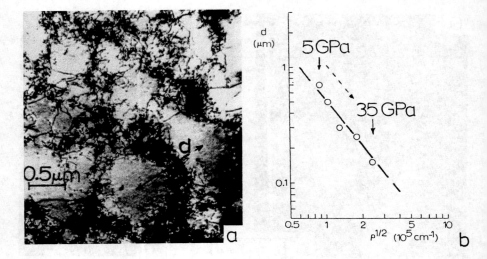

FIG. 4 *Dislocation cell structure in plane-wave shock loaded
copper (a) and a plot of dislocation cell spacing, d, versus
square root of dislocation density ($\sqrt{\rho}$) for shock loaded
copper over a range of pressures (After Murr [14]).*

liner cone. The effective strain associated with the precursor bar in Fig. 2(a) can be
estimated from the dislocation cell spacing (d) in Fig. 5(b), and comparison with the insert
in Fig. 3.

Finally, Fig. 6 shows for comparison dislocation cell structure in a native copper
fragment [insert in Fig. 6(a)] with microstructures typical of the recovered jet fragment in
Fig. 2(a). The insert in Fig. 6(b) shows the electrodeposited copper surrounding the jet
fragment in Fig. 2(a) prior to sectioning and electron transparent disc preparation for TEM.

IV. CONCLUSIONS

We have demonstrated a capability and developed a methodology for comparing the
microstructures of selected elements from the shaped-charge regime by TEM. Similarities
in dislocation cell structure in variously formed and deformed copper have been
demonstrated through TEM observations, and these observations begin to provide a
microstructural context for the comparison of shaped-charge microstructures. A systematic
application of the methodologies demonstrated here will begin to elucidate the fundamental
features of the physical properties of shaped charge jets and their relationship, if any, to the
precursor microstructures - specifically those of the liner.

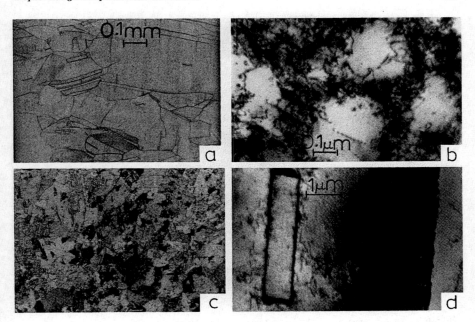

FIG. 5 *Comparison of microstructures in copper precursor bar stock and copper (cone) liner. (a) Optical micrograph of bar stock. (b) TEM image showing dislocation cell structure in a grain of (a). (c) Optical micrograph of liner cone midsection region. (d) TEM image showing annealing twin boundaries and dispersed dislocations.*

ACKNOWLEDGMENTS

We are especially grateful for the provision of the shaped charge components [illustrated in Fig. 2(a)] by Dr. Louis Zernow, Zernow Technical Services, Inc. This work was supported by a Mr. and Mrs. MacIntosh Murchison endowed chair (L.E.M.) and the Phelps-Dodge Scholars Program at the University of Texas - El Paso. (A.G. and S.T.).

REFERENCES

1. J. Plateau, Statique experimentale et theoretique des liquides soumis aux seules forces moleculaires, Paris, 1873.

2. J. S. W. Rayleigh, Theory of Sound, Vol. II, Dover, New York, 1945.

3. F. A. Nichols and W. W. Mullins, *J. Appl. Phys. 36:* 1826(1965).

4. K. T. Miller and F. F. Lange, *Acta Met. 37:* 1343(1989).

5. R. E. Green, Jr., *Rev. Sci. Instrum. 46*(9): 1257(1975).

6. F. Jamet and G. Thomer, *C. R. Acad. Sci.* (Paris) *B279:* 501(1974).

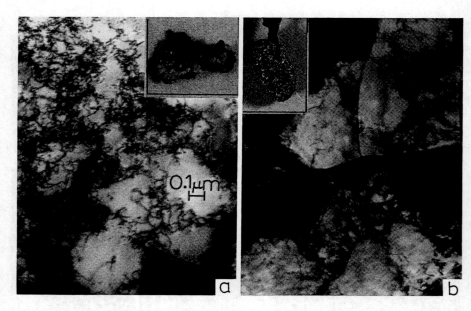

FIG. 6 *Comparison of dislocation cell structures in a native copper fragment (a) with a copper shaped charge jet fragment (b) in the TEM. The insert in (a) shows a native copper cluster in a calcium carbonate/quartzite matrix. The insert in (b) shows the copper electrodeposit on the jet fragment in Fig. 2(a).*

7. F. Jamet, 8th Int. Symp. on Ballistics, Orlando, Florida, 1984.

8. F. Jamet and R. Charon, "A flash x-ray diffraction system for shaped charge jets analysis", Report (0211/86 Franco-German Research Institute, Saint-Louis, France, May 15, 1986.

9. F. I. Grace, These proceedings: Shock-Wave and High-Strain-Rate Phenomena in Materials, M. A. Meyers, L. E. Murr, and K. P. Staudhammer (eds.), Marcel Dekker, Inc., New York, 1991.

10. M. D. Merz and R. W. Moss, "Fine-Grained and Amorphous Metal Liners, Produced for the Ballistic Research Laboratory", 1.0 2333, Pacific Northwest Laboratory Report (Contract No. DE-AC06-76 RLO 1830), March, 1988.

11. L. E. Murr, Electron and Ion Microscopy and Microanalysis, Marcel Dekker, Inc., New York (2nd printing), 1982.

12. L. E. Murr, Industrial Materials Science and Engineering, Chap. 16, Marcel Dekker, Inc., New York, 1984.

13. Louis Zernow and Lynn E. Lowry "High Strain-Rate Deformation of Copper in Shaped Charge Jets", These Proceedings.

14. L. E. Murr, Chap. 37 in Shock-Wave and High-Strain Rate Phenomena in Metals, M. A. Meyers and L. E. Murr (eds.) Plenum Press, New York, 1981, p. 607.

15. L. E. Murr and D. Kuhlmann-Wilsdorf, *Acta Metall. 26: 847(1978)*.

48

High Strain, High-Strain-Rate Deformation of Copper

M. A. MEYERS, L. W. MEYER**,J. BEATTY*, U. ANDRADE, K. S. VECCHIO,
and A. H. CHOKSHI

University of California, San Diego
La Jolla, California 92093, U.S.A.

*U.S. Army Materials Technology Laboratory, Watertown, Massachusetts, U.S.A.

**IFAM, Bremen, West Germany

While copper exhibits total elongations that typically
do not exceed 0.5 at low strain rates, the strains exceed 10
in tension under the special conditions imposed during
shaped-charge deformation. The reasons for this extended
plasticity are poorly understood and the role of
microstructural evolution has not been examined in any
detail. Residual microstructures produced in shaped charges
(jets and slugs) and in special Hopkinson bar tests in which
high strains and strain rates were imposed are characterized
by optical and transmission electron microscopy. The
microstructure was preconditioned by shock loading it to a
pressure of 40 GPa in order to enhance heat generation during
subsequent high-strain-rate plastic deformation. This
preconditioning simulates shock-wave strengthening of the
shaped charge liner prior to its collapse. The residual
structures can exhibit recrystallized regions. By means of
heat transfer calculations, it is shown that the cooling in
the shaped charges slugs and jets takes place over seconds,
while in the especially designed experiments it takes place
over milliseconds. Therefore, the microstructures
investigated under the special testing configurations are
closer to the ones existing during deformation.

The basic equations that govern dynamic recrystallization as they apply to high-strain rate deformation are presented, and the plastic response of the copper after having undergone these microstructural changes is discussed.

I.   OBSERVATIONS ON SHAPED CHARGES

The enhanced ductility exhibited in uniaxial and biaxial tension by metals under dynamic deformation is of great importance. Examples are the large extensions undergone by shaped charge jets and the extensive plastic deformations undergone by space shields when subjected to hypervelocity impact, forming a plume. A number of investigations have treated the problem of an extending jet analytically; Chou and Carleone [1], Walsh [2], and Grady [3] performed stability analyses on stretching plastic jets. The effect of lateral inertia was incorporated by Grady [3], Fressengeas and Molinari [4], and Romero [5], based on observations by Walsh [2] and Frankel and Weihs [6]. It was found that a confining pressure is produced by the gradient in radial velocities resulting from the rapid longitudinal stretching. This inertial confining pressure has the effect of inhibiting failure by internal void formation. Grady [3] estimated that this pressure could be as high as 10 Y at a strain rate of 10^5 s^{-1}, where Y is the yield stress. These studies have successfully predicted times-to-failure and fragment sizes. However, there still remain many unknown aspects in shaped-charge deformation, and microstructural evolution in both liner collapse and jet stretching is virtually unknown. Jamet [7,8] performed flash X-ray diffraction experiments on aluminum jets using transmitted X-rays, and on copper jets using reflected X-rays. For aluminum, he conclusively showed that the jet was solid. For copper jets, the reflected mode could only identify the surface which was shown to be crystalline (i.e. solid).

In this report these microstructural evolution processes will be addressed. First, observations made on slug and jet fragments are described. Second, special experiments in which high strains are applied at a high strain rate are presented. And third, dynamic recrystallization at high strain rates is described analytically by two models, which are developed in detail elsewhere [9, 10].

The recovery of jets and slugs is not simple. Jets propelled at velocities approaching 10^4 ms^{-1} have to be decelerated without subsequent damage. Two slugs were obtained, from D. A. Lassila (Lawrence Livermore National Laboratory) and from C. Wittman (Honeywell). Figure 1 shows micrographs from these two slugs taken from longitudinal and transverse sections. The microstructures exhibit, typically, a recrystallized

FIG. 1 *Typical optical micrographs from Lassila (a,b) and Wittman (c,d) slugs; (a and c) transverse sections: (b and d) longitudinal sections.*

morphology. The absence of grain elongation, in the longitudinal section, in spite of the large strains undergone by the slug, shows that recrystallization occurred. Nevertheless, there is a significant difference in grain size between the two materials: it is equal to 40 μm for the Wittman slug, while it appears to be less than 5 μm for the Lassila slug. A small jet section was also characterized and it is shown in Fig. 2. The optical micrograph shows a recrystallized structure (grains equiaxed), as evidenced in Fig. 2(a), with a grain size of ~30 μm. Transmission electron microscopy of this jet fragment shows some evidence of dislocation activity, 2b, and deformation twins, 2c; most of the regions were, however, simply recrystallized.

On the other hand, preliminary transmission electron microscopy of the slug provided by Lassila showed a microcrystalline structure with low dislocation density [9]. Other investigators (Jamet [11]; Hirsch [12]; Buchar [13]) observed columnar grains surrounding voids in slugs. This is strongly suggestive of melting pockets forming within the slug. Upon solidification, shrinkage produces the void; the solidification grains are elongated, and their longitudinal axes converge towards the center of the void. The

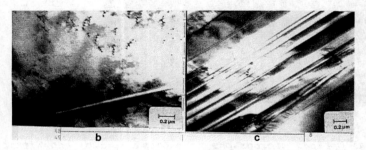

FIG. 2 *Jet section: (a) optical micrograph; (b) and (c) transmission electron micrographs.*

conclusions that can be drawn from the above observations are: (a) considerable variation in microstructure exists; (b) recrystallization (static) can destroy a great extent of the deformation structure; and (c) some melt pockets probably exist, but the jet is at least partially solid.

The temperature rises experienced by the slug and jet are a direct consequence of the adiabaticity of the plastic deformation process. The adiabatic temperature rise in copper can be calculated based on the plastic deformation energy. It is known that ~90% of deformation energy is converted to heat. The temperature rise, ΔT is

$$\Delta T = \frac{0.9}{\rho c_p} \int_0^\varepsilon \sigma \, d\varepsilon \tag{1}$$

where ε is the plastic strain, σ is the flow stress, ρ is the density of the material, and c_p is the heat capacity.

The following simple constitutive model for the stress as a function of homologous temperature, T_h ($=T/T_m$, where T is absolute temperature and T_m is the absolute melting temperature), strain, and strain rate was proposed by Johnson and Cook [14]:

$$\sigma = \left(\sigma_0 + B\varepsilon^N\right)\left(1 + C\ln\dot{\varepsilon}^*\right)\left(1 - T_h^M\right) \tag{2}$$

where σ_0 is the yield stress, B, C, N, and M are constants and $\dot{\varepsilon}^*$ ($= \dot{\varepsilon}/\dot{\varepsilon}_0$) is the normalized strain rate ($\dot{\varepsilon}$ is the imposed strain rate and $\dot{\varepsilon}_0 = 1$ s^{-1}). Equations 1 and 2 may be combined as follows by assuming a constant strain rate:

$$\int_{T_0^*}^{T_f^*} \frac{dT}{1 - T_h^M} = \frac{0.9\left(1 + C\ln\dot{\varepsilon}^*\right)}{c_p\,\rho} \int_0^{\varepsilon_f} \left(\sigma_0 + B\varepsilon^N\right) d\varepsilon \tag{3}$$

where T_0^* and T_f^* are the initial and final homologous temperature, respectively. The solution to Eqn. 3 gives the temperatures rise as a function of plastic deformation.

Calculations for a typical Cu shaped charge reveal that the strain rates achieved are of the order of $>10^4$ s^{-1}. Johnson and Cook [14] reported a value of M = 1.09 for Cu; for the purposes of the present approximate calculations, it is assumed that M = 1 in Eqn. 3. The temperature rise occurring during the dynamic deformation of Cu at room temperature and at a strain rate of 10^4 s^{-1} was estimated using values of the other parameters reported by Johnson and Cook [14].

An analysis based on Equation 3 reveals that temperatures of $>0.4\,T_m$ may be achieved by plastic strains of ≈ 3 [9]. It is also noted that an increase in the yield stress, by shock hardening, decreases the strain necessary to achieve a given temperature. Recrystallization generally occurs at temperatures of $>0.4\,T_m$. Approximate calculations show that shaped charges experience strains of ≈ 5 during the initial collapse. Consequently, the compressive deformation of shaped charges satisfies the experimental conditions necessary for recrystallization.

Jamet [11] measured the temperature attained in copper jets produced from 81.3 mm cones. He found values ranging from 428 to 537° C (0.5 to 0.6 T_m). These values are consistent with the above calculations and well within the realm of recrystallization.

II. CONTROLLED RECOVERY EXPERIMENTS

The temperature history of slugs and jets determines the residual microstructures. The total deformation occurs in approximately 0.3 ms. This is computed approximately by

$$t = t_D + t_t \tag{4}$$

where t_D is the time taken for the detonation front to propagate through the charge and t_t is the travel time of the jet. These values are obviously dependent on geometry, and a cone with apex height of 10 cm was used; the travel distance of the jet was assumed to be 20 cm. The total cooling of the jet and slug are dependent on their initial temperature, mass, shape, and cooling medium. Carslaw and Jaeger [15] provide a formalism for the calculations of heat transfer from a perfect cylindrical conductor into a surrounding medium. Assuming that the slug and jet are cooled in sand, a common decelerating medium for shaped charges, one can estimate the cooling rate for the center along the axis of cylinders with 1 cm (simulating slug) and 1 mm (simulating jet) diameters. Figure 3(a)

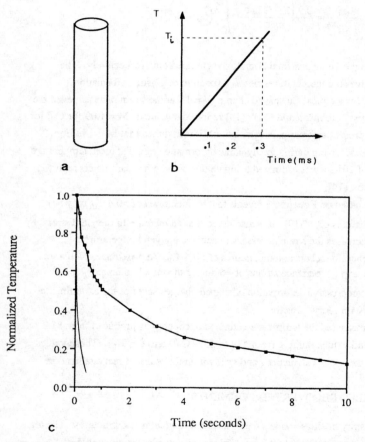

FIG. 3(a) *Geometry used for calculating cooling of slug and jet;*
        (b) heating as a function of time (deformation);
        (c) cooling in sand; of slug (1 cm diameter) and jet (1 mm
        diameter).

shows the idealized geometry (infinitely long cylinder), while Fig. 3(b) schematically shows a linear heating within 0.3 ms. By using appropriate heat capacities and thermal diffusivities, one obtains the curves displayed in Fig. 3(c). The normalized temperature, equal to the ratio: $(T - T_0)/(T_i - T_0)$ is plotted as a function of time after deformation stops. T is the current temperature, T_i the initial temperature of the slug (or jet) and T_0 the surrounding temperature. It can be seen that the cooling takes place over a number of seconds for the slug and over a fraction of a second for a jet. This order of magnitude calculation indicates that thermal recovery processes are most likely to alter the deformation substructure developed during deformation if the temperature reaches a level of 0.4-0.5 T_m. The discussion in Section I shows that this is the case. The exposure to temperature during cooling is approximately one thousand times longer than during deformation (.3ms). For this reason, it is felt that special recovery experiments need to be devised to:

(a)   produce high-strain, high-strain-rate deformation with real-time diagnostics, under controlled conditions,

(b)   provide rapid cooling after deformation to reduce, as much as possible, post-deformation structural recovery processes (recovery, recrystallization, and grain growth.)

Two experimental geometries were tested as described below. They utilize the Hopkinson bar and well-controlled, well-characterized stress pulses. The two geometries are shown in Fig. 4. The disk-shaped specimens have been used directly between the striker and the incident bars; they can also be placed between the incident and transmitter bars. The reductions in thickness accomplished were as high as 80 pct.,which corresponds to a true strain of 1.6. By comparison, a tensile elongation of 400% is necessary to obtain a true strain of 1.6. Copper disks with an initial thickness of 0.5 mm were impacted at ~ 100 m/s; the resulting microstructures are shown in Fig. 5. Different areas exhibited different structures; while a large proportion of the deformed material exhibited a high density of dislocations (Fig. 5(a)), there were some recrystallized areas present (Fig. 5(b)). The grain boundaries are irregular, indicating a high mobility. Figure 5(c) shows a small recrystallized region within which one can see some dislocations. The presence of dislocations inside recrystallized regions is usually considered evidence for dynamic recrystallization. However, in the present case the results are too preliminary to allow a positive identification.

The hat-shaped configuration, developed by Hartman, Kunze, and Meyer [16], and shown in Fig. 4(b), provides shear concentration in the regions circled. The stress pulse, produced by impacting the incident bar, generates controlled plastic shear deformation in the regions indicated, while the specimen itself undergoes mostly elastic deformation. The spacer ring establishes the limit for the plastic deformation within the shear zone. The gap between the incident bar and the ring is exaggerated in Fig. 4(b). A typical longitudinal

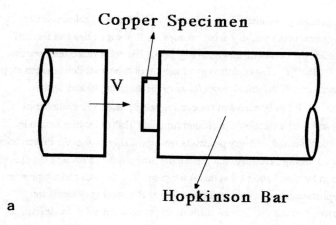

Copper Specimen

a

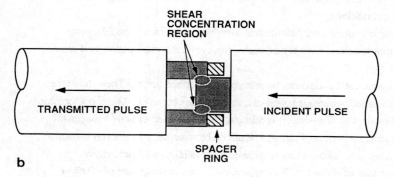

b

FIG. 4 *Experimental geometries for controlled high strain, high strain-rate experiments; (a) disc-shaped specimen; (b) hat-shaped specimen.*

section of specimen after impact is shown in Fig. 6(a). Copper was tested in the Hopkinson bar under two different conditions : 1) annealed, and 2) shock-hardened to a pressure of ~40 GPa by impact with a copper plate. The shock hardening procedure is described in detail in, e.g., DeCarli and Meyers [17]. An explosive lens initiated a 2.54-cm thick slab of PBX 9404 explosive, which accelerated a 4.7 mm thick stainless steel plate, impacting a properly momentum-trapped copper block. Velocity pins were used to determine the impact velocity. The average of three readings was 2,200 m/s. Shock hardening of the copper simulates the microstructural hardening process undergone by a shaped charge when the explosive detonates in contact with it.

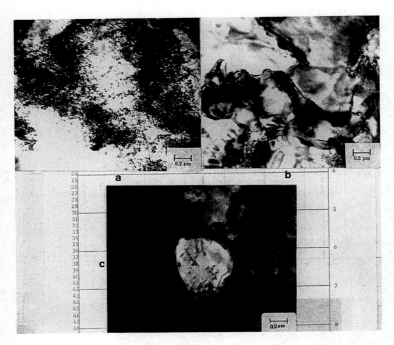

FIG. 5 *Copper disk specimens impacted in Hopkinson bar (ε_{eng} ~1.7); (a) dislocated region; (b) fully recrystallized region; (c) partially recrystallized region.*

Optical micrographs of the shear concentration regions for both annealed and pre-shocked specimens are shown in Fig. 6(b) and 6(c). There are significant differences between the two shear localization regions. The width of the band is larger for the annealed (~400 μm) than for the pre-shocked (~200 μm) specimen. This increased localization of the shocked specimen is the direct result of a decreased work hardening during deformation and is a well-known characteristic of the material. Within the band in the pre-shocked specimen, most microstructural features are annihilated. The microstructure of the shocked material exhibits a large incidence of deformation twins, which are known to be produced by shock loading. Within the band in the pre-shocked material (which had undergone a localized shear strain of 0.8 at a strain rate of ~ $10^5 s^{-1}$), the individual grain boundaries are not visible any longer under optical observation. Transmission electron microscopy was performed and the preliminary results are described herein. Figure 7 shows one of the features identified within the high strain region. Both

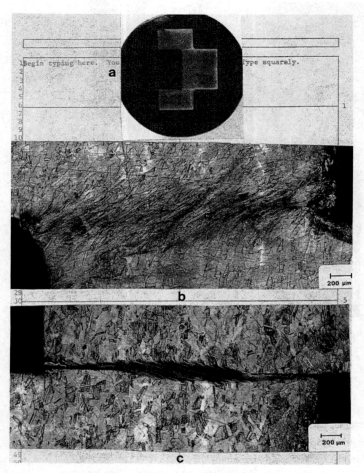

FIG. 6 *(a) Longitudinal section of copper hat specimen after impact; (b) and (c) shear concentration regions for annealed and shock-hardened specimens, respectively.*

the bright- and dark-field electron micrographs and the diffraction pattern give evidence of a fine microstructure. The sizes of the micrograins, as shown in the dark-field image, are approximately 0.05 μm. This microcrystalline structure is similar to the one observed for the copper shaped charge slug [9]. The cooling rates for the disk- and hat-shaped specimens, after high-strain rate deformation, are much higher than the ones encountered in shaped charges, and therefore the deformation structure is better retained. Preliminary heat transfer calculations can be conducted assuming a semi-infinite body. The thicknesses are assumed to be equal to the disk thickness and the shear localization width. Figure 8 shows

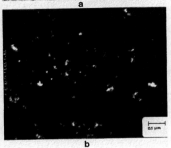

FIG. 7

Transmission electron micrograph of shear localization region $\left(\gamma = 8; \ \dot{\gamma} \cong 10^5 \ sec^{-1}\right)$ *in pre-shocked copper hat-shaped specimen; (a) bright field; (b) dark field.*

the post-deformation temperature histories for the two cases. The temperature distribution at time t = 0 is assumed to be a step function; the temperature is equal to T_0 in the surrounding material and equal to T_i within the band. The band has a thickness of 200 μm for both the disk- and the hat-shaped specimens (see Fig. 6a). The solution to the problem is provided by an error function (Carslaw and Jaeger [15]). Because of the higher diffusivity of copper (10x) than that of steel, the cooling rate is greater in the hat-shaped specimen. In contrast with the cooling undergone by a slug or jet in a normal recovery experiment, the cooling rate is on the order of one thousand times higher in the controlled recovery experiments described here. The cooling time is of the same order of magnitude as the deformation time (0.3 ms).

III.   ANALYSIS OF PLASTIC DEFORMATION

The high strain rates, large strains, and large adiabatic temperature rises experienced by the material create a unique thermomechanical environment. It is premature to develop quantitative predictive analyses; nevertheless, based on the observations, some general principles can be outlined. The existing constitutive models for plastic deformation are based on dislocation motion. This dislocation motion is controlled by overcoming obstacles, viscous drag, or relativistic effects, as the imposed strain rate is increased. The model developed by Follansbee and Kocks [18,19] represents the most advanced formulation of these concepts. It also incorporates dislocation recovery as a softening

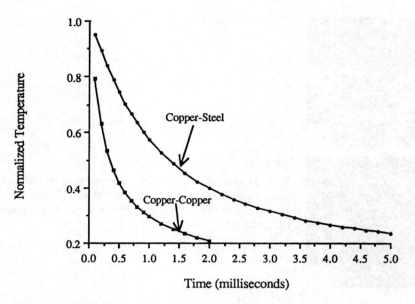

FIG. 8   *Computed normalized temperature (Tᵢ − T) / (Tᵢ − T₀) for*
slab of 200 μm thickness surrounded by (1) copper
(simulating hat-shaped specimen) and (2) steel
(stimulating discoid specimen).

process. These constitutive models are restricted to relatively low strains and we believe
that the large imposed strains (in shaped charges, shear bands, and other hypervelocity
events) require the incorporation of additional microstructural evolution mechanisms. We
further propose that dynamic recrystallization plays a key role in this regime.

Dynamic recrystallization plays an important role in many plastic deformation
processes. The experimental results and theoretical aspects of this phenomenon have been
reviewed by McQueen and Baudelet [20], Sakai and Jonas [21], and Ueki et al. [22].
Dynamic recrystallization essentially involves the development of a dislocation cell and
sub-grain structure and the transformation of low angle grain boundaries to high angle
grain boundaries during plastic deformation. The process is repeated continuously during
deformation, and leads eventually to the development of a steady-state recrystallized grain
size, d_s. Dynamic recrystallization is a thermally activated process, which is important at
temperatures greater than $0.4\,T_m$. From the theory for dynamic recrystallization developed
by Sandstrom and Lagneborg [23] it is possible to obtain the recrystallized steady-state
grain size. It is important to note that, in spite of the different approaches used by

Sandstrom and Lagneborg [23] and Derby and Ashby [24], both models predict the proportionality between the recrystallized grain size and $\dot{\varepsilon}^{0.5}$, where $\dot{\varepsilon}$ is the strain rate. Thus, at high strain rates one should expect small recrystallized grain sizes.

The mechanical response of this recrystallizing microstructure is not well understood yet. It is proposed that the inhibition of tensile instability in shaped charges is a direct consequence of this microstructural evolution. Both hypotheses lead to responses that are stable in tension. The first hypothesis is that classical superplasticity sets in; this occurs by grain boundary sliding. This is explained in detail by Chokshi and Meyers [9]. The second hypothesis is that dynamic recrystallization *per se* can lead to the desired response mechanism; a detailed account will appear shortly [10].

ACKNOWLEDGMENTS

This research was supported by the U. S. Army Research Office through Contracts DAAL03-88-K-0194, DAAL03-89-17-0396, and DAAL03-86-K0169. The use of the facilities of the Center of Excellence for Advanced Materials and of the Electron Optics and Microanalysis Facility at UCSD is gratefully acknowledged. Mr. U. Andrade was supported by the National Research Council (CNP), Brazil. The help of Mr. Jon Isaacs in carrying out the dynamic deformation experiments is gratefully acknowledged. Mr. Wittman (Honeywell) and Dr. D. Lassila (Lawrence Livermore National Lab) generously provided shaped charge specimens.

REFERENCES

1. P. C. Chou and J. Carleone, *J. Appl. Phys.*, 48: 4187 (1977).

2. J. M. Walsh, *J. Appl. Phys.*, 56: 1997 (1984).

3. D. E. Grady, *J. Impact Eng.*, 5: 285 (1987).

4. C. Fressengeas and A. Molinari, *Proc. Int. Conf. Mech. Prop. Materials at High Rates of Strain,* Inst. Phys. Conf. Ser. No. 102, p. 57, (1989).

5. L. A. Romero, *J. Appl. Phys.*. 65: 3006 (1989).

6. I. Frankel and D. Weihs, *J. Fluid Mech.,* 155: 289 (1985).

7. F. Jamet, "La Diffraction Instantanée," Report CO 227/84, Institut St. Louis, France, August 1984.

8. F. Jamet and R. Charon, "A Flash X-Ray Diffraction System for Shaped Charge Jets Analysis," Report CO 211/86, Institut St. Louis, France, June 1986.

9. A. H. Chokshi and M. A. Meyers, *Scripta Met.*, 24: 605 (1990).

10. M. A. Meyers, K. S. Vecchio, U. Andrade and A. H. Chokshi, unpublished results (1991).

11. F. Jamet, "Methoden zur Untersuchung der Physikalischen Eigenschaften eines Hohlladungsstrahles," Report CO 227/82, Institut St. Louis, France, December 1982.

12. E. Hirsch, *Propellants and Explosives*, 6: 11 (1981).

13. J. Buchar, Institute of Physical Metallurgy, Czechoslovakia, private communication, 1990.

14. G. R. Johnson and W. H. Cook, *Proc. 7th Intern Symp. Ballistics*, Netherlands, 1983.

15. H. S. Carslaw and J. C. Jaeger, *Conduction of Heat in Solids*, Oxford, 2nd ed., 1959, pp. 55, 343.

16. K.-H. Hartman, H.-D. Kunze, and L. W. Meyer, in *Shock Waves and High-Strain Rate Phenomena in Metals*, eds. M. A. Meyers, and L. E. Murr, Plenum, N.Y. 1981, p. 325.

17. P. S. DeCarli and M. A. Meyers, in *Shock Waves and High-Strain Rate Phenomena in Metals*, eds. M. A. Meyers, and L. E. Murr, Plenum, N.Y. 1981, p. 341.

18. P. S. Follansbee, in *Metallurgical Applications at Shock-Wave and High Strain Rate Phenomena*, L. E. Murr, K. P. Staudhammer, and L. E. Murr eds., M. Dekker, 1986, p. 451.

20. H. J. McQueen and B. Baudelet, *Proc. ICSMA 5*, P. Haasen, V. Gerold and G. Kostorz, eds., Pergamon Press, Oxford, 1979, p. 329.

21. T. Sakai and J. J. Jonas, *Acta Met.*, 32: 189 (1984).

22. M. Ueki , S. Horie, and T. Nakamura, *Mater. Sci. Tech.* 3: 329, (1987).

23. R. Sandstrom and R. Lagneborg, *Acta. Met.*, 23: 387 (1975).

24. B. Derby and M. F. Ashby, *Scripta Met.*, 21: 879 (1987).

49

Material Characteristics Related to the Fracture and Particulation of Electrodeposited-Copper Shaped Charge Jets

DAVID H. LASSILA

Materials Test and Evaluation Section
Engineering Sciences Division
Lawrence Livermore National Laboratory
Livermore, California 94550, U. S. A.

Shaped charges with two different types of electro-deposited copper liners have been found to produce fragmented and particulated jets of poor quality. Results of tensile testing and chemical analysis of fracture surfaces of these materials show that segregated impurities cause grain boundary embrittlement (loss of ductility) under tensile loading. Increase in the embrittlement occurs with increase in test temperature over some temperature ranges. Good correlation between the fracture behavior and chemical analysis of the fracture surfaces of the materials and the degree of jet fracture and particulation is observed. This correlation suggests fracture and particulation of the shaped charge jets is primarily due to the segregated impurities in the electrodeposited materials. Results of the study suggest tensile testing at elevated temperature can be used to screen liner materials for optimum performance.

I. INTRODUCTION

Copper is a material commonly used for shaped charge liners for performance and econo-mical reasons. For some experimental and complex designs it is desirable to use a electro-deposition process for liner fabrication, for example, small apex angle cone designs and multi-material liners. There are many options regarding electrodeposition processes in terms of the chemical content of the plating solution and physical conditions of the plating process, such as current density and temperature of the plating solution. All of these parameters can affect the final product's impurity content and microstructure, and are often varied to obtain a desired product. Many studies have shown that these types of variabilities, i.e., microstructure and impurity content in a liner material, can cause changes in the performance of the shaped charge, primarily due to fracture and particulation of the jet (1–5).

Two electrodeposited-copper liners, with distinctly different microstructures, have been tested for jetting performance using 42 degree, 81 mm diameter BRL-type shaped charges at Lawrence Livermore National Laboratory (LLNL). X-ray photographs indicated brittle breakup and particulation of the jets during jet stretching. In this report, embrittlement of these electrodeposited-copper materials caused by segregated impurities is investigated as a primary reason for the brittle breakup. High temperature tensile tests were performed to assess the propensity of the materials to particulate during shaped charge jet elongation.

II.   EXPERIMENTAL

A.   TEST MATERIALS

Two electrodeposited-copper materials are examined in this work and are designated UBAC and LEA, which are aconyms for the plating solutions used for their deposition. The UBAC copper was produced at LLNL utilizing UBAC No. 1 plating bath additive obtained from OMI International Corp. The LEA copper was produced by Nickel Electroform, Inc. The electrodeposited materials for analysis were received in the form of plates approximately 3 mm in thickness and were produced using the same electrodeposition conditions used to form the shaped charge cones. All materials were given a stress relief heat treatment in an argon atmosphere for one hour at 250° C prior to testing. The bulk chemical analyses of UBAC and LEA coppers are given in Table 1.

TABLE 1 *Bulk Chemical Analysis of the UBAC and LEA Electrodeposited Coppers (PPM)*

	C	O	H	N	Si	Ca	Al	Ti	Fe	Mg
UBAC	30	600	40	<100	<6	1	1	–	–	<1
UBAC	10	400	10	<100	<6	2	4	4	<4	<1

FIG 1 *Micrograph showing the grain structure of the UBAC copper. The grain size is approximately 3 μm diameter and is equiaxed. (Magnification = 100×)*

The two electrodeposited materials had distinctly different microstructures. The UBAC microstructure, shown in Figure 1,[*] is a very fine grain and equiaxed structure with an average grain diameter of approximately 1 μm to 3 μm. The LEA microstructure, shown in Figure 2, is a columnar structure with an average grain diameter of 50 μm.

B. TENSILE TESTING

Tensile testing was performed using a screw-driven test machine at a strain rate of 1.3×10^{-3} s^{-1}. Testing was performed in a flowing Argon environment, which was

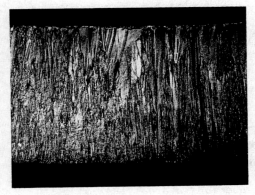

FIG 2 *Micrographs showing the through-thickness and in-plane microstructure of the LEA copper indicating a relatively large grain columnar structure. (Magnification = 100×)*

---

*Throughout, magnification is that of original photo, not as reproduced.

evacuated and back-filled several times prior to testing. Test temperatures ranged from 21° C to 300° C. Flat tensile bars were used with a gage length of 12.7 mm and a cross section of 3.82 mm by 1.27 mm in the gage section.

Scanning electron microscope (SEM) photos of the fracture surfaces were used to determine the fracture surface area of each sample which was then used to calculate the percent reduction of area (%RA). The %RA of the UBAC and LEA coppers are plotted versus the test temperature in Figure 3. The %RA of the UBAC copper at failure drops abruptly from 90% at 21° C to 22% at 180° C. Examination of the fracture surfaces show this to be caused by a change in fracture mode from ductile shear rupture at low test temperature (Figure 4) to a brittle fracture mode at high test temperature (Figure 5). A high magnification SEM micrograph of a UBAC copper brittle fracture is shown in Figure 6. The predominate fracture surface features are approximately the same size as the grain size in this material, which suggests the brittle fracture occurred along grain boundaries.

The LEA copper exhibited brittle fracture at all test temperatures (room temperature and above). For a given test temperature, the values %RA are all substantially below those of the UBAC copper indicating that the LEA copper is a more embrittled material. An SEM photo

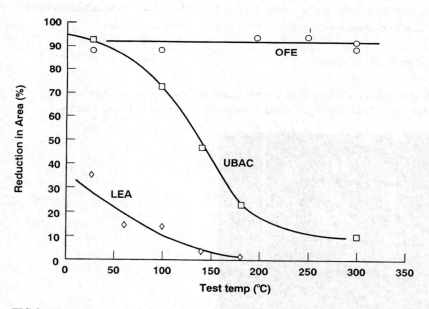

FIG 3 *Dependence of percent reduction in area (%RA) on test temperature. The UBAC copper data shows a ductile-to-brittle transition. The LEA copper is brittle at room temperature and becomes more brittle with increasing test temperature.*

FIG 4 *SEM micrograph of a ductile UBAC copper tensile failure (room temperature test temperature). The necking that occurred during failure is extensive, yielding a %RA close to 100%.*

FIG 5 *SEM micrograph of a brittle UBAC copper fracture surface (180° C test temperature). Essentially no necking occurred prior to failure.*

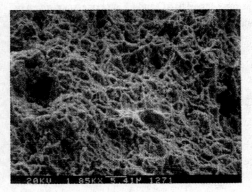

FIG 6 *High magnification SEM micrograph showing details of the brittle UBAC copper fracture surface. The size of the predominate features suggests the failure was intergranular (occurred along grain boundaries).*

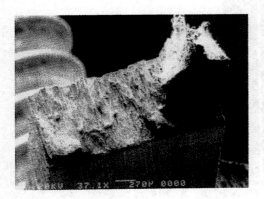

FIG 7 *SEM micrograph of the brittle LEA copper fracture surface*
(180° C test temperature). Considerable amounts of secondary
cracking occurred prior to and/or during failure which
indicates an extremely brittle material. The columnar features
on the fracture surface indicate the fracture occurred along
grain boundaries.

of a LEA copper fracture surface is shown in Figure 7. The main fracture occurred at the
grain boundaries in this material with considerable amounts of secondary cracking also
occurring along grain boundaries.

For comparison, oxygen-free electronic (OFE) copper in form of cold-rolled sheet stock
was also tested. The OFE copper showed no indications of embrittlement over the range of
test temperatures used in this study. At all test temperatures, the OFE copper underwent
extensive necking prior to fracture, similar to the UBAC copper tested at room temperature
(Figure 4).

The ultimate strength and percent elongation of the UBAC and LEA coppers are given in
Table 2 for several test temperatures. In general, there is a trend towards a decrease in these

TABLE 2 *Ultimate Strength and Elongation of UBAC and LEA*
Coppers

		Test Temperatures		
		22°C	100°C	180°C
Ultimate Stress (MPa)	UBAC	303	273	229
	LEA	210	169	139
Elongation (%)	UBAC	68	72	38
	LEA	36	16	11

measurements with increasing test temperature for each of the materials. Caution should be used, however, in exercising a relationship between embrittlement and these measurements because of the interrelationships that exist between test temperature, constitutive behavior, and the embrittlement behavior. For example, the elongation of the UBAC copper at 100° C is greater than that at 22° C, while the fracture surface observations indicate the material is embrittled at 100° C and is not embrittled at 22° C. Thus, it would be erroneous to conclude that the UBAC copper is not embrittled at the test temperature of 100° C because the elongation is greater than that at 22° C. Clearly, fracture surface observations and %RA measurements are needed to determine if the material is undergoing a transition in fracture mode.

III.   CHEMICAL ANALYSIS OF FRACTURE SURFACES

Auger Electron Spectroscopy (AES) was used to investigate the segregation of impurities at the brittle fracture surfaces of the UBAC and LEA coppers. A Perkin-Elmer Auger electron spectrometer equipped with a hot stage and a fracture stage was used for the analysis. Chevron notched samples were fractured at approximately 250° C either insitu in the Auger instrument under UHV conditions (vacuum of 10^{-8} torr or less) or in a glove box filled with an ultra pure argon environment and then transferred to the Auger instrument without exposure to air.

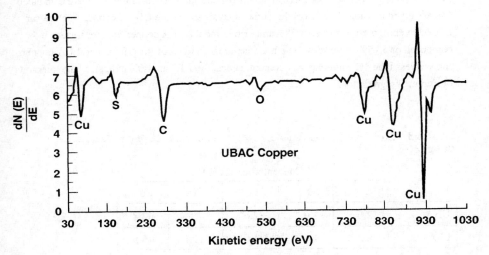

FIG 8 *AES spectra of brittle UBAC copper fracture surface (250° C fracture temperature).*

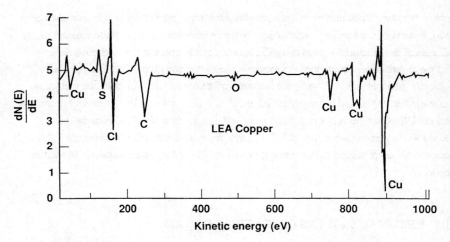

FIG 9 *AES spectra of a brittle LEA copper fracture surface (250° C fracture temperature).*

The AES spectra for the UBAC copper is shown in Figure 8 and indicates the presence of three major impurities; carbon, oxygen, and sulfur. The AES spectra for the LEA copper, shown in Figure 9, indicates the presence of carbon, oxygen, sulfur, and chlorine. Using the AES sensitivity factors for these elements (6), the surface concentrations were calculated and are given in Table 3. The results of the analysis shows the UBAC copper fracture surface to be composed of a total of 38% impurities and the LEA copper fracture surface to be composed of 53.5% impurities. The bulk chemical analysis of the UBAC and LEA coppers showed less than 1% total impurity content; hence, the AES analysis indicates large amounts of impurity segregation.

TABLE 3 *Chemical Analysis of UBAC and LEA Copper Fracture Surfaces*

<table>
<tr><th colspan="6" style="text-align:center">Concentration (at %)</th></tr>
<tr><th></th><th>S</th><th>O</th><th>C</th><th>Cl</th><th>Cu</th></tr>
<tr><td>UBAC</td><td>2.6</td><td>2.0</td><td>32.6</td><td>–</td><td>62.8</td></tr>
<tr><td>LEA</td><td>6.2</td><td>1.4</td><td>34.6</td><td>11.1</td><td>46.5</td></tr>
</table>

IV. DISCUSSION

A. *EMBRITTLEMENT OF ELECTRODEPOSITED COPPER*

The embrittlement of electrodeposited materials is a well established phenomenon which is generally attributed to impurities occluded in the material during deposition (7,8). The observation of a thermally induced ductile-brittle transition in electrodeposited copper has been reported by Zakraysek (9) who found that embrittlement was a result of fracture along grain boundaries. It is generally believed that segregation at grain boundaries is driven by a reduction in grain boundary interfacial energy which, in turn, leads to a reduction in the grain boundary cohesive strength (10,11). The exact nature of the embrittlement mechanism and the observed increase in embrittlement with increasing temperature, i.e., the ductile-brittle transition, are not understood.

The results of the tensile testing and analysis of the fracture surfaces of the UBAC and LEA coppers are consistent with previously mentioned works. The impurities detected by the Auger analysis on the grain boundary fractures are believed to cause decohesion of the boundaries. What segregated impurity or impurities promote the decohesion is not known. Of the impurities detected, sulfur and oxygen have both been found to promote brittle fracture, together and by themselves in copper (12–14). Carbon, which was found in large amounts on the fractures, has not been found to cause brittle fracture, although this does not exclude this possibility.

Hydrogen, which cannot be detected by Auger analysis, was detected in the bulk chemical analysis of the UBAC and LEA coppers at significant levels (40 PPM and 20 PPM, respectively). Hydrogen related embrittlement in copper is generally associated with the formation of bubbles at grain boundaries on the order of 1 μm in diameter (15). Since metallographic observations of the materials did not detect porosity of this nature, it is not likely that this particular hydrogen embrittlement mechanism was responsible for the brittle fractures. Sulfur and hydrogen, cosegregated at grain boundaries, has been shown to cause severe embrittlement of nickel (16). This may also be a potent embrittlement combination in copper.

Although the nature of the embrittlement in terms of the embrittling elements and mechanisms is not known, a general correlation between the total impurity coverage on the fracture surfaces and the degree of embrittlement exists. For a given temperature the UBAC copper is less embrittled than the LEA copper, and also has less total impurity coverage at the fracture surface, as indicated in Table 3. Additional studies are needed to determine what impurity or combination of impurities causes embrittlement.

B. CORRELATION OF DATA WITH JET BREAKUP

The electrodeposited materials examined in this study were used as liner materials in 42 degree, 81 mm diameter BRL shaped charges. These charges were fired at LLNL's Site 300. A flash x-ray photograph of the jet produced by the UBAC copper liner, shown in Figure 10, indicates brittle breakup of the jet with an average length to diameter ratio of about 1.0. A flash x-ray photograph of the LEA copper jet (Figure 11) indicates a more chaotic breakup of the jet with many more fragments and particles than the UBAC copper jet. For

FIG 10 *X-ray photograph showing the breakup of a UBAC copper shaped charge jet. The fracture mode appears to be brittle with little or no %RA prior to fracture, which is similar to the high temperature tensile test failures.*

FIG 11 *X-ray photograph of a LEA copper shaped charge jet which shows the fracture and particulation of the material. The particulation of the jet is somewhat analogues to the large amounts of secondary cracking shown in the micrograph of the high temperature tensile failure of this material (Figure 7).*

FIG 12 *X-ray photograph of a ductile type failure of a copper shaped charge jet which shows extensive necking in the failure areas. The length-to-diameter ratio of the fragments of this jet is on the order of 5. (Courtesy of Dr. Glen Randers-Pehrson, Dept. of the Army, BRL)*

the purpose of comparison, an example of a ductile necking type breakup of a copper jet is shown in Figure 12. The length to diameter ratio of the ductile jet is approximately 10, which is due in part to the extensive necking which occurs prior to fracture.

In a shaped charge application, an exceedingly large amount of deformation of the liner material occurs over a very short period of time; consequently, the deformation is adiabatic and the thermal energy of the jet material increases with increasing strain (17). The temperature of the jet is estimated to be about 450° C at the time of breakup (18). Although the deformation history of the copper in the shaped charge application is radically different than that of the tensile tests performed herein, it is believed that the funda-mental material characteristic of high temperature embrittlement under tensile loading is com-mon to both cases; the presence of segregated impurities is believed to be the primary cause of the brittle fracture and particulation of the UBAC and LEA copper jets.

The degree of embrittlement measured by %RA as a function of test temperature (Figure 3) correlates in a qualitative fashion with the degree of brittle fracture and particulation indicated in the radiographs of the jet breakups. Both the tensile test results and the radiographs of the jet breakup show the LEA copper to be inferior to the UBAC copper. Because of this qualitative correlation, it is believed that the elevated temperature fracture mode data can be used to screen liner materials for optimum performance in terms of shaped charge jet breakup behavior. Materials which retain a ductile fracture mode at elevated temperature, such as the OFE copper tested in this study (Figure 3), are expected to produce jets which undergo a great deal of stretching and necking prior to breakup.

V.   CONCLUSIONS

1.   Results of this study show that the UBAC and LEA electrodeposited- coppers are embrittled by segregated impurities which are occluded in the coppers during electro-deposition. This embrittlement has been shown to increase with increasing temperature.

2.   The brittle fracture and particulation of the UBAC and LEA copper shaped charge jets is believed to be due to the embrittlement caused by the segregated impurities in these materials.

3.   High temperature tensile testing appears to have some utility in selecting copper materials for shaped charge liners. Materials that retain a ductile fracture mode at elevated test temperatures are expected to produce ductile jet breakup.

ACKNOWLEDGEMENTS

The author would like to thank Ms. Mary Leblanc and Mr. Mike Vardanega for their help with the tensile testing. Mr. Todd Fatheree, Ms. Allison Connor and Dr. David Harris are acknowledged for their efforts in obtaining the AES data. The author would also like to thank Mr. J. W. Dini for supplying the UBAC copper. This work was performed under the DARPA funded Joint Armor Anti-Armor Project at LLNL. The work was performed under the auspices of the U. S. Department of Energy by Lawrence Livermore National Laboratory under Contract No. W-7405-Eng-48.

REFERENCES

1.   A. Lichtenberger, *Proceedings of the 11th International Symposium on Ballistics, Vol. II*, Brussels Belgium (1989).

2.   M. L. Duffy and S. T. Golaski, Technical Report BRL-TR-2800, U.S. Army Ballistic Research Laboratory, Aberdeen Proving Ground, Maryland (1987).

3.   A. Lichtenberger, M. Scharf, and A. Bohmann, *6th International Symposium on Ballistics*, Orlando (1981).

4.   D. Dorfman and S. T. Golaski, Contract Report ARBRL-CR-00474, U.S. Army Ballistic Research Laboratory, Aberdeen Proving Ground, Maryland (1981).

5.   M. Held, *Propellants, Explosives, Pyrotechnics 10*, 125–128 (1985).

6.   Physical Electroncis AES Handbook.

7.   J. W. Dini, H. R. Johnson, and L. A. West, *Plating and Surface Finishing*, 36–40 (1978).

8.   C. E. Moeller and F. T. Schuler, Rocketdyne Division of Rockwell Int., paper presented at ASM Metals Show, Cleveland, OH (October 1972).

9.   L. Zakraysek, Fractography and Material Science, ASTM STP 733, L. N. Gilbertson and R. D. Zipp, Eds., *American Society for Testing and Materials*, 428–439 (1981).

10.   D. Mclean and H. R. Tipler, *Metal Science Journal 4*: 103–107 (1970).

11.   D. L. Wood, *Transactions TMS-AIME 209*: 209 (1957).

12.   M. Meyers and E. A. Blythe, *Metals Technology* 165–170 (May 1981).

13.   S. P. Clough and D. F. Stein, *Scripta Metallurgica 9*: 1163–1166 (1975).

14.   T. G. Neih and W. D. Nix, *Mettallurgical Transactions A 12A*: 893–901 (May 1981).

15.   S. Nakahara, *Acta Metallurgica 36*: 1669–1681 (1988).

16.   D. H. Lassila and H. K. Birnbaum, *Acta Metallurgica 35*: 1815–1822, (1987).

17.   D. H. Lassila and T. McAbee, UCID in preparation, University of California, Lawrence Livermore National Laboratory.

18.   W. G. Von Holle and J. J. Trimble, *6th International Symposium on Detonation, Naval Surface Weapons Center* ACR-221 ONR (1976).

50

RHA Plate Perforation by a Shaped-Charge Jet: Experiment and Hydrocode Simulation

MARTIN N. RAFTENBERG and CLAIRE D. KRAUSE

U. S. Army Ballistic Research Laboratory
Terminal Ballistics Division
Aberdeen Proving Ground, Maryland 21005, U.S.A.

Models for metal failure by tensile voids and shear banding are presented for insertion into a hydrocode. Each model consists of an onset criterion and a post-onset prescription on stresses. These have been added to EPIC-2 and applied to the particular case of a BRL3.2" shaped charge fired into a 13.0-mm-thick RHA plate. Use of the two failure models in conjunction with the code's slideline erosion algorithm allows for a predicted hole geometry that agrees closely with experiment both in size and shape.

I. INTRODUCTION

The present study develops models for metal failure by tensile voids and shear bands. These models are then applied by means of hydrocode EPIC-2 to the case of a BRL3.2" shaped-charge warhead fired at normal incidence and a standoff of 1238 mm (15.23 cone diameters) into a 13-mm-thick plate. The warhead has a conical copper liner with a 42° apex angle, a 81.28 mm diameter base, and a 1.91 mm wall thickness and is filled with Comp-B explosive. A Brinnell Hardness number of 364 for the front and back surfaces of the target plate is measured. The plate has a square cross-section, with an edge length of 197 mm.

Section II presents experimental evidence from this range firing for the failure mechanisms of tensile voids and shear bands. Section III describes a model that has been installed into EPIC-2 for each of these two mechanisms. Computational results are compared with measurements in Section IV. Some concluding remarks are made in Section V.

II. OBSERVATIONS OF TENSILE VOIDS AND SHEAR BANDS

RHA specimens are extracted from the perforated plate in the vicinity of the
hole and are examined using scanning electron microscopy. Fig. 1 displays a
crack that has been formed by coalescing voids. Fig. 2 shows a shear band, with
a characteristic 6 μm width, while Fig. 3 displays a crack that runs along part of
a shear band's length. From these figures it is concluded that cracks, and
therefore fragmentation of RHA, are associated with the two phenomena of
voids and shear bands that result from interaction with the jet.

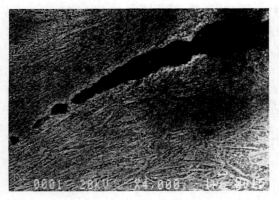

FIG. 1 *A crack formed by void coalescence in the RHA target plate. Scanning
electron micrograph, x4000, 2% Nital etch.*

FIG. 2 *A shear band in the RHA target plate. Scanning electron micrograph,
x3700, 2% Nital etch.*

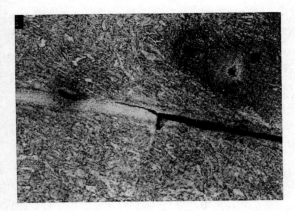

FIG. 3 *A crack that extends along a shear band in RHA. Optical micrograph, x800, 2% Nital etch.*

III. MODELS OF MATERIAL FAILURE FROM TENSILE VOIDS AND SHEAR BANDS

The above range firing will be modeled using the 1986 version of Lagrangian hydrocode EPIC-2 [1]. The two metals, copper and steel, are both handled in the same fashion, with the single exception that the shear band model is applied only to the steel. Their dilatational deformation is governed by the Mie-Gruneisen equation of state. The constant parameters that appear are assigned values from Kohn's handbook [2].

The distortional deformation of copper and steel is treated by means of a plasticity model, which applies the von Mises yield condition. Flow stress σ_y is computed using the Johnson-Cook strength model [3]. According to this model, σ_y is a function of equivalent plastic strain ε^p, equivalent plastic strain rate $\dot{\varepsilon}^p$, and homologous temperature T^*. The functional form is

$$\sigma_y = \left[A + B \, (\varepsilon^p)^N \right] \left[1 + C \ln \left(\frac{\dot{\varepsilon}^p}{1.0 s^{-1}} \right) \right] \left[1 - (T^*)^M \right] \tag{1}$$

Associated material constants are assigned values presented for OFHC copper and for 4340 steel by Johnson and Cook [3].

A. TENSILE VOID MODEL

The voids in the RHA plate that are seen in Fig. 1 are assumed to have been caused by tensile failure. It is further assumed that a similar mechanism is operative in the copper jet. A tensile failure model applicable to individual elements is therefore added to EPIC-2. The model consists of two parts: a failure onset criterion and a prescription for post-onset behavior. The criterion is a negative-pressure cutoff. Pressure is defined as the negative hydrostatic stress, or

$$p = -\frac{1}{3}\left[\sigma_{rr} + \sigma_{zz} + \sigma_{\theta\theta}\right] \tag{2}$$

Thus failure occurs when

$$p \leq p_{fail} < 0 \tag{3}$$

Once this tensile failure criterion is met in an element, at that time step and at all time steps thereafter, hydrostatic tensile stress and all deviatoric stresses are set to zero. Only hydrostatic compression is allowed. The model introduces a single additional input parmeter for each material, p_{fail}.

B. SHEAR BAND MODEL

The algorithm inserted into EPIC-2 to model material failure from shear bands also involves the two stages of onset criterion and a prescription for post-onset behavior. The criterion adopted is called "Zener-Hollomon", after its apparent originators [4]. In a given time step, the product $\left[A + B(\varepsilon^p)^N\right]\left[1 - (T^*)^M\right]$ is evaluated for each RHA element that has not yet failed or eroded and that has undergone further plastic flow during that time step. The above product includes contributions to the flow stress by plastic straining and by temperature. If the element also underwent plastic flow during the immediately preceding time step and if the above product has decreased from its level during that preceding time step, then the element is deemed to have failed by shear banding. In effect, the criterion compares the strength increase due to plastic straining (work hardening) to its decrease due to temperature rise (thermal softening). Failure occurs when the latter exceeds the former.

   Once this criterion has been satisfied in an element, at that and all subsequent time steps, deviatoric stresses are reduced from their computed values. However, hydrostatic tension is not altered, in contrast to the post-onset modeling for the case of tensile failure. Another difference is that the deviatoric stresses are not instantaneously set to zero, as with the previous model. Instead, their gradual reduction is imposed, qualitatively consistent with results from shear band modeling studies reported in Reference 5. If s_{rr}^*, s_{zz}^*, $s_{\theta\theta}^*$, and s_{rz}^* are the four components of deviatoric stress computed by EPIC-2 in such an element for a given time step, then each of these is altered to the values given by

$$s_{ij} = s_{ij}^* f(\varepsilon^p) \tag{4}$$

where

$$f(\varepsilon^p) = \begin{cases} 1 & ; \quad \varepsilon^p \in [0, \varepsilon_{onset}^p] \\ 1 - \dfrac{\varepsilon^p - \varepsilon_{onset}^p}{\Delta \varepsilon_{fail}^p} & ; \quad \varepsilon^p \in (\varepsilon_{onset}^p, \varepsilon_{onset}^p + \Delta \varepsilon_{fail}^p) \\ 0 & ; \quad \varepsilon^p \in [\varepsilon_{onset}^p + \Delta \varepsilon_{fail}^p, \infty) \end{cases} \tag{5}$$

Here, ε_{onset}^p is the value of equivalent plastic strain ε^p that existed in the element at the time step in which the Zener-Hollomon criterion was first met. ε_{onset}^p is therefore computed by the code. On the other hand, $\Delta \varepsilon_{fail}^p$ is a user-prescribed material property. $\left[\varepsilon_{onset}^p + \Delta \varepsilon_{fail}^p \right]$ is the level of ε^p beyond which the element no longer supports deviatoric stresses.

The post-onset conditions imposed on an element that has failed according to the tension model are more stringent than those that have failed by the shear band model. In the former case hydrostatic tensile stress as well as deviatoric stresses are affected, and stress reduction to zero is instantaneous. For this reason, elements that have failed by shear banding continue to be checked for tensile failure.

C. APPLICATION TO THE RANGE FIRING

These tensile void and shear band models are inserted into EPIC-2. The mesh used in the axisymmetric calculations consists of three-node triangular elements. The copper projectile's mesh closely approximates the geometry of the leading particle. The projectile's mesh is assigned the uniform initial velocity of 7.73 mm/μs, as determined from pre-impact radiographs from the experiment. The target's mesh approximates the steel plate by a disk with a thickness of 13.0 mm and a 104.0 mm radius. Individual elements in the target have an edge length of 1.3 mm.

The mesh contains two slidelines, both coincident with the projectile/target interface. One slideline has the projectile's surface nodes serving as master nodes and the target's surface nodes as slaves. In the other slideline this assignment is reversed.

The erosion feature is operative for both slidelines. This allows an element having one or more corner node on the master surface of the slideline to be discarded from the problem, in the sense that all stresses in the element are thereafter set to zero. Since both copper projectile nodes and steel target nodes alternately serve as master nodes for one of the two slidelines, the procedure provides a means for modeling both projectile erosion and target hole formation. EPIC-2 triggers erosion of a given surface element when a user-supplied cutoff value for equivalent plastic strain, ε^p_{erode}, is reached in that element. EPIC-2's eroding slideline algorithm is described further in Reference 6. The tensile void and shear band models are applied to elements throughout the domain, both on and off the slideline.

IV. RESULTS COMPARED WITH EXPERIMENT

Fig. 4 displays the plate's final experimental hole geometry, determined by removing a radial section. $\bar{r}_{hole}$, the hole radius averaged over the plate's thickness, is 18.2 mm based on this figure.

The computed hole geometry is dependent on values assigned to ε^p_{erode}, p_{fail}, and $\Delta\varepsilon^p_{fail}$. These dependencies are studied through three series of numerical experiments. First, EPIC-2 is applied to the firing with both tensile and shear band models inoperative. In each run the same value for ε^p_{erode} is assigned to both the copper and the steel. For ε^p_{erode} varied over the range of 0.25 to 2.00, the largest $\bar{r}_{hole}$ obtained is 13.9 mm, substantially smaller than the experimental value. The second series employs the tensile void failure model with p_{fail} set to -3.0 GPa for both steel and copper. The shear band model is still inoperative. $\bar{r}_{hole}$ values increase slightly from the previous series, but the largest value observed, 15.0 mm, is still smaller than the experimental result.

Finally, runs are performed with both failure models operative. p_{fail} for the tensile model is set to -3.0 GPa for both steel and copper throughout. ε^p_{erode} and $\Delta\varepsilon^p_{fail}$ are both varied. In a given run the same value for ε^p_{erode} is assigned to both copper and steel, while $\Delta\varepsilon^p_{fail}$ is of course only applied to steel. Fig. 5 plots

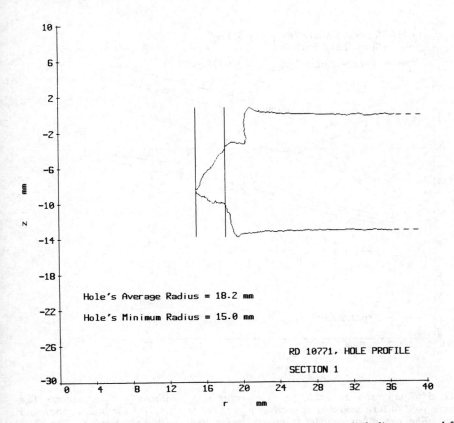

FIG. 4 *The hole's profile in the experiment, based on a radial slice removed from the perforated target.*

$\bar{r}_{hole}$ versus ε^p_{erode} with $\Delta\varepsilon^p_{fail}$ as a parameter. The three curves corresponding to $\Delta\varepsilon^p_{fail}$ = 0.0, 0.10, and 0.25 all intersect the experimental result of $\bar{r}_{hole}$ = 18.2 mm. Thus, inclusion of the shear band model allows a sufficiently large hole to be obtained. Fig. 6 shows the mesh at 500 μs after initial impact for the case of ε^p_{erode} = 0.50 and $\Delta\varepsilon^p_{fail}$ = 0.0 (instantaneous reduction to zero of deviatoric stresses.) The $\bar{r}_{hole}$ value of 18.1 mm and the general shape of the hole's profile agree quite closely with the experimental results in Fig. 4. Shaded elements in Fig. 6 indicate those that have failed by the shear band model and have not eroded from the plate.

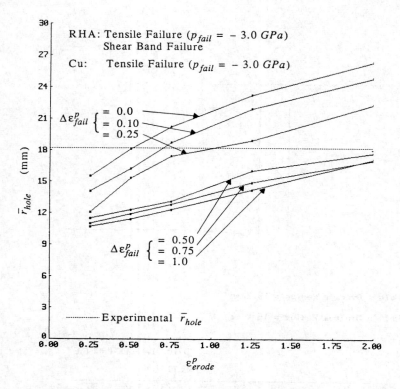

FIG. 5 *Computational results for* $\bar{r}_{hole}$ *versus* ε^p_{erode}, *with* $\Delta\varepsilon^p_{fail}$ *a parameter. The tensile void failure model is operative;* p_{fail} *is set to -3.0 GPA for both copper and steel. The shear band model is operative for steel, with* $\Delta\varepsilon^p_{fail}$ *varied. The* ε^p_{erode} *value applies to both slidelines.*

V. CONCLUDING REMARKS

Models for material failure by tensile voids and by shear banding have been added to EPIC-2. These models introduce two new material parameters, p_{fail} and $\Delta\varepsilon^p_{fail}$. When combined with the code's slideline erosion algorithm, these failure models provide a means for investigating plate perforation by a shaped-charge jet.

EPIC-2 has been applied to the case of a BRL3.2" shaped charge fired at long standoff into a 13.0-mm-thick RHA plate. Without both failure models active, the code has been shown to predict a plate hole that is smaller than the

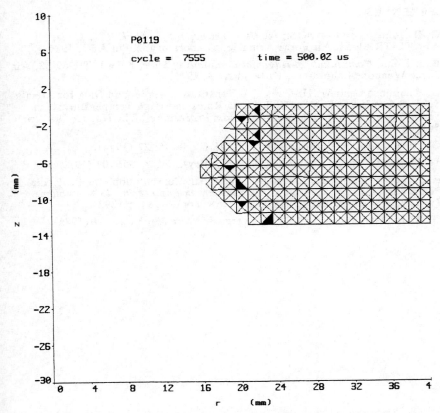

FIG. 6 *The deformed mesh at 500 μs after impact for the case when* pfail = − 3.0 GPa, ε_{erode}^{p} = 0.50, *and* $\Delta\varepsilon_{fail}^{p}$ = 0. *Shaded elements have failed by shear banding.*

experimental result.[*] Inclusion of the two failure models allows for prediction of a hole that agrees closely with experiment both in size and shape.

ACKNOWLEDGEMENTS

It is a pleasure to thank Thomas W. Wright, John W. Walter, and Michael J. Scheidler of the Ballistic Research Laboratory and Howard Tz. Chen of the Naval Surface Warfare Center for assistance with the shear band model.

[*] *EPIC-2 contains a failure model that is phenomenological and was derived to fit data from Hopkinson bar tests and quasi-static tensile tests [7]. Such tests involve relatively low strain rates, so the model was judged inappropriate for the present application. However, its use would presumably have led to a larger predicted hole size.*

REFERENCES

1. G. R. Johnson and R. A. Stryk, *User Instructions for the EPIC-2 Code*, AFATL-TR-86-51, Air Force Armament Laboratory, Eglin AFB, 1986.

2. B. J. Kohn, *Compilation of Hugoniot Equations of State*, AFWL-TR-69-38, Air Force Weapons Laboratory, Kirtland AFB, 1965.

3. G. R. Johnson and W. H. Cook, "A Constitutive Model and Data for Metals Subjected to Large Strains, High Strain Rates and High Temperatures", in *Seventh International Ballistics Symposium Proceedings*, The Hague, 1983, p. 541.

4. C. Zener and J. H. J. Hollomon, *J. Appl. Phys.*, 15: 22 (1944).

5. T. W. Wright and J. W. Walter, *J. Mech. Phys. Solids*, 35: 701 (1987).

6. F. P. Stecher and G. R. Johnson, "Lagrangian Computations for Projectile Penetration into Thick Plates", in *Computers in Engineering 1984 - Volume Two*, W. A. Gruver (ed.), A.S.M.E., New York, 1984, p. 292.

7. G. R. Johnson and W. H. Cook, *Engrg. Fracture Mech.*, 21: 31 (1985).

51

On the Relationship Between the Microstructural Condition of the Liner and the Performance of a Shaped Charge

C. S. DA COSTA VIANA and C. N. ELIAS

Seção de Engenharia Mecânica e de Materiais
Instituto Militar de Engenharia
Praça General Tibúrcio 80 - Urca
22290 Rio de Janeiro, RJ - Brasil

An analysis is made of the material of a shaped charge metal liner during collapse. The nature of the stress and strain states imposed by the detonation on the collapsing surface and its property requirements for a uniform contraction are discussed. Texture development during collapse is predicted and some resulting plastic properties calculated.

I. INTRODUCTION

The collapse of a shaped charge metal liner is a phenomenon still to be understood throughly. The nature of the microstructural and textural conditions of the collapsed material, its properties to guarantee uniform contraction and structural stability under the high detonation pressures can only be guessed. Typical deformation figures include true strains of about 3 and strain rates near 10^5 s^{-1}. The deformation is essentially adiabatic and the plastic deformation realized under these conditions is also assisted by a sharp temperature rise.

Certain characteristics are however known from observation of shocked materials. Before deformation the material is hardened by the detonation shock wave [1]. The dislocation density increases, forming a cellular substructure or a uniformly distributed

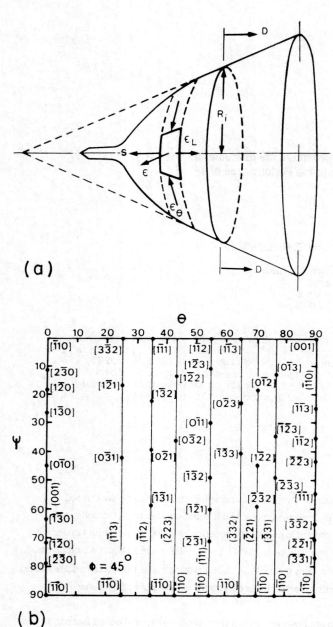

FIG. 1(a) *Schematic diagram of a liner collapse, (b)* $\phi = 45°$ *indexing chart for CODF.*

array with practically no tendency for cell formation [1,2]. Local crystal misorientations of less than 1° are observed with no general texture alteration [3]. In the case of a conical liner the collapsed material flows along a curved surface of revolution towards a stagnation point, as shown schematically in Figure 1a. This figure also shows the reigning strain state involving both hoop compressive and longitudinal tensile components. Under the detonation pressure the material converges to the stagnation point without wrinkling and, except near that region, with little thickness variation

In the present work the development of texture and plastic properties in the collapsed material will be studied in an attempt to help explain the collapse of metal liners.

II. TEXTURE AND PROPERTY PREDICTION

In the present model the liner collapse is imagined to occur according to Figure 1a. The collapsed surface is smooth and with no thickness variation ($\varepsilon_z = 0$). The strain state is supposed to be one of pure shear ($\varepsilon_\theta = -\varepsilon_L$, $\varepsilon_{ij} = 0$ for i≠j) everywhere. The material is assumed to be face centered cubic with [111] <1$\overline{1}$0> restricted glide slip systems. Texture development simulation was done according to the methods of Kallend [4], using full constraints and the maximum work principle of Bishop and Hill [5]. The simulation starts from a random distribution of orientations and predicts the final distribution according to the imposed strain tensor. The final texture is represented by the crystallite orientation distribution function (CODF) as described by Roe [6], using a 20th order series expansion in Euler angles (ψ, θ, $\varnothing$). Plastic properties were predicted according to a method described elsewhere [7]. During collapse, the shock wave is supposed to increase the density of defects to a value where a mechanism such as Coble creep [8] can assist the deformation.

III. RESULTS

The texture simulation in the liner was performed for total hoop strains equal to -0.5, -0.75 and -1. This corresponds to an initial radius R_i being reduced respectively to 0.60 R_i, 0.47 R_i and 0.36 R_i. These conditions were named A, B and C. Figure 1b is an indexing chart for $\phi = 45°$ for the CODF plot. This section contains most of the important orientations for cubic materials. Vertical lines represent {hkl} planes parallel to the cone's surface and the dots represent <uvw> directions parallel to the L direction. The textures are plotted in times random units and interpreted in terms of {hkl} <uvw> components. Figure 2 shows the predicted texture for conditions A and C. The main components are X - {110} <3$\overline{3}$2>,

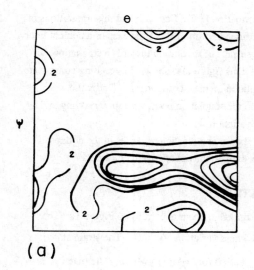

(a)

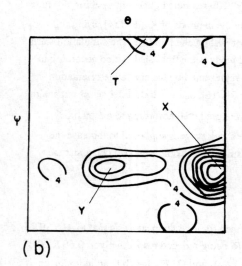

(b)

FIG. 2 *CODF plots of predicted textures for conditions (a) A;*
contour levels incremented by 2, (b) C; contour levels
incremented by 4.

TABLE 1

CODF intensity values for the X, Y and T components in the predicted textures, in times random units.

Cond/Comp	X	Y	T
A	14	8	8
B	21	12	9
C	28	12	8

Y - {221} <11$\bar{3}$> and T - {111} <11$\bar{2}$> but gradually the material tends to become X - single componented, as shown in Table 1. Components Y and T become stable but component X increases continuously as collapse proceeds.

The importance of this can only be appreciated from an analysis of the predicted plastic properties. Since the collapsed surface was assumed to behave like a shrinking deep-drawing blank, the same forming parameters can be used. Thus, the plastic strain ratio, R, in the L direction, which measures the resistance to thinning (or thickening), and the resistance to pure shear deformation, λ, were calculated from restricted upper bound yield loci predicted from the texture data [7]. Figure 3a shows a schematic yield locus where the

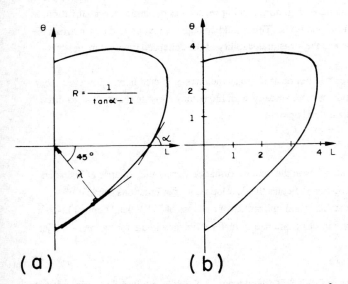

(a) **(b)**

FIG. 3(a) *Schematic yield locus showing R and* λ, *(b) predicted yield for material in condition C.*

TABLE 2

R and resistance to pure shear deformation predicted form the texture data.

Cond.	A	B	C	Random
R	1.94	2.30	3.30	1
λ	2.335	2.309	2.305	2.806

methods of measuring R and λ can be seen. According to the maximum work principle,

the smaller λ the smaller the resistance to pure shear deformation. Figure 3b shows the predicted locus for condition C. The coordinates are measured in Taylor factor units.

Table 2 shows the R and λ values also measured in Taylor factor units, for the textures and for a randomly oriented material. As can be seen R increases and λ decreases as collapse proceeds. Therefore, in-plane deformation becomes easier and the material flows towards the stagnation point with decreasing crystallographic restriction. the easier hoop contaction also helps Coble creep by reducing the effective grain size for diffusion. It is important to say that other phenomena may set in during collapse, such as dynamic recrystallization, as a result of the deformation conditions. This would change the texture and might affect the homogeneity of collapse and the structural stability of the collapsed material in an unpredictable manner.

Real collapse textures will not be as sharp as the ones predicted here. However shear spun or mechanically spun or mechanically spun liners may show textures close to these which would facilitate their development.

IV. CONCLUSION

If no other mechanism interferes in the texture formation during the collapse of a shaped charge liner, texture simulation indicates the development of orientations capable of increasing the structural stability and uniformity of collapse of FCC metal conical liners. These are important properties to guarantee a greater efficiency in the formation of the jet.

ACKNOWLEDGEMENTS

The authors are indebted to Dr. J. S. Kallend from I.I.T. for supplying the original version of the computed programs.

REFERENCES

1. M. F. Rose and T. L. Berger, *Phil. Mag.* 17: 1121 (1968).

2. O. Johari and G. Thomas, *Acta Met* 12: 1153 (1964).

3. A. G. Dhere, M.Sc. Thesis, IME, (1987) Brasil.

4. J. S. Kallend, Ph.D. Thesis, Cambridge, (1970).

5. J. F. W. Bishop and R. Hill, *Phil Mag.* 42: 414 (1951).

6. R. J. Roe, *J. Appl. Phys.* 36: 2024 (1065).

7. C. S. da Costa Viana, J. S. Kallend and G. J. Dovies, *Int. J. Mech. Sci.* 21: 355 (1979).

8. R. L. Coble, *J. Appl. Phys.* 34: 1679 (1963).

52

Transformation and Structural Changes in Metal Under Shock and Dynamic Loading

C. FENG

Metallic Materials Branch
Armaments Technology Division, AED
ARDEC, Picatinny Arsenal, NJ 07805-5000

   Atomic movement in shock-induced metals and alloys differs significantly from conventional plastic flow by slip and glide operations. When materials are shock loaded, adiabatic shear and mechanical twins are produced by the rapidly moving atoms under the high strain rate. Dynamic recrystallization may also take place under certain favorable conditions. An analysis is given of the expected physical and structural changes in terms of atomic movement for these processes. Descriptions are given of the observation for a number of metals and alloys subjected to shock loading and high strain rate deformation. The microstructural features associated with adiabatic shear and twinning are analyzed. A modified atomistic model is suggested for the mechanical twinning process in which a mirror image is produced on the opposite side of the symmetry plane. The mechanisms and the governing parameters are discussed.

I. INTRODUCTION

From the results of a large number of ultradynamic tests and experiments, we have witnessed the significant stress increase in metals subjected to strain rates of 10^3 sec^{-1} and greater. Much attention has been focused on the exploitation of this stress increase because of its potential for

penetration applications. The increase in stress in metals
may reach a level exceeding its fracture strength under
conventional strain rate tests as shown in Fig. 1, and is
frequently followed by a stress attenuation. Many workers
attributed this to the thermal softening effect [1-3] and
empirical models and constitutive equations have been
proposed to evaluate the parameters involved [4-8]. The
materials response under high strain rate loading is complex
and many mechanical test experiments do not always yield
fruitful results. Another approach to reach a better understanding of the dynamic response of materials is to examine
the basic mechanisms involved in dynamic plastic flow.

It is well known that in addition to the conventional
plastic flow by the slip operation, metals also undergo
adiabatic shear and mechanical twinning when subjected to
shock and dynamic loading. Of these, shock induced adiabatic shear and mechanical twinning have been reported for many
metals and alloys. More recently, attention has been given
to the phenomena of shock induced recrystallization [9]. In
the present study, a general review of the three phenomena
is made in terms of the basic characteristics and the atomic
movements involved. The controlling parameters for each of
the three processes will be examined and compared.

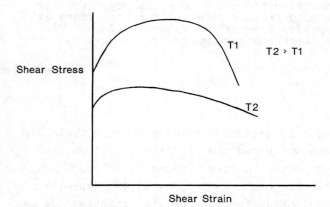

FIG. 1 *Generic illustration of stress/strain relationship
of the shock and high strain rate ($\dot{\varepsilon} > 10^3$ sec^{-1}) test at two
different temperatures, T_1 and T_2.*

II. EXPERIMENTAL STUDY AND ANALYSIS

A. *DYNAMIC RECRYSTALLIZATION*

It is well known that heat is generated when metals undergo plastic flow. Under dynamic loading, the amount of heat generated is significantly increased. We are familiar with examples concerning the use of this heat in shock and dynamic synthesis of materials. The heat associated with dynamic flow can also induce recrystallization in plastically deformed metals.

Dynamic recrystallization is not a new phenomenon and is believed to follow the general rules of nucleation and growth governing the conventional recrystallization process. It is a metallurgical phenomenon in which new grains are formed and developed at the expense of the prior structure and atomic movement is involved. Of importance is the fact that it can take place rapidly at a temperature lower than the conventional recrystallization temperature T_r, which is considered approximately half the melting temperature of the metal.

Figures 2 and 3a are two photomicrographs showing the microstructure for an unfired copper shaped charge liner and the recovered slug which never impacted the target. The

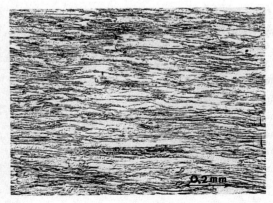

FIG. 2 *Typical microstructure of a deep drawn copper shaded liner showing elongated fibrous grains.*

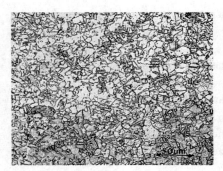

FIG. 3a *Microstructure of the trailing slug which did not impact the target, recovered after firing, showing fine equi-axed grains.*

FIG. 3b *TEM microstructure of the same slug in Fig. 3a.*

liner, as shown in Fig. 2, consists of long fibrous grains whereas in Fig. 3a, equi-axed recrystallized grains are present in the recovered slug. Figure 3b is a transmission electron micrograph showing the internal defect structure in the newly recrystallized grains. The temperature of the slug has not reached 405°C, equivalent to half the melting temperature for copper. Since it did not impact on the target, its temperature decreased rapidly to the ambient temperature of the surrounding area, probably within a few minutes. Another example is taken from a tantalum expanding ring experiment [10]. The maximum temperature of the specimen was considerably lower than 1364°C which is half the melting temperature for tantalum. The total duration of the experiment was less than a few seconds. An examination of

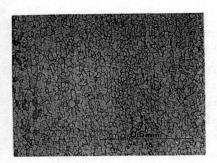

FIG. 4 *Fine tantalum grains, average diameter 25 microns, are present in the starting material for the expanding ring experiment.*

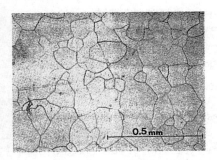

FIG. 5 *Coarse tantalum grains, diameter 50-120 microns present at the completion of the expanding ring experiment.*

the grains before and after the experiment, Figs. 4 and 5 respectively, showed that the grains size has increased from an average of 25 microns to greater than 100 microns. There was no indication of cold working in the microstructure. It may be concluded from the above experiments that more than a temperature rise is involved in dynamic recrystallization. Certainly, additional research is needed.

III. ADIABATIC SHEAR

Large numbers of experiments have been conducted where adiabatic shear was found to take place in different steels of martensitic, pearlitic and bainitic structures and in aluminum, copper and titanium and their alloys. Only the highlights of some of these experiments were discussed here.

Generally speaking, the adiabatic shear process is instantaneous and is influenced by the strain rate and the direction and the intensity of stress [4-7]. Meyer and Manwaring [5] identified the existence of a critical stress above which the structure is instable, leading to the formation of the adiabatic shear. Zerilli and Armstrong [6] proposed a relationship between the "maximum load strain" and the production of adiabatic shear. The structural instability as a function of localized heating and thermal effects have been analyzed by Mescall et al. [1,2] and other workers [3,11]. Cho et al. [12] have correlated the work hardening effect on crack development in association with the adiabatic shear.

It is well known that grain boundaries and other forms of barriers such as inclusions in the structure, do not appear to be effective in preventing the propagation and growth of the shear bands. A clearly defined demarcation line may appear between the shear band and the matrix in some metals but not others. Because of the great amount of heat generated, local melting can take place at the tip or the boundary layer of the shear band in metals of low thermal conductivity, e.g., titanium alloys [3,11] but not

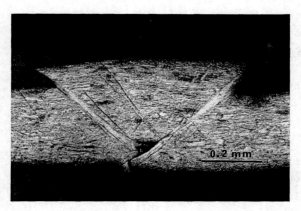

FIG. 6 *Shock induced adiabatic shear which led to failure of the brass cartridge case at the time of firing.*

necessarily in other metals. It is strongly stress dependent and is often seen to propagate in the direction of the
maximum shear. As shown in Fig. 6 for a brass cartridge
case which has failed under tension, the adiabatic shear
line was parallel to the maximum resolved shear stress,
i.e., making a 45 degree angle with the tensile axis.

IV. MECHANICAL TWINNING

Generally speaking, there are similarities between mechanical twinning and adiabatic shear. Both processes involve
the atomic shear and both are related to instability in the
matrix. But differences also exist between the two processes in terms of the mechanics in formation and the basic
characteristics.

In contrast to adiabatic shear, twins are without
exception confined within the individual grains. They are
found either to terminate at the grain boundaries or to make
the necessary adjustment to accommodate the change in orientation as they proceed across the neighboring grain. While
adiabatic shear is strongly dependent on the intensity of
stress, mechanical twinning is strongly dependent on the
physical properties of the materials. It is generally
believed that metals with low stacking fault energy are more
susceptible to twinning.

Twinning is defined as the shifting many blocks of
atoms in the lattice with the end result that a mirror image
is achieved in the opposite side of the twin plane. For a
given system, there is usually a preferred twinning plane in
which the atoms move at certain crystallographic direction.
In polycrystals, twinning may simultaneously take place in
many different grains and it is possible that more than one
set of mechanical twins are found in a single grain. When
multiple phase structures are present, twinning may be found
to take place in one but not the other phases. As shown in
Fig. 7 for a sample of mild steel, mechanical twins were
found in the ferrite grains but not in the pearlite region.
Twinning is seen to take place selectively in some but not
all ferrite grains. There is no detectable difference in
the amount of plastic flow between the grains where twins

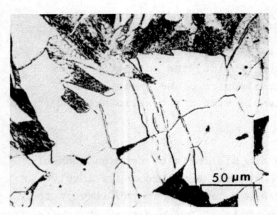

FIG. 7 *Shock induced twins in mild steel are found in the*
ferrite grains but not in the pearlite region.

are observed and those without twins. The microhardness
measurements indicated only negligible difference in hardness between the grains with twins and those without twins.
For other metals, an appreciable increase in hardness resulting from twinning has been reported.

The twinning planes may be determined from x-ray back
reflection experiments and stereographic analyses. For
polycrystals, the crystallographic orientations of the twins
may be determined by TEM selective area diffraction. The
twinning planes in different grains can be effectively
identified and a determination can be made to identify
whether or not they belong to the same family of planes. It
has been established that different family twinning planes
exist in titanium and other close packed hexagonal (c.p.h.)
metals. With rapid advance in transmission electron microscopy in recent years, mechanical twins from different family
planes were also identified in the b.c.c. and some other
systems. For the three common metallic systems the preferred twinning plane and direction are shown below.

Crystallographic System	Twinning Plane	Twinning Direction	Reference Sources
f.c.c.	{111}	<211>	
b.c.c.	{211}	<111>	
c.p.h.	{1012}	<1011>	14,15
	{1122}	<1123>	14

An early model, proposed by Barrett [15], describes the atomic shear taking place in a set of crystallographic family planes in a specific direction, e.g., the {211} planes and the <111> direction for the b.c.c. metals. The amount of shear for a given row of atoms in the (211) plane is proportional to the distance from the row to the symmetry plane. Other models for twinning have been proposed [16,17] recently. In these models, a large part of the restrictions in the previous model on atomic shear was removed and the idea of lattice periodicity and "reshuffling" of atoms in association with the twinning operation was introduced. As discussed below, the new models succeeded in greatly short-ening the atomic movement needed for the twinning operation.

Mechanical twinning involves the movement of 10^3 and more layers of atoms. This may be easily verified by divid-ing the width of the twin by the d-spacing of the twinning plane. In the early model [15], long distance atomic move-ments are required for the atoms occupying the positions hundreds of atomic layers away from the symmetry plane. Applying the concept of lattice periodicity and atom reshuf-fling, a schematic representation for the shear movement in twinning is shown in Fig. 8. It is a reciprocal lattice plot of two consecutive layers of the ($0\bar{1}1$) plane for the b.c.c. system. Atoms on the surface include those occupying the corners of the rectangle and those in the middle of the long sides. The atoms in the layer beneath the top surface include those occupying the middle of the rectangle short sides and those at the center of the rectangle. The atoms in the 3rd layer occupy positions similar to those in the first layer. All atoms in the b.c.c. lattice are, thus, taken into account. From the definition of the reciprocal lattice, the distance between the lattice points (000) and (hkl) is equivalent to the d-spacing for the {hkl} planes.

The atoms are divided into three groups according to their distance to the (211) plane. On the left side of the (211) plane which is also the twinning plane, group (a) atoms in the first layer from the twin plane move in the

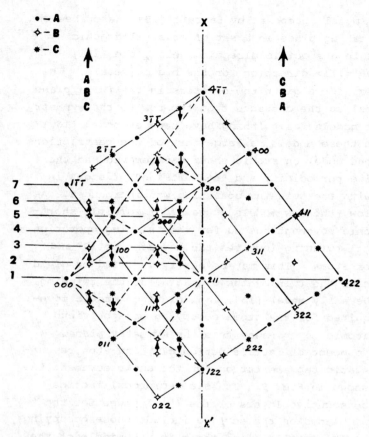

FIG. 8 *A reciprocal lattice with two consecutive (0̄11) la-yers the b.c.c. system. Atomic shear in the left side of the X-X' axis produces a mirror image of the right side. The X-X' axis is also the projection of the (211) plane.*

[1̄11] direction and travel a distance 1/6 of the d-spacing for the {111} planes or 1/6 (√3 a_o), where a_o is the lattice constant. Group (b) atoms which occupy the second layer from the twin plane travel the same distance as (a) atoms but in an opposite direction [111]. Finally, group (c) atoms which occupy the third layer from the twinned plane will remain stationary. The atoms in the next three layers and every three layers thereafter will repeat the same movements as described for the first three layers. The required shear for any atom in the twinned region never exceeds 1/6 of the d-spacing for the {111} planes. The force required to accomplish the twinning operation is, thus, significantly reduced compared to the Barrett model.

It may be added that in the untwinned portion, i.e., right side of X-X', the atoms are arranged following the order A-B-C while in the left side of X-X' (the twinned portion), the order of atoms changes into C-B-A, a typical arrangement of the stacking faults. The new model is applicable to other lattice systems, e.g., the f.c.c. which is twinned in the {111} planes at the <211> directions. A more detailed description of the twinning operation is given in a separate article [18].

It is of interest to note that in f.c.c. metals, {111} serves for both twinning and slip operations. However, the slip direction is <110>. As mentioned earlier, the twin and the slip planes may be identified from x-ray or electron diffraction experiments. But neither techniques are effective in determining the direction of the atomic movement in the plane. It remains difficult for many researchers to differentiate between the twins and the slips, especially in the latter stages of plastic flow when multiple and cross slip is operative. Many workers believe that twinning and slip, especially cross-slip, are interrelated. The interactions between the twinning and slip operations was analyzed by Mahajan [19].

V. SUMMARY AND DISCUSSION

The present study found that the thermal effect and thermal softening are important parameters in dynamic plastic flow. While dynamic recrystallization and adiabatic shear are associated with the heat generated by the plastic flow, mechanical twinning is probably less strongly affected by the heat generation. The latter may be viewed as a process of atomic shear to relieve the excessive stress built up in the metal matrix; probably at a relatively early stage of the flow process before a tremendous amount of heat is generated. This is in agreement with the findings by Mahajan [19] and Brusso, et al. [20] that twinning is preceded by the slip operation.

The exact moment when the slip operation terminates and twinning initiates or whether the slip operation continues after twinning has begun is a matter of conjecture. Mahajan [19] outlined the conditions required for the twinning operation due to the interactions of different slip systems in Mo-35Re alloy. Feng [18] considered that at least in the f.c.c. system when the [110] slip in the (111) plane is exhausted due to dislocation barriers or other obstacles, atomic shear in the <211> direction may occur in the same manner as described earlier and plastic flow will continue by mechanical twinning. It is believed that twinning by itself does not necessarily work harden the metal. As discussed previously, twinning is closely related to stacking faults. Atoms in separate rows are allowed to move in opposite directions, a certain degree of instability in the metal matrix is prerequisite. Shock and dynamic loading are believed to induce instability in the lattice leading to mechanical twinning.

Various workers at times have expressed their view point on adiabatic shear. Shockey [13] chose the terms "crystallographic" and "noncrystallographic" to differentiate the conventional slip bands and the adiabatic shear because of the large amount of heat generated in the latter. It is unfortunate that this rather restrictive definition was chosen for the adiabatic shear process which is, without exception, preceded by the slip operation. The two are closely related but the term "noncrystallographic" suggest termination of the slip process which is certainly not true.

ACKNOWLEDGMENT

The author gratefully acknowledges the valuable discussions with James Beetle of the Metallic Materials Branch at ARDEC in the course of preparing this paper.

REFERENCES

1. J. F. Mescall and H. Rogers, "Role of Shear Instability in Ballistic Penetration," MTL Tech. Rep't 89-104, 1989, Watertown, MA 02171-1145.

2. J. F. Mescall, "Metallurgical Applications of Shock-Wave and High-Strain-Rate Phenomena," Eds. Murr, Staudhammer and Meyers, Marcel Dekker, Inc., N.Y., 1986, Ch. 36, p. 689.

3. R. L. Woodward, *Metall. Trans. Vol. 10A*, 1979, p. 569.

4. J. E. Dunn, "Shock Compression of Condensed Matter," Eds. Schmidt, Johnson and Davison, APS Topical Conf., 1989, Albuquerque, NM, p. 21.

5. Meyer and Manwaring, "Metallurgical Applications of Shock Wave and High Strain Rate Phenomena," Eds. Murr, Staudhammer and Meyers, Marcel Dekker, Inc., N.Y., 1986, Ch. 34, p. 657.

6. F. J. Zerilli and R. W. Armstrong, *J. Appl. Phys. Vol. 61* (5), March 1987, p. 1816.

7. F. J. Zerilli and R. W. Armstrong, "Shock Compression of Condensed Matter," Eds. Schmidt, Johnson and Davison, APS Topical Conf. 1989, Albuquerque, NM, p. 357.

8. P. S. Follansbee, "Metallurgical Applications of Shock-Wave and High-Strain-Rate Phenomena," Eds. Murr, Staudhammer and Meyers, Marcel Dekker, Inc., NY, 1986, Ch. 24, p. 451.

9. C. Feng, "Dynamic Recrystallization," presented at the ADPA Bomb and Warhead Meeting, Walton Beach, FL, May, 1990.

10. Private communication.

11. J. D. Bryant, D. D. Makel and H. G. F. Wilsdorf, "Metallurgical Applications of Shock-Wave and High Strain Rate Phenomena," Eds. Murr, Staudhammer and Meyers, Marcel Dekker, Inc., NY, 1986, Ch. 38, p.723.

12. K. Cho, Y. C. Chi and J. Duffy, *Metall. Trans. Vol. 21A,* 1990, p. 1161.

13. D. A. Shockey, "Metallurgical Applications of Shock-Wave and High-Strain-Rate Phenomena," Eds. Murr, Staudhammer and Meyers, Marcel Dekker, Inc., NY, 1986, Ch. 33, p. 633.

14. S. K. Schiferl, "Shock Compression of Condensed Matter," Eds. Schmidt, Johnson and Davison, APS Topical Conf., 1989, Albuquerque, NM, p. 353.

15. C. S. Barrett, "Structure of Metals," McGraw-Hill, NY, 1952, p. 376-84.
    p. 621, Ibid, Vol. 10, 1960, p. 14.

17. J. W. Christian, "The Theory of Transformation in

Metals and Alloys," Ch. XX, p. 743-801, Pergamon Press, Oxford, 1965.

18. C. Feng, to be submitted for publication.

19. S. Mahajan, *Acta Metallurgica, Vol. 23,* 1975, p. 671.

20. J. A. Brusso, R. N. Wright and Mikola, "Metallurgical Applications of Shock-Wave and High-Strain-Rate Phenomena," Eds. Murr, Staudhammer and Meyers, Marcel Dekker, Inc., NY, 1986, Ch. 21, p. 403.

53

High-Strain-Rate Deformation Behavior of Shocked Copper

D. H. LASSILA and M. LEBLANC

University of California
Lawrence Livermore National Laboratory
Livermore, Ca 94551 U. S. A.

G. T. GRAY III

University of California
Los Alamos National Laboratory
Los Alamos, NM 87545 U. S. A.

    Uniaxial mechanical testing of shock-prestrained copper was performed over a range of strain rates from 10^{-3} to 7×10^3 s^{-1} in tension and compression to study the effects of a shock-induced substructure on constitutive behavior. For comparison, copper in the preshocked condition (annealed, 15 μm grain size) was also tested. Results of the study show that the post-shock mechanical behavior of copper at high strain rates is similar to that previously reported at low strain rates, i.e., the shocked copper had an increase in yield strength and a decrease in work hardening rate. The strain rate sensitivity of the shocked copper was found to be similar to that of the unshocked material. Tensile test results for both high and low strain rate tests are presented and show the elongation of the shocked material to be considerably less than that of the unshocked material. The effects of the constitutive behavior of the shocked and unshocked materials on elongation are discussed.

I.   INTRODUCTION

Deformation which accompanies the shock loading of annealed copper occurs at strain rates on the order of 10^6 and results in unique dislocation substructures [1]. The shock-induced-substructure is known to be responsible for an increase in the yield stress and a decrease in the work hardening rate of shock recovered copper deformed in compression at quasi-static strain rates [2]. In this work the deformation behavior of copper that had been shock loaded at 10 GPa for a duration of 1 μs, using shock recovery techniques described in Refs. 3 and 4, is examined over a wide range of loading conditions including high strain rate loading in both tension and compression. This work is important because the deformation of shocked material, as that which occurs during explosive forming, proceeds at high strain rates.

II.   EXPERIMENTAL PROCEDURES

Shock-recovery test samples were fabricated from cross-rolled oxygen-free-electronic (OFE) copper that had been annealed at 375°C for one hour, which resulted in a final grain size of 15 μm. The test samples were shocked at 10 GPa for a 1 μs pulse duration by impacting the test sample, which was mounted in shock assembly with a copper flyer plate at 518 m/s. A

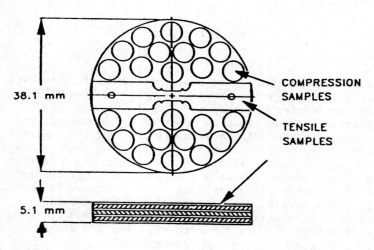

FIG. 1. *Compression and tensile samples taken from a shock-recovered disk. Four sheet tensile samples are made from each disk.*

detailed description of the shock recovery experiment and microstructural and substructural examination of shock-recovered OFE copper is given in Refs. 2 and 3.

Compression samples 5.0 mm in diameter and 5.0 mm in length were electro-discharge machined (EDM) from the shocked disks with the compression axis parallel to the shock direction, as shown in Fig. 1. Samples were also EDM from unshocked cross-rolled OFE copper, with the compression axis perpendicular to the rolling plane. Unshocked compression samples were annealed in a vacuum of 10^{-5} torr for one hour at 375°C to produce a microstructure essentially identical to the shocked material prior to shock loading. Compression testing of the samples was performed at room temperature at strain rates ranging from 10^{-3} to 7×10^3 s^{-1}. Testing up to a strain rate of 10 s^{-1} was performed using a servo-hydraulic test machine. A split Hopkinson pressure bar (SHPB) was used to perform tests up to strain rates of 7×10^3 s^{-1}. Test samples, 2.5 mm in diameter and 2.5 mm in length, were used for the SHPB tests.

High and low rate tensile testing was performed at room temperature using a sheet test sample with a gage length of 5.0 mm. The tensile samples were EDM from the shocked disk with the tensile axis perpendicular to the shock direction,* as shown in Fig. 1. Low rate tests were performed using a screw driven test machine at an initial nominal strain rate of 10^{-3} s^{-1}. Two opposing strain gage extensometers were used to measure axial extension of the gage length, which was subsequently used to calculate strain. High strain rate tensile tests were performed using a tensile SHBP apparatus, which is described in Ref. 5. A high-speed framing camera was used to photograph the test sample as it deformed. The resulting images were used to establish the strain history of the sample and determine the point at which necking initiated. The average strain rate of the tensile SHBP tests was 6×10^3 s^{-1}.

III. RESULTS AND DISCUSSION

A. STRESS–STRAIN RESPONSE AND STRAIN RATE SENSITIVITY

The true stress-strain response of shocked and unshocked copper in tension and compression at a strain rate of 10^{-3} s^{-1} is shown in Fig. 2 and shows the copper in the shocked condition to have a higher yield stress and lower work hardening rate than the unshocked material. This result is consistent with previous works[2,4,6]. The stress-strain

---

*Testing performed on compression samples taken from a similar shocked copper disk indicated the orientation of the compression axis had no effect on the stress-strain behavior. This result allows the direct comparison of the constitutive behavior determined from the compression and tensile tests.

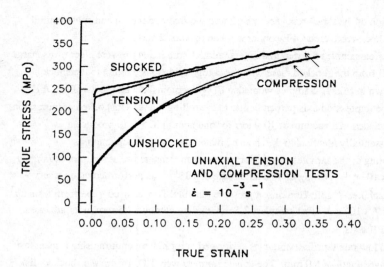

FIG. 2.   *True stress–strain behavior of shocked and unshocked copper.*

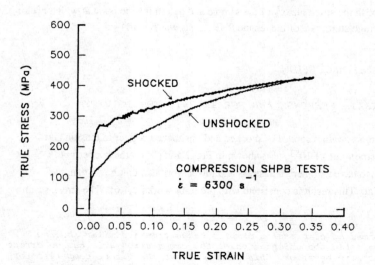

FIG. 3.   *Compression SHBP results of shocked and unshocked copper.*

response of the tensile and compression tests is similar which suggests that the work hardening and texture development of copper, in either the shocked and unshocked conditions, is similar during tensile and compressive deformation. The high rate compression tests performed using the SHPB technique (Fig. 3) produced results similar to the low rate test result.

Engineering stress-strain plots determined from low rate tensile testing and tensile SHBP testing of shocked and unshocked materials are shown in Fig. 4. The strains at which necking and failure occurred are significantly different as a function of deformation rate or the pretest condition of the copper. This behavior is discussed in the following section.

The compressive flow stress at a strain of 15% versus strain rate is plotted for all of the compression tests in Fig. 5 and shows that the strain rate sensitivity of the copper in the shocked condition is, in a global sense, similar to that of annealed copper, i.e. the strain rate sensitivity has exponential dependence at strain rates below approximately 10^2 s^{-1}, and a deviation from this at higher strain rates. The data presented in Fig. 5, may suggest that the rate dependence of the shocked material at strain rates greater than 100 s^{-1} is less than that of the unshocked material. This result is in good agreement with previous work [3].

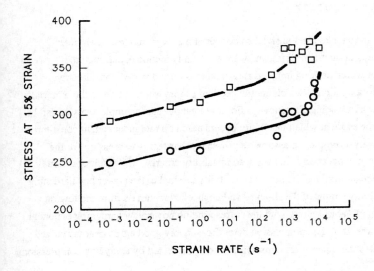

FIG. 4. *Engineering stress-strain response of shocked and unshocked materials deformed in tension at strain rates of 10^{-3} s^{-1} and 6×10^3 s^{-1}.*

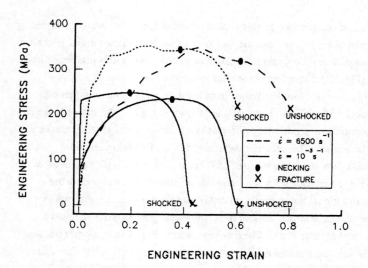

FIG. 5. *Plot of flow stress (15% strain) as a function of strain rate.*

B. DEFORMATION STABILITY

The failure mode of the shocked and unshocked copper at the high and low strain rates involved large amounts of ductile necking and were essentially identical, although the engineering strains at which necking initiated and fracture occurred varied considerably (Fig. 6). The tensile elongation at which ductile necking occurs is a function of the material's constitutive behavior and loading conditions. For low loading rates a simple analysis can be used to predict the true strain at which instability occurs based on rate independent constitutive behavior [8]. The analysis employed assumes a power law hardening expression for the constitutive behavior of copper* and a tensile load instability criteria. The result obtained is simply that the true tensile strain at the point of instability is equal to the power law hardening exponent. The instability strains of the shocked and unshocked materials from the low rate tensile tests are given in Table 1 along with the corresponding power law hardening exponent based on a least squares fit of the compression data. Reasonably good agreement is realized indicating that, at low strain rates, tensile instability can be predicted by analyzing compression

---

*For this analysis we are assuming copper to have constitutive behavior which is strain rate independent. This is a good assumption because both the shocked and unshocked materials have low strain rate sensitivities at low strain rates, as shown in Fig. 6.

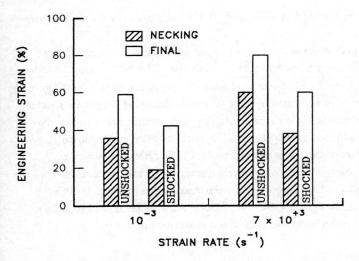

FIG. 6. *Bar chart showing the values of engineering strain at which necking initiated and fracture occurred. Neck initiation during the low strain rate tests was taken to be the strain at which maximum load occurred. The strain at which necking initiated during the SHBP tests was determined by inspections of the framing camera images.*

TABLE 1.

	n	ε_n
UNSHOCKED	0.32	31%
SHOCKED	0.14	17%

data and that the decrease in elongation of the shocked material is primarily due to the decreased work hardening.

In comparison to the low rate results, the shocked and unshocked materials exhibit increased extents of elongation when deformed at a strain rate of 7×10^3 s^{-1} using the tensile SHBP (86% and 54%, respectively). Several phenomenon may be responsible for the increase in elongation at high strain rates. Inertial forces in the rapidly deforming sample can result in uniform elongation past the point of deformation stability; in essence, inertial effects in the sample retard the nucleation and growth of a neck [9]. Also, the strain rate sensitivity of a material has been shown to retard the growth of a neck [10]. This may be a particularly important factor in the observed increase in elongation at high strain rates because the strain rate sensitivity of both the shocked and unshocked material at strain rates greater than 10^2 s^{-1} is much greater than that at low rates (Fig. 6). The results of the study presented herein are in good qualitative agreement with previous studies on high rate deformation stability.

IV.   SUMMARY

•   The shock-recovered copper was found to have an increased yield stress and lower work hardening rate than the unshocked material under all test conditions. The true stress-strain behavior of the materials was found to be similar in tension and compression.

•   The shocked copper was found to have considerably less elongation at the onset of necking and at failure when deformed in tension at high and low strain rates than the unshocked material. At low strain rates this is primarily due to the differences in work hardening behavior of the shocked and unshocked materials.

•   Both the shocked and unshocked materials had greater elongations under high strain rate loading conditions than under low rates. This is believed to be due to inertia effects that are present at high strain rates and the high strain rate sensitivities of the shocked and unshocked materials at strain rates greater than 10^2 s^{-1}.

V.   ACKNOWLEDGEMENTS

The efforts of Mr. James Thournir in carrying out much of the compression testing are greatly appreciated. Discusssions with Prof. Oleg Sherby, who also reviewed the manuscript, are gratefully acknowledged. This work was performed for the Joint Department of Defense/Department of Energy Munitions Development Program under the auspices for the U.S. Department of Energy by the Lawrence Livermore National Laboratory under Contract No. W-7405-Eng-48.

REFERENCES

1. L. E. Murr, *Shock Waves and High-Strain-Rate Phenomena in Metals*, (Plenum Press, New York, NY 1981), p. 607.

2. G. T. Gray III and P. S. Follansbee, *Impact Loading and Dynamic Behavior of Materials*, DFG, 1988, p. 541.

3. P. S. Follansbee and G. T. Gray, III, *Materials Science and Engineering*, submitted for publication, May 1990.

4. J. C. Huang and G. T. Gray, III, *Met. Trans.* 20A: 1061 (1989).

5. D. H. Lassila and M. LeBlanc, in preparation for submittal to *Rev. Sci. Inst.*

6. E. A. Ripperger, in N. J. Huffington (ed.), *Am. Soc. Mech. Engr.*, 62 (1965).

7. P. S. Follansbee and U. F. Kocks, *Acta Metall.*, 36: 81 (1988).

8. A. Considere, *Annales des Ponts et Chaussees*, 9: 574 (1885).

9. P. Perzyna, *Advances in Applied Mechanics*, 9: 243 (1966).

10. A. K. Ghosh, *Acta Metall.*, 25: 1413 (1977).

54

Characterization of Copper Shaped-Charge Liner Materials at Tensile Strain Rates of 10⁴s⁻¹

W. H. GOURDIN

University of California
Lawrence Livermore National Laboratory
Livermore, California 94551, U. S. A.

We have used electromagnetic ring expansion to study the mechanical behavior of 81-mm shaped-charge liners made from oxygen-free electronic (OFE) and electrolytic tough pitch (ETP) copper at tensile strain rates of 10^4 s^{-1}. The OFE copper was processed to yield a uniform grain size of approximately 25 μm, whereas the ETP material was reportedly processed in such a way as to encourage excessive grain growth and a broad distribution of grain sizes. However, the microstructures of the materials studied are similar, and we find no evidence of gross secondary grain growth in the ETP liners, although they do contain oxide inclusions. The OFE liners are characterized by reproducible stress-strain relationships nearly identical to independently processed OFE copper of comparable grain size. The flow stress of the ETP specimens, in contrast, is both lower and generally more erratic than that of the OFE specimens. Elongations at failure for the OFE liner materials are consistently large (0.55 ± 0.01) and are significantly larger than values observed for annealed 10-μm OFE (0.49 ± 0.04). The ETP materials appear to show somewhat less elongation at failure, although their erratic behavior makes comparisons difficult. We suggest that the erratic behavior of ETP shaped-charge liners under test and their poor performance relative to OFE copper are the result of chemical impurities and related microstructural nonuniformities, rather than differences in grain size alone.

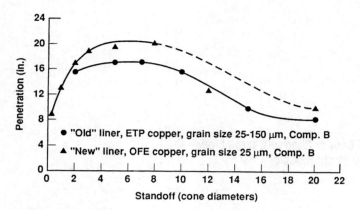

FIG. 1. *Penetration versus standoff distance (in cone diameters) for liners of ETP (25- to 150-μm grain size) and OFE (25-μm grain size) copper.*

I. INTRODUCTION

It has been shown recently [1] that the performance of shear-formed ETP M456/Ballistic Research Laboratory (BRL) 81-mm copper shaped-charge liners can be improved significantly by refining the initial liner grain size. A decrease of the average liner grain size from 120 to 20 μm produces a 25% increase in the liner's penetration of rolled homogeneous armor (RHA) from a standoff of five cone diameters (405 mm) [1]. Although the reasons for this improvement are not well understood, and although there is some indication that further grain refinement yields diminishing returns, Golaski [2] has suggested that the same effect applies to conventionally drawn and coined ETP and OFE* copper liners. Data from a limited number of experiments [2], as shown in Fig. 1, suggest an improvement between 10 and 20% in the penetration at standoff distances of four to eight cone diameters.

To better understand this phenomenon and to quantify differences in material behavior under controlled conditions, we used electromagnetic ring expansion to study liner materials. In using this technique [3], not only is it possible to achieve tensile strain rates comparable to those found in a stretching shaped-charge jet, but ring specimens can easily be fabricated directly from as-formed liners. This paper describes the results of this study. In brief, we find that, with the exception of the presence of characteristic oxide inclusions in the ETP materials, the microstructures of the liners are similar, with comparable grain sizes and mor-

---

* Also referred to as "oxygen-free high-conductivity" (OFHC) copper, a registered trademark of Amax Copper Incorporated.

phology. The ETP copper, in particular, shows no evidence of the wide range of grain size noted by Golaski [2]. We find that the stress-strain behavior of the OFE liner materials is nearly identical to that of independently processed 10-μm annealed OFE copper and that sample-to-sample variations in the flow stress are small. The flow stress of the ETP copper, in contrast, generally falls below that of the OFE and shows a much wider variation. Significantly, the strains at failure of the OFE liner material (0.55 ± 0.01) are consistently the highest we have observed. Ultimate elongations of the ETP material appear to be somewhat less, although the erratic behavior of this material and the limited number of ring specimens make comparisons difficult.

I first describe briefly the experimental techniques and the liner materials used in this study. I then present our measurements of the stress-strain behavior and strains at failure and compare them with the behavior of comparable materials studied elsewhere. I conclude with a discussion of the results and some speculation as to the origin of the effect of grain size on ultimate tensile elongation.

II. EXPERIMENTAL

The electromagnetically launched expanding ring experiment is described in detail elsewhere [3,4], and only a brief summary is given here. In this technique, the specimen, in the form of a thin ring, is the moving secondary winding of a transformer in which the primary is a six-turn solenoid of 18-gauge magnet wire wound on a polycarbonate mandrel. This solenoid is pulsed with a capacitor bank charged to 4 kV, thereby inducing counterrotating currents in the solenoid and the specimen. The current in the specimen interacts with both its own field and that of the solenoid, producing a large outward force that expands the ring rapidly and uniformly. The oscillations of the solenoid-capacitor circuit are terminated by rapid switching, which removes the capacitor from the circuit shortly after maximum expansion speed is reached. The currents in both solenoid and specimen then decay monotonically. When the resistance of the specimen is low enough (as is the case with the experiments considered here), the specimen and solenoid currents are nearly in phase, and the switching time can be chosen such that both are small. Thereafter, residual magnetic forces are likewise small, and the ring is essentially in free expansion.

Measured specimen and solenoid currents are used both to remove the effects of residual magnetic fields and to calculate the rise in specimen temperature. The current induced in a low-resistivity material such as copper (1.7 μW-cm) may be quite large (on the order of tens of kiloamperes), and resistive heating can be substantial, thus limiting the strain rates that can be achieved practically [3]. The expansion speed is determined as a function of time by using a velocity interferometer system for any reflector (VISAR) [3,5], and the flow stress

Table 1: Forming schedules for 81-mm copper shaped-charge liners.

ETP copper: "Old" schedule (L22 and L26)	OFE copper: "New" schedule (L1 and L2)
Cup.	Cup.
Deep draw in 4 stages to near final shape.	Deep draw in 4 stages to near final shape.
Heat treat at 468°C (875°F) for 1 h.	Heat treat at 371°C (700°F) for 1/2 h.
Coin.	Coin.
Heat treat at 468°C (875°F) for 1 h.	Heat treat at 204°C (400°F) for 1/2 h.
Coin.	Finish machine.
Heat treat at 468°C (875°F) for 1 h.	
Finish machine.	

is inferred from the derivative of this data by using the equation of motion for a thin ring [3]. The ring radius at failure is determined independently from measurements of records from framing cameras, the VISAR, and Rogowski probes, which are used to measure the specimen and solenoid currents [4]. These values generally agree closely.

The conical liners (81-mm base diameter, 42° apex angle) used in this study were provided by the U.S. Army Ballistic Research Laboratory [2] and were manufactured by using different procedures, as summarized in Table 1 [2]. The ETP liners* were fabricated according to an "old" schedule that is reported to encourage secondary grain growth and a wide range (25–150 μm) of grain sizes [2]. This was modified [2] into a "new" schedule designed to produce refined (25-μm) and more uniform grains in OFE copper liners.† The performance of liners of both types is given in Fig. 1.

Rings 3.2 cm in mean diameter with a square cross section 0.1 cm on a side were cut from the liners near the apex of the cone, as illustrated in Fig. 2. Two rings were cut from liners L2 and L26, but the second rings from L1 and L22 were lost during machining.

Figure 3 shows the microstructures of the as-received liners, and Fig. 3(a) shows that the OFE material has a uniform 10- to 15-μm average grain size. Surprisingly, the ETP copper was not greatly different, with an average grain size of 25 to 30 μm. Examination of a complete cross section of each cone revealed no substantial variation in the microstructures. In particular, I found no evidence for excessive grain growth in the ETP liners processed using the "old" schedule, in contrast to what has been found by others [2]. The ETP material does, however, contain oxide inclusions that do not occur in the OFE specimens, as shown more clearly in the highly magnified micrographs of expanded-ring cross sections in Fig. 4. The initial hardnesses‡ of the OFE and ETP liners were both 70 kg/mm^2, somewhat higher than for annealed 10-μm OFE copper (60 kg/mm^2) [6].

---

* Firestone Tire and Rubber Co., Defense Research Division, ca. 1976–1977.
† Physics International Company, Contract DAAD05-84-C-8052. Drawing C-3001, November 1984.
‡ Diamond pyramid (Vickers) hardness, 100-g load, 15-s dwell time.

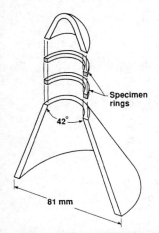

FIG. 2. *Locations of ring specimens cut from the conical liners.*

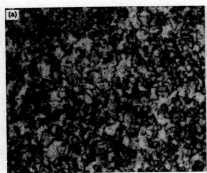

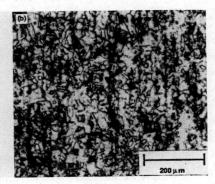

FIG. 3. *Microstructures of liners (a) L1 (OFE copper, Physics International) and (b) L22 (ETP copper, Firestone Tire and Rubber Co.). The latter shows striae and oxide inclusions, but is otherwise comparable to (a). The striae are parallel to the side of the cone.*

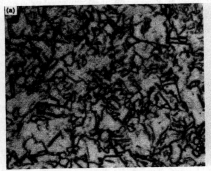

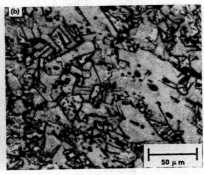

FIG. 4. *Microstructure of the cross sections of rings after expansion. The ETP copper (b) clearly shows the presence of oxide inclusions, which do not occur in the OFE material (a).*

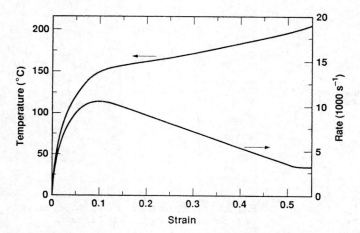

FIG. 5. *Strain-rate and temperature histories for experiment 606, liner L22. These curves are typical of all the experiments discussed here.*

III. RESULTS

The temperature and strain-rate histories typical of all the experiments conducted here are shown as a function of strain in Fig. 5. It should be noted that the heating produced by the large specimen currents is substantial and that the strain rate decreases during free expansion. The stress-strain curves derived from the VISAR records are presented in Fig. 6, along with a flow curve for annealed 10-μm OFE copper [6,7]. Only three rings of each material were available for testing, and the low-frequency "noise"[*] was particularly severe and difficult to remove from the ETP data. I cannot, therefore, easily define a "typical" flow curve, and I have simply elected to show results derived from the best VISAR data. An estimate of the sample-to-sample variation is indicated by the "error bars" at the end of each curve. The stress-strain behavior of the OFE liner materials is virtually identical to that of independently processed OFE copper of comparable grain size [6,7], and the variation among the specimens is quite small. In contrast, although the highest stresses observed for ETP copper (liner L22) overlap those of the OFE, the variations from experiment to experiment are large. Note that the flow curves from ETP liner L26 are substantially below that of L22.

The strains at failure[†] for the OFE liner materials, summarized in Table 2, are consistently high, ranging from 0.54 to 0.56. The ETP copper displays somewhat lower

---

[*] Unlike the high-frequency noise associated with the diagnostic electronics, low-frequency "noise" imay be physical in origin, associated with local motion of the surface observed by the VISAR and caused by nonuniform deformation.

[†] Defined here as $\ln(r_f/r_0)$, where r_f and r_0 are the final (at failure) and initial mean radii of the ring.

Table 2: Strains at failure.

Material	Specimen	Uniform strain	Overall strain at failure	Remarks
OFE	L1	0.50	0.54	Excellent VISAR data.
	L2-1	0.46	0.56	Good VISAR data.
	L2-2	0.47	0.54	Good VISAR data, some low-frequency "noise."
ETP	L22	0.46	0.54	Good VISAR data, excellent σ-ε results.
	L26-1	0.36	0.48	Good VISAR data, erratic σ-ε results, much low-frequency "noise."
	L26-2	0.37	0.49	Fair VISAR data, σ-ε results difficult to interpret, much low-frequency "noise."

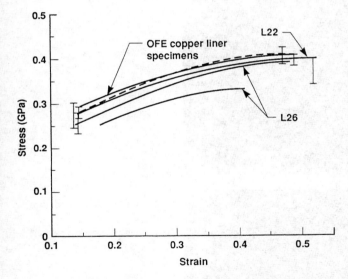

FIG. 6. *Stress-strain curves obtained from VISAR records of ring-expansion experiments. To emphasize the variability of the ETP results, the flow curves for the ETP specimens are shown individually: The highest flow stresses are for ETP liner L22 (experiment 606), while the lower flow stresses are for ETP liner L26 (experiments 630 and 632). The error bars on the L22 curve refer to all of the ETP experiments. The flow curve for 10-μm OFE copper (dashed) is also shown for comparison. The VISAR data could not be analyzed to failure in all cases; hence, some of the curves terminate at smaller strains than others.*

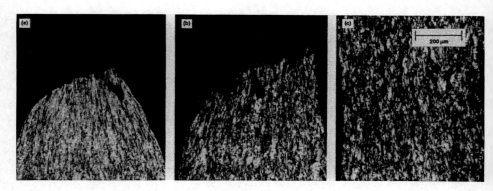

FIG. 7. *Optical micrographs of ring fragments: (a) OFE liner L2, (b) ETP liner L26 (experiment 630), and (c) a necked but unbroken region of a specimen of ETP liner L26 (experiment 632) that appears to have a higher concentration of voids than the surrounding material.*

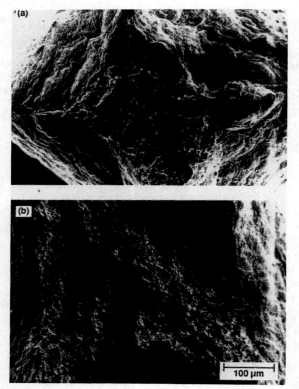

FIG. 8. *Scanning electron micrographs of fracture surfaces: (a) OFE liner L2 and (b) ETP liner L26. The ETP copper shows a higher areal density of voids and an apparently smaller void diameter than the OFE.*

values, and the spread is significantly larger (0.48 to 0.54). The lowest values, furthermore, are associated with "noisy" stress-strain data (Table 2). Annealed 10-µm OFE copper has an average strain at failure under the same expansion conditions of 0.49 ± 0.04 [6]. Also included in Table 2 are estimates of the uniform strain, determined by measuring the ring cross-sectional areas away from any necks. These are substantially smaller than the ultimate strains, indicating that significant deformation occurs after neck formation. The low values of uniform strain observed for the specimens taken from ETP liner L26 are consistent with the smaller ultimate elongations observed for this material.

Optical micrographs of the fragmented rings (Fig. 7) reveal pronounced necking and ductile failure. Although voids are associated with the fractures in both materials, they are more numerous for ETP [Fig. 7(b)] than OFE copper [Fig. 7(a)]. Examination of a necked but unbroken section of an ETP specimen [Fig. 7(c)] suggests a concentration of voids within the neck. Scanning electron micrographs of the fracture surfaces (Fig. 8) also indicate a higher areal density of voids in ETP than in OFE copper. Examination at higher magnification (Fig. 9) reveals oxide particles within many of the voids in the ETP copper.

IV. DISCUSSION

The similarity of the microstructures and initial hardnesses of the OFE and ETP materials (the oxide particles in the latter notwithstanding) suggest at the outset that their mechanical properties should be similar. The behavior of the specimen cut from ETP liner L22 supports this assertion. It is important to note, however, that the response of the OFE specimens studied here is much more reproducible than that of the ETP specimens. The OFE stress-strain results show a variation consistent with previous work [6,7] on carefully prepared materials, and the strains at failure fall within a very narrow range. The ETP copper, in contrast, consistently demonstates a large variability in both flow stress and elongation at failure. The differences between ETP liners L22 and L26, both in their stress-strain behavior and their strains at failure, are particularly striking: although their microstructures are comparable, their properties are noticeably different. Lichtenberger [8] recently found that an ETP copper with an oxygen content of 110 ppm but an otherwise low impurity content yielded a shaped-charge jet with a longer time to breakup than either ETP copper of lesser overall purity or OFE copper. For a single material (OFE), however, performance clearly improved as the grain size decreased [8]. These observations suggest that variability in the chemistry of the starting material, perhaps amplified by the processing schedule, may contribute as much to the relatively "poor" performance of the ETP liners and their erratic behavior under test as does grain size. Inclusions, for example, can affect not only the constitutive relationships, and hence the plastic instability strain, but also encourage early

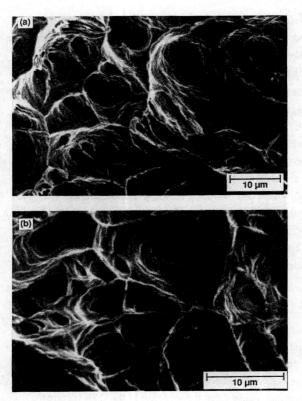

FIG. 9. *Scanning electron micrographs of ring fracture surfaces of OFE liner L1 (a) and ETP liner L22 (b) materials. Second-phase inclusions, presumably oxide, are readily visible within surface voids in the ETP copper but are absent in the OFE.*

failure by providing nucleation sites for the voids (Fig. 9) that ultimately lead to rupture. Hence, their size and distribution, which are functions of processing as well as oxygen content, are important material variables. Other impurities, which may concentrate at and therefore weaken grain boundaries, can produce comparable effects. A refined and uniform microstructure, achieved through a combination of material chemistry and process control, appears to be a prerequisite for superior liner performance.

If nonuniformities in the microstructure or cross-sectional area are large enough, the criterion for unstable plastic flow, defined by

$$d\sigma/d\varepsilon \leq \sigma \tag{1}$$

where σ is the true stress and ε is the true (or logarithmic) strain, can be satisfied in a local region at smaller strains than in the rest of the material. Although stable deformation could continue elsewhere, stresses will concentrate within these locally unstable regions, producing a neck and, ultimately, failure. Conversely, if the structure is very homogeneous, metastable deformation may continue well beyond the formal onset of plastic instability until small perturbations appear and eventually develop into failures. At a given temperature and strain rate, the hardening of OFE copper decreases monotonically with increasing strain and is apparently independent of grain size [7]. For such a material, it is interesting to note that, because the increase in the flow stress with decreasing grain size is additive and independent of either strain rate or strain [9–11], Eq. (1) implies that the instability strain should increase with increasing grain size for fixed deformation conditions. Not only is this at odds with the behavior of stretching metal jets [1,2], but the strains at failure observed in ring specimens of OFE copper also show the opposite trend, with the smallest grain sizes clearly demonstrating the largest deformations [6]. These observations suggest that plastic instability alone does not determine ultimate elongation.

Numerical studies of stretching perfectly plastic jets [12] indicate that instabilities with a wavelength of the same order of magnitude as the diameter of the jet grow most rapidly. Large wavelengths grow very slowly, and very short wavelengths, which are initially stable but become unstable as deformation proceeds, begin to grow rapidly too late to overtake instabilities at longer wavelengths. It seems plausible that as the grain size is refined, the minimum wavelength, and hence the growth rate, of perturbations in an otherwise homogeneous material will decrease, resulting in longer times to failure and larger ultimate elongations, as observed in ring specimens of OFE copper [6]. Furthermore, if the grain size decreases below some critical value associated (for example) with perturbations introduced during the formation process of a shaped-charge jet or during the machining of a ring specimen, the effects of grain size on the elongation at failure will be diminished. The reduced effect on shaped-charge penetration performance as the grain size of the liner is decreased below 10 to 30 μm has been noted by Duffy and Golaski [1].

V. SUMMARY AND CONCLUSIONS

We have used electromagnetic ring expansion to study the mechanical behavior of specimens taken directly from OFE copper shaped-charge liners formed by using a "new" processing schedule and from ETP copper liners formed by using an "old" processing schedule. Although the ETP copper contains oxide inclusions, the microstructures of these materials are otherwise very similar. In particular, we find no evidence of gross secondary grain growth in liners made with the "old" processing schedule. The mechanical behavior of the OFE liner materials is reproducible, whereas the ETP specimens show a much larger variability. From this work, I conclude that

- The OFE liners materials have a stress-strain relationship nearly identical to that of independently processed OFE copper of comparable grain size, and they show a small sample-to-sample variation in the flow stress, demonstrating that consistent processing of pure material will yield reproducible behavior.

- The strains at failure of OFE materials are consistently high (0.55), and are consistent with the improvements in jet performance associated with these liners. The strains at failure of the ETP materials are generally lower and more variable (0.48–0.54).
- The content and distribution of impurities are important in determining the behavior of ETP liner materials. Variations from lot to lot of starting material or introduced by the processing schedule may be the primary cause of both the erratic behavior of this material observed under test and its relatively poor performance as a liner.
- Refinement of the grain size in a uniform mircostructure may reduce the growth rate of plastic instabilities, producing an increase in the total elongation at failure, even though the strain at which the plastic instability condition is satisfied has decreased.

VI. ACKNOWLEDGMENTS

I would like to thank Stan Golaski and Clarence W. Kitchens, Jr. of the U.S. Army Ballistic Research Laboratory for supplying the liners and for permission to use Fig. 1. I am grateful to Glenn Randers-Pehrson and Mary Lou Duffy, also of BRL, as well as Clarence Kitchens for reading and commenting on the manuscript. Stuart Weinland and Larry Crouch performed the experiments, as always, with efficiency and care. This work was performed for the Joint Department of Defense/Department of Energy Munitions Development Program under the auspices of the U.S. Department of Energy by the Lawrence Livermore National Laboratory under Contract No. W-7405-Eng-48.

REFERENCES

1. M. L. Duffy and S. K. Golaski, BRL Report BRL-TR-2800 (1987).

2. S. K. Golaski, BRL, Aberdeen, Md., private communication (1985).

3. W. H. Gourdin, *J. Appl. Phys.* 65, 411 (1989).

4. W. H. Gourdin, S. L. Weinland, and R. M. Boling, *Rev. Sci. Instrum.* 60, 427 (1989).

5. W. H. Gourdin, *Rev. Sci. Instrum.* 60, 754 (1989).

6. W. H. Gourdin, "Metallurgical Effects on the Constitutive and Fragmentation Behavior of OFHC Copper Rings," in *Shock Waves in Condensed Matter 1987*, S. C. Schmidt and N. C. Holmes (eds.), Elsevier, New York, 1988, p. 351.

7. W. H. Gourdin and D. H. Lassila, *Flow Stress of OFE Copper at Strain Rates from 10^{-3} to 10^4 s^{-1}: Grain Size Effects and Comparison to the Mechanical Threshold Stress Model*, Lawrence Livermore National Laboratory, Livermore, Calif., to be published.

8. A. Lichtenberger, "Some Criteria for the Choice of Shaped Charge Copper Liners," *Proc. 11th Internat. Ballistics Symp.*, Vol II, Brussels, Belgium, 1989.

9. R. Armstrong, I. Codd, R. M. Douthwaite, and N. J. Petch, *Philos. Mag. 8*, 45 (1962).

10. D. H. Lassila, *AIP Conf. Proc. 102*, 323 (1989).

11. D. J. Parry and A. G. Walker, *IOP Conf. Ser. 102*, 329 (1989).

12. L. A. Romero, *J. Appl. Phys. 65*, 3006 (1989).

55

Correlation Between the Ultimate Elongations of Rapidly Expanding Rings and Stretching Metal Jets

W. H. GOURDIN

University of California
Lawrence Livermore National Laboratory
Livermore, California 94551, U. S. A.

I suggest that a correlation exists between the ultimate elongation observed for rapidly expanding rings of various copper materials under fixed launch conditions and that of stretching jets produced by shaped-charge liners of the same or similar materials.

I. INTRODUCTION

The recent work of Duffy and Golaski [1] and Lichtenberger [2] demonstrated that the initial metallurgical condition of a material can significantly affect its performance when used as a shaped-charge liner. Duffy and Golaski [1] showed that the penetration of copper shaped charges increases by 10–20% when the grain size of the liner is reduced from 150 to 25 μm and the uniformity of the microstructure is improved. Lichtenberger [2] confirmed this behavior in liners made from oxygen-free electronic (OFE) grade copper and also showed that the purity of the liner material can produce similar effects. These observations suggest that some characteristic of the liner material itself can be correlated with shaped-charge performance.

Such a characteristic, along with a corresponding laboratory test procedure, would be very useful as a means of identifying promising candidate materials prior to the expensive testing of prototype munitions. Lichtenberger [2] has suggested that the "recrystallization temperature" is one such characteristic, and he has successfully demonstrated a correlation with jet behavior.

In this paper, I suggest that the ultimate elongation of specimen rings expanded by means of the electromagnetic launch technique is another material characteristic that appears to correlate with the breakup behavior of stretching metal jets. I briefly discuss in general terms why such a correlation may exist and then offer some limited experimental evidence for it.

II. THE ELECTROMAGNETIC EXPANDING RING

The electromagnetically launched, expanding ring is discussed elsewhere [3–5], and I will not repeat the details here. Strain-rate and temperature histories typical of the expansion of a copper ring 3.2 cm in mean diameter with a square cross section 0.1 cm on a side are shown in Fig. 1. Two important features should be noted: (1) the strain rate has a maximum value of approximately 10^4 s^{-1}, after which it decreases monotonically as the material flow stress slows the ring; and (2) the temperature rises in only 10 to 15 μs to a value of approxmately 180°C. Temperatures as high as 400°C can be achieved with rings of smaller cross section. These conditions, while they do not encompass the entire range that obtains during the

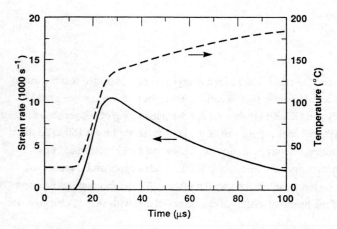

FIG. 1 *Strain-rate and temperature histories typical of the expansion of an OFE copper ring with a 3.2-cm mean diameter and a 0.1-cm square cross section.*

formation of a shaped-charge jet, nevertheless are comparable to those that occur during the stretching phase [6,7].

The combination of an appropriately high strain rate with a rise in temperature of several hundred degrees over only tens of microseconds, as found in the electromagnetic expanding ring, thus appears to be ideally suited to reproducing at least the behavior of the jet itself, if not its formation. Furthermore, except for its curvature (which is small), the expanding ring represents a stretching rod with a periodic boundary condition at a length corresponding to the ring's circumference. Hence, the geometry by itself suggests a close correspondence between an expanding ring and a stretching metal jet.

III. EXPERIMENTAL

Data for the ultimate elongations (given as failure strains) of OFE copper rings with grain sizes between 10 μm and 150–200 μm [8,9] are summarized in Table 1. Although the uncertainties are substantial, the elongation at failure clearly decreases with increasing grain size. Because penetration of a jet decreases when the jet breaks up, this is likewise in qualitative agreement with the observations of Duffy and Golaski [1]. Lichtenberger [10] has determined the times to breakup for OFE copper liners of various grain sizes. These are plotted in Fig. 2 along with ring data, both normalized by the times extrapolated to a grain size of zero. In this plot, the time of formation of the jet is implicitly assumed to be small compared with the time over which it stretches. The times to failure for the rings decrease with increasing grain size, in generally good agreement with the breakup times for jets.

I have reported [11] the properties of ring specimens cut directly from Ballistic Research Laboratory liners made from 25-μm-grain-sized OFE copper. The ultimate

Table 1. Failure characteristics for copper rings expanded at 10^4 s^{-1}.

Material	Grain size (μm)	Failure strain[a] $(\ln (r_f/r_0))$
OFE	10	0.49 ± 0.04
	30–50	0.47 ± 0.04
	90–120	0.44 ± 0.09
	150–200	0.40 ± 0.06
OFE (BRL liner)	25	0.55 ± 0.01
UBAC	<10	0.42 ± 0.03
ETP[b]	50	0.51 ± 0.05

[a] Ranges indicated are 95% confidence intervals.
[b] ETP material provided by Institut Saint-Louis (A. Lichtenberger).

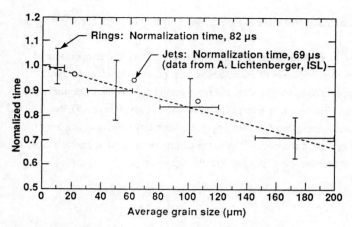

FIG. 2 *Times to breakup for shaped charges and rings.*

elongation of this material (also given in Table 1) is larger than even the 10-μm material [8], consistent with its known good performance as a shaped charge.

Udylite Bright Acid electrodeposited copper (UBAC) has been studied as both a shaped-charge liner and a ring specimen [12]. The jet produced by the prototype liner was very poor, breaking up early into many small fragments [12,13]. Consistent with this poor behavior, ring tests indicated an elongation to failure of only 42% (Table 1).

Lichtenberger [2] found that an ETP copper with a reduced impurity content produced the best jets, better even than OFE copper. His measurements of the recrystallization temperature for this material were consistent with this good performance. Ring experiments on a comparable ETP material showed more elongation at failure than OFE copper (Table 1), in agreement with the recrystallization-temperature measurements [10], which place it intermediate between OFE and the best-performing ETP materials.

IV. DISCUSSION

Although the general arguments and the experimental evidence are suggestive, they are not conclusive. In only one case (UBAC) was a direct comparison made between ring tests and the performance of a prototype liner, and the poor performance of the latter suggests only that large elongation at failure is a necessary condition for a good jet. However, assuming that the BRL liner would have performed as well as it is reputed to [1], I think it is reasonable to infer that large elongation to failure is a sufficient condition as well. The other evidence is unsatisfying because it is circumstantial.

Romero [14] studied the growth of plastic instabilities in stretching metal jets and found that perturbations with wavelengths comparable to the radius of the jet grew fastest. Differences in rates of growth suggest that the kinetics of neck development may be a dominant factor in the breakup of shaped-charge jets. In this regard, the observation [8] that work hardening at a fixed strain rate is independent of the grain size indicates that, for OFE copper, the strain at which necks can begin to form decreases with decreasing grain size. The apparently contradictory observation that failure elongation increases with decreasing grain size thus suggests that, in rings as in jets, the kinetics of neck growth are critical.

V. CONCLUSION

In this paper, I have argued that electromagnetically launched expanding rings offer a unique means of assessing, with a controlled and relatively inexpensive experiment, the jet characteristics of metals used as shaped-charge liners. Although the experimental data is limited, I believe it is compelling enough to warrant further, more systematic study. What is necessary now is a series of comparisons between ring experiments and prototype tests for a variety of materials that perform both well and poorly. Once the plausibility of the correlation has been more firmly established, it can be put to use in screening a much wider field of materials than would otherwise be experimentally or economically feasible.

VI. ACKNOWLEDGMENTS

I would like to thank André Lichtenberger of Institut Saint Louis for permission to use his data for OFE jets in Fig. 1. This work was performed under the auspices of the Joint DOD/DOE Munitions Development Program and the U.S. Department of Energy by the Lawrence Livermore National Laboratory under Contract No. W-7405-Eng-48.

REFERENCES

1. M. L. Duffy and S. K. Golaski, *Effect of Liner Grain Size on Shaped-Charge Jet Performance and Characteristics*, U.S. Army Ballistics Research Laboratory, Aberdeen Proving Ground, MD, BRL-TR-2800 (1987).

2. A. Lichtenberger, "Some Criteria for the Choice of Shaped-Charge Copper Liners," in *Proc. of the 11th Internat. Ballistics Symp., Vol II: Warhead Mechanisms*, Brussels, Belgium, 1989.

3. W. H. Gourdin, *J. Appl. Phys.* 65, 411 (1989).

4.   W. H. Gourdin, S. L. Weinland, and R. M. Boling, *Rev. Sci. Instrum.. 60*, 427 (1989).

5.   W. H. Gourdin, *Rev. Sci. Instrum. 60*, 754 (1989).

6.   W. P. Walters and J. A. Zukas, *Fundamentals of Shaped Charges*, Wiley, New York, 1989, p. 2.

7.   W. G. von Holle and J. J. Trimble, *J. Appl. Phys. 47*, 2391 (1976).

8.   W. H. Gourdin and D. H. Lassila, *Flow Stress of OFE copper at Strain Rates from 10^{-3} to 10^4 s^{-1}: Grain-Size Effects and Comparison to the Mechanical Threshold Stress Model*, Lawrence Livermore National Laboratory, Livermore, CA, UCRL-JC-104679 (1990).

9.   W. H. Gourdin, in *Shock Waves in Condensed Matter 1987*, Elsevier, New York, 1988, pp. 351–354.

10.   A. Lichtenberger, private communication, Institut Franco-Allemand de Recherches de Saint-Louis, France (1990).

11.   W. H. Gourdin, "Characterization of copper shaped-charge liner materials at tensile strain rates of 10^4 s^{-1}," in these *Proceedings*.

12.   J. W. Dini and W. H. Gourdin, "Evaluation of electroformed copper for shaped charge applications," to appear in *Plating and Surface Finishing*, Vol. 77, Aug. 1990.

13.   D. H. Lassila, "Material characteristics related to the fracture and particulation of electrodeposited copper shaped charge jets," in these *Proceedings*.

14.   L. A. Romero, *J. Appl. Phys. 65*, 3006 (1989).

Section V
Shear Localization

56

On the Instability of the Uniform Shear Deformation in Viscoplastic Materials: The Material Instability and the Dynamic Instability

C. FRESSENGEAS and A. MOLINARI

Laboratoire de Physique et Mécanique des Matériaux, U.R.A. C.N.R.S. 1215
Université de Metz, Ile du Saulcy
57045- Metz Cedex 1, France.

The instability of the uniform shear flow of a viscoplastic hardening material and the localization of the deformation into shear bands is investigated by using a two-dimensional linear Lagrangian perturbation method.The solution reveals two different modes of instability. One mode is the material instability due to the material softening; since its destabilizing mechanism is inertia, the second mode can be dubbed a dynamic instability. The prominent two-dimensional features of both instabilities, such as shape, orientation and propagation effects are discussed. Comparisons are made with existing non-linear exact results and the classical hydrodynamic instability theories.

I- INTRODUCTION

The instability of the uniform shear flow, and the localization of the plastic deformation into shear bands is commonly observed at large strains in métals and polymers. The destabilizing factor is either the "adiabatic" material thermal softening or some isothermal strain softening, such as material damage or texture developement. Shear banding may thus be qualified as a material instability. As such, it has recently received considerable theoretical attention, with most studies being confined to the one dimensional approximation. In this paper, we are interested in multi-dimensional features, such as shape and orientation effects or band "propagation". In previous related papers, Freund et al. [3] and Kuriyama and Meyers [4] investigate the band propagation in a time-independent hyperelastic material with a numerical approach. Anand et al. [1] devise a linear perturbation method restricted to initial tendencies leading in particular to the prediction of a possible onset of shear bands in the cross-stream direction. Hereafter, an incompressible viscoplastic strain-hardening material model is used; elasticity is neglected. The time-evolution of initial perturbations to the basic shear motion is described with a two-dimensional Lagrangian linear method; the disturbance is solution for a partial differential equation with time-dependent coefficients. The solution reveals an additional mode of instability for which the destabilizing mechanism is not the material softening, but rather inertia.

II - BASIC FORMULATION.

Let us consider an infinite medium submitted to simple shear. The governing equations are written in non dimensional quantities; for this purpose, we introduce an arbitrary length H and a velocity V. H is the distance between two parallel planes. The lower plane is assumed to be at rest, the upper one being moved at the constant velocity V. H and V scale all of the lengths and velocities used hereinafter. The scaling time and strain rate are $T=H/V$ and $1/T$. As a characteristic stress, we shall use the shear stress τ_0 associated with the linear velocity profile. All quantities are considered to be functions of the Lagrangian coordinates $\mathbf{a}=(a_1,a_2)$, and of the time t. The position relative to a fixed cartesian frame of the particle $\mathbf{a}$ at time t is denoted by the Eulerian coordinates $\mathbf{x} = (x_1, x_2)$. The velocities on the reference planes are   (2-1)

$$v_1 (a_1, 0, t) = 0 \ , \quad v_2 (a_1, 0, t) = 0 \ , \quad v_1 (a_1, 1, t) = 1 \ , \quad v_2 (a_1, 1, t) = 0$$

Let us denote by $\mathbf{D}$ the strain-rate tensor. Let us use in addition the equivalent strain-rate d_e, the equivalent strain γ_e and the equivalent stress τ_e defined by   (2-2)

$$d_e^2 = 2\mathbf{D} : \mathbf{D} \ , \quad \gamma_e = \int_0^t d_e d\tau, \ 2\tau_e^2 = \mathbf{s} : \mathbf{s}$$

The material time-dependent behavior is specified in terms of the symmetric Cauchy stress tensor $\underline{\sigma}$ and of its deviator $\underline{s}$ by :   (2-3)

$$\underline{s} = \Lambda \underline{D} \ , \quad \Lambda = 2 \tau_e/d_e \ , \quad \tau_e = f(\gamma_e)d_e^m \ , \quad 0<m\leq 1$$

where m denotes the rate-sensitivity. The function $f(\gamma_e)$ takes account of strain hardening; it is not monotonically increasing in order to account for any softening. The material is assumed incompressible, and the momentum equations are expressed in terms of the non symmetric Boussinesq stress tensor $\mathbf{n}$ (the transpose of the nominal stress tensor).   (2-4)

$$R\frac{\partial v_1}{\partial t} = \frac{\partial n_{11}}{\partial a_1} + \frac{\partial n_{12}}{\partial a_2} \ , \quad R\frac{\partial v_2}{\partial t} = \frac{\partial n_{21}}{\partial a_1} + \frac{\partial n_{22}}{\partial a_2}$$

R is the dimensionless parameter $\rho V^2/\tau_o$. It is straightforward to show that a solution of the problem equations is the uniform shearing motion: (2-5)

$$x_1^0 = a_1 + t\, a_2, \; x_2^0 = a_2, \; v_1^0 = a_2, \; v_2^0 = 0, \; d_e^0 = 1 \; \gamma_e^0 = t$$

$$\sigma_{11}^0 = \sigma_{22}^0 = \sigma_{33}^0 = -p, \; \sigma_{12}^0 = \tau_e^0 = f(t), \; \Lambda^0 = 2f(t)$$

A stream function $\psi^0 = -a_2^2$ of the uniform flow can be defined; note that ψ^0 is time-independent. We are now concerned with the stability of that fundamental solution.

III - LINEAR PERTURBATION ANALYSIS

To determine whether the basic solution $(x_i^0, v_i^0, \sigma_{ij}^0, n_{ij}^0, \psi^0)$ is stable we add on small disturbance variables $(\delta x_i, \delta v_i, \delta\sigma_{ij}, \delta n_{ij}, \delta\psi)$ under the form (3-1)

$$(x_i, v_i, \sigma_{ij}, n_{ij}, \psi) = (x_i^0 + \delta x_i, \, ...)$$

substitute the variables $(x_i^0 + \delta x_i, \, ...)$ into the problem equations, and select the first-order terms in the disturbance variables. Using the perturbed stream function $\delta\psi$, the analysis results in a single fourth-order partial differential equation (3-2)

$$R\frac{\partial}{\partial t}\left(\left(\frac{\partial^2}{\partial a_1^2} + \left(\frac{\partial}{\partial a_2} - t\frac{\partial}{\partial a_1}\right)^2\right)\delta\psi\right) = \; ....$$

$$4\frac{\partial^2}{\partial a_1^2}\left(\frac{\partial}{\partial a_2} - t\frac{\partial}{\partial a_1}\right)^2\delta\psi + m\left(\frac{\partial^2}{\partial a_1^2} - \left(\frac{\partial}{\partial a_2} - t\frac{\partial}{\partial a_1}\right)^2\right)\delta\psi + ...$$

$$f'(t)\left(\frac{\partial^2}{\partial a_1^2} - \left(\frac{\partial}{\partial a_2} - t\frac{\partial}{\partial a_1}\right)^2\right)\int_0^t\left(\frac{\partial^2}{\partial a_1^2} - \left(\frac{\partial}{\partial a_2} - \tau\frac{\partial}{\partial a_1}\right)^2\right)\delta\psi\,d\tau$$

Since the medium is infinite in both the a_1 and a_2 directions, $\delta\psi$ does not have to satisfy any boundary condition which proves satisfactory if the material edges are far from the instability phenomenon under scrutiny. In addition, the initial conditions are arbitrary. One can build up the solution to general initial conditions by combining the plane waves (3-3)

$$\psi_+^+\exp(i(k_1a_1+k_2a_2)), \; \psi_-^+\exp(-i(k_1a_1+k_2a_2)), \; \psi_+^-\exp(i(k_1a_1-k_2a_2)), \; \psi_-^-\exp(-i(k_1a_1-k_2a_2))$$

where the ψs are functions of time only. Each wave represents a simple shear parallel to the phase line $k_1a_1\pm k_2a_2$= constant. The wavenumbers (k_1, k_2) are real positive; the wave vectors are respectively $\pm\mathbf{k}^+ = \pm(k_1, k_2)$ and $\pm\mathbf{k}^- = \pm(k_1, -k_2)$. A linear combination of these waves displays an initial cellular structure, as sketched in figure 1. The shape factor of the rectangular perturbation cell is defined as $\xi = k_1/k_2$; small shape factors indicate long flat disturbances parallel to the shearing direction. Substituting (3-3) into the equation (3-2), one obtains two integro-

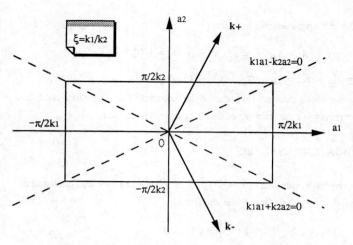

FIG. 1   The initial disturbance cell. Assuming an equal amplitude
1/4 of all the waves (3-3), $\delta\psi(0)=1/2(\cos(k_1a_1+k_2a_2)+\cos(k_1a_1-k_2a_2))$
$\delta\psi$ is zero along the rectangular cell boundaries.

differential equations. The first one describes the time evolution of the waves (w^+) of wave vectors $\pm k^+$ (3-4)

$$R\,(k_1^2(1+t^2)+k_2^2-2tk_1k_2)\frac{d\psi}{dt}=(2Rk_1(k_2-k_1t)-4k_1^2(tk_1-k_2)^2-m(k_1^2(1-t^2)-k_2^2+2tk_1k_2)^2)\,\psi-\cdots$$

$$\cdots f'(t)\,((1-t^2)k_1^2+2tk_1k_2-k_2^2)\int_0^t((1-\tau^2)k_1^2+2\tau k_1k_2-k_2^2)\psi(\tau)\,d\tau$$

the second equation (3-5) pertains to the waves (w^-) of wave vectors $\pm k^-$; it is obtained by changing k_2 into $-k_2$ in equation (3-4). A closed form integration of the equations (3-4, 3-5) is possible in the quasi-static approximation ($R=0$), as amplified in the next section, and in the dynamic case $1/R=0$, as will be seen in Section V.

IV- THE MATERIAL INSTABILITY.

It is straightforward to show that the solution of equation (3-4) for $R=0$ is   (4-1)

$$\frac{\psi(t)}{\psi(0)}=\frac{4\xi^2+mf(0)(1-\xi^2)^2}{(1-\xi^2)f'(0)}\frac{((1-\xi t)^2-\xi^2)\,f(t)}{4\xi^2(1-\xi t)^2+mf(t)(\xi^2-(1-\xi t)^2)^2}\exp(-\int_0^t\frac{(\xi^2-(1-\xi\tau)^2)^2\,f(\tau)\,d\tau}{4\xi^2(1-\xi\tau)^2+mf(\tau)(\xi^2-(1-\xi\tau)^2)^2})$$

The evolution (4-2) of the waves (w^-) is obtained by changing ξ into $-\xi$ in (4-1). It is interesting to compare the present linear results to the predictions of a fully non-linear theory. This can be done in the one dimensional approximation ($\xi=0$), where such a theory is available from the work of Molinari and Clifton [6]. These authors define the asymptotic L_∞ localization as follows: If for every point A different from B, the ratio γ_B/γ_A tends to infinity with increasing time, then L_∞ localization of the deformation is said to occur at point B. Using the constitutive law (2-3-c), we have L_∞ localization if and only if the function $(f(t))^{1/m}$ is integrable at infinity. Assuming that $f(t)$ has a power law behavior at infinity of the form $f(t) \sim a t^{-p}$ as $t \to \infty$ where a and p are positive constants, L_∞ localization occurs if and only if : $-p+m<0$. The one dimensional reduction of both relations (4-1) and (4-2) is (4-3)

$$\frac{\psi(t)}{\psi(0)} = \frac{f'(t)}{f'(0)} \left(\frac{f(0)}{f(t)}\right)^{(1+m)/m}$$

It is seen that $\psi(t) \to \infty$ as $t \to \infty$ for $-p+m<0$, and $\psi(t) \to 0$ for $-p+m>0$; these conditions are the same as the critical conditions for L_∞ localization to occur or not. Therefore, the present perturbation analysis provides results about localization as defined above.

For $-p+m>0$, the asymptotic L_∞ localization does not occur. However, when m-p is small enough, the amplification $\psi(t)/\psi(0)$ may take large values, say larger than P, before turning down to zero; then the practical L_p localization of the deformation is said to occur. An example is given in figure 2, where the amplification (4-3) is plotted for the function f used by Freund et al. [3] to model the material strain hardening / softening (4-4)

$$f(t) = t \left(1 + \frac{b}{n} t^2\right)^{n-1}$$

A maximum exists in the stress/strain curve if $2n<1$, and the L_∞ localization condition turns out to be $2n+m<1$. Using $b=8000$, $n=0.4995$ and $m=0.0117$, as provided by Marchand and Duffy [5] for the low structural steel (HY-100), $2n+m= 1.017$, slightly larger than 1. Thus there is no L_∞ localization; however shear bands are being observed. Considering the figure 2, we see that the practical L_p localization of the deformation occurs with $P \cong 400$. When, like in that case, P is large enough, irreversible events may take place inside the band and lead to failure instead of an asymptotic decrease; in such cases, where non linear destabilizing effects are triggered, L_p localization is of more practical importance than L_∞ localization. Otherwise L_p localization is "reversible", which illustrates the fact that a maximum in the stress/strain curve does not necessarily lead to the localization of the plastic deformation.

We now turn to the investigation of the disturbance shape effects. To illustrate the issue, we select a material obeying the constitutive law (4-4), with parameters slightly different from those of the (HY-100) steel so as to make possible a L_∞ localization for $\xi=0$: $n=0.49$, $b=8000$, $m=0.0117$ ($2n+m=0.9917<1$). The amplifications (4-1) and (4-2) of disturbances initiated immediately after the maximum in $f(t)$ are plotted in figure 3 for various values of the shape factor ξ. The figure suggests that the most amplified perturbation is obtained for $\xi=0$, for which the L_∞ localization occurs, but that the L_∞ localization is precluded for $\xi>0$, whatever the

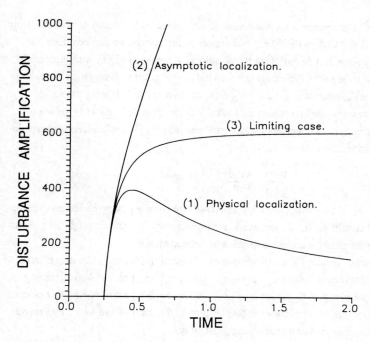

FIG. 2   *One-dimensional disturbance amplification; Freund's hardening/softening law.*
1) Marchand and Duffy (HY-100): b=8000, n=0.4995 and m=0.0117. 2n+m>1, L_p localization.
2) b=8000, n=0.4995 and m=0.0005: 2n+m<1, L_∞ localization.
3) b=8000, n=0.4995 and m=0.001: 2n+m=1, limiting case.

value of the material parameters n and m; this is readily seen from an asymptotic expansion of (4-1, 4-2) when t→∞. Only the L_p localization can hold for $\xi>0$; the function $P(\xi)$ describing the evolution of the disturbances extremum vs. the shape factor is seen to tend to zero as $\xi\to\infty$. For the (w^+) waves however, the maximum amplification $P(\xi)$ does not decrease monotonically: when ξ is large enough, a maximum occurs for $t\cong1/\xi$ prior to any significant amplification of the small-ξ waves. Thus the amplitude $\psi(t)$ of the large-ξ (w^+) waves may become very large before its eventual decrease; therefore it is concluded that an instantaneous early tendency for the occurence of cross-stream bands exists, as predicted by Anand et al. [1]. However their time-evolution reveals that they may be interrupted before undergoing completion.

The wave amplifications (4-1), (4-2) are real. Thus there is no propagation of the waves (3-3), with respect to the material, despite the word "propagation" used to designate the process by which the bands extend at the tip; they are standing waves to an observer moving with

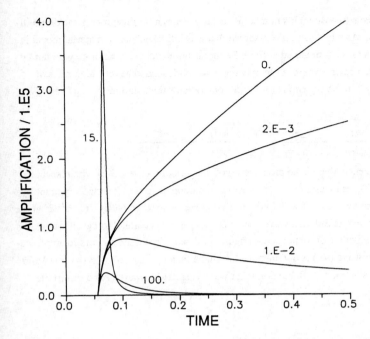

FIG 3 *Two-dimensional disturbance amplification; b=8000,
n=0.49, m=0.0117; Shape factor $\xi=0.,2.10^{-2},10^{-2},10^{2},15$. L_{∞} localization occurs only to one dimensional disturbances.For $\xi=15$, a
significant temporary amplification occurs*

the material. In the present framework, the mechanism by which shear bands propagate is as
follows: within the general disturbance to the basic shearing motion, which is formed as a linear combination of waves like (3-3), with varying wave numbers (k_1, k_2), the waves with
small ξ become dominant as time goes on (see figure 3). Thus the general disturbance becomes more and more a combination of waves with long streamwise wavelengths. As t→∞, all
the waves but the one-dimensional wave ($\xi=0$) die away, since there is no L_{∞} localization for
$\xi\neq0$. Therefore the process ends up with a band of infinite length in the shearing direction, unless failure triggered by the L_p localization occurs before.

V- THE DYNAMIC INSTABILITY.

Inertia is known to have stabilizing effects on the one-dimensional shear band material instability (Fressengeas and Molinari, [2]); it is not attempted hereafter to document further the issue.
Instead we focus on a two-dimensional dynamic instability arising when the ratio of inertial ef-

fects to viscous effects (the Reynolds number R) is large enough; in this process the destabilizing factor is inertia. As a first step, it is assumed that R is infinite; thus the material model is an inviscid fluid. The results obtained at finite Reynolds numbers, using a non-newtonian viscous fluid, are published elsewhere. Both may apply to solids loaded at very high rates of strain. Making $1/R=0$ in the equations (3-4, 3-5), one obtains the solutions for (w^+) and (w^-), respectively

$$\frac{\psi(t)}{\psi(0)} = \frac{1+\xi^2}{\xi^2+(1-\xi t)^2} \quad \text{and} \quad \frac{\psi(t)}{\psi(0)} = \frac{1+\xi^2}{\xi^2+(1+\xi t)^2}$$

(5-1)

Again $\psi(t)$ is real; hence, there is no propagation of the periodic waves (3-3) with respect to the material, and $\psi(t)$ is a measure of their temporal evolution. It reveals a temporary amplification phenomenon: any wave (w^+) with $\xi \neq 0$ will first increase up to $\psi_m/\psi(0) = (1+\xi^2)/\xi^2$ at the critical time $t=1/\xi$ when its Eulerian wave vector $\pm(k_1,(k_1-k_2t))$ is parallel to the shearing direction, then decay with a $1/t^2$ dependence (figure 4). At any time t, the maximum amplification of the dominant wave (w^+) is obtained for the shape factor $\xi_m = 1/2 \ ((t^2+4)^{1/2} -t)$ and its value is $1/\xi_m^2$. As $t \to \infty$, $\xi_m \to 0$: the dominant wave is shifted toward a band of infinite

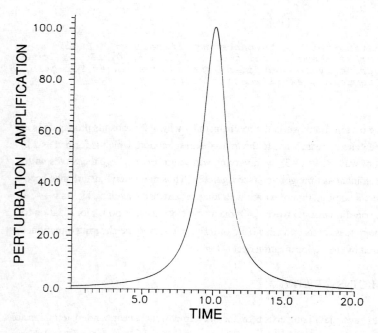

FIG 4 The dynamic instability; amplification of the disturbance of shape factor $\xi=1/10$. Note the critical time $t=1/\xi$.

length and its amplification tends to infinity. As for the waves (w⁻), they monotonically tend to zero as t→∞. A similar temporary amplification phenomenon was exhibited as early as 1907 by Orr in his Eulerian account [7] of the instability of the uniform shear flow of an inviscid fluid between moving planes. A general initial disturbance, formed as a linear combination of waves (3-3) with varying wave numbers (k_1, k_2), evolves toward a single wave of both infinite length and amplitude parallel to the Lagrangian a_1 direction. Thus, an infinite band forms as t→∞; in the Eulerian frame, it is rotated so as to become parallel to the cross-stream axis.

The mechanism responsible for the instability is a transfer of kinetic energy from the basic uniform shear to the perturbed deformation: the (w⁺) waves raise their kinetic energy from their interactions with the mean flow. The rate of change of the (w⁺) waves kinetic energy is   (5-2)

$$\frac{\partial \delta T}{\partial t} = 2k_2 a_2 \frac{(1+\xi^2)\xi^3}{(\xi^2+(1-\xi t)^2)^2} \psi(0) \, i \, \exp(i(k_1 a_1 + k_2 a_2))$$

Assuming a small shape factor ξ, it is found that $\partial \delta T/\partial t$ is small either when t is small or when t→∞. At the critical time $t=1/\xi$, $\partial \delta T/\partial t$ is large and the interactions with the mean flow are at their maximum efficiency. This never occurs to the (w⁻) waves, which therefore do not exhibit any amplification.

VI- SUMMARY AND CONCLUDING REMARKS.

In the present paper, the evolution of initial two-dimensional disturbances to the uniform shear deformation is described by using a linear Lagrangian method; the analysis is not limited to initial tendencies, and it may yield the whole disturbance evolution, provided that the linear assumption remains valid throughout the process. The time dependence of the perturbations is not merely exponential, but rather the solution for an integro-differential equation. The parallelism between the predictions of the present analysis and the exact results provided by a one-dimensional non linear theory suggests that the critical conditions for unbounded growth of the perturbations are the same as the critical conditions for L_∞ localization; it is emphasized that such a parallelism cannot be expected in general for any perturbation analysis. The practical L_p localization may occur in material conditions such that the asymptotic L_∞ localization is precluded; when the asymptotic linear tendency to stability is overshadowed by non-linear destabilizing effects, the L_p localization is of more significance than the L_∞ localization. The examination of their time-evolution, as provided by the present linear analysis, suggests that L_∞ localization does not occur to general two-dimensional disturbances, and is limited to one-dimensional disturbances; yet, L_p localization is likely to be more severe at small shape factors. The onset of cross-stream bands as predicted by Anand et al. [1] is viewed as being temporary; whether those bands actually form depends on the eventual activation of non-linear destabilizing effects.

No wave propagation occurs in this model; the "propagation" of the band at its tip is viewed as the progressive prominence of the waves with the smaller shape factors. The process ends up with a single wave of infinite wavelength in the shearing direction.

In addition to the material instability due to strain softening, a two-dimensional dynamic instability is revealed by the analysis when inertial effects are strong enough; in metals at very high strain rates with a fluid-like behavior, it might result in shear bands oriented in the cross-stream direction. Experimental and material conditions such that both the material instability and the dynamic instability coexist remain to be precisely defined; in such conditions, shear bands oriented in both the streamwise and cross-stream direction would appear.

REFERENCES

1.   L. Anand, K. H. Kim and T. G. Shawki, *J. Mech. Phys. Solids*, 35; 4: 407, (1987).

2.   C. Fressengeas and A. Molinari, *J. Mech. Phys. Solids*, 35; 2: 185, (1987).

3.   L. B. Freund, F. H. Wu, and M. Toulios, Initiation and Propagation of Shear Band in Anti-plane Shear Deformation, in *Proceedings of an International Symposium on Plastic Instability,* Considère Memorial, 125-134, Paris (1985).

4.   S. Kuriyama and M. A. Meyers, *Met. Trans. A.,* 17A: 443, (1986).

5.   A. Marchand and J. Duffy, *J. Mech. Phys. Solids,* 36, 3: 251 (1988).

6.   A. Molinari and R. J. Clifton, *J. Appl. Mech.,* 87-WA/APM-26, (1987).

7.   W. M. F. Orr, *Proc. Roy. Irish Acad.* A27: 9 and 69, (1907).

57

Reverse-Ballisitic Impact Study of Shear Plug Formation and Displacement in Ti6Al4V

W. H. HOLT, W. MOCK, JR., W. G. SOPER , and C. S. COFFEY
V. RAMACHANDRAN and R. W. ARMSTRONG

Naval Surface Warfare Center
Dahlgren, VA 22448-5000 and Silver Spring, MD 20903-5000

Department of Mechanical Engineering
University of Maryland
College Park, MD 20742

Gas-gun-accelerated disks of Ti6Al4V alloy have been impacted onto smaller-diameter flat-end hardened steel rods to push out shear plugs of disk material. The range of impact velocities was 219-456 m/s. For velocities of 290 m/s and below, the plugs were pushed partway through the disks. Optical and scanning electron microscopies were used to determine shear band widths and to describe microstructural details associated with the primary shear zones, including evidence for thermal softening and melting. A simple model is used to relate observed plug displacements to impact velocities and to provide estimates of the shear zone strength as well as threshold energies for plug displacement and for plug separation. The appreciable localized temperature rises evident during the shear plugging of this alloy are discussed in the context of a dislocation pile-up avalanche model.

I. INTRODUCTION

Titanium and titanium-based alloys are known to be shear band prone. Meyers and Pak [1] have studied impact-induced shear bands in commercially pure titanium by high-voltage transmission electron microscopy and by scanning electron microscopy. Shear banding has been reported in microstructural studies of impacted plates by Woodward [2], Me-Bar and Shechtman [3], and Timothy and Hutchings [4]. This paper is concerned with shear banding in reverse-ballistic experiments on Ti6Al4V alloy over a range of impact velocities to achieve differing amounts of material response. In each experiment, a disk of the alloy is accelerated in a gas gun and impacted onto an initially stationary, smaller-diameter, hardened steel rod to push out a plug of disk material. The experiments include impact velocities for which the plug is pushed partway through the disk, as well as higher velocities for which the plug is separated from the disk. The observed microstructural features associated with the main shear zone that forms the interface between the disk and plug are described and discussed in the context of macroscopic and microscale models.

II. IMPACT EXPERIMENTS

A 40-mm-bore gas gun [5] was used for the experiments. The reverse-ballistic impact configuration has the advantage of permitting precise control of the impact geometry. Figure 1 is a schematic of the muzzle region of the gas gun showing an impactor disk in a sabot and a steel rod supported in a frangible target holder, just prior to impact. The impact occurs in vacuum and electrical contact pins in the barrel wall are used to measure impact velocity. After impact, the sabot, impactor disk, and rod are soft recovered in a box of cloth rags. For each experiment, the disk and plug were diametrically sectioned and a toolmaker's microscope was used to measure the plug displacement relative to the corresponding undamaged region of the impact surface of the disk.

All of the disks were machined to be 9.52 mm thick and 34.0 mm in diameter. The disks were machine-lapped with 600 grit SiC, and polished with 6 micron diamond paste.

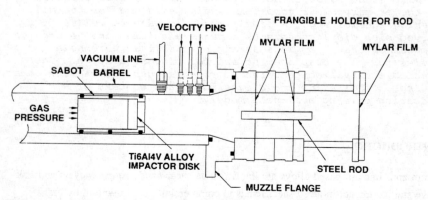

FIG. 1 *Schematic of muzzle region of the gas gun showing a titanium alloy impactor disk in a sabot and a steel rod supported in a frangible holder, just prior to impact.*

The supplier's mechanical property determination gives average values of 840 MPa 0.2% offset yield strength, 10% elongation, and 941 MPa ultimate strength [6]. The hardened steel rods were fabricated from dowel pins having a supplier-specified core hardness range of RC50-58 [7]. The 7.94 mm diameter rods were machined to be 71.9 mm long and the flat ends were polished with 6 micron diamond paste.

The range of impact velocities was from 219 m/s to 456 m/s. For velocities at or below 290 m/s (six experiments), the plug was pushed only partway through the disk. For experiments at 362 and 456 m/s, the plug separated from the disk.

III. MICROSCOPIC OBSERVATIONS

Figure 2 shows a sectioned disk and displaced (but not separated) shear plug from an experiment for which the impact velocity was 283 m/s and the plug displacement was 5.70 mm.

Figure 3 shows the main shear band between the disk and plug for an experiment at 290 m/s. The sectioned surfaces were etched with a 5 ml HNO3, 10 ml HF, 85 ml H2O solution. The observed 60 micron width of the main shear band is consistent with the 50 micron width calculated by Coffey [8] using a dislocation dynamics model and with the same approximate width estimated on a continuum adiabatic shear band model basis by Dodd and Bai [9]. Scanning electron microscopy of this shear band showed it to be

5 mm

FIG. 2 *Sectioned disk and displaced shear plug. Impact velocity was 283 m/s and plug displacement was 5.70 mm. The plug contains additional shear fractures.*

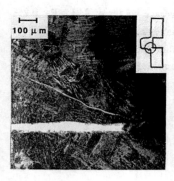

FIG. 3 *Optical microscopy of the main shear band between the disk and the plug for an experiment at 290 m/s. The width of the main shear band is approximately 60 microns. A narrower satellite shear band can be seen in the plug material.*

relatively free of finer microstructural detail, even when studied at magnifications up to 4000X.

Figure 4 is scanning electron micrograph of a central portion of the shear surface of the separated plug from the experiment at 456 m/s. Clear evidence is observed of melted material, including a solidified metal spray thrown onto the plug, suggesting that melting had occurred prior to plug separation. This would indicate the generation of temperatures at least as high as the 1660°C melt temperature for Ti6Al4V [10]. This melting observation is further supported by EDAX analysis of a metallic deposit distributed over the periphery of the impact end of the steel rod, indicating the presence of titanium and aluminum, thus demonstrating that disk material was deposited onto the rod. In experiments with hardened steel conically-tipped projectiles impacting Ti6Al4V targets,

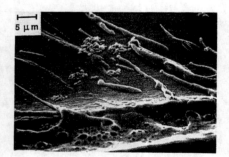

FIG. 4 *Scanning electron micrograph of a central portion of the shear surface of the separated plug from the experiment at 456 m/s. There is clear evidence of melted material, including a solidified metal spray thrown on the plug, suggesting that melting had occurred prior to plug separation. This would indicate the generation of temperatures at least as high as the 1660°C melt temperature for Ti6Al4V.*

Woodward [2] has also reported evidence of melted material on ejected shear plugs as well as deposits of titanium on the eroded projectile tips.

IV. MODEL CONSIDERATIONS

A. MACROSCOPIC DISK–ROD INTERACTION

Soper [11] has developed a macroscopic model for the disk-rod interaction by considering the conservation of momentum and energy for the case where the rod does not punch through the disk (the rod and disk move together as a single mass after impact). A linear relationship is assumed between the plug relative displacement and the force that opposes that displacement. The force is assumed to arise from a constant shear stress τ acting at the interface between the plug and disk. Elastic vibrations are assumed to absorb available kinetic energy until a threshold value E_T is reached and the additional energy goes into plastic work associated with the plug displacement. For the velocity range which gives plug displacements h that are less than the disk thickness H, energy and momentum considerations yield for the assumptions stated an elliptical relationship between h and the impact velocity V :

$$\frac{MV^2}{2} = E_T + \pi Dh \left(H - \frac{h}{2}\right)\tau, \quad \text{where} \quad M = \frac{M_D M_R}{M_D + M_R}. \tag{1}$$

M_D and M_R are the masses of the disk and rod, respectively, and D is the rod diameter.

This model contains two parameters, E_T and τ, which can be chosen to provide a best fit to the experimental data. Figure 5 shows a least-squares fit of Equation (1) to data from six impact experiments. The fit provides $E_T = 312$ Joules and $\tau = 335$ MPa. From these values, one may calculate the threshold velocity for plug displacement, $V_T = \sqrt{2E_T/M} = 197$ m/s, the threshold velocity for plug separation, $V_S = \sqrt{2\,E_T + \pi D\tau H^2/M} = 293$ m/s, and the threshold energy for plug separation, $E_S = MV_S^2 / 2 = 691$ Joules. $E_S - E_T = 379$ Joules is the energy expended during plug separation. The calculated shear strength τ can be compared to an estimated shear strength for the bulk disk material based on one-half of the average room temperature 0.2% offset yield strength. This gives $\tau_{BULK} = 420$ MPa, and the fact that $\tau < \tau_{BULK}$ is consistent with the concept of thermal softening (or even melting) of the material in the shear zone.

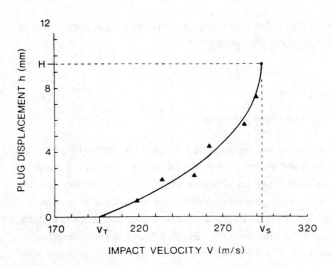

FIG. 5 *Least-squares fit of Equation (1) to data from six impact experiments. H is the disk thickness. V_T and V_S are the threshold velocities for plug displacement and plug separation, respectively.*

B. MICROSCALE ΔT FOR A DISLOCATION PILE-UP AVALANCHE

Armstrong, Coffey, and Elban [12] have proposed that dynamic shear banding is explained by individual dislocation pile-ups overcoming internal obstructions during the initiation of plastic yielding or fracturing. The discontinuous nature of this crystal deformation process is critical to producing very localized and substantial temperature rises. On a micromechanics model basis, the obstacle strength determining the energy available to be released when pile-up breakthrough occurs is measured in terms of the microstructural stress intensity, k, that is directly analogous to the fracture mechanics stress intensity for pre-cracked materials [13]. The microstructural stress intensity has been measured for a wide range of crack-free metal alloys from the predicted yield strength or fracture strength dependencies on the inverse
square root of the polycrystal grain size [14]. The magnitude of k is large for materials showing a pronounced discontinuous yield point behavior and such behavior, for example, in steel, is also associated with the occurrence of shear banding. A k value of 24 MPa·mm$^{1/2}$ has been reported for the yield stress of steel as compared with 1.3 MPA·mm$^{1/2}$ for the more gradual yielding behavior of aluminum [14]. A k value

approximately four times greater than that for yielding of steel is attributed to the cleavage fracture stress dependence on grain size.

The upper-limiting temperature rise at the breakthrough of a circular dislocation pile-up of diameter l has been estimated for a number of metals from the expression

$$\Delta T \leq \left(\frac{kl^{1/2}\upsilon}{16m\pi K}\right) \ln\left(\frac{2K}{c^*\upsilon b}\right) \tag{2}$$

where υ is the dislocation velocity, m is a numerical orientation factor, K is the thermal conductivity, c^* is the specific heat at constant volume, and b is the dislocation Burgers vector [15]. An upper-limiting value of the shear stress intensity, $k_S = k/m$ also was estimated for cleavage crack initiation from the relation

$$k_s = \frac{\pi Gb}{4\alpha\Delta x^{1/2}} \leq \frac{\pi Gb^{1/2}}{4\alpha} \tag{3}$$

where G is the shear modulus, Δx is the separation of the dislocations at the pile-up tip, and $\alpha = 2(1 - \nu)/(2 - \nu)$, with ν being Poisson's ratio.

Armstrong and Elban [15] used the limiting value for k_S to determine a relative temperature rise 1.64 times greater for α-titanium than for steel. An even greater temperature rise is predicted for Ti6A14V material. With $k_S = 25$ MPa.mm$^{1/2}$, υ taken as the shear wave velocity of 3.06×10^3 m/s, $K = 6.66$ J/m·s·°K, $c^* = 2.57$ MJ/m^3·°K and $b = 3.5 \times 10^{-10}$ m, ΔT is obtained as 1.57 times greater than that for α-titanium. Thus, the dislocation pile-up avalanche model is shown to be consistent with the experimentally demonstrated importance of shear banding in this alloy.

V. SUMMARY

Experiments on shear plugging of Ti6A14V material have provided a range of plug displacements relative to the disk surfaces, including separation from the disk. The experiments have led to a model evaluation of plug displacement versus impact velocity. The analysis has led to estimates of the shear zone strength, the threshold energy for the shear plug displacement, and the threshold energy for the plug separation. Large temperature rises calculated on the basis of a dislocation pile-up avalanche model of shear banding in the disk material are consistent with clear experimental evidence for molten material having been produced as part of the plug separation process.

VI. ACKNOWLEDGMENTS

The authors would like to acknowledge the assistance of X. J. Zhang of the University of Maryland in obtaining scanning electron micrographs of the disk indentation. Also acknowledged is M. K. Norr of NAVSWC for providing the scanning electron micrograph of the melted material on the separated shear plug. R. L. Lumpkin and G. R. Silkensen of NAVSWC, and E. Nelson of Southeastern Center for Electrical Engineering Education are acknowledged for assistance in the preparation of the manuscript. The work has been supported at NAVSWC by the Independent Research Fund and at the University of Maryland by Contract N60921-89-C-0002.

REFERENCES

1. M. A. Meyers and H-R. Pak, *Acta. Metall. 34*: 2493 (1986).
2. R. L. Woodward, *Metall. Trans. 10A*: 569 (1979).
3. Y. Me-Bar and D. Shechtman, *Mater. Sci. and Eng. 58*: 181 (1983).
4. S. P. Timothy and I. M. Hutchings, *J. Mat. Sci. Lett. 5*: 453 (1986).
5. W. Mock, Jr., and W. H. Holt, Naval Surface Weapons Center Report NSWC TR-3473, (1976).
6. RMI Company, Niles, OH.
7. SPS, Inc., Jenkintown, PA.
8. C. S. Coffey, *J. Appl. Phys. 62* : 2727 (1987).
9. B. Dodd and Yilong Bai, *Mater. Sci. and Tech. 5*: 557 (1989).
10. *Metals Handbook 3, 9th ed., ASM International, Metals Park, OH, 389*: (1980).
11. W. H. Holt, W Mock, Jr., W. G. Soper, C. S. Coffey, V. Ramachandran, and R. W. Armstrong, in *Shock Compression of Condensed Matter-1989*, S. C. Schmidt, J. N. Johnson, and L. W. Davison, Ed. (North-Holland, 1990), p. 915.
12. R.W. Armstrong, C. S. Coffey, and W. L. Elban, *Acta metall. 30*: 2111 (1982).
13. Q. Gao and H. W. Liu, *Metall. Trans. 21A*: 2087 (1990).
14. R. W. Armstrong, in *The Yield, Flow and Fracture of Polycrystals*, T. N. Baker, Ed., Applied Science Publishers, London, p.1, (1983).
15. R.W. Armstrong and W. L. Elban, *Mater. Sci. Eng. A122:* L1 (1989).

58

Survey of Adiabatic Shear Phenomena in Armor Steels with Perforation

Y. MEUNIER, R. ROUX, J. MOUREAUD

Creusot-Loire Industrie
BP 56 71202 Le Creusot Cedex France

This paper shows how, during the penetration of a projectile crossing an armour steel target, the adiabatic shearing damage represents quite a detrimental factor on its ballistic behaviour. Relationships between metallurgical parameters (microstructure and microstructural banding, hardness) and adiabatic shearing are discussed. In-depth microstructural investigations, carried out in adiabatic shear banding, allow us to give a first approach of the structure morphology.

I - INTRODUCTION

Adiabatic shearing damage of materials occurs in loading conditions with fast strain rates and sometimes high shock wave pressure in such events as armour penetration, explosive fragmentation, high velocity forming and machining, and leads to a local plastic instability of the structure [1].
    The purpose of this article is to show :
- on the one hand, the inter-relations between the ballistic behaviour of armour steels, some of their metallurgical parameters and their damage by adiabatic shearing AS,
- on the other hand, the in-depth microstructural investigations explaining the structure morphology of the shear banding and giving a first approach of the origins of AS,
in order to better understand the occurrence of this damage mechanism, with the aim to control it and to improve the dynamic and ballistic behaviours of armour steels.

II - RELATIONSHIP BETWEEN BALLISTIC BEHAVIOUR AND ADIABATIC SHEARING

During the normal attack of armour steels by perforating ogival-head projectiles, the general progression of the protection ballistic limit - VLP - versus the hardness - Δ - of the material

allows us to distinguish three areas (Fig. 1) :
- in area I, the armour perforation is achieved by plastic flow of the metal, without AS phenomenon ;
- in area II, the perforation mode is carried out with formation of backward armour fragments, parts of discing or plugging with AS phenomenon ;
- in area III, the projectile breaks up during impact and loses a part of its perforating force, thus leading to an increase of VLP versus Δ ; in this case, the perforation mechanism is achieved especially by discing or plugging mode with AS phenomenon.

For instance, the discing failure mode can be described with a physical and mechanical model based on the perforation work of the projectile crossing the target [2].

Roughly speaking, in the beginning of the penetration, due to the projectile pressure, the target metal tends to be drived back towards the lower strength areas, i.e. the plate front, in the opposite course of the projectile progress. Then, metal compressing which takes place forward and along the projectile nose, forms maximum shear zones where AS occurs. Finally, shear stresses, located in a radial direction, normal to the impact direction, are developed by the bending of metal forward the projectile, leading to separate a disc along a dynamic failure path following adiabatic shear bands (Fig. 2).

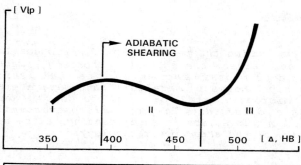

FIG. 1   *General progression of protection ballistic limit versus hardness and associated perforation mechanisms.*

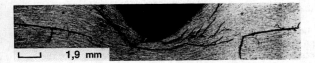

FIG. 2   *Initiation of a separating plane along adiabatic shear bands during the discing mode (magnification x 5,2).*

On a polished and etched section, AS is seen as white bands, about ten micrometers wide, that are the sectional traces of shear surfaces, to which shear cracking is sometimes associated [3].

III - RELATIONSHIPS BETWEEN METALLURGICAL PARAMETERS AND ADIABATIC SHEARING

The local formation of AS represents quite a detrimental factor to the thermo-mechanical and metallurgical properties of the material when the attempt is to develop armour steels whose ballistic performances will result from an optimization of these properties.

A - MICROSTRUCTURE - AS INTER-RELATIONS

The microstructure of the material is a parameter having an effect upon the AS occurrence. For instance, Figure 3 shows the progression of the number of adiabatic shear bands - n - (roughly representative of the phenomenon initiation phase) versus the hardness - Δ - of a 28CND8 type material, for two types of microstructure, martensite and bainite + martensite, after normal ballistic tests with a perforating projectile, for comparable penetration velocities of the projectile.

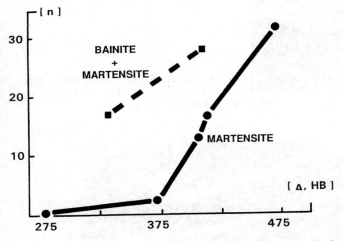

FIG. 3   *Sensitivity of the microstructure of the material to adiabatic shearing.*

From this graph, it is seen that the mixed microstructure - bainite + martensite - is more sensitive to AS occurrence than the martensitic microstructure, in relationship with its less microstructural stability due to the micro-heterogeneities of hardness. The n parameter actually increases for Δ values upper than 380 BHN, which confirms the decrease observed on Figure 2, beyond this value, of VLP versus Δ, relating to the initiation of the AS damage for the material.

B - MICROSTRUCTURAL BANDING - AS INTER-RELATIONS

An inherent feature of rolled steel products is a marked anisotropy in mechanical properties which derives from the directionality of the microstructure in rolled plate. Chemical segregation produced during casting frequently persists in rolled steel plate in the form of longitudinal banding while many types of non-metallic inclusions are deformed by the rolling process into elongated stringers aligned in the rolling direction. These microstructural inhomogeneities in the steel are considered important in the fracture of armour plate by projectile impact. In particular, the mode of failure classified as discing, which is evident in medium to high hardness wrought materials, occurs by single or multiple crack propagation along planes parallel to the rolling direction with separation of material along these planes [4].
   Initiation and the subsequent sustenance of a shear band requires a number of criteria to be satisfied [5]. First, the deformation should become unstable so that the localization of strain within a narrow shear band is energetically favourable. The second requirement is that the material should contain pre-existing strain inhomogeneities like inclusions, second phase particles and cracks. Without these structural defects or inhomogeneities, no shear bands will form even above the critical strain for localization ; only a general softening of the material close to the projectile-target interface will occur. Thus, the structural factors are quite important. Finally, if the shear band width is such that a significant fraction of the heat generated within the band is lost to the surroundings, the critical condition for shear band initiation will also depend on kinetic factors like strain rate of deformation and thermal diffusivity. So, a complete treatment of the shear band initiation problem requires the

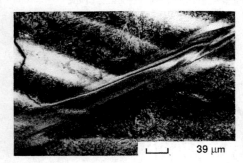

FIG. 4   *Adiabatic shear band crossing the microstructural banding* (magnification x 255).

⌐ ⌐ 39 μm

⌐ ⌐ 40 μm

FIG. 5   *Adiabatic shear band following the microstructural banding (magnification x 250).*

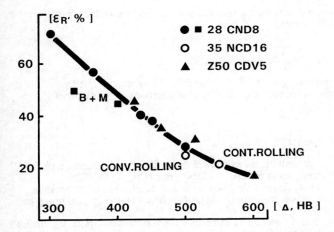

FIG. 6 *Progression of dynamic rupture deformation versus hardness.*

consideration of thermodynamic, structural and kinetic factors.

Figures 4 and 5 exhibit details of AS for a 35NCD16 type material - Δ = 513 BHN - in terms of the directionality of microstructure in rolled steel plate. These attempts to determine from metallographic sections whether the planes on which the shear bands propagated are associated with the microstructural banding or stringer type inclusions are inconclusive. These micrographies show that the shear bands and accompanying crack propagation either cross or follow the microstructural banding, not allowing us to support that the structural factors play a prominent role in determining the probability of shear band formation for this material.

C - HARDNESS - AS INTER-RELATIONS

The dynamic behaviour of the armour steel, operating under high strain rates, can be determined through the split Hopkinson pressure bar, by compressive loading, leading to a specific range of strain rates from 2.2 to 3.10^3 s^{-1}.

Figure 6 shows the progression of the rupture dynamic deformation - ε_R - obtained through stopped dynamic compressive tests versus the hardness - Δ - for different martensitic armour steels. From this graph, it is seen that :
- the higher the hardness is, the less the material buckles ;
- the mixed microstructure - bainite + martensite - is more sensitive to the rupture deformation by AS than the martensitic microstructure, for comparable hardness levels of a 28CND8 type material ;
- a controlled rolling process applied to a 35NCD16 type armour plate leads to a significant increase of hardness with only a low decrease of the rupture deformation.

Microscopic observations allow us to compare the strain at fracture and the strain at the initiation of adiabatic shear bands. The higher the initial hardness of the material is, the lower the difference in the deformation is, indicating that the time required for the total phenomenon - AS and crack propagation - is shorter for the higher hardness than for the lower.

IV - METALLURGICAL ASPECTS OF ADIABATIC SHEAR BANDS

Microscopic investigations have been carried out on a 35NCD16 type armour steel - Δ = 490 BHN - after ballistic tests by a perforating projectile. On this material damaged according to the discing mode, thin foils have been sampled from an armour disc, and made thinner by mechanical and electrolytic polishings.

A – AREA OUT OF ADIABATIC SHEAR BANDING

The transmission electronic microscopy scanning of thin foils shows an austenite free microstructure with martensitic laths, very high density of dislocations and sometimes precipitates (Fig. 7).

B – ADIABATIC SHEAR BANDING

The material structure is micro-crystalline, the as-formed grains shown off against a dark field microscopy have very small sizes, below 50 nm for visible grains (Fig. 8) [6]. The unorganized pattern grains, the very dense dislocation state, and the electron beam crossing through several grain layers contribute to standing-out problems, not allowing us to bring the images into focus.

The investigations of thin foils by electronic diffraction show diffraction spots, significant of a micro-crystalline phase, and a ring diagram, giving evidence of a equi-axed structure (Fig. 9) [7].

The characterization of diffraction patterns and the determination of crystalline structures show the simultaneous presence of body-centred cubic bcc - ferrite - and face - centred cubic

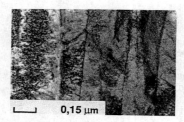

FIG. 7 *Microstructure out of adiabatic shear banding.*

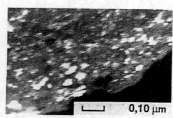

FIG. 8 *Micro-crystalline structure in adiabatic shear banding.*

FIG. 9 *Diffraction pattern in adiabatic shear band.*

FIG. 10 *Diffraction pattern in adiabatic shear band after tempering.*

fcc - austenite - crystallographic lattices [8]. It is not possible to quantify a relative ratio of these two phases, though the diffraction spots associated to the first phase are more numerous.

The fine grain microstructure and the diffraction scanning do not exhibit precipitates such as carbides. Probably the original precipitates were put back into solution.

The investigations of thin foils, sampled in adiabatic shear bands after tempering at 150°C during 1 hour, followed by an air cooling, show no modification of the microstructure morphology. The diffraction pattern displays spotted rings, with characterization only of the bcc crystallographic lattice phase : the low temperature tempering destabilizes the fcc crystallographic lattice phase (Fig. 10).

C – INTERPRETATION – DISCUSSION

The previous metallographic investigations allow us to give the following conclusions :
- the adiabatic shear banding structure, micro-crystalline with fine grains, and the lack of self-tempering due to a very fast cooling-rate after transformation, contribute to the increase of hardness, as noted in the adiabatic shear bands compared to the area out of AS [9] ;
- the adiabatic area involves the simultaneous presence of a bcc phase and a fcc phase ;
- the heating temperature reached in the AS area is higher than the transformation temperature of the bcc into fcc phase.

Two assumptions can be formulated in regard to the thermo-mechanical cycle which has led to the structure noted in adiabatic zone :
- *assumption 1* : the maximum temperature has reached the melting point of the liquid + δ ferrite phase in the given strain state, with carbides back into solution. During the fast cooling rate, a micro-crystallization of the steel occurs from the liquid + δ ferrite phase, which changes into δ ferrite + γ austenite + α' martensite phase. The distinction between δ ferrite and α ferrite or α' martensite by transmission electronic microscopy is not possible.
- *assumption 2* : the maximum temperature belongs to the microstructural equilibrium field of α + γ or γ phases with incomplete or total recrystallization into γ phase. During the fast cooling rate, the γ phase changes into γ + α' phase.

On the one hand, estimates of the local heating in adiabatic shear banding [10, 11], on the other hand, the joint effect of both the shock wave pressure (about 1 GPa) and the strain which promotes, by the dislocation motion, the carbide dissolution in a solid phase becoming hyper-saturated in carbon, do not allow us to select any conclusion from these two assumptions. A better knowledge of carbide dissolution kinetics and phase diagrams under high pressure is essential to the determination of mechanisms leading to the adiabatic shear phenomenon and to those which control it, in order to improve the dynamic behaviour of materials.

V - CONCLUSIONS

The survey of adiabatic shear phenomenon in armour steels with perforation has shown how this plastic instability process represents quite a detrimental factor on the armour ballistic behaviour through the progression of the protection ballistic limit versus the hardness of the material and associated perforation mechanisms. For instance, the discing failure mode, which arises for high hardness armour steels, results from the initiation of a separating plane along a dynamic failure path following adiabatic shear bands.

Relationships between metallurgical parameters and adiabatic shearing have been discussed :
- the mixed microstructure - martensite + bainite - is more sensitive to adiabatic shearing damage than the martensitic microstructure ;
- microscopic investigations do not allow us to determine whether the planes on which the shear bands propagated are associated with the microstructural banding or stringer type inclusions ;
- the higher the hardness is, the less the material buckles after dynamic compressive loading, the rupture deformation occurring as soon as adiabatic shearing arises for very hard steels.

In-depth metallographic investigations carried out in adiabatic shear bands have allowed us to give a first approach in the structure morphology. More thorough knowledges of carbide dissolution kinetics and phase diagrams under high pressure are still essential to understand the origins of the adiabatic shear phenomenon.

REFERENCES

1.   H. C. Rogers, *Ann. Rev. Mater. Sci.*, *9*: 283 (1979).

2.   R. L. Woodward, *Met. Technol*, March; 106. (1979).

3.   Y. Tirupataiah, V. M. Radhakrishnan, and K. R. Raju, Adiabatic Shearing and Associated Cracks in Ballistic Steel Targets, Proceedings of the 6th International Conference on Fracture, New Delhi, India, 4-10 Dec. 1984, *Advances In Fracture Research*, 5; 3127 (1984).

4.   P. W. Leach, R. L. Woodward, *J. of Mater. Sci.*, *20*: 854 (1985).

5.   G. Sundararajan, *ibid.*, 3: 3119.

6.   G. Coquerelle, Contribution à la compréhension des mécanismes de perforation, *Rappor ETCA 85 R 066*, 11 Oct. (1985).

7.   J. L. Derep, *Acta Met.*, *35;* 1245 (1987).

8.   J. L. Derep, Cisaillement adiabatique, Journée thématique sur la perforation - le blindage, Paris, *DGA/DRET*, 14 Jan. 1987.

9.   R. C. Glenn andW. C. Leslie, *Metall. Trans.*, 2: 2945 (1971).

10.   T. A. C. Stock and K. R. L. Thompson, *Metall. Trans.*, *1*: 219 (1970).

11.   G. L. Moss, Shear Strains, Strain Rates and Temperature Changes in Adiabatic Shear Bands, *Technical Report ARBL-TR-02242*, May 1980.

59

Formation of Controlled Adiabatic Shear Bands in AISI 4340 High Strength Steel

J. H. BEATTY, L. W. MEYER*+, M. A. MEYERS+, and S. NEMAT-NASSER+

Metals Research Branch
US Army Materials Technology Laboratory
Watertown, MA 02172 U.S.A.

* Fraunhofer-Institut für angewandte Materialforschung(IFAM), D-287,
Bremen 77, West Germany

+Department of Applied Mechanics and Engineering Sciences
University of California, San Diego
La Jolla, CA 92093 U.S.A.

Adiabatic shear banding is one of the predominant failure modes of Ultra-High Strength (UHS) steels under high rates of deformation. Though these bands have been previously studied, the specific factors governing when, how, or if a certain material fails by shear band localization are still relatively unknown. This article examines these questions in terms of the microstructure using a controlled shear strain. Four microstructures of equal hardness (Rc52), but with different carbide distributions were produced in a VAR 4340 steel by varying the normalizing temperature (1hr at 845, 925, 1010, or 1090°C). A short duration re-austenitizing treatment was subsequently used to produce similar prior austenite grain sizes; this was followed by a 200°C temper. The controlled shear strain was introduced by the use of a hat-shaped specimen loaded in a split Hopkinson bar rig, producing a shear zone at local strain rates up to $5 \times 10^5 s^{-1}$. The use of mechanical stops to arrest the deformation process allowed the development of the shear bands to be studied under controlled stress conditions. TEM studies revealed a *microcrystalline structure within the shear band with a crystallite size of 8-20nm. A gradual change from the microcrystalline structure within the band to lath martensite was observed. Tensile unloading cracks were observed in the sheared regions. A normalizing temperature of 925°C produced the microstructure with the greatest resistance to unstable shearing.*

I. INTRODUCTION

Adiabatic shear bands have been previously studied because of their importance both in metal working operations (such as turning, reaming, punching, and forging) and in other processes at high strain rates (such as armor and projectile failure, and explosive forming). An adiabatic shear band consists of a narrow region of concentrated strain where the strain rate is sufficient to preclude significant heat transfer away from the sheared region. The material within the band experiences both significant increases in temperature and a large accumulated strain. In steels, two types of adiabatic bands have been distinguished using optical metallography: the "transformed" type which etches white using nital, and the "deformed" type which etches dark. The term "transformed" was originally used to suggest that the phase transformation from α (ferrite) to γ (austenite) had taken place within the band during shearing, though many investigations have failed to provide conclusive evidence that a phase transformation is necessary to produce a white etching or "transformed" band. However, because of common usage, this notation will be adhered to until conclusive evidence to the contrary is found.

This research program is aimed at studying the microstructural and mechanistic evolution of adiabatic shear bands in high-strength steels. Specifically, the deformation processes and their relation to controlled microstructural constituents (carbides) have been examined through the use of optical and both scanning and transmission electron microscopy (SEM and TEM). This paper examines the applicability of our testing techniques to study initiation mechanisms, and presents some preliminary observations.

II. EXPERIMENTAL PROCEDURE

In order to study the effects of microstructure on adiabatic shear initiation, both the microstructure and the shear deformation must be controlled. The microstructures of VAR 4340 steel used in the present study have been previously characterized by Cowie et al[1,2]. The controlled variable is the size and distribution of the carbides present in the tempered martensitic microstructure, without changing the material's hardness (52Rc). This is accomplished by using different normalizing temperatures (845, 925, 1010, or 1090°C for 2 hours) to produce each microstructure. The range of possible grain refining carbide phases are redissolved and/or coarsened to differing degrees at each normalizing temperature. These carbide distributions do not change during the final, short duration austenitization treatment at 845°C for 15 minutes (oil quenched). This is followed by a 200°C temper for 2 hours. This produces microstructures with constant prior austenite grain size (ASTM 12) and hardness, but with different carbide distributions.

The controlled initiation and propagation of shear bands occurred during split Hopkinson compression bar testing using the experimental technique of Meyer et al[3,4] with a hat-shaped specimen. The experimental arrangement, schematic stress history, and sample geometry are shown in Figure 1. Using this type of test, it is possible to study microstructural changes just before and after the initiation of the bands. A 19mm (0.750

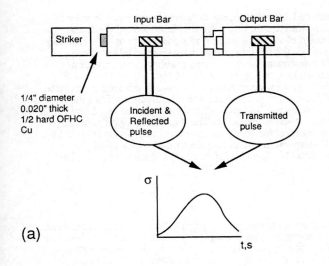

(a)

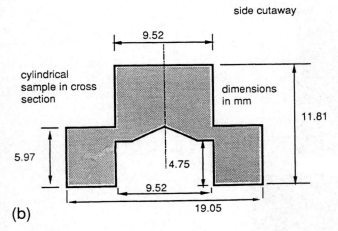

(b)

FIG. 1 *Hopkinson bar with experimental configuration and schematic stress history(a) and "hat" specimen geometry(b) developed by Meyer et al (refs 3,4).*

in) diameter Hopkinson bar was utilized, with a 152mm (6 in) long striker traveling at a velocity of 18.3 m/s. Shear strain rates from 10^3 to 3×10^5 s^{-1} are obtained.

Initial tests using this apparatus showed that ringing effects altered results during sample failure. To eliminate this problem, a small copper disc (dia 6.35mm, thickness 0.5mm) is placed in between the striker bar and the input bar. This creates a trapezoidal (ramped) shaped input wave, and eliminates the ringing. Pulse shaping of this kind has been utilized for dynamic testing of ceramics.[5] Stress and displacement measurements were calculated from the transmitted and reflected elastic waves in the bars. In order to control the degree of deformation introduced, stop rings are placed around the small cylinder of the "hat". With a range of stop ring lengths, it is possible to halt the shearing process at different stages of plastic deformation. From these interrupted samples, metallographic, SEM, and TEM specimens were produced enabling the observation of microstructure-deformation interactions and development of bands.

TEM samples of the sheared regions were made by mechanically thinning the material to a thickness of less than 0.125mm. Three mm discs were abrasively cut from the thinned material using a slurry disc cutter. The foils were electropolished in a 5% perchloric 95% methanol solution at -20°C and 50 V. The shear band region polished preferentially, but produced insufficient electron transparent regions. Ion milling at 15 degrees impact angle and 4 kV was used to further thin the specimens for TEM examination. Both a 300kV Philips CM-30 and a JEOL 200 CX electron microscopes were used.

III. RESULTS

Shear bands are easily produced in the hat specimens. Figures 2a and b show shear band development using optical microscopy in typical samples. The shear bands etch white in nital, typical of the "transformed" band type. Figure 2b shows the deformation zone before the "transformed" bands have formed in a sample with interrupted deformation.

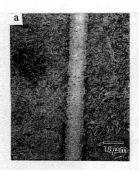

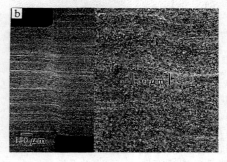

FIG. 2 *Optical micrographs of shear bands produced using this sample geometry. Samples were from the microstructure normalized at 845°C. 2b shows a sample "stopped" prior to the development of a "transformed" band.*

FIG. 3 *Tension cracks produced in the specimens upon unloading. Original loading direction is noted. Normalized at 925°C.*

Some specimens undergoing larger displacements with full shear band formation often exhibited cracks along the shear bands. These cracks along shear bands are common. Since their orientation with the original imposed shearing direction (Figure 3) would *close* the cracks, and not open them, it is concluded that the cracks must have formed after the removal of the imposed shear stresses. It is also indicative of the increased hardness usually associated with the shear band material after testing .

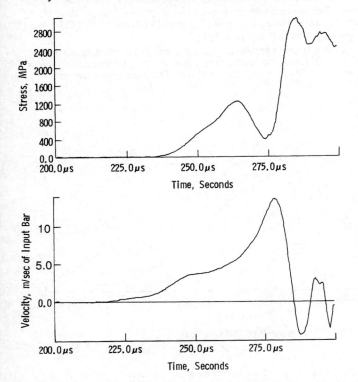

FIG. 4 *Typical a)shear stress/time and b)input bar velocity/time curves.*

A typical stress-time diagram obtained by these tests is shown in Figure 4a, along with the velocity of the input bar in Figure 4b. (This sample was normalized at 845°C) After the sample reached its ultimate shearing strength (instability stress), the load drops until the stopper ring is impacted at about 275 μs. Note also that the velocity of the input bar remains smoothly increasing at a moderate rate (from 4.4 to 5.6 m/s) throughout the regime of interest (260 to 265 μs), during which the sample transitions from elastic deformation to stable plastic flow to unstable plastic flow. An average displacement rate of 5.1 m/s was used to calculate the average strain rate of these tests.

Using the above-mentioned displacement rate and the uniform deformation zone width estimated from Figure 2b (50 μm), a strain rate before shear band formation of $\dot{\gamma}=1.01\times10^5 s^{-1}$ can be derived. The strain rate after shear band formation can be approximated if one assumes that only the material within the shear band continues to deform, and remains at constant width. (However, we have observed some widening of shear bands as shearing has progressed.) Using a shear band width of 9μm (Figure 2b), a strain rate of $\dot{\gamma}=5.64\times10^5 s^{-1}$ in the shear band is calculated.

From the stress-time and velocity-time curves, a stress displacement curve may be generated. Both a typical engineering shear stress and the corresponding true shear stress vs. displacement curve are shown in Figure 5. (The true stress values are valid only up to the instability stress). These curves have a large linear region, and the stresses fall smoothly after instability. Earlier studies by Meyer et. al [3,4] displayed a steeper drop after the maximum shear stresses were reached for a low alloy steel (VHN 480) and a CrMoV steel of medium strength ($\sigma_{ys}=1300$ MPa at $\dot{\gamma}=10^{-4}s^{-1}$).

Figure 6 is a plot of the maximum engineering and true stresses reached as a function of normalizing temperature. Virtually no change is apparent, for the maximum shear stress attained was independent of the four microstructures tested. The same microstructures

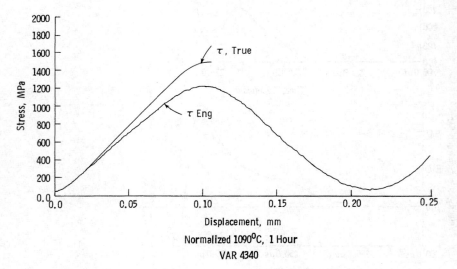

FIG. 5 *Typical engineering shear stress-displacement and true shear stress-displacement curves (true stress curve is only accurate up to the point of instability).*

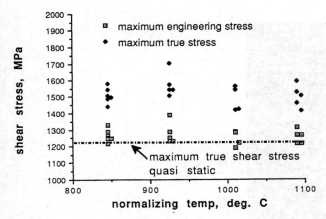

FIG. 6 *Maximum engineering shear stresses and true stresses vs. normalizing temperature. Quasi-static data is from Cowie et. al[1,2].*

were previously studied[1,2] at low strain rates in pure shear, and here too, no effect on the instability stress was found. However, significant differences between the microstructures were noted when energy absorption was considered. Figure 7 plots the energy absorbed before instability (E_i) vs. normalizing temperature relationship for a strain rate of $\dot{\gamma}=10^5 s^{-1}$; note the energy peak for the 925°C normalizing temperature.

Electron microscopy was used to examine the transition between the matrix and the shear bands. Figure 8a is a SEM micrograph of a shear band formed in the microstructure normalized at 925°C. Figure 8b shows this band near its tip. Note the alignment of the

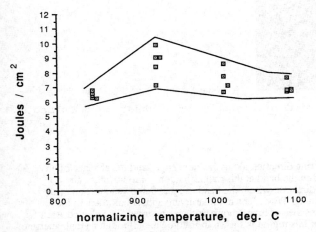

FIG. 7 *Energy absorbed to the point of instability per unit area sheared, E_i, vs. normalizing temperature. $\dot{\gamma}=10^5 sec^{-1}$.*

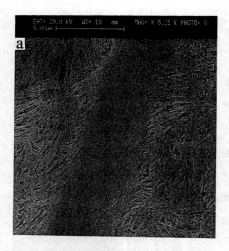

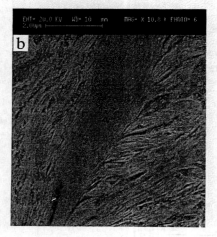

FIG. 8 *Scanning electron micrographs of a shear band formed in a hat specimen.*

martensite laths with the shearing direction along the band edges and the absence of any resolvable grain structure within the band at this magnification. "Flow" lines can be detected parallel to the shear direction within the band. Figure 9 shows a portion of the shear band region examined in the TEM. The microstructure appears mottled, unlike normal martensite, and the presence of many moire' fringes are noted. A Selected Area Diffraction Pattern (SADP) of this region is shown in Figure 9b. Although a small aperture was selected (10μm, which illuminates a 1μm diameter disc at the foil) , a distinct ring pattern is observed, indicating a microcrystalline structure. This ring pattern is typical of

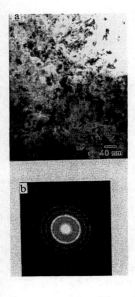

FIG. 9 *TEM bright field micrograph of shear band material in the N-925 microstructure. b) SADP using a 10 μm aperture.*

"transformed" bands [6-10]. The rings do appear somewhat broadened with the pattern indexing to be α–Fe(bcc). No fcc reflections or any individual carbide reflections can be seen..

Using dark field microscopy and centering the first bright ring along the optical axis, the microcrystals can be individually illuminated. Figure 10 is such a dark field micrograph where the crystallite size can be determined to range from 8 to 20nm. The change from a true microcrystalline structure to "normal" heavily deformed martensite away from the band was gradual, agreeing with the works of Wittman et al[8,9]. SADPs as a function of distance away from the center of the band (Figures 11 a-d and 9b) illustrate the gradual transition from a ring pattern to a typical spot pattern .

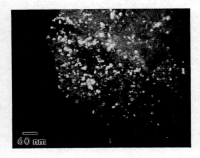

FIG. 10 *TEM dark field centering the first strong ring, which illuminates the microcrystals. Their sizes range from 80-200nm in diameter.*

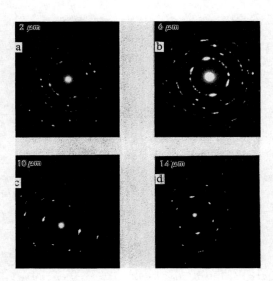

FIG. 11 *Variation of selected area diffraction patterns as a function of distance from the center of the shear band. Distance from the band center is noted on each pattern. Also compare to Figure 9b from near the center of the band.*

IV. DISCUSSION / CONCLUSIONS

The primary purpose of this paper is to examine for a VAR 4340 steel at Rc52 the applicability of using the hat shaped shear specimens to study shear band initiation mechanisms, and present preliminary observations on their formation . The hat specimens with tailored stress-pulse profiles have controlled the deformation, producing unfractured specimens in various stages of shear band formation (prior to band formation, shear band propagation, and shear band widening). The technique provides a simple way to produce shear bands in a controlled manner, at relatively high strain rates; thus, this method permits experimental correlation between metallurgical variables and shear band formation in high strength steels.

The relative resistance to unstable shear of the four microstructures at high strain rates is summarized in Figure 7. It shows the influence of microstructure can be important in this regime of strain rates, despite the intense heating that accompanies shear band formation. Since the principal variable studied, the carbide distribution, probably has little if any effect on the material behavior after destabilization it must affect the processes leading to the destabilization of the deformation process. The fact that one condition had higher energy resistance even though all microstructures had equal hardness at this high strain rate is significant.

These same microstructures have been examined by Cowie et al.[1,2] at low to moderate strain rates. In their study, the 925°C normalizing treatment also produced a peak in the resistance to unstable shear, but the resistance to unstable shear had a minimum for 1010°C , suggesting the presence of a second peak in the instability strain at low strain rates for higher normalizing temperatures. Charpy tests over this same range of normalizing temperatures at room temperature and -40°C correlated in the same manner; two peaks in absorbed energy at room temperature, while the -40°C Charpy results showed a single peak in toughness. This suggests that the increase in strain rate in the "hat" shear tests is analogous to the decrease in temperature in the Charpy tests.

Examination by optical and electron microscopy of the hat specimens in various stages of shear band formation and propagation has not yet shown any significant difference between the four microstructures tested. The tips of the shear bands show the martensite laths gradually aligning with the shearing direction, agreeing with the observations of Wittman et. al [9]. Note the absence on any observable grain structure within the shear band, and the gradual change from the band to the matrix.

The TEM examination of the shear band showed the change from the internal shear band structure to the matrix structure to be gradual. This study has clearly resolved that the interior of the shear band has an extremely fine microcrystalline structure, with grain diameters ranging from 8 to 20nm. Previous studies had suggested that the crystallite size in "transformed" bands was between 100 to 1,000 nm[10]. However, this new evidence clearly shows that microcrystals exist in these shear bands which are an order of magnitude smaller than this. This accounts for the difficult resolution of the band in earlier SEM and TEM studies. Foils that are many crystallites in thickness produce repeated scattering by subsequent crystallites and mask their true structure. No austenite or carbide reflections could be detected in the diffraction patterns taken from the regions showing the extremely fine microcrystals. Since a small crystallite size present within the shear band would enhance the stability of austenite that was present, the complete absence of austenite reflections is significant. It suggests that no transformation has occurred, and that the shear band is simply very heavily deformed martensite, in which both the extremely fine grain size combined with the additional carbon in solution (from dissolved carbides) have increased the hardness of the residual (after testing) microstructure. The absence of carbides combined with the microcrystalline structure would prevent preferential etching, creating the white etching typical of "transformed" bands.

Acknowledgements
The work was performed at the University of California, San Diego (UCSD), and at the US Army Materials Technology Laboratory, Watertown, MA. The work performed at UCSD was supported by the U.S. Army Research Office (ARO) under Contract No. DAAL-03-86-K-0169, as part of ARO's URI center for the Dynamic Performance of Materials. The authors are grateful to Mr. Jon Isaacs for valuable contributions in Hopkinson bar testing.

REFERENCES

1. J. G. Cowie, "The Influence of Second Phase Dispersions on Shear Instability and Fracture Toughness of Ultrahigh Strength 4340 Steel," US Government Report MTL TR 89-20, March 1989.

2. M. Azrin, J. G. Cowie, and G. B. Olson, "Shear Instability Mechanisms in High Hardness Steel, "US Government Report MTL-TR 87-2, January 1987.

3. K.-H. Hartman, H. D. Kunze, and L. W. Meyer, in *Shock Waves and High-Strain-Rate Phenomena in Metals*, M. A. Meyers and L. E. Murr, eds.,Plenum Press, New York, 1981, p. 325.

4. L. W. Meyer and S. Manwaring, in *Metallurgical Applications of Shock-Wave and High-Strain-Rate Phenomena*, L. E. Murr, K. P. Staudhammer, M. A. Meyers, eds., Marcel Dekker, New York, 1986, p. 657.

5.   G. Ravichandran and S. Nemat-Nasser, "Micromechanics of Dynamic Fracturing of Ceramic Composites: Experiments and Observations," Proceedings of the ICF-7, Texas, March 20-24, 1989, *Advances in Fracture Research*, 1: 4 (1989).

6.   R. C. Glenn and W. C. Leslie, *Met. Trans.*, 2: 2954 (1971).

7.   A. L. Wingrove, "A Note on the Structure of Adiabatic Shear Bands in Steel," Australian Defense Standard Laboratories, *Technical Memo, 33*, 1971.

8.   C. L. Wittman and M. A. Meyers, "Effects of Metallurgical Parameters on Shear Band Formation in Low Carbon Steels," *Metall. Trans.*, Dec. 1990.

9.   C. L. Wittman, M. A. Meyers, and H.-r Pak, *Metall. Trans.*, *21A*; 707 (1990).

10.   Y. Meunier, L. Sangoy, G. Pont, in *Impact Loading and Dynamic Behaviour of Materials*, C. Y. Chiem, H.-D. Kunze, L. W. Meyer, eds, DGM Informationsgesellschatt, Oberursel/Frankfurt, 1988, v.2, p. 711.

60

Adiabatic Shear-Band Formation in Explosives Due to Impact

P. C. CHOU, W. J. FLIS, AND K. L. KONOPATSKI

Dyna East Corporation
Philadelphia, Pennsylvania 19104, U.S.A.

Finite-element simulations are performed for shear-band and hot-spot formation in explosives due to fragment impact. The hot spots appear in regions where there are adiabatic shear bands. Formation of shear bands requires thermal softening of the material. The traditional study of shear-band formation in metals requires thermal softening of plastic flow stress. It is shown in this paper that the plastic work generated under fragment impact in plastics and explosives is not sufficient to cause shear bands. Instead of the thermal softening of the flow stress, a temperature-dependent viscosity is used. For this viscous model of constitutive relation, the "hardening" is dependent on the strain rate, and the "softening" is due to a temperature-dependent coefficient of viscosity. Three types of impact are simulated; all show high temperature in the shear band region.

I. INTRODUCTION

This paper presents the hydrocode modeling of shear bands in energetic materials, including explosives and propellants. Shear bands in energetic materials are a source of high temperatures (hot spots) that can lead to undesired initiation even at low impact velocities (less than 100 m/s). We have applied our modeling to several sensitivity problems, including drop-weight indentation, high-speed impact, and shock compression of a void inclusion.

Many authors such as Frey [1] and Howe [2] have hypothesized that localized hot regions are due to two mechanisms, uniaxial shock compression and shear bands. Shear bands could exist whenever there is a large velocity gradient in the specimen. This gradient could be caused by micro-defects such as voids and inclusions, or by grain boundaries in multi-phase explosives. The velocity gradient could also be caused by larger-scale impact geometry. Experimentally, only the latter has been observed. For instance, shear bands in explosives have been observed by Field et al. [3]. They conducted drop-weight tests with transparent anvils and high-speed photography, and after impact observed localized shear bands in explosives that did not initiate. Howe et al. [4] impacted heavily confined explosive targets with plates and fragments. Examination of the damaged explosive showed shear failures, in the form of either shear cracks or shear bands.

Shear-band formation in metal laboratory specimens due to impact has been studied extensively. Under the impactor, or indentor, a distinct shear band forms and propagates parallel to the impactor motion. A comprehensive review was made by Rogers [5]. Recent work comparing experimental results with computer simulation was given by Chou et al. [6].

Computer-code simulations of shear-band formation have been reported recently by a few investigators, including Nemat-Nasser [7], Needleman [8], and Batra and Liu [9]. Most of these works involved tension or compression of simple specimens and required the introduction of artificial initial defects or perturbations. The finite-element technique of this work simulates the impact phenomenon directly, without introduction of artificial defects.

The mechanism of shear-band formation may be considered a thermo-mechanical instability. A solid under dynamic loading may experience thermo-mechanical instability if sufficient heat is generated during deformation. This heat raises the temperature of the material, which may reduce its strength. As this process continues, localized regions develop where the strain is greater than in the rest of the material. These regions, which are warmer and softer than neighboring material, tend to absorb greater amounts of strain, leading to unstable flow of material. If the flow is primarily plastic, the unstable flow coalesces into the classical type of adiabatic shear band, which may be called thermo-plastic instability. Shear bands in metals are usually of this type. In explosives, these bands may be a source of hot spots, as advocated by Winter and Field [10] and Winter [11].

Note that the material undergoing shear must have an effective stress-strain relation that is concave downward; that is, within the domain of strain involved in the problem, the stress may rise to a maximum value but must then decrease, as shown in Fig. 1.

A Lagrangian finite-element hydrocode, DEFEL, has been used to simulate the formation of shear bands in metal. The plastic flow-stress model of Johnson and Cook [12], which yields a concave-downward adiabatic stress-strain curve, was used. This code was used to model a series of shear-punch experiments of high-strength steel, reported previously (Chou et al. [6]). The code predicted a concentration of shear in the workpiece near the edge of the indentor, as observed in test specimens. Other simulations in which thermal softening was not considered did not show such concentrations. The code predicted well the dependence of the length of the shear band on the depth of penetration by the punch.

The concave-downward stress-strain curve required for shear bands is also observed in many explosives. Field et al. [3] showed such behavior for several pressed, HMX-based

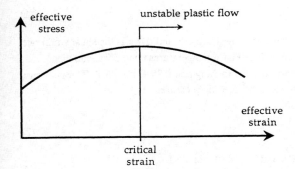

FIG. 1 *Typical plastic stress-strain curve for a shear-banding material.*

explosives. Similar curves for other explosives over a wider range of strain rates have been compiled by Dobratz [13]. Generally, in these curves, the critical strain is only a few percent.

From these data, it may be calculated that the total plastic strain energy (area under the stress-strain curve up to failure) is small, on the order of 10 MJ/m³. This energy can raise the temperature only a few degrees, which is not high enough to allow shear-band formation, nor high enough to be considered "hot spots" for detonation. This conclusion, however, is based on the assumption that the dynamic stress-strain curves are obtained from specimens that have a uniform state of strain throughout. It is possible that shear bands form during these tests and have strains that are much higher than the measured strains which are averaged over the entire specimen. Even with this reservation, we believe that plastic strain is probably not the major mechanism causing thermal softening in explosives.

Thermo-mechanical instability may also be associated with viscosity or friction. If the material has a high coefficient of viscosity, then the resistance to deformation is proportional to its strain rate; this may be called strain-rate hardening. For most such materials, it is known that the coefficient of viscosity decreases with temperature; this is a form of thermal softening. If the thermal softening of the viscous stress is sufficient to overcome the strain-rate hardening, then flow instability may also occur. This phenomenon may be called thermo-viscous instability, another case of thermo-mechanical instability. The model of Frey [1] (see also Boyle et al. [14]) is based on this kind of instability.

As mentioned above, for brittle explosives, the solid strength and fracture strain may be too small to allow enough heat to be generated by plastic work for the plastic type of instability to occur. However, viscous behavior of the material may be important. Viscous stresses are present in liquids and also powdered materials, and come into play after melting or crushing.

II. FREY'S VISCOUS CONSTITUTIVE EQUATIONS

Frey [1] developed an analytical model for initiation of explosives by rapid shear, including the formation of shear bands. In this model, the shear stress is given by

$$\tau = Y + \mu \, d\gamma/dt \tag{1}$$

where Y is the plastic flow stress, μ is the absolute viscosity, and $d\gamma/dt$ is the shear strain rate. The flow stress Y has a constant value of 70 MPa for temperatures up to 55°C and then decreases linearly with temperature to zero at the melting point, 85°C for TNT. The viscosity μ is considered to be dependent on the temperature and pressure,

$$\mu = \mu_0 \exp\left(\frac{P}{P_0}\right) \exp\left(\frac{1}{T_0} - \frac{1}{T}\right) \tag{2}$$

where $\mu_0 = 1.39$ poise, $P_0 = 165$ MPa, and $T_0 = 85$°C for TNT.

Shear bands are possible in a material whose effective stress-strain curve has a negative slope. For this material model, this occurs immediately (at zero strain) because the viscosity μ decreases with temperature, which increases with strain; at larger strains, the plastic term Y also decreases as temperature nears the melt point.

We have incorporated Frey's constitutive equations for explosives into our DEFEL hydrocode. The total stress is the sum of elastic-plastic and viscous parts,

$$\sigma_{ij} = \sigma_{ij}^{e\text{-}p} + \sigma_{ij}^{v} \tag{3}$$

where the elastic-plastic part is computed by the method of Wilkins [15] used in most hydrocodes, with effective stress taken as Y(T), as above; the viscous part is Newtonian,

$$\sigma_{ij}^{v} = 2\mu \dot{e}_{ij} \tag{4}$$

where μ retains the dependency on temperature and pressure used by Frey.

In comparison with Frey's analytical model, the hydrocode of course requires no assumption of a quasi-steady stress state. The hydrocode does have a capability for heat conduction, but we have not yet used it in these types of calculations. Currently, we are adding the capability for heat generation due to reaction.

III. CODE CALCULATIONS

To test this model, hydrocode calculations were performed using Frey's constitutive equation for several problems, including drop-weight indentation, high-speed impact, and shock compression of a specimen with a void inclusion. Figure 2 illustrates these three cases.

The first calculation was a drop-weight indentation test with an explosive specimen. The drop-weight had a mass of 128.5 grams, a velocity of 50 m/s, and an indentor 1.07 mm longer than the thickness of the stopper plate. Results of this calculation are shown in Fig. 3, a close-up of the specimen near the indentor, which has indented from the right of the figure. In this plot, only those finite-element grid lines that were initially vertical are shown. This plot

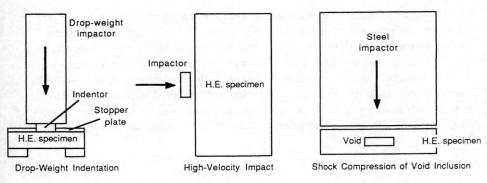

FIG. 2 *Schematic illustration of hydrocode calculations using Frey's constitutive model.*

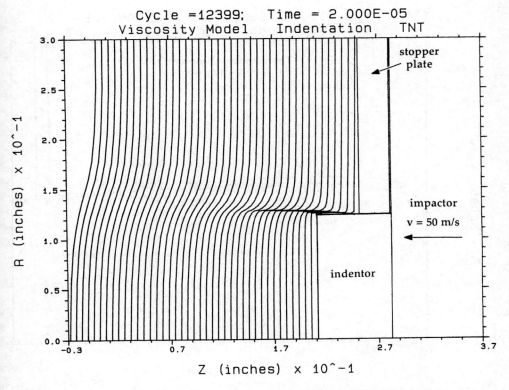

FIG. 3 *Formation of a shear band partially through a TNT specimen in a controlled-depth drop-weight impact simulation.*

clearly shows the concentration of shear in the specimen near the edge of the indentor. A contour plot of temperatures in this region, Fig. 4, shows that this region has been heated beyond the melting temperature of 174°F. These high temperatures are possible because the viscous term continues to contribute to the work of deformation, even after melting.

A second calculation using this model was of a high-velocity impact of an explosive by a very small cylindrical steel projectile having velocity of 200 m/s. The initial grid used in this calculation featured a very fine finite-element grid in the specimen, particularly near the edge of the impactor. Results of this calculation, shown in Fig. 5, indicate the formation of a shear

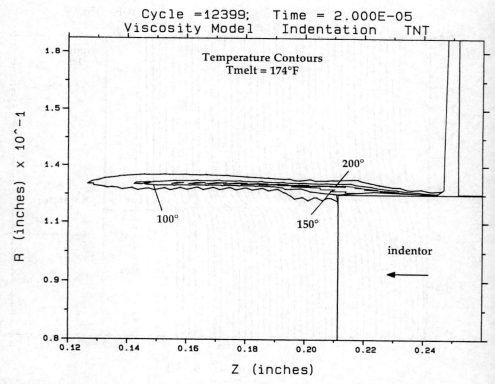

FIG. 4 *Temperature contours near the shear band in a TNT specimen in a controlled-depth drop-weight impact simulation.*

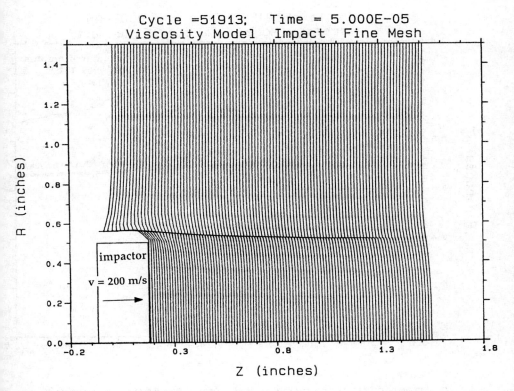

FIG. 5 *Localized shear deformation indicating the formation of a shear band in high-velocity impact of TNT explosive modeled by Frey's constitutive relation.*

band in the specimen near the edge of the impactor. Figure 6 shows the temperature contours in the impact region, with temperatures above melting in the shear-band region.

A third calculation involved plate impact of a layer of explosive containing a cylindrical void. The velocity of the steel impactor plate was 50 m/s. A plot of the vertical finite-element grid lines, Fig. 7, shows the formation of shear bands near the corners of the void. Figure 8, a contour plot of temperature, shows the elevated temperatures in these bands.

In other simulations of all of these types of problems, if the stimulus (impact velocity or shock strength) was not sufficiently strong, shear bands did not occur, even though some gradual shear was evident.

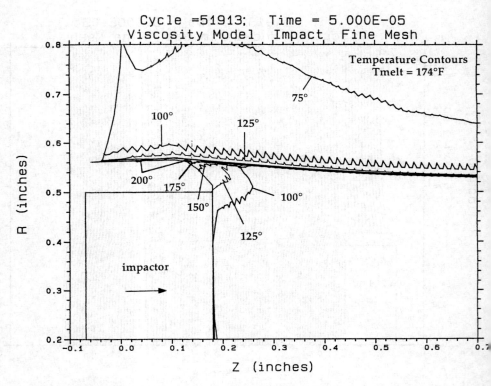

FIG. 6 *Temperature contours near the shear bands in hydrocode simulation of high-velocity impact of TNT explosive modeled by Frey's constitutive relation.*

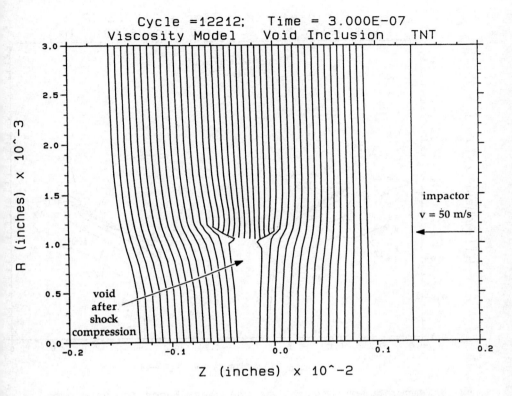

FIG. 7 *Detail of deformed mesh in hydrocode simulation of plate impact of a TNT specimen containing a cylindrical void.*

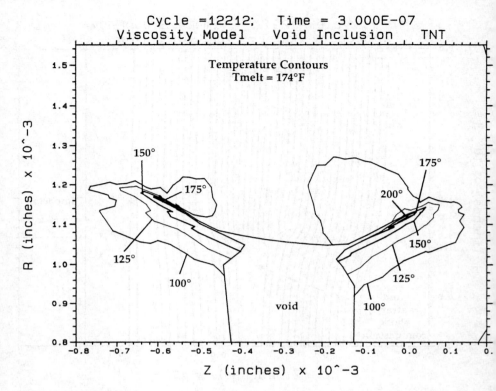

FIG. 8 *Temperature contours near the shear bands in hydrocode simulation of plate impact of a TNT specimen containing a cylindrical void.*

IV. SUMMARY

The code, so modified, can serve as an analytical tool to aid in the design of insensitive munitions. Current work involves the simulation of shear bands with the use of heat transfer; this will allow also prediction of the width of the bands. We are also working on the addition of an algorithm for the release of chemical energy by the explosive or propellant, which will accelerate the formation of shear bands and will serve as a criterion of the degree of reaction. The resulting code will be able to predict initiation by all types of mechanical stimuli, including impact and shock, and will thus be able to handle problems of sympathetic detonation and hostile and accidental impact.

ACKNOWLEDGMENT

This work is supported by the Air Force Armament Laboratory under contract F08635-89-C-0133. We acknowledge Mr. David Wagnon and Dr. Joseph Foster, Jr., for their technical comments and discussions.

REFERENCES

1. R. Frey, "The Ignition of Explosive Charges by Rapid Shear," in *Proc. 7th Symp. Detonation*, Annapolis, 1981.

2. P. Howe, "On the Role of Shock and Shear Mechanism in the Initiation of Detonation by Fragment Impact," in *Proc. 8th Symp. Detonation*, Albuquerque, 1985.

3. J.E. Field, S.J.P. Palmer, P.H. Pope, R. Sundarajan, and G.M. Swallowe, in *Proc. 8th Symp. Detonation*, Albuquerque, 1985.

4. P. Howe, G. Gibbons, and P. Weber, "An Experimental Investigation of the Role of Shear in Initiation of Detonation by Impact," BRL-TR-2718, Aberdeen Proving Ground, 1986.

5. H.C. Rogers, *Ann. Rev. Mater. Sci.*, 9: 283-311 (1979).

6. P.C. Chou, J. Hashemi, H. Rogers, and A. Chou, "Experimentation and Computer Simulation of Adiabatic Shear Bands in Controlled Penetration Impact", in *Eleventh Natl. Congress of Appl. Mech.*, Phoenix, 1990.

7. S. Nemat-Nasser, "Micromechanics of Failure at High Strain Rates: Theory, Experiments, and Computations," BRL-TR-2662, Aberdeen Proving Ground, 1985.

8. A. Needleman, *J. Appl. Mech.*, 56: 1-99 (1989).

9. R.C. Batra and D. Liu, *J. Appl. Mech.*, 56: 527-34 (1989).

10. R.E. Winter and J.E. Field, *Proc. Roy. Soc. (Lond.)*, A343: 399-413 (1975).

11. R.E. Winter, *Phil. Mag.*: 765-73 (1975).

12. G.R. Johnson and W.H. Cook, "A Constitutive Model and Data for Metals Subjected to Large Strains, High Strain Rates, and High Temperatures," in *Proc. 7th Inter. Symp. Ballistics*, The Hague, The Netherlands, 1983.

13. B.M. Dobratz, *LLNL Explosives Handbook, Properties of Chemical Explosives and Explosive Simulants*, Lawrence Livermore Natl. Lab., UCRL-52997, 1981.

14. V.M. Boyle, R.B. Frey, and O.H. Blake, "Combined Pressure-Shear Ignition of Explosives," in *Proc. 9th Symp. Detonation*, Portland, 1989.

15. M.L. Wilkins in *Methods of Computational Physics, V. 3*, B. Adler, S. Fernback, and M. Rotenberg (eds.), Academic Press, New York, 1964.

61

A Dislocation-Microscopic Approach to Shear Band Formation in Crystalline Solids During Shock or Impact

C. S. COFFEY

Naval Surface Warfare Center
White Oak
10901 New Hampshire Avenue
Silver Spring, Maryland 20903-5000

Shear bands have been observed to develop in crystalline solids during shock or impact in time of the order of 10 nanoseconds or less which precludes any form of bulk material motion for the level of forces loading the solid (P<25 GPa). It is reasonable to assume that the same processes occur in metals but that because of their opacity to visible and infrared emissions from the interior of the sample this early shear band formation has not yet been observed.

Here we examine the microscopic processes responsible for shear band formation including the rate at which a dislocation source can create dislocations, the localized confinement of these dislocations to lie on nearly adjacent slip planes, and the very large energy dissipation rates associated with high velocity dislocation motion.

It will be shown that the initial shear band formation occurs as early as a few nanoseconds after shock arrival but because of thermal conduction and, in metals, the absence of internal molecular vibrational states significant temperature increase does not occur until later times (5μs). The width and rate of growth of the shear bands will be determined.

I. INTRODUCTION

The response of crystalline solids to the rapid loading imposed by shock or impact is of general interest. There exists a substantial amount of experimental evidence that indicates that a significant portion of the plastic shear deformation that occurs in a crystalline solid during shock or impact loading is localized in discrete and often almost discontinuous band-like structures. This is a departure from the classical theory that the plastic deformation occurring during these loads is more or less uniformly distributed throughout the crystal in a manner consistent with the applied shear stress. It has also been observed that this localized plastic deformation and associated heating can often determine the response of the crystalline solid to these rapid loads.

The earliest modern observations of localized plastic deformation were on shocked steels reported by Zener and Holloman [1]. Since then numerous others have reported similar observations in shock or impact experiments on steels, titanium and other metals [2,3,4]. Localized plastic deformation and energy concentration during impact experiments on LiF crystals were reported by Johnston and Gilman [5]. Similar results have been reported in other crystals. Observations on strongly shocked quartz reported visible light emitted from localized shear band-like structures in times less than 15 nanoseconds after shock arrival [6,7,8]. Localized deformation has been reported in more complicated molecular crystals such as ammonium perchlorate [9,10].

Here our intent is to draw from both our current and earlier research on this subject to give an understanding of how localized plastic deformation may occur due to shear during shock or impact. It has been shown that this is basically a quantum mechanical problem not totally accessible to classical continuum methods [11]. The energy dissipation rates within the regions of localized plastic deformation can be very large, capable of melting the solid or causing chemical reactions to occur in reactive materials. For large amplitude shocks very rapid, < 1ns, multiphonon processes can occur leading to rapid molecular vibrational and even electronic excitation [12]. Here these will be

assembled along with more recent developments to establish
the behavior of shear bands and energy localization during
shock or impact.

II. LOCALIZED DEFORMATION

The mechanism of plastic deformation in crystalline solids
is the creation and motion of dislocations. Over the past
several years evidence has been accumulated that in crystalline solids the plastic deformation that occurs due to shear
during shock or impact is often localized in narrow band-
like structures here referred to as shear bands. These
bands generally lie along the dislocation slip planes and
have often been observed to contain large numbers of dislocations.

To explain the formation of discrete shear bands it is
necessary to postulate the existence of fast acting dislocation sources. These sources respond to an applied shock or
impact stress to produce copious quantities of dislocations
which are constrained to move along nearly adjacent slip
planes. The energy dissipated by these moving dislocations
generates heating and molecular excitation of the molecules
of the host lattice that lie within the shear bands.

Several models for dislocation sources have been proposed, the most notable of these is that due to Frank and
Read [13]. In actuality, there are likely to be a large
number of different source configurations capable of generating dislocations during a shock or impact. In order to
deal with this problem in a reasonable manner, we propose to
treat these possible sources in a general way. What is most
important for dynamic loading imposed by a shock or impact
is not the actual form of the source, but rather the rate at
which the source generates dislocations. This determines
the response of the host lattice to the applied shear stress
of a shock or impact.

In order to predict the rate at which a source can
generate dislocations during shock or impact, we have proposed a simple model in which the source has a finite size

ℓ_0 [14]. In this picture a pair of newly created but oppo-
sitely orientated edge dislocations have to clear the source
region before the source can function again to create another dislocation pair. The rate at which the source can
create dislocation pairs is

$$\frac{dn}{dt} = \frac{V}{\ell_0} P(\tau)$$

(1)

where $P(\tau)$ is the quantum mechanical probability of breaking
an intercrystalline bond with the applied shear stress, τ,
to form a dislocation pair and V/ℓ_0 is the rate at which the
dislocations leave the source region. V is the dislocation
velocity under the applied shear stress. The expression for
the dislocation velocity developed by Gilman will be used
since it has wide validity, $V = V_0 \exp(-\tau_0/\tau)$, where V_0 is
nearly the shear wave speed and τ_0 is a shear stress characteristic of the material. For high amplitude shock or
impact $\tau \gg \tau_0$ the probability of creating a dislocation
pair approaches unity, $P(\tau) \to 1$, and $dN/dt \approx V_0/\ell_0$. Assuming some nominal values, $V_0 = 3 \times 10^3$ m/s and $\ell_0 = 3 \times 10^{-8}$m,
the maximum rate of creating dislocations during a strong
shock or impact is approximately 10^{11} dislocations/s.

At low or nearly static loading rates the newly created
dislocations move along the slip plane on which they were
created until they encounter an obstacle such as a grain
boundary where upon they stop and form a pile-up. The
number of dislocations in the pile-up will increase until
the back stress that the pile-up exerts on the source increases to the point where it becomes more energy efficient
to create the next dislocations on a nearby slip plane. In
time these dislocations will encounter an obstacle and a new
pile-up will form until the increasing back stress causes
the source to shift its dislocation production to another
nearby slip plane. In this way a shear band is built up
from a collection of dislocation pile-ups on formerly active
slip planes.

For high level shocks or impacts the above picture is
slightly modified because the source will create disloca-

tions at such high rates that they cannot travel far from
the source before the subsequently created dislocations are
piling up behind the earlier created dislocations. In this
way a dynamic pile-up is formed. The back stress from this
dynamic pile-up will temporarily cause the source to shift
the production of dislocations on to other near by slip
planes and in this way form a shear band composed of dynamic
pile-ups on nearly simultaneously active slip planes. For
short duration shock pulses none of these dynamic pile-ups
or associated shear bands need extend entirely across the
crystal.

The rate at which the shear band increases in width, w,
is just

$$\frac{dw}{dt} = \frac{D}{N_0} \frac{dN}{dt} \tag{2}$$

where D is the average separation distance between slip
planes and N_0 is the number of dislocations in the pile-up.
For typical values of $D = 10^{-9}$m, $N_0 \sim 10^2$ to 10^3 dislocations
in a static pile-up and a source rate of $dN/dt \sim 10^{11}$ dislo-
cations/s, the rate of increase in the width of a shear band
during strong shock or impact is approximately .1 to 1 m/s.

III. ENERGY BUILD UP IN A SHEAR BAND

The rate at which energy is accumulated in a shear band
during shock or impact is determined by the rate at which
energy is deposited in the shear band by the moving dislocations minus the rate at which energy is removed from the
band by thermal conduction. It is straight forward to show
that the energy dissipation by a moving dislocation is a
quantum mechanical process not accessible by a classical
continuum mechanical approach. Consider a dislocation
moving at a speed V through a crystal lattice of interatomic
or intermolecular spacing d. To a first approximation the
lattice intercrystalline potential may be represented by the
sinusoidal Peierls-Nabarro potential. As it moves through
the lattice the dislocation encounters this potential at a

rate of V/d times per second. To the moving dislocation and
the surrounding atoms or molecules this represents a time
varying perturbation at a radial frequency of ω = 2 π V/d
rad/s. The maximum velocity at which the dislocation can
travel is nearly the shear wave speed, typically V = 2 to 5
km/s. For most materials of interest the interatomic or
intermolecular spacing is typically about 3 x 10^{-10}m. Thus
the radial frequency near the core of the dislocation for
this simple potential can exceed 10^{13}rad/s. For a more
realistic intercrystalline potential appropriate for molecules with more complicated structures, the above frequency
is just the center frequency of the perturbation frequency
distribution that the molecules near the core of the dislocation experience. These maximum frequencies in excess on
10^{13}rad/s are sufficient to directly excite the internal
molecular vibrational modes of most molecules of interest.
This direct molecular excitation process is not available in
classical physics.

The energy dissipation rate due to the moving dislocations during shock or impact has been calculated by determining the interaction between a moving edge dislocation and
the lattice [11,12]. This was accomplished by constructing
an interaction Hamiltonian from the interaction energy of
the stress field of an idealized edge dislocation and the
local strain field in the lattice and treating the moving
dislocation as a superposition of plane waves. Combining
this expression for the energy dissipation rate due to the
moving dislocations in the shear band with expression for
the loss of energy due to thermal conduction out of the
band, the net rate of energy deposition in a shear band
during shock or impact is

$$\frac{dE}{dt} = 2 \; \Gamma \; R^2 \int \Sigma_k \; (n_q + 1) N_k \; (N_{k-q} + 1) \; dq +$$
$$Higher \; Order \; Terms \; - \; K \; \frac{dT}{dx} \tag{3}$$

Here

$$\Gamma = \frac{1}{64 \; \pi^3 Nm} \; \frac{Gb}{1 \; - \; v}^2 \; \frac{1}{V_0 d^2} \tag{4}$$

and G is the shear modulus, υ is the Poisson ratio, d is the interatomic or intermolecular spacing, V_o is the shear wave speed, K is the thermal conductivity, R is the approximate radius of the dislocation core, and Nm is the mass density, ρ, of the solid. N_k is the number density of moving dislocations with momentum k and N_{k-q} is the number of dislocations moving with momentum k - q after emitting a phonon momentum q. The higher order terms can be obtained from an analysis that follows that of Bebb and Gold [15]. These terms are important in the case of molecular solids subjected to high level shock or impact, $\tau \gg \tau_o$, for which the optical phonons generated by the high velocity dislocations can directly excite the molecular internal vibrational modes, $\omega \sim 10^{13}$ rad/s. This situation has been treated elsewhere [12].

The thermal gradient, dT/dx, is taken to be in the direction normal to the plane of the shear band. For the present purposes it will be assumed that it is adequate to use the classical expression for the thermal conduction process although a more rigorous treatment may be required to describe the conduction process near the edge of a very narrow shear band. It can be inferred directly that the thermal gradient can be very large near the edge of an extremely narrow shear band.

The initial terms arise from a first order perturbation calculation of the energy dissipated by relatively slowly moving dislocations whose velocity is less than the shear wave speed. These low velocity dislocations occur due to low amplitude shock or impact and dominate the energy dissipation process when the phonon frequencies generated by the dislocation motion are inadequate excite the low lying molecular vibrational levels, or as in most metals, there are no internal vibrational levels to excite.

At the tail of a static pile-up the dislocations come to a stop so the k - q = 0. The dislocation number density is also zero at the tail of a static pile-up, $N_{k-q} = 0$. In keeping with this observation it will be assumed that near a

free running dislocation $N_{k-q} \sim 0$, so that $N_{k-q} + 1 \sim 1$.
Further, the number density of phonons moving nearly coincident with dislocation core is very small, $n_q \ll 1$, so that to a good approximation $n_q + 1 \approx 1$. These approximations considerably simplify the first order energy dissipation expression, (3), which reduces to

$$\frac{dE}{dt} \approx 2 \ \Gamma \ R^2 \ \overline{N}_q \ - \ K \ \frac{dT}{dt} \tag{5}$$

where

$$\overline{N} = \sum_k N_k \tag{6}$$

is the number of moving dislocations. The higher order terms will be neglected here. Typically, most dislocations move with the same velocity determined by the applied shear stress so that

$$\overline{N} = \int \frac{dN}{dt} \ dt \tag{7}$$

where dN/dt is the rate at which the source creates dislocations. The phonon wave vector can be written as $q = 2 \ \pi$ $V/(V_o d)$, which when substituted into equation (5) gives the energy dissipation rate the appearance of a velocity dependent dislocation drag expression.

Using the Gilman expression for the dislocation velocity, the first order energy dissipation rate in a shear band simplifies to

$$\frac{dE}{dt} \approx \frac{4 \ \pi \ \Gamma \ R^2 \ \overline{N}}{d} \ e^{-\tau_{0/\tau}} \ - \ K \ \frac{\Delta T}{\Delta x} \tag{8}$$

For experiments in which just the energy dissipation rate per moving dislocation is measured the thermal conduction term can be dropped. With nominal values for the material properties of LiF, Appendix 1, the energy dissipation rate per moving dislocation in impacted and mildly shocked LiF crystals approaches 1 watt/cm of dislocation length in the limit $\tau > \tau_0$. This is in good agreement with the experimental results shown in figure (15) of Johnston and Gilman [5].

IV. SHOCKS AND IMPACTS IN SIMPLE METALS

A simple metal composed primarily of a single atomic species will not have the internal or optical vibrational states that characterize more complicated molecules. Consequently, only the first order terms in the dislocation energy dissipation rate will be important. Early in the development of a shear band during shock or impact the energy loss to thermal conduction will be so great that little energy will accumulate in the band. However, at later times as the band increases in width, the thermal gradient will decrease as Δx increases. To solve this problem exactly would be very difficult. Here, an approximate solution will be examined in which it will be assumed that the width of the thermal gradient is proportional to the width of the shear band

$$\Delta x = \beta V_s t \tag{9}$$

Where β is the proportionality constant and $V_s = dw/dt$ is the velocity at which the shear bands grow in width as given in equation (2).

A limiting condition for stability is obtained by setting to zero the net rate at which energy accumulates in the shear band. After this time the thickness of the thermal gradient zone will increase until the energy accumulated in the band due to energy dissipation by the moving dislocations dominates the energy removed by the conduction processes. Setting the total energy accumulation rate to zero simplifies equation (8) to

$$0 = \Gamma' t' - \frac{K \Delta T}{\beta V_s t} \tag{10}$$

where

$$\Gamma' = \frac{4 \pi \Gamma R^2}{d} \frac{dN}{dt} e^{-\tau_0/\tau} \tag{11}$$

It is necessary to distinguish between the time t' associated with the number of moving dislocations and the total
elapsed time, t, associated with the growth of the shear
band.

Two possible situations may arise, the first occurs
early during the shock or impact loading when all of the
dislocations in the shear band are moving in which case t' =
t. In this situation equation (9) gives

$$t = \left(\frac{K \, \Delta \, T}{\beta V_s \Gamma'} \right)^{1/2}$$

(12)

as the time beyond which the energy in the shear band begins
to increase because thermal conduction ceases to dominate
the energy flow. The second situation arises at later times
when large numbers of dislocations created by the source in
response to the shock or impact have moved to the crystal
boundaries and have either stopped or have moved through the
boundary into adjacent crystals. Currently we do not know
how many dislocations are moving and how many are stopped,
consequently this second case will not be considered further
here.

Using nominal material property values for steel given
in Appendix (1), the energy dissipation rate per moving
dislocation is

$$\frac{1}{N} \frac{dE}{dt} \approx 2.75 \, e^{-\tau_0/\tau} \quad Watts/cm$$

(13)

From equation (11), setting $\beta = 1$, the time beyond which the
energy concentrated in the shear band begins to increase
rapidly is approximately $t = 3 \times 10^{-6}$s. At this time the
width of the shear band is approximately

$$w = V_s t \approx .3 \times 10^{-6} m$$

(14)

The magnitude of the energy dissipation rate combined
with sources capable of generating in excess of 10^{11} dislocations per second implies that after the thermal conduction

process ceases to be important as a means of removing energy from the shear band, energy build up will occur leading eventually to thermal softening and even melting in the shear band regions.

V. CONCLUSION

There is ample experimental evidence that much of the plastic shear deformation and energy dissipation that occurs during shock or impact loading of crystalline solids are concentrated in small local regions within the crystal. The plastic shear deformation in crystalline solids occurs mainly through the creation and motion of dislocation. Here, we have sought to combine the rate of dislocation generation by a generalized source with the energy dissipated by the motion of these dislocations to describe both the localized plastic deformation and the localized energy dissipation.

APPENDIX 1

For lithium fluoride the calculation assumes the following nominal values, $\rho = 2.6 gm/cm^3$, $G = .5 \times 10^{+12} dynes/cm^2$, $b = 2.8 \times 10^{-8} cm$, $d = 4.0 \times 10^{-8} cm$, $\upsilon = .187$, $V_o = 3.6 \times 10^5 cm/s$. The major uncertainty comes in assigning a value for the radius of the dislocation core. Here we take $R = d$. The experimental measurements of Johnston and Gilman [5] shown in their figure (15) exclude the effects of thermal conduction so that will be neglected for the purposes of these calculations. The energy loss per moving dislocation from equation (9) is

$$\frac{1}{N} \frac{dE}{dt} \approx 5\, e^{-\tau_0/\tau} \; Watts/cm$$

In the asymptotic limit of $\tau \gg \tau_o$ for the LiF crystals the energy dissipation rate per moving dislocation approaches 5 watts per cm of dislocation length which is in reasonable agreement with the projected limiting dissipation rate observed experimentally in the "hard" LiF crystals.

For shear bands in steel the following nominal material property data was used, $\rho = 7.86 \text{gm/cm}^3$, $G = 10^{12} \text{dynes/cm}^2$, $b = d = 1.2 \times 10^{-8} \text{cm}$, $\upsilon = .2$, and $V_0 = 5.5 \times 10^5 \text{cm/s}$. As before assume $R = d$. The energy dissipation rate per moving dislocation is approximately $2.75 \exp -\tau_{0/\tau} \text{watt/cm}$ of dislocation length.

REFERENCES

1. C. Zener and J. H. Hollomon, *J. Appl. Phys. 15,* 22(1964).

2. D. B. Mikkola and R. N. Wright, *in "Shock Waves in Condensed Matter 1983"* edited by J. R. Asay, R. A. Graham and G. K. Straub. North Holland, New York, 1984, p. 415.

3. H. C. Rogers and C. B. Shastry, *in "Shock Waves and High Strain Rate Phenomena in Metals,"* edited by M. A. Meyers and L. E. Murr. Plenum Press, New York, 1981, p. 285.

4. R. E. Winter and J. E. Field, *Proc. Roy. Soc. A, 343 London,* 1975, pp. 399-413.

5. W. G. Johnston and J. J. Gilman, *J. Appl. Phys. 30,* 129(1959).

6. W. P. Brooks, *J. Appl. Phys. 36,* 2788(1965).

7. P. J. Brannon, C. H. Konrad, R. W. Morris, E. D. Jones and J. R. Asay, *J. Apply. Phys. 54,* 6374(1983).

8. P. J. Brannon, R. W. Morris and J. R. Asay, *J. Appl. Phys. 57,* 1676(1985).

9. P. J. Herley, P. W. M. Jacobs and P. W. Levy, *J. Chem. Soc. 3(A),* 434(1971).

10. J. O. Williams, J. M. Thomas, Y. P. Savintsev and V. V. Boldyrev, *J. Chem Soc. A, 11,* 1757(1971).

11. C. S. Coffey, *Phys. Rev. B 24,* 6984(1981).

12. C. S. Coffey, *Phys. Rev. B 32,* 5335(1985).

13. F. C. Frank and W. T. Read, *Phys. Rev. 79,* 722(1950).

14. C. S. Coffey, *J. Appl. Phys. 66,* 1654(1989).

15. H. B. Bebb and A. Gold, *Phys. Rev. 143(1),* 1(1966).

62

Mechanical Properties in Shear at Very High Strain Rate of AISI 316 Stainless Steel and of a Pure Iron Comparison with Tensile Properties

C. ALBERTINI, M. MONTAGNANI, E. V. PIZZINATO,
A. RODIS*, S. BERLENGHI*, G. BERRINI*, G. PAZIENZA*, A. PALUFFI*

Commission of the European Communities, Joint Research Centre, Ispra Site, 21020 Ispra (Varese), ITALY

*OTO Melara, P. O. Box N° 71, 54011 Aulla, (Massa Carrara), ITALY

A new double shear specimen is proposed for testing at very large strain and strain rates avoiding the constraints of geometrical instabilities. Strain rates up to $3 \times 10^4 s^{-1}$ have been reached on AISI 316 H and ARMCO iron showing change of deformation and fracture mode with respect to low and medium strain rate.

I. INTRODUCTION

Reliable calculations by finite element codes of rapid metal forming and penetration of metal shields by missile impact require the knowledge of material models which describe the response of the unit metal volume up to large strains, at very high strain rate, under multiaxial loading, taking into account the different deformation modes and straining paths. The nodal point for the experimental investigation of the material response under such extreme loading conditions is the conception and the analysis of a suitable specimen permitting to perform tests in controlled conditions of stress, strain and strain rate. In fact most of the normally used specimen shapes fail to reach large strains due to geometric instabilities as well as to reach very high strain rates due to too large gauge length to obtain homogeneous stress distribution. Uniaxial tension specimens are restricted to small strains by necking. Uniaxial compression specimens are plagued by barrelling and end effects from the plateau. Torsion of solid bars suffers of non-uniform stress state while torsion of thin walled tubes gives rise to torsional buckling [1]. A specimen is here proposed which should avoid the mentioned constraints to tests at large strains and very high strain rates, under simple and double shear stresses.

II. SPECIMEN CONFIGURATION

The proposed specimen is sketched in Fig. 1. It consists of two thick hollow cylindrical parts jointed through a thin circular crown which is the gauge part of the specimen. The thin circular crown is built up by means of a slight difference between the outer diameter of the smaller cylindrical part and the inner diameter of the larger cylindrical part. Due to its resemblance the specimen has been named "bicchierino," the Italian name for a small calice glass.

The deformation mode of the gauge part is simple shear when the specimen is subjected either to axial loading or to a torque as shown in Fig. 1. The simultaneous application of axial and torque loading to the specimen brings the gauge part to deform in a

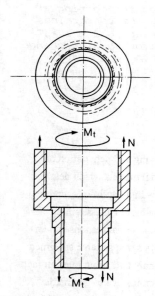

FIG. 1 *Specimen "Bicchierino" (Calice Glass)*.

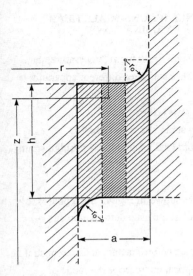

FIG. 2 *Cross section of the gauge part of "Bicchierino" specimen.*

double shear mode. The longitudinal (radial) section of the gauge part is shown in Fig. 2: it consists of a rectangle of the dimensions axh, and of two appendices bounded by the radii of curvature r_0 necessary both for a gradual transition into thicker parts of the specimen, and for manufacturing reasons.

An optimization study by finite element calculation has been performed, by the last two authors of reference [2], of the dimensions of the specimen in order to obtain a state of uniform shearing stresses in the gauge part and to minimize the other stress components. The equations of state assumed for the shape optimization of the specimen are those of linear elasticity because the elastic case has been judged to pose the most severe requirements for obtaining a homogeneous stress distribution. The optimization procedure led to the following optimal dimensions of the gauge part (Fig. 2): $a/h = 0.5$ $r_0/h = 0.2$. The actual specimen has been constructed following the optimization conditions with a short gauge length a = 0.25 mm, which permits to have homogeneous stress distribution along the gauge part also in case of testing at very high strain rate. At the same time the gauge part has a length which contains at least ten grains of the metals under testing and has a cross section of 16 mm^2, both parameters being sufficient to make the gauge part representative of the average mechanical response of the material to external loading [3].

III. VERIFICATION OF SPECIMEN PERFORMANCE BY UNIAXIAL SHEAR EXPERIMENTS AND OPTICAL MICROSCOPE EXAMINATIONS

Experiments in uniaxial shear have been performed by pulling, at room temperature, the two thick hollow cylindrical parts of the specimen by means of three different apparatus:

- a Hounsfield Tensometer for low strain rate tests ($10^{-2}s^{-1}$);
- A hydropneumatic apparatus [4] for medium strain rate tests ($10\ s^{-1}$);
- a modified Hopkinson bar with prestressed bar loading device [4] for high strain rate tests > $10^4\ s^{-1}$), sketched in Fig. 3.

The typical phases of a test with the Hopkinson bar are shown in Fig. 4. Such phases, by eliminating the time, give a stress-strain diagram with elastic-plastic deformation and fracture of the gauge part of the specimen.

    Some tests at low strain rate have been interrupted before fracture at different strain levels of the gauge part in order to observe under the optical microscope the mode of deformation. An example of such observations (Fig. 5), corresponding to a high straining value, demonstrate that the deformation mode is pure shear. The cross section resisting shear remains practically unchanged although the high strain value. Furthermore, the deformation is practically confined inside the nominal gauge length.

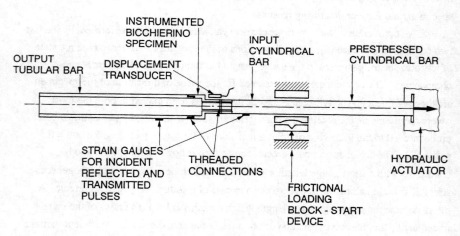

FIG. 3 *Record of a test of the "Bicchierino" specimen with the Hopkinson bar.*

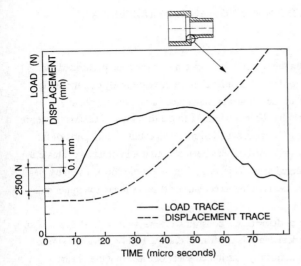

FIG. 4 *Record of a test of the "Bicchierino" specimen with the Hopkinson bar.*

FIG. 5 *Contour of the deformation zone after a shear strain of 1.6 Material AISI 316 H (x 100).*

IV. VERIFICATION OF SPECIMEN PERFORMANCE BY NUMERICAL
SIMULATION

Some numerical investigations on the "bicchierino" specimen have been performed by
means of the NIKE 2D, a vectorized implicit finite element code for analyzing the static
and dynamic response of 2-D solids developed by Lawrence Livermore National
Laboratory (USA). The specimen has been modelled using a mesh with 712 axisymmetric
quadrangular elements connecting 779 nodes, to describe gauge and pulling parts of the
specimen under examination. A static analysis has been carried out in order to evaluate the
plastic deformations in the examined area. This analysis has been performed step by step
by means of incremental displacements of the lower side of the specimen, the upper side
being rigidly clamped.

The material model used to describe the constitutive behaviour of AISI 316 was an
elastic plastic model. The numerical analysis confirmed that the highly prevalent stress
component in the gauge part is a uniformly distributed shear one and that the plastic
deformation is concentrated in the restricted area of the gauge part as desired. Figs. 6 and
7 illustrate such conclusions for the case of interrupted test with large strain whose
microscopical examination has been shown in Fig. 5. The numerical analysis showed that
also at very large strains no geometrical instabilities perturbate the material deformation.
Good agreement has been obtained by comparing the plastic deformations numerically
obtained and those derived from microscopical examinations.

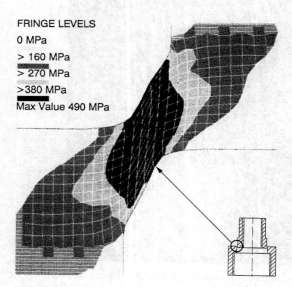

FRINGE LEVELS
0 MPa
> 160 MPa
> 270 MPa
>380 MPa
Max Value 490 MPa

FIG. 6 *Shear stress distribution in the gauge part of the
"Bicchierino" specimen.*

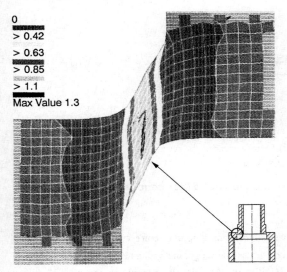

0
> 0.42
> 0.63
> 0.85
> 1.1
Max Value 1.3

FIG. 7 *Effective plastic strain distribution in the gauge part of the "Bicchierino" specimen at displacement of 0.395.*

V. MECHANICAL PROPERTIES IN UNIAXIAL SHEAR OF AISI 316H AND ARMCO IRON

Two materials have been tested, a f.c.c. austenitic stainless steel AISI 316 H and a b.c.c. pure ARMCO iron at room temperature and strain rates ranging between 10^{-2} and 30,000 s^{-1}. The shear stress-strain curves are reported in Figs. 8 and 9. From these figures we observe that at low and medium strain rate the shear stress-strain curves of both materials

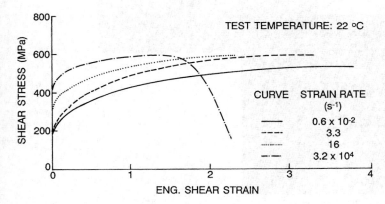

FIG. 8 *Shear stress–strain curves at different strain rates of AISI 316 H stainless steel.*

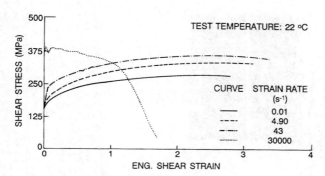

FIG. 9 *Shear stress-strain curves at different strain rates of ARMCO Pure Iron.*

show a stable flow characterized by strain hardening up to fracture which happens at very large values of strain. This fact demonstrates that no geometrical and material instabilities perturbate the flow up to the fracture initiation. Also the microscopic observations of the fracture surfaces (Figure 10) show that at low and medium strain rate the whole material of the gauge part participated to the deformation without particular strain concentrations. At the very high strain rate of 3×10^4 s^{-1} the shear stress-strain curves of the two steels

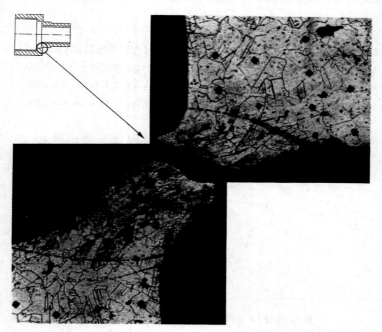

FIG. 10 *Deformation contour and fracture aspect of the gauge part at low and medium strain rate.*

show the following common phenomena, with some differences:

- common is the strong reduction of fracture strain with respect to low and medium strain rate, when these are calculated by taking the nominal initial gauge length of 0.25 mm;
- Different is the flow characterized by strain hardening for AISI 316 H and by initial instabilities and strain softening for ARMCO iron, up to the fracture initiation;
- common seems also the mode of fracture which should take place during the falling branch of the flow curve.

The observation of the fracture surfaces of Fig. 11 in comparison with Fig. 10 shows that at very high strain rate the reason of the strong reduction of fracture strain is probably due to the concentration of the large strain values in a band narrower than the initial nominal length of the gauge part. This last point is particularly important and needs to be confirmed by microscopical observations with higher resolution and by numerical calculations in dynamic range. In fact considering that at the very high strain rate of 3 x 10^4 s^{-1} the ultimate strength of the two materials did not increase with respect to low and medium strain rate, we can make the hypothesis that probably the temperature increase of

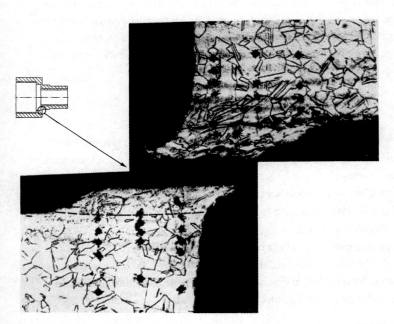

FIG. 11 *Deformation contour and fracture aspect of the gauge part at strain rate 32000 s-1 AISI 316 H.*

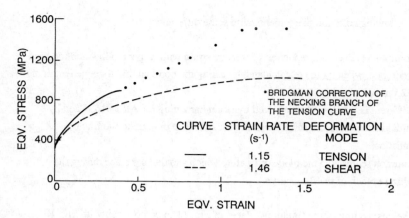

FIG. 12 *Comparison of the equivalent flow curves determined in tension and shear, at medium strain rate, for AISI 316 H.*

the gauge part, due to deformation, causes the concentration of strain in a band narrower than the gauge length.

A comparison of the equivalent flow curves (following Von Mises model) determined in tension and shear at medium strain rate is shown in Fig. 12 for AISI 316, where the experimental tension flow curve after necking has been modified following the Bridgman correction [5]; the flow curves practically coincide at low strain while they diverge at large strains confirming the findings of other authors [1] and the important influence of the deformation mode on the flow curve.

VI. CONCLUSIONS

It has been demonstrated by shear testing, microscopical examinations and numerical calculations that the new double shear specimen named "bicchierino," permits to deform the metal under investigation up to large strains (4) and at very high strain rate (3×10^4 s⁻¹) in a well controlled state of shear stress. The flow curves measured for AISI 316 H and ARMCO iron in this large strain rate range are regular and showed strong reduction of ductility as main strain rate effect.

Nevertheless, microscopical examinations showed that the very high strain rate concentration the deformation in a band narrower than the initial nominal gauge length. This phenomenon demands further examinations to gain more resolution into the real effects of very high strain rate on the mechanical properties.

REFERENCES

1. S. S. Hecker, M. G. Stout, D. T. Eash, "Experiments on Plastic Deformation at Finite Strains, Proceedings of Research Workshop on Plasticity of Metals at Finite Strains," *Stanford University* (1981).

2. C. Albertini, M. Montagnani, M. Zyczkowski, and S. Laczek, "Optimal Design of a Bicchierino Specimen for Double Shear," *Int. J. Mech. Sci.* (in press).

3. J. Lemaitre and J. L. Chaboche, *Mécanique des Materiaux Solides*, Dunod ed., Paris, (1985), pp. 72-73.

4. C. Albertini and M. Montagnani, "Dynamic Material Properties of Several Steels for Fast Breeder Reactor Safety Analysis," (1977) *EUR 5787 EN*.

5. P. W. Bridgman, "Studies in Large Plastic Flow and Fracture," *Harvard University Press*, (1964).

63

Localized Melting During the Separation of High Strength Tensile Samples

D. D. MAKEL and H. G. F. WILSDORF

Department of Materials Science
University of Virginia
Charlottesville, VA, USA, 22903-2442

Examinations of fracture surfaces formed during tensile tests of certain high strength alloys have shown localized regions in which the dimple walls and surrounding surface debris are spheroidized, apparently from melting during separation. Experiments on Ti-8Mn, Ti-6Al-4V, Ti-10V-2Fe-3Al and AISI 4340 ultrahigh strength steel have shown that melting is a result of localized deformation during the rapid separation of shear surfaces, typically of shear lips. The presence of localized melting has been found to rely on a high rate of separation, and it can consequently be suppressed by heat treating to microstructures which decrease the crack propagation rate. Similarly, localized melting in AISI 4340 samples can be suppressed by testing under conditions in which the samples separate in a purely ductile regime, resulting in fracture surfaces which have completely fibrous central crack areas devoid of radial cracking. The model proposed to explain the unusual localized melting and the associated fracture surface features involves temperature increase caused by localized shear followed by a final temperature increase to the melting point during the coalescence of voids through the rupture of the inter-void ligaments.

I. INTRODUCTION

During routine examinations of fracture surfaces of Ti-8Mn quasi-static tensile samples local regions were found which displayed spheroidized dimple walls and surface debris [1-5], an example of which is shown in Figure 1. Since these initial

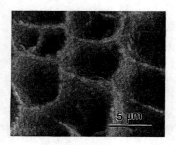

FIG. 1 *Example of spheroidized dimple walls and surface debris on the surface of a Ti-8Mn tensile sample (micrograph courtesy of Steven Y. Yu).*

observations experiments have been performed on Ti-8Mn, Ti-10V-2Fe-3Al [1-4], Ti6Al-4V [3-6] and AISI 4340 steel [7] samples in an attempt to elucidate the mechanisms responsible for the formation of these unusual fracture surface features.

A. FRACTURE SURFACE FEATURES ASSOCIATED WITH SPHEROIDIZED SURFACE STRUCTURES

Investigations of hundreds of fracture surfaces since the initial observations have shown that the unusual spheroidized features are found only on surfaces which lie at high shear orientations to the tensile axis, typically in shear lips. Three general types of fracture surface feature have been identified in conjunction with the spheroidized structures and these have been named Open Surface Shear Zones, Transition Dimples, and Localized Shear Band Dimples [3-6].

Open Surface Shear Zones, an example of which is shown at the right hand side of Figure 2, are dimple free bands found typically at the borders of sample-scale shear surfaces which display striations in the overall direction of shear. An important feature of the Open Surface Shear Zones is that they are <u>never</u> found on matching regions of both fracture surfaces, as opposed to most other fracture surfaces features which always have matching counterparts.

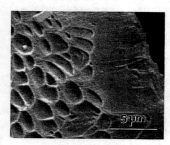

FIG. 2 *Fracture surface region of Ti-8Mn sample containing a dimple-free Open Surface Shear Zone at the right and flat-topped Transition Dimples showing fine surface texturing ("microroughening").*

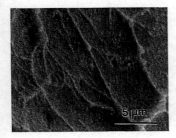

FIG. 3 *Examples of shallow Localized Shear Band Dimples on the fracture surface of a Ti-10V-2Fe-3Al sample showing "microroughening" and short, spheroidized dimple walls (micrograph courtesy of Steven Y. Yu).*

Transition Dimples, shown to the left of the Open Surface Shear Zone in Figure 2, are different from typical ductile dimples in that the individual dimples are not divided by thin ruptured ligaments, rather, they are separated by relatively flat areas which approximately define a single plane.

Localized Shear Band Dimples are shallow dimples divided by short dimple walls which typically are spheroidized rather than sharp at their upper edges. Figure 3 displays a field of Localized Shear Band Dimples and it should be noted that the dimples are generally equiaxed, showing little sign of shear during void growth. Spheroidized debris not connected to dimple walls can typically be found scattered throughout areas of Localized Shear Band Dimples and the occasional groups of large spheroidized dimples such as those shown in Figure 1 are typically found in fields of or directly adjacent to Localized Shear Band Dimples.

Figures 2 and 3 also show a fine texturing of the flat tops of the Transition Dimples and of the areas of Localized Shear Band Dimples. This "microroughening" is common to areas of localized melting in the three titanium alloys investigated [3-5]. On the surfaces of the AISI 4340 samples the areas around the spheroidized features are covered with fine (approximately 0.1 μm or less diameter) bumps [7].

B. PLASTIC DEFORMATION BELOW UNUSUAL SURFACE FEATURES

Selected fracture surfaces were plated with nickel to preserve the fine surface features and sectioned parallel to the tensile axis to expose the underlying microstructures [1-6]. The sections show highly concentrated shear deformation below the unusual features of interest, an example of which is shown in Figure 4. These bands of concentrated shear bear strong resemblance to deformational adiabatic shear bands, although adiabatic shear is not a mechanism typically associated with tensile deformation.

FIG. 4 *Section of a Ti-8Mn tensile sample showing concentrated shear directly below spheroidized fracture surface features.*

C. INVESTIGATING THE CONSEQUENCES OF LOCAL MELTING

Localized melting was the first mechanism proposed for the formation of the spheroidized structures but it was suspected that oxidation might be also partly responsible. The role of oxidation was discounted, however, when it was found that the spheroidized features persist when the samples are fractured in an argon atmosphere [3-5].

Shortly thereafter it was found that initially smooth surfaces could be artificially "microroughened" by flash heating with a hydrogen-oxygen torch [3-5], suggesting that the "microroughening" is a result of high surface temperatures caused by separation through the concentrated, possibly adiabatic shear bands.

It was reasoned that if melting is responsible for spheroidization the necessary temperatures should be high enough to cause visible light emission. In fact, photographs of tensile tests taken using high speed film and an open shutter in a darkened room clearly show light emission at separation and comparisons of these photographs with SEM micrographs of the fracture surfaces show direct correlation between the origins of light emission and the local surface areas containing spheroidized material [3-5].

D. SUPPRESSION OF MELTING

The effect of microstructure on localized melting was investigated by varying the heat treatments of Ti-6Al-4V samples [6]. While melting was present in mill-annealed and equiaxed α samples, no melting was found on the surfaces of samples treated to contain α colonies. Likely causes for the absence of local melting in the colony samples were i.) a lack of smooth, sample-scale shear surfaces and ii.) a decrease in the crack propagation rate caused by in increase in crack path due to crack tip deflection at colony boundaries.

After localized melting was found on the fracture surfaces of quasi-static AISI 4340 steel samples high rate tests were performed at the Fraunhofer-Institut in Bremen, Germany to investigate any possible effects of higher applied strain rates [7]. These tests yielded fracture surfaces which not only had a total lack of spheroidized structures but also had no radial cracking, a feature indicative of high crack propagation rates [8,9] which was clearly present on the quasi-static sample surfaces. Analysis of the strain rates and deformational energy input into the gage and neck regions show firstly, using a general heat flow analysis [10], that the conditions are nearly quasi-static and adiabatic, and secondly that this energy causes the incipient fracture region at the center of the neck to be heated above the ductile/brittle transformation range, resulting in a decrease in the crack propagation rate. This indicates that localized melting is dependent on a high crack propagation rate but not necessarily on a high applied strain rate.

II. DISCUSSION

A model describing the formation of the unusual fracture surface features must include the observed localization of plastic deformation into a thin, concentrated shear plane. Since void growth to coalescence is typically a regular incremental process and localized shear is frequently found to be rapid and unstable, especially in the case of adiabatic shear, the model which will be presented involves a rapid shear event occurring before and during void initiation and growth to coalescence.

A. CONCENTRATED SHEAR DURING VOID INITIATION AND GROWTH

The situation of a band of concentrated shear cutting through a volume of growing voids is depicted in Figure 5. Voids in this figure are shown at a 45 degree angle to indicate that the deformation occurs inside of a volume including an incipient shear surface. Because the voids are linked before inter-void ligament rupture, the void field consist of shallow dish-like dimples connected by a common flat plane. These dimples have been named Transition Dimples because they form by the transition from regular void growth to shear linkage.

The right hand side of the figure also shows the shearing of a pre-existing open surface into the plane of separation, thereby forming an Open Surface Shear Zone. Since the surface making up the Open Surface Shear Zone does not form by separation this explains both the lack of dimples and the lack of matching features on the matching surfaces. Surface areas which have formed by separation through concentrated shear bands display the characteristic "microroughening", and this feature has also been depicted in Figures 5 and 6.

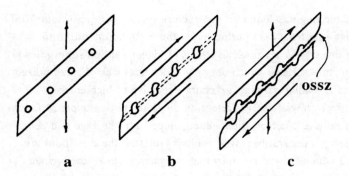

a   **b**   **c**

FIG. 5 *Separation mechanism model proposed for the formation of Open Surface Shear Zones, Transition Dimples and "Microroughened" surfaces. Growing voids (a) are linked by concentrated shear event (b) leaving flat-topped Transition dimples (c). Surfaces which result from separation through concentrated shear display "microroughening" and pre-existing open surfaces sheared into the concentrated shear plane, such as the one depicted at the right of the figure, result in Open Surface Shear Zones.*

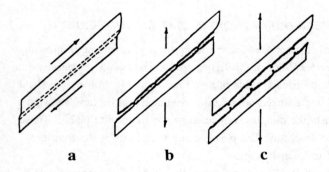

a   **b**   **c**

FIG. 6 *Separation mechanism model proposed for the formation of Localized Shear Band Dimples. After an initial localized shear event (a) void growth is limited to a narrow, concentrated shear band (b) resulting in dimples which are shallow and have spheroidized dimple walls and a generally "microroughened" surface (c). The dimples are not shear*

B. CONCENTRATED SHEAR PRIOR TO VOID NUCLEATION AND GROWTH

In the second separation model the concentrated shear event occurs prior to void initiation and the subsequent growth of the voids is restricted to the localized shear plane. This results in dimples which are very shallow, uniformly "microroughened" and have spheroidized dimple walls and surface debris across their surface. As indicated by the arrows in Figure 6 the primary separation mode is Mode I, since the dimples are essentially equiaxed, which indicates that the major shear event which forms the localized, softened shear band is followed by a return to tensile deformation during final separation. Localized Shear Band Dimples form through material softened by localized shear, not during localized shear.

C. PROPOSED ROLE OF ADIABATIC SHEAR

As discussed in the previous section, a rapid, highly concentrated and locally softening shear event occurring either just prior to or during final separation would explain the Transition Dimples and the Open Surface Shear Zones. Indications of high temperatures such as spheroidized material, visible light emission and "microroughening" strongly suggest that these concentrated shear bands are adiabatic shear bands. This is not to say that all of the shear localization is a consequence of adiabatic shear. In light of recent investigations into isothermal, quasi-static shear localization [11,12] the initial stages of shear localization are quite likely isothermal, with short periods of adiabatic shear occurring only in limited locations.

An important consideration in the analysis of the subsurface shear bands in the titanium alloys is that they are deformational in nature and do not show clear indications of phase transformation. This seems to contradict the idea that the spheroidized material found on the fracture surfaces is a product of localized melting, since the melting point of the alloys is above the alpha-beta phase transformation temperature. Lack of transformation zones, however, actually emphasizes the fact that the final input of strain energy comes from the rupture of the inter-void ligaments. In this regard the adiabatic shear bands may serve multiple purposes, to localized deformation into a narrow plane, to pre-heat the volume of incipient separation, and to form high temperature layers on either side of the deforming ligaments, separating them from the cooler bulk material, reducing the rate of heat dissipation from the ligaments and creating a more nearly adiabatic condition during ligament separation.

CONCLUSIONS

1. Formation of the surface features which have been observed in conjunction with local areas of melting on the surfaces of tensile samples, Open Surface Shear Zones, Localized Shear Band Dimples and Transition Dimples, can be described by a combination of void nucleation and growth with rapid localized shear.
2. Fine surface texturing, "microroughening", results from separation through concentrated shear bands.
3. Where localized melting occurs, the final temperature increase to the melting point comes from the energy involved in the rupture of inter-void ligaments.

ACKNOWLEDGEMENT

This research was supported by the U. S. Office of Naval Research (Grant N00014-88-K-0111) and we would like to thank Dr. George Yoder, program director, for his continuous interest and support. Thanks to Professor Dr. Hans-D. Kunze, Dr. L. W. Meyer, and B. O. Reinders at the Fraunhofer-Institut fur angewandte Materialforshung for conducting the high rate tests on the 4340 steel samples. The authors also acknowledge the award of a travel grant from the NATO Scientific Affairs Division, Brussels, Belgium.

REFERENCES

1.   J. D. Bryant, D. D. Makel, and H.G.F. Wilsdorf, "Metallurgical Applications of Shock-Wave and High-Strain-Rate Phenomena," ed. Murr, Staudhammer and M. A. Meyers, Marcel Dekker, p.723,(1986)

2.   J. D. Bryant, D. D. Makel and J. G. F. Wilsdorf, *Mater. Sci. and Engr.*, 77: p. 85, (1986).

3.   D. D. Makel, "Strain Rate Dependent Process in the Fracture of Ti 8w% Mn," *Ph.D. Dissertation*, University of Virginia (1987).

4.   D. D. Makel and H.G.F. Wilsdorf, "Impact Loading and Dynamic Behaviour of Materials," eds. Chiem, Kunze, and L. W. Meyers, Verlag, 2: p.587, (1988).

5.   D. D. Makel and J.G. F. Wilsdorf, *Scripta Mettal, 21*: p. 1229, (1987).

6.   D. D. Makel and D. Eylon, *Metallurgical Transactions*, in press.

7.   D. D. Makel and J.G. F. Wilsdorf, unpublished research.

8.   F. R. Larson and F. L. Carr, *Metal Progress*, p. 26, (1964).

9.   F. R. Larson and F. L. Carr, *Metal Progress*, p. 75, (1964).

10. G. B. Olson and J. F. Mescall, "Shock Waves and High-Strain-Rate-Phenomena in Metals," eds. M. A. Meyers and L. E. Murr, Plenum Press, p. 221, (1981).

11. J. E. Bird and J. M. Carlson, *Mettal. Trans. A*, 18A: p. 563, (1987).

12. J. M Carlson and J. E. Bird, *Acta. Mettal.*, 35: 7, p. 1675, (1987).

Section VI
Dynamic Fracture

64

Fragmentation Processes for High-Velocity Impacts

S. A. FINNEGAN and J. C. SCHULZ

Research Department
Naval Weapons Center
China Lake, California 93555, U. S. A.

Results from an experimental program to measure and quantify the various projectile and target fragment systems created by normal impact of mild and hard steel spheres against aluminum plate targets at ordnance and ultra-ordnance velocities are described. Results for these systems are compared with previously presented results for steel targets.

I.   INTRODUCTION

For impacts at ordnance to ultra-ordnance speeds (below 3 km/s), and especially against thicker targets, debris and fragmentation characterization becomes complicated. Several distinct fragment systems are often produced, in each of which the fragments have significant size and are relatively few in number [1]. This is in contrast to, for example, hypervelocity impacts which produce roughly spherical clouds of debris with all material concentrated near the surface of the cloud [2].

In a previous paper [3], a portion of the results of an experimental program to measure and quantify the various projectile and target fragment systems created by normal impacts of spheres and cubes against plate targets at ordnance and ultra-ordnance speeds were presented. Spatial mass and velocity distributions for these systems were also given. Emphasis was placed on impacts against thick steel armor targets.

In the present paper, further results are presented from the experimental program described above. Emphasis is placed on impacts against thick aluminum alloy targets by either hard or mild steel spheres. Mild steel results are presented in detail; hard steel results are summarized due to space limitations. Results for these systems are compared with previously-presented results. Reasons for observed differences are discussed.

II. EXPERIMENTAL FIRINGS

Firings of 6.35 mm steel spheres against 6.35 mm, 2024-T4 aluminum plates at normal incidence were conducted. Both hardened (BHN 700) and mild steel (BHN 160) spheres were used as impactors. These were launched using a 50 caliber smooth bore powder gun [4] at velocities between 0.6 and 2.4 km/s. Velocities were measured at the muzzle end of the gun using a pair of photo diodes and a time-interval counter.

A diversity of instrumentation and recovery techniques were employed. A high speed, 6-frame Kerr cell camera was used to observe the impacts and to determine velocity distributions for the fragment debris. A ballistic pendulum was used to measure the momentum transferred to the target and to various fragment systems. Celotex packs positioned downrange from the targets were used to determine spatial distributions of fragment mass and to collect fragments for identification. Wax block target surrounds were also used to collect fragments in order to determine the mass and size distributions in each fragment group and also to make mass balances. A number of post-impact measurements of target damage were also made, including weight loss, perforation hole dimensions and two measures of plastic deformation.

III. FRAGMENT SYSTEMS

A number of fragment systems are produced in a sphere-plate impact process as seen in Fig. 1. A detailed description of the various systems is contained in [1]. Entry side systems include jet and crater-lip fragments from the target. Exit side systems include most of the projectile fragments, fragments from the central region of the target (a single plug fragment at the ballistic limit) and a group of large peripheral fragments (ring or sleeve) which are ejected from the edge of the crater at high impact speeds and whose momentum does not come from direct transfer from the projectile. Of these systems, only the two major exit-side ones (central target region and residual projectile systems) will be discussed in detail in the present paper because most of the residual momentum is contained within these systems.

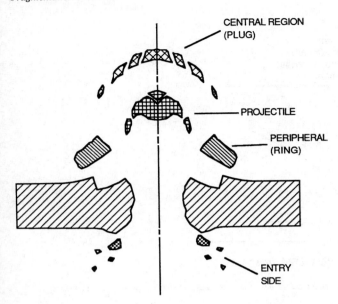

FIG. 1 *Fragment systems for sphere-plate impacts.*

IV. DATA ANALYSIS

For impacts involving steel spheres and aluminum targets, the residual projectile fragment(s) can usually be identified in the high-speed photographs throughout the ordnance and ultra-ordnance velocity regime because of their positions at or near the front of the debris cloud (typical of impacts involving overmatched targets). As a result, the residual momentum and mean velocity of the projectile are easily estimated. This allows both the momentum and mean velocity of the central region system to be calculated without knowledge of the spatial distributions. This turned out to be fortuitous, as an analysis of central region mean velocities showed the presence of large numbers of high-angle fragments that would not have been present in the spatial mass distribution; high-angle fragments are difficult to capture because of their tendency to ricochet.

V. MILD STEEL RESULTS

Velocity trends for the projectile, central region and peripheral (ring) fragment groups are shown in Fig. 2. The abrupt change in projectile system velocity, between 1.1 and 1.3 km/s impact velocity, is caused by the onset of general yielding throughout the mild steel sphere during impact. Below 1 km/s only the frontal portion of the projectile deforms. The change

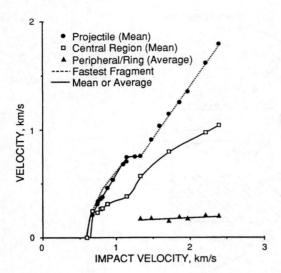

FIG. 2 *Mean system velocities for mild steel sphere/aluminum plate impacts.*

in the trend for the central region system above 1.2 km/s reflects the onset of peripheral (ring) fragmentation, which results in an abrupt drop in the mass of the central region along with an increase in fragmentation. This latter phenomenon was also observed in the previous study [3].

Normalized mean velocity trends for the projectile and central region are shown in Fig. 3. Data were normalized with respect to the fastest fragment in the debris cloud. The normalized mean velocity for the projectile is zero at the ballistic limit (0.6 km/s) and increases asymptotically to 1.0 at about 1.4 km/s. The central region normalized mean velocity is 1.0 at the ballistic limit, decreases sharply to about 0.5 at 1.1 km/s, increases sharply to 0.76 at 1.3 km/s, and then declines gradually to a value of 0.58 at about 2.4 km/s. The two lower values reflect the presence of high-angle fragments, as discussed below.

Fig. 4 shows four high-speed photographs of exit-side fragments at impact velocities where normalized central region mean velocities were extremely large or small. At 0.7 km/s, the central region is just beginning to break up and the normalized mean velocity is very high (0.98). At 0.8 km/s, the mean velocity is sharply lower (0.6) and the central region fragment system has become bimodal. Approximately half of the total mass, consisting of one fragment, travels directly ahead of the projectile while the other half, consisting of approximately 11 fragments, travels at angles between 30 and 60°. At 1.3 km/s, the mean velocity is again sharply higher (0.76) while the central region fragments begin to spread out

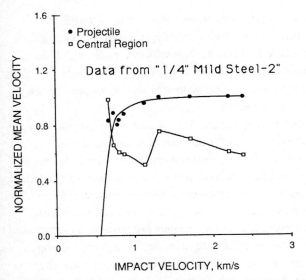

FIG. 3 *Normalized mean system velocities for mild steel sphere/aluminum plate impacts.*

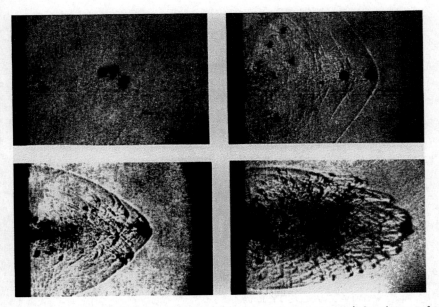

FIG. 4 *Kerr cell photographs of mild steel sphere/aluminum plate impacts. (Upper left, 0.67 km/s; upper right, 0.81 km/s; lower left, 1.33 km/s; lower right, 2.4 km/s.)*

more uniformly with angle. An additional system of penetration tunnel lip (peripheral) fragments is just beginning to emerge at this velocity; these larger, slower-moving fragments cluster about the perforation tunnel. At 2.4 km/s, the mean velocity is again much lower (0.58) and central region fragments are found at angles at high as 80°. High-angle fragmentation of this kind was not observed in other systems studied [3]; neither were the extreme gyrations in central region mean velocity.

Momentum trends for all of the entry and exit side fragment systems are shown in Fig. 5. Two interesting observations can be made from these data. One is the rough equivalency between entry side and exit side peripheral momentum values for similar impact velocities. Another is the momentum transfer between projectile and central region systems at velocities above 1.3 km/s. It is not surprising that the projectile loses momentum when general yielding occurs because of the increased contact area and, hence, increased shear surface for plugging. What is surprising is that the momentum loss shows up as an increase in the central region fragment system momentum rather than, perhaps, as increased target momentum. This phenomena was not observed for any of the impact systems studied earlier. In these systems, general yielding of the projectile occurred at velocities below the ballistic limit.

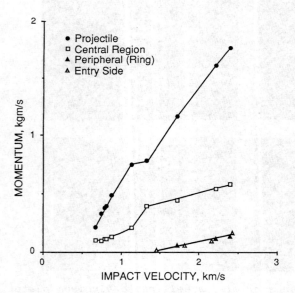

FIG. 5 *System momentum trends for mild steel sphere/aluminum plate impacts.*

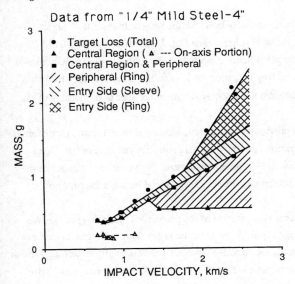

FIG. 6 *System mass trends for mild steel sphere/aluminum plate impacts.*

System mass trends are shown in Fig. 6. The increase in central region mass between 1.0 and 1.3 km/s is caused by general yielding of the projectile, while the abrupt decrease above 1.3 km/s is caused by the emergence of the peripheral fragment system as discussed earlier. The peripheral fragment system is considered a separate system because the momentum comes indirectly from the target plate and not directly from the projectile [1,3]. Above 1.7 km/s, ring fragments are found on the entry side in addition to the sleeve fragments found at lower velocities. A previous study [1] has concluded that fragment shapes for "peripheral" systems (which would include both entry and exit side lip fragment systems) are heavily influenced by material anisotropy.

VI. HARD STEEL RESULTS

Differences in fragmentation behavior for hard steel versus mild steel sphere impacts result from the fact that the hard steel spheres shatter, while the mild steel spheres deform plastically. Shatter occurs at a higher impact velocity than general yielding, and the degree of shatter increases gradually with impact velocity, while general yielding occurs abruptly. Thus, trend line changes associated with shatter/deformation of the projectile tend to shift upward in velocity and be somewhat less pronounced for hard steel impacts.

For example, the change in the velocity trend for the projectile system at about 1.1-1.3 km/s that was observed for mild steel spheres (see Fig. 2) is shifted to about 1.5-1.7 km/s for hard steel spheres. On the other hand, the break in the central region velocity trend (at 1.1 km/s) and onset of peripheral fragmentation (at 1.3 km/s) remain virtually the same. This indicates that these latter events are not associated with either shatter or yielding of the projectile.

For mild steel impacts, as previously mentioned, additional momentum is transferred from the projectile to the central region system when the projectile yields. For hard steel impacts a similar transfer occurs when the projectile shatters. However, the changes in the momentum trend lines for hard steel impacts are less pronounced because the onset of shattering is more gradual.

For mild steel impacts, plastic deformation of the projectile (squashing) causes the effective diameter of material plugged from the target to be larger than the initial diameter of the projectile. For hard steel impacts, the effective diameter increases with shattering, though not so abruptly. Thus, less mass is removed from the target at a given impact velocity for impacts involving hard steel spheres. In spite of this difference in mass, a rough count of exit side fragments (including the projectile) showed similar numbers up to about 1.6 km/s [6]. Above that velocity, shatter of the hard steel spheres resulted in an increase in the number of fragments for projectile system. Whether shatter causes an increase in the number of target fragments was not determined.

REFERENCES

1. M. E. Backman and S. A. Finnegan, *Proc. 8th Int. Symposium on Ballistics, ADPA.* (1984).

2. H. F. Swift, D. D. Preonas and W. C. Turpin, *Rev. Sci. Instr. 41:* 746 (1970).

3. S. A. Finnegan, J. C. Schulz, and O. E. R. Heimdahl, *Proc. 1989 Hypervelocity Impact Symposium, J. Impact Eng:* in press (1990).

4. W. Goldsmith and S. A. Finnegan, *Int. J. Mech. Sci. 13:* 843 (1971).

5. T. W. Ipson and R. F. Recht, *Experimental Mechanics 15:* 249 (1975).

6. M. D. Alexander, S. A. Finnegan, A. J. Lindfors, J. D. Yatteau, M. N. Plooster, and G. F. Kinney, *Proc. 11th Int. Symposium on Ballistics, ADPA.* (1989).

65

Natural Fragmentation of Exploding Cylinders

D. E. GRADY and M. M. HIGHTOWER
Sandia National Laboratories
Albuquerque, New Mexico 87185, U. S. A.

The natural fragmentation of a 4140 steel cylinder fully loaded with RX-35-AN insensitive high explosive is investigated through experiment and analysis. Methods of Taylor and Gurney are used to determine the fracture strain and kinematic state of the expending cylinder. Energy methods based on mechanisms of both tension fracture and adiabatic shear fracture are used to calculate the circumferential fragmentation intensity.

I. INTRODUCTION

The fragmentation of an explosive-filled smooth-wall metal cylindrical case will occur naturally without the aid of fragmentation enhancement techniques. In natural fragmentation the size, velocity and statistical distribution of fragments is determined through a complex interplay among explosive characteristics, geometry of the explosive-case system, and the mechanical properties of the case metal. The objective of the present study is the analysis of natural fragmentation in explosively expanding steel cylinders. Extensive earlier work on this topic has been pursued by Gurney [1], Taylor [2], Mott [3], and numerous others.

The present analysis also applies more recent theoretical developments on dynamic fragmentation to the natural fragmentation of exploding cylinders. This later work is based on an analysis of energy and momentum balance in the dynamic fragmentation process [4,5] and is an extension of an earlier analysis of Mott [3]. The energy balance established within this theory leads to the prediction of the characteristic spacing of fractures. The driving fragmentation energy is a kinetic energy associated with the rate of divergence of neighboring material points in the expanding cylinder. Fracture resistance is determined by energy dissipated in fragmentation and, depending on material and loading conditions, may be associated with tensile fracture or adiabatic shear band enhanced fracture.

In the present report, an analysis for predicting the natural fragmentation of explosively expanding cylinders is developed. In Section II, an explosive fragmentation experiment performed in support of this analysis is described which provides the basis for comparison of theory and experiment. Explosive acceleration of the shell is induced by the detonation and it is necessary to establish the expansion velocity and kinetic energy at the moment of fragmentation. Methods of Taylor [2] and Gurney [1] are used to determine the fracture strain and expansion velocity for the present fragmentation analysis in Section III.

The mechanism of fracture must then be determined and the magnitude of fracture energy dissipation calculated. These issues are addressed in Section V following a derivation of the theory for predicting the characteristic spacing of fractures in Section IV. Unique in this analysis is the development of the characteristics of the process zone in an adiabatic shear band needed to calculate a fracture energy associated with the shear fracture mechanism. Lastly, the analysis of natural fragmentation in explosively expanding cylinders is compared with experimental results in Section VI.

II. NATURAL FRAGMENTATION EXPERIMENT

In the present study, an explosive fragmentation experiment was performed on a 15.2 cm diameter smooth wall metal cylinder. The cylinder was 38.1 cm in length with a wall thickness of 5.7 mm. The cylinder was machined of 4140 steel and heat treated to a Rockwell hardness of 40 ($Y=1.1$ GPa). The cylinder was filled with RX-35-AN explosive and the cylinder ends were confined. The explosive was center detonated at one end.

The insensitive high explosive RX-35-AN [6] used in the present study has been calibrated through instrumented copper cylinder expansion experiments [7] to provide expansion velocity data for purposes of establishing appropriate nonideal explosive equation-of-state parameters. In this study, the measured expansion velocity data are scaled with appropriate Gurney relations to determine expansion velocity behavior for the steel cylinder experiment.

High speed front-lit photography using a CORDIN framing camera with 5 μs frame intervals was used to observe acceleration and breakup of the expanding cylinder. The opening of fractures and emergence of explosive gases were consistent with the 1.2 - 1.25 fracture strain predicted from the Taylor relation described in Section III. Also, an expansion velocity of 1760 - 1830 m/s determined from the photographs compare well with the limiting Gurney velocity of about 1800 m/s calculated for this cylinder.

Multiple flash radiography was used to determine fragment velocity, trajectory and pattern for a 40° sector of the cylinder. Fragments from this sector were captured in fiberboard bundles which were placed approximately 6 meters from the event. From the bundles, 161 fragments were recovered which represents 90% of the weight of the 40° cylinder sector.

From the recovered fragments, it was observed that fracture was predominantly along elongated strips with the fracture parallel to the axis of the cylinder. A number of the fragments were 4 to 5 times longer than they were wide. Both tensile and shear fracture were observed from examination of fracture surfaces. Shear fracture appeared to be the dominant breakup mechanism. Fragment size statistics were determined from the recovered fragments for comparison with the present fragmentation analysis.

III. GURNEY AND TAYLOR ANALYSES OF EXPLOSIVELY ACCELERATED CYLINDRICAL SHELLS

Within the present fragmentation theory, it is necessary to establish the radial expansion velocity of the exploding cylinder at the moment of fracture. An early successful theory on the fracture strain of explosively expanding cylinders is due to Taylor [2]. Later improvements on Taylor's theory have been offered, however their calculated fracture strains do not differ significantly from those of Taylor. Taylor's analysis led to a relation for the circumferential stress in the shell subjected to an internal pressure P given by $\sigma(y) = Y - P(1 - \frac{y}{h})$, where Y is the yield stress in simple tension, h is the shell thickness and y is a coordinate through the thickness, $0 \leq y \leq h$. Thus $\sigma = Y - P$ (compression) at the inner surface and $\sigma = Y$ (tension) at the outer surface. The crossover point occurs at an interior point of the shell. Taylor assumed that failure occurs when the internal pressure within the expanding cylinder decreases to a value such that tension is just achieved at the inner surface — that is when $\sigma = Y - P = 0$.

It is common to assume ideal gas behavior for the explosive products and develop an expression for pressure versus expansion radius to calculate fracture strain with the Taylor method. The RX-35-AN explosive used in the present study is not suited to an ideal gas description of the explosive products, however. Instead, the velocity history data described in Section II [7] was used, through appropriate Gurney expressions, to calculate pressure versus radius behavior [8].

Through this method, an internal pressure of $P = Y = 1.1$ GPa, corresponding to the yield stress of 4140 steel, is calculated at an expansion radius of $R_{crit}/R_o = 1.24$. This fracture strain calculated through the Taylor criterion is compared with through-the-thickness

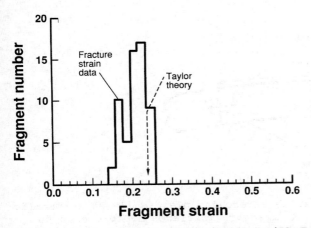

FIG. 1 *Fracture strain date compared with Taylor theory prediction for natural fragmentation experiment on 4140 steel cylinder.*

measurements on a number of fragments recovered from the natural fragmentation experiment on the 4140 steel cylinder in Fig. 1. The agreement is quite satisfactory.

The analyses of Gurney [1] were then applied to determine the radial expansion velocity at the predicted fracture strain [8]. For the present experiment an expansion velocity of 1530 m/s was calculated. This value will be used in the subsequent fragmentation analysis to establish the kinetic energy driving fragmentation.

IV. FRAGMENT SIZE ANALYSIS

The method for predicting the circumferential fracture spacing of an explosively expanding cylindrical shell follows from a study of Kipp and Grady [4] and builds on earlier work of Mott [3]. The method is based on a calculation of momentum diffusion and energy balance in the fragmentation of a rapidly stretching plastic body. Although the study focused on the fragmentation of stretching rods, the method is expected to apply equally well to circumferentially stretching shells.

In the analysis, a body is assumed to be stretching axially (circumferentially in the present application) at a strain rate $\dot{\epsilon}$ and at a flow stress Y. The configuration is illustrated in Fig. 2. A fracture initiates at a point $\xi = 0$ and dissipates an energy Γ as a crack opening coordinate ψ proceeds from $\psi = 0$ to $\psi = \psi_c$. ξ is a coordinate along the stretching axis of the body marking the front of an unloading stress wave propagating away from the opening fracture. Balance of momentum and compatibility leads to this system of ordinary differential equations,

$$\xi\dot{\xi} = \frac{Y^2}{2\rho\dot{\epsilon}\Gamma}\psi \ , \tag{1}$$

$$\dot{\psi} = \dot{\epsilon}\xi \ , \tag{2}$$

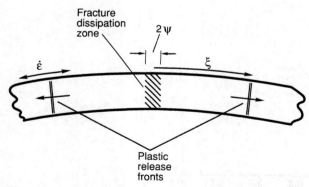

FIG. 2 *A cylindrical surface stretching plastically at a specific strain rate is shown. As fracture initiates and proceeds to completion, tensile stress release propagates away from the point of fracture as moving interfaces separating rigid and plastic material.*

which determines the rate of crack opening and propagation of the unloading wave. Solving Eqs. (1) and (2) provides the time dependent solution for the rigid-plastic wave boundary,

$$\xi = \frac{1}{12}\frac{Y^2}{\rho\Gamma}t^2 \ ,$$

(3)

and the crack opening displacement

$$\psi = \frac{1}{36}\frac{\dot{\epsilon}Y^2}{\rho\Gamma}t^3 \ .$$

(4)

A work-energy balance at the critical crack opening displacement $\psi_c = 2\Gamma/Y$ leads to a fracture time,

$$t_c = \left(\frac{72\rho\Gamma^2}{Y^3\dot{\epsilon}}\right)^{\frac{1}{3}} \ .$$

(5)

Using the fracture time from Eq. (5) in Eq. (3) provides the extent of the plastic unloading at fracture completion. Twice this distance is assumed to determine the nominal fracture spacing,

$$S = \left(\frac{24\Gamma}{\rho\dot{\epsilon}^2}\right)^{\frac{1}{3}} \ .$$

(6)

Eq. (6), with a stretching rate $\dot{\epsilon}$ corresponding to the appropriate fracture strain, is used to calculate the average circumferential fracture spacing of the fragmenting cylinder. Solving for Γ in Eq. (6) illuminates the energy balance nature of the fragmentation process in that the derived expression is a measure of the kinetic energy of divergence of two points separated by a distance S.

V. THE FRAGMENTATION ENERGY

Calculation of the nominal circumferential fracture spacing from Eq. (6) requires knowledge of the fragmentation energy Γ. The fragmentation energy is a material and mechanism dependent property which is determined through experimental measurements and models of the fracture dissipation process. There are two predominant modes of fracture in the breakup of an expanding metal shell which are illustrated in Fig. 3. The first is tensile fracture where failure proceeds by the opening of mode I cracks. Fracture dissipation is governed by the material fracture toughness K_c, and an estimate of the fragmentation energy is provided by,

$$\Gamma = \frac{K_c^2}{2E} \ ,$$

(7)

where E is the elastic modulus of the material. Material properties for the 4140 steel tested in the present study are provided in Table 1, and provide a fragmentation energy of $\Gamma \approx 16,000$ J/m² for tensile fracture.

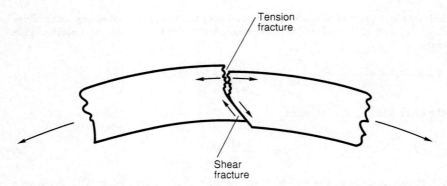

FIG. 3 *Models of tension and shear fracture assumed in calculating fragmentation energy.*

In explosively-expanding cylinders, shear fracture preceded by localized adiabatic shear banding on the planes of fracture is also an important mode of failure. In determining the fragmentation energy associated with shear fracture, we will assume that the energy is principally accounted for by dissipation in the adiabatic shear banding process. Grady and Kipp [5] have analyzed the energy dissipated in adiabatic shear banding and have arrived at the expression,

$$\Gamma = \frac{\rho c}{\alpha} \left(\frac{9\rho^3 c^2 \chi^3}{Y^3 \alpha^2 \dot{\gamma}} \right)^{\frac{1}{4}} . \tag{8}$$

In Eq. (8), $\dot{\gamma}$ is the shear strain rate and is approximately equal to the circumferential stretching rate in the present application. A value of $\dot{\gamma} \approx 16{,}000/s$ is calculated for the experiment described in Section II. The new material properties include the specific heat c, the thermal diffusion coefficient χ, and the thermal softening coefficient α. From properties provided in Table 1 a fragmentation energy of $\Gamma \approx 19{,}000 \text{ J/m}^2$ is obtained for shear fracture.

Table 1

Material Properties for 4140 Steel

ρ	(kg/m³)	7870
Hardness	(HRC)	~40
E	(GPa)	200
K_c	(MN/m²ᐟ³)	80
Y	(GPa)	1.1
χ	(m²/s)	1.5×10^{-5}
c	(J/kg K°)	450
α	(K°⁻¹)	7.5×10^{-4}
Γ Tensile	(J/m²)	16,000
Γ Shear	(J/m²)	19,000

Whether tensile or shear fracture dominates in the present fragmentation test on 4140 steel has not yet been determined. Preliminary metallography on explosively fractured specimens suggest that this steel has a strong tendency to shear band and fracture along shear banded zones. There is also a propensity for shear fracture on planes oriented at approximately 45° to the shell surface. The close numerical values for the tensile and shear fragmentation energies will lead to similar predictions of circumferential fracture spacing. We do not have a ready explanation for this closeness in magnitude.

VI. COMPARISON OF FRAGMENTATION ANALYSIS AND EXPERIMENT

The relations from the preceeding sections are necessary to the calculation of circumferential fracture spacing in an explosively expanding metal cylinder. We focus on the test performed on 4140 steel described in Section II. Eq. (6) is used to calculate the average circumferential fracture spacing, but written in the form,

$$N = 2\pi \left(\frac{\rho R_o V^2}{24\Gamma} \right)^{\frac{1}{3}} . \tag{9}$$

In Eq. (9), $\dot{\epsilon}$ has been replaced by V/R_o where R_o is the initial cylinder radius and V is the expansion velocity at the radius of fracture (fracture strain). The fracture spacing S has been replaced by $N = 2\pi R_o/S$, where N represents the total number of circumferential fractures occurring in the fragmentation process.

For the fragmentation energy, the value of $\Gamma = 19,000$ J/m^2 derived for shear band enhanced fracture is used. A value of $\Gamma = 16,000$ J/m^2 is used for tensile fracture. To calculate the expansion velocity V, the fracture strain is first calculated from the Taylor criterion. A calculated value of $R_{crit}/R_o = 1.24$ is found to be in good agreement with observed strains in recovered fragments. At this fracture strain, an expansion velocity of $V = 1530$ m/s is determined through Gurney analysis from calibration experiments on RX-35-AN explosive.

The present study is focused on predicting the circumferential fragmentation intensity. It does not consider axial breakup of the longitudinal strips. To make a statistical comparison of the fragment size data with the present analysis, the following reduction of the data was performed: Every fragment was weighed and the length of every fragment was measured. An effective rectangle was assumed for a fragment such that the mass is given by $m = \rho wtl$, where ρ is the density and w, t and l are the width, thickness and length, respectively. A circumferential fragment width was then determined from $w = m/\rho tl$, where m and l are the measured values and t is the initial wall thickness of the cylinder. The width w then provides an effective average measure of fracture spacing for that fragment in terms of the initial cylinder dimensions. A fragment of length l was then considered to be a fraction of a strip of length L, given by $n = l/L$, where L is the length of the cylinder. Through these methods, a number versus circumferential width distribution was determined for the fragment data.

This experimental circumferential fragment number distribution is plotted as a histogram in Fig. 4. The average circumferential fragment widths (cylinder circumference divided by fragment number) predicted through the present fragmentation analysis are compared with the data in this figure. The agreement between prediction and data is

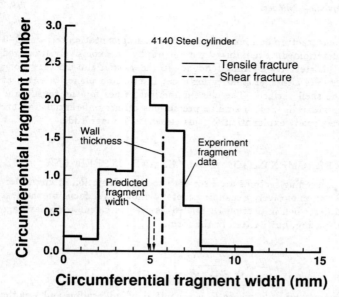

FIG. 4 *Number histogram for the 161 fragments from the natural fragmentation experiment plotted against the circumferential fragment width. The average fracture spacing predicted from the tensile fracture and shear fracture fragmentation models are compared with the data.*

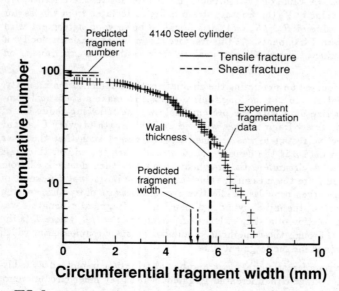

FIG. 5. *Cumulative fragment number plotted against width for the fragmentation experiment. Fragment number and spacing predicted from the analysis are compared with the data.*

remarkably good. As is readily seen, the difference in predicted fragment size between tensile and shear fracture is insufficient to select one mechanism over the other.

The same data are plotted as a cumulative number distribution in Fig. 5. The cumulative number data have been multiplied by 9 to scale from a 40° sector to an equivalent cylinder. Both total number of circumferential fragments and average fragment width predicted from the analysis are compared with the data in this plot. Figure 5 also illustrates the strong central tendency or small variance in the fragment width distribution. The 10 - 90% spread in fragment number lies between about 3 mm and 7 mm. A statistically random distribution in fragment widths as discussed by Grady and Kipp [9], for example, would yield a linear plot in Fig. 5 and would have a significantly larger variance. We speculate that the narrow measured variance in widths is a consequence of a fragmentation process in which the average fragment width is close to the cylinder wall thickness.

VII. CLOSURE

The present study has outlined an analytic method for predicting circumferential fragmentation (circumferential spacing of fractures) resulting from the explosive breakup of smooth wall cylinders. The analysis draws on earlier theoretical work where appropriate, and has also implemented more recent energy-momentum balance theories of dynamic fragmentation. This study has resulted in a first-principles predictive theory in that no experiment specific parameters were required to complete the analysis.

REFERENCES

1. R. W. Gurney, *Army Ballistic Research Laboratory Report BRL* 405 (1943).

2. G. I. Taylor, *Scientific Papers of G. I. Taylor*, Vol. III, No. 44, Cambridge Univ. Press (1963).

3. N. F. Mott, *Proc. Royal Soc.* A189: 300 (1947).

4. M. E. Kipp and D. E. Grady, *J. Mech, Phys.* Sol. 33: 399 (1985).

5. D.E. Grady and M. E. Kipp, *J. Mech. Phys. Sol.* 35: 95 (1987).

6. R. R. McGuire, *Personal Communication* (1989).

7. E. Lee, *Personal Communication* (1990).

8. D. E. Grady, *Sandia National Laboratories Report SAND90-0254* (1990).

9. D.E. Grady and M. E. Kipp, *J. Appl. Phys.* 58: 1210 (1985).

66

Rate-Dependent Modeling of Multidimensional Impact and Post-Spall Behavior

J. A. NEMES and J. EFTIS

Mechanics of Materials Branch
Naval Research Laboratory
Washington, D. C. 20375, U. S. A.

Department of Civil, Mechanical, and Environmental Engineering
George Washington University
Washington, D. C. 20052, U. S. A.

Plate impact spallation is considered using the Perzyna viscoplastic constitutive theory, formulated with a scalar variable description of damage, which for ductile metals is interpreted as the void volume fraction. A spallation criteria based on critical void volume fraction is utilized. The theory is applied to impact of circular plates, where the diameter to thickness ratio is not large, resulting in multidimensional axisymmetric strains. Post-spall behavior, including detached spall, is simulated using element deletion techniques in a Lagrangian finite-element code.

I. INTRODUCTION

Spallation is a particular type of dynamic fracture that occurs, for example, during high velocity impact when superposition of release waves produce tensile stresses of sufficient magnitude to cause material separation along a plane parallel to the wave fronts. Under severe enough conditions complete separation of the spalled section from the remaining target material may occur, which is sometimes referred to as detached spall or scabbing.

Under laboratory conditions impact experiments used to study spall fracture are typically designed so that the lateral dimensions of the plate are many times larger than the

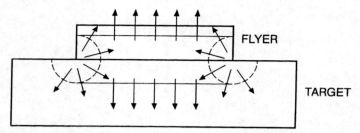

FIG. 1 *Two-dimensional plate impact configuration.*

thickness dimension, thus producing a one-dimensional state of strain that prevails during several wave transits across the plate thickness. However, in cases where large lateral to thickness ratios are not maintained, or impact of unequal diameter plates is involved, a much more complicated multidimensional strain field is produced. Also if consideration is given to studying the post-spall behavior of the material, including determination of detached spall, multidimensional strain must be considered.

A viscoplastic/damage constitutive model developed by Perzyna [1], which previously has been applied to spall fracture under uniaxial strain conditions [2-4], is applied here to impact of circular plates of unequal diameter, illustrated in Fig. 1. A plane compressive wave front propagating across each plate is shown along with the nonplanar wave arising from the circular edge of the flyer. The latter travelling toward the center reduces the region of the target experiencing high intensity tensile mean stress. Nevertheless, as long as the diameters of the plate are several times their thickness, a central region of high tensile stress is produced that can result in spallation. After development of the initial fracture, the momentum remaining in the spalled section causes the opening to enlarge, as shown in Fig. 2, and possibly result in detached spall. Numerical simulation of this phenomena is depicted using the constituive model described below in a Lagrangian finite-element computer code.

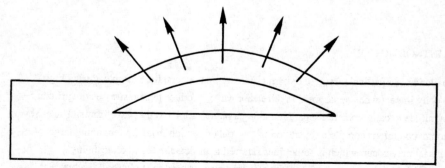

FIG. 2 *Behavior after initial spall.*

II. CONSTITUTIVE MODEL

A. *Constitutive Equations*

The theory assumes a linear decomposition of the rate of deformation into elastic and viscoplastic components

$$\underset{\sim}{D} = \underset{\sim}{D}^e + \underset{\sim}{D}^p \quad , \tag{1}$$

where the elastic deformation is assumed to be of small order. Because of the presence of microvoids, however, the elastic properties of the material are assumed to be degraded according to the model proposed by Mackenzie [5] and Johnson [6]. The elastic shear and bulk moduli and the Poisson ratio are given respectively, by

$$\overline{\mu} = \mu(1-\xi)\left(1 - \frac{6K+12\mu}{9K+8\mu}\xi\right) \qquad \overline{K} = \frac{4\mu K(1-\xi)}{4\mu + 3K\xi} \qquad \overline{v} = \frac{1}{2}\frac{3\overline{K} - 2\overline{\mu}}{3\overline{K} + \overline{\mu}} \quad . \tag{2}$$

In these relations ξ is the void volume fraction of the material and K, μ, v are the value of the moduli for the non-voided solid, which is an idealization for polycrystalline materials which contain some initial volume of voids, ξ_0, of the order $10^{-3} - 10^{-4}$. The constitutive equation for the rate of elastic deformation of the voided solid, then, is given by

$$\underset{\sim}{D}^e = \frac{1}{2\overline{\mu}}\left[\overset{\triangledown}{\underset{\sim}{T}} - \frac{\overline{v}}{1+\overline{v}}\left(\operatorname{tr}\overset{\triangledown}{\underset{\sim}{T}}\right)\underset{\sim}{1}\right] \quad , \tag{3}$$

where $\overset{\triangledown}{\underset{\sim}{T}}$ is an objective stress rate of the Cauchy stress tensor $\underset{\sim}{T}$.

The viscoplastic rate of deformation is given by the overstress form [1]

$$\underset{\sim}{D}^p = \frac{\gamma}{\phi}\Phi(\hat{F})\frac{\partial f}{\partial \underset{\sim}{T}} \qquad \text{for } \hat{F} > 0 \quad , \qquad \hat{F} = \frac{f}{\kappa} - 1 \tag{4}$$

and $\underset{\sim}{D}^p = 0$ for $\hat{F} \leq 0$, where Φ is a material functional of the yield function. The function ϕ is a rate of deformation control function that depends upon the argument $(I_2/I_2^s - 1)$, and is defined to be such that $\phi(0) = 0$ and $\phi(....) \equiv 0$ for $(I_2/I_2^s < 1)$. The invariant $I_2 = \left\|(II_{D'})^{\frac{1}{2}}\right\|$, where D' is the rate of deformation deviator tensor, and I_2^s is the value of I_2 at the quasi-static rate of deformation. Also, $\gamma^* = \frac{\gamma_0}{\kappa_0}$,where γ_0 is the material viscosity constant and κ_0 is a material parameter related to the rate independent (quasi-static) initial yield stress. Because of the presence of the voids, the plastic deformation of the voided solid will be compressible.

Consequently, the yield function f is taken to have the form

$$f = f\left(J_1, J_2', \xi\right) = J_2' + n\xi J_1^2 \quad , \tag{5}$$

where J_1 is the first invariant of the Cauchy stress tensor and J_2' is the second invariant of the Cauchy stress deviator tensor. The coefficient n is a material parameter that serves as a weighting between the deviatoric component J_2' and the dilatational component ξJ_1^2 of the yield function.

The function κ accounts for the isotropic strain hardening and for the softening that accompanies any increase of the void volume and is taken to have the form

$$\kappa = \kappa(\varepsilon^p, \xi) = \left[q + (\kappa_0 - q)\,e^{-\beta\varepsilon^p}\right]^2 \left[1 - n_1\,\xi_2^{\frac{1}{2}}\right]^2 \quad , \qquad \varepsilon^p = \int_0^t \frac{2}{\sqrt{3}} \left\|(II_{D^p})^{\frac{1}{2}}\right\| \, dt' \quad , \tag{6}$$

where q and β and n_1 are material hardening and softening parameters. The material functional Φ and the rate of deformation control function ϕ are taken as simple power functions of their arguments

$$\Phi(\widehat{F}) = \left(\frac{f\left(J_1, J_2', \xi\right)}{\kappa(\varepsilon^p, \xi)} - 1\right)^{m_1} \qquad \phi\left(\frac{I_2}{I_2^s} - 1\right) = \left(\frac{I_2}{I_2^s} - 1\right)^m \quad . \tag{7}$$

Insertion of the expressions for Φ, f, κ and ϕ into (4), leads to the following constitutive equation for the rate of inelastic deformation of the solid with microvoids

$$\underset{\sim}{D}^p = \frac{\gamma_0}{\left(\frac{I_2}{I_2^s} - 1\right)^m} \left[\frac{J_2' + n\xi J_1^2}{\left[q + (\kappa_0 - q)\,e^{-\beta\varepsilon^p}\right]^2 \left[1 - n_1\xi_2^{\frac{1}{2}}\right]^2} - 1\right]^{m_1} \frac{1}{\kappa_0}\left(2n\xi J_1 \underset{\sim}{1} + \underset{\sim}{T'}\right) \quad . \tag{8}$$

B.   Damage Evolution Equations

Increase in void volume can be attributed to the nucleation and growth microprocesses. Under high strain rates the nucleation of new microvoids is attributed principally to thermally activated mechanisms [7] whereby

$$\dot{\xi}_{nucl} = \frac{h(\xi)}{1 - \xi}\left[\exp\left(\frac{m_2|\sigma - \sigma_N|}{k\theta}\right) - 1\right] \quad , \quad \sigma > \sigma_N \tag{9}$$

Here $\sigma = (1/3)J_1$ is the mean stress, σ_N is the threshold mean stress for nucleation, θ is the temperature, k is the Boltzmann constant and m_2 is a material constant. The function $h(\xi)$ was included by Perzyna [1] to account for the effect of microvoid interaction, here however it will be taken as a constant.

The expression for the rate of void growth is determined from a model of the growth of voids in porous ductile solids, which considers a random distribution of microvoids idealized as spherical holes of arbitrary size. The initial hollow sphere model developed by Carroll and Holt [8] has since been generalized to allow for rate dependence and nonlinear isotropic hardening of the material. The void growth rate, derived in detail in [9], for a porous rate sensitive material with nonlinear hardening can be expressed by the set of relations

$$\dot{\xi}_{grow} = \frac{1}{\eta} g(\xi) F(\xi, \xi_0)(\sigma - \sigma_G), \quad \text{for } \sigma > \sigma_G \tag{10}$$

where

$$F(\xi, \xi_0) = \frac{\sqrt{3}}{2} \xi \left(\frac{1-\xi}{1-\xi_0}\right)^{\frac{2}{3}} \left[\left(\frac{\xi}{\xi_0}\right)^{\frac{2}{3}} - \xi\right]^{-1} \qquad \sigma_G = \frac{1}{\sqrt{3}}(1-\xi) \ln\left(\frac{1}{\xi}\right)\left[2q + (\kappa_0 - q) F_1(\xi, \xi_0)\right],$$

and

$$F_1(\xi, \xi_0) = \exp\left[\frac{2}{3}\beta \frac{(\xi_0 - \xi)}{\xi(1-\xi_0)}\left(\frac{1-\xi}{1-\xi_0}\right)^{-\frac{2}{3}}\left(\frac{\xi_0}{\xi}\right)^{-\frac{2}{3}}\right] + \exp\left[\frac{2}{3}\beta \frac{(\xi_0-\xi)}{(1-\xi_0)}\left(\frac{1-\xi}{1-\xi_0}\right)^{-\frac{2}{3}}\right]. \tag{11}$$

In these expressions ξ_0 is the average initial void volume fraction, σ_G is the void growth threshold mean stress, η is a material parameter, and $g(\xi)$ is a material function included to account for void interaction which is here taken to have the form $g(\xi) = e^{\alpha\xi}$.

Thus, the damage evolution equation, determined from (9) and (10), is given by

$$\dot{\xi} = \frac{h(\xi)}{1-\xi}\left[\exp\left(\frac{m_2|\sigma - \sigma_N|}{k\theta}\right) - 1\right] + \frac{1}{\eta} g(\xi) F(\xi, \xi_0)(\sigma - \sigma_G). \tag{12}$$

C. Spallation Criteria

Micrographic examination of spalled plates [10] that exhibit ductile type fracture shows failure to be the final stage of a process of void nucleation, growth, and coalescence. This suggests that a local criterion for ductile spall fracture can be defined as the attainment of a critical value for the void volume fraction. This condition is incorporated directly into the constitutive equation by the requirement that as material separation occurs the isotropic hardening/softening function goes to zero. It follows from Eq. (10), therefore, that

$$\kappa(\epsilon^p, \xi)_{\xi=\xi_F} = \left[q + (\kappa_0 - q) e^{-\beta\epsilon^p}\right]^2 \left[1 - n_1\xi_F^{\frac{1}{2}}\right]^2 = 0, \tag{13}$$

from which the material constant n_1 is defined as $n_1 = 1/(\xi_F)^{\frac{1}{2}}$.

III. NUMERICAL METHOD

Solution of the problem of impact of circular plates of unequal diameter is performed by the computer code PRONTO [10], which is a Lagrangian finite element code utilizing four-noded, uniform strain quadrilateral elements. The balance of mass equation and the equations of motion in cylindrical coordinates are solved incrementally using explicit integration. A modified central difference, in which velocities are integrated with a forward difference and displacements are integrated with a backward difference is employed. The computational procedure used in [10] defines a Cauchy stress tensor and a rate of deformation tensor at each point for a locally unrotated configuration. Thus each material point of the body has its own reference frame which is oriented such that the deformation relative to this frame is a pure stretch. The advantage in defining this frame is that the material derivative of the stress tensor is objective, whereas in general the material derivative in the rotated configuration is not. Use is also made of the element deletion capability in PRONTO to accomodate the large deformations that occur after the onset of spallation. Elements are deleted upon reaching a critical value of the void volume fraction, which is established as a value somewhat less than ξ_F, which is only approached in the limit. In the calculations presented here, elements were deleted upon reaching a void volume fraction of .25.

IV. SIMULATION OF PLATE IMPACT

The constitutive theory and numerical procedure discussed in previous sections were used to simulate the impact of circular plates under varying impact velocities. Due to the axis of circular symmetry only one half of the configuration of Fig. 1 needs to be modelled. A 3 mm thick, 40 mm diameter flyer impacting a 6 mm thick by 80 mm diameter target at 350 m/s is considered for the first simulation. Material parameters for high purity OFHC copper determined in [2,4] and listed in Table 1 were used for the simulations. A uniform finite element mesh containing 49 x 29 elements for the flyer plate and 96 x 59 elements for the target plate was employed. Calculations were performed on a Cray X-MP computer, requiring on the order of 150 seconds CPU time for each 1μs of the simulation.

Figure 3 shows the deformed geometry of the plates at successive instants of time during the impact event. At 5 μs, time is sufficient for several wave transits across the

Table 1 OFHC Copper Material Parameters

μ	= 4.84×10^4 MPa	γ_0	= $337\ s^{-1}$	m	= $1/2$		ξ_0	= .0003
K	= 14.00×10^4 MPa	l_2^s	= $7 \times 10^{-4}\ s^{-1}$	m_1	= 2		ξ_F	= 0.32
q	= 125 MPa	κ_0	= 9.31MPa	$k\theta$	= 4.05×10^{-21}J		n_1	= 1.77
β	= 6.14	n	= 0.25	σ_N	= 500 MPa		h	= 70.37
ρ_R	= 8.93 gm/cm³	η	= 120 Poise	m_2	= 2.025×10^{-23} cm³		g	= $e^{\alpha\xi}$
							α	= 20

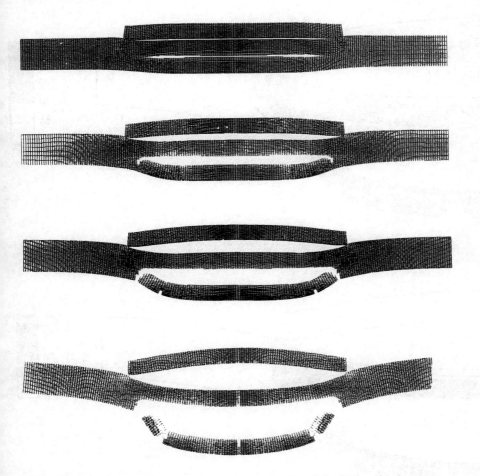

FIG. 3 *Deformed geometry for 3 mm flyer with 350 m/s impact velocity (a) 5 µs, (b) 15 µs, (c) 25 µs (d) 50 µs.*

thickness of the target plate to occur, which results in the formation of the spall plane shown in the figure. By 15 µs the momentum remaining in the rear portion of the spalled target plate results in the more pronounced opening. In addition the tensile stresses that develop at the edges of the spalled portion are sufficient to cause an increase in the void volume fraction, resulting in ductile fracture being formed. This process continues to the point of complete separation or scabbing, shown at 25 µs, including development of some additional fracturing at the rear surface of the scab. Figures 4 and 5 show the distribution of the accumulated

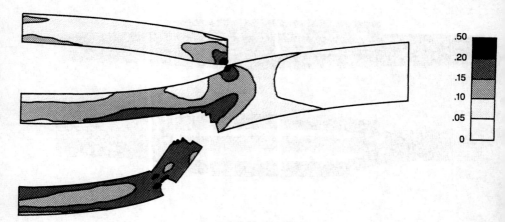

FIG. 4   *Accumulated plastic strain at 50 μs.*

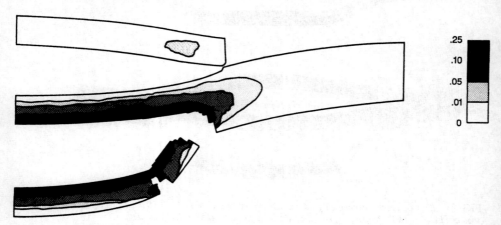

FIG. 5   *Void volume fraction at 50 μs.*

plastic strain and void volume fraction in the final configuration, i.e. after which no significant changes are expected.

The second configuration simulated is identical to the first with the exception that the thickness of the flyer is taken to be 1.5 mm. The deformed geometries are shown in Fig. 6. In this case it can be seen that the no fractures develop at the attachment point of the spalled section indicating that insufficient momentum remains to cause detached spall.

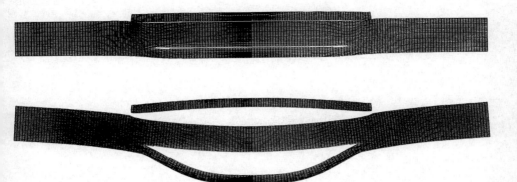

FIG. 6 *Deformed geometry for 1.5 mm flyer with 350 m/s impact velocity (a) 5 μs, (b) 50 μs.*

REFERENCES

1.   P. Perzyna, *Int. J. Solids Structures* , 22: 797 (1986).

2.   J.A. Nemes, J. Eftis, and P.W. Randles, *J. Appl. Mech.*, 57: 282 (1990).

3.   J.A. Nemes and J. Eftis, in *Shock Waves in Condensed Matter -1989*, S.C. Schmidt, J.N. Johnson and L.W. Davidson (eds.), Elsevier Science Publishers, B.V. Amsterdam, 1990.

4.   J. Eftis, J.A. Nemes, and P.W. Randles, (forthcoming in *Int. J. Plasticity*). Also *Advances in Plasticity 1989*, A.S. Kahn and M. Tokuda (eds.), Pergamon Press, Oxford, 1989, p. 381.

5.   J.H. Mackenzie, *Proc. Phys. Soc.*, 63B: 2, (1950).

6.   J.N. Johnson, *J. Appl. Phys.*, 53: 2812 (1981).

7.   D.R. Curran, L. Seamen, and D.A. Shockey, *Physics Today*, 46 (1977).

8.   M.M. Carroll and A.C. Holt, *J. Appl. Phys.*, 43, 1626 (1972).

9.   J. Eftis and J.A. Nemes, (forthcoming in *Int. J. Plasticity*).

10.   L.M. Taylor and D.P. Flanagan, SAND86-0594 (1987).

67

Spall of Differently Treated High-Strength Low-Alloy Steel

C. N. ELIAS, P. R. RIOS AND A. W. ROMERO

Department of Mechanical Engineering and Materials Science
Instituto Militar de Engenharia
Rio de Janeiro, RJ 22290, BRASIL

Department of Metallurgical Engineering
Pontifica Universidade Catolica
Rio de Janeiro, RJ 22453, BRASIL

The resistance of a quenched and tempered high strength low alloy steel was investigated as a function of heat treatment. The flyer-plate technique was used. Fracture mechanism and spalling resistance were found to be significantly affected by the heat treatments.

I. INTRODUCTION

Materials submitted to high strain rates behave very differently from materials deformed slowly. In the former, the strain produced in the material is not instantly transmitted through the specimen, but from atom to atom in the form of elastic, plastic, and shock waves [1]. As a consequence, peculiar phenomena can arise, such as adiabatic shear bands, phase transformations, hardening by shock waves and dynamic fractures [2-5].

Dynamic fracture, in particular, can lead to the failure of the material by spalling, also known as spalling fracture, scabbing and back spalling. In this type of fracture, there is nucleation, propagation and coalescence of a high number of microcracks. These microcracks are distributed in parallel to the surface and are caused by a high tension stress, higher than the dynamic resistance of the material.

The metallurgical factors involved in the nucleation and propagation of dynamic fracture causing spalling are not as yet fully understood. The fracture is dependent on the microstructure and in particular, grain boundaries, substructure and interfaces [4].

Dynamic fracture is often divided into four stages [4]: microcrack nucleation, growth and coalescence and finally spalling or fragmentation. All these stages could in principle depend on the extent of microstructure.

In this work 25mm HSLA steel plates were submitted to different heat treatments and shock loaded using the flyer-plate technique in order to investigate the effect of the microstructure on spalling.

II. EXPERIMENTAL METHODS AND MATERIALS

The flyer-plate technique consists of a plate which makes an angle to the target plate. On top of the flyer plate there is a layer of explosive; with the explosion, the plate is impelled to the target which is placed on soft materials to prevent shock wave propagation above the target. The material of the flyer plate was the same of the target and had half the thickness of the latter.

Velocity and pressure were varied by using different number of layers of PLASTEX-P explosive, each 2.18 mm thick. The properties of the explosive were: density = 1.6 g/cc, caloric constant = 4.14 kJ/g, detonation velocity = 7340 m/s. Three combinations of velocity and pressure were used: i) 289 m/s and 5.42 GPa; ii) 389 m/s and 7.41 GPa; iii) 437 m/s and 8.37 GPa. These velocities and pressures were calculated by Gurney's equation. The composition of the steel used was in wt%: C 0.18 - Mn 1.32 - Cr 1.02 - Ni 1.23 - Mo - 20 - Si 1.25 - V 0.78.

The material was received as a 25 mm thick plate. Samples of 25 x 25 x 25 mm were cut and heat treated. The austenitization temperature was 1173 K followed by furnace cooling or water quenching. Water quenched specimens were subsequently tempered at 473 K and 673 K for 2 hs and at 873 K for 7hs. After each treatment mechanical properties were characterized by tensile and hardness tests, and after ballistic tests the specimens were examined by optical and scanning electron microscopy. The measurement of the size of the cracks was made in an area of 25 x 25 mm with a magnification of 25 times.

III. RESULTS

A. HEAT TREATED SPECIMENS

Results of the tensile tests are shown in Table 1. Optical microscopy showed that all water quenched samples were fully martensitic, for samples tempered at 873 K carbides could be seen in optical microscopy. Furnace cooled specimens exhibited typical ferrite and pearlite "banded" structure.

B. MACROSCOPIC EXAMINATION OF IMPACTED SAMPLES

The surface of impacted specimens parallel to the load directions exhibited a "wavy" appearance as shown in Fig. 1a. It is clear that the thickness of each "fringe" decreases as the bottom of the specimen is approached or as the intensity of the shock wave is "damp."

Table 2 shows the severity of spalling for each treatment and combination of pressure and plate speed. In the present work the following classification is proposed for the level of damage in spalling:
1 - No spalling cracks can be detected. 2 - Some cracks are present, but can only be seen by optical or electron microscopy. 3 - Cracks can be detected in a polished specimen for magnifications up to 20 times. 4 - Occasional external cracks can be detected. 5 - Several external cracks can be seen. 6 - Specimen is separated in two or more pieces.

One can see in Table 2 that the level of damage increases for an increase in flyer-plate pressure and velocity. An exception is the material quenched and tempered at 873 K in which the level of damage remained approximately constant for each velocity and pressure. One can also see decrease in resistance with an increase in tempering temperature. Another aspect is that as the pressure and the velocity of the flyer-plate are increased, the influence of heat treatment appears to decrease and the damage level is high in all specimens. As a consequence one may infer that the effect of microstructure becomes more important at lower pressures and velocities.

TABLE 1. MECHANICAL PROPERTIES AFTER HEAT TREATMENT

	STRENGTH			HARD
	σ_y (MPa)	σ_s (MPa)	elong. %	HRa
Quenched	1299	1489	13.3	69.2
Tempered 473K	1246	1511	12.2	68.9
Tempered 673K	1197	1413	11.3	67.6
Tempered 873K	673	807	21.7	58.4
Annealed	421	658	29.3	50.5

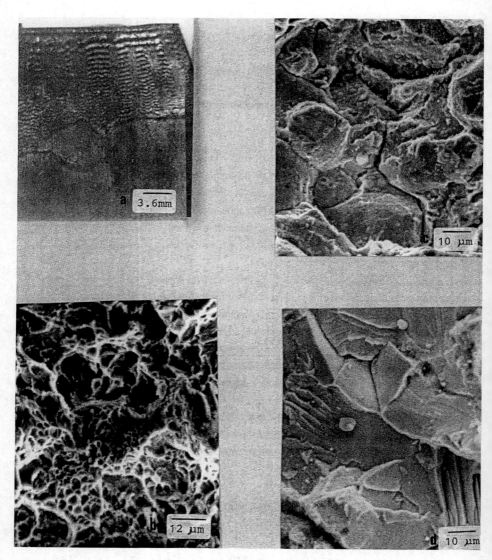

FIG. 1 (a). Specimen surface after ballistic test parallel to the
load direction. Aspects of surface fractures. (b) Dimples in
a sample quenched, P = 8, 37 GPa, (c) intergranular propagation
in a sample quenched and tempered at 673 K, P = 8.37 GPa, (d)
Annealed sample with fracture by cleavage, P = 8.37 GPa.

TABLE 2. SPALLING DAMAGE

	V = 289 m/s P = 5.4 GPa	V = 389 m/s P = 7.4 GPa	V = 457 m/s P = 8.4 GPa	Average
Quenched	3	3	4	3.33
Tempered 473 K	1	2	4	2.33
Tempered 673 K	4	4	6	4.67
Tempered 873 K	6	5	5	5.33
Annealed	2	5	6	4.33
AVERAGE	3.2	3.8	5	

C.   FRACTOGRAPHY AFTER BALLISTIC TEST

Table 3 shows the fracture mechanism of spalling.

One can observe in Tables 1 and 3 that the more ductile samples (annealed) in the tensile tests failed in the ballistic test by a more brittle mechanism, and the less ductile samples (those tempered at 673 K) failed by a more ductile mechanism.

After the ballistic tests different surface fractures could be observed, e.g. dimple (Fig. 1b), intergranular propagation (Fig. 1c) and a surface showing brittle fracture (Fig. 1d).

Crack initiation could be detected by separation of the interfaces (Fig. 2a) between the inclusion and the matrix, segregation bands (Fig. 2b), at grain boundaries (Fig. 2c) and at carbide grain boundary interfaces (Fig. 2d).

The HSLA steel used in this work has a composition that justifies the tempering embrittlement. One can suppose that precipitation during the cooling of the samples tempered at 873 K had occurred, which induced intergranular fracture. This observation agrees with results obtained from Charpy impact test of this steel [5, 6].

TABLE 3. FRACTURE MECHANISM AFTER BALLISTIC TEST

Treatment	Fracture Mechanism	Observation
Quenching	Dimples and Quasi-Cleavage	
Tempered 473 K	Dimples and Quasi-Cleavage	Higher Fracture Resistance
Tempered 673K	Dimples	
Tempered 873 K	Intergranular and Cleavage	Lower Fracture Resistance
Annealed	Cleavage	

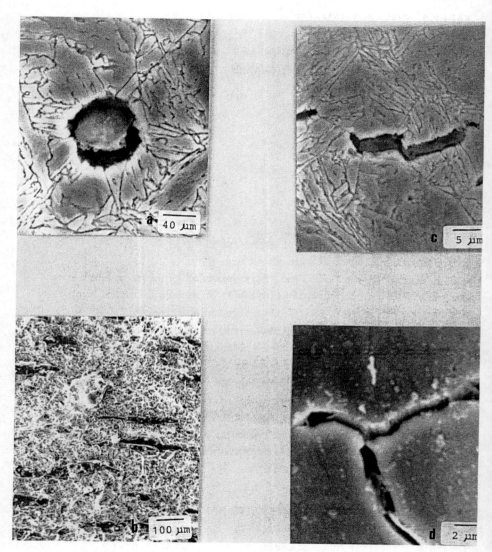

Fig. 2 *Crack initiation in ballistic test. (a) Separation of the interface in a quenched sample, P = 5.42 GPa, (b) microcracks in a sample quenched and tempered at 673 K, P = 8.37 GPa, (c) initiation of microcracks at grain boundary in a sample quenched and tempered at 873 K, P = 7.41 GPa, (d) initiation of fracture at interface of carbide in a grain boundary.*

IV. CONCLUSIONS

The influence of heat treatment was more important for lower pressures and velocities of the flyer plate. For a certain velocity and pressure it was found that tempering at 473 K gave the best spalling resistance of all heat treatments used here.

REFERENCES

1. R. Kinslow, *High-Velocity Impact Phenomena*, Academic Press, New York (1970).
2. A. R. Champion, and R. W.Rhode, *J. Appl. Phys.*, 41: 2213 (1970).
3. W. J. Nellis, *Scripta Metallurgica*, 22: 121 (1988).
4. M. A. Meyers and C. T. Aimone, *Progress in Materials Science*, 28: 1 (1983).
5. H. R. Andrade, Master of Science Thesis, Instituto Militar de Engenharia (1987).
6. A. W. Romero, Master of Science Thesis, Instituto Militar de Engenharia (1988).

68

Spalling of Aluminum and Copper Targets by Laser Shocks

M. BOUSTIE [*], F. COTTET [*], Y. CHAUVEAU [**]

[*] Laboratoire d'Energétique et Détonique (U.R.A. 193)
ENSMA, 20 rue Guillaume VII - 86034 POITIERS (FRANCE)

[**] S.A. MATRA - Branche Défense - DAST
37 A[ue] Louis Bréguet - 78146 VELIZY VILLACOUBLAY (FRANCE)

This work presents the results of an experimental and numerical investigation to determine the ability of very short laser pulses in generating shock waves of sufficient magnitude to cause spallation in metal targets and to test the use of a cumulative damage criterion for simulating scabbing process. The irradiation conditions leading to spallation have been determined by experiments on pure aluminum and copper targets using three different pulse durations: 0.6 ns, 2.5 ns and 27 ns. Consistent results between simulations and experiments are shown. Besides, experiments with stepped targets have been performed and analysed with a 2D code.

I. INTRODUCTION

When a shock wave propagates in a material plate, a fracture process, called *spallation* or *scabbing,* can occur near the rear face. The spallation phenomenon is due to the crossing of two rarefaction waves. One is coming from the front face of the plate when the loading falls down, and the other is sprung from the rear face, when the incident shock reflects back into the material. When these two release waves meet, they stretch the material. If the tensile conditions are sufficient in magnitude and time application, they can lead to the formation of a scab [1]. This spall can remain attached to the material or be ejected .

High-power pulsed laser provides us with new experimental shock configurations to study spallation: we can reach very high pressure levels (a few GPa) during a very short time (a few nanoseconds). Our purpose has been to study scabbing under this kind of particular conditions. By performing experiments by laser-shocks on material foils, we obtain spallation data which are correlated with a numerical study. By comparison between experimental and numerical results, a numerical spallation criterion well-known for aluminum has been checked. Thus, we can simulate correctly the propagation of shock waves initiated by a laser irradiation, taking into account the spallation phenomenon.

Afterwards, an attempt to extend it to other materials (such as copper or iron) has been done. Experiments on targets with a step on the rear face have been performed. We have compared successfully experimental scabbing results with simulation results for this kind of experiments.

II. NUMERICAL STUDY

A. SIMULATION TOOLS

In order to describe the spallation phenomenon, we need tools able to reproduce shock waves propagation through materials. For this purpose, both codes SHYLAC and RADIOSS™† are available.

SHYLAC, fully described elsewhere [2], is a monodimensional code solving conservation equations written in Lagrangian coordinates with a finite-difference method. Together with this resolution, the Mie-Grüneisen equation of state, referenced to the Hugoniot curve with a cubic form, is used. The behavior of materials is described by an elastic perfectly plastic law with the Von-Mises criterion.

The other code, RADIOSS™, is a more sophisticated one. It is a three-dimensional code solving conservation equations written with a finite-element method through an arbitrary Lagrangian or Eulerian mesh. The same equation of state as in SHYLAC is used. A Johnson-Cook law describes the material behavior [3].

In both codes, shock configurations are initial and boundary conditions on the front face of the target. As no matter-laser interaction process is included in the codes, laser-shock is driven by a pressure pulse applied on the front surface. This gaussian-shaped pulse is characterized by the laser energy deposition time τ at medium height and the maximum pressure P_M. This

---

† *RADIOSS is a trade mark of Mecalog SARL, rue Barrault, 75013 PARIS*

latter is estimated according to a scale law [4]:

$$P_M = 12 \left(\frac{I_a(W/cm^2)}{10^{11}} \right)^{2/3} \tag{1}$$

where I_a is the absorbed flux which is about 80% of the incident intensity in the irradiation range of our experiments.

After having checked the consistency of laser induced waves propagation given by SHYLAC and RADIOSS, we have included into them a spallation criterion.

B. SIMULATION OF SCABBING

Many spallation criteria have already been proposed [5,6]. The difficulty consists into choosing among them one which will describe accurately the spallation process over a wide range of stress and strain. For ductile fracture, the simplest one is a cut-off criterion. It presents the drawback of an adaptative pressure level according to the loading conditions. Besides, it does not take into account time effects, what is done in more sophisticated criteria such as N.A.G models. But, very often, these models require many independant parameters. So, we have chosen an intermediate criterion which uses only three parameters: the Tuler-Butcher criterion [7,8]. It assumes that tension must exceed a fixed level σ_R so that small voids appear. This over-stress must act over a finite time for leading to void coalescence. The degree of coalescence is measured by the integral of a damage function $f(\sigma)$. The integral must reach a given level for fracture be effective. Tuler and Butcher model proposes a particular form for this function:

$$f(\sigma,t) = (\sigma - \sigma_R)^A \tag{2}$$

Thus, this cumulative damage criterion is expressed at a defined location x by:
If ($\sigma(x,t) > 0$ and $\sigma > \sigma_R$), then fracture occurs where and when the time integral of $f(\sigma)$ is greater than a constant K.

Therefore, with the knowledge of the stretching history of the material and only three parameters A, K, σ_R, we can predict where and when a spall can occur. The parameters of this criterion being given for pure aluminum, we have validated this criterion by comparison between numerical scabbing data and experimental results obtained with laser irradiation of aluminum targets.

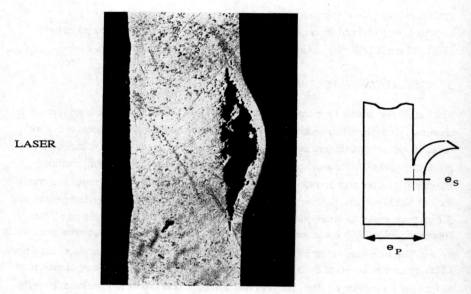

LASER

FIG. 1 *Cross sectional micrography of spalled laser irradiated copper target* – $\Phi_i = 4.10^{12}$ W/cm^2 – $\tau_{FWMH} = 27\ ns$ – *Measurement of the spall thickness vs target thickness*

III. EXPERIMENTAL STUDY

A. EXPERIMENTAL SET-UP

All experiments have been carried out at LULI[++] with a Neodymium glass laser with a 1.06 μm wavelength. The laser beam was focused onto the front face of the target under vacuum. The focused spot had a diameter larger than the thicknesses of the targets used (typically a few millimeters) so that we could consider the experiments as monodimensional. For an average output energy of about 80J and pulse durations at full width medium height of 600 ps, 2.5 ns and 25 ns (with a half gaussian shape), we can reach incident fluxes ranging from $0.5 \ 10^{11}$ W/cm^2 to 10^{12}W/cm^2. The spalled samples are recovered and coated with resin, cut through a plane including the laser axis, polished and observed with an optical microscope. From these observations, we can scan the damage degree and measure the spall thickness e_s (see figure 1).

B. ANALYSIS OF EXPERIMENTAL RESULTS

In our experiments, for a given incident intensity, we vary foil thickness. As we use very short pulse durations, a quick hydrodynamic attenuation of shocks occurs [9]. The thicker the target is, the less important the amplitude of stress reaching the rear face is and so on for the tensile magnitude. For different target thicknesses, we modify the reflexion conditions of the shock at the rear face, and therefore, we have different scabbing data. We show these results on figure 2 where numerical results are in good agreement with experimental data.

For pure aluminum, the Tuler-Butcher coefficients are well known and these results allow us to validate the spallation criterion in our shock conditions. In order to extend it to pure copper, we have performed the same experiments as for aluminum for copper targets. The problem for simulating correctly spallation into this material consists in finding the appropriated values of the Tuler-Butcher criterion parameters. This has been approximately done by fitting the numerical curve to the experimental results. The better agreement we found to reproduce well the spalling evolution and mainly the scabbing thresholds (i. e. the target thickness over which no spall appears for a given flux) is shown on figure 3 with the values of table 1.

[++] *Laboratoire d'Utilisation des Lasers Intenses - Ecole Polytechnique-91128 PALAISEAU - FRANCE*

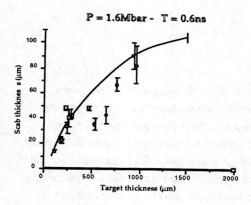

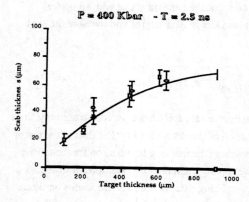

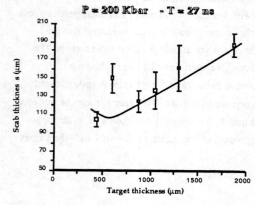

FIG. 2 *Comparison of numerical spallation results (———) and experimental data (⌶) for different laser irradiation conditions with* **aluminum** *targets*

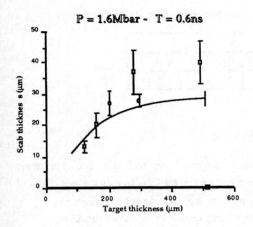

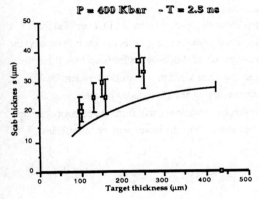

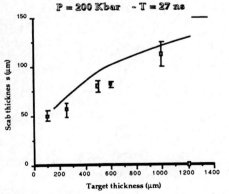

FIG. 3 *Comparison of numerical spallation results (————) and experimental data (⊥) for different laser irradiation conditions with* **copper** *targets*

TABLE 1 *Set of values used in simulations for aluminum and copper*

	$\rho_o (kg/m^3)$	σ_R (Pa)	A	K
Al	2785.	10^9	2.02	$3.8\ 10^{11}$
Cu	8930.	$3.5\ 10^9$	2.02	$6\ 10^{11}$

IV. EXPERIMENTS WITH STEPPED TARGETS

Some experimental results correspond to particular targets. A step at the rear face of the target has been machined and the laser beam has been centered on the front face on the axis of the step (see on fig. 4). This kind of experiments presents the advantage of providing two spallation results for a single laser shot, corresponding to exactly the same irradiation conditions.

In order to integrate this kind of experiments with classical ones, we first verified that spallation process was quite the same as in targets without a step. This has been done with the use of RADIOSS™ code simulating the propagation of shock-waves initiated by a laser irradiation through a stepped target. By examining the simulation results when tension has occured on both steps, we can notice that the stretching zone is reduced on the side of the upper step along the laser axis because of the release wave coming from the edge of this step. But the central tensile zone on this step is not altered. For the lower step, no significant perturbation can be observed.

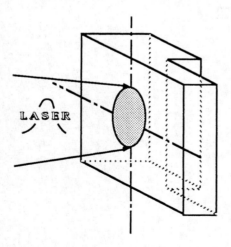

FIG. 4 *Sketch of a stepped target*

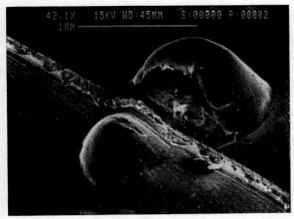

FIG. 5 *Rear view of an aluminum stepped target* $-$ $\Phi_i = 4.10^{12}$ W/cm^2 $-$ $\tau_{FWMH} = 0.6\ ns$

These phenomenons can be seen on the photography of the rear face of a spalled stepped target (see fig. 5). Thus we could consider a stepped target experiment as two classical ones. This has also been experimentally checked by comparing successfully the thicknesses of scabs measured on such stepped targets and those obtained by the same irradiation of two single foils of identical thicknesses.

V. CONCLUSION

We have shown the ability of the Tuler-Butcher cumulative damage criterion to well reproduce laser induced spallation into aluminum foils for laser induced pressures ranging from 200 kbar to 2 Mbar with pulse durations of 0.6 ns, 2.5 ns and 27 ns. Experiments with copper targets have been performed in the same irradiation range: we propose a set of the Tuler-Butcher parameters for simulating spallation into copper for this range.

The study of stepped targets has shown the ability of the RADIOSS™ code to well reproduce laser shock waves in a 2D configuration. Moreover, it helps us to interprete the experimental scabbing results on this kind of targets.

REFERENCES

1.   J. A. Zukas, T. Nicholas, H. F. Swift, L. B. Gresczuk, D. R. Curran, *Impact Dynamics*, Wiley, New York, 1982.

2.   F. Cottet and M. Boustie, *J. Appl. Phys.* 66 (9), 1989.

3.   G. R. Johnson and W. H. Cook, "A constitutive model and data for metals subject to large strains, high strain rates and high temperature" - *7th Int. Symposium on Ballistics*, 1983.

4.   R. Fabbro, E. Fabre, F. Amiranoff, C. Garban-Labaune, J. Virmont, M. Weinfeld and C. E. Max, *Phys. Rev. A 26*, 2289, 1982.

5.   J. N. Johnson, *J. Appl. Phys. 52*, 2812, 1981.

6.   S. Cochran and D. J. Banner, *J. Appl. Phys. 48*, 2729, 1977.

7.   F. R. Tuler and B. M. Butcher, *Int. J. Fract. Mech. 6*, 431, 1968.

8.   J. J. Gilman and F. R. Tuler, *Int. J. Fract. Mech. 6* (2), 1970.

69

On Anomalous Increase of Steel Spall Strength and Its Relationship to Martensitic Transformation

A. N. DREMIN, A. M. MOLODETS, A. I. MELKUMOV AND A. V. KOLESNIKOV

Institute of Chemical Physics
USSR Academy of Sciences
Chernogolovka, Moscow Region,
USSR, 142432

Spall strength measurements have been performed with manganin gauges for austenitic steel of 18/8 type at 17, 140, 180 and 300 K initial temperatures. An anomalous increase of the strength has been revealed in the temperature interval of 140 − 180 K. Some metallographic investigations and α − phase content distribution measurements along the shock wave (6 GPa amplitude) motion within the samples recovered have been done for the steel at 77 K. It has been found that the location of the distribution maximum falls on the spall range. It means that the martensitic transformation proceeds intensively in the region of maximum tensile stress and that it results in stress relaxation. The anomalous increase of the austenite steel spall strength is interpreted in terms of this new type of tensile stress relaxation.

I. INTRODUCTION

In impact experiments the spall fracture occurs in target under the effect of a tensile stress pulse produced by the interaction of the rarefaction waves moving in opposite directions from free surfaces of both projectile and target. It is well known that the spall phenomenon is a kinetic process [1] and that metal spall strength dependence on temperature in general is weak [2-5]. Austenitic steels are an exception; their spall strength increases strongly at low temperatures [6-7]. It has been shown before that the austenite-to-martensite transformation is possible under the shock wave effect still during the compression [8-10]. This means that the shock wave loading history would influence the material spall strength. Moreover, it turns out that the martensitic transformation takes place at low temperatures under the effect of tensile stress pulses. Meyers and others [11, 12] have performed a detailed investigation of the martensitic transformation induced by the tensile stress pulse (2 GPa intensity and microsecond range time duration) a little below that necessary for the spall phenomenon can occur. It has been revealed that the transformation is the kinetic fast isothermal process and that the tensile stress stimulates the process.

So, for austenitic steels two kinetic processes proceed under tensile stress effect: The spall phenomenon and the martensitic transformation and therefore it is not improbable that there is some interdependence between the two. In order to explain the large increase of the austenitic steel spall strength that is experimentally observed it has been already implied that the increase is governed by the martensitic transformation [7]. To make a close study of the problem we have performed two types of the experiments with austenitic steel of 18/8 type: the spall strength measurements at 77, 140, 180 and 300 K temperatures and some microstructural investigations of the samples recovered after the shock effect at 77 K initial temperature.

II. EXPERIMENTAL PROCEDURES

The rectangular samples of a 100 mm size were machined from the plates of 4.9 mm thickness of industrial production. Shock waves of 6 GPa intensity have been generated by aluminum flyer impact of 600 m/s velocity. The flyer diameter and thickness were 70 and 3 mm. The flyer central plane part had about 50 mm diameter. Manganin technique has been employed for the compression wave stress registration at the steel sample-glass interface. A glass of 80% quartz content and 2.3 g/cm^3 density has been used because it has the following advantages. First, it has rather small coefficient of thermal expansion; this property prevents the manganin gauge and insulating films from deformation during cooling of the experimental assembly and thus ensures the gauge efficiency. The sufficiently low acoustic impedance is the second advantage of the glass; the property makes it possible to observe the steel spallation under the experimental conditions. And the third advantage is the glass elastic behaviour for all shocks up to 6 GPa intensity at 300 K as well as at 77 K temperatures [13]. It follows from the above-said that in the investigation the steel spall strength has been calculated using the compression wave stress registration at the steel sample - glass interface. It is obvious that the third property makes the calculation procedure rather simple. The experimental procedures have been presented in greater detail before [14].

III. RESULTS AND DISCUSSION

The compression wave stress profiles at the steel sample-glass interface registered by manganin gauges at different temperatures are presented in Fig. 1. As follows from Fig. 1 the steel spall strength increases sharply in the temperature interval of 140-180 K. Really, both profiles at 300 K and 180 K clearly demonstrate the spall impulse. However, the impulse practically ceased at the following decrease of temperature only by 40 K. Spall strength data σ^* are presented in Fig. 2 (curve 1). The following expression [15] has been employed to calculate the spall strength using experimental values σ_1 and σ_2 shown in Fig. 1.

$$\sigma^* = 0.5 \left\{ \sigma_1 \left[\frac{(\rho c)_s}{(\rho c)_g} - 1 \right] - \sigma_2 \left[\frac{(\rho c)_s}{(\rho c)_g} + 1 \right] \right\}$$

Here ρ and c denote the initial density and sound velocity; indexes s and g refer to steel and glass. It should be mentioned that two assumptions have been introduced to derive the expression. It has been implied that single shock adiabat, isentrope and double shock adiabat coincide and that the materials acoustic impedance does not depend on pressure. The assumptions are valid for steel as well as for glass. It means that the expression has been deduced within the acoustic approximation. However it is well-known that for materials having an appreciable dynamic elastic limit the spall strength value calculated by the

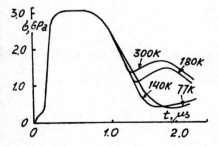

FIG. 1 *The compressive stress profiles at the steel sample-glass interface at four various temperatures.*

expression is to be corrected taking into account the elasto-plastic state of the materials [15]. The correction, in general, increases the value by approximately 20%. Curve 2 in Fig. 2 represents the corrected spall strength data for the steel under investigation. It is evident that according to both uncorrected (curve 1) and corrected (curve 2) data the steel spall strength increases anomalously in the temperature range of 140 - 180 K.

Taking into account the aforementioned data obtained by Meyers and others [11, 12] as well as the data obtained by Golubev and others [7], the metastable austenite-to-martensite transformation seems to be the cause of the increase . However, it should be mentioned that the gasdynamic measurements performed and the steel spall strength data presented in Fig. 2 do not contain any direct information on the transformation. Therefore, in order to substantiate the transformation some metallographic and X-ray diffractometric studies have been performed with the samples recovered after gasdynamic measurements at 77 K temperature. It should be noted that the studies have been carried out with the samples of 10 mm thickness since it was rather difficult to do it with the samples of 4.9 mm thickness. Two blocks of 10 mm size have been cut out of the recovered sample central part for the studies. Thick samples of the same 18/8 steel type had the following composition: Si: 0.57; Mn: 1.27; W: 0.05; Cr: 17.31; Ni: 10.40; V: 0.03; Mo: 0.12; Cu: 0.21; C: 0.09; S: 0.005; N: <0.05. The steel austenite grain size was found to be about 100 μm.

X-ray radiation of 1.5418 Å wavelength has been used for the diffractometric study. Two diffraction peaks have been detected: one is J_γ (111) at a diffraction angle of 2θ = 43.63° and corresponding to austenite and the other - J_α (110) at an angle of 2θ = 44.60° and corresponding to martensite [16].

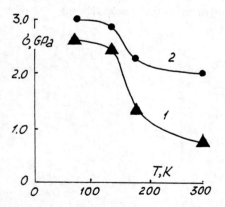

FIG. 2 *Spall strength temperature dependence for 18/8 type steel ([1,2] see in the text).*

The studies have shown that the diffractograms of neither the initial material at 77 K temperature nor the recovered sample shock-treated at room temperature had J_α (110) martensite spike. However, one can observe the spike for "cool" recovered sample shock-treated at 77 K temperature. So, it has been ascertained that for the steel under investigation the austenite-to-martensite transformation under the shock wave effect takes place only at low temperatures.

Naturally, the question arises: what stresses (compressive or tensile) stimulate the transformation? Obviously, to answer the question one has to measure the α phase content distribution within the samples along the shock wave motion. However, it should be noted that it was not possible to use for the measurement the above-mentioned diffractometric studies performed with solid samples since the steel was rather textural. Therefore, it was done by a different way, described below. It was afore-said that two blocks had been cut out the recovered sample central part for the austenite-to-martensite transformation studies. One was used in solid state for the martensite qualitative observations and the other was reduced to powder by layers oriented perpendicular to the shock wave motion and of 0.5 - 0.8 mm size. Each portion of the powder was sifted through a sieve of 100 x 100 µm cell size and then was used for the quantitative diffractometric studies. It is known that the α-phase content is determined by the relation of the martensite diffractometric peak intensity J_α (100) to that of the austenite J_γ (111) [16]. In Fig. 3b the relation data are shown by points for the layers in sequence within the sample along the shock wave motion. Star point in Fig. 3b refers to the initial material, that is, to the sample which has not been treated by the shock and had no α-phase. The point testifies to the fact that some martensite quantity appears at the powder making process during the reduction of block to powder. Obviously, to determine the quantity of martensite originated just due tot he shock wave effect one has to displace the abscissa up by the value corresponding to the star point.

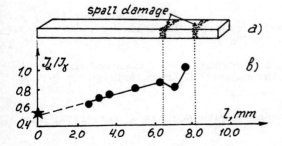

FIG. 3 (a)Sample cross-section and (b) J_α/J_γ distribution inside the sample along the shock wave-motion.

It should be noted that double spall appears inside the 10 mm thick samples. It is evident from the microdamage metallographic observation at sample cross-section schematically shown in Fig. 3a. One can see from Fig. 3b that the α-phase content distribution has two maxima and that they fall on the spall ranges. At the same time the α-phase has not been registered at the impact surface that is at the sample part with the highest compressive stress. It leads to the conclusion that the austenite-to-martensite transformation in the steel under investigation (as well as in the material discussed in [11, 12]) is stimulated not by compressive but by exclusively tensile stresses. Therefore, one can consider that the steel spall strength anomalous increase discovered in the sample initial temperature interval of 140-180 K is governed by the transformation.

The dynamic elastic limit of steel is small in comparison with the shock amplitude employed in our experiments. Therefore, the material tension in the experiments is accompanied by its plastic deformation. At the same time, it is known [17] that the plastic deformation in real heterogeneous materials promotes local stress appearance at some micro-volumes, these stresses considerably exceeding macroscopic external stress. At spall the local stresses relax through the microcracks origin and it leads to the spall fracture [18]. In the case of austenitic steels the local stresses can relax also through formation of the martensite, that is, through the creation of the material with smaller density. The alternative method of tensile stress relaxation does not lead to the microcrack appearance. Thus, as a matter of principle, the martensitic transformation turns out to be the process which depresses the spall damage origin.

REFERENCES

1. L. Davison and R.A. Graham, *Phys. Rep.*, 5: 255 (1979).

2. S. A. Novikov, I. I. Divnov and A. G. Ivanov, *Fizika Metallov i Metallovedenia*, 21: 608 (1966).

3. V. K. Golubev et al., *Zhurnal Prikladnoi Mekhaniki i Teknitsheskoi Fiziki* N6: 108 (1982).

4. S. J. Bless and D. L. Paisley, in *Shock Waves in Condensed Matter, 1983*, eds. J. A. Asay, R. A. Graham, and G. K. Straub, North-Holland, 1984, p. 163.

5. A. M. Molodets and A. N. Dremin, *Fizika Gorenia i Vzriva N4*: 101 (1989).

6. V. K. Golubev et al., *Problemi Protshnosti, N1*: 67 (1981).

7. V. K. Golubev et al., *Problemi Protshnosti, N6*: 28 (1985).

8. P. O. Pashkov and Z. M. Gelunova, *Shock Wave Effect on Hardened Steels*, Volgograd (1968).

9. A. N. Kiselev, *Fizika Gorenia i Vzriva, 11*: 945 (1975).

10. L. E. Murr et al, *Scripta Met., 12*: 425 (1978).

11. M. A. Meyers and J. R. L. Guimaraes, *Mater. Sci. and Eng., 24*: 289 (1976).

12. M. A. Meyers, N. N. Thadani, and S. N. Chang, *J. de Phys., C3-49*: 355 (1988).

13. A. M. Molodets, A. V. Orlov and A. N. Dremin, *Detonation. Proc. IX All-Union Symp. on Combustion and Explosion*, Chernogolovka, USSR: 74 (1989).

14. A. M. Molodets, A. I. Melkumov and A. N. Dremin, *Detonation Proc. 4th All-Union Meeting on Detonation*, Telavi, USSR 1: 251 (1988).

15. G. V. Stepanov, Elastic-plastic Deformation of Metals Under the Impulse Loading Effect, Kiev, Naukova Dumka (1979).

16. S. S. Gorelik, A. N. Rastorguev and V. A. Skakov, *Rentgenographic and Electronographic Analysis*, Moscow, Metallurgia (1970).

17. V. R. Regel, A. I. Slutkser and E. E. Tomashevski, *Solid Strength Kinetic Nature*, Moscow, Nauka (1974).

18. A. M. Molodets and A. N. Dremin, *Fizika Gorenia i Vzriva N1: 88* (1983).

70

The Influence of Shock Pre-Strain and Peak Pressure on Spall Behavior of 4340 Steel

A. K. ZUREK, CH. E. FRANTZ, AND G. T. GRAY, III

Materials Science and Technology Division
Los Alamos National Laboratory
Los Alamos, New Mexico 87545, U.S.A.

A fundamental study of the influence of peak stress amplitude and pre-strain on the spall fracture of pearlitic 4340 steel is presented. Spall tests were performed at projectile velocities to achieve approximately 5, 10 and 15 GPa peak stress amplitudes. Some spall tests were preceded by a pre-shock and recovery test at 10 and 15 GPa. Spall strength measurements suggest that there is a decrease in the spall strength of 4340 with an increase in the shock wave amplitude as the transition pressure of 13.1 GPa is approached. At this transition pressure, a substantial increase in the spall strength, as well as a change in a mode of fracture from brittle to ductile are observed, both attributed to the allotropic phase transformation at this amplitude.

I. INTRODUCTION

The majority of shock-spall studies have concentrated on investigating the dynamic fracture and strengthening of annealed or stress-relieved metals and alloys [1,2]. Few studies have explored the shock response of materials possessing a pre-existing dislocation substructure formed via heat treatment, quasi-static deformation or prior shock loading [3-6].

An additional complication in the dynamic response of iron and steels is the reversible, allotropic phase transformation which occurs at approximately 13 Gpa. Above this pressure the spall morphology [7-13] and the measured spall strength of iron and alloyed steel (this study) changes drastically. The phase transformation, α to ϵ and back to α, has been linked to the change in the mode of fracture from brittle to ductile and to the formation of the smooth fracture topography at spall pressures above 13 GPa [7-13].

The purpose of this study is to correlate the measured changes in the spall strength and the observed changes in fracture morphology with the peak stress amplitude and the changes in deformation substructure caused by the shock preceding the spall test.

II. EXPERIMENTAL PROCEDURE

The material used in this investigation was pearlitic 4340 steel. Shock recovery experiments, preceding spall, were performed using an 80-mm single-stage-gas gun as described elsewhere [6,14]. The flyer plates and all assembly components were fabricated from 4340 steel.

The spall assembly consisted of a 7-mm-thick, 19-mm-dia sample, surrounded by two concentric rings with outside diameters of 25.4 and 57 mm. The sample and the surrounding rings were placed on a plexiglass backing plate (1.37-mm-thick) and then on a plexiglass holder 5.7-mm-thick and 76-mm-dia. A manganin gauge measuring the resistance vs. time response during the spall test, was placed immediately behind the center of the sample and sandwiched between the plexiglass backing plate and plexiglass holder.

Spall tests with and without a shock pre-stresses were performed at 5, 10 and 15 GPa peak stress amplitudes for a 1-μs pulse duration. Some spall samples were pre-shocked in a shock recovery test at 10 and 15 GPa peak stress amplitude for 1-μs. The residual plastic strain in the pre-shocked samples (defined as the change in sample thickness divided by the starting sample thickness) was $\leq 2\%$. All samples were soft-recovered in a water catch chamber positioned behind the impact area. The fracture topography of all the spalled samples was analyzed using scanning electron microscopy (SEM). In addition, samples for optical and transmission electron microscopy (TEM) were cut from selected shock recovered and spalled samples.

III. RESULTS AND DISCUSSION

In order to present the results it is of value to examine the effects of shock pre-stress on the microstructure and stress-strain curves obtained by subsequent reloading. The microstructures of samples subjected to peak stress amplitudes below the level needed for the phase transformation consisted of a lower density of deformation twins and dislocations than observed in samples which were spalled at 15 GPa or pre-shocked at 15 GPa and subsequently spalled at either 10 or 15 GPa. Fig. 1 shows the high density of dislocations and deformation twins observed in 4340 pre-shocked at 15 GPa. This high density of dislocations combined with

FIG.1 *Transmission electron micrograph of 4340 steel shock pre-stressed at 15 GPa showing high dislocation and twin substructure.*

the high density of twins is reflected in the increase in the reload yield stress from 350 MPa for the annealed pearlitic 4340 steel to 500 and 620 MPa for 4340 samples which were pre-shocked at 592 m/s (10 GPa pressure) and 871 m/s (15 GPa pressure), respectively (Fig. 2). The reload stress-strain curves of the shock-recovered 4340 further show that, the rate of strain hardening is higher initially and then quickly saturates. Due to the shock hardening introduced during the pre-shock the sample appears to be saturated by a high density dislocation networks and numerous deformation twins, prior to the spall event. This microstructure may provide the nucleation sites for either brittle or ductile fracture.

Spall experiments were conducted on both as heat treated and shock pre-stressed samples. The spall strength are summarized in Table 1. The numbers reflect an average of two or more tests.

The spall strength measurements suggest that there is a decrease in the spall strength of 4340 associated with an increase in the spall pressure up to the α - ϵ transformation pressure. Samples spalled above this pressure show a significant

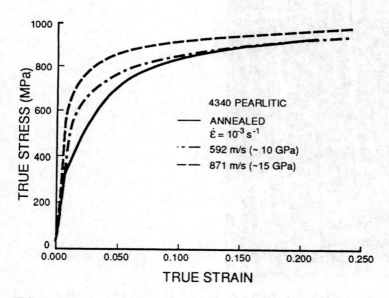

FIG. 2 *The reload true stress – true strain curves of 4340 steel, as annealed, and pre-shocked at 10 and 15 GPa.*

TABLE 1

Pressure Amplitude (GPa)	SPALL STRENGTH (GPa)		
	Spall Test	Pre-shock at 10 GPa and Spall	Pre-shock at 15 GPa and Spall
5	- 3.1(B)	NA	NA
10	- 2.6(B)	- 2.3(B)	- 2.8(B)
15	- 4.8(D)	- 2.9(D)	- 3.9(D)

(B) corresponds to brittle mode of fracture
(D) corresponds to ductile mode of fracture

increase in the spall strength combined with a change in the mode of fracture, from cleavage below, to ductile fracture above the phase transition pressure. Shock pre-stressing prior to the spall is observed to substantially change the spall strength of the sample, decreasing the gap between the spall strength of the sample tested in a simple spall test, (below and above the phase transformation pressure), but no significant change with respect to the fracture mode is observed. Samples spalled below the phase transformation pressure either directly or preceded by a pre-shock regardless of pressure, exhibit a cleavage fracture mode, shown in Fig. 3. Fig. 4 shows the ductile mode of fracture obtained in a spall test performed at a pressure above 13 GPa. This fracture appearance is representative of the fracture obtained in a simple spall test or a spall test preceded by a pre-shock at an amplitude of 10 or 15 GPa.

In discussing the results it is of value to consider three important features: i) the decrease in spall strength with increase shock pressure up to the 13 GPa, ii) the observation that in the brittle spall regime (i.e. below 13 GPa) the spall strength is independent of shock pre-stress, iii) the decrease in spall strength in the ductile regime (i.e. greater than 13 GPa) for shock pre-stress of 10 or 15 GPa.

In spall fracture it is difficult to consider the separation of nucleation and propagation events in the fracture process. However, in the brittle process at lower pressures, cleavage is influenced by the local hydrostatic pressure [15]. Thus in the spall process the fracture occurs under triaxial tension. As the superimposed

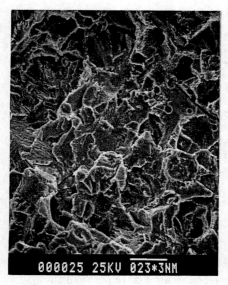

FIG.3 *Scanning electron micrograph of 4340 steel spalled at 10 GPa showing brittle mode of fracture.*

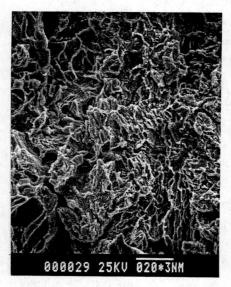

FIG.4 *Scanning electron micrograph of 4340 steel spalled at 15 GPa showing ductile mode of fracture.*

hydrostatic tensile pressure increases with shock pre-stress, the spall strength decreases. It is important to note, that the spall strength is independent of shock pre-stress in the brittle regime. This suggests that the brittle fracture process is dynamic and therefore independent of work hardening and damage induced by pre-shock. If we now consider the change in fracture mode with pressure, the question arises as to why there is such a pronounced change in the mode of fracture in this material associated with the allotropic phase transition, as shown in Figs. 3 and 4. In addition, the fracture surface morphology of the 4340 steel spalled at 15 GPa is very smooth in comparison to the rough surface of the sample spalled at 10 GPa. Several studies have previously investigated this problem [7-13]. Erkman [7], Banks [8] and Barber and Hollenbach [9] have all proposed that formation of a smooth spall is related to the existence of a rarefaction shock wave fan, created during the phase transformation at a high pressure. The phase transformation provides a sudden pressure drop in the wave resulting in a sudden rise in the tensile stress pulse occurring in a very narrow region. Therefore, the fracture region is very localized, as we have also observed by metallographic examination, and consequently the spall is much smoother in appearance. In the case of a spall below the phase transformation pressure, the tensile stress pulse increases slowly and monotonically, allowing for the deformation to occur in a wider region, and initiation of a large number of fracture sites. Consequently, we observe a very irregular deformation zone extending deep into the material. Our metallographic observations of the cross sections of spalled samples are consistent with this hypothesis.

An additional reason for the ductile mode of fracture at a higher stress amplitude, may be a sudden local increase in temperature, prompted by the thermodynamics of the phase transformation and a high deformation strain rate localized in a very narrow region. Recently, Zehnder and Rosakis [16], used an infrared method to measure the local temperature increase at the vicinity of dynamically propagating cracks in 4340 steel. Their experimental data show that a temperature increase of up to 465°C may occur at the tip of the dynamically propagating cracks in 4340 steel, and that the region of intense heating (> 100°C) may extend to as much as a third of the active plastic zone size. Others have shown that such a temperature increase in Armco iron [17] or in low carbon steel [18], deformed at a strain rate of approximately 10^4 s^1, is sufficient to promote ductile fracture even when the spall pressure is far below that necessary for the α - ϵ phase transformation. This temperature is above the dynamic fracture toughness corresponding to the ductile-to-brittle transformation temperature DTBT, promoting void growth and coalescence. Thus at pressures above 13 GPa we propose that either the substructure modifications due to the α - ϵ and ϵ - α transformations or the local temperature increase may raise the stress for brittle fracture above the peak stress level. Hence, ductile rather than brittle fracture is observed. The observation that the spall strength in the ductile regime is decreased by shock pre-stress is consistent with this model because both local deformation events occurring during pre-stress and the decrease in hardening capacity due to shock pre-stress would aid the ductile fracture process and thus lower the observed spall strength. Further experiments are in progress to examine the link between the deformation structure evolution and fracture behavior of 4340 steel.

IV. SUMMARY

In this study we measured the spall strength of 4340 pearlitic steel as a function of peak stress amplitude and pre-stress and correlated it with the mode of fracture and substructure evolution. It is evident from this study that the phase transformation has a major influence on the change in the mode of fracture from brittle (below), to ductile (above the 13 GPa pressure) and the spall morphology from rough (below) to smooth (above 13 GPa) and consequently on a spall strength. Spall strength of this material increases due to the phase transformation. Pre-stress decreases the spall strength of 4340 spalled at above the 13 GPa transition pressure, but does not change the mode of fracture. In addition, we have qualitatively rationalized the reason for the change in the mode of fracture from cleavage to ductile at 13 GPa. Although, we do not yet have conclusive evidence, the possibility exists that the excessive heating due to the localized deformation initiated by the phase transformation, combined with high dislocation and twin density may by itself be sufficient to initiate ductile mode of fracture in this material.

V. ACKNOWLEDGMENTS

The authors would like to thank J. David Embury for his help in preparing this manuscript. This work was performed under the auspices of the U.S. Department of Energy.

VI. REFERENCES

1. W. C. Leslie, in *Metallurgical Effects at High Strain Rates*, (R. W. Rohde, B. M. Butcher, J. R. Holland and C. H. Karners, eds.): Plenum Press, New York, 571 (1981).

2. L. E. Murr in *Shock Waves and High-Strain-Rate Phenomena in Metals*, (M. A. Meyers and L. E. Murr eds.): Plenum Press, New York, 607 (1981).

3. B. Kasmi and L. E. Murr, Ibid, p. 753.

4. T. M. Sobolenko and T. S. Teslenko, in *IX Int. Conf. on High Energy Rate Fabrication*, (I. V. Yakovlev and V. F. Nesterenko eds.): Novosibirsk, 116, (1986).

5. N. V. Gubareva, A. N. Kiselev, T. M. Sobolenko and T. S. Teslenko in *Impact Loading and Dynamic Behavior of Materials*, (C. Y. Chiem, H.-D. Kunze, and L. W. Meyer eds.): DGM Verlag, 801 (1988).

6. P. S. Follansbee and G. T. Gray III, "Dynamic Deformation of Shock Prestrained Copper" submitted to *Materials Science and Engineering*.

7. I. O. Erkman, *J. Appl. Phys. 31*: 939 (1961).

8. E. E. Banks, *J.I.S.I. 206*: 1022 (1968).

9. L. M. Barker and R. E. Hollenbach, *J. Appl. Phys. 45*: 4872 (1974).

10. A. G. Ivanov and S. A. Novocov, *J. Exp. Theor. Phys. (USSR) 40*: 1880 (1961).

11. A. S. Balchan, *J. Appl. Phys. 34*: 241 (1963).

12. M. A. Meyers, C. Sarzeta and C-Y. Hsu, *Met. Trans. 11A*: 1737 (1980).

13. M. A. Meyers and C. T. Aimone, in *Dynamic Fracture (Spalling) of Metals, in Progress in Materials Science* (J. W. Christian, P. Haasen, and T. B. Massalski, eds.): Pergamon Press, vol. 28, No.1: (1983).

14. J. C. Huang and G. T. Gray III, *Met. Trans. 20A*: 1061 (1989).

15. D. Teirlinck, F. Zok, J. D. Embury and M. F. Ashby, *Acta Met. 36*: 1213 (1988).

16. A. T. Zehnder and A. J. Rosakis, *On the Temperature Distribution at the Vicinity of Dynamically Propagating Cracks in 4340 Steel*, California Institute of Technology Report No. SM 89-2: (1989).

17. D. R. Curran, in *Shock Waves and Mechanical Properties of Solids*, (J. J. Burke and V. Weiss, eds.): Syracuse University Press, New York, 121 (1971).

18. A. K. Zurek, P. S. Follansbee and J. Hack, *Met. Trans. 21A*: 431 (1990).

71

Dynamic Fracture (Spalling) of Some Structural Steels

J. BUCHAR, S. ROLC* AND J. PECHÁČEK*

Institute of Physical Metallurgy C.A.S.
Zizkova 22, 616 Brno, Czechoslovakia

*Materials and Technology Research Institute,
Rybkova 2a, 602 00 Brno, Czechoslovakia

Fracture behavior of several structural steels under explosive attack was studied. In the first part the fracture behavior of the given steel was evaluated from the point of crack initiation under high loading rates $\dot{K}_{IC}$ i.e. in terms of fracture toughness K_{IC}. The results on dynamic fracture (spalling) were characterized by the spall strength σ_C. It was found that σ_C was dependent on the fracture toughness and on the strain rate in spall plane $\dot{\varepsilon}$ as well. The dependence σ_C (K_{IC}, $\dot{\varepsilon}$) was nearly the same like in the case of purely brittle materials like rocks, etc.

I. INTRODUCTION

The origin of spall is linked to material microstructure. Considerable research has focused on the identification and development of the microstructural events leading to spall failure of material [1-3]. Even if this procedure is very successful in predicting of material behavior under different dynamic loading conditions, in some applications a full knowledge of material microstructure history is not necessary. Rather, a prediction of spall strength, the time to fracture, and perhaps some measure of the fragment size is needed. This second approach is much less sensitive to microstructural details of the tested material. This procedure, developed in [4-7], enables the prediction of main spall characteristics, especially spall strength in terms of some continuum measure of fracture dissipation such as fracture toughness, flow stress, viscosity, or surface tension.

This paper presents the results of experiments that were performed to generate data connecting the spall behavior with the behavior of single cracks, which were studied under very similar loading conditions.

II. EXPERIMENTAL DETAILS

For the experiments, several steels with different microstructures and tensile properties were used.

The aim of our studies is verified by the relation between spall strength, σ_c , and fracture toughness K_{Ic} [4,6]:

$$\sigma_c \approx K_{Ic}{}^{2/3} \; \dot{\varepsilon}^{\,1/3} \tag{1}$$

where $\dot{\varepsilon}$ is the strain rate in spall plane.

Owing to this fact it is necessary to obtain a proper value of K_{Ic} . K_{Ic} is according to standards evaluated under plane strain Mode I fracture at loading rates $\dot{K}_I$ up to 10^5 MPam$^{1/2}$. As it is discussed e.g. in [7] the stress state under stress wave propagation differs from plane strain conditions. At the same time the strain rate at the crack tip loaded by stress wave can achieve values $\dot{\varepsilon} \approx 10^4 - 10^6 s^{-1}$, corresponds to the values of $\dot{K}_I \varepsilon (10^8 - 10^{10})$MPam$^{1/2}s^{-1}$. We used the method of wedge loaded compact tension (WLCT) specimens developed in [8,9] which enables obtain loading rates 10^7 MPam$^{1/2}s^{-1}$. Even if this value of K_I is more close to loading conditions leading to spall fracture, the first problem mentioned above remains. The dynamic spall strength was measured in spall test using arrangement outlined in Fig. 1. The loading pressure stress pulses p (t) were measured using a manganin gauge pulsed by a Wheatstone bridge circuit conjunction with a delayed triggering system [10]. The numerical simulation of this loading revealed that one-dimensional conditions in specimens were prevailing, i.e. the macroscopic strain was uniaxial [11]. The free surface velocity, v, was measured by the capacitor gauge technique [10].

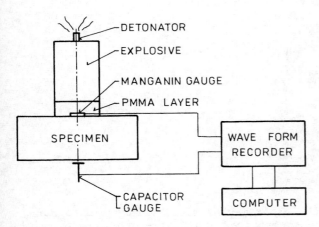

FIG. 1 *Experimental set-up used for the investigation.*

The peak value of pressure pulse, p_m, was nearly the same for all tested specimens, $p_m \approx 12$ GPa.

III. EXPERIMENTAL RESULTS AND THEIR DISCUSSION

In the first step the temperature dependence of fracture toughness K_{Id} was determined. The dependence of K_{Id} vs temperature T can be fitted by the function:

$$K_{Id} = A + B \exp (C \cdot T) \tag{2}$$

where material parameters A,B,C and K_{Id} at room temperatures are given in [11].

In the next step the spall strength measurements were performed. The values of the spall strength, σ_c, were evaluated according to the procedure outlined e.g. in [10], i.e. from the experimental records of the free-surface velocities. The strain rates in the spall plane were estimated using the theory developed in [12] which was also used in [7].

In Fig. 2 the dependence of the spall strength, σ_c, on the fracture toughness, K_{Id}, is displayed. The experimental results are in reasonable agreement with theoretical conclusion for the spall strength of the brittle materials given in [6], i.e.:

$$\sigma_c = (3\rho c_0 K_{Id}^2 \dot{\varepsilon})1/3 \tag{3}$$

where ρ is the material density and c_0 is the longitudinal elastic wave velocity.

FIG. 2 *Comparison of the experimental spall strength data for tested steel with the theoretical spall strength.*

In order to verify these conclusions we performed experiments on a low alloyed high strength steel (0.75 C, 18 Mn, 1.6Si, 0.014P, 0.005S, 1.03Cr, 0.14Ni, 0.08Cu in wt %). Using the same specimen geometry like for the structural steels we obtained instead of spall fracture the fragmentation of the specimen - see Fig. 3. This kind of fracture was prevailing for relatively low values of detonation pressures, $p_m \leq 8$ GPa. For higher pressures the transition of fragmentation to spall fracture was observed. The values of spall strength

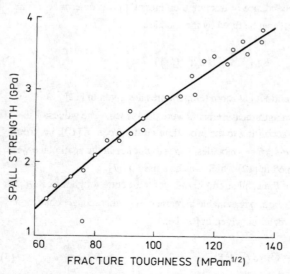

FIG. 3 *Optical micrograph of the explosive loaded specimen of UHSLA steel. Magnification 0.6 times.*

ranged from 2.7 up to 3.1 GPa. The fracture toughness of these steel is very low, at room temperature $K_{Id} \approx 18$ MPam $^{1/2}$. The spall strength σ_c, is thus higher than the prediction from the Equation (3). A more detailed investigation of this steel under plane wave loading which included different heat treatment and thus different mean carbides spacing, 1, led to the conclusion that the transition of "radial" fragmentation to spall fracture was determined by the dimensionless number $K_{Ic}/p_m l^{1/2}$. The critical value of this number for the given steel is about 0.7-0.8.

These results show the influence of the steel microstructure on the resulting fracture behavior under shock loading. The role of the carbides was discussed in [7].

The detailed metallographic and fractographic analysis of the recovered specimens made from structural steels did not show on some role of carbides. For these steels it was found that the fracture was predominantly brittle, both as cleavage and as transgranular. Only small areas of ductile fracture were found.

In order to obtain more information on the validity of the theory of spall behavior we performed a series of experiments on the steel 1OGN2MFA [13] where the maximum detonation pressure p_m was changed from 3 up to 35 GPa. In Fig. 4 the dependence of σ_c, on $\dot{\varepsilon}$, together with values of p_m, is shown. It may be seen that for loading pressures $p_m \geq$ 25GPa which corresponds to the strain rate $\dot{\varepsilon} \approx 3 \times 10^5 s^{-1}$ there is a change in the slope of the dependence $\sigma_c(\dot{\varepsilon})$. This phenomenon may be interpreted as the transition from brittle-to-ductile spall behavior as it is described in [14]. The detailed investigation of the recovered specimens supports this hypothesis [13].

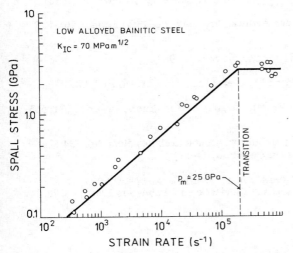

FIG. 4 *Spall strength data for low alloyed bainitic steel (1OGN2MFA) as the function of strain rate.*

IV. CONCLUSIONS

The experimental results obtained in the given paper support the spall theory which was developed in [4-6]. The conclusions of this theory expressed in Eq. (3) are valid at least for the steels where cleavage fracture occurs. The validity of this approach to the spall behavior is dependent on the microstructure of the tested steel as it was approved for one low alloyed high strength steel. The influence of maximum loading pressure can be also expected.

It seems that these spall theories enable to predict the spall behavior on the knowledge of standard fracture properties.

REFERENCES

1. L. Davison and R. A. Graham, *Phys. Rept.*, 5: 257 (1979).

2. M. A. Meyers and C. T. Aimone, *Prog. Mat. Sci.*, 28:1 (1988).

3. D. R. Curran, L. Seaman and D. A. Shockey, *Phys. Rept.*, 147: 255, (1987).

4. D. E. Grady and M.E. Kipp, *Mech. Mat.*, 4: 311 (1985).

5. L. A. Glen and A. Chudnovsky, *J. Appl. Phys.*, 59: 1379 (1986).

6. D. E. Grady, *J. Mech. Phys. Solids*, 36: 353 (1988).

7. A. K. Zurek, P.S. Follansbee and J. Jack, *Metallurgical Transactions*, 21A: 431 (1990).

8. J. R. Klepaczko, *J. Eng. Mat. and Technology*, 104: 29 (1986).

9. M. Holzmann, J. Buchar and Z. Bilek, *Journal de Physique*, 46: 155 (1985).

10. J. Buchar, *Acta Technica CSAV*, 5: 545 (1988).

11. J. Buchar and J. Krejčí, *Metallic Materials*, 23: 477 (1985).

12. V.I. Romanchenko and G.V. Stepanov, *J. Appl. Mech. Tech.*, 21: 555 (1981).

13. J. Buchar, Spall damage in low alloyed bainitic steel, Res. Rept. No. 734/521. *Institute of Metallurgy Brno*, (1989).

14. D.E. Grady, in *Shock Waves in Condensed Matter 1987*, S. C. Schmidt and N.C. Holmes, eds., Elsevier Science Publishers 1988, p. 327.

72

The Dynamic Strength of Copper Single Crystals

G. I. KANEL, S. V. RASORENOV, and V. E. FORTOV

Institute for High Temperatures
USSR Academy of Sciences
IVTAN, Moscow, USSR

Department of Institute for Chemical Physics
USSR Academy of Sciences
Chernogolovka, Moscow Region, USSR

Plate-impact experiments were performed to study the spall strength of copper single crystals and, for comparison, polycrystalline commercial-grade copper. Impacts of aluminum flyer plates 0.2-0.4 mm thick with velocity 620-670 m/s induced shock load pulses in the samples of thickness 0.6-4 mm. The VISAR wave profile measurements at sample rear free surface were carried out. Results of measurements show that spall strength of polycrystalline copper in this condition is 1.36 GPa and single crystal spall strength is between 3.3 and 4.6 GPa. The great excess of single crystal spall strength above polycrystal one is related with a small defect concentration in the monocrystalline material.

I. INTRODUCTION

The objective of this work was to examine the mechanisms responsible for elementary fracture events under dynamic loading. The material tensional strength is susceptible to various factors, such as mechanical inclusions, grain boundaries and other inhomogeneities. Single crystals are largely free from influence of similar factors. It is anticipated that single crystals, which are free from various stress concentrators, have an ultimate dynamic strength for solids.

The dynamic strength of materials in microsecond load durations range is studied using analysis of spall fracture phenomena [1,2]. A comprehensive review of spalling has been written by Meyers and Aimone [3]. The spalling of monocrystalline and polycrystalline copper was investigated by McQueen and Marsh [4], who conducted examination of specimens after shock-wave loading by the plate impact in the pressure range 25-60 GPa. The fracturing stress value was estimated in excess of 15 GPa for both monocrystalline and polycrystalline copper at load duration 1-2 μs. The experimental procedures were not precise: strength was obtained from comparison of recovery experiments with corresponding hydrodynamic calculations. Subsequent measurements (see, for example, [5-9]) do not verify such high values of fracturing stresses for polycrystalline copper.

II. EXPERIMENTAL TECHNIQUE

The most reliable and informative method for the determination of spall stress is based on measurements of free surface velocity profiles during load pulse refraction [2, 6-8] In the work reported here plate-impact experiments were performed to determine the fracturing stress under spalling for copper single crystals and, for comparison, polycrystalline commercial-grade copper M2. A single crystal was grown by Czochralski method. Specimens were cut out of single crystal bar. They were 30 mm in diameter and varied from 0.6 to 4.5 mm in thickness. Shock loading was carried out in directions <111> and <100>. Polycrystalline copper specimens were cut out of rolled sheets 2-4 mm thick.

Shock load pulses were produced by colliding aluminium plate impactors 0.2-0.4 mm thick with samples. Impactors were accelerated by explosive means up to a velocity of 620-670 m/s. The VISAR [10] free surface velocity wave profile measurements were made. Results of measurements are shown in the figures.

III. RESULTS

The elastic-plastic compression wave causes fast rise of the free surface velocity. The unloading wave produces a following velocity decreasing. The interference of this incident unloading wave with a rarefaction wave reflected from the free surface is accompanied by the appearance of tensile stresses in the specimen. A fracture of the material within the specimen (spalling) occurs after tensile stresses reach sufficiently high levels. Tensile stress value decreases during the fracture process and this results in the appearance of a compression wave, so called spall pulse, that propagates in the expanded

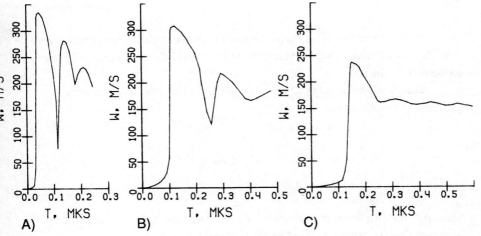

FIG. 1. *Free surface velocity profiles for copper samples.* A
- single crystal, load direction <111>, sample thickness H =
0.63 mm, impactor thickness h = 0.2 mm; B - single crystal,
<111>, H = 1.95 mm, h = 0.4 mm; C - polycrystal, H = 2.73 mm,
h = 0.4 mm.

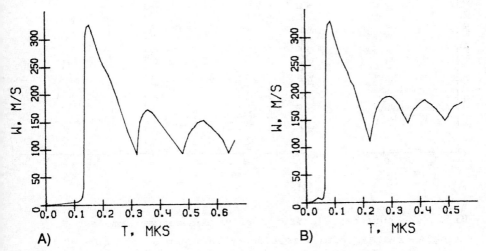

FIG. 2. *Free surface velocity profiles for copper single*
crystals.   A - load direction <100>, sample thickness H =
4.25, impactor thickness h = 0.4; B - annealed sample, load
direction <100>, H = 4.6 mm, h = 0.4 mm.

material to free the surface of the sample. Following attenuated oscillations of the surface and velocity are connected with wave reflections between sample surface and spall plane repeated many times.

The fracture stress values σ* were determined from the free surface velocity profiles w (t) by relationship:

$$\sigma^* = \frac{1}{2}\,\rho_0\,c_0\,\Delta\,w \tag{1}$$

where ρ_0, c_0 are the initial material density and sound velocity, Δ w is the velocity drop from peak value to its value ahead of spall pulse front.

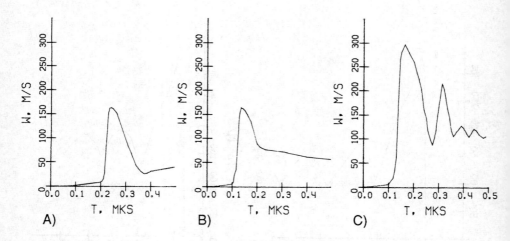

FIG. 3. *Free surface velocity profiles for copper samples. A - single crystal, load direction <111>, sample thickness H = 4.35 mm, impactor thickness h = 0.2 mm; B - polycrystal, H = 3.85 mm; h = 0.2 mm; C - example of fracture within spall plate in result of spall pulse action, single crystal, load direction <111>, H = 1.95 mm, h = 0.4 mm.*

The difference in results of measurements for polycrystalline copper and single crystals with different orientations can be seen immediately from a comparison of wave profiles. The spall strength of single crystals is considerably higher than that of polycrystals. Spall pulses on wave profiles for single crystals are pronounced, have steeper fronts and higher amplitudes than in the case of polycrystalline copper. It is possible to conclude that fracture of single crystals is brittle in nature in that it has more pronounced threshold and grows faster. Wave profiles are also differentiated by the rate of attenuation of velocity oscillations. Evidently the attenuation would be smallest in case of the smooth fracture surface. If this surface is very developed and layers of material near the fracture surface are loosened, a wave reflection would be accompanied by a substantial dissipation. In this count the distinction between polycrystalline copper and single crystal with orientation <100> is most evident. In experiments with single crystals of orientation <111> the first spall pulse is clearly pronounced usually, but after that velocity oscillations rapidly damp. In some cases shock-wave profiles permit to suppose the producing of fracture within spall plate in result of a spall pulse action (Fig. 3C). The spall pulse slope falls off practically to zero as the shock load peak value decreases (Fig. 3A, B).

Results of spall strength measurements for polycrystalline copper and single crystals are presented in the table.

TABLE. SPALL STRENGTH MEASUREMENTS FOR MONOCRYSTALLINE AND POLYCRYSTALLINE COPPER

Sample	Sample thickness, mm	Impactor thickness, mm	Spall strength, GPa
Polycrystal	2.7	0.4	1.35
Polycrystal	3.85	0.2	1.5
Single-crystal <111>	1.9	0.4	3.3-3.9
Single-crystal <111>	0.7	0.2	4.5-4.6
Single-crystal <111>	4.35	0.2	>2.5
Single-crystal <100>	4.3	0.4	4.1-4.2
Single-crystal <100> annealed, 900 C, 2h	4.5	0.4	3.9-4.0

IV. DISCUSSION

The spall strength values obtained in this work for polycrystalline copper are in agreement with data of previous works [7,8]. The value of monocrystalline spall strength is not reproducible exactly (the spread in data 2-4 measurements is shown in the table), but this value exceeds approximately three times as much the strength of polycrystalline samples. The spread in experimental data is rather large, but nevertheless it may be argued that the resistance to fracture increases with a load pulse duration reduction.

It is known that closely packed planes <111> in the f.c.c. structure are weakest not only for slip, but for crack growth and for ideal plane breakage, too. Measurements show some tendency to increasing of a dynamic strength under change in the load direction from <111> to <100>.

A comparison of the initial load pulse duration and period of velocity oscillations after spalling can indicate an additional information about mechanism of the spalling. Such analysis for the single crystal with orientation <100> shows that delay of fracture probably exists. The width of the initial load pulse at the velocity level immediately ahead spall pulse, is higher than the period of velocity oscillations after spalling. The difference between these values is about 20 ns. It is unlikely that this effect is associated with an elastic-plastic behavior of the material because yield stress for copper is small as it can be seen from elastic precursors.

Estimations of the ultimate theoretical tensile strength for copper in terms of a surface energy or cohesive energy [11] indicate the value about 24-39 GPa. The dynamic strength of copper single crystals is about 20% of ultimate theoretical strength and 3-4 times as high as the one for polycrystalline samples. This difference is undoubtedly associated with small defects concentration in the monocrystalline material. This conclusion is verified by measurements of the spall strength for glasses which have a high degree of homogeneity [12] and for commercial alloys in a wide range of shock amplitudes [13]. According to results of measurements, the spall strength of glasses is about 5 GPa or more, and this exceeds strength of most of commercial alloys. Spall strength of alloys does not depend on the peak shock pressure. This indicates that microdefects, generated by high-rate plastic strain in shock front, do not influence the strength. Christy, Pak and Meyers [9] have shown in experiments on copper with various degrees of purity that primary damage nucleation sites are grain boundaries, second-phase particles and other inhomogeneities. Probably the nucleation of fracture at deformation microdefects produced by deformation occurred in experiments with single crystals only. Metallographic examination of single crystals after spalling is desirable for the observation of elementary fracture events.

In routine conditions the largest strength is obtained in defect-free filamental single crystals with a micron cross-section [14]. The strength of filamental copper single crystals can reach to 3.5 GPa; this is at least an order of magnitude greater than the strength of typical size samples. Shock-wave loading is more suitable for obtaining the ultimate strength because deformation is one-dimensional with almost spherical tensor of stresses in this condition; any influence of sample surface as well as isolated and relatively large inhomogeneities on the deformation and fracture is ruled out.

V. CONCLUSION

Measurements of free surface velocity profiles for copper single crystals show that dynamic strength of single crystals is about 20% of ultimate theoretical strength and 3-4 times as high as the one for polycrystalline samples. This fact is associated with a small defect concentration in the monocrystalline material. Metallographic examination of single crystals after spalling is desirable for the subsequent study of mechanisms responsible for the elementary fracture events.

REFERENCES

1. Я. Б. Зельдович, Ю. П. Райзер. Физика ударных волн и высокотемпературных гидродинамических явлений. Москва, Наука (1966).

2. G.I. Kanel, V.E. Fortov. *Advances in Mechanics, 10:* 3 (1987).

3. M.A. Meyers, C.T. Aimone, *Prog. in Mat. Sci. 28:* 1 (1983).

4. R.G. McQueen, S.P. Marsh, *J. Appl. Phys. 33:* 654 (1962).

5. Smith J.H., *ASTM Spec. Techn. Publ. 336:* 264 (1962).

6. С. А. Новиков, И. И. Дивнов, А. Г. Иванов, Физика металлов и металловедение 24: 608 (1966).

7. S. Cochran, D. Banner, *J. Appl. Phys. 48:* 2729 (1977).

8. Г. И. Канель, С. В. Разоренов, В. Е. Фортов, Доклады АН СССР, 275: 369 (1984).

9. S. Christy, H.-R. Pak, M.A. Meyers, *Metallurg. Applications of Shock-Wave and High-Strain-Rate Phenomena:* 835 (1986).

10. L.M. Barker, R.E. Hollenbach, *J. Appl. Phys. 43:* 4669 (1972).

11. B.L. Averbach, *Fracture, ed. by H. Liebowitz, 1.* Academic Press, N.-Y. & London (1968).

12. М. М. Абазехов, Г. И. Канель, С. В. Разоренов, В. Е. Фортов. Исследования свойств вещества в экстремальных условиях: 137 (1990).

13. Г. И. Канель, С. В. Разоренов, В. Е. Фортов, Доклады АН СССР 294: (1987)

14. С. З. Бокштейн, Нитевидные кристаллы и их свойства: Москва, (1966).

73

Compression-Induced High-Strain-Rate Void Collapse

S. N. CHANG and S. NEMAT-NASSER

Center of Excellence for Advanced Materials
University of California, San Diego
La Jolla, CA 92093, U.S.A.

Void collapse is used to prove material response under very large strain and strain rates, employing the Hopkinson bar technique, as well as quasi-static testing methods. Here, we summarize the results of a series of experiments of this kind on single-crystal copper, and polycrystal mild steel, pure iron, and copper. As the void collapses into a crack, the material in the neighborhood of the collapsing void undergoes extremely large deformations at strain rates orders of magnitude larger than the nominal overall ones. It is observed that tensile microcracks are nucleated at the intersections of slip planes in single-crystal copper, growing both along the slip lines, as well as normal to the applied compression. In polycrystals the cracks are generally normal to the applied compression. The fracture surfaces resemble brittle failure, especially in pure iron. Recrystallization in pure single-crystal copper occurs when the void is collapsed at high overall nominal strain rates of $10^4/s$ and greater. Tension cracks seem to grow into newly formed crystals.

I. INTRODUCTION

The possibility of dynamic failure by tensile cracking normal to the applied compression in a very ductile single crystal, was first examined by Nemat-Nasser and Hori [1]. These authors analytically studied dynamic void collapse in single crystals deforming plastically by rate-dependent crystallographic slip, and elastically by lattice distortion. They showed by incremental computations that a void which has been collapsed under a compressive pulse, may grow as a crack in its own plane, normal to the direction of compression, during unloading. This represents a newly identified dynamic failure mode of ductile crystalline solids in overall pure compression, requiring experimental verification. To this end, a series of experiments on void collapse in single-crystal copper, mild steel (AISI 1018 steel), and pure iron have been performed; see, Nemat-Nasser and Chang [2]. Here we summarize some of the results presented by these authors.

II.   SAMPLE PREPARATION AND EXPERIMENTAL PROCEDURES

The experiments are performed in a compression Hopkinson bar, as well as quasi-statically. Sample preparation, experimental setup, and other related details are found in [2]. Here we confine attention to a few selected results. The pure copper (99.99%) samples (1 mm thickness, 7.6 mm width, and 9 mm length) are cut in the {011}-plane, and mechanically and electrolytically polished. A circular hole of about 120-150 μm is produced at the center of the sample by EDM. The sample is then electro-polished, suitably heat treated, sandwiched between two similar copper plates, and loaded in the $<0\bar{1}1>$-direction. All experiments are performed with a predetermined total nominal axial compression strain at pre-assigned strain rates. The samples with collapsed voids are then recovered and characterized, using optical microscopy and SEM. In addition, microhardness tests were performed at various locations, close to as well as away from the crack tips.

III.   RESULTS AND DISCUSSION

A.   *SINGLE-CRYSTAL COPPER*

With stress axis in the $[0\bar{1}1]$-direction, slip occurs on the $(\bar{1}\bar{1}1)$- and $(1\bar{1}1)$-planes. This results in slip on the (011)-plane in the $[\bar{2}1\bar{1}]$-direction in the $(\bar{1}\bar{1}1)$-plane and in the $[21\bar{1}]$-direction in the $(1\bar{1}1)$-plane; see Fig. 1. Figure 2 shows the slip lines near one tip of the partially collapsed void in a single-crystal copper specimen deformed at the nominal 1,100/s strain rate, by the total nominal axial compression strain of 9%. At the intersection of the slip planes, Lomer-Cottrell (L-C) sessile dislocations can develop from partial dislocations. The dislocations on the (100)-plane which is not a slip plane, are immobile. They can serve as obstacles to the movement of mobile dislocations on the $(\bar{1}\bar{1}1)$- and $(1\bar{1}1)$-planes. The density of these immobile dislocations increases during void collapse, resulting in substantial work-hardening. During unloading their presence may hinder reverse plastic deformation, leaving

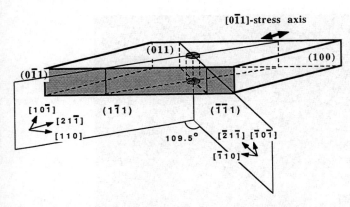

FIG. 1 *Primary slip systems in single crystal copper specimen cut parallel to the (011)-plan, and loaded in the [0$\bar{1}$1]-direction.*

FIG. 2 *Slip lines near the right tip of the partially collapsed void in a single crystal copper deformed at strain of -9% and at 1,100/s strain rate.*

microcracking as a favorable mechanism for stress relief. Note that in the early stages of unloading, additional L-C sessile dislocations may form as a result of reverse plastic flow. Clustering of these lattice defects can serve as a crack initiation mechanism. Crack growth in the <110>- and <112>- directions has been observed in silver and in gold crystals, by Lyles and Wilsdorf [3] and Wilsdorf [4] in *in situ* experiments.

Crack initiation by the above-discussed mechanism is evident in Fig. 3(a) which shows highly deformed regions at the tip of a collapsed void with nominal total strain of -31%, at nominal strain rate of 10^4/s. There are three main cracks, one in the [100]-direction, straight

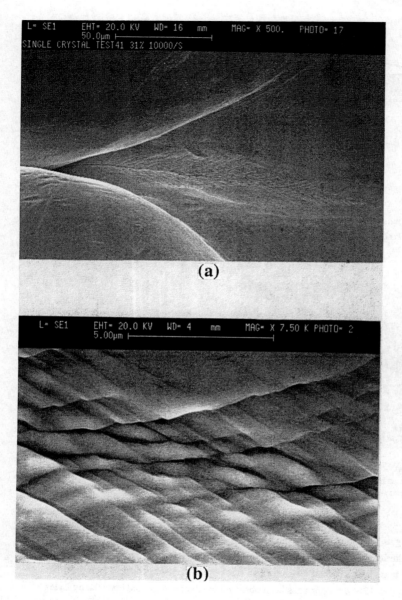

FIG. 3   (a) Highly deformed area ahead of a collapsed void in Cu
single-crystal; (b) microcracks nucleated at L-C locks in the
central region of (a); (c) polished surface of (a).

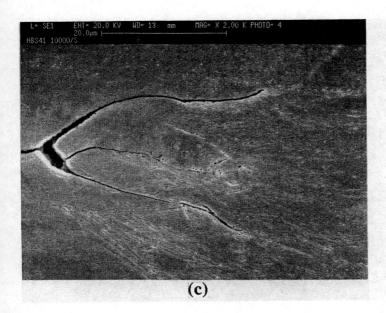

(c)

ahead of the tip of the collapsed void. The surface structure of the specimen around this crack clearly shows extensive plastic deformation on the $(\overline{1}11)$- and $(1\overline{1}1)$-planes; Fig. 3(b). This is a region with high-density L-C locks. Microcracks are then nucleated in unloading, and grow along slip lines. Close to the tip of the collapsed void, extremely high tensile stresses are produced during the removal of compression. This then leads to the formation of macroscopic cracks, straight ahead in the [100]-direction. Figure 3(a) also shows two additional main cracks, located almost symmetrically about the [100]-direction, along highly deformed curved slip-directions. Figure 3(c) is the polished surface of this specimen, showing these cracks.

B. RECRYSTALLIZATION

Void collapse involves strains of several hundred percent, at the vicinity of a collapsing void, even for the overall nominal strains of 20-30%. At a nominal strain rate of 10^4/s, the plastic flow near a collapsing void is almost adiabatic. The temperature close to the collapsed void can become quite high. Extremely high dislocation density and the accompanying plastic heating at very high strain rates can lead to recrystallization. This happens when the overall nominal strain exceeds -25%, at the overall nominal strain rate of 10^4/s, as seen in Fig. 4. Note that recrystallization is accompanied by fracturing, with cracks growing into the newly formed crystals. This experiment shows the recrystallization to occur *prior to unloading*, since tensile cracks which can only occur in unloading, extend into the new grains.

FIG. 4 *Back scattered electron image showing recrystallized grains on the electro-polished surface in Cu single-crystal specimen at 31% strain and at 10⁴/s strain rate.*

C.   *POLYCRYSTALS*

Similar experiments were performed on 1018 mild steel and pure iron (grain size of about 15 μm) specimens. Figure 5 compares void collapse in steel specimens at three different strain rates for the essentially the same total nominal strain. From this figure, the resistance of the material to plastic flow during compression loading and the extent of subsequent tensile cracking are seen to increase with increasing strain rate. The cracks extend through grains normal to the compression axis. These tensile cracks initially run straight ahead for a short distance and then branch out. Crack branching occurs even at the quasi-static strain rate, although their lengths are much shorter than those formed at higher loading rates. Similar results are obtained for void collapse experiment in pure iron and polycrystal copper; see [2].

D.   *FRACTURE SURFACE*

Figure 6 shows the fracture surface for a specimen tested at a 5×10³/s strain rate, indicating cleavage-like cracking during unloading. As is seen, the structure of these surfaces is quite

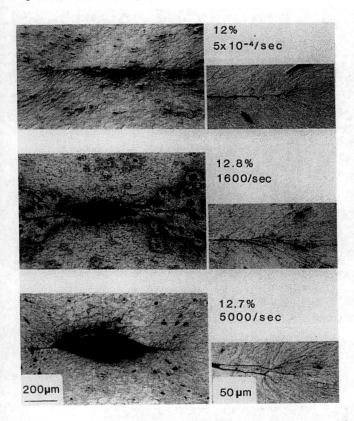

FIG. 5 *Void collapse and subsequent tensile cracking under uniaxial compression in 1018 steel: right-hand figures show magnified tensile cracks, formed at the right ends of collapsed voids.*

different than those produced in tension when the sample was pulled apart (after test) to expose the fracture surface. Figure 7 shows the fracture surface at the edges of a collapsed void in polycrystal pure iron. To the left of the figure is the surface of the crack formed during the *compression* test, indicating brittle fracturing. The right side shows ductile fracturing took place when (after the compression test was completed) the sample was pulled apart to expose the fracture surface.

FIG. 6 *Evidence of a cleavage-like river-shaped fracture surface at the tip of collapsed void in copper single-crystal.*

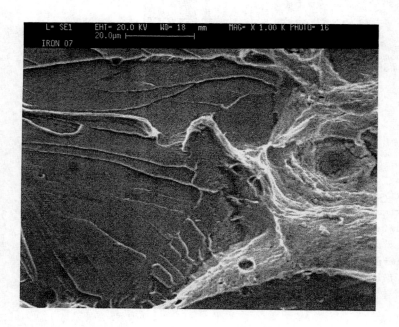

IV. CONCLUSIONS

The resistance to void collapse in ductile metals such as copper, iron, and mild steel, increases as the overall nominal strain rate is increased. Once a void is partially or fully collapsed, tensile cracks can develop at its elongated tips, during the removal of the compression pulse, when sufficient overall straining has taken place. For pure copper single crystals, the cracks grow in a direction normal to the applied compression, as well as in the dominant slip-planes. Recrystallization can occur before unloading, for sufficiently high strains and strain rates, followed by tensile cracking in unloading.

ACKNOWLEDGMENTS

The authors thank Messrs. J. B. Isaacs and J. Schwartz for their assistance in conducting the experiments. This research is supported by Army Research Office under contract No. DAAL-03-86-0169 to the University of California, San Diego.

REFERENCES

1. S. Nemat-Nasser and M. Hori, *J. Appl. Phys.* 62; .2746 (1987).

2. S. Nemat-Nasser and S. N. Chang, *Mech. Maters*: (to appear).

3. R. L. Lyles and H. G. F. Wilsdorf, *Acta Metall.* 23; 269 (1975).

4. H. G. F. Wilsdorf, *Maters. Sci. Eng.* 59; 1 (1983).

FIG. 7 *Fracture surface in pure iron showing brittle fracture (left) which occurred during compression loading at 4,100/s, and unloading; and ductile fracture (right) which occurred in pulling the specimen apart to expose the fracture surface.*

Section VII
Shock Phenomena and Superconductivity

74

Magnetic and Electrical Properties of Shock Compacted High-T$_c$ Superconducters

S.T. WEIR*, W.J. NELLIS*, C.L. SEAMAN[†], E.A. EARLY[†], M.B. MAPLE[†], M.J. KRAMER[††], Y. SYONO[°], M. KIKUCHI[°], P.C. McCANDLESS*, and W.F. BROCIOUS*

*Lawrence Livermore National Laboratory
University of California
Livermore, CA 94550

[†]Department of Physics
University of California, San Diego
La Jolla, CA 92093

[††]Ames Laboratory
Iowa State University
Ames, IA 50011

[°]Tohoku University
Katahira 2-1-1, Sendai 980
JAPAN

We present the results of shock compaction experiments on high-T$_c$ superconductors and describe the ways in which shock consolidation addresses critical problems concerning the fabrication of high J$_c$ bulk superconductors. In particular, shock compaction experiments on YBa$_2$Cu$_3$O$_7$ show that shock-induced defects can greatly increase intragranular critical current densities. The fabrication of crystallographically aligned Bi$_2$Sr$_2$CaCu$_2$O$_8$ samples by shock-compaction is also described. These experiments demonstrate the potential of the shock consolidation method as a means for fabricating bulk high-T$_c$ superconductors having high critical current densities.

I. INTRODUCTION

The shock-consolidation of high-T_c powders has many features which make it an interesting and promising method for fabricating bulk high-T_c samples having high critical current densities[1]. In the shock-consolidation method, the energy and heating are largely localized to the surfaces of the powder particles, the particle interiors remaining relatively cool and stable[2,3]. Thus, if single-crystallite powder particles are used, shock-consolidation processes the grain boundaries in a manner distinct from the grain interiors, and thus enables a degree of control over the bonding process not possible with sintering.

Shock-processing can also enhance the intragranular critical current density of high-T_c superconductors through the introduction of high densities of shock-induced dislocations and other defects which act as flux-pinning sites[4]. Because dislocations perturb the lattice on a length scale comparable to the superconducting coherence lengths of high-T_c superconductors, they are effective at pinning magnetic fluxoids and increasing intragranular critical current densities without reducing the volume of superconducting material.

We describe here the results of our shock experiments on various high-T_c superconductor samples. These experiments were performed with two-stage light-gas guns at Lawrence Livermore National Laboratory (LLNL). Our shock experiments are characterized by small quantity samples ($\approx$ few hundred mg) shocked under carefully controlled conditions, with the starting powder density, the powder particle size and the shock pressure all being precisely measured quantities. The shock-recovered samples were examined by a variety of means including optical microscopy, electrical transport, SEM, TEM, and SQUID magnetometer measurements.

A. THE SHOCK COMPACTION PROCESS

The shock waves in our experiments were generated by means of a two-stage light-gas gun which accelerates a 20 mm diameter, 5 g lexan projectile at a copper capsule containing the high-T_c sample. A schematic of the copper sample capsule and the target area is shown in Figure 1. The impact of the lexan projectile onto the copper capsule generates a planar shock wave which compresses the sample. Further details can be found in Ref. 5.

Because the generation of shock waves by means of a gas gun occurs under well-controlled conditions, computer simulations of these shock experiments are fairly straightforward through the use of a two-dimensional axisymmetric finite-element code such as DYNA2D. These simulations are valuable because they yield information on the stress history and stress distribution of the shocked specimens, and thus allow one to better understand the conditions under which the powders were consolidated and the shock-induced defects generated. Standard Mie-Gruneisen equations of state and consti-

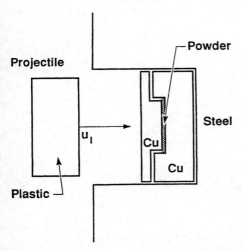

FIG. 1   *Schematic of the shock-recovery fixture for shocked high-*
T_C *powders.*

tutive models were used for all materials. Figure 2 shows the recovery target in the vicinity of the copper sample capsule as calculated by DYNA2D at 30 μsec after impact of a 1.5 km/sec lexan projectile onto the copper capsule. The plastic projectile is seen to have rebounded off the copper sample capsule and is moving away (downward) from the copper capsule. The copper capsule itself has deformed and extruded outward slightly, trapping it in the steel recovery block. Because the sample thickness is small (≈ 0.5 mm), the sam-

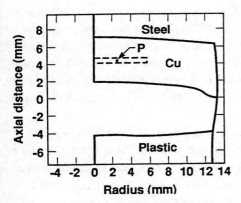

FIG. 2   *DYNA2D cross-section calculation of the shock-recovery fixture 30 μsec after lexan projectile impact at 1.5 km/sec. Axisymmetry about a radius=0 axis is assumed.*

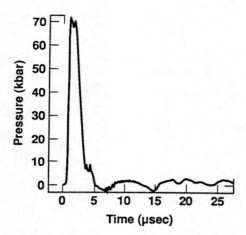

FIG. 3   *Calculated pressure history at point P of Fig. 2.*

ple chamber was not included in these calculations. Figure 3 shows a plot of the calcu-lated pressure history of point P in Figure 2. The peak pressure is 70 kbar and the dura-tion of this pressure is about 2 µsec. This calculation reveals tensile stresses at point P at times of 7 and 15 µsec after impact, the magnitude of these stresses being about a kbar.

II.   SHOCK COMPACTION OF $YBa_2Cu_3O_{7-\delta}$ + Ag POWDER

Shock compaction experiments on $YBa_2Cu_3O_{7-\delta}$ + Ag powder at a number of shock pres-sures were performed to determine the behavior of $YBa_2Cu_3O_{7-\delta}$ as a function of shock pressure. Since the amount of heterogeneous heating of the particle surfaces depends on the magnitude of the shock pulse, it was expected that the electrical transport properties of the shocked compacts might be very sensitive to the shock pressure, and so a full study at a wide variety of shock pressures was performed to explore and understand the effects of shock consolidation on $YBa_2Cu_3O_{7-\delta}$.

The samples were prepared using $YBa_2Cu_3O_{7-\delta}$ powder obtained from Argonne National Laboratory (ANL). The $YBa_2Cu_3O_{7-\delta}$ powder was sifted for a uniform 28-45 µm particle size. The sifted $YBa_2Cu_3O_{7-\delta}$ was then mixed with 20-30 µm Ag powder to form a $YBa_2Cu_3O_{7-\delta}$ + 30% vol. Ag mixture, the silver here serving to act as a binder for the $YBa_2Cu_3O_{7-\delta}$ and to increase the ductility of the shocked compact. In a practical applica-tion the silver would also function as a current shunt in the event that the superconductive component of the composite were to quench.

Shock recovery experiments were prepared with this powder mixture for shock pressures of 29, 58, 87, and 167 kbar[5]. Approximately 190 mg of the powder mixture was used for each shot. The powder was loaded into the 10 mm diameter disk-shaped sample chamber of the copper shock-recovery capsule diagrammed in Figure 1.

Optical micrographs of the shock-recovered samples revealed that fully dense composites were obtained even at the lowest shock pressure (29 kbar). There was considerable shock-induced fracturing, however, with the grain size decreasing with increasing shock pressure. For the 167 kbar specimen, the average grain size had been reduced to ≈ 1 μm from the starting grain size of ≈ 10 μm. Furthermore, twin lamellae were observed throughout the grains at all shock pressures, indicating that the orthorhombic, superconducting phase of $YBa_2Cu_3O_7$ is largely retained in the grain interiors for shock pressures up to at least 167 kbar.

Powder diffraction x-ray spectra on the unshocked starting powder and the 167 kbar shocked compact revealed some amount of transformation from the orthorhombic superconducting phase of $YBa_2Cu_3O_7$ to the tetragonal semiconducting phase. The tetragonal phase is presumably localized on the particle surfaces where heterogeneous shock heating is greatest, since the bulk shock temperature in the 167 kbar shocked sample is estimated to have been below 400 °C. This view is supported by optical micrographs of the 167 kbar shocked specimen, as described in the previous paragraph.

Zero-field-cooled (zfc) and field-cooled (fc) magnetization measurements were made on the shocked composites under a 25 Oe applied magnetic field. The zero-field-cooled magnetic screening signals are shown in Figure 4. First, note that there is no decrease in the bulk critical temperature with shocking. The grain interiors, then, apparently retain their superconducting properties with little loss of oxygen. Secondly, the screening signal decreases with increasing shock pressure, this effect apparently being due to decreasing grain size; as the average grain sizes in the shocked compacts approach the length scale of the superconducting penetration depth, a larger and larger volume fraction of the superconducting material will reside within a penetration depth of a grain surface, and this results in progressively smaller superconducting screening signals with decreasing grain size. Quantitatively, this size effect can be described by the following equation derived by Clem and Kogan[6] for weakly-linked superconducting spheres:

$$\frac{M}{M_o} = -\frac{3}{2} f_o \left(1 - \frac{3\lambda}{R} \right)$$

Here M is the zfc magnetization at T=0, M_o is the ideal Meissner magnetization ($-H/4\pi$), f_o is the volume fraction of superconducting material, λ is the magnetic penetra-

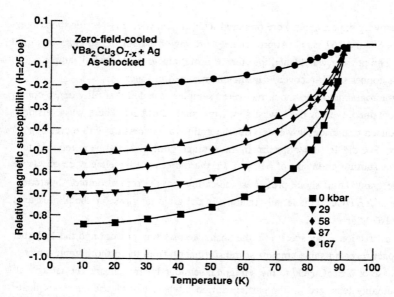

FIG. 4 *Zero-field-cooled magnetic susceptibilities of the shocked $YBa_2Cu_3O_7$ + 30 vol. % Ag samples relative to the ideal Meissner value, $(4\pi)^{-1}$.*

tion depth, and R is the radius of the spheres (assuming $\lambda/R \ll 1$). Indeed, taking the average grain size of the unshocked powder to be R=5 μm and using the measured magnetization ratio of M/M_0=0.84, the above equation yields average grain diameters of 2.8, 1.9, 1.4, and 0.8 μm for the 29, 58, 87, and 167 kbar shocks, respectively. These sizes are consistent with grain sizes observed by optical microscopy, which supports the view that the decreasing screening signals observed in Figure 4 can be largely accounted for by grain size reduction.

Transmission Electron Microscopy (TEM) was performed on the $YBa_2Cu_3O_7$ + 30 vol. % Ag shocked composites. The sample foils were prepared by mechanical polishing and then dimpling the shocked specimen to 40-50 μm, followed by argon-ion milling on a liquid-N_2 cooled stage. These studies revealed a remarkably high density of shock-induced dislocations. The $YBa_2Cu_3O_7$ shocked to 167 kbar, for instance, had a dislocation density in the range of 10^{12} cm^{-2}, which is approximately two orders of magnitude higher than dislocation densities generated in $YBa_2Cu_3O_7$ samples which had been subjected to slow mechanical deformation[7].

The shock-induced dislocations had line vectors of <100> and <110> lying in the (001) planes. These dislocations result from the large shear stresses generated across the (001) planes during shock loading. Indeed, due to the different crystallographic orientations of the individual grains with respect to the shock front, which resulted in a wide range of shear stresses being generated across the (001) planes, the dislocation density was found to vary widely from grain to grain of the polycrystalline sample. According to Schmid's law[8], the shear stresses across the (001) planes and, hence, the dislocation densities should be greatest in those grains oriented with their c-axis at an angle of 45° with respect to the shock direction. A TEM micrograph showing numerous shock-induced dislocations in a single crystal of $YBa_2Cu_3O_7$ shocked to 30 kbar is presented in Fig. 5. As will be seen, these dislocations are an important microstructural feature because they serve as very effective magnetic flux-pinning sites, and result in a considerable increase in the intragranular critical current density. A high density of twin lamellae was also observed in the 167 kbar as-shocked specimen, the average twin spacing being about 0.02 μm. This spacing is several times narrower than what is found in typical as-sintered materials, indicating that some of the shock stress is accommodated by deformation twinning.

The TEM studies of the 167 kbar sample also found evidence of melting around some of the grains due to heterogeneous shock-heating. The melted regions appear to be tetragonal phase material which form a layer ≈1 μm thick around the $YBa_2Cu_3O_7$ grains.

1 μm

FIG. 5   *TEM micrograph of a $YBa_2Cu_3O_7$ crystal shocked to 30 kbar. The crystal was oriented so that its c-axis was at an angle of 45° with respect to the shock direction.*

Low-angle grain boundaries similar to those found in melt-textured material as well as dendritic growth structures were found in these melted regions. Melted regions were not observed in the specimens shocked to lower pressures.

Interestingly, we found that the 167 kbar shocked $YBa_2Cu_3O_7$ retained a high density of defects even after annealing the shocked material at 890 °C for an extended period of time[4] (24 hrs.). Although few shock-induced dislocations remained after this long anneal, a very high density of extrinsic stacking faults was found, these stacking faults consisting of an intercalation of an extra Cu-O plane by b/2. Because of the large average separation between the partial dislocations which bound each stacking fault (≈ 0.5 μm), the stacking fault energy is fairly small (≈ 10 erg/cm^2), and this fact may partially explain the stability of the stacking faults even with extensive high-temperature annealing.

Specimens for electrical transport measurements were cut from the shocked compacts and had dimensions of roughly 8mm x 1mm x 0.3 mm. Resistivity measurements were made from room temperature down to 2 K. The resistivities of the as-shocked specimens tended to be linear with temperature, indicating that the current flows mostly through the Ag matrix rather than through the $YBa_2Cu_3O_7$, even below T_c.

To recover superconducting transport, the resistivity specimens were subjected to various annealing runs in so as to heal microcracks and transform the tetragonal, non-superconducting surface layers on the grains back to the orthorhombic, superconducting phase. Anneals were performed for various times and at various temperatures from 450 °C to 890 °C. Superconducting transport was recovered only after an 890 °C anneal, suggesting that resintering is needed to heal microcracks. The sample shocked to 167 kbar and then annealed at 890 °C exhibited a transport J_c of 320 A/cm^2 at 77 K with zero magnetic field.

III. FLUX-PINNING AND INTRAGRANULAR J_c

The high density of dislocations found in the 167 kbar shocked specimen suggests that the magnetic flux-pinning energy and the intragranular J_c of this sample may be extremely high. In order to determine the flux-pinning energy of this sample, we performed magnetic relaxation experiments with a SQUID magnetometer. The magnetic decay rate of a sample, assuming cylindrical geometry, was calculated by Beasley *et.al.*[9] to be

$$\frac{dM}{d\ln t} = \frac{r\,J_c}{3}\frac{kT}{U_o} \tag{1}$$

where r is the cylinder radius, J_c is the intragranular critical current density, and U_0 is the flux-pinning energy. Within the approximation that the grains of the sample are cylindrical, the above equation can be applied with r now being the characteristic grain size.

To obtain an averaged value for rJ_c, we make use of Bean's equation[10] which relates rJ_c to $\Delta M(H)$, the width of the magnetic hysteresis loop at H:

$$\Delta M(H) = \frac{2}{3} \frac{r J_c}{c} \tag{2}$$

Combining Eqns. (1) and (2) yields the following grain-size-independent expression for U_0:

$$U_0(H) = \left[\frac{\Delta M(H)}{[dM(H)/d\ln t]}\right] \frac{kT}{2} \tag{3}$$

Flux-relaxation experiments were performed at various temperatures after zero-field-cooling to the desired temperature and applying a 10 kOe magnetic field. Magnetic hysteresis runs were also performed to determine $\Delta M(H=10\ kOe)$. Using this data and Eqn. 3, the pinning energies shown in Figure 6 were determined for the starting

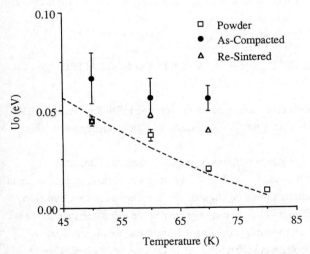

FIG. 6 *Flux-pinning energy U_0 (at H=10 kOe) as a function of temperature for shocked and unshocked $YBa_2Cu_3O_7$ + 30 vol. % Ag samples.*

$YBa_2Cu_3O_7$ + Ag powder mixture, the as-shocked sample, and the shocked + 890 °C annealed sample. Note that the shock process results in a substantial increase in the flux-pinning energy of the sample, apparently because of the large flux-pinning contribution from the high density of shock-induced defects. Also, note that the enhanced pinning is retained to some degree even after an 890 °C anneal. This enhanced pinning is presumably due to flux-pinning by the partial dislocations associated with the extrinsic stacking faults in the annealed material.

The increases in flux-pinning energy shown in Figure 6 have a considerable effect on the intragranular J_c's. Because the characteristic magnetic flux-pinning energies of $YBa_2Cu_3O_7$ and other high-T_c superconductors at 77 K are nearing thermal energies, the intragranular critical current densities of these materials at liquid-N_2 temperature are limited by the process of thermally activated flux-line motion. This results in a strong dependence of the intragranular J_c on the pinning energy U_0, the intragranular J_c having a roughly exponential, Arrhenius-like dependence on U_0 (i.e. $J_c \sim \exp(U_0/kT)$) (Ref. 11). Consequently, modest increases in U_0 can translate into substantial increases in intragranular J_c. Indeed, by using optical micrographs to estimate the average grain size of the shocked compact and applying the magnetic hysteresis data to Bean's equation (Eqn. 2), we find that the intragranular critical current density of the as-shocked material is approximately 10 times greater than that of the starting powder. A similar measurement of the intragranular J_c of the shocked + 890 °C annealed specimen reveals a factor of 5 increase in intragranular J_c.

IV. SHOCK EXPERIMENTS ON $Bi_2Sr_2CaCu_2O_8$ - CRYSTALLOGRAPHIC ALIGNMENT

Shock recovery experiments were also performed on the compound $Bi_2Sr_2CaCu_2O_8$. This compound is much more ductile than $YBa_2Cu_3O_7$ and shows little tendency to fracture under shock.

The crystallographic grain alignment of high-T_c superconductors is important for forming high-J_c bulk materials because (1) the high-J_c directions of high-T_c materials are in the basal plane and these directions should be aligned with the direction of current flow, and (2) the coherence length of high-T_c materials is greater in the a-b plane than in the c-axis direction, so to maximize intergranular coupling it is desirable to align the grains. Additionally, if the grain morphology is platelet-like, as in the case of $Bi_2Sr_2CaCu_2O_8$, alignment of the grains yields a microstructure with large overlap areas between grains, which results in better intergranular coupling simply by virtue of increased intergranular contact areas[12,13].

To form crystallographically aligned $Bi_2Sr_2CaCu_2O_8$ compacts we shock compacted prealigned pressed powders. The starting powder was made by solid state reaction by the DuPont Company. This micaceous powder was sifted for the 30-40 μm size. The sifted powder was observed by SEM and found to consist of platelets 10 μm thick, 30-40 μm wide, and 40-100 μm long. A high degree of alignment was then obtained by simply tapping this powder into a specimen holder, with the planes of the platelets aligning perpendicular to the tapping direction. For all compactions, the shock direction was along the tapping direction, or perpendicular to the *a-b* planes of the aligned powder.

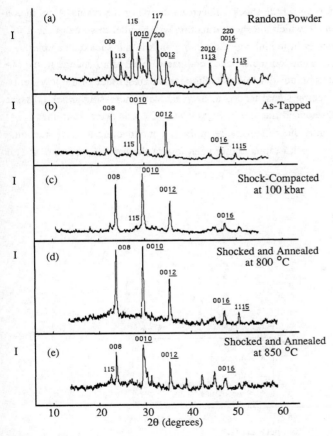

FIG. 7 *Powder diffraction patterns of (a) starting powder with random orientation, (b) as-tapped powder; also, shocked to 100 kbar for (c) as-shocked, (d) shocked and annealed at 800 ˚C in oxygen, and (e) shocked and annealed at 850 ˚C in oxygen. Cu Kα line was used for all patterns.*

Specimens were prepared and shocked at 30, 50, 70, 100, and 140 kbar with 2 μsec duration. Figure 7 shows powder x-ray diffraction patterns taken for a randomly oriented starting powder specimen (Figure 7(a)), the sieved powder after tapping (Fig. 7(b)), and the compact shocked to 100 kbar (Fig. 7(c)). Note the strong ($00l$) reflections in the as-tapped specimen, which indicates a high degree of crystallographic alignment with the c-axes perpendicular to the plane of the compact. This alignment is preserved with shock-consolidation, as can be seen in Fig. 7(c).

Optical microscopy of the shocked specimens reveals a high density (>95% of crystal density), and a high degree of texturing by plastic flow. In contrast to $YBa_2Cu_3O_7$ shock experiments, there is very little shock-induced fracturing of the grains. Also, in contrast to shock experiments in which the shock direction was parallel to some of the a-b planes[14], there was relatively little kinking of the a-b planes in our shocked samples.

Magnetometer experiments on the shocked compacts revealed no change in the on-set T_c from that of the starting powder (≈90 K), although the magnitude of the low-field (20 Oe) diamagnetic screening signal tended to decrease with increasing shock pressure. Since there is little shock-induced fracture in $Bi_2Sr_2CaCu_2O_8$, this decrease in the screening signal is apparently due to a decrease in the superconducting volume fraction because of heterogeneous shock heating and oxygen loss at the particle boundaries. This

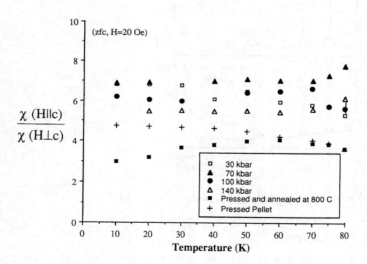

FIG. 8 *Zero-field-cooled magnetic susceptibility anisotropy for magnetic field parallel and perpendicular to the "c-axes" of the pressed or shocked compacts ($\chi(H||c)/\chi(H\perp c)$).*

contrasts with the case of $YBa_2Cu_3O_7$ which showed little decrease in the superconducting volume fraction as a result of shocking but substantial shock fracturing.

An interesting result of the shock compaction of the $Bi_2Sr_2CaCu_2O_8$ powder was the discovery that there was a large increase in the average aspect ratio of the particles due to plastic flow. Low-field magnetic screening measurements were performed both parallel and perpendicular to the c-axis of the shocked compact. Figure 8 is a plot of the magnetic anisotropy for H parallel and perpendicular to the c-axis for various shocked samples as well as for a $Bi_2Sr_2CaCu_2O_8$ pellet pressed and annealed at 800 °C. Note the enhancement of magnetic anisotropy for the shocked samples, indicating high aspect ratios for the grains in these samples. The increase in aspect ratio is apparent from optical micrographs of these samples; the shocked specimens have highly textured microstructures with high aspect ratio grains, while the sintered specimen has low aspect ratio grains and no discernable texture. Shock consolidation of $Bi_2Sr_2CaCu_2O_8$, then, can produce highly textured and dense bulk high-T_c superconductors with heterogeneously processed grain boundaries. Future work will focus on finding the proper shock conditions and post-shock treatments to maximize intergranular J_c's.

VI. CONCLUSIONS

We have seen that shock-consolidation of high-T_c superconductors addresses both the intragranular and intergranular aspects of the fabrication of bulk superconductors having high critical current densities. The dramatic increase in intragranular J_c with shocking is particularly significant since the relatively low intragranular J_c's in high-T_c superconductors are starting to become a limiting factor in the attainment of high J_c's in bulk materials[15]. Currently, there are a limited number of methods for increasing intragranular J_c's. Although neutron[16] and proton[17,18] irradiation methods have been valuable in demonstrating that intragranular J_c's of $\approx 6 \times 10^5$ A/cm^2 are possible at liquid-N_2 temperatures, there are numerous problems associated with irradiation techniques including the destruction of intergranular links[17], the requirement of long-time irradiation periods, and the production of residual radioactivity in the samples. Because of these problems, the applicability of irradiation methods to the large-scale fabrication of bulk superconductors is questionable. Shock-processing with explosives, on the other hand, can quickly and efficiently process large quantities of shock-defected high-T_c material having greatly enhanced intragranular J_c's.

We have also seen that highly textured bulk specimens of $Bi_2Sr_2CaCu_2O_8$ can be fabricated by shock-consolidation. This technique, then, is a promising one for creating superconducting material with good intergranular links and high critical current densities.

Future experiments will include further studies of the properties of shock-consolidated, textured $Bi_2Sr_2CaCu_2O_8$ compacts as a function of pressure and post-shock treatment with the aim of attaining the highest possible transport J_c.

REFERENCES

1. W.J. Nellis and L.D. Woolf, "Novel Preparation Methods for High-T_c Superconductors," *MRS Bulletin, 14:* 63-66 (1989).

2. W.J. Gourdin, *J. Appl. Phys., 55:* 172 (1984).

3. V.F. Nesterenko and A.V. Muzykantov, *Sov. Comb., Expl. and Shock Waves, 21:* 240 (1985).

4. S.T. Weir, W.J. Nellis, M.J. Kramer, C.L. Seaman, E.A. Early, and M.B. Maple, *Appl. Phys. Lett. 56:* 2042 (1990).

5. W.J. Nellis, C.L. Seaman, M.B. Maple, E.A. Early, J.B. Holt, M. Kamegai, G.S. Smith, D.G. Hinks, and D. Dabrowski, *High Temperature Superconducting Compounds: Processing and Related Properties, S. Whang and A. DasGupta eds.,* (TMS Publications, Warrendale, PA 1989), p. 249.

6. J.R. Clem and V.G. Kogan, *Jpn. J. Appl. Phys. 26, Suppl. 26-3:* 1161 (1987).

7. M.J. Kramer, L.S. Chumbley, R.W. McCallum, W.J. Nellis, S. Weir, and E.P. Kvam, *Physica C 166:* 115 (1990).

8. D.R. Askeland, *The Science and Engineering of Materials,* (PWS Publishers, Boston, MA 1984).

9. M.R. Beasley, R. Labusch, and W.W. Webb, *Phys. Rev. 181:* 682 (1969).

10. C.P. Bean, *Rev. Mod. Phys. 36:* 31 (1964).

11. A.P. Malozemoff, T.K. Worthington, R.M. Yandrofski, and Y. Yeshurun, *Towards the Theoretical Understanding of High-T_c Superconductors, S. Lundqvist, E. Tosatti, M. Tosi, Y, Lu, Eds., (World Scientific, Singapore, 1988),* p.757.

12. J. Mannhart and C.C. Tsuei, *Z. Physik B77:* 53 (1989).

13. A.P. Malozemoff, *to be published in High Temperature Superconducting Compounds II, eds. S.H. Whang, A. DasGupta, and R.B. Laibowitz (TMS Publications, Warrendale PA, 1990).*

14. Y. Syono, M. Nagoshi, M. Kikuchi, A. Tokiwa, E. Aoyagi, T. Suzuki, K. Kusuba, and K. Fukuoka, *in the Proceedings of the 1989 APS Topical Conference on the Shock Compression of Condensed Matter, Albuquerque, NM August 14-17, 1989,* p. 579.

15. N. Alford, "Contrasting Critical Currents," *Nature 345:* 292 (1990).

16. R.B. van Dover, E.M. Gyorgy, L.F. Schneemeyer, J.W. Mitchell, K.V. Rao, R. Puzniak, and J.V. Waszczak, *Nature 342:* 55 (1989).

17. J.O. Willis, D.W. Cooke, R.D. Brown, J.R. Cost, J.F. Smith, J.L. Smith, R.M. Aikin, and M. Maez, *Appl. Phys. Lett. 53:* 417 (1988).

18. R.B. van Dover, E.M. Gyorgy, A.E. White, L.F. Schneemeyer, R.J. Felder, and J.V. Waszczak, *Appl. Phys. Lett. 56:* 2681 (1990).

75

Shock Treatment of High T$_C$ Ceramics

V. F. NESTERENKO

Lavrentyev Institute of Hydrodynamics
USSR Academy of Sciences, Siberian Division
Novosibirsk 630090, U.S.S.R.

The properties of high-T$_C$-ceramics Y-123, Bi-1112, Bi-4334, Bi-4457, (Bi-Pb) - 4457 after shock loading and subsequent heat treatment were investigated. The peculiarities of their structure, changes of superconductivity parameters depending on loading conditions at P < 10 GPa, regimes of heat treatment were revealed. The feasibility of obtaining crack-free cylindrical superconducting shields was demonstrated.

I. INTRODUCTION

High-T$_C$ ceramics are very interesting objects for the study under shock loading conditions. Many papers concerning this domain were recently published [1-11]. They are connected, on the one hand, with attempts to solve practical problems - enhance density and strength, superconductivity destruction currents, obtain joints with normal metals, some details of superconductivity devices. On the other hand, high dynamic pressures are promising for synthesis of metastable superconducting phases. They produce unique conditions for creating the structure of crystals extremely saturated by defects that can be used as a precursor for new phases. The last feature is especially important because superconductivity in high-T$_C$ ceramics is in principle connected with ordered defect structure. The study of its management by high dynamic pressures is an important scientific problem closely connected with practical applications of these materials.

II. STRUCTURE CHARACTERISTICS OF Y-123 AFTER EXPLOSIVE LOADING

The structure and properties of compacts from Y-123 powder after explosive compaction (EC) in a cylindrical geometry at detonation pressures P = 2.2 and 6.5 GPa were studied [12]. The initial density (ρ_{oo}) had values of 3 and 3.5 g/cm^3, more than 90% were particles with sizes < 50μm. The lattice parameters of orthorhombic phase YBa$_2$Cu$_3$O$_{7-x}$ calculated from XRD pattern were: a=3.82 A, b=3.89 A, c=11.66 A.

The densities after EC at P = 2.2 and 6.5 GPa were 5 and 6.2 g/cm^3, respectively. Analysis of XRD patterns (the relative positions and heights of 200, 006, and 020 peaks) for compacts after EC with P = 6.5 GPa have shown that the main part of the material is orthorhombic phase with decreasing oxygen content in comparison with the initial powder.

For samples after EC (P = 6.5 GPa, ρ_{oo} = 3 g/cm^3), a decrease of the average grain size was obtained (Fig. 1a, b), as in [13] after plane shock wave loading. Except for the main phase Y = 123, the material after EC contains a small part of phases 1, 2, 3, and 4 with composition CuO, BaCuO$_2$, Y$_2$BaCuO$_5$, BaO, respectively (which are typical for the initial powder) and eutectic from 1 and 2 phases (sometimes 1, 2, 3) [12], Fig. 2. The characteristic feature of Y-123 is a high density of twins (Fig. 1b). The distance between twin boundaries is from 0.1 μm to few microns. The arbitrary relative position of twins testify the absence of

FIG. 1 *Structure of starting material Y- 123 (a) and after EC with P= 6.5 GPa, ρ_{00} = 3 g/cm^3 in polarized light X400 [12].*

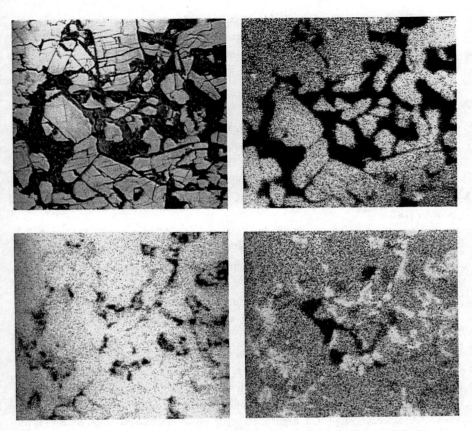

FIG. 2 *Compact structure of Y - 123 after EC with P = 6.5 GPa, X800 [12] a - in the regime of back scattered electrons, b, c, d - in characteristic X-ray radiation YLα, BaLα, CuKα respectively.*

texture of Y-123 phase in a plane perpendicular to the cylinder axis. Its micro-hardness for a load of 0.02 kg is 13 GPa, a few times more than a typical value for the ceramic prepared by the ordinary method [14,15].

The content of eutectic localized in thin layers on particle boundaries and cracks of Y-123 phase decreases with the pressure increase from 2.2 to 6.5 GPa. It can be the consequence of mutual plastic deformation, or rubbing of particles at the shock front that results in heating of boundary layers. That can cause the synthesis of Y-123 from eutectic phases 1, 2 and phase 3 enriched by Y. Moreover, pressure increase promotes the melting of eutectic and subsequent rapid solidification prevents its reverse formation.

The above-mentioned peculiarities of the material structure after EC result in semiconductive behaviour of its transport electrical properties and make necessary postshock heat treatment (HT) for obtaining a proper level of superconductive characteristics [3,6].

III. COMPACT STRUCTURE OF Bi- BASED SUPERCONDUCTORS

It is known that doping by Pb of Bi-based ceramics increases significantly the amount of phase Bi-2223 ($T_C \approx 110K$) [16]. On the other hand, the impulse loading of Bi-4334 system itself promotes the creation of Bi-223 phase [7] as in systems Bi-1112 [7-9], Bi-2212 [17] after post-shock. Combination of Ec (P = 5 GPa) and HT results in essential improvement of semiconductive transition (ST) in Bi-2212 phase in comparison with the same HT of the initial powder [18].

A partial amorphisation of the crystal structure after EC of system Bi-1112 has been obtained [10]. The weak initial peak (002) at $2\theta \approx 6°$, corresponding to superconducting phase Bi-2212 ($T_C \approx 80K$) has revealed abnormal increase of its intensity after postshock HT, as the content of Bi-2223 is increased.

Taking into account all these peculiarities of Bi-based superconductors, it is very interesting to compare the behaviour of this type ceramic with and without Pb doping after impulse loading.

Starting powders Bi-4457 and (Bi-Pb) - 4457 (Bi$_{3,2}$Pb$_{0.8}$Sr$_4$Ca$_5$Cu$_7$O$_{6-x}$) have been obtained by ordinary ceramic technology. XRD pattern analysis has revealed in system Bi-4457 superconducting phase Bi-2212 with c=30.6 A, and in system (Bi-Pb) - 4457 two superconducting phases in comparable quantities: Bi-2212 and Bi-2223 with c=30.6 A and c=37.0 A respectively [19]. On magnetic susceptibility dependence on temperature χ (T) for starting powder of Bi-4457 have been observed ST with $T_{co} \approx 110$ K (Bi - 2223) and $T_{co} \approx 85$ K (Bi - 2212).

EC has performed, as in [10], in cylindrical geometry with detonation pressures of 1.7 to 6.5 GPa. XRD patterns of both systems after EC have shown broadening of peaks with decreasing of its amplitude, as for Bi -1112 [10] and for Bi-2223 system with doping Pb and S [20].

For Bi-4457 the essential decrease of peak intensity on XRD patterns over the pressure range 1.7 - 4 GPa and stabilization of its amplitude at further pressure increase to 6.5 GPa have been observed. In addition, peak intensity (002) at $2\theta \approx 6°$ has been appreciably higher than for Bi-1112 [10] and in the last case the intensity of this peak did not distinguish essentially at 1 and 4 GPa.

For (Bi-Pb) -4457 after EC over the pressure range 1.7 to 6.5 GPa the intensity peaks have been obtained on XRD pattern corresponding to Bi-2212 and Bi-2223 phases with c=30.6 A, 37.0 A respectively, including close spaced peaks (002) at $2\theta \approx 6°$ (Bi-2212) and at $2\theta \approx 5°$ (Bi-2223). Over the pressure ranges 1.7 to 2.3 GPa and 4 to 6.5 GPa the peak intensities have changed slightly unlike the range 2.3 to 4 GPa. Besides these two high-Tc phases a small amount of Ca$_2$PbO$_4$, Ca$_2$CuO$_3$ and others have been found [19,21].

The temperature dependence of magnetic susceptibility for Bi-4457 compacts have been observed to be very broad ($\Delta T \approx 10 - 30$ K) ($T_c \approx 85$ K) with ΔT increase with pressure over the range 1.7 - 4 GPa, according to XRD results [19]. For (Bi-Pb)- 4457 measurements of χ (T) revealed two ST ($T_{co} \approx 110$ and 85 K) at pressures of 1.7 and 2.3 GPa. At high pressures (4 and 6.5 GPa) ST were absent [19].

It is possible to obtain the ratio of total amplitude of ST - A to sample mass-m based on χ(T) curves, which is proportional to superconductive phases in both materials over the pressure interval 1.7 to 2.3 GPa. At further pressure increase the stabilization of the superconductive phase content in Bi-4457 and its destruction in (Bi-Pb) - 4457 have been observed.

XRD pattern analysis has shown that after EC there are other changes other than broadening of peaks and decrease of their amplitudes [19]. With pressure increase the change of relative intensity of peaks (001) types has been revealed, for example, ratio of I$_{008}$/I$_{113}$ (curve 3 in Fig. 4) decreased with P. This behaviour indicated to decreasing of texture after EC with pressure increase.

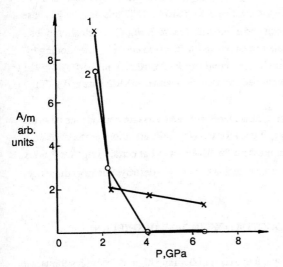

FIG. 3 Dependence of superconductive phases on pressure content [19]: Bi-2212 in compact Bi - 4457 (curve 1); Bi-2212 + Bi-2223 in compact (Bi-Pb) - 4457 (curve 2).

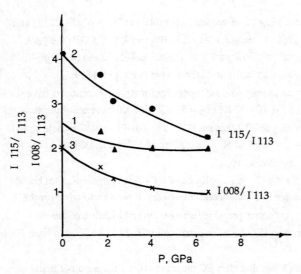

FIG. 4 *Change of intensity ratios after EC [19]: Bi-4457 (curve 1); (Bi - Pb) - 4457 (curves 2, 3).*

Pressure dependence of intensity ratios which are distinguished from (001) differs for Bi-4457 and (Bi-Pb) - 4457. The decreasing of I_{115}/I_{113} for Bi-4457 is characterized by stabilization on level ≈ 2, (curve 1, Fig. 4), whereas for (Bi-Pb)-4457 this tendency is absent and changes of above ratio take place at wider intervals (curve 2, Fig.4). Such alterations testify to essential rearrangement of the crystal lattice for Bi-based ceramic, after doping by Pb, at impulse loading. That can result in superconductivity destruction in (Bi-Pb)-4457 at P > 2.3 GPa and explains the above-mentioned contradiction between XRD data and χ (T) measurements.

In this connection the qualitative coincidence between pressure dependence of A/m (curve, Fig. 3) and I_{115}/I_{113} (curve 1, Fig. 4) for Bi-4457 calls attention on itself. A monotonic decrease if I_{115}/I_{113} with pressure for (Bi-Pb)-4457 at certain stage (at P > 2.3 GPa) evidently results in such structure change that completely destroys superconductivity [19].

IV. OBTAINING OF SUPERCONDUCTING CYLINDRICAL SHIELDS

In [2-9, 11] the possibilities of explosive methods for obtaining high-T_C ceramics joints with normal metals, solenoid type devices, metal-ceramic cylinders were demonstrated. For applications of high-T_C shields in heavy current commutation apparatus [22] it is necessary to develop the method of producing cylinders with proper level of superconductive parameters. With this purpose different technologies are studied. In [23] the properties of high-T_C shields were described. The maximum values of permanent magnetic field had a large scattering - 19;4.3 and 9.5 G. From these data and geometry of shields in [23] were evaluated critical current densities j_C (T=77 K): 150; 63 and 30 A/cm^2, respectively.

Explosive compaction for producing shields have been used in [24]. Special features of the cylindrical scheme developed [6,8,9] provide the possibility of easy removal of central rod after EC, obtaining of smooth surface of high-T_C material without macrocracks. As is well known, the cracking of compacts after EC is the main problem in terms of application. The Y-123 powder with a particle size of $\leq$ 30 μm, starting density $\rho_{oo} \approx 2.8$ g/cm^3 was used for EC. Detonation velocity was 2.5 km/s.

Metal-ceramic cylinders (Cu - Y - 123) were obtained with dimensions: outer diameter - 30 mm; thickness of Y-123 layer - 1.9 mm, its density - 6 g/cm^3, length - 50 mm (Fig. 5a). Their HT was conducted in oxygen atmosphere in standard regimes for Y - 123.

FIG. 5 a - metal - ceramic cylinders after EC and mechanical treatment; b - superconductive shields [24].

As a result, two high-T_C shields were obtained with an outer diameter of 26 mm, wall thickness of 1.8 mm and heights of 13 mm (No. 1) and 28 mm (No. 2), which are shown in the left part of Fig. 5(b). In the same figure, for comparison, a part of the shield produced by the ordinary ceramic technology with the analogous shielding characteristics to cylinder No. 1 and the extreme right - Cu - Y - 123 cylinder after HT are shown. The shield No. 2 had some defects - cracks and laminations. The measurement of ST width at low magnetic fields (H = 0.4 0e, No. 1; H = 0.55 0e, No. 2) gave $\Delta T \approx 1K$ for No. 1 and $\Delta T \approx 5K$ for No. 2 [24]. Shielding characteristics were measured in alternating magnetic field ($\omega = 100$ Gc), as in [22], and consisted of initial linear part in the following fast magnetic field increase in a cavity (Fig. 6). Penetration field H_C was determined in the beginning of deviation from the linear dependence, which is connected with tip effects [23]. For shields No. 1 and No. 2 Hc values were 17.3 Oe and 5.5 Oe which correspond to average $j_c = 42$ A/cm^2 (No. 2) and $j_c = 135$ A/cm^2 (No. 1) in appropriate fields; this is close to the best results in [23].

The dependence of emf inside the cavity (ϵ) on H was observed by Hc increase with temperature decrease; this is depicted in Fig. 7 for shield No. 1. It is worthwhile to note that

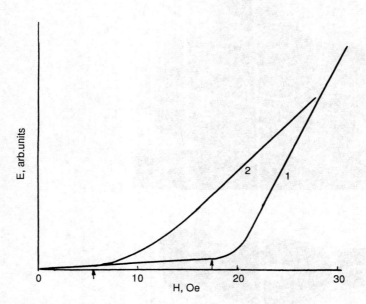

FIG. 6 *Dependence of emf in cavity of shields No. 1 and No. 2 on effective magnetic field at T = 78.4 K [24]. Values of Hc are shown by arrows.*

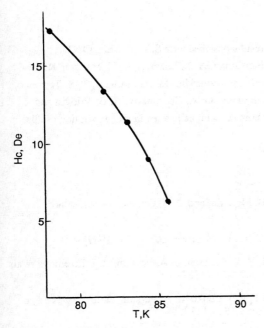

FIG. 7 *Dependence of effective value of H$_C$ on temperature for shield No. 1 [24].*

Hc for cylinder No. 1 is close to Hc for the shield produced by the ordinary ceramic technology with a wall thickness of 20 mm (Fig. 5b) [24].

CONCLUSION

Explosive loading, that retains at some pressure interval the starting phase content of Y - 123 system, drastically changes its structure. First of all, it influences the transport electrical properties (appearance of semi-conductive behaviour, decreasing j$_c$), width of ST; this requires inevitable post-shock HT for obtaining a good level of superconductive properties.

Impulse loading of Bi-based high T$_c$ ceramics can promote the appearance of superconducting phase with T$_c \approx$ 110 K. Doping Bi - 4457 by Pb results in the destruction of superconductivity at P > 2.3 GPa.

Explosive compaction with postshock HT can be used for producing superconductive cylinders which are able to shield alternative magnetic field with effective amplitude H$_m$ = 17.3 Oe at ω = 100 Gc, T = 77 K.

ACKNOWLEDGMENTS

The paper is based on the experimental results obtained with S. A. Pershin, O. G. Epanchintsev, Yu. A. Bashkirov, L. S. Fleishman, A. N. Lazaridi, E. V. Matizen, R. I. Efremova, E. B. Amitin, A. N. Voronin, P. P. Bezverkhii, M. A. Starikov, T. S. Teslenko, D. P. Kolesnikov, A. E. Korneev, B. A. Baterev, G. A. Chesnokov, A. B. Vdovin and others. I would like to thank all my collaborators who helped me in the preparation of this paper.

REFERENCES

1.   R. A. Graham, E. L. Venturini, B. Morosin, and D. S. Ginley, *Phys. Lett.* *A123*: 87 (1987).

2.   L. E. Murr, A. W. Hare, and N. G. Eror, *Nature 329* : 37 (1987).

3.   V. F. Nesterenko, A. N. Lazaridi, E. V. Matizen, S. A. Pershin, R. I. Efremova et al, *SDOHRH Report No. GR186001264* (1987).

4.   A. A. Deribas, E. V. Matizen, V. F. Nesterenko, S. A. Pershin, et al, *Izvestiya SO AN SSSR, Chemical Seria 5*: 84 (1988).

5.   E. V. Matizen, A. A. Deribas, v. F. Nesterenko, S. A. Pershin, P. P. Bezverkhii, A. N. Voronin, R. I. Yefremova, M. A. Starikov, *Intern Journal of Mod. Phys. 3B* 97 (1989).

6.   E. V. Matizen, V. F. Nesterenko, S. A. Pershin, R. I. Efremova, S. I. Afanasenko, A. N. Voronin, M. A. Starikow, P. P. Bezverkhii, A. N. Lazaridi, M. M. Goldshtein, T. S. Teslenko, E. D. Tabachnikova, *Preprint IIC SO AN SSSR,* No. 89 - 13 (1989).

7.   E. V. Matizen, V. F. Nesterenko, S. A. Pershin, R. I. Efremova, E. B. Amitin, A. N. Lazaridi, A. N. Voronin, T. S. Teslenko, P. P. Bezverkhii, M. A. Starikov, *Preprint IIC SO AN SSR,* No. 89-21 (1989).

8.   V. F. Nesterenko, S. A. Pershin, A. N. Lazaridi, S. I. Afanasenko, E. V. Matizen, R. I. Efremova, E. B. Anitin, P. P. Bezverkhii, A. N. Voronin, M. A. Starikov, T. S. Teslenkol *12 AIRAPT Conference: 832* (1990).

9.   V. F. Nesterenko, *in Shock Compression of Condensed Matter,1989,* S. C. Schmidt, J. N. Johnson, and L. W. Davison (eds.), North Holland, Amsterdam, p. 553.

10. O. G. Epanchintsev, D. P. Kolesnikov, A. E. Korneev, V. F. Nesterenko, S. A. Pershin, *Physics of Explosion,Combustion and Shock Waves: No. 3,* 129 (1990).

11. K. Takashima, H. Tonda, M. Nishida, S. Hagino, M. Suzuki, T. Takeshita, in *Shock Compression of Condensed Matter: 591* (1990).

12.   B. A. Baterev, O. G. Epanchintsev, V. F. Nesterenko, S. A. Pershin, N. S.
      Tzekunov, O. K. Bushueva, E. A. Orlova, *Physics of Explosion,
      Combustion and Shock Waves: No. 1* (1991).

13.   C. L. Seaman, E. A. Early, M. B. Maple, W. J. Nellis, J. B. Holt, M. Kamegai, G.
      S. Smith, *in Shock Compression of Condensed Matter*, S. C. Schmidt,
      J. N. Johnson, and L. W. Davison (eds.), North Holland, Amsterdam, p. 571
      (1989).

14.   H. C. Ling and M. F. Yan, *J. Appl. Phys ,64* 1307 (1988).

15.   L. B. Harris and F. K. Nyang, *J. Mater. Sci. Lett. 801* (1988).

16.   A. Maddalena, *J. Mat. Sci. Lett. 8:* 983 (1989).

17.   Z. Iqbal, N. N. Thadhani, N. Chawla, K. V. Rao, S. Skumryev, B. L. Ramakrishna,
      R. Sharma, H. Eckhardt and F. J. Owens, in *Shock Compression of
      Condensed Matter:* p. 575 (1990).

18.   Y. Syono, M. Nagoshi, M. Kikuchi, A. Tokiwa, E. Aoyagi, T. Suzuki, K. Kusaba,
      and K. Fukuoka, *in Shock Compression of Condensed Matter*, p. 579
      (1990).

19.   O. G. Epanchintsev, D. P. Kolesnikov, A. E. Korneev, V. F. Nesterenko, S. A.
      Pershin, G. A. Chesnokov, *Physics of Explosion, Combustion and
      Shock Waves. No. 2* (1990).

20.   M. Yoshimoto, H. Yamamoto, H. Koinuma and A. B. Sawaoka, *2nd Workshop
      on Industrial Application Feasibility of Dynamic
      compaction Technology:* p. 89 (1988).

21.   V. A. Baterev, V. K. Bushueva, O. G. Epanchintsev, A. E. Korneev, V. F.
      Nesterenko, I. A. Orlova and S. A. Pershin, *Physics of Explosion,
      Combustion and Shock Waves: No. 2,* (1990).

22.   Yu. A. Bashkirov, R. N. Baranova, N. F. Kerichenko, K. Sh. Lutedze and
      L. S. Fleishman, *Superconductivity: Physics, Chemistry,
      Techniques 6:* 43, (1989).

23.   O. G. Symko, W. J. Yeh, D. J. Zheng and S. Kulkarni, *J. Appl. Phys. 15:*
      2142 (1989).

24.   Yu. A. Bashkirov, A. B. Vdovin, L. S. Fleishman, V. F. Nesterenko, S. A. Pershin,
      Materials Impulse Treatment: 120 (1990).

76

Enhanced Superconducting Properties and Defects in Shock Compacted YBa$_2$Cu$_3$O$_{7-x}$ and Shock-Synthesized Tl$_2$Ba$_2$CuO$_6$ Superconductors

Z. IQBAL,[1] N.N. THADHANI,[2] K.V. RAO,[3] AND B.L. RAMAKRISHNA[4]

[1]Allied Signal Inc., Corporate Research and Technology, Morristown, NJ 07962, USA.

[2]Center for Explosives Technology Research, New Mexico Tech, Socorro, NM 87801, USA.

[3]Department of Solid State Physics, Royal Institute of Technology, Stockholm, SWEDEN.

[4]Dept. of Chemistry, Arizona State Univ., Tempe, AZ 85287 USA.

Commercially obtained polycrystalline YBa$_2$Cu$_3$O$_{7-x}$ powders (~40 micron particle size) were shock consolidated using a cylindrical implosion geometry at peak pressures of ~1-5 GPa and powder packing density of ~50%. The shock compacted powders were then subjected to controlled oxygen annealing treatments to homogenize the microstructural defects produced during shock compaction and optimize the particle size. The resulting shock processed and annealed samples showed significant improvement of flux pinning forces in contrast to conventionally sintered samples. The intergrain critical currents obtained via low-field and AC susceptibility data show an enhancement by a factor of **three** at temperatures between 4K and 78K, and magnetic fields up to 150 Oe. Even higher intergrain critical currents were obtained for samples that were shocked and then melt-processed. Electron microscopic analysis indicates formation of interpenetrating twins which may be the seat of defects in microcrystals. At the atomic scale level, the defects consist of intergrowths of a Y-Ba-Cu 223 phase sequence in bulk 123 structure. Some studies on shock-synthesized Tl$_2$Ba$_2$CuO$_6$ are also presented.

I.INTRODUCTION

An important concern regarding the potential of the new high-T_c ceramic
superconductors [1] for large-scale (bulk) applications is whether there is
an intrinsic limitation in their ability to carry large currents, in a
magnetic field, with little or no loss. Particularly in the highly-layered
copper-oxide based bulk superconducting materials, this has been attributed
to the problem of flux creep [2], which can keep current densities
disappointingly low. However, microscopic defects [3] in the
superconductor can pin the flux lattice in place and prevent creep, thus
allowing the supercurrent to flow unimpeded. In the absence of such
pinning sites, the supercurrent causes the flux lines to shift or creep.
However, during shock-compression loading of powders, the rapid increases
in pressure and subsequent plastic deformation introduce large numbers of
vacancies, dislocations, and other planar defects. Thus, not only can
complete densification and bonding of individual powder particles be
achieved under high pressure, but microstructural defects can also be
locked in (or frozen), due to extremely high quench rates. Subsequent
tailoring by controlled post-shock thermal treatments can result in uniform
distribution of the defects in the grains, oxygen ordering in the grain
boundaries, and optimization of the grain size distribution.

The shock-compression processing technique has been used to fabricate high-
T_c ceramic superconductors because of the unique characteristics of this
dynamic process [4]. Processing at shock pressures less than 10 GPa has
demonstrated the best results. Although the resulting compact is not a
good bulk superconductor in the as-shocked state, the sample recovers its
superconductivity upon subsequent annealing. Oxygen disordering [4] and
fracturing of particles are believed to be the cause of degradation of
superconducting properties in the as-shocked state. The post-shock
annealing treatment causes complete recovery of superconductivity without
loss of the shock-induced microstructural modifications and microscopic
defects.

Efforts to synthesize novel superconducting compounds by using shock-
induced chemical reactions have met with limited success. Morosin et al
[5] used high-pressure shock loading to chemically synthesize a
$La_{1.85}Sr_{0.15}CuO_4$ compound in the K_2NiF_4 structure with yields as high as 90%
from mixed oxide reactants. However, a subsequent thermal treatment of the
shocked material was necessary to render it superconducting with a T_c of
36K.

Our research has focused on three basic areas of dynamic processing of high-T_c oxide ceramic superconductors [6]:

(a)   Controlled shock compaction and post-shock annealing of pre-synthesized $YBa_2Cu_3O_{7-x}$ superconducting powders;

(b)   Shock-induced grain orientation and grain boundary welding of the bismuth-based superconductors;

(c)   Shock-induced chemical reactions for one-step synthesis of $Tl_2Ba_2CuO_6$ ($T_c = 50\text{-}70K$) with atomic-scale defects;

In this paper we will discuss highlights of recent results on enhanced critical currents and defects in shock processed $YBa_2Cu_3O_7$ and in one-step shock-synthesized $Tl_2Ba_2CuO_6$ (T_c ~55-70K).

II.   EXPERIMENTAL PROCEDURE

The basic experimental arrangement adopted for shock processing of the superconducting powders employs a single-tube cylindrical implosion system. The powder to be compacted is placed in a thick-walled copper tube, which is surrounded by the explosive charge. Upon detonation of the explosive, the shock wave is transmitted through the container into the powder. The CETR/Sawaoka 12-capsule plate-impact shock recovery fixture [7] was also used for shock synthesis experiments. In this system, an explosively accelerated flyer plate is used to impact the powders contained in cylindrical steel capsules at peak pressures of 5 to 100 GPa. For shock synthesis of the Tl-based superconductor, a mixture of Tl_2O_3, BaO_2, and a pre-synthesized $Ba_2Cu_3O_x$ powder (obtained by firing $BaCO_3$ and CuO at 930°C in air for 30 hours with an intermediate grinding) was used. The powders were blended to yield three different mixtures in which Tl:Ba:Cu ratios were 1:2:3, 2:2:3, and 2:2:1.

III.   MICROSTRUCTURAL CHARACTERISTICS

A.   *SHOCK PROCESSED $YBa_2Cu_3O_7$*

SEM analysis showed that the average particle size was reduced to 1-10 microns from the average size of 30-40 microns of the starting powder. XRD showed no evidence of phase degradation and textured grain alignment in the as-shocked material. TEM analysis showed the formation of a regular array of twin boundaries in the as-shocked material which was transformed to a checkerboard pattern of interpenetrating boundaries upon annealing at 890°C in oxygen. Melt processing the shocked material at 1150°C for 15 minutes however provided sizable grain orientation in the material.

With high-resolution TEM both the as-shocked and the 890°C annealed materials showed the presence of a novel atomic scale defect in the form of stacking faults [8]. Computer image simulations and measured interlayer spacings suggest that the defect region may have a layer sequence -Ba-Cu-Y-Y-Cu-Ba- in which an extra Y-layer is introduced in the layer sequence -Ba-Cu-Y-Cu-Ba- of $YBa_2Cu_3O_7$. The presence of this defect suggests the likely occurrence of a metastable $Y_2Ba_2Cu_3O_8$ phase. Attempts are currently underway to synthesize this phase. Such atomic scale defects are likely to provide centers for flux pinning in these extreme type II superconductors.

B.   SHOCK-SYNTHESIZED $Tl_2Ba_2CuO_6$

XRD analysis indicated nearly 100% conversion to the $Tl_2Ba_2CuO_6$ structure for the Tl:Ba:Cu = 2:2:1 precursor, 80-85% conversion for the 1:2:3 precursor, and only 10% conversion of the 2:2:3 precursor. However, only in the case of the 1:2:3 and the 2:2:3 precursors was the pseudo-tetragonal structure (expected to be superconducting) formed. SEM analysis showed the particle sizes of the shock-synthesized materials to be in the range of 1 to 10 microns, and EDX analysis showed the pseudo-tetragonal $Tl_2Ba_2Cu_6$ particles to be Tl deficient and Cu-rich.

Consistent with the above observations, high resolution electron microscopy showed the presence of two types of defects in the $Tl_2Ba_2CuO_6$ phase microcrystals. One defect shows a sequence -Cu-Ba-Tl-Ba-Cu- in the bulk phase of sequence -Cu-Ba-Tl-Tl-Ba-Cu-. The former corresponds to a known non-superconducting single Tl-O layered $TlBa_2CuO_5$ phase. The second defect is rather novel, and simulations suggest that it has a sequence -Tl-Ba-Cu-Tl-Cu-Ba-Tl- where the Tl occupancy between the two copper layers is 80% and the inter-layer Cu-Cu distance suggests that the Tl should be present largely as Tl^+. This defect corresponds to a phase of ideal composition $TlBa_2TlCu_2O_x$ (equivalent to the Tl 1212 phase containing Tl instead of the Ca between the CuO_2 layers). Preliminary synthesis experiments on this metastable phase, which was found to be non-superconducting, have been reported [9].

IV. SUPERCONDUCTING PROPERTIES, CRITICAL CURRENTS, AND FLUX PINNING

A. *SHOCK PROCESSED* $YBa_2Cu_3O_7$

DC resistivity measurements as a function of temperature, on the as-shock-processed material, showed semiconducting behavior with resistance drops at 30K and 90K, although Meissner fractions were indicative of bulk superconductivity. The results suggest the occurrence of disorder in the intergranular regions of the as-shocked material. Metallic behavior and a sharp drop (transition width less than 1K) to zero resistance at 90K was restored by annealing under oxygen at 890°C for 12 hours followed by slow cooling. Similar transport behavior was observed in samples that were subjected to a melt-annealing protocol. Both types of annealed samples showed complete shielding in zero field cooled DC SQUID magnetometry, but rather high (of the order of 10) ratios of the zero-field to field-cooled diamagnetic fractions at 5K, indicating enhanced flux pinning energies in these samples.

Enhanced flux pinning is also indicated in the as-shocked material via the grain surface sensitive technique of field modulated microwave absorption [10]. Here one measures the hysteresis in the peak magnetic field position of the low field absorption signal after subjecting the sample to various preparation fields. A factor of ten enhancement in hysteresis at a preparation field of 1600 Oe was observed, consistent with the bulk enhancement of flux pinning seen in the SQUID data.

Hysteresis loops were measured for the sintered Y123 (from the starting powder) and for both types of shock-annealed samples, at 5K and for a few intermediate temperatures to 70K. Hysteresis values at 5K and 70K at 2-3 kOe are tabulated below:

		M/cm^3 (at 2-3 kOe)	
	SAMPLE	5K	70K
Y 123	sintered	94	4.2
Y 123	shock + annealed	62	--
Y 123	shock + melt	125	4.6

In the Bean state where M is independent of the field, the above values of the hysteresis translate to an intra-grain critical current enhancement for the shock-melt annealed sample relative to the sintered specimen. The increase is by a factor of at least 10 at 5K when the decrease in particle size on shock treatment is taken into account. However, due to flux creep at temperatures above 40K, the improvement in intra-grain critical current is at best only a factor of 2 to 3 for shock-melt-annealed samples.

The intergrain critical currents were obtained using three
techniques: (a) low field DC susceptibility hysteresis loops, (b) AC real
and imaginary (loss) data, and (c) DC transport measurement using a 0.1
microvolt criterion and plasma sprayed Ag contacts at 77K in zero field.
In method (a) magnetic moment versus field loops were obtained at low
fields (lower than the limit for field penetration of the grains).
Intergrain J_c is then determined using the Bean formula and the sample
thickness. Data obtained by method (a) for the shock-annealed and sintered
Y123 samples at 5K are depicted in Figure 1 for fields up to 120 Oe. A
clear improvement in the intergrain J_c field dependence is seen at 5K.
Method (b) involves the measurement of the real and imaginary (loss)
components of the AC susceptibility at various temperatures and applied
fields. Using the analytical techniques discussed by Chen et al [11] it is
possible to computer simulate for each temperature and field a low field
hysteresis loop similar to that obtained directly by method (a). Using
this technique near zero-field, inter-grain J_c values are obtained as shown
for the shock-melt-annealed, shock-annealed and sintered samples as a
function of temperature in Figure 2. Some results obtained by method (a)
are also included in this figure, showing good agreement with the AC
values. At 5K a factor of 3.5 and 8 enhancement of the intergrain J_c's are
seen for the shock + annealed and shock + melt-annealed samples,

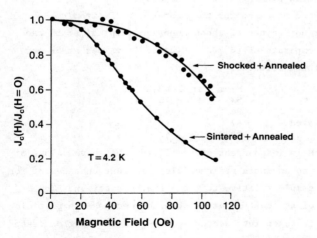

FIG. 1 *Normalized inter-grain J_c's for indicated samples
obtained by method (a) as a function of magnetic field.*

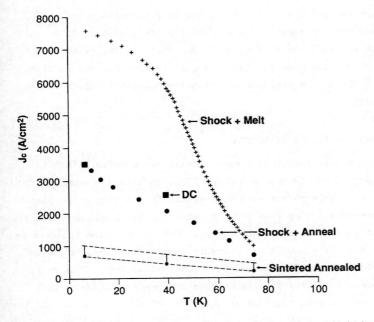

FIG. 2 *Intergrain J_c's obtained by AC susceptibility as a function of temperature.*

respectively. However, due to flux creep the numbers at 75K, though better than the sintered values, do not approach the values obtained by conventional melt-textured and melt-quenched Y123 materials [12]. The shock + annealed samples were measured by method (c) also, and were found to have values consistent with, but below, those obtained by the AC susceptibility method. It is, however, worth noting here that the AC method senses the maximum critical current in the sample whereas the DC transport method obtains an average value for the entire ~1 cm long sample at, in our case, a criterion of 0.1 microvolt.

B. *SHOCK-SYNTHESIZED $Tl_2Ba_2CuO_6$*

The shock-synthesized material from the Tl:Ba:Cu = 1:2:3 precursor showed zero resistance at 55K with a transition width of ~5K. Susceptibility data showed a diamagnetic onset at 70K and a bulk Meissner fraction of >10%. AC susceptibility data showed the presence of a large fraction of voids and hence intergranular critical currents were not measured. The relative flux pinning parameter was determined on the synthesized material using the microwave absorption technique. At a preparation field of 1500 Oe a hysteresis, that was a factor of 5 greater than that for the furnace synthesized material, was measured indicating a sizable increase in flux pinning possibly due to the presence of extensive defects in the shock-synthesized material.

VI.   CONCLUSIONS AND FUTURE DIRECTIONS

We have established the formation of novel atomic scale defects in both $YBa_2Cu_3O_7$ and $Tl_2Ba_2CuO_6$ by shock processing. These defects are considered to be remnants of metastable phases with potentially interesting properties. The defects themselves act as flux pinning centers. Clear enhancements are seen in the critical currents in $YBa_2Cu_3O_7$ upon shock processing followed by oxygen and melt annealing. Critical currents approach 10^4 amps/cm² below 30K in zero field but drop down to below 10^3 amps/cm² at 77K. This suggests that a higher density of flux pinning defects and fewer weak links are required to obtain critical currents of the order of 10^4 amps/cm² at 77K in fields above 1T.

ACKNOWLEDGMENTS

This work at New Mexico Tech was funded in part by Allied-Signal, Inc., Morristown, and NSF/MPI Grant No. DMR 8713258. Mr. Yates Coulter of ERDC at Los Alamos National Lab. performed transport property measurements.

REFERENCES

1.   J.G. Bednorz and K.A. Mueller, *Z. Phys. B.*, 64: 189 (1986).

2.   Y. Yeshurun and A.P. Malozemoff, *Phys. Rev. Lett.*, 60: 202 (1988).

3.   H.W. Zandbergen et al., *Nature* 331: 596 (1988).

4.   S.C. Schmidt, J.N. Johnson, and L. Davison (editors), *Proc. of the Americal Physical Society Topical Conference on "Shock Compression of Condensed Matter,"* North Holland, Chapter IX, 1990, pp. 548-610.

5.   B. Morosin, E.L. Venturini, R.A. Graham, and D.S. Ginley, *Synth. Metals* 33: 185 (1989).

6.   Z. Iqbal, N. Thadhani et al, source cited in Ref. 4, p.575.

7.   N.N. Thadhani, *Adv. Mater. and Manuf. Proc.* 3: 493 (1988).

8. R.Sharma, B.L. RamaKrishna, Z. Iqbal, N.N. Thadhani, and N. Chawla, *Proc. MRS Symposium on High Resolution Electron Microscopy of Materials*, 1990, (in press).

9. Z. Iqbal and N.N. Thadhani et al, *Proc. Intl. Conf. on Superconductivity*, Bangalore, India (in press).

10. F.J. Owens and Z. Iqbal, source cited in Ref. 4, p. 599.

11. D.X. Chen, J. Nogues, and K.V. Rao, *Cryogenics* 29: 75 (1989).

12. S. Gin et al., *Appl. Phys. Lets.*, 54: 584 (1989).

77

Low Peak Shock Presssure Effects on Superconductivity in $Bi_7Pb_3Sr_{10}Ca_{10}Cu_{15}O_x$

M.A. SRIRAM, L.E. MURR and C.S. NIOU

Department of Metallurgical and Materials Engineering,
The University of Texas at El Paso,
El Paso, Texas 79968.

The effects of peak shock pressure on the resistance-temperature signatures is more drastic in the case of Bi-Pb-Sr-Ca-Cu-O superconductors than in the case of Y-Ba-Cu-O superconductors. In this work, a transition has been observed for the first time in Bi-Pb-Sr-Ca-Cu-O presintered coupons loaded at a low peak shock pressure of 1.5 GPa by a plane wave. A comparison has been made of the variously shock loaded and annealed samples with reference to their resistance temperature signatures. The results indicate that a predictable resistance temperature behavior in the normal state and the superconducting transition can be obtained by variously altering the peak shock pressure and the annealing treatment.

I. INTRODUCTION

Recent developments on shock wave fabrication of superconducting monoliths from precursor powders and plane-wave shock loading of sintered superconducting Y-Ba-Cu-O and Bi-Pb-Sr-Ca-Cu-O has shown the dramatic effect of peak shock pressure on the magnitude of change of the resistance-temperature (R-T) behavior of the material [1-4]. It has indeed been shown by Murr, et al. [5] that the R-T response of plane-wave shock-loaded and shock-wave fabricated superconducting materials is the same.

In the case of $YBa_2Cu_3O_{7-\delta}$ [1-3], the transition is broadened and the $T_C(1/2)$ decreases with increasing peak shock pressure. The normal state behavior of the shocked material switches to semiconducting from metallic. Above around 15 GPa, the semiconducting behavior continues and no transition is observed up to liquid nitrogen temperatures. In Bi-Pb-Sr-Ca-Cu-O [4] on the other hand, no transition had been observed up to a peak shock pressure as low as 4 GPa. The resistance of the as shocked samples varied from 200 to 500Ω in the range of ambient to liquid nitrogen temperatures. Yoshimoto, et al. [6] have also confirmed that Bi-Pb-Sr-Ca-Cu-O does not show a transition after it has been shock fabricated at 20 to 40 GPa. The Bi-Sr-Ca-Cu-O (without PbO) material did show a transition, thus showing that doping with PbO to stabilize the high temperature phase could have altered the shock wave response of the material.

The recovery of the superconducting properties of both Y-Ba-Cu-O and Bi-Pb-Sr-Ca-Cu-O in annealing experiments were observed to be rather systematic [3,4,7,8]. In Y-Ba-Cu-O, the transition became sharper and the normal state behavior turned more and more metallic with time and temperature. In the Bi-Pb-Sr-Ca-Cu-O [4,7,8] samples, the transition was recovered at annealing temperatures as low as 650°C .The annealed material showed a two phase R-T signature with a low T_C phase at 80°K which disappeared at annealing temperatures above 750°C. The rest of the recovery in terms of the superconducting transition width and the normal state slope was systematic as in the case of Y-Ba-Cu-O.

In our experiments of plane wave shock loading of sintered $Bi_3Pb_7Sr_{10}Ca_{10}Cu_{15}O_x$ specimens at 1.5 GPa, we have observed a superconducting transition at a $T_C(onset)$ of about 100°K. Therefore Bi-Pb-Sr-Ca-Cu-O followed the same trends as Y-Ba-Cu-O in terms of increased degradation of superconductivity with increased peak shock pressure, only more drastically. Annealing showed similar changes as were observed with samples subjected to higher peak shock pressures.

II. EXPERIMENTAL

The experimental sandwich design for shock loading Bi-Pb-Sr-Ca-Cu-O and Y-

Ba-Cu-O coupons consisted of a base plate with channels and the cover plate as shown in Fig. 1(a). The material used was Zn-5%Al-2%Cu alloy with a density nearly compatible with the density of the superconducting material placed in the channels(ρ = 6.7 g/cm^3). The alloy was cast into plates of thickness 25 mm. and 12.5 mm. respectively. The base plate (20 mm. thick) and the cover plate (5 mm. thick) were machined out of these cast plates.The diameter of the base and cover plates was about 11.5 mm. The sintered superconducting bars of Y-Ba-Cu-O, Y-Ba-Ca-O + Ag and Bi-Pb-Sr-Ca-Cu-O were hand polished to the shape of little coupons to fit well into the channels (the coupons were 28 mm. long, 2.8 mm. wide and 1.5 mm. thick.). As shown in Fig. 1(a), Bi-Pb-Sr-Ca-Cu-O coupons were placed into three of the five channels, Y-Ba-Cu-O with silver was placed into one and a coupon without silver into another. The surface of the base plate was then lapped flat. The cover plate was pinned onto the base plate with dowel pins (which were cut from the cast alloy stock) and lapped flat on the outer surface. This assembly was set up in a gas gun tube as shown, in Fig. 1(b) within an outer momentum ring. A 5 cm. diameter and 4 mm.thick Al projectile was fired at it at a velocity of 150 m/s to develop a pressure of 1.5 GPa in the superconducting coupons. The detailed calculations are discussed by Murr [9]. The sandwich assembly was collected in front of the gun tube in a catcher stuffed with cloth. Although parts of the samples within the channels had spalled after the shock, pieces large enough to measure the R versus T curves were recovered from the sandwich.

Samples recovered from the sandwich were also annealed in flowing air with the carbon dioxide filtered out. They were annealed at 650° and 750°C for 90 hours each. Resistance versus temperature measurements were made with the four probe method.

III. RESULTS AND DISCUSSIONS

The R versus T signature for the Bi-Pb-Sr-Ca-Cu-O shocked at 1.5 GPa is shown in Fig 2, in comparison with the unshocked sample and one shocked at 4 GPa. (At pressures at and above 4 GPa the Bi-Pb-Sr-Ca-Cu-O showed no transition). It is interesting to note that the material shows a two-phase signature with the high T_c phase shifted slightly to the left ($\approx$100K) and the low T_c phase existing at about 80K. The normal state behavior is semiconducting.

Fig 3 on the other hand shows that Y-Ba-Cu-O shows transitions up to

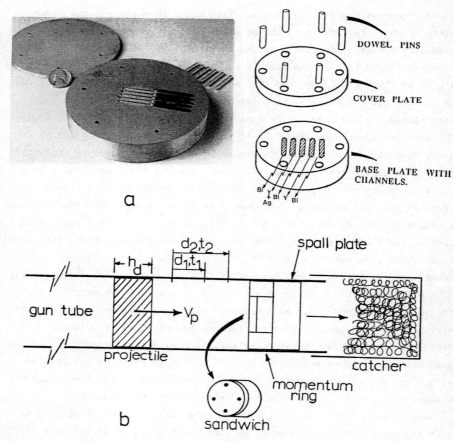

FIG. 1 *(a) The experimental sandwich used in the shock loading of the superconducting coupons shows the base plate with the channels milled out for the coupons, the cover plate and the dowel pins. The schematic shows the arrangement for assembly and the filling sequence of the channels. (b) A schematic of the gas gun equipment where the sandwich was shock loaded to a peak shock pressure of 1.5 GPa.*

much higher peak shock pressures ($\approx$13 GPa). The T_c(onset) is observed to be nearly constant at 90K up to a peak shock pressure of 7 GPa where it begins to decline, and at 19 GPa the transition is lost and the R-T signature becomes semiconducting like Bi-Pb-Sr-Ca-Cu-O shocked at 4 GPa (as shown in Fig. 2). The resistances of Y-Ba-Cu-O material shocked at 19 GPa is about a hundred times lower than in Bi-Pb-Sr-Ca-Cu-O shocked at 4 GPa.

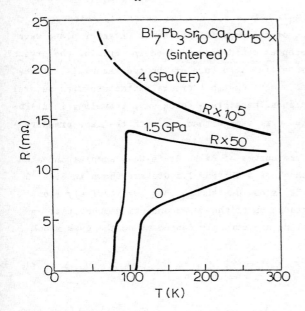

FIG. 2 A comparison of the R-T signatures of Bi-Pb-Sr-Ca-Cu-O: unshocked and shocked at 4 and 1.5 GPa.

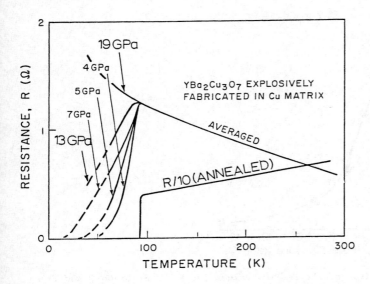

FIG. 3 A comparison of the as-shocked R-T signatures of Y-Ba-Cu-O samples shocked at peak shock pressures from 4 to 19 GPa. The superconducting transition shows complete recovery on annealing as shown (950°C for 60 hours).

Fig. 4 compares the $T_c(R=0)$ values for the two materials which shows the differences in the response of the materials to shock waves or plane wave shock loading. These differences could be explained in terms of the crystal structure or the micro-crystalline structure of the two materials. Bi-Pb-Sr-Ca-Cu-O has a taller and a more unstable crystal structure with several easily cleavable crystal planes. Therefore a shock wave traveling in Bi-Pb-Sr-Ca-Cu-O could have caused more damage than one with the same pressure traveling in Y-Ba-Cu-O.

The resistance-temperature curves of Bi-Pb-Sr-Ca-Cu-O samples annealed at 650 and 850°C after a shock of 5 GPa and 1.5 GPa are shown in Figs. 5 (a) and (b) respectively. It is obvious though only qualitatively from these figures that the degradation of the superconducting properties and their recovery after annealing are strongly dependent on the peak shock

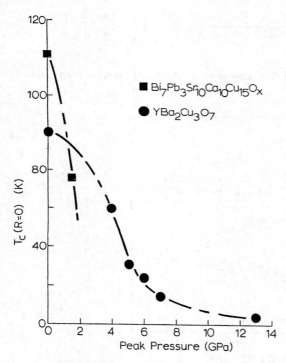

FIG. 4   *A comparison of the Tc(R=0) values for Bi-Pb-Sr-Ca-Cu-O and Y-Ba-Cu-O at different peak shock pressures. Bi superconductors show a more drastic degradation than Y superconductors with increasing pressure. At pressures higher than 1.5 GPa no transition was found in the as shocked Bi-Pb-Sr-Ca-Cu-O.*

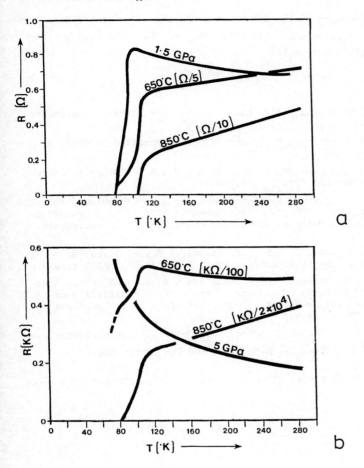

FIG. 5 *Resistance-Temperature curves of annealed Bi-Pb-Sr-Ca-Cu-O samples after shock loading at (a) 1.5 GPa and (b) 5 GPa respectively. The annealing time is 90 Hrs. The degradation is very strongly dependant on the peak shock pressure, as both the as-shocked and the annealed R-T signatures show.*

pressure. Therefore by altering the peak shock pressure and the post shock annealing treatment, properties of the material can be controlled to suit particular applications.

ACKNOWLEDGEMENTS

This research was supported in part by the Materials Research Center for

Excellence, a component of the NSF Minority Research Center For Excellence (NSF Cooperative Agreement No. RII-8802973), and the DARPA HTSC program (Contract ONR-N00014-88-C-0684)

REFERENCES

1. L.E. Murr, N.G. Eror and A.W. Hare, *J. Soc. Adv. Mater. Process Eng. 24(15);* (1988).

2. L.E. Murr, C.S. Niou and M. Pradhan-Advani, submitted to *Physica Status Solidi (A)*.

3. L.E. Murr, M. Pradhan-Advani, C.S. Niou and L.H. Schoenlein, *Solid State Communications 73(10);* 695(1990).

4. L.E. Murr, C.S. Niou, M. Pradhan-Advani and L.H. Schoenlein, Annual AIME Meeting, Feb. 1990, to be published in *HTSC (Second Symposium)*, ed. S.H. Wang, A. Dasgupta and R. Laibowitz, TMS-AIME, Warrendale, PA(1990).

5. L.E. Murr, C.S. Niou, S. Jin, T.H. Tiefel, A.C.W.P. James, R. Sherwood and T. Siegrist, *Applied Phys. Lett. 55(15);* 1575 (1989).

6. M. Yoshimoto, H. Yamamoto, H. Koinuma and A.B. Sawaoka, *Proceedings of the Second Workshop on Industrial Application Feasibility of Dynamic Compaction Technology, Tokyo;* 83(1988); Tokyo Institute of Technology, Tokyo, 1988.

7. L.E. Murr and C.S. Niou, *J. Mater. Sci. Lett.*, in press (1990).

8. C.S. Niou, M.A. Sriram, R. Birudavolu and L.E. Murr, these proceedings.

9. L.E. Murr, "Novel Applications of Shock Recovery Experiments", these proceedings.

78

Thermal Recovery and Kinetic Studies of Degraded High-T$_c$ Superconductivity on Explosively Fabricated (Shock-Loaded) YBa$_2$Cu$_3$O$_7$ and Bi$_7$Pb$_3$Sr$_{10}$Ca$_{10}$Cu$_{15}$O$_x$

C. S. NIOU, M. A. SRIRAM, R. BIRUDAVOLU, and L. E. MURR

Department of Metallurgical and Materials Engineering
The University of Texas at El Paso
El Paso, Texas 79968-0520

The residual superconductivity of explosively fabricated (consolidated) powder of superconducting YBa$_2$Cu$_3$O$_7$ (T$_c$~90K) and Bi$_7$Pb$_3$Sr$_{10}$Ca$_{10}$Cu$_{15}$O$_x$ (T$_c$~110K) and sintered (98% dense) shock-loaded YBa$_2$Cu$_3$O$_7$ and Bi$_7$Pb$_3$Sr$_{10}$-Ca$_{10}$Cu$_{15}$O$_x$ exhibit identical systematic change with peak shock pressure. Thermal recovery and kinetic studies show that degraded superconductivity can be fully recovered. X-ray diffraction peaks at $2\theta \approx 32°$ and 58° and resistance-temperature (R-T) signature changes in a systematic way. In addition, the manipulation of fabrication pressure and thermal treatment can allow for considerable manipulation of normal state and superconducting state behavior.

I. INTRODUCTION

It has been recently shown that explosively fabricated superconducting YBa$_2$Cu$_3$O$_7$ and Bi$_7$Pb$_3$Sr$_{10}$Ca$_{10}$Cu$_{15}$O$_x$ powder and sintered coupons cause the resistance-temperature signature to be changed rather dramatically: the normal state resistance after shock wave consolidation exhibits a semiconducting behavior, increasing with temperature up to T$_c$ start (~90°K) and then exhibiting a broad superconducting transition temperature [1-5]. The a.c. susceptibility and x-ray split peaks at $2\theta \approx 32°$ and 58° were also correspondingly altered [6]. It has been demonstrated by Murr, et al [2, 4] that shock-wave induced degradation can not be avoided even at peak shock pressures as low as 1.5 GPa [3]. In addition, Morosin, et al [7] observed a significant drop in the shielding fraction at pressure above 20 GPa and $(-4\pi\chi)$ dropping from 0.98 in the initial $YBa_2Cu_3O_7$ powder to 0.32 at 20 GPa and 0.17 at 27 GPa. Yoshimoto [8] also noted a reversion to semiconducting behavior (resistance increasing with decreasing tempera- ture) accompanied by significant structural alteration for Bi-Pb-Sr-Ca- Cu-O and Tl-Ba-Ca-Cu-O powders shock consolidated at peak pressures of 30 GPa.

The annealing of the residual materials at 750°C in oxygen caused the resistance-temperature signature to recover: the normal-state resistivity began to revert to a metallic behavior while the broadening of the super- conducting transition was correspondingly reduced after sufficient time. It was also observed that shock peak pressure < 8 GPa did not cause a reduction in the oxygen content, and the T_c start was unchanged after shock consolidation or heat treatment from that observed (90°K) for the superconducting $YBa_2Cu_3O_7$ powder and sintered coupons. $Bi_7Pb_3Sr_{10}Ca_{10}Cu_{15}O_x$ showed semiconductor behavior after shock wave consolidation. Recently, Sriram, et al. [3] demonstrated that peak pressure of 1.5 GPa would show the transition at 110°C.

II. EXPERIMENTS

The explosive or shock fabrication (consolidation) of superconducting powder and plane-wave shock loading of sintered superconductors has been reported by Murr, et al. [9]. Samples of these materials consolidated at peak pressure 4, 5, and 7 GPa in a copper tooling matrix were extracted by milling the metal away from the consolidated channels to produce experi- mental samples. These samples (~ 0.5 cm in length and 0.1 cm x 0.5 cm in cross-section) were fitted with four contacts in a four-probe configura- tion for resistance-temperature measurements.

The heat treatment of the $YBa_2Cu_3O_7$ at different conditions were always followed by a final annealing at 400°C for 12 hours in flowing oxygen and slow cooling to room temperature to stabilize the orthorhombic phase. The recovered $Bi_7Pb_3Sr_{10}Ca_{10}Cu_{15}O_x$ was also heat treated under flowing air at different temperatures to revert the superconductivity. This process does not need low temperature oxygen annealing.

Crystal structures of the initial $YBa_2Cu_3O_7$ and $Bi_7Pb_3Sr_{10}Ca_{10}Cu_{15}O_x$ powders and representative sections following shock-wave fabrication were examined by x-ray diffraction using CuKα radiation. All specimens were ground to roughly the same size fraction (and size distribution) to eliminate any possible texture or artifactual ordering phenomena, and orthorhombic peak splitting [6] near $2\theta \approx 32°$ and $58°$ were also compared in $YBa_2Cu_3O_7$.

III. RESULTS AND CONCLUSION

The shock peak pressure induced degradation on high-T_c superconducting $YBa_2Cu_3O_7$ and $Bi_7Pb_3Sr_{10}Ca_{10}Cu_{15}O_x$ have been shown as Fig. 1. Figure 1(a) shows the resistance-temperature (R-T) signatures changing after shock-wave consolidation, in which the normal state resistance exhibits a semiconducting behavior, increasing with temperature up to the transition temperature ($YBa_2Cu_3O_7 \sim 90$ K) and then exhibiting a broad superconducting transition temperature corresponding with the peak shock pressure. Figure 1(b) shows that the semiconducting behavior of $Bi_7Pb_3Sr_{10}Ca_{10}Cu_{15}O_x$ increases with temperature to a very low temperature (77 K). Fig. 1(c, d) compares the resistance-temperature (R-T) signatures for the explosively fabricated and annealed $YBa_2Cu_3O_7$ and $Bi_7Pb_3Sr_{10}Ca_{10}Cu_{15}O_x$ corresponding to a peak shock pressure during fabrication of 7 GPa and 5 GPa. The results shown in Fig. 1(c, d) are very similar to those obtained for plane-wave shock loaded $YBa_2Cu_3O_7$ [5]. The variations in the R-T signatures with temperature are in fact identical to those previously observed, including the stability of the T_c start.

By comparison with the $YBa_2Cu_3O_7$ there is no detectable superconducting transition in the $Bi_7Pb_3Sr_{10}Ca_{10}Cu_{15}O_x$ at flowing shock fabrication, and this is consistent with observations of shock loaded $Bi_7Pb_3Sr_{10}Ca_{10}Cu_{15}O_x$ at much higher pressures (~ 30 GPa) by Yoshimoto, et al.[8]. However superconducting transitions become apparent after annealing at 650°C, and the higher-temperature transition (T_c start 110 K) remains relatively stable up to the annealing temperatures where a sharp transition occurs. This is even true of the low-temperature (second phase) transition observed around 80°K (T_c start) in Fig. 1(d).

The x-ray diffraction split peaks at $2\theta \approx 32°$ and $58°$ have been used as a diagnosis for the quality control of "good" superconducting $YBa_2Cu_3O_7$ powder by Murr and Eror [6]. Fig. 2(a) shows the degradation of the high T_c superconductors with increasing pressure is implicit in the orthorhombic x-ray split-peak signatures and the line broadening which increases

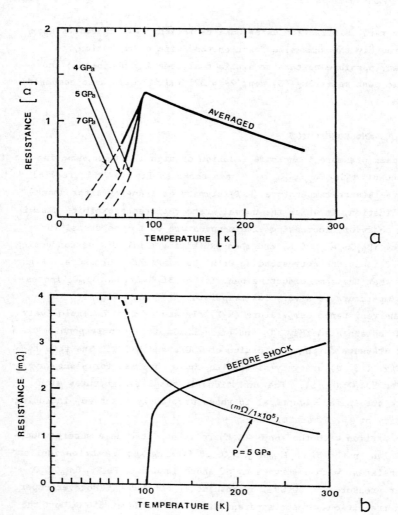

FIG. 1 *(a) Resistance-temperature (R-T) signatures for explosively fabricated $YBa_2Cu_3O_7$ at different shock pressures. (b) comparison of resistance-temperature singatures for sintered $Bi_7Pb_3Sr_{10}Ca_{10}Cu_{15}O_x$ ($Tc\sim110°C$) and explosively fabricated (5 GPa) $Bi_7Pb_3Sr_{10}Ca_{10}Cu_{15}O_x$ powder. (no transition temperature). (c) Thermal recovery of $YBa_2Cu_3O_7$ at different temperature for 24 hours shows the normal state resistance change from semiconducting behavior to metal-like behavior above 700°C. (d) Thermal recovery of $Bi_7Pb_3Sr_{10}Ca_{10}Cu_{15}O_x$ at different temperature for 90 hours shows the resistance change from semiconducting behavior to metallic behavior between 650-750°C.*

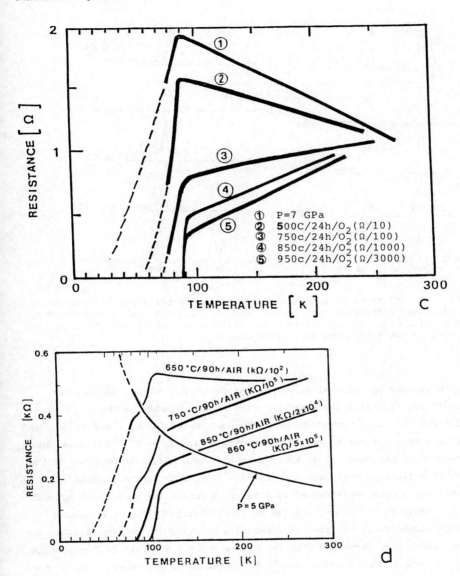

FIG. 1 *(continued)*

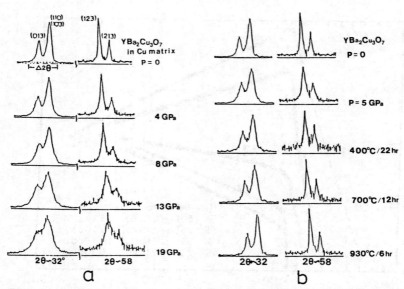

FIG. 2 *Variations in characteristic (orthorhombic) X-ray diffractometer split-peak signatures near $2\theta = 32°$ and $58°$ (a), and the component ratio variation with shock pressure. (b) the recovery of split-peak after annealing at high temperature ($YBa_2Cu_3O_7$).*

with increasing residual strain in the fabricated copper matrix. In addition, Fig. 2(b) shows the recovery of the superconductivity at $2\theta \approx 32°$ and $58°$. It is possible that defects created in the shock front contribute in a systematic way to broadening of the $2\theta \approx 32°$ split peak while related microstructural alterations contribute to the suppression of the (013) reflection peak amplitude at $2\theta \approx 32°$ and the corresponding (213) reflection peak amplitude at $2\theta \approx 58°$ with increasing peak shock pressure.

While superconductivity can be re-established through annealing at high temperature, the fact that metallic behavior in the normal state can be established above 700°C in $YBa_2Cu_3O_7$ and above 600°C in $Bi_7Pb_3Sr_{10}Ca_{10}Cu_{1-5}O_x$ indicates that massive atomic displacement in the shock front created significantly different microstructure. In order to understand the fundamental nature of shock-wave induced degradation and the mechanism of the recovery, a systematic high temperature heat treatment has been applied at different temperature (500°C to 950°C) and different time (6 to 60 hours). Figure 3 shows the activation energy increasing with peak shock pressure,

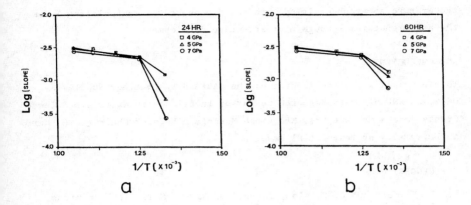

FIG. 3 *(a) Arrhenius plots for log(slope) versus (1/T), which slope = [(R₃₀₀ - Rₜ)/(R₃₀₀•ΔT)], the activation energy increasing when the shock pressure increasing at low annealing temperature. Annealing temperature between 800°C and 950°C for 24 hours, the pressure effect are very small. (b) comparison of 60 hours heat treatment results with Fig. 3(a) shows the similar activation energy change.*

and also shows the activation energy changing with the annealing time. The Arrhenius plot shows higher activation energy between 750°C and 800°C and lower activation energy above 800°C. From the activation energy slope change on Fig. 3, we may conclude that more defects will be created at higher pressure than the lower peak pressure. There is some indication from transmission electron microscope observations that unique defects are created in the shock front which show the fine microtwin structure characteristic of the orthorhombic superconducting $YBa_2Cu_3O_7$ to increase with explosive fabrication or shock loading [2, 9]. It is not known how these defects are created in the shock front or what their structural or chemical details involve. However, they have been observed to increase significantly in density with peak shock pressure, and apparently require temperature above 900°C to completely anneal out or reorder (Fig. 1).

IV. SUMMARY

It would appear that more detailed observation must be made to observe the microstructures which may be associated with the individual thermal recovery curves shown for example in Fig. 1. It may be possible to understand some of the fundamental aspects of microstructure in high-temperature superconductivity, and the prospects for manipulating T_c or the R-T signatures through microstructure development.

ACKNOWLEDGMENTS

This work was supported in part by the DARPA-HTSC Contract N00014-88-C-0684, an NSF Minority Research Center of Excellence in Materials Science Grants, and a Mr. and Mrs. MacIntosh Murchison Endowed Chair (L. E. Murr) at University of Texas at El Paso.

REFERENCES

1. L. E. Murr and C. S. Niou, *Journal of Mater. Sci. Lett.* in press, (1990).

2. L. E. Murr, C. S. Niou, and M. A. Sriram, to be published in *Proceedings of the 7th CIMTEC-World Ceramic Congress & Satellite Symposia: Symposium 4:* High temperature Superconductors, P. Vincenzini, editor, Elsevier Science Publishers, The Netherlands, in press, 1990.

3. M. A. Sriram, C. S. Niou, and L. E. Murr, *these Proceedings.*

4. L. E. Murr, M. Pradhan-Advani, C. S. Niou, and L. H. Schoenlein, *Solid St. Comm. 73(10):* 695 (1990).

5. L. E. Murr, C. S. Niou, S. Jin, T. H. Tiefel, A.C.W.P. James, R. C. Sherwood, and T. Siegrist, *J. Appl. Phys. Lett. 55 (15):* 1575 (1989)

6. L. E. Murr and N. G. Eror, *Mater. & Manuf. Processes 4(2):* 177 (1988).

7. B. Morosin, R. A. Graham, E. L. Venturini, and D. S. Ginley, *Proce. of. 2nd Workshop in Industrial Application Feasibility of Dynamic Compaction Technology,* Tokyo Institute of Technology, Yokohama, Japan, 1988, p. 96.

8. M. Yoshimoto, H. Yamamoto, H. Koinumai, and A. B. Sawaoka, *ibid,* p. 84.

9. L. E. Murr, *these Proceedings.*

79

Microstructural Modifications and Critical Current Densities of Explosively Compacted Oxide Superconductors

K. TAKASHIMA*, H. TONDA*, M. NISHIDA*, S. HAGINO**, M. SUZUKI**
and T. TAKESHITA**

* Department of Materials Science and Resource Engineering
  Kumamoto University
  Kumamoto 860, JAPAN

** Central Research Institute
   Mitsubishi Metal Corporation
   Oomiya, Saitama 330, JAPAN

   Explosive compaction experiments were performed on silver-sheathed Bi-Sr-Ca-Cu oxide superconducting tapes, and effects of the compaction on microstructural modifications and superconducting properties of Bi-Sr-Ca-Cu oxides were investigated. Explosive loading was applied on as-rolled and pre-sintered tapes by the flyer plate method. X-ray diffraction analysis showed that the amount of high-T_c phase in both as-rolled and pre-sintered tapes was increased by the compaction. Although a zero resistance was not obtained above 77K for as-compacted tapes, the tapes showed $T_c(\rho=0)$ of 100K by the post heat treatment at 1118K in air. The critical current density (J_c) for rolled, shock compacted and heat-treated tape was $3 \times 10^2 A/cm^2$, while the J_c for pre-sintered, shock compacted and heat-treated tape was increased to $3 \times 10^3 A/cm^2$ at 77K under zero magnetic field. The latter compaction was performed with explosive of low detonation velocity. Core of pre-sintered and shock compacted tape consists of the highly oriented grains which appear to be effective for increase of a critical current density.

I. INTRODUCTION

The improvement of critical current density (J_c) of high-T_c oxide superconductor is the most important subject for their practical applications. J_c value of oxide superconductors depends on their bulk densities and microstructures (crystal defects, grain boundary structure, crystalline orientation of grains etc.), which are mainly controlled by fabrication method. Shock compaction technique is capable of densifying various powders. In addition, many crystal defects and grain boundaries having unique structure are formed in shock compacted materials by the passage of shock wave [1]. Murr et al. [2-5] have conducted the shock compaction of Y-Ba-Cu oxide superconductor, and found that the superconducting properties were very erratic after shock compaction. However, they also suggested that the atomic scale defects created during shock compaction could act as flux pinning centers. In our previous studies [6-8], we have applied the explosive compaction technique to prepare Y-Ba-Cu oxide superconducting coils. After the explosive compaction and adequate post heat treatment, the J_c was reached to higher than 10^4 A/cm^2. Transmission electron microscopy observation of the coil core revealed that the high J_c value was strongly related with the introduction of many dense microtwins acting as flux pinning center and also the formation of clear grain boundaries without any impurity phase. Thus , the shock compaction technique seems to be an effective fabrication process to increase J_c of oxide superconductors

The oxide superconductor of Bi-Sr-Ca-Cu system has two superconducting phases of T_c of 110K and 80K, called high-T_c phase ($Bi_2Sr_2Ca_2Cu_3O_y$) and low-T_c phase ($Bi_2Sr_2Ca_1Cu_2O_y$), respectively. Usually the high-T_c phase is thermodynamically less stable than the low T_c phase, so it is difficult to produce a high-T_c single phase. Recently, partial substitution of Pb for Bi [9] and sintering in reduced oxygen atmosphere [10] were found to increase the amount of the high-T_c phase in a sample. Furthermore, an intermediate pressing between two sintering treatments enhances the growth of oriented grains, which causes the J_c to increase [11]. However, since large grains and intergrain structures are broken during the intermediate pressing, it is difficult to restore the connection between grains entirely. Shock compaction technique is an effective means not only to improve the connectivity between grains by the densification of powder but to produce a non-equilibrium material. In this study, explosive compaction experiments were performed on silver-sheathed Bi-Sr-Ca-Cu oxide superconducting tapes, and the microstructural modifications and superconducting properties were investigated.

II. EXPERIMENTAL

Powder with nominal composition of $Bi_1Pb_{0.25}Sr_1Ca_{1.25}Cu_{1.75}O_y$ was prepared from high-purity (99.9%) powders of Bi_2O_3, PbO, $SrCO_3$, $CaCO_3$ and CuO. The mixed powder was calcined at 1083K for 72ks in air. After grinding, the powder was put into silver tubes, which were drawn and rolled into tapes with thickness of 0.4mm and width of 1.2mm.

Figure 1 shows an experimental arrangement for explosive compaction used in this study. Tapes of 20mm in length were placed on a steel base plate of 50mm x 50mm x 20mm. A flyer plate was put on the tapes and explosive was set on it. The explosive used in the first experiment was manufactured by Asahi Chemical Industry Co., Ltd. under the trade name "SEP", which consists mainly of ammonium nitrate. The detonation velocity, V_D, of the explosive was about 7.0km/s. The weight of the explosive on the flyer plate was 35g, since this condition was found to give the best compactibility of tape cores in our previous experiments. In the second experiment, the explosive used was "PAVEX" manufactured by the same company, V_D of which was about 2.3 km/s.

T_c and J_c were measured by the four-probe dc method. The J_c was defined as the transport current which induces an electric voltage of $1\mu V/cm$ in the tape. X-ray diffraction analysis using Cu-Kα radiation was performed on tape core, and fractured surface of tape core was examined by scanning electron microscopy.

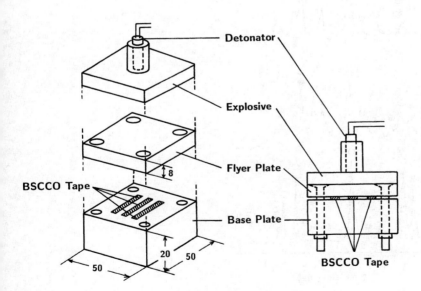

FIG. 1 *Experimental arrangement for explosive compaction (dimensions in mm).*

III. RESULTS AND DISCUSSION

A.   *Effects of shock compaction on as-rolled tape*

First, the results of the shock compaction with SEP explosive are given. Figures 2(a)-(c) show the variation of X-ray diffraction patterns for as-rolled, as-compacted and post heat-treated tape cores, respectively. The intensity of (00$\underline{10}$) line of high-T_c phase, $I_{H(00\underline{10})}$, and that of (008) line of low-T_c phase, $I_{L(008)}$, were measured, and a volume fraction of high-T_c phase was roughly estimated as $I_{H(00\underline{10})}/[I_{H(00\underline{10})} + I_{L(008)}]$ [12]. Table 1 gives the fraction of the high-T_c phase and superconducting properties. The as-rolled tape core mainly consists of low-T_c phase, and the fraction of high-T_c phase is 14%. The fraction of the high-T_c phase is markedly increased to 53% after the explosive compaction, although the diffraction peaks are broad because of the lattice distortion induced by explosive com-

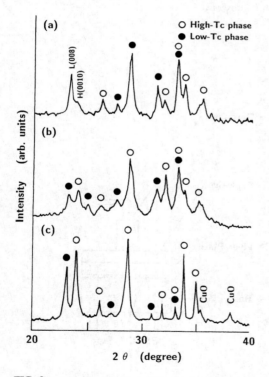

FIG. 2 *X-ray diffraction patterns of rolled(a), compacted(b) and post heat-treated(c) tape cores.*

TABLE 1 *Influence of fabrication process on high-T_c phase content and superconducting properties.*

Process*	Fraction of high-T_c phase (%)	T_c (onset) K	T_c ($\rho=0$) K	J_c (77K, 0T) A/cm^2
R	14	-----	-----	-----
R + S	53	-----	-----	-----
R + S + H	58	104	80	230
P	50	102	80	210
P + S	71	-----	-----	-----
P + S + H	75	104	95	330
P + S2 + H	92	106	102	3000

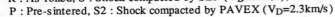

* R : As-rolled, S : Shock compacted by SEP (V_D=7km/s), H : Post heat-treated,
P : Pre-sintered, S2 : Shock compacted by PAVEX (V_D=2.3km/s)

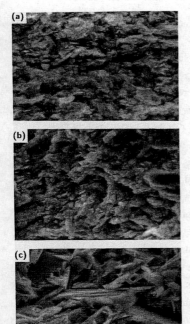

FIG. 3 *Scanning electron micrographs of fractured cross sections of (a)as-rolled tape, (b)as-compacted one and (c)post heat-treated one.*

paction. It is well known that high-T_c phase is formed via low-T_c phase and the growth of high-T_c phase is enhanced by the partial melting of low-T_c phase [13]. Therefore, the increase of the high-T_c phase in our samples is considered to be caused by local temperature rise or local melting at particle interfaces during shock compaction. Although the tape cores were found to be extremely dense, zero resistance was not measured above 77K for as-compacted tapes. This result may be due to insufficient interparticle bonding, micro-cracks or impurity phases formed during the compaction.

Then the tapes were heat-treated at 1118K for 360ks in air. The fraction of high-T_c phase increased slightly after heat treatment. Figures 3(a)-(c) show the scanning electron micrographs of the fracture surfaces of transverse cross sections for as-rolled, as-shocked and post heat-treated tapes, respectively. The grain growth was not observed for as-compacted tape core. The grain growth and the penetrations of plate-like and large grain into small grains were observed after post heat treatment. The T_c(onset) and $T_c(\rho=0)$ were 104K and 80K, respectively, and the J_c of 230 A/cm^2 was obtained at 77K in zero external magnetic field.

B.   Effects of shock compaction on pre-sintered tape

Recently, an intermediate statically pressing between sintering treatments has been found to be very effective for both improving J_c and making high T_c phase to grow [14]. We conducted the shock compaction using SEP on our pre-sintered tapes. Figures 4(a)-(c) show the variation of X-ray diffraction patterns, and Figures 5(a)-(c) show the scanning electron micrographs of the fractured cross sections for pre-sintered, shock compacted and post heat-treated tape cores, respectively. The high-T_c phase grew during pre-sintering at 1118K for 720ks in air, and the fraction of high-T_c phase was 50% (Table 1). The plate-like grains were formed by pre-sintering, but the oriented structure was not observed (Fig. 5(a)). The $T_c(\rho=0)$ was 80K, and J_c at 77K was 210A/cm^2 which is slightly low in comparison with that of the rolled, shock compacted and post heat-treated tape. The high-T_c phase after pre-sintering grew markedly during the shock compaction, which is similar to the result described in the previous section, and the fraction of high-T_c phase was reached to 71% (Table 1). The plate-like grains formed by the pre-sintering were crashed during shock compaction (Fig. 5(b)). The crashed grains would collide and rub each other so that the surface layer of grain melts locally and the generated heat diffuses. It is reasonably assumed from this consideration that the generated heat promotes the transformation to high-T_c phase in addition to the effect of shock pressure. The tapes after shock compaction, however, did not show zero resistance above 77K. This will be due to the imperfect connection between grains and the formation of microcracks by the shock loading. Then the tapes were post heat-treated at 1118K for 360ks. After the heat-treatment, the fraction of high-T_c phase was

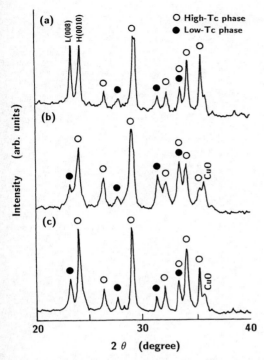

FIG. 4 *X-ray diffraction patterns of (a) pre-sintered, (b) compacted after pre-sintering, and (c) post heat-treated tape core.*

increased slightly (Table 1), and the highly oriented and large plate-like grains were stacked densely as shown in Fig. 5(c). The $T_c(\rho=0)$ and $J_c(77K)$ are 95K and 330A/cm^2, respectively. The J_c was not so high in spite of the increase of high-T_c phase and the oriented grain structure. This may be attributed to that the microcracks formed by the shock compaction were not entirely removed by the post heat-treatment. The broad superconducting transition shown in Table 1 also suggests the presence of microcracks as weak links.

C. *Effects of shock pressure on microstructures and J_c*

In the previous section, it is shown that explosive compaction technique is effective to produce oriented microstructures but microcracks will be still remained in the tape core even after post heat treatment. This may be due to the extremely high shock pressure of SEP. Then, in the second experiment, "PAVEX" having low detonation velocity was used. The weight of the explosive on flyer plate was 100g.

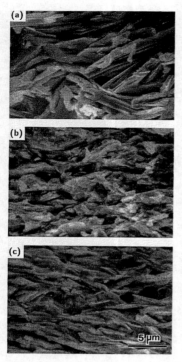

FIG. 5 *Scanning electron micrographs of fractured cross
section of (a) pre-sintered, (b) compacted after
pre-sintering and (c) post heat-treated tape.*

Figure 6 shows X-ray diffraction pattern of pre-sintered, shock compacted and post heat-treated tape core. Many peaks of low T_c phase in Figs. 2 and 4 are not seen in Fig. 6. This is accordance with the result that the fraction of high-T_c phase was increased to 92% (Table 1). Table 1 also shows that the tape compacted using PAVEX exhibits a narrow superconducting transition in comparison with those for tapes compacted using SEP. The tape core consists of the plate-like and oriented grains which are compacted densely, similarly to the structure shown in Fig. 5(c). These results are expected to provide a superior superconducting property. In fact, Table 1 shows that the compaction using PAVEX makes the T_c to rise and the J_c to increase up to extremely large value of 3000A/cm^2 at 77K in zero external magnetic field. That is, the use of explosive with low shock pressure is preferable for improving J_c. The long duration time of shock wave generated by the explosive may have also strong influence on the microstructure of grain boundary and the growth of high-T_c phase although the detail of mechanism is not known.

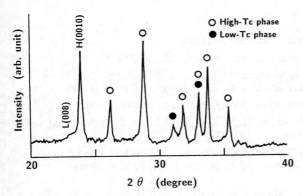

FIG. 6 *X-ray diffraction pattern of a shock compacted and post heat treated tape core. The tape was pre-sintered before the compaction by low detonation velocity explosive.*

IV. CONCLUSION

The explosive compactions were conducted on silver-sheathed Bi-Sr-Ca-Cu oxide superconducting tapes, and the microstructural modifications and superconducting properties were investigated. Shock compaction promoted the growth of high-T_c phase. This appears to be caused by local melting at the particle interfaces and/or heat diffusion from the interfaces during the shock compaction. Although the as-compacted tapes did not show zero resistance above 77K, the T_c was restored by the post heat treatment at 1118K in air. The amount of the high-T_c phase increased by the shock compaction was much more for pre-sintered tape than for as-rolled one. The core of shock compacted and heat-treated tapes consists of the plate-like and highly oriented grains. J_c was increased to 3000A/cm^2 by the shock compaction using the explosive with lower detonation velocity as compared with the explosive with high detonation velocity (~300A/cm^2). The long duration time of shock wave generated by the explosive helps to increase J_c because the amount of microcracks could be reduced and the high-T_c phase is increased.

ACKNOWLEDGMENTS

This work was partly supported by Grant-in-Aid for Scientific Research on Priority Areas from the Ministry of Education, Science and Culture, Japan. The authors would like to express their thanks to Mr. N. Tsukimata and Mr. I. Shimizu of Kumamoto University for assistance in the explosive experiment.

REFERENCES

1.   W. H. Gourdin, *Prog. Mater. Sci.*, 30 : 39 (1986).

2.   L. E. Murr, A. W. Hare, and N. G. Eror, *Nature*, 329 : 37 (1987).

3.   L. E. Murr, T. Monson, J. Javadpour, M. Strasik, U. Sudansan, N. G. Eror,
     A. W. Hare, D. G. Brasher, and D. J. Butler, *J. Metals*, 40(1) : 19 (1988).

4.   L. E. Murr, and N. G. Eror, *Mater. Manuf. Processes*, 4 : 177 (1989).

5.   L. E. Murr, C. S. Niou, S. Jin, H. Tiefel, A. C. W. P. James, R. C. Sherwood, and
     T. Siegrist, *Appl. Phys. Lett.*, 55 : 1575 (1989).

6.   H. Hagino, M. Suzuki, T. Takeshita, K. Takashima, and H. Tonda, in *Advances in
     Superconductivity*, K. Kitazawa, and T. Ishiguro (eds.), Springer-Verag,
     1989, p.365.

7.   H. Hagino, M. Suzuki, T. Takeshita, K. Takashima, and H. Tonda, in *Processing &
     Applications of high T_c Superconductors*, M. E. Mayo (ed.), in press.

8.   K. Takashima, H. Tonda, M. Nishida, H. Hagino, M. Suzuki, and T. Takashita, in
     Shock Waves in Condensed Matter 1990, S. C. Schmidt, J. N. Johnson, and L. W.
     Davison (eds.), North-Holland, 1990, p.591.

9.   M. Takano, J. Takada, K. Oda, H. Kitaguchi, Y.Miura, Y. Ikeda, Y. Tomii, and
     H. Mazaki, *Jpn. J. Appl. Phys.*, 27 : L1041 (1988).

10. U. Endo, S. Koyama, and T. Kawai, *Jpn. J. Appl. Phys.*, 27 : L1476 (1988).

11. T. Asano, Y. Tanaka, M. Fukutomi, K. Jikihara, and H. Maeda, *Jpn. J. Appl. Phys.*,
     28 : L55 (1989).

12. Y. Hayashi, H. Kogure, and Y. Gondo, *Jpn. J. Appl. Phys.*, 28 : L2182 (1989).

13. J. Tsuchiya, H. Endo, N. Kijima, A. Sumiya, M. Mizuno, and Y. Oguri, *Jpn. J.
     Appl. Phys.*, 28 : 11918 (1989).

14. T. Asano, Y. Tanaka, M. Fukutomi, K. Jikihara, J. Machida, and H. Maeda, *Jpn. J.
     Appl. Phys.*, 27 : L1652 (1988).

80

Structural Changes in Metallic and Superconducting (YBa$_2$Cu$_3$O$_7$) Powders Induced by Laser-Driven Shocks

P. DARQUEY, J. C. KIEFFER, J. GAUTHIER, H. PEPIN, M. CHAKER,
B. CHAMPAGNE*, H. BALDIS*, and D. VILLENEUVE*

INRS-Energie
Varennes, Québec, Canada, J3X 1S2

*Institut des Matériaux Industriels, Conceil National de Recherche Canada
Boucherville, Québec, Canada, J4B 6Y4

*Division of Physics, National Research Council Canada
Ottawa, Ontario, Canada, K1A OR6

A new method based on the use of laser driven shocks
to density Cu powders and YBCO superconductors was studied.
Pressures of 10 GPa on a nanosecond time scale were
generated by producing a confined plasma with a high
intensity laser pulse. Significant consolidation was
generated 600 μm and 100 μm below the irradiated surface
for Cu and YBCO respectively.
Magnetic susceptibility, X-ray diffraction and
magnetic relaxation were used to characterize the induced
changes of Cu and YBCO before and after laser shocks. Even
though laser shocks led to lower susceptibility signal, the
decay rate of the remanent magnetization is slightly
decreased. Laser shocks of about 10 GPa do not
significantly modify the properties of YBCO superconductor.

I. INTRODUCTION

Structural changes have been induced in various materials by conventional shock techniques [1]. In these experiments large pressures are maintained during a long time, microseconds up to milliseconds, by impact of flyer plates or explosives. In the present study a strongly different regime was investigated: shock waves were induced on a nanosecond time scale by the generation of a confined plasma from a laser pulse.

The consolidation process resulting from laser driven shocks on powders still needs to be better understood [2]. Consolidation of large volumes depends on the properties of the materials. Two different types (metallic and ceramic) were selected in this study. Moreover, structural modifications are also generated by shocks. It has been shown [3] that conventional shock largely modify the properties of YBCO super-conductors. The purpose of this study was to characterize the dynamics of consolidation and assess the structural effects of nanosecond laser driven shocks.

II. EXPERIMENTAL

A. *METHOD*

Very high pressures are generated from a plasma produced by a laser at the surface of a material. In the ablative regime, the pressure is maintained only during the laser pulse and high peak pressures require high laser intensity, and therefore small focal

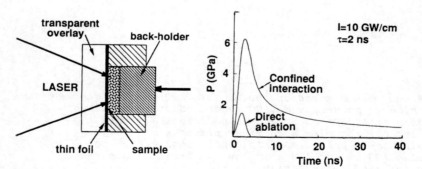

FIG. 1 *(a) Schematic view of experimental set up for producing laser shocks in confined interaction. (b) Pressure-time profile for a laser intensity I = 10 GW/cm² and a laser pulse duration τ = 2 ns compared to direct ablation.*

spots. High pressures on large material surfaces with a reasonable laser energy can be achieved by the confinement plasma technique developed by Anderholm [4]. Figure 1a schematically illustrates the method used in the present study to generate shock waves. A plasma is produced by a laser radiation at the interface between a thin foil (typically a few microns thick) and an overlay transparent to the laser radiation. The shock wave then propagates through the sample which is thermally isolated from the plasma by the foil. A back-holder is used to avoid reflection of any rarefaction wave from the rear sample surface. With a confined plasma, the peak pressure is higher and the temporal profile is longer compared to the direct ablation [5].

B. LASER AND PRESSURE LOADING

The experiments were carried out with the LP2 laser of the National Research Council Canada in Ottawa. This Nd laser (wavelength 1.06 μm) delivers up to 200 J in a 4 ns pulse (FWHM) with a 300 ps rise time. The beam is focused with an f/8 spherical lens in a 10 mm diameter spot.

Figures 2a and 2b show the behavior of the impulse and the pressure as a function of the laser intensity used in this study. The transparent windows used to confine the plasma are 6 mm thick glass. The impulse and the pressure increase with the laser intensity and then saturate at about 20 GW/cm^2 for 1 μm light. This saturation is produced by the breakdown of the transparent window when the laser intensity is too high. Maximum impulse and pressure of 1 GPa.ns and 10 GPa have been obtained at 1.06 μm in the saturation regime.

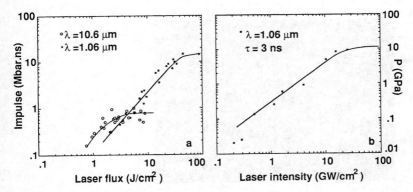

FIG. 2 (a) Impulse and (b) pressure versus laser intensity induced in confined interaction.

C. CU AND YBCO POWDERS

The particle size distribution of Cu (99.99%) ranged between 5 and 20 μm. Cu was statically pressed at 4 t/cm² and 5 t/cm² to cylindrical samples (2mm thick and 15 mm in diameter) resulting in 75% and 80% of the theoretical density (T.D.) respectively.

Superconducting samples were prepared by mixing high purity (> 99.99%) Y_2O_3, $BaCO_3$ and CuO powders. Solid state synthesis was carried out in air for 10 hours at 950° with intermediate grinding. The annealing steps was carried out from 900° to 500° at a rate of 100°C per hour in air for 12 hours. Ground pellets were thin sieved (-120 mesh), statically pressed at 5 t/cm² sintered at 950° and annealed for 12 hours at 500° in O_2. The resulting cylinders had a 80% theoretical density, a critical temperature of 92 K with a transition less than 10 K, an orthorombic structure with c = 11.69 Å corresponding to an oxygen content of about 6.87 [6].

III. COMPACTION OF Cu POWDERS

Figure 3 shows the physical aspect of irradiated samples with a 1 μm laser intensity of 40 GW/cm² equivalent to an applied pressure of about 10 GPa. It illustrates the extent of the compaction along the shock propagation axis and indicates a complete densification up to a depth of 150 μm. However as the shock propagates inside the material the pressure decreases and a compaction gradient appears perpendicularly to the irradiated surface. Figure 4a shows the density gradients induced by the propagation of the shock through Cu of different initial density. It is apparent from this figure that the attenuation of the shock is faster when the initial density is lower.

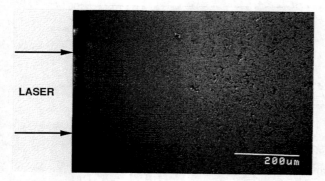

FIG. 3 *S.E.M. micrograph of Cu powders with 70% T.D. shocked with a laser intensity I = 40 GW/cm² (P ~ 10 GPa).*

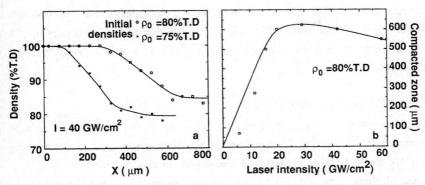

FIG.4 *(a) Density gradient of Cu having initial density of 70% T.D. and 80% T.D. for laser intensity I = 40 GW/cm2. (b) Thickness of the density gradient region versus laser intensity.*

The typical density gradient thickness is about 600 μm for Cu with an initial density ρ_0 = 80% T.D. Figure 4b shows that the compaction zone initially increases with the laser intensity and then saturates at about 600 μm for laser intensities greater than 20 GW/cm².

IV. EFFECTS OF LASER DRIVEN SHOCKS ON $YBa_2Cu_3O_4$

A. COMPACTION

Figure 5 shows that the compacted zones are lower in YBCO than in Cu indicating that the shock does not propagate in the same way since their mecanical properties

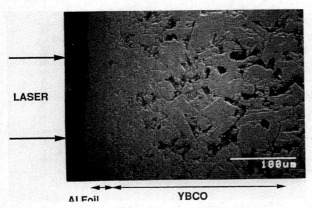

FIG.5 *SEM micrograph of YBCO having initial density 80% T.D. shocked with a laser intensity I = 30 GW/cm2.*

are different (hardness, hugoniot curve). One notes a good densification in a 50 μm thick layer below the irradiated surface, and a very fast decrease of the density in 150 μm. It is worthy to note the dimensions of the grains appear to be smaller in the shocked zones.

B. SUPERCONDUCTING PROPERTIES

Magnetic susceptibility: Figure 6 shows the out of phase and the in phase susceptibility signals which respectively correspond to the diamagnetism and to the resistive losses of the material for equal masses of loaded and unloaded materials.

The out of phase signal (fig. 6a), obtained at zero magnetic field and measured with an a.c. amplitude of 3.2×10^{-5} T indicates a T_c onset of 92 K for both loaded and unloaded YBCO. However, a decrease of the magnetic susceptibility signal with a broadening of the transition occurs after shock. An increase of the a.c. field up to 6.4×10^{-4} T led the in phase signal (fig. 6b) to exhibit two peaks corresponding to intragrain and intergrain currents [5]. The position of the intragrain peak is not changed by the shock but the intergrain peak is broadened and shifted to lower temperatures. This can be related to a reduction of grain size and a destruction of their links.

Magnetic relaxation measurements: We performed magnetic relaxation measurements on loaded and unloaded YBCO using a SQUID magnetometer. The samples were zero field cooled to 50 K, a field of 1 Tesla was applied and removed and the

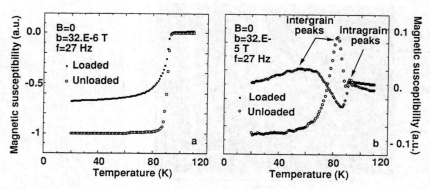

FIG.6 *(a) Real and (b) imaginary part of the a.c. susceptibility versus temperature for shocked and unshocked samples.*

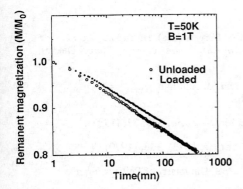

FIG. 7 *Relaxation of remanent magnetization as a function of time normalized to its initial value M_O for loaded and unloaded YBCO.*

magnetic moment was measured each minute for 250 minutes. Figure 7 shows the time dependence of the relaxation of the remanent magnetization on a logarithmic scale. The slope of the shocked material is slightly lower and may indicate an increase of the flux pinning. The pinning energy U_0 can be calculated using the thermal activated model [8] and the order of magnitude of U_0 for shocked YBCO is about 0.11 eV which is similar to reported U_0 values for YBCO [8,9].

V. DISCUSSION

Nanosecond laser driven shocks have been successfully used to compact Cu and superconducting powders. The characteristics properties of YBCO ceramics are slightly modified by laser shocks decreasing the intergrain coupling and broadening of the magnetic susceptibility transition temperature. Furthermore, preliminary magnetic relaxation measurements do not indicate a significant increase in the flux pinning of loaded YBCO.

Thus, laser driven shocks constituted an interesting technique, with a high repetition rate, for consolidating YBCO without significantly altering the superconducting properties.

ACKNOWLEDGMENTS

The authors wish to thank J. Cave (IREQ, Varennes, Québec) for helpful discussions on the characterization of YBCO and W.R. McKinnon (NRC, Ottawa, Ontario) for magnetic relaxation measurements.

REFERENCES

1. L. E. Murr, K. P. Staudhammer, M. A. Meyers (eds), *Metall. Appl. of Shock Waves and High Strain Rate Phenomena*, Marcel Dekker, New York, (1986).

2. P. Darquey, J. P. Romain, F. Cottet, and M. Hallouin, *Journal de Physique*, Colloque C3, 49, (1988).

3. L. E. Murr, A. W. Hare, and N. G. Eror, *Nature*, 329, 37, (1987).

4. N. C. Anderholm, *Appl. Phys. Lett.*, 16 (3), 113, (1970).

5. P. Darquey, *Ph.D. Thesis No. 256*, University of Poitiers, France, (1989).

6. Küpfer et al., *Cryogenics*, 28, 659, (1988).

7. P. W. Anderson, *Phys. Rev. Lett.*, 9, 309, (1962).

8. S. T. Weir et al., *Appl. Phys. Lett.*, 56 (20), 2042, (1990).

9. Y. Yeshurun, A. P. Malezemoff, *Phys. Rev. Lett.*, 60 (21), 2202, (1988).

81

A Comparison of Residual Superconductivity in Shock Processed, Oxygen Deficient, and Irradiated Y-Ba-Cu-O

L. E. MURR

Department of Metallurgical and Materials Engineering
The University of Texas at El Paso
El Paso, Texas 79968 USA

Shock processed $YBa_2Cu_3O_7$ exhibits variations in both the normal state resistance-temperature (R-T) behavior and in the nature of the superconducting transition. However these variations are different from those exhibited in the R-T signatures for oxygen dificient and ion and electron irradiated Y-Ba-Cu-O. A comparison of these characteristic R-T signatures led to the conclusion that manipulating the microstructure of the initial superconducting powder could alter the shock instabilities, and change the residual superconductivity. The addition of silver to the Y-Ba-Cu-O resulted in dramatic variations in the R-T signature, and provided evidence in support of strategies which may allow for the explosive processing of useful bulk prototypes of Y-Ba-Cu-O.

I. INTRODUCTION

It is now well established that explosive or shock processing of $YBa_2Cu_3O_7$ and other superconducting powders [1,2] as well as the plane wave shock loading of dense, sintered superconductors induces degradation [3-6]. This degradation manifests itself in variations in both the normal state resistance-temperature behavior, and in the nature of the superconducting transition. As shown in the experimental data reproduced in Fig. 1(a), peak shock pressures in excess of about 13 GPa destroy the superconducting transition and produce a continuous semiconducting behavior -- with resistance increasing with decreasing temperature. In

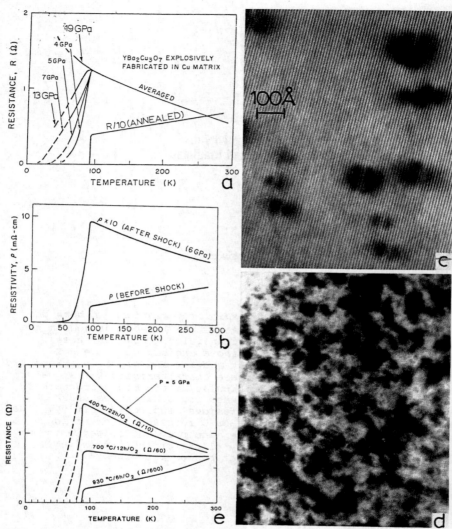

FIG. 1 *Explosively fabricated Y-Ba-Cu-O.* (a) *Resistance -temperature (R-T) signatures for explosively fabricated Y-Ba-Cu-O powders to produce bulk monoliths at peak pressures noted.* (b) *Plane-wave shock loading of a sintered coupon for comparison (Data from [4]).* (c) *Shock-induced defect clusters in samples corresponding to fabrication pressure of 7 GPa.* (d) *Shock-induced defect clusters in samples corresponding to fabrication pressures of 19 GPa.* (e) *Thermal recovery of shock fabricated Y-Ba-Cu-O. (After [5]).*

addition, the broadening of the superconducting transition, indicated by: $[T_c$ (onset) - T_c (R=0)], decreases with decreasing pressure.

While the curves in Fig. 1(a) are characteristic of the residual properties of explosively fabricated, superconducting powder, the curves in Fig. 1(b) illustrate the behavior of sintered and plane-wave shock loaded Y-Ba-Cu-O, and attest to the overwhelming role that the shock wave plays in this process. Figure 1(c) and (d) show some comparative examples of shock-induced microstructures in explosively fabricated Y-Ba-Cu-O powders which, because of the apparent increase in density with increasing pressure, can be at least phenomenologically associated with the resistance-temperature (R-T) signature changes shown in Fig. 1(a).

Finally, as shown in Fig. 1(e), thermal annealing of explosively fabricated Y-Ba-Cu-O at increasing temperatures exhibits some features similar to reducing the peak pressure in comparing the R-T signatures of Fig. 1(a). The arrows in Figs. 1(a) and 1(e) in fact illustrate the shocking-"up" of residual R-T behavior or the annealing-"down" of the R-T behavior. Another feature of interest in comparing Fig. 1(a) and 1(e) is the fact that the T_c (onset) at roughly 90K is not changed unless the pressure exceeds about 13 GPa and is unchanged at various annealing temperatures as well. Thermogravimetric analysis has indicated that the oxygen content is not changed appreciably in the explosively fabricated material until 13 GPa, and the downward shift in the T_c (onset) shown in Fig. 1(a) at 13 GPa is indicative of this feature.

The interesting features about the resistance-temperature signatures shown in Fig. 1 are not only the variations in both the normal-state behavior with pressure and annealing treatment, but also the superconducting-state behavior. In addition, there is ample evidence that while shock-induced defects (illustrated in Fig. 1(c) and (d)) are involved in creating the residual R-T behavior noted, they also promote strong flux pinning which can have a dramatic effect on the transport supercurrent density [7,8].

II. RESISTANCE-TEMPERATURE SIGNATURES FOR OXYGEN DEFICIENT AND IRRADIATED Y-Ba-Cu-O

The implications of Fig. 1 are that if the defects (microstructure) induced by explosive fabrication can be elucidated and manipulated either by manipulating the microstructure prior to fabrication to alter the way in which the instabilities in the shock front are accomodated by post fabrication annealing, or both; it may be possible to predictably control the residual R-T signature. In effect, it may be possible to manipulate the R-T signature while optimizing the flux pinning if that is an important issue to be considered.

While there is a dearth of direct observations of microstructures associated with various R-T signatures in Y-Ba-Cu-O, there are some interesting comparisons to be made for variously treated Y-Ba-Cu-O which may provide at least some basis for developing an appropriate or desirable fabrication strategy. The data reproduced in Fig. 2 is intended for comparison with Fig. 1 while providing some simple features from which an experimental strategy might be developed. Figure 2(a) shows that decreasing the oxygen concentration has a dramatic if not irregular influence on the R-T signature for $YBa_2Cu_3O_X$ ($7 < X < 6$) [9]. This effect is due to variations in oxygen ordering in the Cu-O chains as illustrated schematically in Fig. 3, which

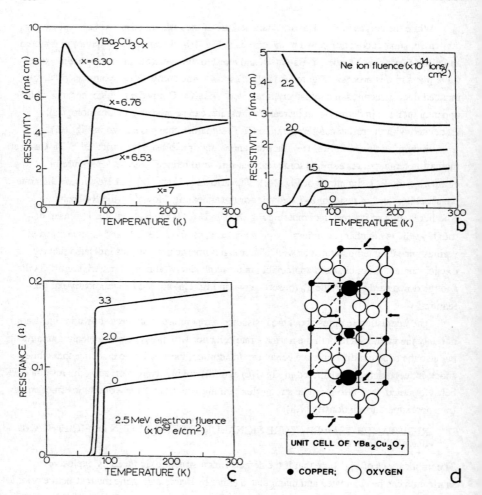

FIG. 2 *Comparison of R-T signatures for oxygen deficient (a) ion irradiated (b) and electron irradiated (c) Y-Ba-Cu-O. Data in (a), (b), and (c) are from [9], [10], and [11] respectively. (d) shows the idealized unit cell for $YBa_2Cu_3O_7$.*

also shows the simple, elemental alterations in the Y-Ba-Cu-O unit cell. By comparison with Fig. 1(a), (b) and (e), Fig. 2(a) shows a consistent decline in both the T_c (onset) and zero-resistance intercept with decreasing oxygen content. In Fig. 2(b), the effect of increasing ion fluence [10] is observed to exhibit some similarities to both the shock-wave effects shown in Fig. 1(a) and the oxygen loss effects shown in Fig. 2(a). However, the signature features are

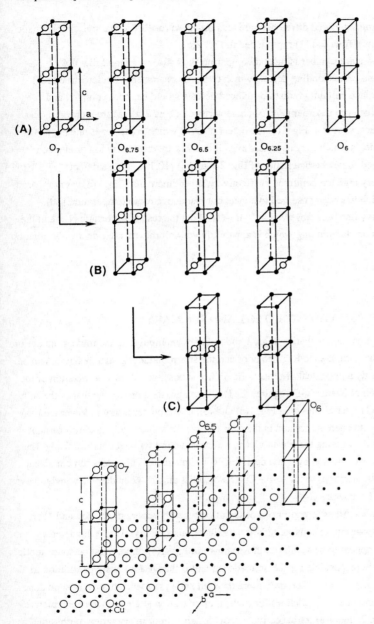

FIG. 3 *Schematic representations of the double-unit cell structures for* YBa$_2$Cu$_3$O$_x$ *showing basal-plane Cu-O chains and oxygen order-disorder. See Fig. 2(d) for reference unit cell.*

different. Both similarities and differences are also observed for R-T signatures characteristic of increasing electron fluence [11] reproduced in Fig. 2(c).

It should be apparent in perusing and comparing Figs. 1(a) and Fig. 2 that the microstructures (and corresponding defects which compose characteristic microstructures) would be different, and certainly radiation-induced defects would be more complex and extensive than the oxygen disorder (Fig. 3) characteristic of Fig. 2(a). Moreover the shock-wave induced defects shown in Fig. 1(c) are defect clusters which create large and even interacting strain fields which may be absent even if clusters are created by ion or electron damage as described to be characteristics of Fig. 2(b and c) [10,11]. These defects (Fig. 1(c) and (d)) require very high temperatures to effectively anneal them out (Fig. 1(d)) as compared to ion damage which has been observed to anneal out just above room temperature [10]. Neutron damage has also been observed in Y-Ba-Cu-O but the results are erratic [12] and the emphasis has been on flux pinning which is apparently very prominent after neutron irradiation [13].

III. PROCESS OPTIMIZATION STRATEGY AND SUMMARY

It seems rather apparent on comparing Figs. 1 and 2 that by manipulating the microstructure of Y-Ba-Cu-O both the normal-state R-T behavior and the superconducting-state behavior can be altered. Consequently a great deal of process flexibility is possible. If the optimization of the explosive fabrication of Y-Ba-Cu-O requires the R-T signature to approximate that of the fully annealed or sintered material shown in Fig. 1(a) and (e), it will be necessary to lower the peak pressure and/or alter the defects created in the shock front. To achieve this at some constant pressure would require altering the pre-fabrication microstructure without compromising the superconductivity (as expressed by some desirable R-T signature). This alteration could start by specific elemental additions to the Y-Ba-Cu-O which has already been demonstrated to have an effect on the R-T signature [14].

Figure 4 shows a rather serendipitous observation recently reported by Niou and Murr [15] which demonstrates quite convincingly the merits of this strategy. While the defects shown in Fig. 4(b) appear to be similar to those shown in Fig. 1(c), there must be some subtle differences which are responsible for the dramatic signature change for the silver addition to Y-Ba-Cu-O as shown in Fig. 4(a). Certainly more detailed and systematic studies must be done to establish more fundamental guidelines for such an optimization strategy, but a comparative glance at Figs. 1(a), 2, and 4(a) retrospectively provides an extraordinary range of possible electronic behavior in Y-Ba-Cu-O which might be selectively attained in explosively fabricated bulk devices.

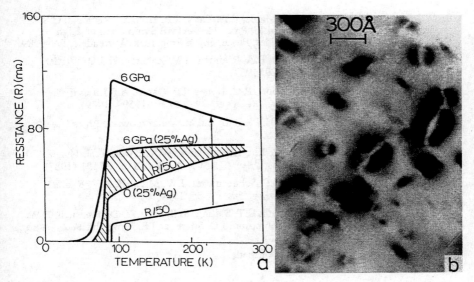

FIG. 4 *Comparison of R–T signatures in plane-wave shock loaded YBa₂Cu₃O₇ and YBa₂Cu₃O₇ + 25% Ag (a) and characteristic defect clusters observed in the silver-doped, shock-loaded material (b).*

ACKNOWLEDGMENT

This research was supported in part by a DARPA High-Temperature Superconductivity Program under Contract ONR-N00014-88-C-0684, an NSF Minority Research Center of Excellence in Materials Science Grant, and a Mr. and Mrs. MacIntosh Murchison Endowed Chair.

REFERENCES

1. L.E. Murr, A.W. Hare, and N.G. Eror, *Nature 329:* 37 (1987).

2. L.E. Murr, T. Monson, J. Javadpour, M. Strasik, U. Sudarsan, N.G. Eror, A.W. Hare, D.G. Brasher, and D.J. Butler, *J. Metals 40(1):* 19 (1988).

3. L.E. Murr, N.G. Eror, and A.W. Hare, *SAMPE Journal 24:* 15 (1988).

4. L.E. Murr, C.S. Niou, S. Jin, T.H. Tiefel, A.C.W.P. James, R.C. Sherwood, and T. Siegrist, *Appl. Phys. Lett. 55(15):* 1575 (1989).

5. L.E. Murr, M. Pradhan-Advani, C.S. Niou, and L.H. Schoenlein, *Sol. State Comm. 73:* 695 (1990).

6. L.E. Murr and C.S. Niou, *J. Mater Sci. Lett.* in press (1990).

7. B. Morosin, R.A. Graham, E.L. Venturini, and D.S. Ginley, Proc. 2nd Workshop in Industrial Application Feasibility of Dynamic Compaction Technology, Tokyo Institute of Technology, Yokohama, Japan, 1988, p. 96.

8. W.J. Nellis, C.L. Seaman, M.B. Maple, E.A. Early, J.B. Holt, M. Kamegai, G.S. Smith, D.G. Hinks, and D. Pabrowski, Proc. TMS-AIME Symposium on High Temperature Superconducting Oxides, Processing & Properties, Warrendale, PA, 1989.

9. R.J. Cava, B. Batlogg, C.H. Chen, E.A. Riefman, S.M. Zahurak, and D. Werder, *Phys. Rev. B 36(10):* 5729 (1987).

10. J.M. Valles, A.E. White, K.T. Short, R.C. Dynes, J.P. Garno, A.F.J. Levi, M. Anzlowar, and K. Baldwin, *Phys. Rev. B 39(16):* 11599 (1989).

11. H. Vichery, F. Rullier-Albenque, H. Pascard, and M. Kanczykowski, *Physica C 159:* 697 (1989).

12. A. Umezawa, G.W. Crabtree, J.Z. Liu, H.W. Weber, W.K. Kwok, L.H. Nunez, T.J. Moran, C.H. Sowers, and H. Claus, *Phys. Rev. B 36(13):* 7151 (1987).

13. R.B. Van Dover, E.M. Gyorgy, L.F. Schneemeyer, J.W. Mitchell, K.V. Rao, R. Puzniak, and J.V. Waszczak, *Nature 342:* 55 (1989).

14. S.B. Oseroff, D.C. Vier, J.F. Smyth, C.T. Salling, S. Schultz, Y. Dalichasuch, B.W. Lee, M.B. Maple, Z. Fisk, J.D. Thompson, J.C. Smith, and E. Zirngiebl, *Sol. State Comm. 64(2):* 241 (1987).

15. C.S. Niou and L.E. Murr, to be published.

Section VIII
Shock Waves and Shock Loading

82

Defect Nucleation Under Shock Loading

M. A. MOGILEVSKY

Lavrentyev Institute of Hydrodynamics
Siberian Division of the USSR Academy of Sciences
Novosibirsk 630090, USSR

The shock loading of crystalline bodies is accompanied by extremely high stresses, attaining or even exceeding the theoretical shear strength and strain rates up to 10^7 - $10^9 s^{-1}$. Under these conditions, the significant changes of deformation mechanisms are possible. The present paper considers the peculiarities of nucleation under shock loading of different defects, such as dislocations, point defects, twins, and adiabatic shear bands.

I. INTRODUCTION

In 1990 we celebrate the 100th anniversary of Sir Lawrence Bragg who is one of the authors of investigation of dislocations on a model of soap bubbles. This investigation happened to be a brilliant confirmation of hypothetic concepts about the crystalline lattice defects by Taylor, Polany and Frenkel. An analysis of structures [1] observed on a layer of soap bubbles in steady state or under deformation allowed not only the confirmation of successive character of shear propagation in a crystal to be made and also nucleation of a couple of dislocations to be observed under the deformation of a defect-free crystal as well as generation of dislocations from grain boundaries and free surface.

The present paper deals with consideration of peculiarities of defect nucleation under the extreme conditions of shock loading. Elementary events of defect nucleation are of cooperative character, for this reason their theoretical study and experimental observation is quite difficult. Most of our results on defect nucleation have been obtained from computer experiments with a plane crystalline lattice. These experiments are thus further developments of the Bragg study.

II. NUCLEATION OF DISLOCATIONS ON SHOCK FRONT

A. THEORETICAL STRENGTH

The well-known Cowan method [2] allows the value of shear stresses on a plane stationary shock front to be associated with deformation. The level of shear stresses increases rapidly with the wave intensity, and by the Cowan estimate, attains the value corresponding to the theoretical shear strength G/30 in copper in a 26 GPa wave. However, the catastrophic changes in copper, predicted by Cowan, are not observed under more intense loading. With an increase in the shock loading intensity up to 100 GPa, gradual strength and structural changes take place as well as gradual decrease in the shock front width. Our study [3] of a model plane crystalline lattice with the interatomic copper potential [4] showed that the Cowan paradox is attributed to a dependence of the theoretical strength on the loading form. The calculation results for critical deformation and the corresponding stresses under which the lattice losses its stability are given in the Table for different forms of loading. In fact, due to deformation uniformity, a calculation was made for interaction forces of one of atoms, placed in the origin of coordinates, including all the neighbours being under the action of the potential. At pure shear with zero normal stresses on the faces of a deformed element the lattice for which the interatomic interaction potential was used in our calculations looses its stability when $\tau = 6.55$ GPa accounting for 1/17 of the shear modulus G. This value is between the estimates of Frenkel and Mackenzie. Application of the hydrostatic pressure results in an increase in the critical value of τ and change of the relation τ/G up to 2/15 when P = 60 GPa. Tension with zero deformation in the lateral direction $\varepsilon_t = 0$ corresponds to spallation. Uniaxial elastic compression with $\varepsilon_t = 0$ is the stressed state characteristic for the shock front. For

TABLE 1

Loading	tg γ^*	ε_n^*	P_n	τ^*, GPa	τ^*/G
Shear $\varepsilon_n = \varepsilon_t = 0$	0.229	0	35.5	18.2	1/6
$P_n = P_t = 0$	0.102	−0.024	0	6.55	1/17
Shear+pressure $P = 60$ GPa	0.194	0	60	23.8	2/15
Elongation $\varepsilon_t = 0$ $[110]$	–	−0.13	−28.2	4.58 *	1/24
$P_t = 0$ $[110]$	–	−0.06	−9.72*	4.38	$P_{n/E} = 1/28$
Compression $\varepsilon_t = 0$ $[112]$	–	0.25	208	54.9	1/2

the lattice with interatomic potential under consideration the stability loss takes place under pressure $P_n = 208$ GPa in the compression direction and $\tau = 55$ GPa. Thus, even under megabar shock loading the copper lattice does not loose its stability (this is in agreement with experiment [5]) (Table 1).

B. "HOT SPOTS" IN A DEFECT-FREE CRYSTAL

With the non-zero temperature the value of one-dimensional crystal compression, under which the lattice looses its stability decreases. For example, in our calculations the temperature increase from 0 to 300K results in decrease in the critical level of shear stresses from 55 to 43 GPa with the non-zero temperature (theoretical strength), the corresponding pressure value P_n decreases from 208 to 150 GPa. Several our papers [4, 6] are devoted to investigation of peculiarities of nucleation and evolution of plastic deformation in a plane copper lattice of 500 atoms under a 18.6% compression corresponding to 100 GPa pressure. A cold lattice under such one-dimensional compression is stable and compressed elastically. During gradual heating of the lattice under uniaxial 18.6% compression, shear nucleation was observed in it at 424K. At first, the bending of planes {011} took place in a certain place. Then, one of planes broke and the half-planes moved apart forming a stacking fault bounded by partial dislocations (Fig. 1). The following stage is the stacking fault breakdown and formation of two split dislocations.

In observation of shear evolution from a certain crystal point, the question arises that concerns the nature of a "hot spot" in which shear nucleation occurs.

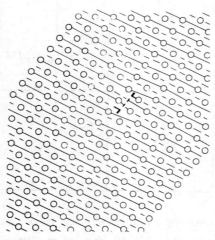

FIG. 1 *Shear nucleation in a defect-free lattice; 18.6% compression in a vertical direction [112].*

Unlike the model Bragg experiments, computer experiments allow the states preceding the lattice stability loss to be studied in detail, since the whole process history is in computer's memory. Additional computer experiments and theoretical analysis of thermal vibrations in a plane lattice, that is analogous to that one from [7] for the three-dimensional crystal, were performed. It has been possible to show the role of correlated displacements of atoms from crystallographic nodes in defect formation [8]. The analytical dependences and construction of displacement fields have shown that in each moments there exist the regions in crystal, where the atoms are shifted from the nodes in one predominant direction with deviation up to 20°. This is an inherent property of the thermal motion in a condensed medium, providing the capability for self-organization of the system in changing conditions. The transversal dimension of such regions is approximately 4-5 atoms. It turned out to be independent of temperature and pressure. The analogous theoretical calculations and observations over the lattice did not show the existence of similar correlation for velocities of atoms. The correlated velocity blocks comprising more than 3-4 atoms are unlikely.

Figure 2 presents the field of atomic displacements from the crystallographic nodes for the lattice under uniform 18.6% compression at the moment of heating

FIG. 2 *Regions of correlated displacements in a lattice at the moment of shear generation (Fig. 1).*

up to 424K. The arrows on each atom point to the displacement from the lattice node. Full arrow means the displacement > 0.25 **A**. Predominant direction of displacements in each block is marked by a large arrow. In the "hot spot", where the shear shown in Fig. 1 arises. the interatomic bond is broken as a consequence of the shift of two large blocks of correlated displacements in the opposite directions along the plane with high shear stresses. Calculations of behaviour of the crystalline lattice with the free boundaries at premelting point show that interaction of blocks of correlated displacements play a leading role also in formation of vacancies, bivacancies and stacking faults near the free surface [8].

C. SHEAR NUCLEATION ON POINT DEFECTS

A defect-free crystalline lattice is a very rough model of real material for studying the strength. So, behaviour of our model copper lattice is elastic under the one-dimensional 150 GPa compression at 300K. A series of papers concerning decay of an elastic precursor [see 9, 10] made a proposal that the shear nucleation takes place on impurity atoms even under this comparatively low pressure level. This suggestion was studied in detail in our computer experiments. In this section the main calculation results [4, 6] are presented. In calculation performed for gradual one-dimensional compression of a lattice of 500 atoms at constant temperature of 300K, the drastic fall of critical pressure P_n a shear stress τ, for defect nucleation took place in the vicinity of a substitution atom with increase in the impurity atom radius (Fig. 3). Inflection on curves of P_n and τ with $r = 1.3r_0$ is due to changes of a number of the nearest neighbours in the plane (111) from 6 to 7 and 8 with $r = 1.4; 1.5r_0$. With the impurity atom in size range $r = 1.5r_0$, the shear nucleation on it would take place in a 10 GPa shock wave if the calculation model is close to the real material characteristics. Calculation for the vacancy with $r = 0$ gave the criticel pressure for shear nucleation 40 GPa. Earlier [11] it was shown, but with other interatomic potential, that efficiency of the interstitial atom in rearrangement occurence under the uniaxial compression is higher than that of the vacancy. The complexes of point defects can naturally be more efficient centers of shear nucleation. The calculation was made for gradual compression of the lattice with three different defects: a single impurity atom with $r = 1.2r_0$, a couple of

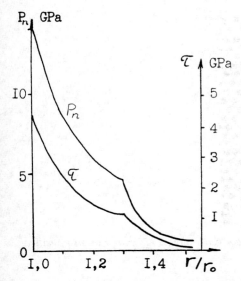

FIG. 3 *Influence of substitution atom radius on critical stresses for lattice rearrangement.* $\varepsilon = 18.6\%$, $T = 300K$.

such atoms and a triangle complex of similar atoms [6]. First of all shear nucleation on a pair of atoms took place, then it occured on a single substitution atom, the impurity complex of three atoms turned out to be the least efficient (Fig.4). This computer experiment allows a conclusion about the leading role of inclusion form in shear nucleation to be made, i.e. the needles and plates in close packed planes are more efficient, less efficiency of spherical particles is due to radial character of distortions around them and their resistance to shear going through the particle. The difference in precursor decay in LiF mentioned in [10] as compared to measurements from [9] at similar parameters is deemed to be a result of the difference in dimension and the form of precipitate that are strongly dependent on the regime of thermal treatment of specimens.

An abrupt decrease in the critical stresses for shear nucleation on point defects and their complexes requires critical consideration of the models of plastic deformation and fracture at the non-zero temperature, in which use is made of the numerical value of theoretical strength since the vacancies, interstitial atoms, and their complexes are the thermodynamic equilibrium defects.

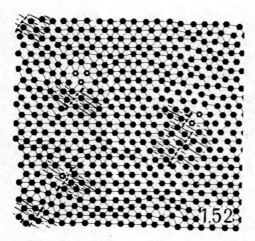

FIG. 4 *Influence of inclusion form on shear generation, 14% compression in vertical direction.*

III. NUCLEATION OF POINT DEFECTS

The shock loading results in very high concentrations of interstitial atoms and vacancies up to 10^{-4} [12, 13]. The main mechanism of nucleation of vacancies and interstitial atoms under ordinary conditions is the motion of dislocation with jogs formed during dislocation intersections (Fig. 5a). This mechanism is very active on the shock front under critically high stresses. As is shown by the crystallographic

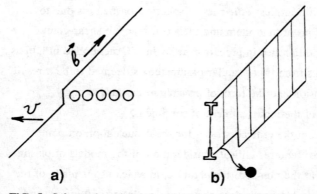

FIG. 5 *Schematic point defects generation: (a) on the moving dislocation jog; (b) from the dislocation dipole.*

analysis of formation of different jogs [14], in this case a larger number of interstitial atoms than vacancies form on the front of a sufficiently strong shock wave. The second mechanism of generation of point defects under the shock loading is the formation of dislocation dipoles when contrary-sign dislocations meet on neighbouring parallel planes (Fig. 5b). With the density of moving dislocations ranging from 10^{11} to 10^{12} cm^{-2}, probability of such meetings is very high. The dipoles are the long-living sources of point defects with decreased activation energy. In the megabar loading range the specific rapid deformation mechanism observed in the computer experiment [15] is probably possible. The relaxation of stresses can be accomplished by means of the formation of interstitial type clusters in close-packed planes nearly perpendicular to the shock front while in the planes parallel to the front vacancy-type clusters appear from which the material was taken. However, such mechanism was not verified yet by structural investigations.

IV. TWINNING

The active twinning under the shock loading was observed on a lot of metals including copper and nickel that do not usually exhibit a tendency to this deformation mechanism. It is known that the twinning formation requires considerably higher shear stresses than its subsequent propagation in a shear plane. The first stage of twinning formation is the appearance of a stacking fault. As is shown in Section II (see Fig. 1), in sufficiently strong shock waves the stacking faults can appear in the dislocation-free crystal parts, the threshold stress decreases abruptly on the impurity atoms. Under considerably less stresses appearance of the primary stacking fault and primary twinning dislocations can take place on dislocations that have already been in the crystal. This question was discussed in detail in literature, see, for example, the analysis for the BCC lattice [16].

The twinning formation is controlled in fact by the second state of the process, i.e. growth of the twin in thickness. The pole mechanism, at least, under the shock loading cannot provide formation of twins of observed dimensions for short loading times. According to our calculation [4], after the stacking fault appearance its growth in thickness is not observed. The subsequent stage is the stacking fault and breakdown accompanied by formation of complete splitted

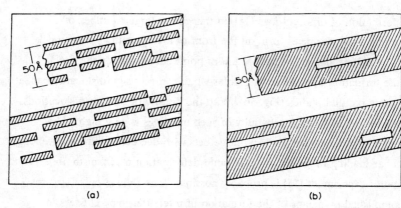

FIG. 6 *Schemes of development of deformation twins as a function of pulse duration:
(a) 0.017 μsec and (b) 0.3 μsec [18].*

dislocations. Probably the absence of stacking fault growth in thickness is associated
with the features of interatomic potential or with the edge type of dislocation. In
calculation [17] an increase in the twin thickness was observed in the case of
propagation of a screw twinning dislocation in the BBC lattice. Dynamic formation
of twinning dislocation pairs was assumed to occur on the surface of the stacking
fault.

The interesting results were obtained in studying the deformation structure
of a Cu - 8.7% Ge alloy after loading by thin plates [18] (see Fig. 6). At the very
short pulse durations the twins were very thin. Approximately 8 A in thickness and
occur in bundles having an overall thickness of about 50 A. With an increase in the
loading duration, formation of more thick twins took place. Twins can grow in
thickness not only by means of producing pairs of twinning dislocations at the
boundary surface but with help of forest dislocations having a screw component. It
is likely that new isolating stacking faults in parallel near planes are generated on
impurity active centers under the action of stress field from the moving primary twin.

V. ADIABATIC SHEAR UNDER RAREFACTION

One more specific form of defects was observed in a number of metals after the
extremely strong shock loading. In aluminum single crystals loaded by 30-50 GPa

shock waves at 77K and copper crystals loaded by 100 GPa the unusual band structures 0.3 mm and 1.6 mm in length, respectively were discovered. The experimental procedure and structural analysis that was the basis to restore the history of deformation evolution during shock wave loading, are described in details in [5]. It was shown that the bands observed are the traces of the intense localized shear that took place under rarefaction. In a 100 GPa loaded copper single crystal the high-rate deformation took place with $\varepsilon \sim 10^5 - 10^6$ s^{-1} and a strain of $\varepsilon = \Delta V / V_0 = 26.3\%$.

At the beginning of rarefaction the temperature was 1620K and at the end of it, it was 785K. The density of moving dislocations on the front was more than 10^{11} cm^{-2}. After structure rearrangement on the final part of the front and during exposure at the amplitude compression the cell structure with misorientation and high density of dislocations inside the cells was formed. Usually the path of dislocations during deformation is restricted by the cell size. However, the stresses under the conditions given turned out to be sufficiently high for the shear to break the boundaries of cells in some places. The intense deformation in the band results in heating, local temperature increase and further shear growth in the band. Thus, the forming band is a variety of the adiabatic shear bands. Due to high thermal conductivity, copper does not exhibit a tendency to the adiabatic shear band formation even at such high-rate processes as target perforation and tube shock loading. However, under 100 GPa loading temperature around the band is 1620K

FIG. 7 *Shear bands in Cu single crystal shock loaded 100 GPa at 77K axis [110]: (a) section (001), bar = 25 μm: (b) section (110), bar = 0.5 μm*

and heat removal is not effective. Ultimately, dislocations in the band are annealed and a thermal influence zone (Fig. 7) is observed near the band. The analogous structures appear also in Al under 30-50 GPa and probably, in stainless steel under 120 GPa [19].

REFERENCES

1.   L. Bragg and J. F. Nye, *Proc. Roy. Soc. A190:* 474 (1947).

2.   G. E. Cowan, Trans. Metallurg. *Soc. AIME 233:* 1120 (1965).

3.   M. A. Mogilevsky and I. O. Mynkin, *Combust. Explos. Shock Waves 24:* 6 (1988).

4.   M. A. Mogilevsky and I. O. Mynkin, *Inst. Phys. Conf. Ser. No.70* (1984).

5.   M. A. Mogilevsky and L. S. Bushnev, *Inst. Phys. Conf. Ser. No.102* (1989).

6.   M. A. Mogilevsky, *Impact Loading and Dynamic Behaviour of Materials,* C. Y. Chiem, H.-D. Kunze and L. W. Meyer, eds., Informationsgesellschaft, VERLAG V2, 957, (1988).

7.   G. Leibfried, *Handbuch der Physik Band VII, Teil 2* (1955).

8.   I. O. Mynkin and M. A. Mogilevsky, *Dinamika sploshnoi sredy 82:* 142 (1987).

9.   Y. M. Gupta. G. E. Duvall and G. R. Fowles, *J. Appl. Phys. 46:* 532 (1975).

10.   J. J. Dick, G. E. Duvall and J. E. Vortman, *J. Appl. Phys. 47:* 3987 (1976).

11.   M. A. Mogilevsky and I. O. Mynkin, *Combust Explos. Shock Waves 14:* 680 (1978).

12.   M. A. Mogilevsky, *Physics Reports 97:* 357 (1983).

13.   L. E. Murr, O. T. Inal and A. A. Morales, *Acta Met. 24:* 261 (1976).

14.   V. L. Indenbom, *Pisma v ZHETF 12:* 526 (1970).

15.   M. A. Mogilevsky, V. V. Efremov and I. O. Mynkin, *Combust. Explos. Shock Waves 13:* 637 (1977).

16.   A. W. Sleeswyk, *Phylos. Mag. 8:* 1467 (1963).

17.   S. Ishioka, *J. Appl. Phys. 46:* 4271 (1975).

18.   R. N. Wright, D.E.Mikkola and S.La Rouche, *Shock Waves and High-Strain-Rate Phenomena,* L. E. Murr and M. A. Meyers, eds., Plenum Press (1980).

19.   L. E. Murr, ibid (1980).

83

Shock Processing Research in the People's Republic of China

J. DING

Department of Engineering Mechanics   Center for Explosives Technology Research
Beijing Institute of Technology   New Mexico Institute of Mining & Technology
Beijing 100081, P.R.C.   Socorro, NM 87801, U.S.A.

China has made headway in modern research on physics of explosion and shock waves since the late fifties. In this paper, research in the following fields during the past decade is reviewed: explosive metal forming and hardening, explosive welding and cladding, dynamic fracture and spalling, shock consolidation of powders, and shock-induced reactions.

I. INTRODUCTION

Modern research on the physics of detonation, explosion, and shock waves in the People's Republic of China was started in the late fifties. With the understanding that the study of shock-wave and high-strain-rate phenomena should be an important part of modern mechanics, a research laboratory on explosion mechanics, headed by C.M. Cheng, was established in 1960 at the Institute of Mechanics, Chinese Academy of Sciences. The first conference on explosive working was held in 1963 in Beijing.

Explosive synthesis of diamond and cubic boron nitride was successfully accomplished in China in 1971 and 1975, respectively. Explosive welding and cladding became industrial practices in the early seventies. Theoretical studies on the mechanism of explosive welding and interfacial wave formation were made in the period 1979 to 1985. In recent years, the shock consolidation of powders and shock-induced reaction in materials become a focus of research interests.

China's first national conference on explosion mechanics was held in 1977. The first International Symposium on Intense Dynamic Loading and Its Effects [1] was convened in Beijing on July 3-7, 1986. A quarterly journal, *Explosion and Shock Waves* [2], has been published in Chinese since 1981; the *Chinese Journal of High Pressure Physics* [3], also quarterly, has been published since 1987. Both of these journals cover the shock-wave and high-strain-rate phenomena in materials.

In this paper, research in the following fields in the past decade is reviewed: explosive metal forming and hardening; explosive welding and cladding; dynamic fracture and spalling; shock consolidation of powders; and shock-induced reactions.

II. EXPLOSIVE METAL FORMING AND HARDENING

The research on explosive metal working was started in China in the late fifties. The book *Explosive Working* [4] by C.M. Cheng, Z.S. Yang *et al.*, provides the first comprehensive review in Chinese of the theory and practice of explosive forming in China. It also illustrates the principles of explosive welding. Based on the understanding of the impact process of deformed flank with the die in explosive forming, a concept of separated inertia die was developed and an inertia die was successfully applied to produce a 3m diameter hemisphere. The theoretical model of inertia die was discussed by Z.S. Yang [5].

In the aviation and aerospace industries in China, explosive metal forming has been used for nearly thirty years, and the first explosion press machine was used in the production of metal parts in 1966 [6,7]. The first explosion container for metal working was designed and fabricated in the early eighties [8]. The 5BR-1 model explosion container, with a maximum capacity of five kilograms of high explosives was tested in 1986 [9]. This container is equipped with high-speed photography, electronics, and flash X-ray photonics. One of the interesting new developments of explosive metal forming is the fabrication of spherical pressure vessels [10] with diameters up to 4 m, thicknesses from 3 to 20 mm of high strength steel, for water or oil storage, paper mill, chemical industries, and others.

Shock loading effects on mechanical properties of structural materials have also been studied in China. The explosive hardening process of Hadfield steel is the most investigated one, and is also the most extensively used in practice. Research on explosive hardening of Hadfield steel railway frogs was carried out in the sixties. The best results reported by S.D. Zhao and W.B. Chen [11] were a depth of hardening for $R_c \geq 38$ of about 16 mm, achieved by firing twice with a plastic sheet explosive 4 mm thick and having a detonation velocity of about 6500 m/s. In contrast, by ordinary means the depth of hardening ($R_c \geq 32$) is only 3-6 mm. In a recent paper by Y.F. Chen and Y.Q. Hong [12] the hardness distribution in Hadfield steel after explosive hardening is calculated analytically. The calculated results are in good accord with the experimental data obtained for different loading conditions.

F.Q. Jing and J.W. Han [13] measured the elastic moduli and yield strength of iron at 1.026 GPa and 11.16 GPa with a shock loading-unloading technique. Combining these data with that reported by L.V. Al'tshuler *et al.* at 111 GPa and 185 GPa [14], they found that the elastic moduli and yield strength increase with pressure over the range 1.0–185 GPa, even though the phase transformation from α to ϵ has occurred. This implies that the shock-hardening effect is significant for iron in the above pressure range. Experimental HEL data for armco-iron and stainless steel [15] also indicate significant deformation-hardening effects in the stress range 1–5 GPa. In the process of explosive hardening of high-manganese steel, no martensitic transformation was found in the metallographic and TEM observations of M.S. Li [16].

III. EXPLOSIVE WELDING AND CLADDING

The research on explosive welding in China was started in the sixties and the first explosive cladding plate was made in 1968. The book *Principles of Explosive Welding and Its Engineering Applications* [17] in Chinese, authored by B.H. Shao and K. Zhang, was published in 1987. This monograph, comprehensive and unique, contains most new developments in China. The industrial applications of explosive welding and cladding in China have entered an era of prosperity. There are now two industrial production bases, one in Dalian and another in Baoji, for the large-scale manufacturing of composite plates, tubes, junctions, etc. Both have a yearly production capacity of over one thousand tons. Aluminum-steel composite plates are being used in naval vessel construction. An aluminum-clad copper wire factory in Shanghai has started mass production recently.

A. MECHANICS OF WAVE FORMATION

In theoretical study of explosive welding the problem of wave formation at the interface of colliding surfaces is of particular importance. It has been studied by many people using different models, such as the indentation model, Helmholtz instability model, von Karmen vortex streets model, and others. C.M. Cheng and his colleagues [18] realized in the sixties, the necessity of deriving a constitutive relation which permits the material to behave like a fluid, when subjected to a pressure far in excess of its strength, and deform like an elasto-plastic solid otherwise. Hydro-elastic-plastic dynamics has been developed as a branch of continuum mechanics for the description of motion and deformation of a body under intense dynamic loading. A new theory for the mechanism of wave formation in explosive welding has been developed by Cheng and Tan [19] and published recently.

Utkin [20] showed that, according to the Helmholtz instability theory of parallel inviscid and incompressible flow, the most preferred wave length of unstable sinusoidal disturbances is given by

$$\frac{\lambda}{H} = 128 \, \sin^2\frac{\alpha}{2}$$

where λ is the wavelength, H the plate thickness, and α the dynamic angle of collision. While this formula fitted well with the data he quoted, it does not predict the dependence on material properties. Another formula was obtained instead by C.M. Cheng and his colleagues [21, 22]:

$$\frac{\lambda}{H} = K\frac{U_c}{U_o} \, \sin^2\frac{\alpha}{2}$$

where K is a non-dimensional constant (in the order of 10^2) independent of material properties; U_o is the impact velocity; U_c is the velocity at $P/\sigma_b = C$, where the transition from fluid-like to solid-like behavior takes place; P is the pressure; and σ_b is the ultimate static strength of the material. The effect of material property is contained in U_c/U_o.

The validity of the hydro-elastic-plastic model of wave formation was further proved by the following experiments.

Geometrical Magnification Test [23] For a symmetrical impact of incompressible flow, the following relation holds:

$$\frac{\lambda}{H} = f\left(\frac{\rho_o V_f^2}{\sigma_b}, \beta, \frac{V_f \rho_o H}{\mu}\right)$$

where ρ_o is the density of plate, V_f, the incoming flow rate of the welding plate, σ_b, the static strength limit of the material, β, the collision angle, and μ, the viscosity coefficient. With a geometric magnification of 5-8, and $V_f = 1800, 2450, 3110$ m/s, it is found that the Re number plays only a minor effect on λ/H; therefore, the effect of viscosity on the wave formation may be neglected. It was also found that the non-dimensional λ/H increases with the specific strength or non-dimensional pressure $\rho_o V^2_f/2\sigma_b$ monotonously in the range of 22-50. That is, in this range the strength of material exerts a certain negative effect on the wave formation. On the other hand, if the specific strength varies in the range of 50-120, the λ/H almost stays constant, which means that in this range the effect of material strength may be neglected, and the model of non-compressible fluid is applicable.

Periodic Change of the Collision Point and the Angle of Collision [24] In a well-designed experiment, it has been proven that the collision point is floating on the interfacial wave and the collision angle changes periodically. However, for the Helmholtz instability and the von Karman vortex street mechanism, the stagnation point is fixed, and the development of instability needs a space of several (say 3-4) wavelengths. The results of this experiment tell us that the collision is not at the stagnant point in any sense and even the collision point itself is unstable.

Freezing Experiments of Wavy Interface near the Collision Point [24] A wedge with an angle of $2\beta_o$, the same as the collision angle, is placed at some distance up-stream. The collision of plates is prevented when the point of collision reaches the tip of the wedge. At the same time, the wave formation process is stopped and the wave phenomena are frozen. Photos of the welded interface show that the interface waves are formed near the tip of the wedge. Obviously, these experimental results are in contrast to the ordinary instability models.

Thermo-plastic Shear Instability In the process of explosive welding, a high-velocity adiabatic shearing process occurs in the vicinity of the collision point, with a very fast temperature increase, approximately 10^{8-9} K/s. In the experiments by G.H. Li *et al.* [25], it is clearly shown in the micrograph that along the interface of explosive welding there are a number of about 45° inclined lines, characteristic for thermo-plastic shear instability, with a micro-hardness H_v of 560 which is more than three times higher than the neighboring ferrite. A mathematical analysis by Y.L. Bai [26] results in a criterion for thermo-plastic shear instability with application to titanium, mild steel, and an aluminum alloy. B.H. Shao and G.H. Li [27] pointed out that a cast texture appeared on the welding interface and the vortex did not result from the penetration of reentrant jet, but from the thermo-plastic instability under a pressure of several tens of GPa.

B. EXPLOSIVE WELDING AND CLADDING TECHNOLOGY

Flyer Motion Study and Impact Pressure Calculation An approximate analytical solution, based upon the theory of Prandlt-Meyer flow and Lighthill's piston theory, was obtained by B.H. Shao *et al.* [28]. The calculated bending angle and the configuration of the flyer plate are in good accord with the flash X-ray photographs with an error below 5%. This procedure is also useful for explosive hardening, shock consolidation of powder, and shock synthesis of new materials. K. Zhang *et al.* [29] investigated the effect of material strength on the velocity and attitude of the flyer plate, and a better agreement with experimental results was achieved. H.J. Chen [30] introduced a procedure for calculating the pressure at the stagnation point in explosive welding, using both incompressible and compressible fluid models, for an arbitrary combination of metals in symmetric or non-symmetric impact.

Design of Multilayer Cladding Using the calculation method for the motion of a flyer plate [28] B.H. Shao *et al.* [31] developed a computational procedure for the design of multi-layer cladding. Welding parameters, the incoming flow velocity V_f, and the dynamic impact angle β, for each layer are determined. The calculated results are in good agreement with experiments even in the case of ten layers of aluminum plate. The calculated β angle for each layer is in good accord with the flash X-radiograph. A 1.2 m long by 90 mm wide soft electric cable of fifteen layers of 1 mm-thick copper sheet has been explosively clad by K. Zhang *et al.* [32]. If symmetrical explosive loading is used, the large amplitude interfacial wave formed will cause a marked drop in welding quality. Therefore, it is important to eliminate these waves.

Experimental Determination of the Polytropic Constant of Explosive For explosive metal working in general, explosives of low detonation velocity are used. The polytropic constant γ_0 of the detonation products is important for explosive performance. B.H. Shao *et al.* [33] used the Prandtl-Meyer flow model to study the motion of detonation products. A relationship between the lateral flying angle, which can be measured directly by flash X-ray photography, and the effective polytropic constant under the working conditions was established. The measured value of γ_0 is accurate within 2.5%.

Experimental Determination of Weldability Domain Several experimental methods [34-36] were used by Chinese scientists to determine the explosive weldability window. These methods and weldability windows for systems like aluminum alloys, copper-stainless steel, aluminum alloy-stainless steel, and stainless steel-low carbon steel were discussed in the book by Shao and Zhang [17]. An upper limit analysis of the weldability domain was proposed by K. Zhang [37] on the basis of an approximate calculation of the temperature at the welding interface using the numerically computed strain field by the stagnation point. By using the experimental method of small inclination angle [28] on a specimen 300mm long, more than 80 data can be collected for different points with different binding strengths. Only two or three tests are needed to get the best welding parameters. A measurement technique for the dynamic welding parameters and the pressure field in tube-to-tube plate welding was developed by C.H. Wang and X.S. Zhang [38].

IV. DYNAMIC FRACTURE AND SPALLATION

Dynamic fracture has traditionally been studied in China for various purposes. The dynamic fracture of expanding shells has been studied by D.N. Chen and his colleagues [39]. One of their experimental apparatus is a spherical shell loaded with high explosive that is initiated at the center. Another one is a loaded cylindrical shell, initiated at one end. Finite difference calculation of elastic-plastic flow was used to study the dynamic response of the shell and the influence of stress waves on the fracture pattern. Both flash X-ray and framing experiments were done for several geometrical configurations. The experimental data agree quite closely with the computational results. According to J.W. Taylor *et al.* [40], when total stress at the internal boundary of the shell becomes tensile, complete fracture of the shell will take place. The computational results indicate that the complete fracture of the shell occurs 14 μs after the initiation of the explosive, while the experimental results show complete fracture at 13 μs. Since thermal softening can lead to the formation of adiabatic shear bands, the material is weakened in the shear zone.

D.N. Chen *et al.* [41] distinguished the spalling from the fracture during shell expansion as two different kinds of dynamic fracture, having two different mechanisms and, therefore, two different criteria. Spallation criteria and numerical modelling of spallation processes were reviewed by Chen and Wang in 1982 [42]. Spalling of aluminum (LY-12) and stainless steel (1Cr18Ni9Ti) was studied by W.J. Zhang and Y.S. Zhang experimentally [43]. They found the instantaneous fracture strength of both materials to be 1.4-1.6 GPa for LY-12 and 2.7 GPa for stainless steel. G.D. Wu *et al.* [44] found the thickness of spall and the cumulative criterion of fracture with embedded manganin gauge measurements. In the paper of G.R. Zhang [45] the rarefaction shock and spalling in iron was discussed for an iron target impacted by a copper flyer plate. He found that only one spallation could occur in this case and only at the place where two rarefaction shocks interact.

L.T. Shen *et al.* [46] reported a study of spallation in which the residual strength of the spalled specimen is adopted to define a damage function relevant to microcracks formed in the specimen. Experiments were carried out with a single-stage gas gun in symmetric impact. They defined a damage function $\alpha = 1 - \sigma_r/\sigma_b$, where σ_r is the residual strength of spalled specimen, σ_b the ultimate tensile strength of the test material. The spallation criterion proposed by Tuler and Butcher [47] is rewritten

$$\left(\frac{\sigma}{\sigma_o} - 1 \right)^n \Delta t - K'$$

where σ is the nominal tensile stress, Δt is the duration of σ, and K', n, σ_o are parameters dependent on damage function. They found in their experiments that the larger the α, the larger and wider are the cracks. Their data also show that the difference between the first maximum stress and the first minimum stress is nearly a constant, independent of impact velocity and stress duration.

The fracturing process of an expanding thin cylindrical shell under the impulsive expansion of detonation products has been studied by J.B. Feng *et al.* [48]. A damage level function $\alpha(t) = A_d(t)/A(t)$ is introduced, in which $A(t)$ is the area of a cross section at the time t, and $A_d(t)$ is the damaged area at the same time. It is assumed that σ_d may be approximated with the yield strength of the material σ_o, $d\alpha/dt$ increases with the increase of $(\sigma - \sigma_d)$ or $(\sigma - \sigma_o)$, the local stress on a microcracked surface is zero, and the effective stress on the surface S is $(1 - \alpha)(\sigma - \sigma_o)$. By the method of dimensional analysis, the following equation of damage level is obtained:

$$\frac{d\alpha}{dt} - \frac{[(1-\alpha)(\sigma-\sigma_o)]^2}{\eta K}$$

where K and η are the hardening modulus and effective viscosity of the material, respectively. Analytical results demonstrate that there exists a maximum complete-fracture strain at an appropriate strain rate. For mild steel, the complete-fracture "plastic peak" maximum strain is 60~80% at a strain rate of $4 \cdot 10^{-4} s^{-1}$.

Y.B. Dong *et al.* [49] extended the model given in previous work [48] to cases of plane geometry and performed numerical simulation on the dynamic damage process under uniaxial strain. The material studied is LY-12 aluminum. The strain-rate dependence of the effective viscosity by D.E. Grady [50], $\eta = \eta_0 \, \dot{\epsilon}^{-1/2}$ is used, where $\dot{\epsilon}$ is the strain rate. When the damage level function α reaches the critical value α_c at a certain section, spallation will occur there. Assuming the initial damage level is zero, then at time t, the damage level is

$$\frac{\alpha(t)}{1-\alpha(t)} - \int_0^t \frac{(\sigma-\sigma_o)^2}{\eta K} dt$$

They performed numerical simulation of the dynamic damage processes under uniaxial strain for LY-12 and 2024 aluminum. The calculated free-surface velocity histories are consistent with experimental results, if the critical damage level is chosen appropriately. In the following table, the critical damage level, α_c, critical strain rate, $\dot{\epsilon}_c$, the calculated and experimental critical spall strengths, σ_c, and spall thickness, δ, for six experiments are summarized.

Shot No.	α_c Calc.	$\dot{\epsilon}_c$ $(10^5 s^{-1})$ Calc.	σ_c (GPa) Calc.	σ_c (GPa) Expt.	δ (cm) Calc.	δ (cm) Expt.
1	0.493	7.4	1.87	1.65	0.094	0.105
2	0.522	7.7	1.94	1.61	0.094	0.090
3	0.551	8.2	2.02	1.65	0.094	0.094
4	0.451	3.8	1.64	1.26	0.173	0.191
5	0.455	3.9	1.65	1.48	0.173	0.181
6	0.346	1.0	1.23	----	0.280	-----

From the above data, a new criterion of spallation, valid in the strain-rate range of $10^5 - 10^6 s^{-1}$, is obtained: $\alpha_c = -0.0671 + 0.202 \lg \dot{\epsilon}$.

Recently, Y.B. Dong and his colleagues [51] used the same approach as shown in [49] to simulate numerically the spallation of a steel cylindrical shell imploded under slipping detonation. The calculated result of spall locations are in good accord with the experimental result of A.G. Ivanov [52] at $\alpha_c = 0.101$, as shown in the following table.

Explosive Thickness cm	Fracture Instant μs	Spalled Layer (cm) Outside Radius Calc.	Spalled Layer (cm) Outside Radius Expt.	Spalled Layer (cm) Thickness Calc.	Spalled Layer (cm) Thickness Expt.	Remnant Layer (cm) Outside Radius Calc.	Remnant Layer (cm) Outside Radius Expt.	Remnant Layer (cm) Thickness Calc.	Remnant Layer (cm) Thickness Expt.
0.75	2.97	1.343	1.40	1.023	0.9-1.0	1.998	2.05	0.494	0.49
0.50	2.60	1.070	1.15	0.833	0.8-0.85	2.258	2.285	0.596	0.58

V. SHOCK CONSOLIDATION OF POWDERS

Research on shock consolidation of powders in China was started early in the last decade, and has received more attention in recent years [53-57]. Efforts are concentrated in the consolidation of amorphous compounds, microcrystalline materials, and ceramics. Compact densities of 99.6-99.7% TMD for amorphous materials are obtained.

To get a specific, mechanistic description of the way in which energy is deposited in the powder during shock consolidation, B.H. Shao and his colleagues [56] suggested recently an adiabatic friction and collision-welding model.

The powder is assumed to be composed of spherical grains between 10-100 μm, in close-packed layers. Since the powder is not homogeneous, the shock front in the powder is not a flat surface. The pores are closed under the compression of shock waves. The thickness of the shock-wave front is about the same size as the diameter of powder grains. The time needed for the shock wave to cross the powder grains is short in comparison with the time duration of the shock wave itself; therefore, the detailed wave motion may be neglected.

Under shock loading, every spherical grain can collide with and weld to six neighboring grains in the same plane. The collision angles are $2\beta = 0\text{-}72°$. The grains will also collide with three grains of the next row with a collision angle of $0\text{-}36°$. During the process of collision welding, the outside layer of the grain, including the oxidation product, will be ejected into the pores as a reentrant jet. An important requirement for explosive welding is no cohesive shock wave at the point of impact, i.e. the condition of subsonic regime. Therefore, the detonation velocity of the explosive used for powder consolidation should be lower than the elastic wave velocity of the powder material. If the collision angle between two iron particles is smaller than the critical angle, the strain rate of the shear layer is about $10^6\text{-}10^7 s^{-1}$, and the shear stress produced by its viscosity can reach a magnitude of 10^2 GPa. In the adiabatic shear layer with a thickness of 1-10 μm, a temperature rise of 10^3 K will occur in a time interval of 10-100 ns.

When the shock wave reaches the free surface of a powder grain, the grain itself will move at a velocity twice that of the particle velocity. There will be a very strong adiabatic friction with the neighboring grain and interfacial melting will occur. In the risetime of the shock wave, the powder is compressed and consolidated. In the process of void collapse, the surface temperature reaches a maximum. When the shock wave is reflected from the lower boundary, the compacted powder is shocked a second time. A pressure of several tens of GPa will be achieved, while the temperature rise in the reshock is negligible. The adiabatic friction is able to raise the local temperature $10^3\text{-}10^4$K in a time duration of 10-100 ns. For shock consolidation, adiabatic friction is more important than collision welding. This adiabatic friction and collision-welding model has been used to calculate the volumetric percentage of melt as a function of shock pressure for the iron powder. The calculated result is in good agreement with the experimental data by D.G. Morris [58].

The planar compaction of tungsten powder was studied by Xue and Lu [57]. The powder was composed of over 90% tungsten and the rest was Ni, Fe, Co. The powder grain sizes were 3 and 20μm, and were precompacted to 55% and 64% of theoretical density, respectively. For coarse powder, good consolidation to 94% TMD is achieved with impact velocities of 850 and 1100 m/s. For the 3 μm powder at 1200 m/s a density of only 90.4% TMD is obtained. Following the spherical shell model of Carroll and Holt, they calculated the change in porosity during the process of consolidation for both coarse and fine powders. The temperature distribution within the spherical unit-shell during consolidation was also calculated. In their calculation for small particle size Ni powder, in mixture with coarse tungsten powder, the Ni powder melts and plays the role of binder. On the other hand, if the fine Ni powder is mixed with fine tungsten powder, it does not play the same role. These results of the numerical study have been confirmed by experiments.

VI. SHOCK-INDUCED REACTIONS

Here, *reaction* means either physical change like a phase transition or chemical reaction. Research on shock synthesis of diamond was started in China in the late fifties. A report "Production of Artificial Diamond by the Method of Explosion" [59], authored by B.H. Shao and his colleagues was published in 1986. In this paper a highly efficient set-up for diamond production is introduced, where an inward-collapsing cylindrical flyer is used for compaction. By this procedure, a graphite-diamond transformation ratio of 8-10% and a yield of 7-8 carat diamond per kilogram of TNT were reached. The mechanism of the phase change, including the critical size of nuclei and the kinetics of the transformation, has been discussed in detail. Parameters of the shock-reaction process are calculated analytically.

The phase transformation of potassium chloride induced by shock loading has been studied by J.W. Han *et al.* [60] using Lagrangian technique with manganin gauges. From the b_1 phase of lower density to the b_2 phase of higher density, the phase transition occurs at 1.935 GPa, and the reverse transition at 1.044 GPa. The sample completes its dynamic martensitic recrystallization from b_1 to b_2 within 0.1 μs. The phase transition of PZT 95/5 ceramics was reported by W.Z. Yuan [61]. For the 40 x 10 x 10 mm sample of PZT 95/5 with a residual polarization of 31 $\mu C/cm^2$, the ferroelectric to anti-ferroelectric phase transition occurs under shock loading of 0.9 GPa, with a charge release efficiency of 98% in a relaxation time of less than 0.15 μs.

In recent years, some interest in shock-wave chemistry is noticeable in several institutions in China. Hugoniot data for a mixture of $BaCO_3$ and TiO_2 powders under a shock pressure range of 10 to 100 GPa were obtained [62]. There are two obvious discontinuities at pressures around 30 and 45 GPa, respectively. The first discontinuity is related to the phase change of TiO_2 from anatase to β-TiO_2, whereas the second one is a manifestation of the chemical reaction between $BaCO_3$ and TiO_2 which forms $BaTiO_3$ and CO_2. These results were confirmed by recovery experiments.

The shock-induced crystallization of amorphous $Fe_{40}Ni_{40}P_{12}B_2$ alloy was studied by H.L. He *et al.* [63]. They found that crystallization is induced under the shock loading in the order of μs with threshold pressure of 30-50 GPa. The precipitated phases depend upon the pressure applied. The shock-crystallization of amorphous materials opens a new way for synthesizing new materials.

CONCLUDING REMARKS AND ACKNOWLEDGMENTS

In this review an outline is given for shock-processing research in China in the past decade. Coverage, however, is limited. Very interesting areas such as shaped charge phenomena, shear localization, and related experimental techniques are not included. References given are only a small fraction of what have and have not been published. It is hoped that this paper may give the reader a general view and arouse interest in knowing more about current and past work in China. The author gratefully acknowledges the support provided by P.-A. Persson. Special thanks are due to B.H. Shao for his helpful communication. Thanks are also due to A.R. Miller and N.N. Thadhani for a critical review of this manuscript.

REFERENCES

1. *Proceedings of the International Symposium on Intense Dynamic Loading and Its Effects (Proc. ISIDLE)*, June 3-7, 1986, Beijing, China, eds.: C.M. Cheng and J. Ding, H. Jing and Y.H. Li, Science Press, Beijing, China, 1986, pp. 1078 + x.

2. *Explosion and Shock Waves (E&SW)*, Quarterly, in Chinese, editor-in-chief: J. Ding (1981-1987), G.R. Zhang (1988-present). The title, author's name and affiliation, abstract, keywords, caption of figures and tables, are shown both in Chinese and English.

3. *Chinese Journal of High Pressure Physics (CJHPP)*, Quarterly, in Chinese, editors-in-chief: Q.Q. Gou and F.Q. Jing (1987-present). The title, etc., are shown both in Chinese and English.

4. C.M. Cheng, Z.S. Yang *et al.*, *Explosive Working*, in Chinese, 1980, Defense Industry Publishers, Beijing, China, pp. 721.

5. Z.S. Yang, "Theoretical Model of Inertia Die in Explosive Forming," *Proc. ISIDLE*, 1986, p. 1032.

6. S.B. Zhang, "Explosion Press Machine, Its Development and Prospect," *E&SW* 2(1): 1 (1982).

7. H.M. Zhou, "Explosive Buldging of Multi-Hole Parts," *Proc. ISIDLE*, 1986, p. 1073.

8. S.B. Zhang, "Explosion Container," *E&SW* 4(2): 83 (1984).

9. G.R. He, "Peak pressure versus distance measurement in 5BR-1 explosion container," *CJHPP* 1: 142, (1987).

10. T.S. Zhang, private communication, 1990.

11. S.D. Zhao and W.B. Chen, "Investigation on Explosive Hardening of Hadfield Steel Railway Frogs," *E&SW* 2(1): 11 (1982).

12. Y.F. Chen and Y.Q. Hong, "A Theoretical Analysis of Hardening of Hadfield Steel by Explosion," *E&SW* 8: 344 (1988).

13. F.Q. Jing and J.W. Han, "Elastic Moduli and Yield Strength of Shock-Loaded Iron," *E&SW* 2(4): 19 (1982).

14. L.V. Al'tshuler *et al.*, *Appl. Mech. and Tech. Phys. USSR* 2: 159 (1971).

15. J.D. Lu, W.J. Zhang and Y.S. Zhang, "Deformation Hardening Effect of Armco-Iron and Stainless Steel," *E&SW* 4(2): 75 (1984).

16. M.S. Li, "Microscopic Study on the Explosive Hardening of High-Manganese Steel," in Chinese, Master Thesis, Institute of Mechanics, Academy Sinica, September 1989.

17. B.H. Shao and K. Zhang, *Principles of Explosive Welding and Its Engineering Applications*, in Chinese, 1987, Dalian Institute of Technology Publishers, Dalian, China, pp. 387 + iv.

18. C.M. Cheng, "Several Problems in Hydro-Elastic-Plastic Dynamics," *Proc. ISIDLE*, 1986, p. 1.

19. C.M. Cheng and Q.M. Tang, "Mechanism of Wave Formation at the Interface in Explosive Welding," *Acta Mechanica Sinica*, 21: 129 (1989).

20. A.V. Utkin *et al., Phys. Comb. & Explo.* 16(4), (1980), in Russian.

21. C.M. Cheng, "Mechanics of Explosive Welding," *Proc. ISIDLE,* 1986, p. 848.

22. C.M. Cheng and G.H. Li, "Effects of Strength and Compressibility of Materials on Wave Formation at Interface in Explosive Welding," *Proc. ISIDLE,* 1986, p. 854.

23. G.H. Li, D.Y. Zhang, Z.H. Zhou and B.H. Shao, "An Experimental Study of the Modeling Law in Symmetric Welding," *Proc. 4th Nat. Conf. on Explosive Working,* Fuzhou, 1982, in Chinese.

24. B.H. Shao, G.H. Li and Z.H. Zhou, "Experimental Research on Process of Instability at the Interface in Explosive Welding," *Proc. ISIDLE,* 1986, p. 974.

25. Cf. Ref. 17, p. 376, Figure 7-5-2.

26. Y.L. Bai, "A Criterion for Thermo-Plastic Shear Instability," *Shock Waves and High-Strain-Rate Phenomena in Metals, Concepts and Applications,* eds.: M.A. Meyers and L.E. Murr, Plenum Press, 1981, p. 277.

27. B.H. Shao and G.H. Li, "Hydro-Elastic-Plastic Instability Model for Wave Formation at the Interface in Explosive Welding," *Proc. 4th Nat. Conf. on Explosive Working,* Fuzhou, 1982, in Chinese.

28. B.H. Shao, D.X, Zhang, W.B. Chen and G.H. Li, "Motion of Flyer Plate under Glancing Detonation," *II Meeting on Explosive Working of Materials,* Novosihirsk, USSR, 1981, p. 63.

29. K. Zhang, L.Y. Li and W.B. Yang, "The Study of Effect of Material Strength on the Motion Curve of Flyer Plate under Glancing Detonation," *VI International Symposium on Use of Explosive Energy in Manufacturing Metallic Materials of New Properties,* Gottwaldov, 1985.

30. H.J. Chen, "Calculation of Impact Pressure in Explosive Welding Interface," *E&SW* 4(3): 10 (1984).

31. B.H. Shao, Z.H. Zhou and G.H. Li, "Calculation Method of the Explosive Cladding Parameters of Multilaminates under Slipping Detonation," *E&SW* 6: 143 (1986).

32. K. Zhang, J.Y. Xi and J.B. Gao, "Research on Wave Formation in Multilayer Cladding," *Shock Compression of Condensed Matter-1989,* eds.: S.C. Schmidt, J.N. Johnson, L.W. Davison, North-Holland Elsevier Science Publishers, 1990, p. 237.

33. B.H. Shao, W.B. Chen and Y.Y. Zhou, *et al.,* "Determination of Effective Adiabatic Exponential of Explosive Products under Lateral Slipping Detonation," *E&SW* 1(2): 30 (1981).

34. H.L. Xue, S.Y. Chen and X.S. Zhang, "Experimental Study on the Weldability Window for the Explosive Welding of Aluminum Alloy Plate," Journal, National Defense University of Science and Technology, (1979).

35. Z.K. Zhang and S.Y. Wu, *Proc. 7th Intern. Conf. of HERF,* Leeds, 1981.

36. H.J. Chen, "Stepping Method for Experimental Determination of Explosive Welding Parameters," *Proc. 4th Nat. Conf. on Explosive Working,* Fuzhou, China 1982, in Chinese.

37. K. Zhang, "An Approximate Method for Calculating the Temperature at the Explosive Welding Interface and Upper Limit Analysis of the Weldable Window," *E&SW* 7: 235 (1987).

38. C.H. Wang and X.S. Zhang, "The Measurement of the Dynamic Welding Parameters and the Pressure Field in Bonding Region," *Proc. ISIDLE,* 1986, p. 997.

39. D.S. Wang, S.H. Ma, G.X. Shi, Y.N. Li and D.N. Chen, "Dynamic Fracture in Solids," *Proc. ISIDLE,* 1986, p. 777.

40. J.W. Taylor and F.H. Harlow, *et al., J. Appl. Mech.* 45: 108 (1978).

41. D.N. Chen, D.S. Wang, S. H. Ma, G.X. Shi and Y.N. Li, "Two Kinds of Dynamic Fracture," *E&SW* 7:27 (1987).

42.   D.N. Chen and D.S. Wang, "Spallation Criteria and Computer Modeling of Spallation Process," *E&SW* 2(4): 50 (1982).

43.   W.J. Zhang and Y.S. Zhang, "Experimental Studies of Spall-Fracture of Aluminum Alloy and Stainless Steel under Impulsive Loading," *E&SW* 3(1): 73 (1983).

44.   G.D. Wu, Y. Li, H.Q. Liu, J.C. Chi and Y.H. Zhao, "Determination of Scab Thickness and Cumulative Criterion of Fracture Using Manganin Gauge Technique," *E&SW* 3(1): 79 (1983).

45.   G.R. Zhang, "Characteristics of Shock Loading in Iron and Its Fracture," *E&SW* 7: 9 (1987).

46.   L.T. Shen, Y.L. Bai and S.D. Zhao, "Experimental Study of Spall Damage in an Aluminum Alloy," *Proc. ISIDLE,* 1986, p. 753.

47.   F.R. Tuler and B.M. Butcher, *Intern. J. Fract. Mech.,* 4: 431 (1968).

48.   J.B. Feng, F.Q. Jing, L.X. Su and J.W. Han, "Studies of the Explosion Expanding Fracture Process of a Thin Cylindrical Shell," *CJHPP* 2: 97 (1988).

49.   Y.B. Dong, W.J. Zhang, F.Q. Jing, J.W. Han, D.N. Chen, L.X. Su and J.B. Feng, "Numerical Analysis for Dynamic Damage Processes and LY-12 Aluminum Spallations," *CJHPP* 2:305 (1988).

50.   D.E. Grady, *Appl. Phys. Lett.,* 38: 825 (1981).

51.   Y.B. Dong, L.X. Su, D.N. Chen, F.Q. Jing, J.W. Han and J.B. Feng, "Numerical Simulation on the Spallation of a Steel Cylindrical Shell Imploded under Slipping Detonation," *CJHPP* 3: 1 (1989).

52.   A.G. Ivanov *et al., Appl. Mech. and Tech. Phys. USSR,* 3: 125 (1984).

53.   R. Yang, Z.Z. Xie, X.P. Wang, B.H. Shao, Z.H. Zhou and G.H. Li, "An Investigation on Sintering Consolidation Mechanism During Dynamic Powder Compaction of High Speed Steel Powders by Explosive Means," *Proc. ISIDLE,* 1986, p. 1025.

54.   K. Zhang, N.S. Wang, X.S. Jing and L. Wang, "Explosion Compaction of Tungsten Powder," *Proc. 6th Nat. Conf. on Metal Working,* Zhangjiaje, China, 1988, in Chinese.

55.   X.S. Jing and K. Zhang, "Computer Simulation of the Dynamic Process of Explosive Compaction of Powder into Tubes," presented at the Intern. Conf. of Computational Physics, Beijing, 1988.

56.   B.H. Shao, J.X. Gao and G.H. Li, "The Mechanism of Energy Deposition at the Interface of Metal Powder in Explosive Consolidation," *E&SW* 9: 17 (1989).

57.   H.L. Xue and F.Y. Lu, "Planar Dynamic Compaction of Tungsten Powder," *CJHPP* 3: 115 (1989).

58.   D.G. Morris, *Metal Sci.,* 15: 116 (1981), 16: 457 (1982).

59.   B.H. Shao, Z.H. Zhou, J.T. Wang and Y.Y. Zhou, "Production of Artificial Diamond by the Method of Explosion," *E&SW* 6: 198 (1986).

60.   J.W. Han, Z.K. Li and Y.S. Zhang, "Experimental Investigation of the Potassium $b_1 \rightarrow b_2$ Phase Transition under Shock Loading, *Proc. ISIDLE,* 1986, p. 632.

61.   W.Z. Yuan, "Electric Charge Release under Shock Loading, *CJHPP* 1: 145 (1987).

62.   X. Chen, X.G. Jin and M.S. Yang, "Experimental Studies of Hugoniot for Powder Mixture of $BaCO_3$ and TiO_2 and Shock Synthesis of $BaTiO_3$," *CJHPP* 3: 67 (1989).

63.   H.L. He, X.G. Jin, P.S. Chen and W.K. Wang, "Experimental Studies of the Crystallization of Amorphous $Fe_{40}Ni_{40}P_{12}B_8$ Alloy under Shock Loading," *CJHPP* 3: 211 (1989).

84

Shock Experiments in Metals and Ceramics

G.T. GRAY III

Materials Science and Technology Division
Physical Metallurgy Group
Los Alamos National Laboratory
Los Alamos, New Mexico 87545, U.S.A.

Shock recovery and spallation experiments, in which material structure / property effects are systematically varied and characterized quantitatively, offer two important experimental techniques to probe the physical mechanisms controlling shock processes and dynamic fracture. This paper highlights the current state of knowledge of the structure / property effects of shock-wave deformation on metals and ceramics. Recent shock-recovery and spallation experimental results on post-mortem material properties and fracture behavior in metals and ceramics are reviewed. Finally, the influence of shock-wave deformation on several inter-metallics and a recent experiment examining the Bauschinger effect in Al-4%Cu during shock loading are presented.

I. INTRODUCTION

Since the late 1950's shock recovery and spallation experiments have been used to probe the structure / property effects and physical mechanisms controlling the deformation processes and dynamic fracture of materials subjected to impulse or shock loading. Several papers have reviewed the structure / property response of numerous metals, alloys, ceramics, and non-metallic materials to shock-wave deformation [1-9] and spallation [10]. Overall, the extreme conditions of strain rate and temperature imposed by a shock are known to induce a high density of defects in crystalline materials which in turn affect the post-shock mechanical properties. Accommodation of the applied high-rate-uniaxial-strain pulse during the shock is observed to cause the generation of dislocations, deformation twins, point defects, stacking faults, or in some materials may activate a pressure-induced phase transition. As a result of this high defect density, most metals and alloys are observed to exhibit a greater degree of hardening , measured via post-shock hardness or stress-strain response, due to shock loading than quasi-static deformation to the same total strain, particularly if the metal passes through a allotropic phase transition, such as the α–ϵ in iron or the α–ω in titanium [1-6,8].

The exact type and arrangement of these defects, whether dislocation cells, planar faults, twins, or phase transformation products in the post-shock substructure is dependent on both chemically and microstructurally controlled components within the material as well as the externally imposed variables such as the temperature and shock loading parameters. Chemical components include the influence of material chemistry (due to the alloying and interstitial content) on the starting crystal structure, stacking fault energy, phase constitution (influencing the precipitation of second phases), and phase stability of the material undergoing shock deformation or dynamic fracture. Independent of the chemistry of the starting material variables such as the grain size, second phase size and distribution, and crystallographic texture can be altered through the manipulation of heat-treatment. Due to some of the recent developments in the growing area of composite materials it is also possible to introduce chemically inert dispersoids, such as Al_2O_3 or SiC in aluminum alloys, which can significantly influence the structure property response of these materials.

Since the last Explomet conference held in 1985[11], numerous studies have investigated the influence of a variety of microstructural and externally applied variables on the "real-time" and post-shock structure / property responses of a broad class of materials. Many of these studies have been reported in recent conference proceedings including: the two American Physical Society conferences on condensed matter held in 1987 [12] and 1989 [13], IMPACT '87 [14], the Oxford conference on materials at high rate [15], DYMAT 88 [16], and a conference on dynamic compaction technology[17]. In addition, two recent books edited by Blazynski [18] and Murr [19] summarize various shock studies on materials. The intent of this paper is to highlight some of the recent experimental results and research trends of shock effects on materials with particular emphasis on the shock deformation behavior of conventional metals, alloys, and ceramic materials. Due to space limitations only studies on existing materials are included. Also, the areas of shock-wave synthesis and dynamic compaction of materials are not included. Based upon a review of the literature over the past five years, the stronger than usual influence of funding priorities from the government and military sector, both in the USA and abroad, on shock-wave and hyper-velocity research are readily apparent. Of particular significance were the formation in 1983 of the Strategic Defense Initiative and, more important for materials research, the report on armor/antiarmor by the Defense Science Board in 1985 [20].

The general area of armor/antiarmor, funded heavily by several DARPA and DOD initiatives to universities, federal labs, and the private sector, started a broad range of studies on materials research in the areas of penetration failure mechanics, constitutive behavior, and spallation particularly in the area of ceramics. This research emphasis has in turn created a heightened level of activity in the materials/physics communities in the areas of shear band formation, microstructural effects on spallation, shock-wave damage mechanisms in ceramics, and constitutive data on armor/antiarmor materials. This paper is divided into sections covering metallic and ceramic materials. Recent experimental results are presented on three emerging materials shock-wave research areas: the effects of laser-induced shocks on materials, the influence of shock-wave deformation on several intermetallic materials, and some recent experimental results demonstrating how the strain-path change inherent during a shock can lead to a Bauschinger effect in two-phase materials.

II. METAL AND ALLOYS STUDIES

Numerous aspects of the deformation and spall behavior of metals and alloys subjected to shock loading have been examined experimentally and correlated with materials models in the past 5 years[21-43]. Systematic soft recovery experiments showed that studies of the influence of shock wave parameters and strengthening mechanisms on the structure / property behavior of shock-recovered materials are only relevant if the effects of additional plastic work due to lateral release or poor recovery techniques are minimized as much as possible[40]. Consistent with previous work the increased strain rate in the shock increases the propensity for deformation twinning in shock-loaded metals[38,39] including in Al-4.8 Mg[21] and 6061-T6 Al[22]. Microstructure is still observed to significantly affect the shock hardening and deformation response of materials[21-23,41]; the shock hardening of solution or underaged JBK-75 stainless steel and 6061 Al alloys exceed that of the peak aged materials[23]. Several studies have also examined the kinetics of the martensite transformation induced by controlled tensile pulses produced by shock loading[29,30]. The effect of dynamic prestraining[34] and repeated shock loading[26] on subsequent mechanical properties have shown that prior dynamic loading reduces the subsequent rate of strain hardening during shock or quasi-static reloading. High temperature preheating prior to shock loading was found to enhance the shock hardening and impact strength of steel [28]. In addition to the broad range of experimental studies, Mogilevsky [27,31,32] and

Follansbee[33] have conducted modeling studies considering the nature of defect generation during shock loading and compared their predictions with experimental results.

Finally, lasers represent a new method for assessing the shock-wave and dynamic fracture behavior of materials. Lasers offer the possibility of producing shock pressures in targets of one to more than 10^2 GPa with pulse durations related to the laser pulse length, usually in the nanosecond range. The limitations of the laser shock pulse are principally its diameter owing to the small diameters of typical lasers and total pulse duration. The laser pulses however can be equivalent to the impacts produced by microparticles at high velocity and are therefore of great interest to studies such as micrometeroite impacts in outer space. Several recent studies have assessed the influence of laser shocks on the deformation substructure of metals[44,45]. These studies have shown that the substructure evolution in iron and stainless steel targets are very similar to those obtained by conventional shock methods.

III. CERAMIC STUDIES

The study of shock-wave and dynamic fracture effects on ceramics has been one of the fastest growing fields in materials science during the past five years[46-62]. The catalyst for this activity has been the need for data on the dynamic deformation and fracture behavior of monolithic ceramics and cermets for potential armor applications. Investigations of the Hugoniot Elastic Limit (HEL) of alumina found that the HEL in Al_2O_3 is not strain rate dependent[46] and that it is higher than the static yield strength [47]. The question of whether plastic flow and microcracking are activated above or below the HEL of a ceramic has been the subject of a number of studies which have failed to define a consistent pattern.

Studies have shown that while alumina retains its shear strength to 15 GPa[49] in some cases alumina is seen to start cracking below the HEL[50,58,60,62] while in another case no microcracking was observed, but only pore closure, up to twice the HEL [51,56,59,61]. In addition to the post-shock sample analysis for cracks, in one study the observation of a decaying stress pulse was attributed to the attenuation of the incident compressive pulse due to compressive damage[60] while Figure 1 shows a spall trace of symmetrically shocked TiB_2 above the HEL displaying a stable shock pulse. Analysis of shock recovered polyphase alumina showed that one reason for the cracking below the HEL was due to the impedance mismatch and residual stresses associated with the glassy phase[53,57]. Studies of recovered alumina[52] and an $Al-B_4C$ cermet [54] showed dislocation activation above the HEL and in the cermet none below the HEL. Several studies have also measured the spall strengths of TiB_2[55] and Al_2O_3[51,56]; the spall strength of TiB_2 has been observed to decrease to zero above the HEL[55] while the spall strength of Al_2O_3 remains non-zero up to twice the HEL [51]. Given the difficulties in recovering ceramic samples after shock loading[9], the resolution of the pressure and material dependencies on the deformation and microcracking in

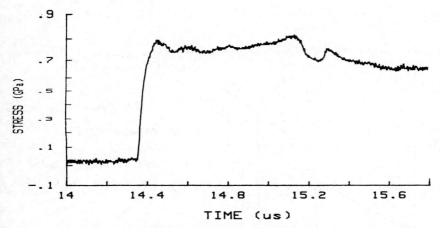

FIG. 1   *Stress vs. time spall trace measured in the plexiglas backing for a symmetric TiB$_2$ test with an impact velocity of 596 m/sec (15.2 GPa). Spall strength = 0.63 GPa.*

shock-loaded ceramics will probably remain a difficult area to obtain definitive research findings.

IV. RECENT SHOCK STUDIES

A. *INTERMETALLICS*

Intermetallics and some composites are receiving increasing attention due to their high specific strengths, stiffnesses, and potential high temperature properties. Within the last few years high-strain-rate and shock-loading experiments have begun to probe the dynamic deformation response of a range of intermetallic and composite materials. To assess the substructure evolution and mechanical response of Ni$_3$Al as a function of peak pressure, we have "soft" shock recovered samples and then sectioned the recovered disks for mechanical testing and TEM samples. Figure 2 shows the reload compressive stress-strain response of the Ni$_3$Al following shock loading as compared to the annealed starting material. The shock-loaded stress strain curves are plotted with offsets starting at the approximate total transient shock strains [calculated as 4/3 ln(V/V$_0$) where V and V$_0$ are the final and initial volumes] for the two shocks. Peak shock pressures are approximated using the EOS of pure nickel in the absence of an EOS for Ni$_3$Al. The reload yield strength of Ni$_3$Al increases from 250 MPa to 750 and 1250 MPa following the ~ 14.0 and 23.5 GPa shocks, respectively. As in the case of most metals and alloys the effective hardening in the shock-loaded Ni$_3$Al exceeds that quasi-statically obtained when deformed to roughly the equivalent strain[4].

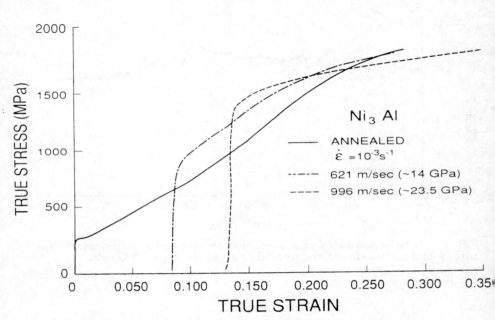

FIG. 2   *Stress-strain behavior of shock-loaded Ni₃Al as a function of peak pressure.*

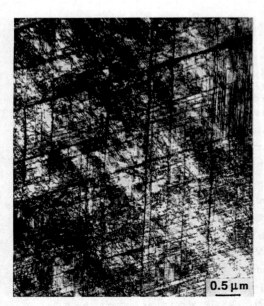

FIG. 3   *Brightfield electron micrograph of dislocations, stacking faults, and deformation twins in Ni₃Al shock loaded at 996 m/sec (~23.5 GPa).*

The substructure evolution in the shock-loaded Ni$_3$Al is observed to depend on the peak shock pressure consistent with the reload data. Increasing the peak pressure is observed to significantly increase the density of stacking faults and deformation twins. Figure 3 shows the substructure of the Ni$_3$Al shock-loaded at 996 m/sec (~ 23.5 GPa) consisting of a high density of dislocations on octahedral planes, planar stacking faults, and deformation twins. The substructure evolution in shock-loaded TiAl and Ti$_3$Al have also been investigated. Similar to Ni$_3$Al the substructure of both Ti-aluminides is dependent on the peak shock pressure. While the TiAl readily deforms via deformation twinning when subjected to shock loading, the Ti$_3$Al exhibits a substructure consisting of solely coarse planar slip. Figure 4 shows the high density of deformation twins formed in TiAl shock loaded to 7.0 GPa. Finally for intermetallics, as mentioned earlier, lasers offer an alternative method for applying shocks to materials. Figure 5 shows the high density of dislocations formed in Ti$_3$Al due to a laser-driven miniature flyer plate impacted at 4.3 km/sec [63]. The deformation mechanisms activated, while formed by a very short pulse duration due to the ~10μm thick flyer plate, are similar to that observed in Ti$_3$Al conventionally shock loaded.

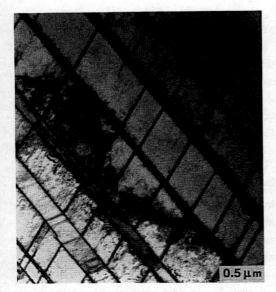

FIG.4   *Brightfield electron micrograph of deformation twins in TiAl containing Ti$_3$Al shock loaded at 580 m/sec (~7 GPa).*

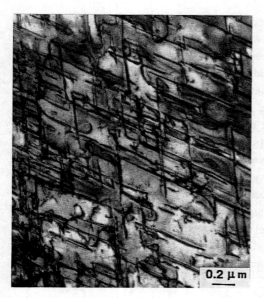

FIG. 5   *Brightfield electron micrograph of dislocation debris in Ti₃Al shock loaded by a*
laser-driven flyer plate accelerated to 4.3 km/sec.

B. *BAUSCHINGER EFFECT DURING SHOCK LOADING*

Due to the intrinsic nature of the shock process the structure/property response of a material is
a result of the total shock excursion comprised of the compressive loading regime occurring at
a very high ($\sim 10^5$ - 10^8 s^{-1}) strain rate (shock rise), a time of reasonable stable stress (pulse
duration), and finally a tensile release of the applied compressive load returning the sample to
ambient pressure at a lower strain rate.   Collectively the loading sequence in a shock amounts
to a single cycle stress/strain path change excursion with elastic and plastic deformation
operative in two directions.   In this regard the shock process may be compared to a single
high-amplitude "fatigue-type" cycle with a dwell time representing the pulse duration [25].

Some materials after reversing the direction of stressing quasi-statically, exhibit an offset
yield upon stress reversal due to directional kinematic and isotropic hardening[64,65].   In
most instances of 2-phase materials, given the existence of a back stress acting in the matrix
due to the presence of the unrelaxed plastic strain in the particle vicinity, yielding in the
reverse direction occurs at a reduced stress level compared with the forward flow stress level.
To directly assess if the strain path reversal inherent to the shock contains a Bauschinger
effect for a shock-loaded two-phase material, two samples of an Al-4wt.% Cu alloy

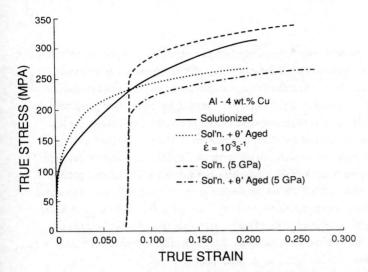

FIG. 6   *Stress-strain response of shock-loaded Al-4wt.% Cu as a function of heat-treatment illustrating the Bauschinger effect inherent to the shock process between the solution-treated and θ′ conditions.*

exhibiting a well known Bauschinger effect as a function of microstructure were shock loaded to 5.0 GPa and "soft" recovered in the same shock assembly to assure identical shock-loading conditions.

Figure 6 shows our results of the stress-strain response of the starting microstructures and quasi-statically-reloaded shock-loaded samples. The reload shock-loaded sample curves starting points are shifted to a strain of 7.5% equal to the calculated total shock transient strain. While the as-heat-treated yield strengths of the starting microstructures are the same, the reload behavior of the shock-loaded samples is quite different between the solution-treated and θ′ aged microstructures. While the solutionized sample shock hardens above the quasi-static response to equivalent strain levels the aged sample flow behavior is considerably below that of the unshocked aged sample response. This data is consistent with the operation of a Bauschinger effect, i.e. reverse strain cycle, occurring during the shock process in the two-phase Al-Cu alloy studied. This effect offers a consistent explanation of the lack of significant shock hardening and/or softening compared to an equivalent quasi-static strain level in 2-phase materials as a whole. Finally, while the Bauschinger effect will be manifested in two-phase materials, the proven existence of the strain reversal contribution occurring during the shock process supports the previous model including reversibility as the controlling variable of pulse duration effects[25].

V. SUMMARY

Shock recovery and spallation experiments, in which material structure / property effects are systematically varied and characterized quantitatively, offer two important experimental techniques to probe the physical mechanisms controlling shock processes and dynamic fracture. In conventional metals, alloys, and also in cermets and ceramics experimental studies continue to find that microstructure significantly affects the shock hardening and deformation response. In ceramic materials, the question of when plastic flow and microcracking are activated, in particular above or below the HEL, of a ceramic has been the subject of numerous studies and still the results have not define a self-consistent pattern. A shock loading study on Ni_3Al has shown that similar to most disordered metals and alloys the effective hardening in shock-loaded Ni_3Al exceeds that quasi-statically obtained when deformed to roughly the equivalent strain. Finally, experiments on Al-4wt.% Cu as a function of heat-treatment have demonstrated that the shock process behaves as a Bauschinger effect test to two-phase materials sensitive to strain-path reversals.

ACKNOWLEDGMENTS

The author acknowledges the collaborative contributions of P.S. Follansbee, C.E. Morris, W.R. Blumenthal, and C.E. Frantz to various of the findings presented in this paper. This work was performed under the auspices of the U.S. Department of Energy.

REFERENCES

1.   E.G. Zukas, *Metals Eng. Quart.*, 6: 16 (1966).

2.   S. Mahajan, *Phys. Stat. Sol.*(a), 2: 187 (1970).

3.   W.C. Leslie, in *Metallurgical Effects at High Strain Rates*, R.W. Rohde, B.M. Butcher, J.R. Holland, and C.H. Karnes (eds.), Plenum Press, N. Y., 1973, p. 571.

4.   L.E. Murr, in *Shock Waves and High Strain Rate Phenomena in Metals*, M.A. Meyers and L.E. Murr (ed.), Plenum Press, N.Y., 1981, p. 607.

5.   L.E. Murr, in *Materials at High Strain Rates* , T.Z. Blazynski (ed.), Elsevier Applied Science, London, 1987, p. 1.

6.   L.E. Murr, in *Shock Waves for Industrial Applications* , L.E. Murr (ed.), Noyes Publications, Park Ridge, New Jersey, 1988, p. 60.

7.   D. Raybould and T.Z. Blazynski, in *Materials at High Strain Rates* , T.Z. Blazynski (ed.), Elsevier Applied Science, London, 1987, p. 71.

8. K.P. Staudhammer, in *IMPACT 1987* , C.Y. Chiem, H.-D. Kunze, and L.W. Meyer (eds.), Duetsche Gesellschaft fur Metallkunde, Oberursel, West Germany, 1988, p. 93.

9. G.T. Gray III, in *Shock Compression of Condensed Matter - 1989*, S.C. Schmidt, J.N. Johnson, and L.W. Davison (eds.), North-Holland Press, 1990, p. 407.

10. M.A. Meyers and C.T. Aimone, Progress in Materials Science, 28: 1 (1983).

11. *Metallurgical Applications of Shock Wave and High-Strain-Rate Phenomena*, L.E. Murr,   K.P. Staudhammer, and M.A. Meyers (eds.), Marcel Dekkar Inc., 1986.

12. *Shock Waves in Condensed Matter 1987* , S.C. Schmidt and N.C. Holmes (eds.), Elsevier Science Publishers, New York, 1988.

13. *Shock Compression of Condensed Matter - 1989*, S.C. Schmidt, J.N. Johnson, and L.W. Davison (eds.), North-Holland Press, 1990.

14. *IMPACT 1987* , C.Y. Chiem, H.-D. Kunze, and L.W. Meyer (eds.), Duetsche Gesellschaft fur Metallkunde, Oberursel, West Germany, 1988.

15. *Mechanical Properties of Materials at High Rates of Strain 1989* , J. Harding (ed.), Institute of Physics Conference Series #102, J.W. Arrowsmith Ltd., Bristol, England, 1989.

16. *DYMAT 88, International Conference on Mechanical and Physical Behavior of Materials under Dynamic Loading*, Journal De Physique, Tome 49, Colloque 3, #9, 1988.

17. *2nd Workshop on Industrial Application Feasibility of Dynamic Compaction Technology* , Tokyo Institute of Technology, 1988.

18. *Materials at High Strain Rates* , T.Z. Blazynski (ed.), Elsevier Applied Science, London, 1987.

19. *Shock Waves for Industrial Applications* , L.E. Murr (ed.), Noyes Publications, Park Ridge, New Jersey, 1988.

20. H. Fair, *Int. J. Impact Engng.,* 5: 1 (1987).

21. G.T. Gray III, *Acta Metall.,* 36: 1745 (1988).

22. G.T. Gray III and P.S. Follansbee, in *Shock Waves in Condensed Matter 1987* , S.C. Schmidt and N.C. Holmes (eds.), Elsevier Science Publishers, New York, 1988, p. 339.

23. M.J. Carr, C.R. Hills, R.A. Graham, J.L. Wise, Ibid, p. 335.

24. J.A. Brusso, D.E. Mikkola, G. Bloom, R.S. Lee, and W. Vonholle, Ibid, p. 375.

25. G.T. Gray III and P.S. Follansbee, in *IMPACT 1987* , C.Y. Chiem, H.-D. Kunze, and L.W. Meyer (eds.), Duetsche Gesellschaft fur Metallkunde, Oberursel, West Germany, 1988, p. 541.

26. N.V. Gubareva, A.N. Kiselev, T.M. Sobolenko and T.S. Teslenko, Ibid, p. 801.

27. M.A. Mogilevsky, Ibid, p. 957.

28. F. Tauadze, E. Kutelia, A. Peikrishvili, T. Eterashvili, and G. Gotsiridze, Ibid, p. 993.

29. N.N. Thadhani and M.A. Meyers, *Acta Metall.*, 24: 1625 (1986).

30. S.N. Chang and M.A. Meyers, *Acta Metall.*, 36: 1085 (1988).

31. M.A. Mogilevsky, in *DYMAT 88, International Conference on Mechanical and Physical Behavior of Materials under Dynamic Loading,* Journal De Physique, Tome 49, Colloque 3, #9, 1988., p. 467.

32. M.A. Mogilevsky and L.S. Bushnev, in *Mechanical Properties of Materials at High Rates of Strain 1989* , J. Harding (ed.), Institute of Physics Conference Series #102, J.W. Arrowsmith Ltd., Bristol, England, 1989, p. 307.

33. P.S. Follansbee, in *Shock Compression of Condensed Matter - 1989*, S.C. Schmidt, J.N. Johnson, and L.W. Davison (eds.), North-Holland Press, 1990, p. 349.

34. A.M. Rajendran, N.S. Brar, and M. Khobaib, Ibid, p. 401.

35. W. Arnold, M. Held, and A.J. Stilp, Ibid, p. 421.

36. L.C. Chhabildas, L.M. Barker, J.R. Asay, and T.G. Trucano, Ibid, p. 429.

37. A.K. Zurek and P.S. Follansbee, Ibid, p. 433.

38. J.C. Huang and G.T. Gray III, *Mat. Sci. Eng.*, A103: 241 (1988).

39. J.C. Huang and G.T. Gray III, *Scripta Metall.*, 22: 545 (1988).

40. G.T. Gray III, P.S. Follansbee, and C.E. Frantz, *Mat. Sci. Eng.*, A111: 9 (1989).

41. J.C. Huang and G.T. Gray III, *Metall. Trans.*, 20A: 1061 (1989).

42. J.C. Huang and G.T. Gray III, *Acta Metall.*, 37: 3335 (1989).

43. P.S. Follansbee and G.T. Gray III, "The Response of Single Crystal and Polycrystalline Nickel to Quasi-Static and Shock Deformation," in *Advances in Plasticity 1989*, A.S. Khan and M. Tokuda (eds.), Pergamon Press, Oxford, 1989, p. 385.

44. M. Hallouin, F. Cottet, J.P. Romain, L. Marty and M. Gerland, in *IMPACT 1987* , C.Y. Chiem, H.-D. Kunze, and L.W. Meyer (eds.), Duetsche Gesellschaft fur Metallkunde, Oberursel, West Germany, 1988, p. 1051.

45. M. Hallouin, M. Gerland, J.P. Romain, and F. Cottel, in *DYMAT 88, International Conference on Mechanical and Physical Behavior of Materials under Dynamic Loading,* Journal De Physique, Tome 49, Colloque 3, #9, 1988, p. 413.

46. J. Cagnoux and F. Longy, in *Shock Waves in Condensed Matter 1987* , S.C. Schmidt and N.C. Holmes (eds.), Elsevier Science Publishers, New York, 1988 , p. 293.

47. D. Yaziv, Y. Yeshurun, Y. Partom, and Z. Rosenberg, Ibid, p. 297.

48. S.J. Bless, N.S. Brar, and A. Rosenberg, Ibid, p. 309.

49. Z. Rosenberg, D. Yaziv, Y. Yeshurun, and S.J. Bless, in *IMPACT 1987* , C.Y. Chiem, H.-D. Kunze, and L.W. Meyer (eds.), Duetsche Gesellschaft fur Metallkunde, Oberursel, West Germany, 1988, p. 393.

50. Y. Yeshurun, D.G. Brandon, and Z. Rosenberg, Ibid, p. 399.

51. F. Longy and J. Cagnoux, Ibid, p. 1001.

52. D.G. Howitt and P.V. Kelsey, Ibid, p. 249.

53. Y. Yeshurun, Z. Rosenberg and D.G. Brandon, in *Mechanical Properties of Materials at High Rates of Strain 1989* , J. Harding (ed.), Institute of Physics Conference Series #102, J.W. Arrowsmith Ltd., Bristol, England, 1989, p. 379.

54. W.R. Blumenthal and G.T. Gray III, Ibid, p. 363.

55. D. Yaziv and N.S. Brar, in *DYMAT 88, International Conference on Mechanical and Physical Behavior of Materials under Dynamic Loading,* Journal De Physique, Tome 49, Colloque 3, #9, 1988, p. 683.

56. J. Cagnoux and F. Longy, Ibid, p. 3.

57. Y. Yeshurun, D.G. Brandon, A. Venkert, and Z. Rosenberg, Ibid, p. 57.

58. L.H.L. Louro and M.A. Meyers, *J. Mat. Sci.,* 24: 2516 (1989).

59. F. Longy and J. Cagnoux, *J. Amer. Cer. Soc.,* 72: 971 (1989).

60. R.J Clifton, G. Raiser, M. Ortiz, and H. Espinosa, in *Shock Compression of Condensed Matter - 1989*, S.C. Schmidt, J.N. Johnson, and L.W. Davison (eds.), North-Holland Press, 1990, p. 437.

61. F. Longy and J. Cagnoux, Ibid, p. 441.

62. LH.L. Louro and M.A. Meyers, Ibid, p. 465.

63. D.L. Paisley, "Laser-Driven Miniature Plates for 1-D Impacts at 0.5-6 km/sec", this conference.

64. G.D. Moan and J.D. Embury, *Acta Metall.,* 27: 903 (1979).

65. M.G. Stout and A.D. Rollett, *Metall. Trans. A*, (1990) in press.

85

Novel Applications of Shock Recovery Experiments

L.E. MURR

Department of Metallurgical and Materials Engineering
The University of Texas at El Paso
El Paso, Texas 79968 USA

The development of density-compatible sandwich arrangements to subject a wide variety of materials to a plane, compressive shock wave at known peak pressures and pulse durations and their recovery is reviewed with the use of several novel examples. It is demonstrated that shock recovery experiments can be used as a powerful diagnostic tool to elucidate the response of materials to shock wave effects and shock loading.

I. INTRODUCTION

The concept of shock recovery experiments can be traced to a number of researchers depending upon the nature of the experiment. Here we will focus on subjecting matter to conditions of high pressure and recovering the matter after the event. Typical of this concept are the experiments of Sir Charles Parsons [1] following the discovery of diamond in meterorites in 1878. Parsons bored a flat-bottomed hole in an iron or steel block, placed carbon-bearing material into the hole, and fired a projectile from a high-powered rifle into the hole. While Parsons actually recovered small diamonds, the conditions were not optimized to synthesize significant quantities of diamond [2].

During World War II, intense interest developed in the effect of shock waves on materials and experiments initiated at Los Alamos National Laboratory paved the way for conducting simple, plane shock wave recovery tests where a material could be subjected to a plane compression shock accompanied by a uniaxial strain [3]. C. S. Smith [4] was one of the first to examine the microstructure of a recovered metal subjected to a shock wave, and this work catalyzed considerable interest in the shock deformation of metals [5].

The flying plate or "mouse-trap" technique originally described by Duvall [6] uses a line-wave generator of a suitable explosive (originated by DuPont [7]) to initiate detonation simultaneously over the surface of a main explosive charge which accelerates a suitable plate to impact a parallel, plane target. If the impacting plate (or flyer) and the target material are the same, simple impact physics can accurately describe the propagation of the shock wave into the target material. Furthermore, there is not much attenuation of the shock pulse in the first several millimeters of the target [8], and it is possible to sandwich density-compatible materials within a target, recover the sandwich, and systematically examine the effects of shock-wave propagation. Inman, et al. [9] and Rose and Grace [10] were among the first to place thin, density compatible foils or coupons of metals and alloys within a sandwich target, subject the sandwich to a controlled shock pulse, and recover the sandwich in order to examine the microstructure of the foils by transmission electron microscopy. This sandwich recovery technique has also been utilized in other impact arrangements including gun-fired projectiles impacting sandwiched targets, and these features are illustrated schematically in Fig. 1.

The pressure, P, imparted on the system by the impact of a flyer plate (Fig. 1(a)) or a projectile (Fig. 1(b)) can be calculated by a technique called impedance matching [11] using the Rankine-Hugoniot equation relating the pressure, P, and particle velocity, U_p. The physical basis for this technique is the requirement that particle velocity and pressure are continuous across the collision interface: conservation of momentum, and energy in Fig. 1(c) and (d). The impedance matching technique has a simple graphical solution [11] after impact when

$$P = P^F = P^T \tag{1}$$

and

$$U_p^F + U_p^T = V_p = U_d \tag{2}$$

where P^F and U_p^F and P^T and U_p^T are the flyer (F) or projectile and target (T) pressures and particle velocities respectively, and V_p and U_d are the corresponding projectile, flyer, or driver plate velocity (the impact velocity).

From conservation of momentum it follows that for two dissimilar flyer (or projectile) and target materials impacting at a velocity V_p, the impact pressure (Eqn. (1)) is given by:

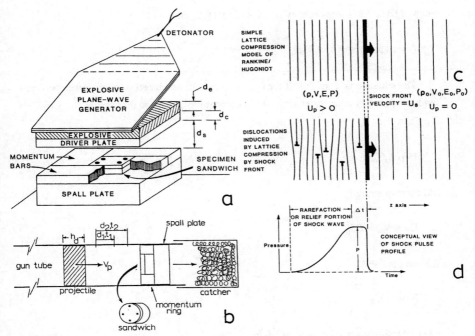

FIG. 1 *Principal features for plane wave shock loading and experimental sandwich recovery. (a) Plane wave generation by explosively initiated flyer plate. The sandwich is recovered in a water tank below the target arrangement. (b) Projectile impact in a gun tube. (c) Idealized view of lattice compression. (d) Shock-induced lattice instabilities and shock pulse schematic.*

$$P = P^F = P^T = \rho_F V_p C_F / (1 + \rho_F C_F / {}_T C_T) \qquad (3)$$

where $_F$ and C_F and ρ_T and C_T are the flyer (F) and target (T) material density and longitudinal (bulk) sound velocities respectively. Of course if the bulk sound velocities are known, the pressure can be measured by measuring the velocity, V_p, from d_1/t_1 or d_2/t_2 (Fig. 1(a) and (b)) using electrical pin contacts, laser ranging, or other suitable measurements. For explosively accelerated flyer plate arrangements illustrated in Fig. 1(a), Gurney [12] developed an equation for the plate velocity based on equating the chemical energy of the explosive to the kinetic energy of the flyer:

$$V_P = \sqrt{2E} \ \sqrt{3/(1 + 5/\zeta + 4/\zeta^2)} \qquad (4)$$

where $\sqrt{2E}$ is the Gurney energy in units of m/s [13] and

$$\zeta = h_e \, \rho_e/h_d \, \rho_d = d_e \, \rho_e/d_c\rho_c \tag{5}$$

The situation is simplified if the flyer or projectile and target (including the specimen sandwich) are the same material or have a compatible density so that $\rho_F \cong \rho_T = \rho$; $C_F \cong C_T = C_o$. Then

$$P \cong \rho V_p C_o/2 \tag{6}$$

and correspondingly

$$U_d = V_p = 2U_p \tag{7}$$

We can also write

$$P = \rho U_s U_p \tag{8}$$

where the shock velocity, U_s is given by

$$U_s = C_o + S_i U_p \tag{9}$$

where S_i is an empirical parameter for the particular material (see Appendix C of reference [13]).

We can calculate the pulse duration (Δt in Fig. 1(d)) from the thickness of the flyer plate or projectile:

$$\Delta t = 2h_d/U_s = 2d_c/U_s \tag{10}$$

Rose and Grace [10] have developed some design criteria for sandwich arrangements illustrated in Fig. 1(a). As a rule of thumb, the momentum bars should be as wide as the maximum sandwich dimension (width) and the spall plate should be twice the sandwich thickness. These conditions generally hold for pressures in excess of 50 GPa. For smaller pressures, these design conditions can be relaxed in some simple proportion.

For energetic explosives having uniform properties such as plastice and high-density (low porosity) explosives with detonation velocities above about 4km/s, the Gurney

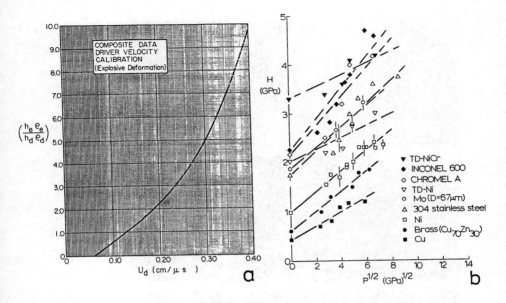

FIG. 2 *(a)* *Zeta* *(ζ)* *versus* *driven* *plate* *velocity.* *(b)* *Residual*
hardness *versus* *rust* *peak* *pressure* *for* *a* *range* *of* *metals* *and*
alloys *shock* *loaded* *and* *recovered* *in* *sandwich* *arrrays* *as* *shown*
in *Fig.* *1(c).* *The* *pulse* *duration* *was* *constant* *at* *2µs.*

equation (Eqn. (4)) can be expressed in terms of ζ (Eqn. (5)). Figure 2(a) shows a plot of
driver (or flyer) velocity versus ζ for a wide range of metal and alloy systems employing
sandwich arrangements where the flyer plates and the target materials were the same, and the
explosives ranged from C-2 to C-4 designations [13].

Figure 2(b) illustrates some typical examples of residual hardness of a host of metals
and alloys subjected to plane shock waves at a constant pulse duration of 2 µs. Figure 3
shows some typical microstructures which illustrate the basis for the shock hardening
shown in Fig. 2(b). By recovering density compatible sandwiches containing foils of
metals and alloys, the microstructures produced by plane shock wave propagation can be
observed and compared without inducing artifacts which could result by sawing or cutting
samples from a bulk target. By maintaining the pulse duration essentially constant, the
effects of peak pressure can be conveniently and accurately compared as illustrated in Fig.
2(b) and Fig. 3. These features have been extensively described for a wide range of metals
and alloys [14].

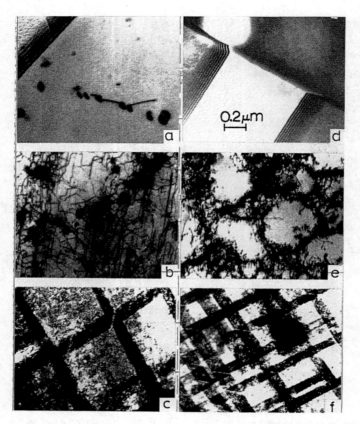

FIG. 3 *Examples of residual (TEM) microstructures in shock loaded, sandwiched materials. (a) Annealed, unshocked Inconel 600. (b) Inconel 600 after shock at 8 GPa. (c) Inconel 600 after shock at 25 GPa. (d) Annealed, unshocked nickel. (e) Nickel after shock at 10 GPa. (f) Nickel after shock at 45 GPa. Pulse durations were constant at 2µs.*

II. NOVEL APPLICATIONS OF SHOCK RECOVERY EXPERIMENTS: SOME EXAMPLES

Although the shock loading of metals and alloys in recoverable sandwich arrangements (Fig. 1(a) and (b)) continues to provide useful microstructural comparisons and access to fundamental materials behavior as exemplified in Figs. 2(b) and Fig. 3 [15], there are situations where this experimental approach can provide truly novel diagnostics. Figure 4 illustrates an example of such a situation. In the example illustrated in Fig. 4, it was

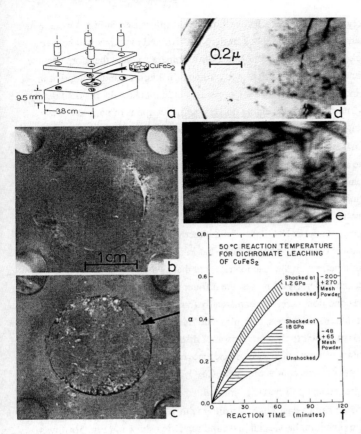

FIG. 4 *Shock-recovery of chalcopyrite mineral (CuFeS₂). (a) Sandwich schematic. (b) Lid of recovered sandwich. (c) Sandwich base with chalcopyrite sample (arrow) after 1.2 GPa shock. (d) Initial TEM microstructures. (e) After 1.2 GPa shock. (f) Comparison of leaching kinetics for shocked and granulated chalcopyrite. Powder size is indicated by mesh size. α - is the fraction reacted in potassium dichromate solution - at temperature.*

necessary to subject a large sample of natural chalcopyrite mineral (CuFeS₂) to a plane shock wave in order to induce crystal defects (dislocations) into the material and recover it largely intact so that it could be ground to various size fractions and reacted at constant temperatures to determine the effect of dislocations on the reaction kinetics [16]. Prior to this experiment, the effects of crystal defects on reaction kinetics was largely speculative since no direct evidence had been obtained relating defect density, particle size (or size distribution) and reaction rate. The basis for the experiments illustrated in Fig. 4 were that, as implicit in the schematic view of the shock front in Fig. 1(d), all materials will behave plastically in the shock front, and even brittle materials or minerals would therefore become dislocated when a plane shock wave were passed through. In addition, and as implicit in Fig. 2(b) and Fig. 3, increasing the peak shock pressure would increase the residual defect density. By carefully machining and grinding polycrystalline coupons of natural chalcopyrite, and fitting them into a density-compatible sandwich (Fig. 4(a)) (a titanium-6Al-4V alloy; $\rho = 4.2$ g/cm^3) and using a plane-wave arrangement as shown schematically in Fig. 1(a), chalcopyrite samples were recovered largely intact as shown in Fig. 4(b) and (c) using a water tank below the shock wave assembly shown in Fig. 1(a). Figure 4(d) and (e) show corresponding initial (unshocked) and shocked (1.2 GPa peak pressure) dislocation structures observed by transmission electron microscopy. A second experiment was also performed at higher peak pressure (18 GPa) and these recovered regimes were compared in kinetic experiments shown typically in Fig. 4(f). What Fig. 4(f) shows is that small particle size has an overwhelming effect on the reaction rate, but that as the peak shock pressure is increased and the dislocation density increases, the difference in fraction reacted (α) between the initial and shocked condition increases dramatically (compare the shaded regions in Fig. 4(f)). Indeed, the fact that metals, alloys, ceramics, and minerals behave plastically in the shock front (Fig. 1(d)) is implicit in the comparison of dislocation in Figs. 3 and 4.

A more contemporary example of a novel shock recovery experiment is presented in Fig. 5. Here sintered bars of superconducting YBa$_2$Cu$_3$O$_7$ having a density of roughly 6.4 g/cm^3 were carefully ground and polished to fit into milled channels in a density compatible sandwich (Fig. 5(d)) made from a cast Zn-Al-Cu alloy with $\rho = 6.7$ g/cm^3. The entire assembly (flyer plate, momentum bars, spall plate and sandwich shown in Fig. 5(a)) was machined from the cast alloy, shocked, and recovered as illustrated in the schematic arrangement of Fig. 5(a), and shown in Fig.5(b). The shocked superconductor bars were recovered essentially intact (Fig. 5(b)), similar to the arrangement shown in Fig. 4(a)-(c) for shocked CuFeS$_2$, after being subjected to a peak pressure of 6 GPa at 4 μs pulse duration, and four-probe resistance-temperature measurements were performed as illustrated in Fig. 5(c) [17]. These signatures show that the shock wave causes both the normal-state resistance-temperature response and the superconducting transition to be altered (degraded) and this same response had been previously observed for explosively consolidated powders of YBa$_2$Cu$_3$O$_7$ as shown for comparison in Fig. 5 [18]. The unique feature of the shock recovery experiment (Fig. 5) was that the shock wave effect could be unambiguously demonstrated with a simple, plane shock wave. The essentially "dry" recovery of the shock loaded sandwich in Fig. 5 was necessary because the superconductor was degraded by moisture, and a water recovery was too risky.

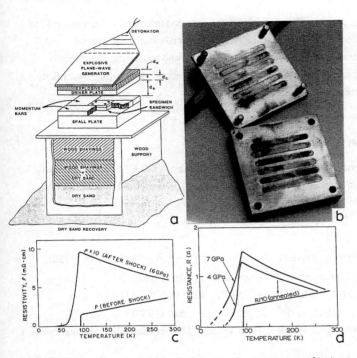

FIG. 5 *Shock-recovery of YBa₂Cu₃O₇ superconductor. (a) Schematic arrangement for shocked-sandwich recovery. (b) Recovered Sandwich with superconductor coupons. (c) Residual resistivity-temperature signature for Y-Ba-Cu-O coupons before and after shock. (d) Resistance-temperature signatures for explosively consolidated Y-Ba-Cu-O powders at pressures shown. The arrow indicates the signature when annealed at 930°C in oxygen.*

In a similar shock recovery experiment to be described by Sriram, et al. [19], the gun impact arrangement shown in Fig. 1(b) was employed to subject another superconductor to a much lower peak shock pressure which was required since no superconducting transition was observed at the lowest fabrication pressure of 4 GPa, and 5 GPa was the lowest pressure which could be attained by explosive shock loading (Fig. 1(a)). This experiment revealed a superconducting transition in a Bi-Pb-Sr-Ca-Cu-O superconductor at a peak shock pressure of 1.5 GPa.

III. SUMMARY

The few examples of shock recovery presented in Figs. 3, 4 and 5 illustrate the applications of a very simple, unambiguous, diagnostic physics tool to a host of materials areas over a period of more that two decades. These examples are intended to illustrate the powerful and elegant comparisons which can be made by subjecting materials to plane shock waves of known pressure and pulse duration, recovering them intact, and measuring specific and varied properties or observing shock-induced microstructures uncompromised by artifacts induced by specimen preparation or handling.

ACKNOWLEDGMENTS

This research was supported in part by DARPA-HTSC Contract N00014-88-C-0684 and a Mr. and Mrs. MacIntosh Murchison endowed chair at the University of Texas at El Paso.

REFERENCES

1. C. A. Parsons, *Philos. Trans. Roy Soc. (London) A220:* 78 (1920).

2. P. S. DeCarli and J. C. Jamieson, *Science 133:* 821 (1961).

3. J. M. Walsh and R. H. Christian, *Phys. Rev. 97:* 1544 (1955).

4. C. S. Smith, *Trans. Met. Soc. AIME 212:* 574 (1958).

5. G. E. Deiter, in "Response of Metals to High Velocity Deformation", Wiley-Interscience (New York), vol. 9 (1961).

6. G. E. Duvall, *ibid.*

7. G. B. Huber, "DuPont Line-Wave Generators", Stanford Research Institute Report No. 040-59, 1959.

8. C-Y. Hsu, K-C. Hsu, L. E. Murr, and M. A. Meyers, Chap. 27 in "Shock Waves and High-Strain-Rate Phenomena in Metals", M. A. Meyers and L. E. Murr (eds.), Plenum Press, New York 1981, p. 433.

9. M. C. Inman, L. E. Murr, and M. F. Rose, in "Advances in Electron Metallography, ASTM-STP 396: *6* , 39 (1966).

10. M. F. Rose and F. I. Grace, *Brit. J. Appl. Phys. 18:* 671 (1967).

11. P. S. DeCarli and M. A. Meyers, Chap. 22 in "Shock Waves and High-Strain Rate Phenomena in Metals", M. A. Meyers and L. E. Murr (eds.), Plenum Press, New York, 1981, p. 341.

12. R. W. Gurney, "The Initial Velocities of Fragments from Bombs, Shells, and Grenades", Report No. 405, Ballistics Research Laboratory (BRC), Aberdeen Proving Grounds, Maryland, Sept. 14, 1943.

13. M. A. Meyers and L. E. Murr (eds.) "Shock Waves and High-Strain-Rate Phenomena in Metals", Plenum Press, New York, 1981.

14. L. E. Murr, Chap. 37 in "Shock Waves and High-Strain-Rate Phenomena in Metals", M. A. Meyers and L. E. Murr (eds.), Plenum Press, New York, 1981, p. 607.

15. G. T. Gray III, these Proceedings.

16. L. E. Murr and J. B. Hiskey, *Met. Trans. 12B:* 255 (1981).

17. L. E. Murr, C. S. Niou, S. Jin, T. H. Tiefel, A. C. W. P. James, R. L. Sherwood, and T. Siegrist, *Appl. Phys. Lett.* *55(15)*: 1575 (1989).

18. L. E. Murr, M. Pradhan-Advani, C. S. Niou, and L. H. Schoenlein, *Sol. St. Comm.* *73*: 695 (1990).

19. M. A. Sriram, L. E. Murr, and C. S. Niou, these Proceedings.

86

Defect Structures of Shocked Tantalum

C.L. WITTMAN,* R.K. GARRETT, Jr.,** J.B. CLARK,** and C.M. LOPATIN*

*   Armament Systems Division
    Honeywell Inc.
    Brooklyn Park, Minnesota 55428, U.S.A.

** Metallic Materials Branch
    Naval Surface Warfare Center
    Silver Spring, Maryland 20903, U.S.A.

For metals deforming at high strain rates under explosive loading conditions, it is difficult to separate the shock-generated damage from that of plastic strain. Pure tantalum was subjected to shock loading by grazing incidence of direct contact explosive in a configuration allowing ~0.02 total plastic strain. Microstructural investigations revealed spall, spall initiation sites, twinning, textural changes, and texture regression upon annealing.

I.   INTRODUCTION

Shock-thermal processing of metals offers an alternative to standard rolling in obtaining conditions for hardening and reduction of grain size in iron and its alloys [1]. It was thought that this technique could be applied to tantalum plate material from which prior conventional processing could not produce fine grain size. Also, if approached from a microstructural sense, such a study could identify defects generated under shock conditions with low plastic strain. This metallurgical interest—identifying the defect structures of tantalum exposed to shock waves without significant plastic strain—is the first step in differentiating the deformation sequences of materials being exposed to explosive forming conditions.

In this study tantalum plate was shocked by grazing incidence detonation on each face using two thicknesses of detasheet. Preliminary microstructural studies showed that repeated exposure to shock was required to increase the material's hardness. No attempt was made to prevent reflection of the shock wave at free surfaces; thus spall was readily evident. Recovered material was analyzed in the as-shocked and post-shock-annealed conditions using standard metallography and SEM techniques, TEM, and X-ray pole figure Orientation Distribution Functions (ODF) to identify the nature of defects present in the material.

II.   EXPERIMENTAL

Commercially pure (99.95%), arc cast, annealed tantalum of 7.62 mm thickness (55 μm grain size) was shocked by a grazing incidence plane wave using detasheet. This resulted in an estimated 24 GPA pressure pulse. A tantalum momentum trap of two 7.62 mm-thick plates backed the sample (Fig. 1). No attempt was made to maintain a coherent interface between the sample and momentum trap, thus allowing a reflected (tensile) wave to be generated creating the conditions for spall.

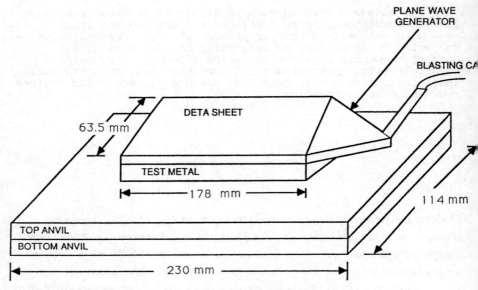

FIG. 1 *Schematic of shock loading configuration.*

Table 1. Shock Exposure Conditions and Resulting Void Measurements

Sample Number	Explosive Thickness	Number of Explosive Exposures for Each Surface	Number Void per mm^2	Void Diameter μm
1	2 mm	1	.24	46
2	2 mm	2	.10	43
3	2 mm	3	.182	63
4	2.8 mm	1	.65	106
5	2.8 mm	2	.54	122

Plate surfaces were individually shocked up to three times alternating exposed surfaces. The conditions are reported in Table 1. Recovered samples were examined using standard metallography techniques with a $1HNO_3:2HCL:2HF$ etchant, also used in preparation of the X-ray samples. A solution of $12H_2SO_4:1HF:12$ methanol was used in preparation of the TEM foils. Shocked samples were also annealed in vacuum at 1065 C for one hour to produce a recrystallized microstructure.

III.   RESULTS

Recovered samples averaged 7.46 mm thick, resulting in a plastic strain of 0.02. Optical and SEM metallography revealed that Neumann bands (deformation twins) and spall voids were present (Fig. 2 and 3). Previous reports of twinning in tantalum were the result of tensile testing at 77K [2], impact testing at room temperature and below [3,4], and cold rolling [5]. For our study, in many areas the twins traversed grain boundaries, and some grains contained twins of more than one direction (Fig. 2). Twins were randomly located through the thickness of the samples, which indicates a uniform exposure to shock.

Spall and spall initiation sites revealed themselves as highly strained areas indicative of the strong etching artifact (Fig. 3). As expected, spall occurred in the mid-thickness region of the plate. Typically, spall voids were present at multiple (>2) grain boundary intersection points. The highly strained regions are thought to be either spall initiation sites or the top/bottom of a spall void region. These sites, when observed perpendicular to the rolling plane, could only be found in the core area of the plate. Average void size generally increased with increased exposures to shock and increased explosive thickness (pulse length); see

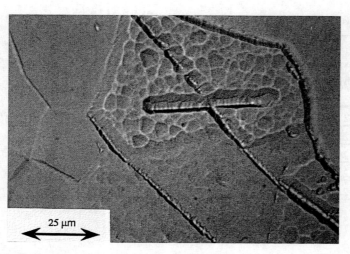

FIG. 2 *Scanning electron micrograph of a twinned grain resulting from shock condition number 1. Note that the twin crosses the grain boundary and that twins are oriented in two directions.*

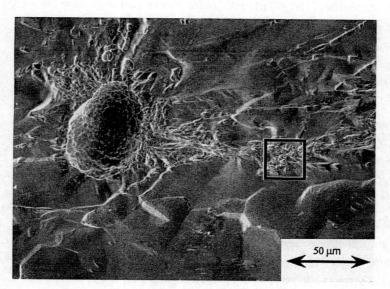

FIG. 3 *Spall void produced by shock loading condition number 1. The scanning electron micrograph indicates that the region around the spall is strained as indicative of the etching conditions. The highlighted area is thought to be either a spall initiation sight or the top or bottom edge of a void.*

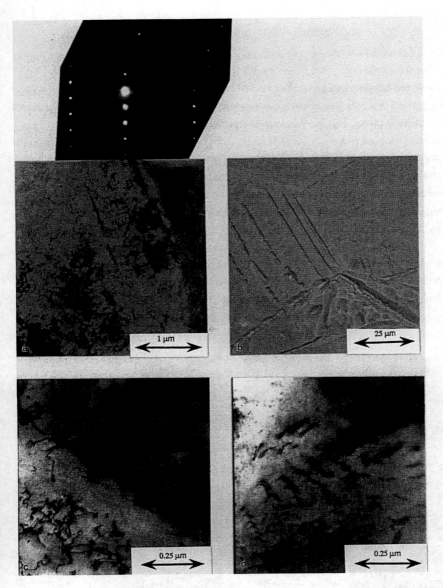

FIG. 4 *These electron micrographs of shocked condition number 4 show a resulting Neumann band and dislocation arrays. The TEM images of (a) and (c) show the dislocation network associated with this grain. Comparison of the SEM image of (c) with (a) confirms the segmented nature and indicates the bands are 0.1 μm wide. In other regions of this same grain, dislocation loops were found to be ordered similar to that expected by a slip trace (d).*

Table 1. Measured void density per unit area varied with the multiple exposure experiments, but a noted increase occurred with the greater explosive thickness.

In test number 4, material from the plate's mid-thickness, parallel to the plate surface, was examined using the transmission electron microscope and revealed what is thought to be a Neumann band (twin), Fig. 4a. Its nature and particle size is similar to that of the noncontinuous band present in the SEM image in Fig. 4b. The extended array with a width of 0.1 μm traverses the grain in a noncontinuous pattern. At higher magnification (Fig. 4c), these array segments appear to be high-angle subgrain boundaries. The area surrounding the defect shows that a dislocation-free region on one side and tangles on the other exist (Fig. 4c). Within this grain there were many other dislocation loops and tangles that were aligned in arrays (Fig. 4d).

Further analysis indicated that the internal damage was strongly dependent on grain orientation to the shock wave—parallel to this image (Fig. 5a). At this multiple-grain boundary intersection region, it may be observed that a varied response to damage accumulation has resulted from the different orientations. Dislocation networks with pileups at the boundaries only occur in grains that have a high dislocation density. The hexagonal dislocation tangles

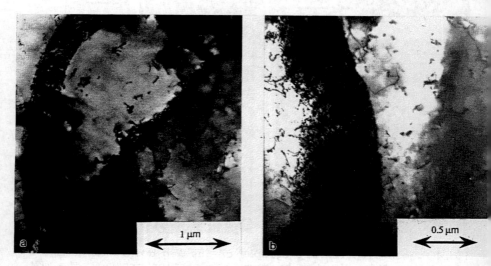

FIG. 5 *The in-plane (parallel to rolled surface) damage accumulation effects, from shock condition number 4, varies based upon grain orientation 5a. This multiple boundary intersection region indicates varying amounts of damage. The effects are shown across a grain boundary, 5b, where dislocation tangles exist in the left grain while individual loops occupy the right grain.*

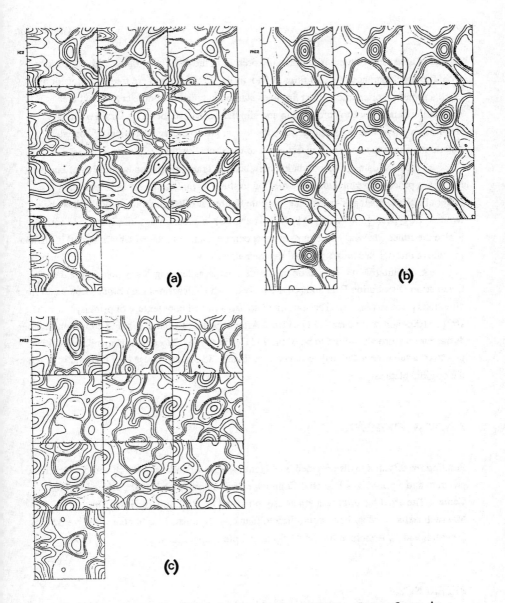

FIG. 6 *Orientation Distribution Function (ODF) plots from in-plane mid-thickness samples. Each box represents the texture present at 10 increments of orientation so the texture fiber and exact direction may be identified. The starting material (a) has a mixed texture of primary {001}<100>/secondary {111}<110> which has been altered by shocking to become a {111} fiber type and, upon annealing, reverted back to the mixed primary {001}<100>/secondary {111}<110>.*

typical of annealed tantalum [6] are readily present in the damaged grains, with added loops from the shock damage. Other grains are not as defect-ridden and are free of dislocations. Such a preferred orientation effect has been studied in tantalum single crystals where orientation dependence to deformation and slip has been described [7]. Grain boundary dislocation pileup and cells of Fig. 5b in the low-angle boundary is atypical of annealed tantalum. This appears to be a direct result of the shock loading.

This local damage accumulation correlated to grain orientation is one area of current interest in processing tantalum. One method to study its bulk nature is to measure the material's texture using X-ray diffraction pole figures. Rolled tantalum typically has one of two texture types: {100} and {111} [8]. The {100} is undesirable since it leads to earring. It is also the rolled (as-worked) texture of most commercially processed tantalum, and is difficult to remove through secondary thermal treatments.

A quantitative representation of texture can be made using X-ray pole figure data for Orientation Distribution Function (ODF) analysis, which describes the primary fibers and fiber directions present (Fig. 6). The starting plate had a mixed core texture of primary {001}<100> and secondary {111}<110>. After exposure to shock loading condition number 4, the core texture was found to be of the {111} fiber type. Subsequent annealing at 1065 C produced a fully recrystallized structure of grain size 32 μm and caused the material to revert to the original plate texture.

ACKNOWLEDGMENTS

Funding for this study was provided by a Honeywell independent research and development program and a joint Navy Potential Contractor Program with the Naval Surface Warfare Center. The anvil material was graciously provided by NRC Inc. of Newwton, Massachusetts. The authors would like to thank M. Nyquist, M. Shephard, J. Swenson, B. Oswood, and C. Anderson for their help in sample preparations.

REFERENCES

[1]   E.G. Zukas and R.G. McQueen, *Trans. AIME*, 221: 412 (1961).

[2]   R.C. Koo, *J. Less Common Metals*, 4: 138 (1962).

[3]   C.S. Barrett and R. Bakish, *Trans. AIME*, 212: 122 (1958).

[4]   R.W. Anderson and S.E. Bronisz, *Acta Met.*, 7: 645 (1959).

[5]   T.K. Chatterjee and C. Feng, "Defect Structure and Mechanical Twinning in Rolled and Annealed Tantalum," in *Proc. of the 45th Annual Meeting of the Electron Microscopy Society*, San Francisco Press, San Francisco, 1987.

[6]   A. Gillbert, D. Hull, W.S. Owen, and C.N. Reid, *J. Less Common Met.*, 4: 399 (1962).

[7]   D.P. Ferriss, R.M. Rose, and J. Wulff, *Trans. AIME*, 224: 975 (1962).

[8]   R.A. Vandermeer and W.B. Synder, *Met. Trans.*, 10A: 1031 (1979).

87

Shock Characterization of Epoxy – 42 Volume Percent Glass Microballons

L. J. WEIRICK

Explosive Projects & Diagnostics Division 2514
Sandia National Laboratories
Albuquerque, New Mexico, USA 87185

In the late 1960's, a series of shock experiments were done on ceramic-filled epoxies. Representative of these mixtures were epoxy-40% corundum (Al_2O_3) with initial density, ρ_o of 2.31 g/cm³; epoxy-40% forsterite (Mg_2SiO_4) with ρ_o = 2.0 g/cm³; and epoxy-quartz (SiO_2) with ρ_o = 1.66 g/cm³. The results from the shock characterization of these mixtures were compared to unfilled epoxy with ρ_o = 1.185 g/cm³. In that the volume percent fill was kept essentially constant, the shock Hugoniots formed a consistent family of curves relative to their initial densities. The present study examined the shock characteristics of an anhydride-cured epoxy (Epon 828 base) filled with approximately 42 volume percent microballoons (GMB) with a density of 0.94 g/cm³. The Hugoniot relationship was determined between the shock pressures of 0.5 and 4.2 GPa. A light-gas gun was used to impact samples of the epoxy-GMB material onto targets of known impedance and known Hugoniot equation. A VISAR laser interferometer system was used to measure the particle velocity in the known target material. From the measured projectile velocity and the Hugoniot of the known material, the particle velocity, shock velocity and pressure in the EPOXY-GMB was determined. The Hugoniot relationship of the epoxy-GMB material is compared to those of the previous mixtures. A comparison of the epoxy-GMB material to a porous epoxy is also made.

Shock attenuation studies were also done on this material. It was found that the shock wave profile consisted of an elastic and a plastic wave due to the initial strength of the glass microballoons, 0.24 GPa, the crushing of the glass microballoons, and the final state of dense epoxy containing a small amount of glass. This material fits the P-α model very well.

I.   INTRODUCTION

Filled epoxy systems are used extensively for potting electronics in the Department of Energy (DOE) complex in general and Sandia National Laboratories (SNL) in particular. Components, and their potting materials, are subjected to shock loading environments. Understanding the shock wave response of these potting materials is necessary both for proper data interpretation of shock events as well as computer modeling of these events. Early in the 1960's, a series of studies were done on the shock response of epoxy and epoxy filled systems[1]. These systems contained an Epon 828 resin, a Z hardener and, if appropriate, a filler.

However, the Z hardener is being removed from the commercial market due to suspected carcinogenic properties. Thus, the first aspect of the present study was to investigate the effect of a replacement curing agent on the Epon 828 resin system without a filler [2]. The curing agent used was a methyl nadic anhydride. The second part of the present study was to investigate the shock response of this anhydride-cured epoxy system filled with 42 volume percent glass microballoons. Hugoniot data were generated between the shock pressures of 0.5 and 4.2 GPa using a gas gun system [3] and a VISAR (Veocity Interferometer System for Any Reflector) [4] [5]. Shock attenuation experiments were done as a function of specimen thickness at shock pressures of 3.1 and 2.2 GPa.

II.   EXPERIMENTAL PROCEDURE

Epoxy Materials. The mix ratios for both the unfilled and GMB-filled, anhydride-cured epoxies are given in Table I[2]. The GMB filler is 27.7 percent of the system by weight or 42 percent by volume.

Projectiles/Targets for Hugoniot Measurements. Epoxy or epoxy-GMB specimens were mounted on the front of either nylon/foam or aluminum sabots. The specimens acted as flyer plates in impacts with the target materials. The target materials of either polymethyl methacrylate (Polycast), or quartz, or sapphire had known Hugoniot relationships. The targets consisted of both a thin, 1-mm, buffer plate and thicker, 12.7-mm, window of the same material. The window had a 0.1-μm thick layer of vapor deposited aluminum on the front which reflected the laser beam for the VISAR.

Projectiles/Targets for Attenuation Measurements. The projectiles consisted of either nylon/foam or aluminum sabots with a layer of carbon foam, 3.81-mm thick, glued to

TABLE I

Mix Ratios for Epoxy and GMB-Epoxy

EPOXY

ppw*	Material
50	Epon 828 resin
42	Methyl Nadic Anhydride
20	Nyax 10-25
3	Curie ATC-3

GMB-EPOXY

Mix ratio for Epoxy plus:

44	Glass Microballoons (GMB)

*ppw - part per million by weight

the front and a 0.5-mm thick disc of tantalum glued onto the carbon foam. The thin tantalum disc acted as the flyer plate imparting a short impulse of known magnitude into the target material at a given projectile velocity. The carbon foam acted as a non-structural spacer between the tantalum flyer plate and the sabot. The targets consisted of discs, 25.4-cm diameter, of epoxy or epoxy-GMB of varying thicknesses. A thin, 0.0254-mm thick, foil of aluminum was glued to the back of the target specimen to act as the laser reflector for the VISAR system. By varying the target thickness and measuring the free-surface velocity at the back of the target, a plot of velocity versus thickness is obtained.

III. HUGONIOT RESULTS

Unfilled Epoxy. Specifics of shot conditions and results for the Hugoniot tests on the unfilled, anhydride-cured epoxy are given in Table II. The predicted values for the epoxy are taken from the Hugoniot data from Marsh[1] for Epon 828 epoxy with a Z hardener. It can be seen that the Epon 828 epoxy, anhydride-cured has the same, within experimental error, Hugoniot properties as the previous Epon 828 epoxy with the Z hardener.

TABLE II

Hugoniot Test Specifications and Results for Unfilled Epoxy

Shot No.	Target	Pressure (GPa)	Project Velocity (km/s)	Particle Target Calc. (mm/μs)	Velocity Epoxy Meas. (mm/μs)	Particle Target Calc. (mm/μs)	Velocity Epoxy Meas. (mm/μs)
310	Polycast	0.79	0.446	0.227	0.222	0.219	0.224
312	Polycast	2.06	1.015	0.515	0.515	0.500	0.500
314	quartz	3.00	0.937	0.260	0.250	0.677	0.687
315	quartz	4.03	1.200	0.350	0.340	0.850	0.860
311	sapphire	5.25	1.15	0.115	0.125	1.035	1.025

TABLE III

Hugoniot Test Specifications and Results for GMB-Epoxy

Shot No.	Target Material	Projectile Velocity (km/s)	Particle Target (mm/μs)	Velocity GMB-Epoxy (mm/μs)	Pressure (GPa)
255	Polycast	0.524	0.210	0.314	0.72
256	Polycast	0.976	0.300	0.676	1.11
254	Polycast	1.353	0.465	0.888	1.82
257	quartz	0.916	0.125	0.791	1.42
259	quartz	1.206	0.190	1.106	2.18
260	quartz	1.422	0.255	1.167	2.92
262	sapphire	1.106	0.0565	1.0495	2.45
263	sapphire	1.266	0.068	1.198	3.20
288	sapphire	1.401	0.088	1.313	4.10
264	WC*	1.384	0.075	1.309	5.80
265	WC	1.443	0.080	1.363	6.20

*WC - tungsten carbide

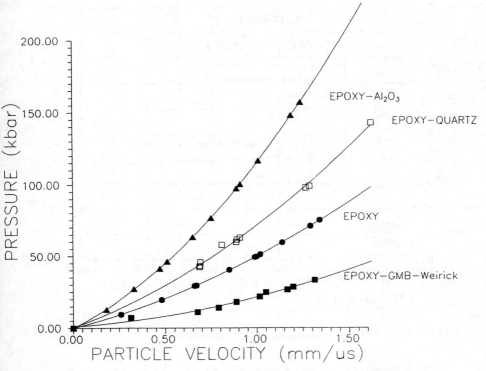

FIG. 1 *Pressure versus particle velocity plots for GMB-epoxy and other filled-epoxy systems.*

GMB-Filled Epoxy. Specifics of shot conditions and results for the Hugoniot tests on the GMB-filled, anhydride-cured epoxy are given in Table III. The particle velocity for this material is plotted as a function of pressure in Figure 1. Also shown in Figure 1 are the Hugoniot plots for epoxy-Al_2O_3, epoxy-SiO_2 and unfilled-epoxy. It can be seen that the quantitative position of the curves are related in an approximately linear manner to the initial density. Thus, the GMB-filled epoxy system behaves in a manner which would be expected of an epoxy with 42 volume percent pores.

IV.   ATTENUATION RESULTS

Unfilled Epoxy. The shock wave profiles for 12.7-mm thick samples of unfilled epoxy impacted by 0.5-mm thick tantalum flyer plates at impact pressures of 3.1 and 2.4 GPa

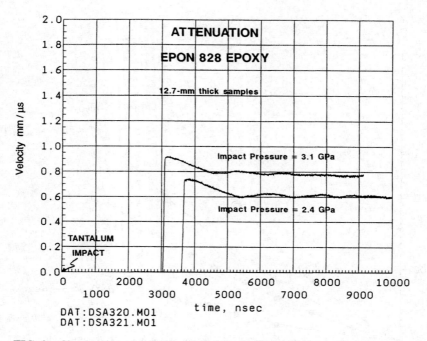

FIG. 2   *Shock wave attenuation profiles for unfilled epoxy.*

are shown in Figure 2. The shock waves attenuated from free surface velocity values at impact of 1.41 and 1.14 mm/μs to ~0.92 and 0.75 mm/μs, respectively, a 35 % decrease in amplitude. These values correspond to drops in pressure from 3.1 to 2.18 GPa and from 2.4 to 1.65 GPa, respectively, a 30 % decrease. The wavefronts were fairly planar and singletary.

GMB-Filled Epoxy. Table IV summarizes the test specifications for attenuation tests on GMB-epoxy samples. The shock wave profiles for GMB-filled epoxy specimens of thicknesses varying from 5.08 mm to 19.0 mm impacted by a 0.5-mm thick tantalum flyer plate at an impact pressure of ~3.1 GPa are shown in Figure 3. Table V summarizes the test results from these attenuation shots.

GMB-filled epoxy specimens of two different thicknesses were impacted at ~2.2 GPa. The profiles have analogous shapes to those shown for impacts at 3.1 GPa. Table V also summarizes the results from these attenuation shots.

Another very interesting aspect of these profiles is their step structure. The first shock raised the free surface velocity to a value of ~0.27 mm/μs. This value remained

TABLE IV
Test Specifications for Attenuation Shots on GMB-Epoxy

Shot No.	Projectile Velocity (km/s)	Initial Particle Velocity (mm/μs)	Initial Pressure (GPa)	GMB-Epoxy Thickness (mm)
316	1.233	1.179	3.12	5.11
317	1.242	1.187	3.16	7.68
318	1.233	1.179	3.12	10.07
290	1.227	1.174	3.09	12.66
289	1.245	1.190	3.18	19.21
319	1.014	0.977	2.15	5.09
303	1.030	0.991	2.22	19.37

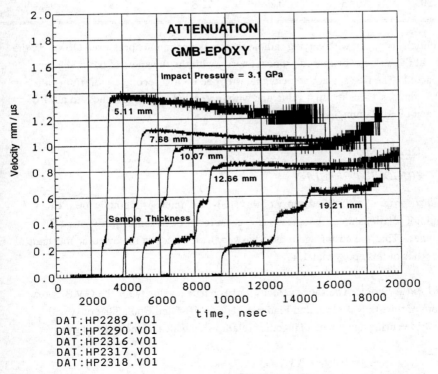

DAT:HP2289.VO1
DAT:HP2290.VO1
DAT:HP2316.VO1
DAT:HP2317.VO1
DAT:HP2318.VO1

FIG. 3 *Shock wave attenuation profiles for GMB-epoxy impacted at ~3.1 GPa.*

TABLE V

Results for Attenuation Tests on GMB-Epoxy

Shot No.	Free Surface Velocity (mm/μs)	Final Particle Velocity (mm/μs)	Percent Decrease Particle Velocity (%)	Final Pressure (GPa)	Percent Decrease Pressure (%)
316	1.38	0.69	41.5	1.09	65.1
317	1.12	0.56	52.8	0.73	76.9
318	0.98	0.49	58.4	0.56	82.1
290	0.86	0.43	63.4	0.44	85.8
289	0.64	0.32	73.1	0.25	92.1
319	1.02	0.51	47.8	0.61	71.6
303	0.55	0.27	72.8	0.19	91.4

essentially constant with varying sample thickness and impact pressure. However, the time at this velocity increased proportionately with the thickness of the specimen impacted. The free surface velocity then stepped up to a second equilibrium value of ~0.54 mm/μs, twice the first value and with increasing sample thickness to a third plateau of ~0.82 mm/μs.

V.   DISCUSSION

A.   HUGONIOT RESULTS

Unfilled epoxy. The unfilled epoxy of this study had an experimentally identical Hugoniot relationship as that previously measured for this same epoxy with a Z hardener. Thus the curing agent used has very little effect on the dynamic mechanical properties of this epoxy material.

GMB-Filled Epoxy. The shock velocity versus particle velocity for the GMB-epoxy system of this study is plotted in Figure 4. The GMB-filled Epon 828 epoxy with anhydride curing agent had a Hugoniot relationship which fit the following curve:

$$U_s = 0.088 + 2.312 \, U_p \tag{1}$$

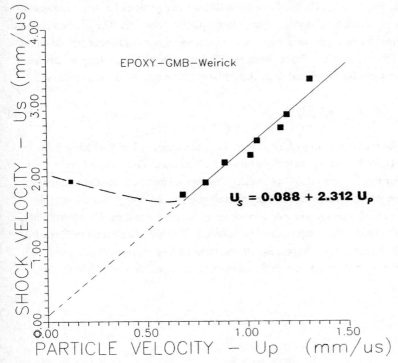

FIG. 4 *Shock velocity versus particle velocity for GMB-epoxy.*

Equation (1) is valid for the initial loading conditions, it is *not* valid as an unloading isentrope as will be discussed in the attenuation section.

In addition, the shock Hugoniot in U_s-U_p space shown in Figure 4 shows a break in the curve at a particle velocity of ~0.7 mm/μs which corresponds to a shock pressure of ~1 GPa. These values will be correlated with subsequent observations in the Discussion section of this paper.

B. *ATTENUATION RESULTS*

Unfilled epoxy. The unfilled Epon 828 epoxy exhibited wave profiles for attenuation tests which were characteristic of single-wave, plastic behavior. The profiles showed a shock jump to the attenuated free surface velocity, a decline as the sample unloaded, and then an approximately steady velocity of the spalled free surface.

GMB-filled epoxy. The GMB-filled epoxy exhibited wave profiles for attenuation tests which were characteristic of double-wave, elastic-plastic behavior. The profiles showed a shock rise to a free surface velocity plateau which was a constant value independent of sample thickness or initial shock pressure. The pressure corresponding to this free surface velocity can be calculated from the equation for momentum conservation

$$P - P_o = \rho_o U_s U_p \tag{2}$$

where P_o is the initial pressure which is zero, U_s is the shock velocity which is 1.95 mm/μs, and U_p is the particle velocity which is 0.27 mm/μs. The value of pressure is 0.24 GPa. For thin samples (2.54-mm thick), the free surface velocity then jumped to the final value which represented the attenuated pulse in the epoxy. As the sample thickness increased, a secondary free surface velocity plateau and finally a third plateau appeared, followed by the jump to the final attenuated value. According to Hayes[6], wave profiles showing these shapes, as reflected from a free surface, are classic for an elastic-plastic material with a cusp in the Hugoniot curve in P-V space, due to the

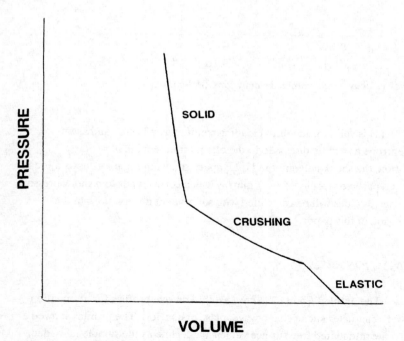

FIG. 5   *Schematic diagram of Hugoniot curve in P-V space for "strong foam".*

presence of a Hugoniot elastic limit (HEL). However, unfilled epoxy does not exhibit this behavior either with respect to the attenuation profiles or the Hugoniot curve in P-V space. Therefore, this behavior is most likely due to the presence of the glass microballoons in the epoxy system. Figure 5 shows a schematic diagram of a Hugoniot curve in P-V space for a "distended" material (foam) which has a finite strength before the pores begin to collapse. According to the P-α model of Herrmann[7], the material loads by moving up the elastic section of the curve until the pores begin to crush. The pores then continue collapsing until a state of solid material is reached which is where the more vertical Hugoniot for the solid material is intersected. The data for the GMB-filled epoxy system is plotted in P-V space in Figure 6 and compared to the Hugoniot for unfilled epoxy. Although more data needs to be acquired at lower shock pressures, the general trend of this curve as compared to the theoretical schematic of Figure 5 can be seen. The GMB-filled epoxy system is analogous to Herrmann's foam system. Upon loading, the material responded in an elastic manner until the dynamic crush strength of the microballoons was reached, a value of 0.24 GPa. The material continued to load

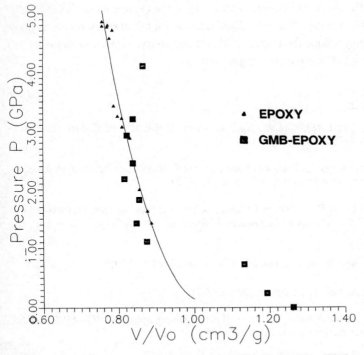

FIG. 6   *Hugoniot curves in P-V space for GMB-epoxy and unfilled epoxy.*

slowly as the microballoons continued to crush. When all the microballoons had crushed (at ~1 GPa), the material reached a solid state and the unfilled epoxy Hugoniot was followed. A final point, the dynamic crush strength of the glass microballoons in this study (3M product no. S 60/10000) was found to be 0.24 GPa or ~35 Ksi. This compares to the "static" crush strength which was given as 10 Ksi.

VI. CONCLUSIONS

The Hugoniot for an unfilled Epon 828 epoxy cured with methyl nadic anhydride exhibited the same material response as a previously developed epoxy with a different curing agent. The Hugoniot relationship in P-U_p space for the GMB-filled epoxy fit the family of curves previously determined for other filled epoxy systems relative to their initial densities Thus, the GMB-filled epoxy was analogous to an epoxy with 42 volume percent pores with an initial density of ~0.94 g/cm³. The Hugoniot relationship in P-V space follows Herrmann's P-α model for foams and shows a dynamic crush strength for the microballoons of 0.24 GPa. All the microballoons have collapsed by a shock wave pressure of ~1 GPa. At shock pressures above 1 GPa, the loading Hugoniot follows closely that of unfilled epoxy. The unloading isentrope for initial shock pressures above 1 GPa approximately follows the Hugoniot for unfilled epoxy. Unloading paths for initial shocks below 1 GPa are more complicated.

VIII. REFERENCES

1.   S. P. Marsh, ed., *LASL Shock Hugoniot Data,* University of California Press, Berkeley, CA (1980).

2.   K.B. Wischmann. Sandia National Laboratories Process Specification SS 390353. Sandia National Laboratories, Albuquerque, N.M.

3.   S. A. Sheffield and D. W. Dugan, "Description of a New 63-mm Diameter Gas Gun Facility," *Shock Waves in Condensed Matter,* ed. Y. M. Gupta, Plenum Press (1986).

4.   L. M. Barker and R. E. Hollenbach, *J. Appl. Phys. 43*:11 (1972).

5.   W. F. Hemsing, *Rev. Sci. Instrum. 50*:1 (1979).

6.   D. B. Hayes, *Introduction to Stress Wave Phenomena,* SNL-73-0801, Sandia National Laboratories, Albuquerque, N. M.

7.   W. Herrmann, *J. Appl. Phys. 40*, 2490 (1969).

88

Dynamic Yield Strength and Spall Strength Measurements Under Quasi-Isentropic Loading

L. C. CHHABILDAS and J. R. ASAY

Sandia National Laboratories
Albuquerque, New Mexico 87185, U. S. A.

In this paper, measurements on the quasi-isentropic compression of tantalum and tungsten to stress levels of 60 GPa and 250 GPa, respectively, are reported. Results of these experiments have been compared to those obtained under shock loading conditions. These experiments have allowed the determination of temperature, pressure, and loading rate effects on the dynamic yield strength of tungsten and tantalum up to 250 GPa and 80 GPa, respectively. The results show that the dynamic yield strength of tungsten is dependent on the loading rate with the strength being higher for the relatively slower rates of loading along the quasi-isentrope. The pressure dependence of the yield strength of tungsten is determined nearly independent of temperature effects from quasi-isentropic loading experiments to 250 GPa, because the temperature rise in an quasi-isentropic loading experiment is much lower than that associated with shock loading experiments. For tantalum, quasi-isentropic loading experiments up to 60 GPa suggest that the dynamic yield strength is comparable to that determined under shock loading conditions up to 50 GPa. There is a significant decrease in the dynamic yield strength for tantalum above 50 GPa under shock loading. A determination of spall strength for tantalum precompressed to 60 GPa under quasi-isentropic loading suggests an increase in spall strength when compared to the strength measurements in a shocked specimen at 20 GPa.

*This work performed at Sandia National Laboratories supported by the U.S. DOE under contract DE-AC04-76DP00789.

I.   INTRODUCTION

A shock Hugoniot is the locus of end states arrived at from an initial state, under the influence of a single shock wave. An isentrope, however, represents a continuous sequence of thermodynamic states that a material undergoes during compression. Under shock loading to a stress of interest, the strain rates induced in the shock front are controlled primarily by the material viscosity[1-2], whereas under plane quasi-isentropic loading conditions to the same stress the loading rates can be varied. Under shock loading conditions, the internal energy change of the material is that of the area under the Rayleigh line in the pressure-volume plane, whereas under plane quasi-isentropic compression the internal energy change of the material is given by the area under the isentropic compression curve. Thus, for large compressions, the internal energy change of a material under quasi-isentropic compression will be substantially less than that for shock loading to the same stress, resulting in energy (temperature) states that are lower than those obtained on the shock Hugoniot. A comparison of shock loading and quasi-isentropic loading experiments to the same stress will, therefore, allow a determination of pressure, temperature, and rate-dependent properties of materials.

II.   EXPERIMENTAL TECHNIQUE

This section briefly describes the experimental techniques employed to determine dynamic material properties. Compressive yield strength and spall strength of candidate plate materials are determined after being subjected to both shock and quasi-isentropic high pressures. These experiments were performed on propellant and two-stage light-gas guns [3]. The impact configuration used to obtain shock loading/release and quasi-isentropic loading is indicated in Fig. 1. By varying the impact assembly used on the projectile nose the stress history obtained at the impact interface can be controlled. The particle velocity histories are then monitored at the specimen/lithium-fluoride window interface [4] using velocity interferometric techniques [5,6].

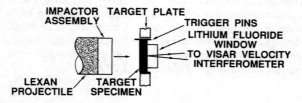

FIG. 1 *Experimental configuration used for impact studies.*

A. Quasi-isentropic Loading Measurements

Quasi-isentropic loading was introduced in the target material either by using graded-density or layered-density impactors. An example of a graded-density material [7] is a "pillow", which is fabricated using powder metallurgical and sedimentation techniques so that a smooth variation in its shock-impedance occurs through its thickness. The shock impedance of the impact surface of the graded density material is that of polyolefin, and the shock impedance of the back surface of the pillow resembles copper. The layered-density impactors [8,9] consist of PMMA/aluminum/titanium/copper. For both material assemblies, the variation in shock impedance through the impactor thickness gives rise to a finite rate of loading at the impact interface.

Measurements of interface particle velocity profiles (Figs. 2 & 3) at two different sample locations are needed to allow the use of Lagrangian wave analysis techniques [8,9] in determining the stress-volume loading path under quasi-isentropic loading. In effect, the pulse measured for the thin sample serves as an input for the extra thickness of the thicker specimen. Impedance matching techniques are used to determine the in-material particle velocity from interface particle velocity profiles. The equations of motion describing the conservation of mass and momentum are then used to determine the stress-volume loading path. A comparison of the experimentally determined quasi-isentrope σ_i to that of a calculated pressure isentrope P_i provides the dynamic yield strength Y_i under quasi-isentropic loading [8], assuming that the material behaves like a von Mises solid,

$$\sigma_i = P_i + 4\tau_i/3, \tag{1}$$

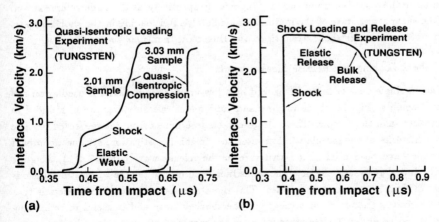

FIG. 2 *Interface velocity measurements in tungsten (a) at two locations in a quasi-isentropic loading experiment at 170 GPa and (b) in a shock loading and release experiment at 200 GPa.*

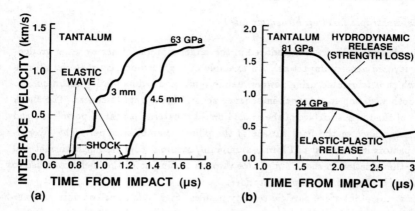

FIG. 3 *interface velocity measurements in tantalum (a) at two locations in a quasi-isentropic loading experiment at 60 GPa and (b) in shock loading and release experiments at peak shock stresses of 32 and 81 GPa respectively at 200 GPa.*

where τ_i is the shear strength ($Y_i = 2\tau_i$) of the material under quasi-isentropic loading. The pressure isentrope is calculated using the shock Hugoniot σ_h as the reference and further assuming a Mie-Grueneisen solid behavior with $\rho\gamma$ being constant,

$$P_i = (\sigma_h - 4\tau_o/3) + \rho\gamma(E_i - E_h). \tag{2}$$

In this relation γ is the Grueneisen parameter, and E_i and E_h are the energy states corresponding to the quasi-isentrope and the Hugoniot, respectively, at the same volume strain. τ_o is the shear stress state of the material immediately after loading in the shocked state. If the material loses its shear strength on the Hugoniot, then τ_o is zero.

B.   Shock Loading and Release Measurements

When a single density material backed by a low-impedance material is used as an impactor assembly (Fig. 1), shock loading and subsequent release states are introduced in the impactor and the sample. These states can be determined from the measured particle velocity histories at the sample-window interface. In this investigation, symmetric impact conditions have been used. Fine structure in the release wave (Figs. 2 & 3) suggests that the initial release is elastic. The transition from elastic release to plastic release is usually evident in release wave profiles. This signature is characteristic of materials that exhibit a finite dynamic shear strength in the shocked state and is useful in estimating the compressive dynamic yield strength Y_h of the material [10] using the relation

$$Y_h = \tau_o + \tau_c = -(3/4)\rho_o \int_{eh}^{ep}(c_l^2 - c_b^2)de, \tag{3}$$

where ρ_o is the density of the material, τ_o is the shear stress of the material in the shocked state, and τ_c is the critical shear strength after shock loading. The velocities c_l and c_b are, respectively, the measured Lagrangian release wave speed and the calculated bulk wave speed estimated from the plastic part of the wave profile at the same strain. The integration is performed over the peak shock strain states e_h to e_p, the plastic strain state to which elastic-plastic effects are apparent.

C. Spall Measurements

The impact configuration shown in Fig. 1, without the window, is used to determine the spall strength of materials that have been pre-compressed by either quasi-isentropic or shock loading. Spall strength measurements in a material pre-compressed by a shock (or a quasi-isentropic wave) is accomplished by using a single density (or a graded density) material, backed by PMMA as an impactor assembly. This introduces a shock (or a quasi-isentropic wave), followed by a rarefaction wave into the target. The interaction of rarefaction waves emanating from the free surfaces of both the target and the impactor, creates tension states in the target material. The target specimen thickness is chosen so that the resulting tension states generate a spall plane near the center of the target. A velocity interferometer measures the free-surface "pull-back" velocity, a signature which is characteristic of a material going into tension. Free-surface "pull-back" velocity measurements on tantalum either shocked initially to 19 GPa or quasi-isentropically precompressed to ~ 60 GPa are shown in Fig 4.

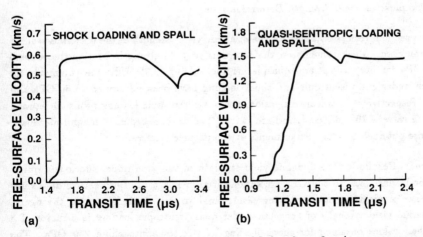

FIG. 4 *Free-surface velocity and pull back velocity measurements in (a) shocked tantalum at 19 GPa and (b) quasi-isentropically loaded tantalum at ~ 60 GPa.*

An estimate for the spall strength S_m from the measured free-surface "pull-back" velocity Δu_{pb} is obtained by using the relation

$$S_m = \frac{1}{2}\rho_o c_l \Delta u_{pb}, \tag{4}$$

where c_l is the elastic wave velocity of the material. As the tensile wave traverses towards the target free-surface, the peak amplitude of the tensile wave (propagating at a bulk wave velocity c_b) is partially attenuated by the elastic wave (travelling faster at an elastic wave velocity c_l) generated at the spall surface [11,12]. This correction ΔS is related to the thickness of the spall plate h [11,12], and is given by

$$\Delta S = \frac{h}{2}\left(\frac{dp}{dt}\right)\left(\frac{1}{c_b} - \frac{1}{c_l}\right), \tag{5}$$

where (dp/dt) is the rate at which the material is led into tension. In this study, (dp/dt) has been approximated by (S_m/t), where t is the duration of the pull-back signal. This correction, when added to Equation 4, yields an estimate for the spall strength S of the material at the spall plane, namely

$$S = S_m + \Delta S. \tag{6}$$

III.   RESULTS AND DISCUSSION

A.   *Compressive Yield Strength Determinations*

In this section, results obtained for the dynamic yield strength of the materials studied are summarized. Measurements on the compressive yield strength for tungsten extend to 250 GPa for both shock and quasi-isentropic loading [8,9], while measurements for tantalum under quasi-isentropic and shock loading have been performed to 60 GPa and 81 GPa, respectively. In this investigation, the quasi-isentropic loading paths correspond to loading rates of 10^5-10^6 sec^{-1}, and are at least two to three orders of magnitude slower than those obtained in shock experiments at comparable stresses.

Tungsten: Results for the dynamic yield strength of tungsten under shock and quasi-isentropic loading [8,9] to 250 GPa are summarized in Fig. 5(a). Both sets of data show an increase in yield strength with increasing final stress. As indicated in the figure, the dynamic yield strength of tungsten under quasi-isentropic loading is a factor of 3 larger than values obtained for shock loading at stresses approaching 250 GPa. The temperature of tungsten shocked to 170 and 250 GPa is calculated to be 2500 K and 5000 K, respectively, whereas under quasi-isentropic compression to 250 GPa it is less

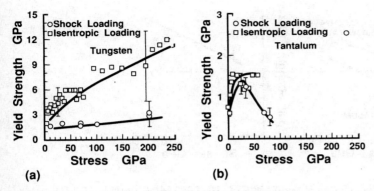

FIG. 5 *Yield strength measurements for tungsten and tantalum under both shock and quasi-isentropic loading.*

than 1000 K. In quasi-loading experiments, the temperature of the material is much lower than that under shock loading. Therefore, the pressure dependence of the yield strength is essentially determined in quasi-isentropic compression experiments, while the temperature dependence of the yield strength is a major effect in shock loading experiments.

Tantalum: The dynamic yield strength for tantalum under shock and quasi-isentropic loading is indicated in Fig. 5(b). The present measurements extend up to 81 GPa for tantalum under shock loading conditions, whereas isentropic compression is limited to 60 GPa. Within the experimental uncertainty, the dynamic yield strength for tantalum is similar under both shock and quasi-isentropic compression over the regime where the strength measurements overlap. This is not surprising, since the temperature rise associated with different rates of stress loading is not appreciably different at these low stresses, and therefore the effect of temperature on yield strength is not observed. Initially, both measurements show an increase from the ambient value of 0.75 GPa. Above 50 GPa, yield strength measurements for tantalum under shock loading, however, indicate a loss in strength. This is not yet understood, but wave speed measurements also support a decrease in shear modulus over this stress regime. Apparently, the material recovers its dynamic yield strength, as is evidenced by a strength measurement of 2.5 GPa at a shock stress of 230 GPa [13]. Additional experiments, both under shock and quasi-isentropic compression are necessary to determine the dynamic yield strength over the range of 60 to 230 GPa to further understand this anomalous behavior.

B. Tensile Strength Determinations

Pull-back velocity measurements indicated in Fig. 4 were used to deduce the spall strength of tantalum precompressed to 19 GPa and 60 GPa under shock and quasi-

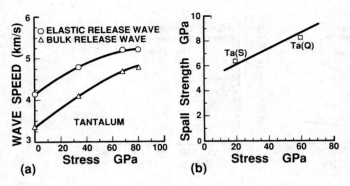

FIG. 6 *(a) Wave speed measurements in tantalum under shock loading up to 81 GPa. (b) Increase in spall strength for tantalum at slower rates of quasi-isentropic loading.*

isentropic loading, respectively. The experiment on tantalum precompressed to 60 GPa quasi-isentropically does indicate an increase in spall strength to 8.1 GPa, which is an increase of 27% when compared to a value of $\sim$ 6.2 GPa after shock compression to 19 GPa. This is indicated in Fig. 6. Although not clearly evident, this increase may be a result of slightly lower temperatures induced as a consequence of quasi-isentropic loading. A lesser concentration of defects (and therefore a lower concentration of nucleation centers for spall initiation) created as a result of slower rates of loading under quasi-isentropic precompression may also have contributed to an increase in its spall strength. Further study is required to determine the physical proceeses responsible for this observed increase.

REFERENCES

1. L. C. Chhabildas and J. R. Asay, *J. Appl. Phys., 50:* 2749 (1979).

2. D. E. Grady, *Appl. Phys. Letter, 38:* 825 (1981).

3. J. R. Asay, L. C. Chhabildas, and L. M. Barker, *Sandia Laboratories Report # SAND85-2009* (1985; unpublished).

4. J. L. Wise and L. C. Chhabildas, *Shock Waves in Condensed Matter - 1985*, Plenum Publishers, New York, 1986.

5. L. M. Barker and R. E. Hollenbach, *J. Appl. Phys., 43:* 4669 (1972).

6. L. C. Chhabildas and R. A. Graham, *AMD Vol. 83*, ASME Publishers, New York, 1987.

7. L. M. Barker, *Shock Waves in Condensed Matter - 1983* , Elsevier Science Publishers B. V., North Holland, 1984.

8. L. C. Chhabildas and L. M. Barker, *Shock Waves in Condensed Matter - 1987*, Elsevier Science Publishers B. V., North Holland, 1988.

9. L. C. Chhabildas, J. R. Asay, and L. M. Barker, *Sandia National Laboratories Report # SAND88-0306*, (1988; unpublished).

10. J. R. Asay and L. C. Chhabildas, *Shock Waves and High-Strain-Rate Phenomena in Metals*, Plenum Publishers, New York 1981.

11. V. I. Romanchenko and G. V. Stepanov, *Zhur. Prik. Mekh. Tekh. Fia.*, *4*: 141 (1980).

12. L. C. Chhabildas, L. M. Barker, J. R. Asay, T. G. Trucano, *Shock Waves in Condensed Matter - 1989*, Elsevier Science Publishers B. V., North Holland, 1990.

13. D. E. Grady, Sandia National Laboratories, Private Communication.

89

Energetics of Nanodefect Structures in Shocked Crystals

F.A. BANDAK, D.H. TSAI and R.W. ARMSTRONG

Research Department
Naval Surface Warfare Center
White Oak Laboratory
Silver Spring, Maryland 20903, U.S.A.

Nat. Inst. of Standards and Technology (Ret.)
10400 Lloyd Road
Potomac, Maryland 20854, U.S.A.

Department of Mechanical Engineering
University of Maryland
College Park, Maryland 20742, U.S.A.

The evolution of a dislocation defect structure from a vacancy cluster subjected to sudden shock compression has been evaluated for a body centered cubic lattice using molecular dynamics. The energetics associated with the transformation process were evaluated. Both $<100>$ and $<111>$ Burgers vector dislocations have been identified and related to a previous model of dislocations proposed to be generated within a shock front. A stress/strain description of the behavior of the crystal has been obtained for pre- and post-shock conditions.

I. INTRODUCTION

Under intense shock loading, a crystalline solid containing defects may undergo structural relaxation which leads to the creation of dislocation structures behind the shock front. Dislocation mechanics models have been proposed in an attempt to describe this process [1-3]. They are based on the idea that the shear stress state occurring within the shock profile will generate large numbers of dislocations [4]. Although the models are ideal in nature, the dislocation numbers they predict give realistic dislocation density values [2]. The distinction between the dislocation density levels occurring during the shock process and those remaining afterwards is made by the model reported by Armstrong (AMS model) [5]. This is based on the proposition that dislocations, emanating from the many points of high shear in the shock profile, react to produce prismatic edge dipoles with Burgers vectors parallel to the shock wave propagation direction (Fig. 1). They are therefore likely to survive unloading forces having become aligned in a direction of low shear stress.

The relation between residual dislocations and work hardening is well established and is a basis for many workhardening theories. According to the Kuhlmann-Wilsdorf mesh length theory [6], workhardening is dependent on the square root of the dislocation density and on the geometric arrangement of these dislocations. A parameter is introduced in the theory to account for this unknown arrangement. The AMS model has been used in an attempt to calculate shock strengthening effects by providing estimates of the dislocation density values along with physically based information on the geometry of the dislocation arrangement [5]. Results from these calculations showed reasonable agreement with strength tests performed on shocked single and polycrystalline nickel [7].

Important issues regarding the transient nature of the events, presented in "snap shots" by these models, remain to be studied under dynamical conditions. The presence of defects in the lattice during shock compression is one such issue. Mogilevsky [8] investigated the influence of the size of substitutional atoms on the development of shear in crystals under gradual compression. Tsai's detonation work on "hot spot" generation [9] gave evidence that vacancy clusters under shock loading collapse into structures containing dislocations.

Here we report initial results of a molecular dynamics study on the effect vacancy clusters have on the production of dislocations in a shocked solid and the consequent strength of that solid after unloading. We also show an observed process by which a vacancy cluster transforms into a structure containing dislocations.

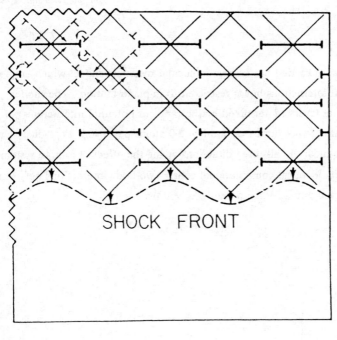

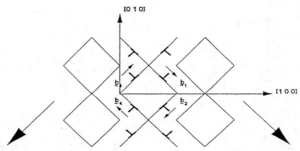

$$\mathbf{b}_3 + \mathbf{b}_1 = \frac{a}{2}[1\,1\,0] + \frac{a}{2}[\bar{1}\,1\,0] = a[0\,1\,0]$$

$$\mathbf{b}_1 + \mathbf{b}_2 = \frac{a}{2}[1\,\bar{1}\,0] + \frac{a}{2}[\bar{1}\,\bar{1}\,0] = a[0\,\bar{1}\,0]$$

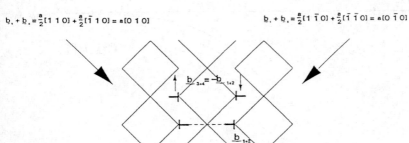

FIG. 1 *Prismatic dipole formation in a crystal under [0Ī0] compression.*

II. METHOD

Simulations were carried out using a molecular dynamics model which employs three dimensional interactions but is restricted to in-plane motion. The model consisted of several layers (determined by the range of potential interactions) of {001} planes of a monatomic bcc lattice containing 200 atoms per layer. Periodicity was invoked at the boundary in all three directions giving the effect of a large system comprised of many identical computational cells. Two-body interactions between atoms were assumed and a Morse potential, given by:

$$\phi(r) = e\left\{e^{-2a(r-r_1)} - 2e^{-a(r-r_1)}\right\}$$

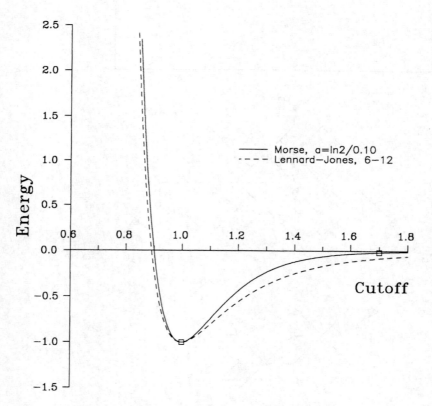

FIG. 2 *Morse potential vs. 6-12 Lennard-Jones potential.*

and shown in comparison with a Lennard-Jones potential in Fig. 2, was used for this purpose. Here $r = r_{ij} = |r_i - r_j|$, r_1 is the equilibrium distance, $a = \ln 2/0.1$, and e is the well depth. The system was assumed to be classical and the positions of the atoms were obtained by solution of the equations

$$m_i \frac{d^2 r_i}{dt^2} = \sum_{j=1}^{n} \frac{\partial \phi(r_{ij})}{\partial r_{ij}} \quad i = 1, ..., N$$

where n is the number of neighbors and N is the number of particles in the system. Units of mass, distance and energy were normalized to unity. Numerical solution of the equations of motion was performed using a multi-step algorithm developed by Beeman [10] and Schofield [11]. An integration time step of 0.002 was used and typical cases took 8000-14000 time steps during which energy conservation was satisfied to a few parts in 10^{-5}. Initial temperature was simulated by assigning a Maxwell-Boltzmann distribution of velocities to all the atoms. Heating and cooling of the system were performed by scaling the velocity components by a common factor. Calculation of the stresses was performed by summing the force components and the momentum flux contributions across an area, and averaging over that area [12].

III. VACANCY CLUSTER UNDER COMPRESSION

A vacancy cluster was formed in the lattice by removing ten atoms from their sites. The system was allowed to relax and equilibrate at an initially assigned temperature of 0.04 and at a constant volume. After relaxation the volume of the system was adjusted until a minimum energy configuration at zero temperature was found. The system was then heated incrementally to a temperature of 0.12 which was still low enough for the vacancy cluster to remain stable. The atomic configuration and the average potential energy distribution are shown in Fig. 3a. One of the equilibrated configurations at this temperature was used as the initial condition for subsequent cases.

In the first case, the crystal was compressed incrementally in six steps to 5.3% strain in the [$\bar{1}$00] direction. At each step, the system was allowed to equilibrate and the average potential energy, temperature, and the atomic configuration were obtained.

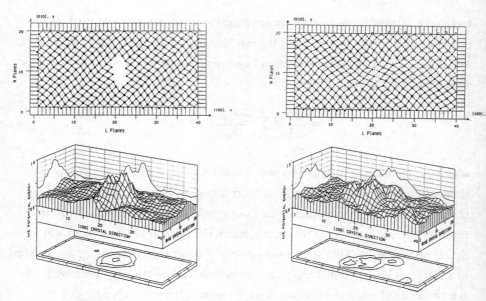

FIG. 3 *Equilibrated states for the vacancy cluster before and after incremental compression.*

The system was then unloaded following the same procedure as the loading until a minimum energy configuration was found. The final configuration of the unloaded state is shown in Fig. 3b and the loading and unloading paths are shown in Fig. 4.

Case two was started from a state with different initial conditions. Here the vacancy cluster was compressed in a single step to 5.3% strain in the [$\bar{1}$00] direction, and then unloaded. Case three involved a 10% compression of the vacancy cluster in the [0$\bar{1}$0] direction followed by unloading. The fourth and last case was a single step 15% compression in the [$\bar{1}$00] direction starting with the same initial conditions as the first case. The initial and final configurations of the first three cases were brought to zero temperature and statically tested at different volumes in the reversible region. This was done to provide constant temperature data for the calculations of the elastic constants.

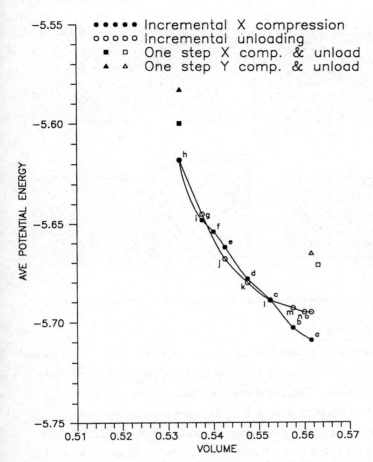

FIG. 4 *Average potential energy per atom for the various compressions.*

IV. RESULTS

A. *ENERGETICS OF THE TRANSITION FROM A VACANCY CLUSTER TO A DISLOCATION STRUCTURE*

The system average potential energy states were determined as a function of volume as shown in Fig. 4. The labelled points (A,B,C,...,N,O) show the incremental loading and unloading paths for case one. An abrupt change in the slope of the loading path curve occurred just past point C thus giving an indication that the beginning of an

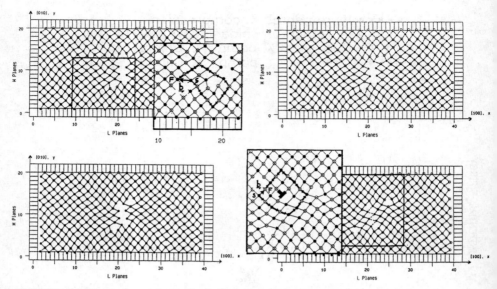

FIG. 5 *Sequence of atomic configurations for the collapse of the vacancy cluster.*

irreversible yield-like structural transition had begun. Unloading from point C showed that the process up to that point was reversible in the sense that the vacancy cluster reverted to its original shape. Upon loading beyond point C, however, the ten vacancy defect structure began to collapse towards a new structure containing dislocations. The beginning of this transformation process, shown in Fig. 5, will be discussed in detail in the next section. Beyond point C, the lattice continued to show readjustment until a fully dislocated structure was obtained at point G. At this point of compression, the vacancy disk had fully collapsed into a microstructure containing a pair of ±[100] and ±[111] dislocations as confirmed by the Burgers circuit definition shown in Fig. 5.

The other cases, indicated in fig. 4, were all one-step compressions starting from the initial state A or near to it. For these cases the filled points represent the compressed states and the open points are for unloaded states. The same dislocation microstructure was observed for each [$\bar{1}$00] compression. The case of compression in the [0$\bar{1}$0] direction has been investigated also to show that the vacancy cluster is resistant to collapse even to a compression of 5.3%.

B. STRESS/STRAIN CALCULATIONS

The stress, σ_{11}, was calculated directly from the atomic forces using the molecular dynamical stress calculation method described by Tsai [12]. The corresponding strain, ε_{11}, defined as:

$$\varepsilon_{11} = \ln\frac{V}{V_0}$$

where V and V_0 are the current and initial volumes, is shown for each stress point in Fig. 6. The labelled points in Fig. 6 correspond to those shown for incremental

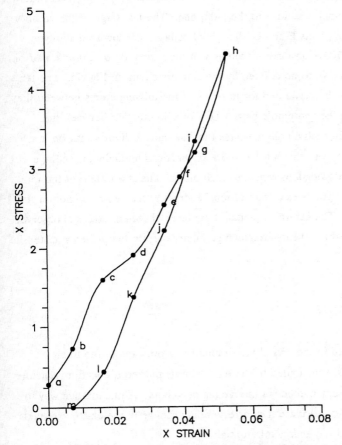

FIG. 6 *Uniaxial stress vs. uniaxial strain for the incremental compression of the vacancy cluster.*

compression steps in Fig. 4. The structural transition described at point C in Fig. 4 is manifested in Fig. 6 as a "yield" phenomenon directly producing the first [$\bar{1}$00] dislocation that is shown in Fig. 5a. The Burgers circuit definition employed in the figure, is denoted as SF/RH, that is, a start-to-finish (SF) closure vector for a right hand (RH) circuit, taking the dislocation line vector in this case as [001]. The dislocation has been formed in the first stage of the vacancy cluster collapse, as revealed in the enlarged portion of Fig. 5a, by the shear-type displacement of the pair of atoms at the core of the dislocation. Fig. 5b shows the second stage of the vacancy collapse which, in a similar fashion, produces a second, in this case [100], dislocation in a sheared dipole geometry relative to the [$\bar{1}$00] one. The last stage of the vacancy collapse occurs between points F and G. The final collapse has involved slip-type displacement of the <100> dislocations along with the formation of a dipole pair of <111> dislocations, shown to be defined by the Burgers circuit in Fig. 5d. The ten collapsed vacancies may be accounted for in terms of the missing atoms between the dislocations comprising the two dipole pairs. The force interaction between the individual dislocations in each dipole indicates that the pair of dipoles may be in a metastable configuration with the outer <100> dislocations mutually attracting each other and the <111> dislocations repelling each other. The strain field of the total dislocation arrangement produces a state of tensile stress at the central region of the collapsed vacancy disk. The central coplanar (001) layer of (filled) atoms is observed in Fig. 5d to be arranged in a near-hexagonal geometry despite the presence of tensile strains.

C. ELASTIC CONSTANTS

A stiffening was observed in the crystal after unloading from each of the final compression cases. This is attributed to the more closely packed dislocation structure which increases the average number of interacting neighbors. A quantitative way to investigate this effect is to calculate the corresponding change in the elastic constants. To do this we first recall the energy differential:

$$dU = dW + TdS$$

Here dU is the internal energy per unit mass, $dW = \sigma_{ij}\,d\varepsilon_{ij}/\rho$ is the work done per unit mass, ρ is the density, σ_{ij} is the stress tensor, ε_{ij} is the strain tensor, dS is the entropy per unit mass and T is the temperature. This equation combined with the Helmholtz free energy (per unit mass) defined as $\Psi = U - TS$ gives:

$$d\Psi = \frac{1}{\rho}\sigma_{ij}d\varepsilon_{ij} - SdT$$

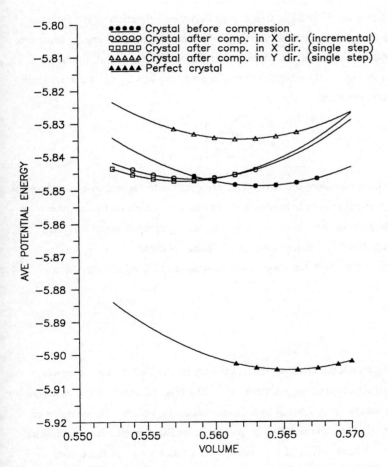

FIG. 7 *Potential energy vs. volume at absolute zero in the elastic region.*

Now from the total differential:

$$d\Psi = \frac{\partial \Psi}{\partial \varepsilon_{ij}}\bigg)_T d\varepsilon_{ij} + \frac{\partial \Psi}{\partial T}\bigg)_{\varepsilon_{ij}} dT$$

we can define as the stress, for an isothermal reversible process:

$$\sigma_{ij} = \rho \left(\frac{\partial \Psi}{\partial \varepsilon_{ij}}\right)_T$$

At zero temperature, the Helmholtz free energy becomes a function of the internal energy only and so equals the potential energy. Isothermal potential energy functions for the initial and final states of the crystal at zero temperatures and in the linear range are shown in Fig. 7. By substitution of Hooke's law, $\sigma_{ij} = c_{ijkl}\varepsilon_{kl}$, it can be seen that the elastic constants are:

$$c_{ijkl} = \rho \left(\frac{\partial^2 \Psi}{\partial \varepsilon_{kl}\partial \varepsilon_{ij}}\right)_T$$

Here, as in the stress equation, the derivatives are taken with respect to each strain component while all others are held constant. In this way the elasticity tensor can be constructed. The purpose here is to consider only the c_{11} compressional modulus component of the elastic constant tensor. An increase of about 8% in c_{11} was calculated for the cases where the shock compression was in the [$\bar{1}00$] direction.

V. SUMMARY

Molecular dynamics calculations show that a vacancy cluster under shock compression collapses into a structure containing dislocations. The level of detail of the calculation reveals atomistic mechanisms associated with dislocation formation. Dislocations of type <100> and <111>, as identified by their specific Burgers circuit vectors, were observed to form. A comparison of the pre- and post-shock states of the crystal showed an 8% increase in the c_{11} compressional modulus.

VI. ACKNOWLEDGMENTS

Appreciation and thanks are extended to Professor A. S. Douglas for helpful discussions, to Messrs L. Brown and R. Thrun for support in producing the graphics, and to Messrs D.E. Phillips, K.W. Reed and R.A. Kavetsky of the Naval Surface Warfare Center for encouraging this work that is intended to be submitted by F. A. Bandak in partial fulfillment of the Ph.D. thesis requirement at the Johns Hopkins University.

VII. REFERENCES

1. C. S. Smith, *Trans. TMS-AIME, 212*: 574 (1958).

2. M. A. Meyers, *Scripta Metallurgica, 12*: 21-26 (1978).

3. R. W. Armstrong, R. S. Miller and H. W. Sandusky, In R. W. Armstrong, "Indentation Hardness, Defect Structure and Shock Model for RDX Explosive Crystals", *ONR Workshop on Dynamic Deformation, Fracture and Transient Combustion, Chestertown Md., Chemical Propulsion Information Agency (CPIA) Publication 474*: 77-89 (1987).

4. F. A. Bandak, A. S. Douglas, R. W. Armstrong, and D. H. Tsai, Dislocation Nanostructures in Shocked Crystals, Poster presented at The Second Int. Symp. on Plasticity and Its applications, Tsu, Japan, (1989).

5. F. A. Bandak, R. W. Armstrong and A. S. Douglas, in preparation.

6. D. Kuhlmann-Wilsdorf, *Material Science and Engineering, A113*: 1-41 (1989).

7. P. S. Follansbee and G. T. Gray, The Response of Single Crystal and Polycrystal Nickel to Quasistatic and Shock Deformation, *The Second Int. Symp. on Plasticity and Its Applications*, Tsu, Japan (1989).

8. M. A. Mogilevsky, Journal de Physique, 49: C3-467 (1988).

9. D. H. Tsai, "Chemistry and Physics of Energetic Materials", ed. S. N. Bulusu, *NATO ASI Series Vol. 307, Proc. NATO Adv. Study Inst. on "Chemistry and Physics of Molecular Processes in Energetic Materials"*, Sicily, Italy (1989).

10. D. Beeman, *J. Computational Physics, 20*: 130-139 (1976).

11. P. Schofield, *Computer Physics Communications, 5*: 17-23 (1973).

12. D. H. Tsai, *J. Chem. Phys., 70*: 1375-1382 (1979).

90

Measurement of Residual Temperatures in Shock Loaded Cylindrical Samples of 304 Stainless Steel

K. P. STAUDHAMMER*

Los Alamos National Laboratory
Material Science and Technology Division
Los Alamos, New Mexico 87545 U.S.A.

Determination of residual temperature effects in cylindrical samples has been difficult due to the non-uniform implosive radial pressure pulse generated by the shock event. This paper discusses the technique used to experimentally determine the residual temperature rise (ΔT_r) and compare them to calculated values. In the present investigation, calorimetry on shocked samples provided the heat input from the shock event. This heat quantity was then used to calculate the residual temperatures via radial thermal conduction in the sample, from the high pressure region of the central axis, to the lower pressure outer surface. The slope of the measured dT/dt extrapolated rather well to the residual temperature values reported for 304 stainless steel.

I. INTRODUCTION

For thermal analysis, the hydrocode data profiles the regions of pressure throughout the sample. These regions can then be defined in terms of their contribution to the overall residual heat content which results from their associated temperature rise as a consequence of

*This work was performed in part at the Fraunhofer Institut fur Angewandte Materialforschung Bremen, West Germany, while on sabbatical leave as an Alexander von Humboldt US Senior Scientist Awardee.

the shock pressure experienced by that region. Thus the total heat input arising from the shock event can be treated as a contribution from each of these regions and subsequently be measured by calorimetry methods. By measuring the surface temperature as a function of time, the inner higher temperature resulting from a greater pressure state could be calculated as well as back extrapolated to the shock event. These values, to a first approximation compare rather well to those calculated by McQueen et al [1] for 304 stainless steel.

These observations are also analyzed in terms of the known shock wave geometry in the cylindrical sample and the static heat conduction states after the passage of the shock wave. The implications of the observations and analysis are thus discussed, with particular regard to the influence of overall strain in the sample.

II. EXPERIMENTAL

A. *Material* - All work was performed on 304 stainless steel (18.12% Cr, 8.60% Ni, 0.055% C, 0.20% Co, 0.14% each of Cu, Mo, W, balance Fe, all in weight percents) purchased from a commercial supplier as stress relieved rod. The as received material was annealed at 1040°C for 1 hour. The material had a grain size of 70 μm and a preshock hardness of 190 DPH.

B. *Strain Considerations* - Strain has a very profound effect on residual temperatures in a shock event. Thus, the knowledge of the strain magnitude and/or its control is essential in the understanding of residual effects. Strain effects are discussed elsewhere [2-8], and for the purposes of this paper, the lowest overall strained (ε_T) sample was used to approximate the conditions used to compare the data published by McQueen et al [1]. The lowest strain achieved for these experiments which had pressures up to 1.7 Mbars, was $\varepsilon_T = 1.9\%$. Having established this low strain base line for the conditions cited here, the residual temperature increase primarily due to the shock event can be obtained.

C. *Shock Loading Geometry and Hydrocode Pressure Calculations* - Figure 1 illustrates the cylindrical arrangement for subjecting a solid bar specimen to a continuously varying pressure along the specimen length from 12 to 170 GPA. In this arrangement, detonation of the main charge (composition C-4 explosive) begins radially at the top edge of the specimen and as the detonation wave moves radially outward and downward, the pressure is incrementally increased at a rapid rate. The key to accurately determining the pressure profile along the specimen, and the shock wave profile within the specimen, is a two-dimensional Eulerian computer code in use at Los Alamos National Laboratory. It should be noted from the shock wave profile which travels through the cylindrical specimen, that it is by comparison to a plane compressive wave a shear wave, making an angle of 53° with the specimen axis. The actual strain rate, which can be deduced from the computer plots were

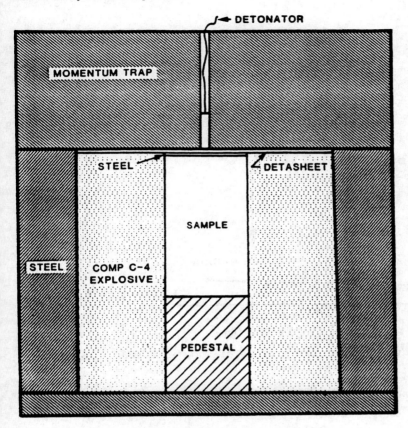

FIG. 1 *Schematic of the shock loading assembly.*

in the range of 10^6/s and the shock pulse durations were calculated to be less than 1 μs. The shock wave completely transversed the specimen in roughly 10 μs. This code can produce shock wave profiles at any point of time (and distance) from the initiation of detonation so that maximum pressure at any point along an axial reference of the specimen can be determined [3,5]. This is shown in fig.2. The pressure profile shown in this figure is only for the inside and outside diameter along the axial length of the 304 SS sample. The radial pressure distribution is shown in fig.3 at a constant axial position.

III. RESULTS and DISCUSSIONS

The residual distribution throughout the cylindrical volume is described by the hydrocode, it can be divided into smaller regions (volumes) of near constant temperature or where greatly

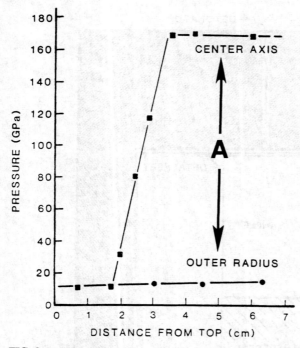

FIG. 2 *Hydrocode pressure profile of shocked 304SS.*

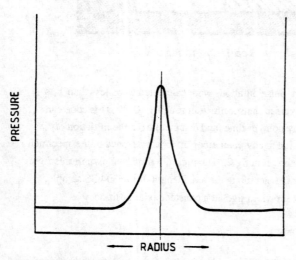

FIG. 3 *Radial pressure profile at a fixed axial length position (shown in fig. 2 marked A).*

varying temperature, the average temperature. Thus knowing the mass within each volume a total heat input can be calculated by summing all the volumes. Consequently, each shock event of given pressure which has an associated temperature results in a given amount of heat input Q. Using the adiabatic and residual temperatures as a function of pressure based on the data of McQueen et al. [1] a similar profile such as that shown in fig.3 can be obtained with temperature being substituted for the pressure axis. By considering the total heat input ie. the associated mass at each incremental temperature position, a profile of heat input can be obtained and be depicted as in fig.4. Each axial position would have an equivalent profile such as that shown in fig.4 The temperature profile radially and axially would look as that shown in fig. 5. Calorimetry and heat conduction as experimental techniques can be used because heat generation can be considered instantaneous (ie. shock event, 10^6/s) relative to the diffusion time to remove/equilibrate that heat.

A. Calorimetry   The experimental measurement of the quantity of heat deposited in a sample resulting from a shock event is a simple matter. However, because the shock event does not produce a constant pressure throughout the sample, the resulting temperature rise is not constant throughout the sample.

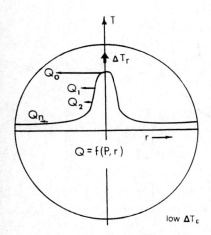

FIG. 4 *Schematic of the heat input Q ($Q_0 \cdots Q_n$) resulting from a shock pressure at a fixed axial length (fig. 2 marked A), showing the radial distribution with the maximum Q at the center (highest pressure) to the outside diameter, minimum Q (lowest pressure).*

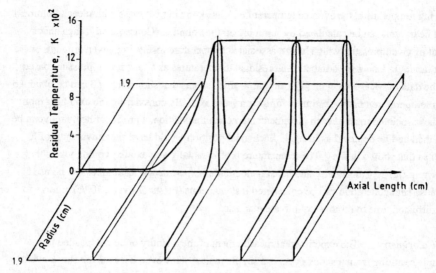

FIG. 5 *Radial and axial temperature profile after fig. 2.*

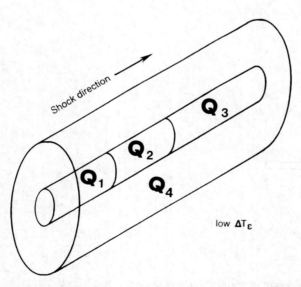

FIG. 6 *Schematic of shock sample showing regions (volumes) undergoing the pressure (average temperature) regimes depicted in fig.2 and 5, with their concomitant heat values listed as $Q_1 \cdots 4$.*

Thus, an approximation based on the hydrocode calculations must be done. This is shown in fig.7 where linear approximations are applied to regions of known pressure and subsequent temperature rises following those pressure regions as illustrated in fig. 2, and 3. Thus, the total heat content which is given by $Q = mc\,(T_2-T_1)$ is the sum of the contributing regions noted Q_1, Q_2, Q_3, Q_4 each with its own mass and associated temperature rise. For 304SS, the value of c was taken as 0.11 cal/gm. deg. and assumed not to be altered by the shock event. In dealing with the residual temperature, this is a valid assumption as it occurs after the shock event. The values for m, Q, and T_1, are measured. This allows for the calculation of T_2, which is the temperature of interest resulting from the shock event, particularly for the higher pressures (lower pressures < 15 GPa have a very small $\Delta T \sim 4$ K and can, for now, be neglected). For T_2 a value of 915 K was obtained and is within 40 % of that of the temperature calculated (1500 K) using the data of [1] at an equivalent pressure of 170 GPa.

B. Heat Conduction (1) Calculations. The heat conduction problem implicit in the cylindrically shocked 304SS samples, can to a first approximation be simplified to that of a series of cylinders and hollow cylinders each with its own mass and peak pressure profile.

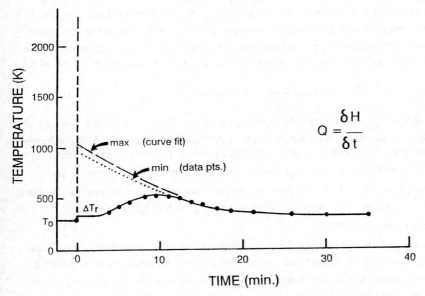

FIG. 7 *Measured temperature profile as a function of time for cylindrically shocked 304 Stainless Steel* .

This was shown in Fig.6. Inherent in this illustration are the regions of the shocked sample containing rather well defined constant or peak temperatures based on the hydrocode calculations for residual temperature (ΔT_r) arising from the pressure-shock event. Thus one can calculate the quantity of heat in each section using the data of [1]. This gives relative values to each contributing section.

(2) Experimental. To experimentally determine T_2 and T_3, heat conduction measurements have been performed on post shocked samples. Temperature measurements were made at 1 cm intervals along the outside diameter of the shocked 304 SS sample. In the high pressure region d = 5.0 cm (center of the Q_3 region), a time-temperature profile as shown in fig. 7 is obtained. The sample shown here contained a 1.9% overall strain. Illustrated in this figure is the initial temperature (T_0=297K) and the residual temperature (ΔT_r) due to the 12 GPa shock event near the outside diameter before the heat dissipation from the center due to the much higher pressure event. Temperature measurements are started 45-60 seconds after the shock event due to safety requirements. The sample temperature at this axial position remains constant for approximately 360 seconds before increasing due to radial conduction from the central position (Q_3). The temperature rises from 297K to 460K in about 780 to 840 seconds after the shock event. At this point in time the temperature exponentially decreases to 340K after 2100 seconds. The portion of the cooling curve between 900 to 2100 seconds can now be used to back extrapolate T_3, the residual temperature immediately after the passage of the high pressure shock wave. Also, T_3 can be obtained by calculating the heat conduction from Q_3 (treated as a heat source) through a 'hollow' cylindrical portion Q_4. The form of that equation was previously worked out [9-10] and used by [11] is of the form:

$$Q = 2\pi kL \, (T_{ID}\text{-}T_{OD}) \; / \; \ln(r_{OD}/r_{ID})$$

In this experiment, the effects of the boundary conditions are neglected, however, to a first approximation, this is not unreasonable. The temperature obtained by heat conduction also agree reasonable well, i.e. within 40%. Similar data at T_2 (not shown) resulted in values of T_2 = 358 K. Having obtained the values for T_2 and T_3, we can compare the calorimetry data, Q_T and back calculate the residual shock temperature for a maximum shock pressure of 170 GPa. This resulted in a value of 1023 K.and is with in 32 % of the calculated value [1] of 1500 K

The temperature obtained from heat conduction and calorimetry agree rather well, each resulting in a ΔT_r of 1023 K and 915 K respectively. However, they are low by about 50 % for those calculated using the Rankine - Hugoniot equation and the Gruneisen parameter [1]. Obviously further experiments are needed to address the conformity of the experimental and calculated values.

IV. CONCLUSIONS

The results of these experiments have revealed a technique-methodology for experimentally obtaining the residual temperature of shocked cylindrical samples, though limited here for 304SS. Calorimetry and temperature - time profiles are viable techniques for determining the residual temperatures of shock driven events on homogeneous monolithic samples. The results obtained, for the shock conditions cited, agree rather well, within 50% of the published data. The technique of curve fitting and back extrapolation of the time-temperature profile represents a powerful technique that has even greater experimental potential if used in conjunction with materials having lower specific heats, thermal conduction and higher residual temperatures resulting from the shock event.

V. ACKNOWLEDGMENTS

The author would like to acknowledge the helpful contributions of W. J. Medina, LANL for the experimental support and Franz-Josef Behler, Fraunhofer-Institut fur Angewandte Materialforschung, for the numerous constructive discussions. This work, in part, was supported by the DOE under contract W-7405-ENG-36.

VI. REFERENCES

1. R. G. McQueen, S. P. Marsh, J. W. Taylor, J. N. Fritz, and W. J. Carter, "The Equation of State of Solids from Shock-Wave Studies," in *High Velocity Impact Phenomena*, R. Kinslow, ed., Ch. 6, Academic Press, NY, 1970, p. 293.

2. K. A. Johnson and K. P. Staudhammer, "High-Strain-Rate ~10^6/s Response of 304 Stainless Steel at Various Strains" in Metallurgical Applications of Shock-Wave and High-Strain-Rate Phenomena, L. E. Murr, K. P. Staudhammer, and M. A. Meyers, eds., Ch. 27, Marcel Dekker, Inc., New York, 1986, p. 525.

3. K. P. Staudhammer and K. A. Johnson, "Technique for Megabar Controlled Strain Experiments," *International Symposium on Intense Dynamic Loading and Its Effects*, Science Press, Beijing, China, June 1986, p. 759.

4. K. P. Staudhammer, *International Conference on Impact Loading and Dynamic Behavior of Materials-Impact 1987*, Informationsgesellshaft, Verlag, 1988, p. 93.

5. K. P. Staudhammer and K. A. Johnson, Ibid., p. 839.

6. K. A. Johnson, K. P. Staudhammer, N. E. Elliott, and W. J. Medina, *Fifth APS Topical Conference: Shock Waves in Condensed Matter*, 1987.

7.   C. Albertini, A. M. Eleiche, and M. Montagnani, "Strain-Rate History Effects on the Mechanical Properties of AISI 316 Stainless Steel," in *Metallurgical Applications of Shock-Wave and High-Strain-Rate Phenomena*, L. E. Murr, K. P. Staudhammer, and M. A. Meyers, eds., Ch. 31, Marcel Dekker, Inc., New York, 1986, p. 583.

8.   M. A. Mogilevsky and L. A. Teplyakova, "Methodical Aspects of Investigation of Structural Changes Under Shock Loading," Ibid., Ch. 22, p. 419.

9.   *Thermal Engineering*, C. C. Dillio and E. P. Nye, eds., International Textbook Co., Scranton, Pennsylvania, 1963, p. 220.

10. *Conduction of Heat in Solids*, H. S. Carslaw and J. C. Jaeger, Clarendon Press, Oxford, England, 1963, p. 188.

11. P. L. Nichols, Jr., and A. G. Presson, *J. of Appl. Phys.*, 25, 12: 1469 (1954).

91

Pressure-Temperature History of Thin Films Recovered from Mbar Shock Pressures

D. J. BENSON*, W. J. NELLIS+, and J. A. MORIARTY+

*Department of Applied Mechanics and Engineering Sciences
University of California, San Diego
La Jolla, California 92093

+Lawrence Livermore National Laboratory
University of California
Livermore, California 94550

Micron-thin films of Nb embedded in a Cu capsule have recently been recovered intact from Mbar shock pressures. A two-dimensional hydrocode was used to calculate the pressure and temperature histories of the Cu capsule. A Cu equation of state calculated with first principles condensed matter physics theory was used.

I. INTRODUCTION

At 1 Mbar, Cu is compressed to 1.3 times its initial density and to 2000 K. Specimens subjected to such high dynamic pressures undergo strain rates greater than 10^8/s on loading and remain at high pressures for about 500 ns. Bulk thermodynamic quench rates during pressure release are $\sim 10^{12}$ bar/s and $\sim 10^9$ K/s. Such a specimen history is appropriate for synthesizing metastable high-pressure phases [1-3]. Micron-thin films have essentially the same pressure-temperature history as Cu because the relaxation times of the film pressure and temperature to the adjacent Cu are about 1 ns. Micron-thin films of Nb embedded in a Cu capsule have recently been recovered intact from Mbar shock pressures [3]. Since thin films of all materials embedded in Cu will have the same dynamic history, we have calibrated computationally the pressure and temperature history of Cu for a range of impact velocities.

One-dimensional calculations are sufficient to calculate early-time loading and unloading. At late times, however, the Cu capsule is extruded laterally, which results in plastic flow and higher residual temperatures than in the one-dimensional case. It is important to know the actual residual temperature in order to calculate late-time quenching by thermal diffusion from the film to the surroundings. Thus, a two-dimensional hydrocode was used to calculate the complete pressure and temperature of the Cu capsule. A Cu equation of state (EOS) calculated with first principles condensed matter physics theory was used [4].

II. THE EXPERIMENT

The thin film specimens (1 to 10 μm thick) are embedded at a depth of 3 mm in a Cu capsule that is 6 mm thick and 20 mm in diameter. The thin film is sputter-deposited on a 3 mm thick Cu substrate, protected with a sputter-deposited Cu film about 5 μm thick, and the remaining 3 mm of Cu is electroplated over the protective layer. The capsule is then machined to its final dimensions.

The recovery fixture is made from Vasco Max 250 maraging steel to minimize the deformation of the Cu capsule. It fractures during the experiment, but only after the Mbar pressures are released. A 25 mm thick Pb ring is placed on the front of the fixture to tamp the surface during the divergence of the impact shock. A stainless steel chill block, cooled with liquid N_2, supports the recovery fixture. The ~100 K initial temperature minimizes the residual shock temperature for quenching shock-induced structures.

Projectiles were accelerated to velocities of 2.7 and 3.36 km/s by a two-stage light-gas gun in the experiment described in [3]. The impact velocities correspond to shock pressures of 0.72 and .97 Mbar. The Cu impact plate is 1.5 mm thick and 20 mm in diameter, the Lexan sabot is 22.3 mm in diameter, and the total projectile length is 9.5 mm.

III. THE EQUATION OF STATE

The calculations of pressure histories are typically performed using a Gruneisen EOS. The temperature histories are calculated by assuming that the internal energy,

$e(\rho, T)$ is separable into cold compression, $e_0(\rho)$, and thermal, $e_T(\rho, T)$, terms. In most codes, such as Dyna2d [5], the thermal term is assumed to have the simple form given in Eq. (1), and the temperature is calculated by inverting the relation. The cold compression curve is calculated within the code by numerically integrating along the 0 K isotherm.

$$
\begin{aligned}
e(\rho, T) &= e_0(\rho) + e_T(\rho, T) \\
&= e_0(\rho) + CT
\end{aligned}
\tag{1}
$$

In this paper we use for Cu a first-principles EOS developed by Moriarty [4]. Since we knew that the Cu did not melt, no phase transformations are included. For units of Mbar and Ry, the Cu coefficients in Eq. (2) for P_0, e_0 and η are 4.1133, 6.6838, and 6.6685 respectively.

$$
\begin{aligned}
P(V, T) &= P_{300}(V) + P_{therm}(V, T) - P_{therm}(V, 300) \\
e(V, T) &= e_{300}(V) + e_{therm}(V, T) - e_{therm}(V, 300) \\
P_{300}(X) &= P_0[(1 - X)/X^2]\exp[\eta(1 - X)] \\
e_{300}(X) &= e_0\{[(1 - X)/\eta - 1/\eta^2]\exp[\eta(1 - X)] + 1/\eta^2\} \\
X &= (V/V_0)^{1/3}
\end{aligned}
\tag{2}
$$

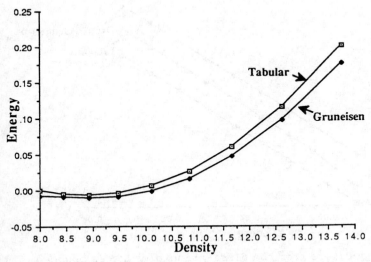

FIG. 1 *The cold compression curve of the first-principles EOS is always above the numerically evaluated curve for the Gruneisen EOS. The reference state is zero energy at 0 Mbar and 300 K.*

The functions $P_{therm}(V, T)$ and $e_{therm}(V, T)$ were provided in tabular form by Moriarty. The pressure is calculated from the current density and internal energy by inverting the internal energy function at a fixed density, $e(V(t), T)$.

The primary difference between calculations using a Gruneisen EOS with Eq. (1) and the first-principles EOS is the calculated temperature. The pressures calculated as functions of density and internal energy from the two equations of state agree within a few percent over the entire range of densities and energies encountered in the calculations. The cold compression curve for the Gruneisen EOS always lies below the first-principles EOS, see Fig. 1, with the departure increasing with density. At low temperatures, the first-principles EOS has a nonlinear relationship between the internal energy and temperature at a fixed density, see Fig. 2. The temperature from the first-principles EOS is bounded up to $\sim$ 3500 K by the temperatures calculated from substituting the numerically-evaluated Gruneisen and first-principles EOS cold compression curves into Eq. 1.

IV. CALCULATIONS

All calculations were performed using Dyna2d [5] with the first-principles EOS and the Steinberg-Guinan model for strength [6]. The two-dimensional calculations

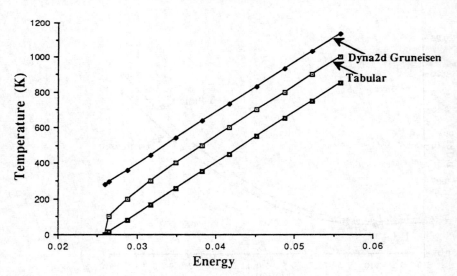

FIG. 2 *The temperature from the first-principles EOS is bounded by the temperatures calculated from Eq. 1 using the numerical and first-principles cold compression curves. The example curve is for a density of 10.837 g/cc.*

have ~6000 elements, with the Cu capsule meshed with 30 elements in the axial direction and 20 elements in the radial direction. Rezoning was performed every 0.1 μs after 2.0 μs to avoid shock diffusion within the target. The impact plate has an initial velocity of 3.36 km/s and 10 elements through the thickenss. The Lexan sabot was deleted from the calculation after it was no longer in contact with the impact plate at 4.5 μs. The calculation was carried out to 10 μs and the effects of thermal diffusion were not included.

The calculations show that the initial shock wave is uniform in the thin film specimen plane out to a radius of ~ 6 mm and that the release wave associated with the Lexan sabot causes oscillations in the plane from the axis out to ~ 3 mm. Within the band from 3 mm to 6 mm, the specimens have pressure and temperature histories that are essentially identical to the one-dimensional calculations. At an impact velocity of 3.36 km/s, plastic strains of ~.4 occur on axis, increase to .5 at a radius of 5 mm and to ~.8 at the outer radius. A cross section of the target is shown at 10 μs in Fig. 3.

A series of one-dimensional calculations, with initial temperatures of 100 and 300 K and initial velocities from 1 to 3.36 km/s, were performed to determine the ef-

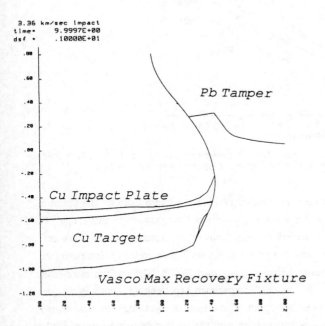

FIG. 3 *The cross section of the target is shown at 10 μs. It exhibits plastic strains from 0.4 to 0.8.*

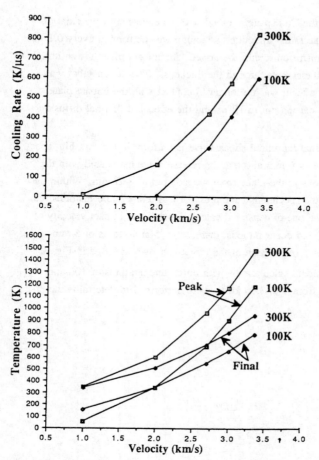

FIG. 4 *The peak temperature, final temperature, and cooling rates for 100 K and 300 K as a function of the impact velocity.*

fect of the initial temperature on the pressure and temperature histories. The pressure history was found to be largely insensitive to the initial temperature in this range. The 200 K quench is translated into a 300 K peak temperature decrease at 3.36 km/s but only a 150 K decrease in the final temperature. As the impact velocity decreases, the peak and post shock temperature differences approach 200 K. The cooling rates reflect the temperature differences, with the chilled specimen experiencing a cooling rate $\sim$ 175×10^{6} K/s less than the other specimen. The effects of the initial temperature on the peak temperature, post shock temperature, and the cooling rates are summarized in Fig. 4.

REFERENCES

1. W. J. Nellis, W. H. Gourdin, and M. B. Maple, "Shock-Induced Melting and Rapid Solidification," in *Shock Waves in Condensed Matter 1987*, S. C. Schmidt, N. C. Holmes (eds.), Elsevier Science Publishers, B. V., 1988, p. 407.

2. J. J. Neumeier, W. J. Nellis, M. B. Maple, M. S. Torikachvili, and B. C. Sales, "Superconductivity of A15-Phase Nb_3Si Synthesized by Mbar Shock Pressure," in *Shock Waves in Condensed Matter 1987*, S. C. Schmidt, N. C. Holmes (eds.), Elsevier Science Publishers, B. V., 1988, p. 447.

3. W. J. Nellis, R. Koch, H. Davidson, J. W. Hunter, W. F. Brocius, A. Marshall, and T. H. Geballe, "Micron-Thin Nb Films Recovered From Mbar Shock Pressures" in *Shock Waves in Condensed Matter 1989*, S. C. Schmidt, N. C. Holmes (eds.), Elsevier Science Publishers, B. V., 1990.

4. J. A. Moriarty, "First-Principles Equations of State for Al, Cu, and Pb in the Pressure Range 1-10 Mbar" in *Shock Waves in Condensed Matter 1987*, S. C. Schmidt, N. C. Holmes (eds.), Elsevier Science Publishers, B. V., 1988, p. 57.

5. J. O. Hallquist, *LS-DYNA2D An Explicit Two-Dimensional Hydrodynamic Finite Element Code with Interactive Rezoning and Graphical Display*, Livermore Software Technology Corporation, Livermore, 1990.

6. D. J. Steinberg and M. W. Guinan, Lawrence Livermore National Laboratory Rept. UCRL-80465, 1978.

92

Numerical Simulation of a Sample Recovery Fixture for High Velocity Impact

F. R. NORWOOD and R. A. GRAHAM

Sandia National Laboratories
Albuquerque, New Mexico 87185, U. S. A.

Numerical results are presented for a model of the sample recovery fixture developed by Sawaoka to preserve powder samples subjected to high velocity impact. These results were obtained with the hydrocode CSQ, for an axisymmetric stainless steel recovery fixture which approximates the actual fixture. Results are presented for stainless steel impactors with impact velocities of 0.9, 1.3, 1.6, 1.9 and 2.5 km/sec, corresponding to typical velocities for a series of experiments. The results obtained for an impact velocity of 0.9 km/sec are used to illustrate focusing effects in rutile powder compacts. After impact, a planar shock wave propagates through the fixture until it reaches the rutile sample. Because the shock velocity is slower in the powder compact, the wave fronts outside the sample radius propagate at a faster speed than those inside the sample. This results in a radial pressure gradient at the edge of the rutile sample. For the 0.9 km/sec impact velocity case, the radial pressure gradient generates wave fronts at the edge of the sample that are almost parallel to the axial direction. The radial effects are the dominant features and account for pressures and temperatures much higher than those predicted by a one dimensional analysis.

I. INTRODUCTION

The present report is a detailed numerical analysis of the overall shock process affecting a porous powder sample embedded within a recovery fixture. This study is needed for an understanding of the experimental conditions to which samples have been subjected during high pressure shock loading. The recovery fixture allows the sample to be preserved for post-shock analysis. The range of complex processes encountered within the sample makes it difficult to provide an interpretation of the characteristics of the recovered samples. Prior to recent numerical work, one-dimensional analysis had been presented as a description of the shock conditions. However, in earlier work [1,3], in which realistic, quantitative, two-dimensional simulations were carried out on recovery fixtures subjected to planar explosive loading or to planar impact, it was discovered that a wave-trapping phenomenon dominates the loading of a porous powder and converts the initially planar loading to a radial-mode loading in the sample. The current numerical work has allowed for a detailed analysis of the mode conversion from planar to radial for different impact velocities. It will be shown that the radial effects are dominant in all cases and therefore that a one-dimensional analysis is incorrect and leads to substantial errors.

The recovery fixture, which has been used in different versions, consists of powder compacts embedded with parallel axes within a larger cylindrical recovery fixture [2-4]. In a previous study [2], this three-dimensional geometry was analyzed and the results were compared with the corresponding ones obtained from an axisymmetric simulation. The good agreement between the 3D and the 2D results showed that, for the dimensions of the fixture given below, the sample to sample interaction and also the interaction with the outside boundary were second-order effects. Thus, the axisymmetric conditions modelled in CSQ provide a good simulation of the 3D geometry since the shock conditions in the sample are controlled by the wave-focusing effects.

The axisymmetric geometry modelled in the present work (see also Ref. 2) considered only one powder compact sample. The powder sample was located 3 mm from the impact surface, with a thickness of 5 mm and a radius of 6 mm. The impacting plate was 3.2 mm thick and had a radius of 60 mm. The initial velocity of the flyer plate ranged through the values 0.9, 1.3, 1.6, 1.9 and 2.5 km/sec. For the calculations, the sample consisted of rutile powder at a packing density of 2.6 Mg/m^3 (61% solid density) and all other parts, including the flyer plate, were of stainless steel. Rutile powder was chosen as the model material because, as has been indicated by Graham and Webb[1], rutile has been studied quite extensively [5] and therefore the numerical results may be compared with existing experimental data with confidence. It should be noted that prior work [1] indicates that the radial focusing in powders creates a condition in which peak pressures are nearly independent of the powder and its equation of state. For this reason, the peak presssure values calculated for rutile represent a good approximation to conditions for other powders.

II. RESULTS FOR 0.9 KM/SEC IMPACT VELOCITY

A careful analysis of this low velocity case will reveal the main effects occurring within the sample. Figure 1 shows pressure contours in the neighborhood of the sample for various times. The solid line represents the boundary of the sample. It is clear that, for early times, plane wave fronts propagate through the fixture. However, propagation through the rutile is substantially slower than propagation through the recovery fixture. This causes the pressure contours within the sample to lag the corresponding contours outside the sample. At t=0.8 microseconds, this effect has become more pronounced and contour A, which is horizontal away from the neighborhood of the edge of the sample at radius=0.6 cm, forms a step of about 0.15 cm in height. Clearly, since particle acceleration

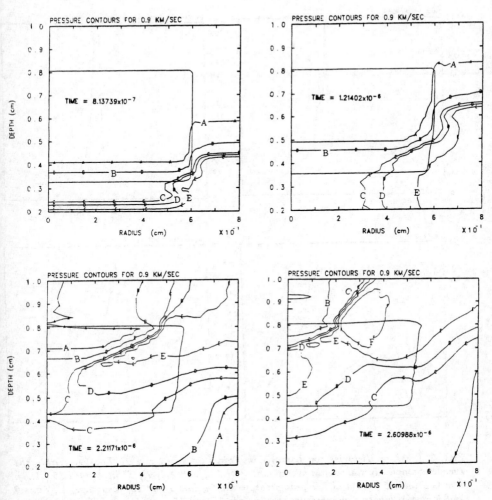

FIG. 1 *Pressure contours for impact velocity of 0.9 km/sec showing that the contours are not planar once the powder sample is reached. The pressure gradient produces inward acceleration, thus focusing the waves. This focusing is the dominant effect. Time is in seconds measured from the time of impact, impactor moves upward and strikes the surface depth=0; contour A : .05GPa, contour B: 1GPa, contour C: 5GPa, contour D: 7GPa, contour E: 10GPa, contour F: 20GPa, contour G: 50GPa, contour H: 70 GPa and contour I: 100GPa. Pressure contour E, 10 GPa, is present at 2.2 microseconds and grows larger along the back region of the sample. Pressure contour F, 20 GPa, is present at 2.6 microseconds.*

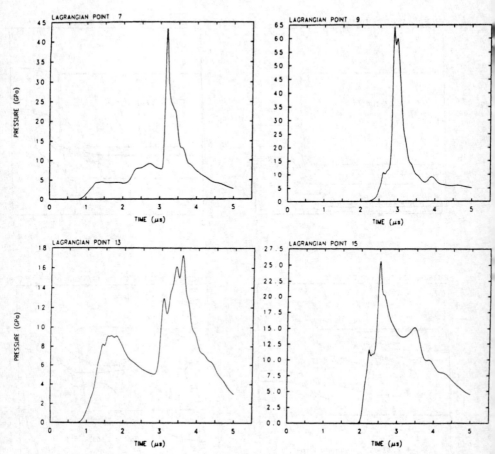

FIG. 2 *Pressure Histories for Lagrangian Points 7, 9, 13 and 15 Located Within the Powder Sample. Initial coordinates in cm. P7:(.1,.4); P9:(.1,.7); P13:(.4,.4); P15:(.4,.7). Points 9 and 15 have a higher peak pressure than points 7 and 13, respectively. Points 7 and 9 have a higher peak pressure than points 13 and 15, respectively. Higher pressures are calculated for the back region of the sample than for the front region. Also, higher pressures are calculated for the central region (radius=0) than for the outer region of the sample.*

is proportional to the negative of pressure gradient, for rutile in the neighborhood of the outer surface of the cylindrical sample, there is a radial component of acceleration. At points close to the center of the sample, the contours retain a one-dimensional character. At t=1.2 microseconds, the pressure contours have reached the top surface of the sample and contour A now forms a step about 0.5 cm high. The character of the pressure contours will change for later times as illustrated in the bottom two plots of Figure 1. At t=2.2

microseconds, the wave fronts have enveloped a substantial part of the sample and a high pressure region has developed on the back side. The region $0 < radius < 0.3cm$ and $0.7cm < depth < 0.8cm$ is an area toward which particles are accelerated by the pressure gradients. As a result, at 2.6 microseconds, high pressures result at the back (depth=0.8 cm) and center (radius=0) of the sample.

Figure 2 shows the pressure histories at Lagrangian points 7, 9, 13 and 15 used in the calculation. The initial (radius,depth) coordinates, of these points were (0.1,0.4), (0.1,0.7), (0.4,0.4) and (0.4,0.7), respectively. Point 13 is the first to feel the focusing effect. The first peak results from this focusing effect. The second peak is from the wave

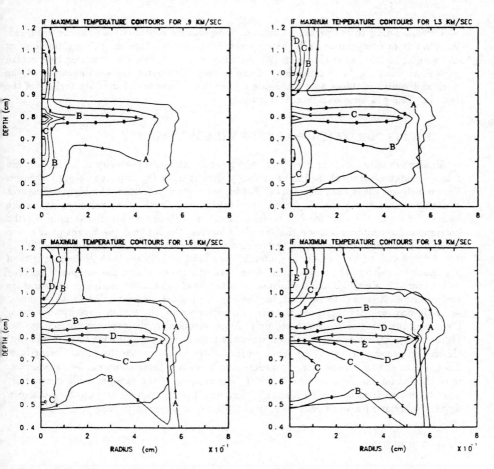

FIG. 3 Maximum Temperature Contours for Impact Velocities of 0.9, 1.3, 1.6 and 1.9 km/sec. Contour A: 887°C, contour B: 1468°C, contour C: 2040°C, contour D: 2628°C and contour E: 3209°C.

which has reflected from the symmetry axis. The maximum pressure calculated at this point is about 18 GPa. Point 15 is the next point to be affected. By the pressure gradients shown in the previous figures, and as has been indicated earlier [2], this point lies in the region where peak pressures and temperatures are higher than at points closer to the impact surface. Note a maximum pressure of about 26 GPa calculated for this point. Point 7 is the point which experiences planar effects longer than the other points. Thus, the first plateau does not reflect the focusing effect. The next portion begins to reflect the influence of the focusing effect. The sharp rise at 3 microseconds is a shock wave which was first felt at the point 9 and reaches a peak of about 43 GPa. Point 9 is in the region where the focusing effect is the greatest. The maximum pressure calculated at this point is about 64 GPa.

Corresponding to the pressure contours, temperature contours may also be analyzed. As a result of compression, a high temperature region develops in the neighborhood of the point (0.6,0.3), as is shown in [2]. As time increases, this high temperature region expands toward the back and center of the sample. Temperatures are developed at the back and center of the sample which are higher than those found in other regions of the sample. This will be shown later in Figure 3.

III. RESULTS FOR OTHER VALUES OF IMPACT VELOCITY

Similar pressure effects as have been calculated for an impact velocity of 0.9 km/sec are found for higher values of the impact velocity. However, as the impact velocity increases, the intensity of these effects increases. For higher impact velocities, the higher pressures available compact the sample faster, thus raising the shock wave velocity within the sample. As was the case for 0.9 km/sec, pressure contours within the sample lag the corresponding contours outside the sample; however, the lag and the height of the step formed by contour A decrease as the impact velocity increases. For impact velocities of 1.3, 1.6, 1.9 and 2.5 km/sec, the main feature is a gradual increase both in the intensity of the shock and also in the temperature developed at a point within the sample. This may be compared in Figures 3 and 4 which show contours of calculated maximum temperature and pressure. It is easy to see that, for the higher pressure contour values in the figures, as the impact velocity increases, the size of a region within the sample encompassed by a given contour value increases accordingly. More specifically, the pressure contour for 90 GPa, contour C, covers a small region when the impact velocity is 0.9 km/sec. Gradually, this region grows to cover a significant portion of the sample when the impact velocity is 2.5 km/sec. For the temperature contours, the process is more complex. For simplicity, note that at 0.9 km/sec, contour A, $887°C$, covers most of the sample, while contour B, $1468°C$, covers most of the sample for 2.5 km/sec. The corresponding results, and also a detailed analysis, for an impact velocity of 2.5 km/sec are given in Ref. 2.

IV. DISCUSSION

As observed previously [4, Chapter 4], in the case of powder compacts which are placed in metal holders, the low shock-compression impedance of the sample relative to the impedance of the sample holder causes relatively different loading patterns than

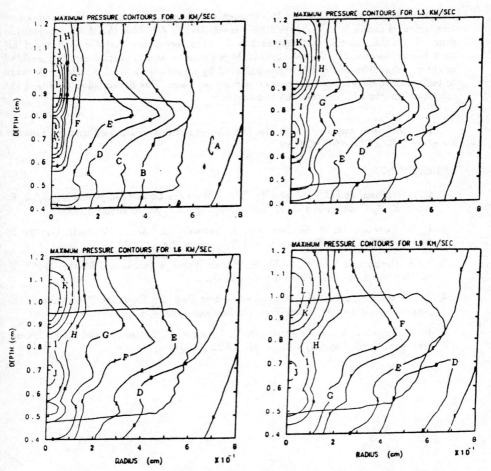

FIG. 4 *Maximum Pressure Contours for Impact Velocities of 0.9, 1.3, 1.6 and 1.9 km/sec. Contour A: 10 GPa, contour B: 15 GPa, contour C: 20 GPa, contour D: 30GPa, contour E: 40GPa, contour F: 60GPa, contour G: 70 GPa, contour H: 90GPa, contour I: 100GPa, contour J: 120GPa, contour K: 140 GPa and contour L: 160 GPa.*

predicted by one-dimensional analysis. Further, one-dimensional and two-dimensional results were compared. The present work extends our previous results [1-3] by showing the effect of varying the impact velocity. This made it possible to identify quite clearly the radial focusing effect which is absent in layered (one-dimensional) geometries.

The present numerical calculations were made for a stainless steel fixture and a flyer plate. This is a material of known shock properties which can be accurately described in numerical simulations with computer codes. For the rutile, the material model has been used in previous work [1], where it has been noted that the peak shock pressures and temperatures in the samples are relatively insensitive to details of the shock compression properties of the material. In the present work, it has been shown that, for all values of impact velocity assumed in the calculations, high pressures and temperatures are generated in the powder. The size of the region enclosed by a given contour of maximum pressure or temperature increases as the impact velocity increases. The figures show details of the calculated maxima in pressure and temperature.

This work was supported by the U.S. Department of Energy under contract number DE-AC04-76DP00789.

REFERENCES

1. R. A. Graham and D. M. Webb, in "Shock Waves in Condensed Matter-1983," J. R. Asay, R. A. Graham and G. K. Straub, eds., p. 211, North Holland (1984).

2. F. R. Norwood, R. A. Graham and A. Sawaoka, in "Shock Waves in Condensed Matter-1985," Y. M. Gupta, ed., p. 837, Plenum Press (1986).

3. R. A. Graham and D. M. Webb, in "Shock Waves in Condensed Matter-1985," Y. M. Gupta, ed., p. 831, Plenum Press (1986).

4. R. A. Graham and A. B. Sawaoka, eds., High Pressure Explosive Processing of Ceramics, pp. 35 and 92, Trans Tech Publications (1987).

5. V. A. Bugaeva, M. A. Podurets, G. V. Simakov, G. S. Telegin and R. F. Trunin, Izvestiya, Earth Physics, Vol. 15, pp. 19-25 (1979).

93

Phase Transition and Hugoniot Data in Shock Loaded Bismuth

R. DORMEVAL, J. PERRAUD and C. REMIOT

Commissariat à l'Energie Atomique
Centre d'Etudes de Vaujours-Moronvilliers
B.P. 7, 77181, COURTRY, FRANCE.

Plate impact experiments were performed on bismuth samples in the pressure range 2 to 5,5 GPa, in which polymorphic transitions occur. In this paper, we present a measurement technique using simultaneously manganin gauges, ferroelectric probes and Doppler Laser Interferometry, which allow to get a better knowledge of these mechanisms. The experimental results are compared to the theoretical Hugoniot curves.

I. INTRODUCTION

Solid-to-solid phase transformations under shock loading have been largely studied since the firts Bancroft's experiments on iron [1]. However, for a better knowledge of these mechanisms, accurate measurement methods are needed to reveal the shock-wave profiles. In this goal, we developed simultaneously, on the same experimental set-up, three different diagnostic methods :

- manganin gauges, to measure the pressure variation versus time,
- ferroelectric probes, to know the shock wave velocity,
- Doppler Laser Interferometry, to record the free surface velocity versus time.

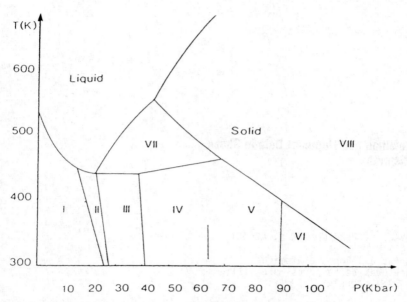

FIG.1   *Pressure-Temperature phase diagram for bismuth.*

Shock experiments were performed on bismuth targets. Bismuth was chosen because many studies have been done on this material [2 to 5], and by the fact that in the studied pressure range, several phase transitions can be observed, as seen on figure 1. This complex pressure-phase diagram is interesting to valid the measurement methods.

II. EXPERIMENTAL PROCEDURE

A. *SET UP DESCRIPTION*

Shock experiments are performed using a 60 mm bore powder gun.

The target is schematically presented on figure 2. It is composed by a 10 mm thick copper plate (transmitter) behind which are positionned two bismuth discs, 3 mm thick each. The impactor is a copper projectile of about 40 mm in thickness.

The target device is fixed at the end of the launch tube : so, the projectile is guided up to the impact and the tilt angle is less than 10^{-3} radian.

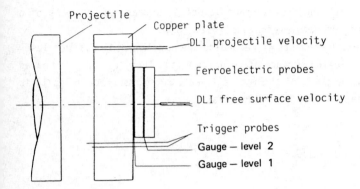

FIG.2   *Schematic view of the target.*

The studied material is a 99,997% purity bismuth. Samples are machined from cast ingots, inducing a large grain size (closely 1 mm). Nevertheless, works done by Larson [6] have shown that grain size had no marked influence on the phase change pressure. The specific mass is $(9,824 \pm 0,007)$ g/cm³.

B.   MEASUREMENTS METHODS

As said in introduction, three measurement methods are used simultaneously.
Manganin gauges are used to get the pressure versus time profile. They have a 150 μm thickness and the rise-time is 150 ns for an initial pressure pulse of 10 GPa. In the pressure range 0-20 GPa, these gauges are calibrated using a classical law :

$$\frac{\Delta R}{R} = kP$$

with $k = 0.27 \ 10^{-3}$ GPa^{-1}.
Manganin gauges are located at two levels in the target [7].
First level is the interface between the copper transmitter and the first bismuth sample. The gauges give the incident pressure at this interface.
The second level is the bismuth-bismuth interface, where is recorded the transmitted pressure through the first bismuth sample, eventually taking into account the solid-solid phase changes during the shock.

On an other way, the free surface velocity is measured by means of a Doppler Laser Interferometry technique (DLI) [8].

This measurement consists in recording the wavelength λ_v (t) of a laser beam reflected by a moving surface at velocity v (t). Then, λ_v (t) is compared to the initial wavelength λ_i of the incident beam. So, the surface velocity is directly determined by the following relation :

$$\frac{\lambda_v(t) - \lambda_i}{\lambda_i} = \frac{-2 \, v \, (t)}{c}$$

where c is the light celerity.

In fact, the variations of λ_v (t) is deduced from an interferogram obtained on a Fabry Perot interferometer : this interferogram is recorded on a streak camera. Its evolution in function of time gives the curve surface velocity versus time. The accuracy of a such method is usually 0.01. In the described experiments, two DLI channels are working : one for the bismuth free surface velocity measurement, the second for the projectile velocity measurement.

Finally, the shock wave celerity is obtained by chronometry, using ferroelectric probes located at different levels in the target. For pressures less than 40 GPa, the rise time is closely 5 ns, and the accuracy of the shock celerity is about 1 %.

III. EXPERIMENTAL RESULTS

Figure 3 shows a schematic diagram of the shock wave propagation in copper transmitter and bismuth specimens, and the associated pressure-time and free surface velocity-time signals. In the case of a phase change, two shock waves appear in bismuth with two different celerities D_1 and D_2 ($D_2 < D_1$). When these shock waves reach the gauge levels or the free surface, they induce the advent of plateaus on pressure and velocity records. An example of such experimental records is presented on figures 4 and 5.

Table I gives the values of pressure P_i, shock celerity D_i and particle velocity u_i obtained with gauges, each value corresponding to the observed plateaus. These results seem to indicate that two phase changes occur in the

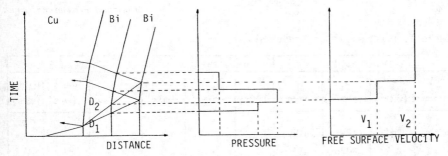

FIG.3   *Schematic diagram of the shock wave propagation, pressure and free surface velocity profiles.*

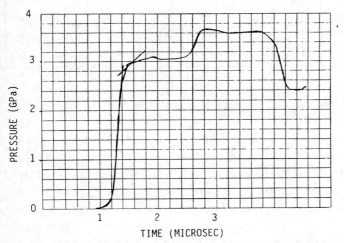

FIG.4   *Experimental pressure-time record (incident pressure : 3.7 GPa).*

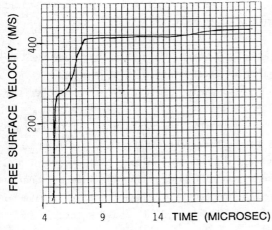

FIG.5   *Experimental free surface velocity record (incident pressure : 3.8 GPa*

Table I. Pressure gauges results.

Shot n°	Plateau 1			Plateau 2			Plateau 3		
	P_1 (GPa)	D_1 (m/s)	u_1 (m/s)	P_2 (GPa)	D_2 (m/s)	u_2 (m/s)	P_3 (GPa)	D_3 (m/s)	u_3 (m/s)
562	2.1	1996	107						
543	2.2	2075	108						
542	2.76	2165	131	3.7	1055	210			
564	2.8	2141	134	4.0	1040	232			
544	2.76	2162	131	4.0	1180	237	4.25	935	253
639	2.76	2152	132	4.0	1190	240	4.8	1030	275
563	2.74	2118	132				5.3	1400	300

studied range pressure. However, the type of transition is quite difficult to know.

Clearly, the two phase transition pressures are about 2.7 - 2.8 GPa and 4 GPa. In reference to the figure 1, between 0 and 5 GPa, it would be observed three phase changes (BiI→ BiII, BiII→ BiIII, BiIII→ BiIV). Plateau 2 can be related to the BiIII→ BiIV phase transition, for which the pressure ($\simeq$ 4 GPa) is in agreement with figure 1. The uncertainty comes from the first plateau. The transition pressures between the states I → II and II → III are of a same order (see figure 1) and the gauge signals are not able to reveal each of these phase changes. Thus, one can think that plateau 1 corresponds to BiI → BiIII transition, maybe directly (this case is observed under quasistatic conditions), or by the existence of a transient state BiII :
BiI → (BiII) → BiIII.

Because of the high accuracy of the free surface velocity measurement by means of DLI, it was hoped that this uncertainty will be removed.

Table II presents the DLI measurement results, where V_i is the free surface velocity.

More than the values of the velocity, the ratio V_i/u_i (free surface velocity/particle velocity) is interesting. The first value is closed to 2, the second about 1.8. This last value seems to indicate that a part of the energy is consumed by the phase transition, increasing the internal energy and decreasing the kinetic energy. Unfortunately, these DLI results give no more informations concerning the BiI → BiII → BiIII phase change.

Table II. Doppler Laser Interferometry results.

Shot n°	P_1 (GPa)	V_1 (m/s)	V_1/u_1	P_2 (GPa)	V_2 (m/s)	V_2/u_2	P_3 (GPa)	V_3 (m/s)	V_3/u_3
562	2.1	218	2.04						
543	2.2	215	1.99						
542	2.76	270	2.06	3.7	380	1.80			
564	2.8	275	2.05	4.0	420	1.81			
544	2.76	270	2.06	4.0	415	1.75			
639	2.76	275	2.08	4.0	440	1.83	4.8	495	1.80
563	2.74	275	2.08				5.3	540	1.80

$$V_1/u_1 \simeq 2.0 \qquad\qquad V_2/u_2 \simeq 1.80$$

On an other hand, we have compared these results with the theoretical datas in the pressure-particle velocity diagram. In respect to the pressure range, two D(u) relations are available :

$$D = 1771 + 2.21\ u\ (m/s)\ \text{for low pressures}$$
$$D = 1113 + 2.14\ u\ (m/s)\ \text{for high pressures}$$

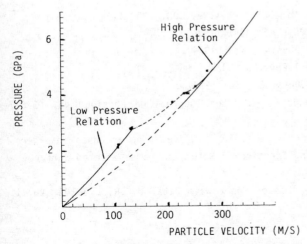

FIG.6   *Pressure versus particle velocity diagram.*

It can be seen on figure 6 that for low and high pressures, there is a good agreement between theory and experiments, but that discrepancy appears at the first phase change level ($\simeq 2.8$ Gpa). More experimental datas are needed to get a better accuracy, but it seems that, from this type of results, it is possible to precise fruitfully the bismuth behaviour in the 2.8-5 GPa pressure range.

IV. CONCLUSION

The results obtained with the different measurement methods are consistent and show the interest of increasing and varying the number of measurements. However, in the case of bismuth, the phase diagram is so complex that an uncertainty still remains concerning the kind of cristallographic transitions. This point has to be studied with additional investigations, and it may be interesting to use the described method with other types of pressure gauges, like polymeric ones.

REFERENCES

1. D.Bancroft, E.L. Peterson and F.S. Minshall, J.Appl.Phys.27, 291 (1956).
2. R.E. Duff and F.S. Minshall, Phys.Rev.108, 1207 (1957).
3. J.R. Asay, J.Appl.Phys.45, 4441 (1974).
4. J.P.Romain, J.Appl.Phys.45, 135 (1974).
5. Z. Rosenberg, J.Appl.Phys.56, 3328 (1984).
6. D.B.Larson, J.Appl.Phys.38, 1541 (1967).
7. J.M.Servas, J.Perraud et G.Le Don, Journal de Physique 49, C3-627 (1988).
8. P.Andriot, P.Chapron, C.Le Dréan, F.Olive, Doppler Laser Interferometer for Measuring Ejection of Material from a Shocked Surface, CLEO'81, Washington.
9. M.Van Thiel, Compendium of Shock Wave Data. U.C.R.L. 50108, Vol.1, Rev.1 (1977).

94

The Prescribed Minimum of Impacted Point Velocity in Multilayer Explosive Welding of Amorphous Foils

KAI ZHANG

Research Institute of Engineering Mechanics
Dalian University of Technology
Dalian, P.R. China

In multilayer explosive welding of amorphous foils, the surface of the foil may appear as a regular ripple shape geometry under certain glancing detonation velocities. In order to eliminate this phenomenon, the velocity of the impact point must be greater than C_g. The term C_g is defined as the prescribed minimum of the impact point.

I. EXPERIMENTAL

In multilayer explosive welding of amorphous foils, on the surface suffice of the top foil may appear a regular ripple shape geometry under certain detonation velocities. Figure 1 demonstrates the surface appearance of such amorphous foils (Fe—Si—B-C), recovered by multilayer explosive welding. The experimental assembly is shown in Fig. 2, where the detonation velocity of explosive was V_d=2109 m/s, the impacting plate 2, impacted the amorphous foils 4, through plate 3 which had a thickness of 0.5 mm. The calculated velocity of plate 3 is V_p=356 m/s with a bending angle of b=58°, the measured average wave length of the surface ripples were 1_a=(0.8 -1) mm.

II. THEORETICAL ANALYSIS

When the one end of foil is impacted by the plate, the foil which is very thin can be subjected the composite action of shear and bending by the shock wave. There are bending–shear waves propagating along the foil, making the foil sheet located in the front of the impact point to

FIG.1   The surface appearance of the foils recovered by multilayer exposive welding.

appear ripple shapes, thus the recovered foil sheet has a distinct ripple appearance under a
squeezing action by the steel plate.   Moreover there are dented vestiges that appear in the
lowest concave point of the ripple.   In order to investigate the shock bending–shear wave in the
foil, the Timoshiko beam can be used to set up the differential wave equation.   Taking a
differential element of foil sheet of width unit, the differential equation can be shown as follows

$$\text{EIW}_{xxxx} + \rho \text{AW}_{tt} - J_p \left[W_{xxtt} - \frac{\rho A}{R'AG} W_{tttt} \right] - \frac{EI\rho A}{R'AG} W_{xxtt} = 0 \tag{1}$$

in which W–vertical displacement, x–the coordinate along the longitudinal direction, θ–Slant
angle, $\partial W / \partial x$–Turning angle due to bending, γ–Shear angle, ρ–density, A–area of cross
section, $R' = 2 / 3$(for rectangle cross section), $I = Ah^2 / 12$–moment of inertia of section,
E–Young's modulus, G–shear modulus $G = E / 2(1+\nu)$.

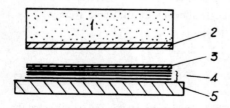

FIG.2   The assembly of explosive welding.   1–explosive,   2–impacting   plate,   3–steel
plate,   4–foils(4layers),   5–bed plate.

For this amorphous material we used the following $\rho = 7.18$ g $/$ cm^3, E $= 16.4 \times 10^{10}$ Pa, $v = 0.37$, substituting in the fomula (1), we have the form

$$C_o^2 k^2 W_{xxxx} - k^2 \left[1 + \frac{2(1+v)}{R'} \right] W_{xxtt} + \frac{2(1+v)}{C_o^2 R'} k^2 W_{tttt} + W_{tt} = 0 \tag{2}$$

in which $C_o = \sqrt{E / \rho} = 4779$ m / s, $k^2 = I / A$. Assuming $W = A\exp[i(pt - fx)]$ and substituing into (2), the frequency equation can be obtained

$$C_o^2 k^2 f^4 - k^4 \left[1 + \frac{2(1+v)}{R'} \right] f^2 p^2 + \frac{2(1+v)}{R'} \frac{k^2}{C_o^2} p^4 - p^2 = 0 \tag{3}$$

p—frequency, f—wave number, $\lambda = 2\pi / f$—wave length, $C_p = p / f$—phase wave velocity, $C_g = dp / df$—the velocity of wave train

$$\left(\frac{C_p}{C_o} \right)^2 = \left\{ \left[\frac{1}{12} \left(1 + \frac{2(1+v)}{R'} \right) \frac{h^2}{\lambda^2} + \frac{1}{4\pi^2} \right] - \right.$$

$$\left. \sqrt{ \left[\frac{1}{12} \left(1 + \frac{2(1+v)}{R'} \right) \frac{h^2}{\lambda^2} + \frac{1}{4\pi^2} \right]^2 - \frac{(1+v)}{18R'} \frac{h^4}{\lambda^4} } \right\} \bigg/ \frac{1+v}{3R'} \frac{h^2}{\lambda^2} \tag{4}$$

$$\frac{C_g}{C_o} = \frac{\left(\frac{C_p}{C_o} \right) \left[(1+v)(\frac{C_p}{Co})^2 (\frac{h}{\lambda})^2 - \frac{4+3v}{6} (\frac{h}{\lambda})^2 - \frac{1}{\pi^2} \right]}{\left[(1+v)(\frac{C_p}{Co})^2 (\frac{h}{\lambda})^2 - \frac{4+3v}{6} (\frac{h}{\lambda})^2 - \frac{1}{2\pi^2} \right]} \tag{5}$$

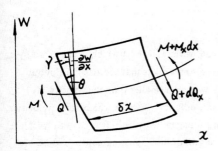

FIG3.   Timoshiko beam.

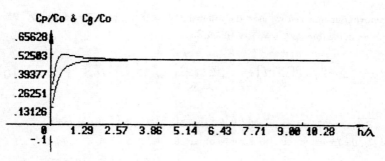

FIG4.   The plot of C_p / C_o and C_g / C_o.

The curve of C_p / C_o and C_g / C_o against h / λ is shown in Fig. 4, from these curves, we can see that the wave velocities traveling along a foil sheet relate to wave length. As the wave length decreases, the greater the wave velocity, thus the wave is dispersed. After impact at $x = x_0$, there are instantaneous reflected waves propagated along the foil sheet, and the initial condition can be written as

$$W(x,0) = 0$$

$$W_t(x,0) = -v_0 \int_{-\infty}^{\infty} \delta(x - x_0)dx \qquad (6)$$

from (2) and (6), the general solution of equation (2) can be obtained [1]

$$W(x,t) = -\frac{v_0}{i} e^{if_k x_0} \int_{-\infty}^{\infty} \frac{e^{if(x-x_0)}}{fC_p} [e^{if(-x+C_p t)} - e^{if(-x-C_p t)}] \delta(f - f_k)df$$

$$= -\frac{v_0}{f_k C_p} \sin(f_k C_p t) \qquad (7)$$

In aforementioned experiment, the impacted end of the foil sheet which is subjected by a continuous impact will also send out shock bending–shear wavelets that are also continuous. In these wavelets all of the wavelets which have wave velocities smaller than the velocity of impacting point are restrained by the impacting point, only the wavelets which wave velocities greater than the velocity of the impacting point can propagate forward. It is easy to calculate the wavelength of this wavelet with a velocity equal to that of the impacting point. At the impacting time, $C_g = V_d = 2109$ m / s, and $C_g / C_o = 0.4413$, when $h / \lambda = 0.176$, from (4) and (5) we can calculate $c_p / C_o = 0.2614$, $C_g / C_o = 0.4400$, this is almost as same as 0.4413 from Fig. 4, so the wavelength corresponding to the wavelet with wave velocity of $C_g = 2109$ m / s is $\lambda = = 0.142$ mm. All of wavelengths corresponding to greater wave velocities than $V_d = 2109$ m / s are smaller than 0.142 mm. Apparently, all these wavelets do not result in a ripple surface of $\lambda = (0.8 \sim 1)$ mm, this has been observed experimently. However in the experimental assembly presented here, the multilayer foils are stacked together, the interspace letween each layer is about $0.4h = 0.01$mm. Supposing that at time t, the upper layer (1) is bent over to AC and impact the point A of lower layer (2). When point A is impacted, all of wavelets which have wave velocities greater than $C_g = 2109$ m / s will be traveling from A towards B rightforward along AB, however wavelets with wavevelocities equal to $C_g = 2109$ m / s will reach point B after $\Delta t = 0.028$ μs. But at this time, point B will turn into a new impact point and send out a new wavelet traveling to the right side of point B, these two wavelets, i.e. one is sent out from A and another from B, superimpose at point B. In order that the wavelet sent out from point A is not restrained by the impacting point, and letting the wave velocity of the wavelet sent out from point A as 10% greater than $C_g = 2109$ m / s, i.e. $C'_g = 2320$ m / s, thus after $\Delta t = 0.028$ μs, this wevelet will reach the position ahead of point B by 0.0057 mm, this value is only 4% of wavelength $\lambda = 0.142$mm corresponding to the wavelet which travels with the impacting point. Now, we can see the superimposed results of the wavelet sent out from A with another wavelet which is sent out from B. From (4), (5), the wavelength corresponding to wave velocity $C'_g = 2320$ m / s is $\lambda' = 0.109$ mm. As is known to all, the two superimposed wavelets of different wavelength with same frequency can construct a new standing wave with variations in amplitude. In fact we have

$$a[\cos(pt + f_1 x) + \cos(pt + f_2 x)] = 2a\cos\frac{f_1 - f_2}{2} x \cdot \cos(pt + \frac{f_1 + f_2}{2} x) \tag{8}$$

the ampitude of the superimposed wave is $2a\cos(f_1 - f_2)x / 2$ which is periodically variated and double amplitude as a function of x and thus becomes standing wave. The wave number of this standing wave is $(f_1 - f_2) / 2$, and the wavelength is $\lambda = 2\pi / [(f_1 - f_2) / 2] = 0.938$ mm.

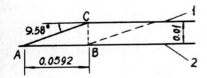

FIG5.   Impacting schema of foils.

Consquently there is a standing wave of 0.938mm wavelength existing ahead of point B and having double the amplitude. At the same time, as a result of the continued foward motion of the impact point, the flying plate forcefully compresses the front ripple foil with a standing wave shape on the baseplate. Apparently, the lowest concave points of the ripple (the lowest point of sine or cosine curve) are compressed dented vestige, this is what was observed in these experiments.

III. CONCLUSION

When the velocity of the impact point is smaller than $0.5286C_0$ (C_0–Sound velocity of foil material), it will result in the apperance of a ripple shape due to the traveling wave in the foil and results in failure of the weld. In order to eleminate this phenomenon, the velocity of the impact point must be greater than the $C_{g_{max}}$. The term $C_{g_{max}}$ is defined as the prescribed minimum of velocity of the impacted point.

IV. ACKNOWLEDGEMENT

The experiments of this paper is done by my assistent lecturer Li Xiao Jie, here allow me to express my gratitude to him. This work was performed under the auspices of the National Advanced Materials Committee of China.

REFERENCES

1. Zhang Kai, Xi–Jingyi, Yan Wenbin, The theory of effect of bending wave in continuous slanted resistance–wire on the measured results of velocity of flyer plate. The 6th Inter. Symposium use of explosive energy. Czechoslovak, 1985.

Section IX
Shock and Dynamic Phenomena in Ceramics

95

Crack Behavior and Dynamic Response of Alumina Under Impact Loading

H. SENF and H. ROTHENHÄUSLER

Fraunhofer-Institut für Kurzzeitdynamik, Ernst-Mach-Institut
Weil am Rhein, D-7858, W-Germany

Two different methods of impact experiments were performed against alumina targets. Firstly, the ballistic performance of a protection system consisting of alumina tiles glued on a thick steel backup was investigated at impact velocities of 1050 m/s and 1500 m/s, respectively. The thickness of the tiles was varied. Differing terminal ballistic behaviour was observed at both velocities. Secondly, an attempt is made to understand the differing results with the aid of additionally performed edge-on impact experiments against thin alumina tiles in which the fracture behaviour of this ceramic was investigated.

I. INTRODUCTION

Non-metallic materials like ceramics gain more and more importance as modern protection materials against fragment and projectile impact [1]. These materials are used as homogeneous plates or in combination with metallic or non-metallic components in so-called structured or layered armor plates. They are lighter in weight than steel plates and their performance against certain threats is superior to existing materials. However, terminal ballistic experiments against such armor show that the interaction processes between projectile and target not only depend on the striking velocity of the impinging projectile and the type of target, but also on the spatial arrangement inside the target configuration.

In order to get a deeper and more fundamental knowledge of the terminal ballistic resistivity of brittle materials, two different methods of impact experiments were performed with alumina targets. Firstly, the tiles were glued onto a semi-infinite steel backup and centrally impacted normal to their surface; and the penetration depth of the projectiles into the steel block was determined. Secondly, by loading the alumina tiles edge-on the propagation of cracks was observed. An attempt is made to explain the differences in penetration resistance observed with the central impact experiments with the aid of the latter results.

II. EXPERIMENTAL SET-UP

The impact experiments are performed by means either of a gas gun or a powder gun, depending on the range of the striking velocity. The different loading configurations (Fig. 1) differ in the impact and in the observation mode. The experimental set-up for central impact is illustrated in Fig. 1a [2]. The target consists of a 100 mm x 100 mm ceramic tile glued on a back-up block of steel (>100 mm thickness). Pointed projectiles were fired at two velocities (v = 1050 m/s and 1500 m/s) against the structured target and the residual penetration depth into the steel block after perforating the tile was measured. The thickness of the ceramic tile was varied from 0 to 30 mm. The ceramic material designated alumina A 18 98 used in both test series was produced by Hoechst CeramTec: (98 % Al_2O_3; ρ = 3793 kg/m^3; c_L = 10.44 km/s; E = 360 GPa).

The experimental set-up for investigating the propagation of waves and cracks in the tiles is shown in Fig. 1b. The projectile hits the edge of the tile of dimensions 10 mm x 100 mm x 100 mm. The impact velocity of the blunt or pointed projectiles was varied between 100 m/s and 1500 m/s. The

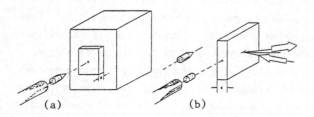

(a)   (b)

FIG. 1 *Loading arrangement for central (a) and edge-on (b) impact experiments.*

crack propagation and the damage processes in the target plates were directly photographed using a Cranz-Schardin 24-spark high-speed camera using
shadow optical method in reflection [3-5]. This photographic method applied
in reflection requires that one of the large surfaces of the opaque target
plates has to be plane and optically polished and coated with an aluminum
layer.

III. EXPERIMENTAL RESULTS

A.   *CENTRAL IMPACT EXPERIMENTS*

The experimental results of two series of central impact investigations
(Fig. 1a) at the nominal test velocities 1050 m/s and 1500 m/s are summarized in Fig. 2. The terminal ballistic resistance of the target configurations is characterized by the areal density ρ_F (mass/area) of the material used. Therefore, the ratio of the areal densities of the totally penetrated materials ρ_{Ftotal} and of steel ρ_{Fsteel} (Y-axis) is plotted versus
the areal density ratio of perforated ceramic ρ_{Fcer} to ρ_{Ftotal} (X-axis),
where

$$\rho_{Fcer} = d \cdot \rho_{cer}$$

$$\rho_{Fsteel} = P \cdot \rho_{steel}$$

d   – thickness of perforated alumina tile

P   – penetration depth into steel without
       front plate

ρ_{cer}; ρ_{steel} – material density of alumina; steel

$$\rho_{Ftotal} = \rho_{Fcer} + P_{res} \cdot \rho_{steel}$$

P_{res} – residual penetration depth into steel
             block after perforating alumina tile.

The value "0" on the X-axis means that the target consists of steel
only without ceramic covering and "1" means that the required ceramic quantity which stops the impacting projectile without penetrating the steel
block, respectively. The penetration performance P into steel only without
alumina covering was defined "1" (on Y-axis) in case of the striking velocity v = 1500 m/s. The experiments performed show that at this velocity the
ballistic performance of the alumina increases with increasing tile thickness d or quantity of alumina. That means the (normalized) total areal density ρ_{Ftotal}, which may be equalized with the total mass of the packed-up
material (alumina and steel) needed for stopping the penetrator, decreases
with increasing amount of alumina in this special semi-infinite target configuration, as is expected. But this tendency changes if the ratio $\rho_{Fcer}/$

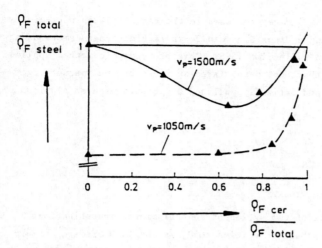

FIG. 2 *Terminal ballistic performance of alumina-steel target configurations. Normalized areal density of penetrated material* ρ_{Ftotal} *vs. ratio of the areal density of perforated alumina and penetrated material.*

ρ_{Ftotal} exceeds values higher than 0.7. Now with more increasing ceramic tile thickness, the ballistic resistance of alumina decreases. The extrapolation of the fitted curve to the point representing the case of ratio $\rho_{Fcer}/\rho_{Ftotal} = 1$ where the penetrator is stopped by alumina only without penetrating the steel backup shows that a higher mass of alumina is necessary than for steel only.

    At low impact velocities (1050 m/s) the resistance behaviour of the discussed target configuration is not significantly influenced by the portion of alumina within a broad regime up to 0.8 which is in contrast to the 1500 m/s experiments. That means that the alumina behaves similar to steel; the mass of steel may be replaced by an equal mass of alumina material without changing the ballistic stopping performance of the layered target. Above this areal density ratio of 0.8 the resistance behaviour of alumina decreases in a drastic way and the experiments performed showed that at this low impact velocity (1050 m/s) tiles of about the same thicknesses are required to stop the impactor without penetrating into the steel backup as in the case of high velocity impact (1500 m/s).

Discussion of Central Impact Experiments The above described target response under central impact conditions may be discussed by splitting up the observed phenomena into five regimes:

1. At low impact loadings (v = 1050 m/s), mainly the target mass is responsible for the stopping mechanism (dashed curve in Fig. 2) and alumina behaves like steel with regard to mass.

2. Above a certain tile thickness ($\rho_{Fcer}/\rho_{Ftotal} > 0.7$) with an increasing amount of ceramic, the stopping performance of alumina decreases.

3. At high impact loadings (v = 1500 m/s) with increasing tile thickness the resistance of alumina increases in cases $0 < \rho_{Fcer}/\rho_{Ftotal} < 0.7$. In opposition to low loading conditions, the stopping response of total mass of this target configuration is influenced by the portion of inserted alumina.

4. Above this ratio of 0.7, the resistance strength of alumina decreases similar to case 2.

5. The experiments showed that under the impact conditions discussed here nearly equal tile thicknesses are necessary to prevent residual penetration into the steel backup independently from the striking velocity.

From these results it is concluded: In case of *thin* alumina tiles and at *low impact velocities*, steel behaves like alumina of equal mass although the two materials suffer differing damage processes and interact in different manner with the penetrator. At *high impact velocities* the mass effectiveness of alumina is higher than that of steel from a terminal ballistic viewpoint. That means that there is a strong influence of loading velocity on the dynamic ballistic strength of thin alumina tiles. In case of *thick* alumina tiles, a strong reduction of ballistic mass effectiveness of ceramic material is observed in relation to steel not only at low but also at high loading rates.

B. *EDGE-ON IMPACT RESULTS*

Two typical examples of high-speed photographic series of edge-on experiments are given in Figs. 3a and 3b. Fig. 3a shows the results obtained from a conically-nosed projectile impinging the tile edge-on at a velocity of v_p = 519 m/s, and Fig. 3b for a blunt projectile impact at a velocity of v_p = 1062 m/s. Both experiments show the development of an extended and irregularly shaped damage zone (black area, no details can be identified) at the impact site. In case of point impact (Fig. 3a) 4 µs after the collision cracks leave the black area at high velocity. The crack path is rough and fuzzy and furcations can be seen. This leads to the assumption

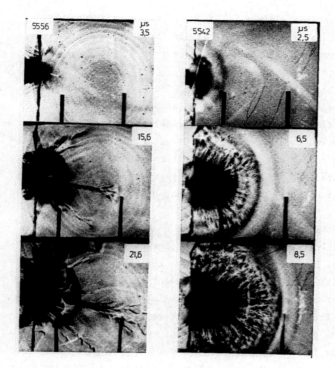

(a)   (b)

FIG. 3 *High-speed photographic series of wave and fracture processes in edge-on impacted alumina tiles, (a) pointed projectile $v_P = 519$ m/s, (b) blunt projectile $v_P = 1062$ m/s.*

that these cracks propagate with terminal velocity. The blunt impact experiment (Fig. 3b) differs considerably from the previous one. A crack field propagates very fast in radial direction from the area of impact into the target. Figs. 3a and 3b are supplemented by space-time diagrams of moving events taken from the pictures (Figs. 4a and 4b). In these diagrams the time axis is given in terms of microseconds. The space axis is representing the propagation direction of a moving crack tip. Both diagrams show the longitudinal wave front and the rarefaction wave reflected at the rear edge of tiles are plotted. The symbols consistently used in both dia-

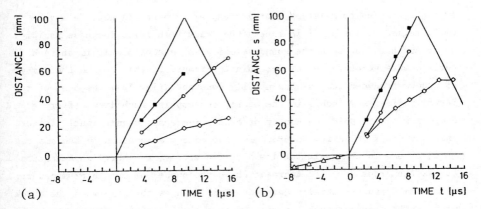

FIG. 4 *Distance-time diagrams of wave and crack propagation.*

grams are (■) characterizes the observed wave, (O) the crack tip position, (◇) the extension of the black area and (△) the movement of the projectile prior to impact. The velocity of the longitudinal wave is determined to be 10.47 km/s ± 0.06 km/s from Fig. 4b which is in good agreement with the previously determined wave speed c_L = 10.44 km/s [6]. In Fig. 5 the crack velocity and the number of cracks are plotted versus energy density data assuming that the energy transported by a shock wave is responsible for the wave and fracture phenomena observed in the pictures.

Discussion of the Edge-on Impact Results Fig. 5a shows at same energy den- sity larger crack velocities for the blunt projectiles than for the pointed

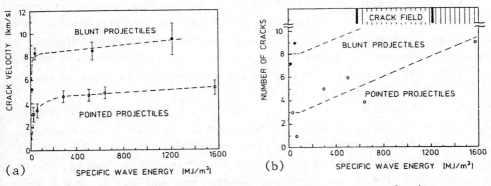

FIG. 5 *Crack velocity (a) and number of cracks (b) vs. energy density of the wave.*

projectiles, due to a larger energy transfer into the target. In a very
small regime of low specific energy the crack velocities increase rapidly
up to about 80% of both the sound velocity in case of planar impact and
surface wave velocity in case of pointed impact. In this regime the number
of cracks produced remains essentially constant (Fig. 5b). The bent of the
curves indicates a change in the energy consumption mechanism. After
exceeding a threshold the number of cracks increases significantly faster
than the crack velocity. Accordingly, two energy absorption mechanisms can
be observed. In the regime of low specific energy, an energy augmentation
causes the crack speed to increase while the number of cracks produced remains small and essentially constant. This changes abruptly when the crack
propagation speed exceeds a threshold at about 80% of the correlated wave
speed. The number of cracks becomes large and a crack field is developed.
The expansion velocity of this field increases only slightly and seems to
approach the wave speed asymptotically.

From energy considerations based on the fracture mechanics concept the
results presented here are not yet understood. Crack speeds higher than 80%
of wave speed were measured and furthermore no distinct terminal crack
velocity could be found as had been expected.

IV.   DISCUSSION AND CONCLUSION

The results of edge-on and central impact investigations show that in the
very beginning of projectile-target collision the energy transfer occurs
mainly in form of shock wave energy. This energy is used to produce cracks
and crack fields in the target material. (Only damages in the target material but not in the projectile are discussed.) The initiated waves and the
crack phenomena propagate with nearly same velocities into the alumina.
That means in this primary phase of impact a strong consumption of kinetic
energy of the impacting projectile and consequently a reduction of penetration speed will take place.

If the alumina tiles are very thick, then a few microseconds after
impact the target material will be completely permeated by cracks and
fragmented into powdery particles due to the high crack velocity. The
strain conditions within the material break down with increasing particle
generation and the particles start moving from their original places because of the absorbed energy and momentum. That means the projectile is

perforating a material with new physical properties during this second penetration phase. The projectile which is mainly decelerated by material displacement then will pass low resistance material in front of its tip. Much higher amount of ceramic in comparison to steel is necessary to stop the penetration process.

The discussed phenomena are not yet fully understood and the proposed hypothesis has to be examined very carefully before generalizing the statements. It is speculated [7] that new material properties have to be correlated to ceramic materials which provide evidence of strange material behaviour not known from other materials. It is discussed that the fracture toughness of ceramics apparently reduces "to negligible quantities once the crack is rapidly propagating" [7]. Therefore, crack and damage propagation are to be investigated in further impact experiments supported by ceramography and fractography. The results of these experiments have to be correlated to results of ballistic performance investigations of armor ceramics under well defined experimental conditions.

V.   REFERENCES

1.   D.J. Viechnicki, A.A. Anctil, D.J. Papetti, J.J. Prifti, Proc. 28th DRG Seminar on Novel Materials for Impact Loading, DS/A/DR(88)245: 125 (1988).

2.   P. Woolsey, S. Mariano, D. Kokodko, 5th TACOM Armor Conf. (1989).

3.   U. Hornemann, H. Rothenhäusler, H. Senf, J.F. Kalthoff, S. Winkler, Phys. Conf. Ser. No. 70, 3rd Conf. Mech. Prop. High Strain-Rates: 291 (1984).

4.   S. Winkler, H. Senf, H. Rothenhäusler, Fraunhofer-Institut für Kurzzeitdynamik, Weil am Rhein, Rept. V 5/89 (1989).

5.   S. Winkler, H. Senf, H. Rothenhäusler, High Velocity Fracture Investigation in Alumina, to be published by THE AMERICAN CERAMIC SOCIETY.

6.   S. Winkler, Fraunhofer-Institut für Werkstoffmechanik, Freiburg, IWM Rept. V 15/88 (1988).

7.   A. S. Kobayashi, K.H. Yang, the 1989 ASME Pressure Vessels and Piping Conf., Honolulu, Hawaii, July 23-27 (1989).

96

Response of Shock-Loaded AlN Ceramics Determined with In-Material Manganin Gauges

N.S. BRAR, S.J. BLESS, and Z. ROSENBERG

University of Dayton Research Institute
Dayton, Ohio 45469-0180, U.S.A.

The dynamic response of aluminum nitride ceramics was determined in a series of plate impact experiments using in-material manganin gauges both in the longitudinal and lateral orientations to the shock direction. The HEL of HP AlN was determined to be 94 ± 2 kbar and that of sintered AlN 70 ± 4 kbar. An anomalous second HEL-like signal was observed in the hot pressed material. The shear strength of the shocked material beyond its Hugoniot Elastic Limit (HEL) was determined as half the difference between the two principal stresses. The present work demonstrates the usefulness of the transverse gauge technique in determining this important parameter, which for the hot pressed AlN specimens was found to be almost constant and 10 % greater than that at the HEL for shock amplitudes between 1 and 2 x HEL.

I. INTRODUCTION

Current interest in the dynamic strength of ceramics is driven by their potential applications as armor and in internal combustion engines. Aluminum nitride has recently emerged as a very effective armor material [1]. It has been shown that the ballistic performance of other ceramics is related to Hugoniot elastic limit (HEL) values and shear strength above the HEL [2,3]. Therefore, we have conducted a shock characterization study of armor-grade AlN materials. Stress in the shock direction (σ_1) and in the transverse direction (σ_2) were measured with embedded in-material piezoresistive gauges. The shear stress is proportional to ($\sigma_1 - \sigma_2$). This technique has been previously used to measure strength of other ceramics [4].

II. EXPERIMENTAL TECHNIQUES

A. *MATERIALS*

Hot Pressed (HP) and Sintered (S) Aluminum nitride (AlN) plates (about 4"x4"x0.5"), were supplied by Dow Chemical Company, Midland, MI 48667. Table 1 lists physical properties of these materials measured by Dow Chemical.

B. *EXPERIMENTAL CONFIGURATION*

Plate impact experiments were performed using the 50 mm gas/powder gun at the University of Dayton Research Institute (UDRI). Three types of Micro-Measurements manganin gauges were used. For longitudinal stress measurements we used the standard gauge (LM-SS-125CH-048) and a somewhat larger gauge with integral leads encapsulated in epoxy sheets (LM-SS-110FD-050). For the lateral or transverse stress measurements we used a narrow gauge (C-880113-B), which is about 2 mm wide, in order to reduce the time for the shock wave to sweep the width of the gauge.

  Three different target configurations were used, each providing certain advantages for data recording, as shown in Figure 1. Configuration A results in the best time resolution for the initial shock and is thus best suited for HEL measurements. Configuration B is used for transverse gauges; the longitudinal gauge shown in Figure 2 (B) is optional. Configuration C provides the

Table 1. Physical Properties of AlN

Property	Hot Pressed	Sintered (Typical)
Density (g/cm^3)	3.226	3.20
Porosity (%)	1	1-2
Grain Size (microns)	2.0	4-6
Hardness (GPa)	11.0	10
Longitudinal Wave Velocity (km/s)	10.72	10.45
Shear Wave Velocity (km/s)	6.27	
Poisson's Ratio	0.238	
Young's Modulus (GPa)	314	307-319
Bulk Modulus (GPa)	·201	
Shear Modulus (GPa)	127	
Compressive Strength (GPa)	4.0	
Flexure Strength (MPa)	319	214-298

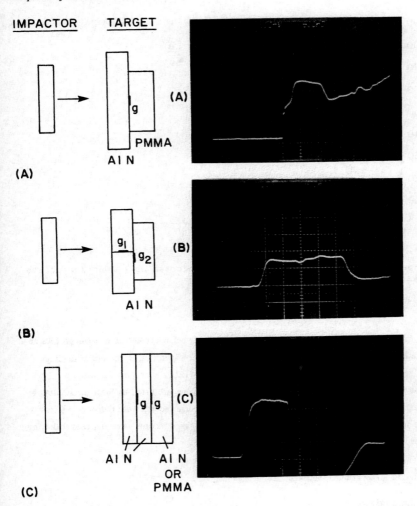

FIG. 1 *Schematic representation of different experimental configurations.*
(A) HEL measurement and manganin gauge profile from experiment 7-1403
(scales 0.5 v/div and 0.5 μs/div),(B) longitudinal (g_1) and transverse (g_2)
stress measuring gauges, and transverse gauge profile from experiment 7-
1345 (scales 0.5 v/div and 0.5 μs/div), and (C) longitudinal stress
measurement at two depths g_1 and g_2 from the impact surface of the target
and longitudinal gauge (g_1) profile from 7-1362 (scales 2 v/div and 0.5
μs/div)

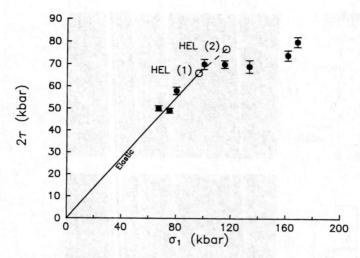

FIG. 2 *Measured shear stress (or strength) τ plotted as a function of the shock stress in HP AlN.*

best data for Hugoniot points, shock velocities and unloading behavior, since the stress profiles at two locations can be compared. In some experiments with Configuration B, the initial loading wave was reasonably resolved to permit identification of the HEL.

Seventeen experiments were performed. The parameters are listed in Table 2. A catalog of all measured profiles can be found in [4]. Longitudinal gauge records were reduced to stress-time histories using the calibration curve in [5]. Transverse gauge records were analyzed following the procedure outlined in [6].

III. RESULTS AND DISCUSSION

A. *HEL MEASUREMENTS*

As detailed in Table 2, HEL values were obtained in five experiments on hot pressed material and three tests on sintered materials. When analyzing the records, we associated the HEL with the first wave arrival, in accordance with the usual definition of the HEL stress as the amplitude of the elastic stress wave in the target. The HEL of hot pressed material was 94 ± 2 kbar. The sintered material was significantly weaker, HEL = 70 ± 4 kbar.

The gauge record from shot 7-1403, in hot pressed material, contained an anomalous second arrival just after the HEL at a peak stress of about 170 kbar, as shown in Figure 1 (A). This phenomenon was absent in the sintered material. However, similar wave forms have been

Table 2. Summary of Impact Experiments and Results

Shot No	Flyer Thick.(mm)/ Vel. (m/s)	Target/Thick. (mm)	Config.	HEL (kbar)	Long. Stress (kbar)	Trans. Stress (kbar)
7-1343	4/590	HP AlN/6.96	A	92±2		
7-1346	8/739	HP AlN/6.96	A	93±2		
7-1360	8/1115	Sint AlN/9.14	A	70±4		
7-1403	3/984	HP AlN/13.00	A	96±2		
7-1345	8/550	HP AlN/10.29	B		100±3[1]	30±1
7-1347	8/760	HP AlN/10.29	B		133±3[1]	64±2
7-1350	8/990	HP AlN/10.29	B	93±4	168±3[1]	88±1
7-1352	8/930	HP AlN/10.29	B	94±4	161±3[1]	87±1
7-1358	8/870	Sint AlN/9.14	B	71±5		
7-1361	8/770	Sint AlN/9.14	B	69±4		73±1
7-1367	8/626	HP AlN/9.10	B		115±2[2]	45±1
7-1393	10/378	HP AlN/8.00	B		67±1[1]	17±1
7-1406	3/410	HP AlN/13.00	B		75±1[1]	26±1
7-1407	3/433	HP AlN/12.95	B		80±1[1]	22±2
7-1348	8/763	HP AlN/6.96	C		134±2[2]	
7-1362	8/1010	HP AlN/9.20	C		170±3[2]	
7-1363	8/1178	HP AlN/9.20	C		185±5[2]	

[1] Longitudinal stress inferred from Hugoniot
[2] Longitudinal stress measured directly

seen in hot pressed TiB_2 [8]. We also reported (but did not comment on) similar signals in fully dense sintered alumina [9]. The second jump in the stress record (Figure 1a) corresponds to a wave velocity very close to that of the first wave (longitudinal velocity), and its amplitude, assuming an elastic behavior, is equal to 116 ± 2 kbar. The "plastic" wave velocity is much lower than either of these, which suggests a much lower slope of the Hugoniot curve at stresses above the HEL-like signal.

We do not fully understand the cause of the second HEL-like signal. It may arise from operation of more than one slip system, as in LiF [10], or from a brittle-ductile transition as occurs in static testing [11].

B. HUGONIOT MEASUREMENT OF HP AlN

The Hugoniot of AlN in the stress (σ_1) - particle velocity (u_p) plane was determined using impedance matching techniques based on the impact velocity and the known copper Hugoniot. The data on the Hugoniot of HP AlN are summarized in Table 2.

We found that when the shock (longitudinal) stress was in the 130-180 kbar regime the standard manganin gauge (LM-SS-125CH-048) showed signs of failure due to conduction through epoxy around the gauge grid lines, or the epoxy around the gauge leads. In order to overcome this difficulty we used a slightly larger and thicker gauge package (LM-SS-210FD-050) with integral leads fully encapsulated between two epoxy sheets. We performed two shots (7-1362 and 7-1363) with these gauges at shock stresses of about 170 and 185 kbar, respectively. The gauge records from these two highest pressure shots still indicate some short-circuiting near the maximum stress levels. The increased uncertainty in peak stress is reflected in large error bounds in Table 2.

C. TRANSVERSE STRESS IN HP AlN

Figure 1 (B) shows a typical record from a transverse gauge. The relatively long rise time is due to the sweeping of the gauge by the shock. The downward dip in the middle of the plateau of the signal is the result of a small amplitude release wave originating from the epoxy layer at the longitudinal gauge plane.

The transverse stress in hot pressed AlN was measured in eight experiments. The shear stress, τ, was computed as half the difference of the principal stresses, and is plotted in Figure 2. Values of σ_1 were either used as measured, or (where no longitudinal gauge was embedded in the target or the gauge failed) computed by impedance matching using the AlN and the copper flyer plate Hugoniots.

In the elastic region, the transverse stress (σ_2) should be correlated with the longitudinal stress (σ_1) through Poisson's ratio (v) by

$$\sigma_2 = \frac{v}{1-v}\,\sigma_1$$

(1)

As seen in Figure 2, below the HEL, measured values of the shear stress, $(\sigma_1 - \sigma_2)/2$, are within about 4% of the theoretical value.

Above the HEL, the measured shear strength is constant within the uncertainty of the data. This behavior resembles that seen previously in alumina [4]. These preliminary observations are consistent with findings in [2] that the shear strength of ceramics after initial failure is of primary importance in determining their usefulness as armor materials.

The value of shear strength above the HEL is about 10 % higher than the theoretical value at the HEL computed from equation (1) (65 kbar). We can estimate the shear stress at the second HEL-like signal by assuming that the bulk modulus is constant. This gives $\sigma_1 - \sigma_2 = 77$ kbar; the measured value is 10 % lower than this. Whether this discrepancy is due to measurement error or changes in the elastic constants has not yet been determined.

We have only one experiment (shot 7-1361) with a transverse gauge in sintered material, $\sigma_2 = 73 \pm 1$ kbar. A close examination of Table 2 shows that in shot 7-1347, on a HP specimen, the impact velocity was very close to that in shot 7-1361. The transverse stress of the HP specimen in this shot was 64 ± 2 kbar, about 10 kbar lower than that of the sintered material. Taking into account the fact that σ_1 is higher by about 5-10 kbar for the HP material (due to its higher HEL), we may expect a difference of 15-20 kbar in the $\sigma_1 - \sigma_2$ values of these materials. Therefore, the shear strength of hot pressed material is estimated to be 7-10 kbar higher than that of sintered specimens, a fact which is likely to have direct consequences on their ballistic performances as armor materials.

IV. CONCLUSIONS

The principal stresses resulting from shock loading hot pressed and sintered AIN were determined using manganin gauges. The HELs of these materials are, respectively, 94 ± 2 and 70 ± 4 kbar. Hot pressed AIN continues to strengthen above the HEL. At shock stress of 100 to 185 kbar, the shear strength is constant. This behavior may help explain the superior armor performance of hot pressed AIN.

ACKNOWLEDGEMENTS

This work was primarily sponsored by the U.S. Army Research Office under contract DAAL03-88-K-0203. Additional sponsorship was provided by Dow Chemical Company (P.O. 903798) and Sandia National Laboratories (P.O. 40-5226)

REFERENCES

1. W. Rafaniello, F.P. Khoury, 92nd Annual mtg. Am. Ceramic Soc., Dallas TX April 22-26, (1990).

2. Z. Rosenberg, S.J. Bless, and N.S. Brar, Int. J. Impact Eng. 9:45 (1990).

3. Z. Rosenberg and J. Tsaliah, Int. J. Impact Eng. 9:247 (1990).

4. Z. Rosenberg, D. Yaziv, Y. Yeshurun, and S.J. Bless, J. Appl. Phys., 62:1120 (1987).

5. Z. Rosenberg, N.S. Brar and S.J. Bless, University of Dayton Research Institute report UDR-TR-89-74 (1989).

6. Z. Rosenberg, D. Yaziv, and Y. Partom, J. Appl. Phys. 51:3702-3705 (1980).

7. Z. Rosenberg and Y. Partom, J. Appl. Phys. 58:3072 (1985).

8. M.E. Kip and D.E. Grady, Sandia National Laboratory report SAND89-1461 (1989).

9. Z. Rosenberg, N.S. Brar, and S.J. Bless, J. de Physique, Colloque C3, 49:707 (1988).

10. N. Kidron, Ph.D. Thesis, Washington State University, Pullman, WA (1983).

11. H.C. Heard and C.F. Cline, J. Mat. Sci., 15:1889 (1980).

97

Response of Alpha-Aluminum Oxide to Shock Impact

Y. WANG and D. E. MIKKOLA

Department of Metallurgical and Materials Engineering
Michigan Technological University
Houghton, Michigan 49931, U.S.A.

The response of $\alpha\text{-}Al_2O_3$ to shock impact has been investigated in terms of substructure development, and the relationship between plastic flow and fracture, including microcracking. Both alumina single crystals and polycrystals were explosively shock loaded permitting the influence of crystallographic orientation and grain boundaries on the defect substructures and fracture to be examined. Several active slip systems, along with basal twinning, were observed. Crystallographic orientation and grain boundaries had significant influences on plastic deformation and fracture. Shock loading with (10$\bar{1}$2) perpendicular to the shock wave direction, introduced the most intense defect substructure. The observations indicate a close relationship between plastic flow and fracture.

I. INTRODUCTION

The response of materials to shock loading involves both: the defect substructures formed by plastic deformation, including dislocations, twins, and stacking faults; and fracture, including microcracking. Although there have been many studies of shock impact effects on the structure of $\alpha\text{-}Al_2O_3$ powders, and powder compacts, as well as some observations on bulk materials [1-12], more work is needed to define the microscopic plastic deformation mechanisms for shocked alumina, and other ceramic materials. Also, although the crystallographic orientation with respect to the shock wave direction should have significant influences on the substructures developed, there is apparently no detailed information about the effects of crystallographic orientation on defect substructures. The purpose here is to comment on studies in progress on alumina of both the plastic deformation effects and the influences of crystallographic orientations and grain boundaries on plastic deformation and fracture under different shock loading conditions.

II. EXPERIMENTAL METHODS

Rectangular specimens of both high purity single crystals and polycrystals of α-Al_2O_3 were shock loaded at the Idaho National Engineering Laboratory using the explosive-driven plate impact method under three conditions: 23 GPa, 5 μs; 12 GPa, 10 μs; and 5 GPa, 23 μs. The polycrystals had an average grain size of 25 μm. Single crystals were shock loaded with (0001), (10$\bar{1}$2), or (11$\bar{2}$3) planes perpendicular to the shock wave direction. Details concerning the experiments are given elsewhere [11,13]. It should be noted that the nature of the confinement permitted numerous reflected waves so that fragmentation of the specimens occurred. In addition to studies of the defects with transmission electron microscopy (TEM), active slip systems and crystallographic crack planes were determined through trace analysis by TEM and from optical observations on different surfaces.

III. RESULTS AND DISCUSSION

A. PLASTIC DEFORMATION

Features of Dislocations and Twins Fig. 1 shows an example of a fragmented single crystal specimen resulting from shocking at 23 GPa and 5 μs with the basal plane perpendicular to the shock wave direction. There are a significant number of slip bands in different crystallographic directions, coupled with surface relief, suggesting that a large amount of plastic deformation has occurred. Also, there are a lot of cracks, many of which are oriented in specific directions, about 70° apart from one another, suggesting a higher index crack plane. However, these crystallographic cracks are apparently on slip planes.

Fig. 2 is a transmission electron image of a single crystal shocked at 23 GPa, 5 μs, perpendicular to (10$\bar{1}$2) plane. A large number of band-like

500 μm

Fig. 1 Fragmented single crystal specimen shocked at 23 GPa, 5μs, perpendicular to the basal plane. Basal plane surface.

defect structures occur that are aligned in one direction. Diffraction analysis shows that the bands consist of basal twins, with arrays of dislocations occurring along the twin boundaries. There are also large numbers of dislocations that are not associated with the twin boundaries.

Overlapping dislocation configurations on successive basal planes are shown in Fig. 3(a). Most of these dislocations are not associated with the twin boundaries, but most lie along specific crystallographic orientations, such as <10$\bar{1}$0> and <11$\bar{2}$0> and have Burgers vectors <10$\bar{1}$0> and $\frac{1}{3}$<11$\bar{2}$0>. There are also some nonbasal dislocations, which glide on planes such as {10$\bar{1}$0} and {11$\bar{2}$0}. Generally, there are more dislocations within twins than in the matrix. The closely-spaced parallel curved dislocations (as shown by arrows) are those associated with the twin boundaries. These dislocations glide on the basal twin interfaces with Burgers vectors of $\frac{1}{3}$<10$\bar{1}$0>, and are emitted from sources on different planes. A new model for basal twinning has been formulated based on the TEM observations [13].

Fig. 2 Transmission electron image of a single crystal shocked at 23 (;
GPa, 5 µs, perpendicular to (1012). Foil plane: (1012)

Basal twins are shown "edge-on" in Fig. 3(b), which shows a bright
field image as well as two dark field images formed with a $(1\bar{1}02)_T$ twin
spot and a $(1\bar{1}04)_M$ matrix spot, respectively. This single crystal was
shocked at 12 GPa, 10 µs, perpendicular to (0001). Generally, basal twins
were uniformly distributed, and very few rhombohedral twins were observed
in the shocked single crystals. Twin lamellae usually extended through
whole grains in polycrystals or entirely through the single crystals. Most
twins appeared to nucleate at grain boundaries in polycrystals, or at
specimen surfaces in single crystals, with few of them nucleating at other
heterogeneous sites in the bulk. It appears that the development of slip
bands leads to the formation of twins, and the thin twin lamellae formed
initially coalesce to form thicker twins as the deformation proceeds.
Various slip systems can be activated through the suppression of fracture
by large hydrostatic stresses during shock loading [4,13]. The slip

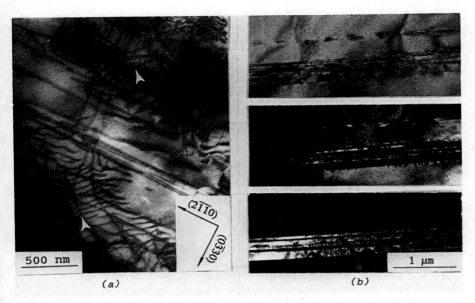

Fig. 3 Single crystal shocked normal to basal plane: (a) overlapping dislocation configurations on successive planes. 23 GPa, 5μs, basal foil. (b) edge-on view of basal twins. 12 GPa, 10μs, perpendicular to (0001). [1120] beam direction. Top - bright field, middle - dark field with $(110\overline{4})_m$, and bottom - dark field with $(110\overline{2})_t$.

systems observed here were $\{0001\}\frac{1}{3}<11\overline{2}0>$, $\{0001\}<10\overline{1}0>$, $\{10\overline{1}0\}\frac{1}{3}<1\overline{2}10>$, $\{11\overline{2}3\}<1\overline{1}00>$, and $\{\overline{1}2\overline{3}0\}<\overline{5}410>$. The first three systems have also been observed in shocked alumina by Yust and Harris [4]. For all of these systems (except the basal plane systems), the slip planes were confirmed in the current work by the analysis of traces of the slip bands through optical microscopy, and the slip directions were arrived at based on crystallography. Most of the slip systems were confirmed by TEM, including $\{0001\}\frac{1}{3}<11\overline{2}0>$, $\{0001\}<10\overline{1}0>$, $\{10\overline{1}0\}\frac{1}{3}<1\overline{2}10>$, and $\{11\overline{2}3\}<1\overline{1}00>$.

Influences of Crystal Orientation and Grain Boundaries Fig. 4 shows the volume fraction of twins vs. shock pressure for single crystals of different orientations. Crystals with a $(10\overline{1}2)$ orientation have the largest value of Schmid factor for basal twinning; $(11\overline{2}3)$ crystals next; and Schmid factor is zero for (0001) crystals. The observations were consistent with this order. In the ideal case, no twins are expected for

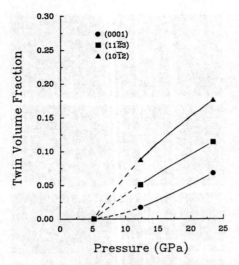

Fig. 4 Volume fraction of twins vs. shock loading pressure

the basal plane crystal so that the observed twins apparently result from
slight crystal misorientations and/or contributions from reflected waves of
various orientations. The behavior of the dislocation densities with
pressure and orientation was generally similar to that for twins.

The fragments from polycrystals, especially at the highest pressure,
appeared to have lower defect densities, which implies less plastic
deformation than the single crystals. This can be attributed to the
influence of grain boundaries on fracture initiation through slip
incompatibility between grains, as well as particular defect/grain boundary
interactions. Also, the numbers of defects generated in different grains
varied, but the fraction with high defect densities increased with
pressure.

B. FRACTURE

Both crystallographic and randomly oriented microcracks were observed. It
is concluded that the random cracks formed after the generation of the
basal dislocations and the formation of the crystallographic cracks, and
then branched during the propagation. There was no evidence for
microplastic deformation, or dislocations, at the tips of most random
cracks even though a few of them indicated interactions with dislocations.

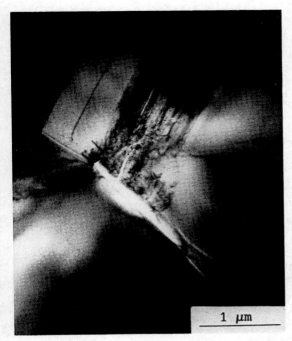

1 μm

Fig. 5 Microcrack nucleated at the intersection of twins with a grain boundary, and propagated along the grain boundary. 23 GPa, 5μs.

Fig. 5 gives evidence for nucleation of cracks at the intersection of bundles of twins with a grain boundary in a polycrystal shocked at 23 GPa.

The mean fragment sizes and fracture-created area per unit volume, determined by screen analyses for different shock loading conditions, are shown in Fig. 6. The differences among the shocked single crystals of different orientations are not large; however, the polycrystals show a strong dependence of fragmentation on pressure. At 23 GPa, the polycrystals suffered considerably more fragmentation than any of the single crystals. This is attributed to the more favorable conditions for nucleating cracks, such as through interactions of twins and dislocations with grain boundaries.

Crystallographic cracks were observed on planes such as $\{11\bar{2}3\}$, $\{10\bar{1}0\}$ and $\{\bar{1}\bar{2}30\}$ [13]. With single crystals, the ratio of crystallographic cracks to randomly oriented cracks increased with increased pressure, indicating a close relationship between microplastic deformation and the

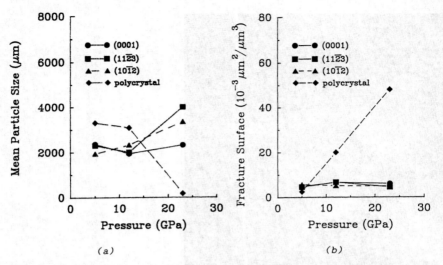

Fig. 6 (a) Mean fragment size vs. shock pressure, (b) fracture surface area per unit volume vs. shock loading pressure

crystallographic cracking. However, the total surface area created by fracture did not increase significantly with pressure. Most cracking in polycrystals was along grain boundaries, with some intragranular cracks.

C. RELATIONSHIP BETWEEN PLASTIC FLOW AND FRACTURE

At high shock pressures, single crystals show considerably less fragmentation than the polycrystals studied. A related phenomenon, that is, high defect density corresponding to less fragmentation, was also found in comparing different shock loaded polycrystalline aluminas [11]. These observations argue that the relief of high local stresses depends directly on the near-balanced competition between plastic flow and fracture [11]. If more of the input energy for given shock loading conditions is involved in relief of stresses through generation of defects, such as dislocations and twins, which contribute to plastic deformation, fracture is inhibited, resulting in less fragmentation. For single crystals, the amount of plastic flow increased with shock pressure, so that the degree of fragmentation did not change significantly. With polycrystals, the grain boundaries served as important crack initiation sites, and crack propagation paths, thus contributing to fragmentation, especially at high pressures.

III. SUMMARY

Slip systems, such as $\{0001\}\frac{1}{3}<11\bar{2}0>$, $\{0001\}<10\bar{1}0>$, $\{10\bar{1}0\}\frac{1}{3}<1\bar{2}10>$, $\{11\bar{2}3\}<1\bar{1}00>$, and $\{\bar{1}230\}<\bar{5}410>$, have been activated in sapphire single crystals through the suppression of fracture by large hydrostatic pressures during shock loading. Changing the crystallographic orientation with respect to the shock wave direction has a significant influence on the substructures developed [13]. Shock loading at 23 GPa, with $(10\bar{1}2)$ perpendicular to the shock wave direction, introduced the most intense defect substructures, e.g., a volume fraction of twins of 0.176 was observed, which suggests that the plastic strain arising from the twinning shear can be as large as 9 pct. Other orientations had less twinning.

The degree of fragmentation among the single crystals of different orientations was similar. With increasing shock pressure (5 to 23 GPa), the relative ratio of crystalline cracks to random cracks increased, however, the total new surface created by fracture was almost unchanged. On the other hand, fragmentation of shocked polycrystals was strongly dependent on pressure; at lower pressures the fragmentation was slightly less than for single crystals, but for 23 GPa the fragments were almost an order of magnitude smaller than for all other specimens. In the polycrystals, the importance of grain boundaries was shown by the fact that cracks usually nucleated at the intersections of bundles of twins with grain boundaries, with most cracks propagating along the grain boundaries. Generally, the observations support the concept that increased plastic deformation can inhibit fragmentation.

The authors gratefully acknowledge the assistance and advice provided by P. V. Kelsey, formerly with the Alcoa Technical Center, and G. E. Korth, R. L. Williamson, and J. E. Flinn of the Idaho National Engineering Laboratory.

REFERENCES

1. O. R. Bergmann and J. Barrington, *J. Am. Ceram. Soc.* 49: 502 (1966).

2. R. W. Heckel and J. L. Youngblood, *J. Am. Ceram. Soc.* 51: 398 (1968).

3. R. A. Pruemmer and G. Ziegler, *Powder Metall. Inter.* 9: 11 (1977).

4. C. S. Yust and L. A. Harris in *Shock Waves and High-Strain-Rate Phenomena in Metals*, M. Meyers and L. Murr (eds.), Plenum Press, New York, 1981, p. 881.

5.   A. Sawaoka, K. Kondo and T. Akashi in *Report of Res.Lab. of Eng.
     Matls., Tokyo Inst. of Tech.* V. 4 (1979).

6.   B. Morosin and R. A. Graham, *Mater. Sci. Eng.* 66: 73 (1984).

7.   C. L. Hoeing and C. S. Yust, *Am. Ceram. Soc. Bull.* 60: 1175 (1981).

8.   E. K. Beauchamp, R. A. Graham and M. J. Carr, *Mat. Res. Soc. Symp.
     Proc.* 24: 281 (1984).

9.   E. K. Beauchamp, M. J. Carr and R. A. Graham, *J. Am. Ceram. Soc.* 68:
     696 (1985).

10. D. G. Howitt, J. P. Heuer, P. V. Kelsey and J. E. Flinn in *Microbeam
     Analysis - 1987,* R. H. Geiss (ed.), San Francisco Press, Inc., Box
     6800, San Francisco, CA 94101-6800, p. 227.

11. J. A. Brusso, D. E. Mikkola, J. E. Flinn and P. V. Kelsey, *Scripta
     Metall.* 22: 47 (1988).

12. L. H. L. Louro and M. A. Meyers, *J. Mater. Sci.*, 24: 2516 (1989).

13. Y. Wang and D. E. Mikkola: unpublished research, Mich. Tech. Univ.,
     Houghton, MI 49931.

98

Dynamic Fracture and Failure Mechanisms of Ceramic Bars

N. S. BRAR and S. J. BLESS

University of Dayton Research Institute
Dayton, Ohio 45469-0180, U.S.A.

Failure of ceramics can be studied with instrumented bar impacts and high speed photography. When the impact stress is below the ceramic strength, there is no fracture of the bar, the stress wave in the bar has a relatively long duration, and the measured peak stress in the bar almost equals the impact stress. When the bar fails, fractures propagate away from the impact interface, the stress pulse is short, and the measured peak stress is less than the impact stress. Failure in alumina bars is by axial splitting; in glass, TiB_2, and SiC it is by a propagating destruction wave.

I. INTRODUCTION

The bar impact experiment on brittle materials (glass and ceramics) is a useful technique for measuring the yield strength of ductile and brittle materials at strain rates in the range 10^2-10^4 s^{-1}. This test can also be used to quantify dilatant behavior in brittle materials.

For ductile materials, one uses an impact velocity that causes plastic flow on the impact face of the bar. An elastic wave propagates down the bar, whose amplitude is measured by embedding a thin manganin gauge in the target 10 diameters away from the impact end, using Rosenberg et al's [1] calibration. The amplitude of the stress wave is the largest compressive elastic stress that can be sustained in the bar material, σ_c. Less than about 2 diameters from the impact end the stress state is 1-d strain, so $\sigma_c = \sigma_{HEL}$. At larger distances $\sigma_c \rightarrow Y_0$, the yield stress for 1-d stress deformation. We have reported the yield stress of a number of ductile materials using this technique [2].

In tests on brittle aluminas with thick steel plate impactors, measured stress (σ_m) at 10 diameters was almost equal to the impact stress (σ_i) when $\sigma_i \leq \sigma_c$ [3,4]. In these cases, a stable compressive wave was recorded at the gauge location. When $\sigma_i > \sigma_c$, gauge profiles were of very short duration (1-2 μs), and $\sigma_m << \sigma_i$. A similar finding was reported by Janach [5] for sandstone and granite bars. Following Janach's simple dilatational model, we were able to explain the observed behavior of the alumina bars [3].

In the present paper we emphasize observations regarding failure processes in brittle bars. Tests were conducted on glass and several ceramics.

II. EXPERIMENTAL METHOD

A. *MATERIALS*

Titanium diboride (TiB$_2$), silicon carbide (SiC), boron carbide (B$_4$C) and hot pressed alumina (Al$_2$O$_3$) bars in the form of 10 mm x 10 mm square prisms were supplied by GTE Products Corporation, Towanda, PA. TiB$_2$, SiC, and HP Al$_2$O$_3$ bars were ground on all four edges so that the finished surfaces of the bars were octagons. Sintered alumina round bars of 12.7 mm diameter, AD-94 (94 % Al$_2$O$_3$) and AD-998 (99.8 % Al$_2$O$_3$), were purchased from Coors Porcelain, Golden CO. Pyrex bars of 12.7 mm diameter were supplied by Corning Glass Works, Corning, NY. Manganin gauges (Type C-880113-B) were purchased from Micro-Measurements, Raleigh, NC.

B. *EXPERIMENTAL CONFIGURATION*

Two types of projectiles were used as shown in Figure 1. For the octagon bar targets of TiB$_2$, SiC, and HP Al$_2$O$_3$ plates of similar materials were used as impactors. The sintered alumina and pyrex bars

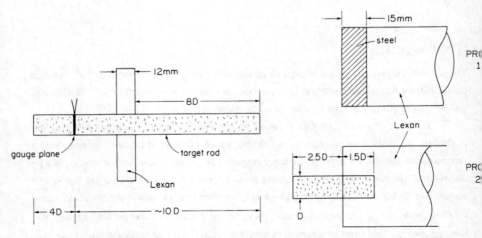

FIG. 1 *Schematic representation of the impactors and the target in the bar impact experiment.*

were impacted with bars with L/D=4-5 of the same material. In the case of plate impact, the impact scenario is similar to a Taylor anvil test. With bar impactors, the scenario is similar to the Hopkinson bar configuration at strain rates in the range of 10^3 - 10^4 s^{-1}. Manganin gauges were embedded at 10 diameters away from the impact end for most of the experiments.

Projectiles (plate or bar) were launched using lexan sabots in a 50 mm gas/powder gun at the University of Dayton Research Institute. The bar targets were aligned for a planar impact using a special fixture [6]. An Imacon 790 high speed camera was used to photograph the fracture pattern in most experiments. Alumina bars were painted black so that the cracks and faults could be distinguished. TiB$_2$, SiC, B$_4$C, and HP Al$_2$O$_3$ bars were painted white for the same reason. The camera speed was 10^5/s, except for two experiment on pyrex at 10^6/s.

III. RESULTS and DISCUSSION

A total of twenty experiments were conducted. The parameters are listed in Table 1. Typical photographs of fracture patterns obtained on alumina, pyrex, and TiB$_2$ bars are shown in Figure 2a-c. Manganin gauge (stress-time) profiles obtained in two shots on alumina, one shot on pyrex, and one shot on TiB$_2$ are shown in Figures 3a-d.

In general, three types of fracture were observed, which are labelled I, S, and D in Table 1. Type I refers to pulverization of the bar on the impact plane. The impact face of the rod becomes

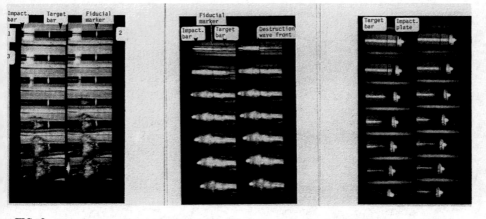

FIG. 2 *Imaging camera photographs of bar impact. (a) AD-998 bar/bar experiment 7-1385, (b) pyrex bar/bar experiment 7-1386, and (c) TiB$_2$ plate/bar experiment 7-1420. Frames are 10 μs apart. The fiducial on target bars in all the photographs is 25.4 mm from the impact face.*

Table 1. Summary of Impact Experiments and Results

Shot No 7-	Imp. Type/ Dim. (mm)	Imp. Vel. (m/s)	Target Material/ Config.	Target Tilt (deg)	Strain Rate (s^{-1})	Gauge Profile Width at 75 % of Peak (μs)	Measured Stress (kbar)	Fract. Type
1274	steel plate T=15	563	AD-998 12.7mm+g_1 +17mm+g_2 +50mm	-	12000 5900	0.4±0.02 1.2±0.1	σ_1=67±2 σ_2=42±2	S
1276	steel plate T=15	595	AD-998 12.7mm+g_1 +17mm+g_2 +50mm	-	14450 11560	0.5±0.02 0.8±0.1	σ_1=58±2 σ_2=43±2	S
1385	AD-998 bar L=50 D=12.7	306	AD-998 127mm+ g+50mm	0±0.5	17700	1.3±0.1	σ=39.5±1	S
1389	AD-998 bar L=63.5 D=12.7	99	AD-998 127mm+ g+50mm	0±0.5	1970	8.5±0.1	σ=21±1	S
1390	AD-998 bar T=12.7	133	AD-998 127mm+ g+50mm	0±0.5	-	-	-	S
1392	AD-998 bar L=63.5 D=12.7	154	AD-998 127mm+ g+50mm	0±0.5	-	-	-	S
1314	steel plate T=15	228	AD-94 127mm+ g+50mm	-	3140	1.6±0.1	σ=29±1	S
1329	AD-998 bar L=50 D=12.7	230	AD-94 127mm+ g+50mm	0±0.5	4410	2±0.1	σ=41±1	S
1331	AD-94 bar L=50 D=12.7	297	AD-94 127mm+ g+50mm	0±0.5	5610	1±0.1	σ=27±1	S
1312	steel plate T=15	210	pyrex 150mm+g +46.4mm	-	5500	2±0.2	σ=15±1	D
1313	steel plate T=15	145	pyrex 152mm+ g+47mm	-	3968	2.4±0.1	σ=15±1	D

Shot No 7-	Imp. Type/ Dim. (mm)	Imp. Vel. (m/s)	Target Material/ Config.	Target Tilt (deg)	Strain Rate (s^{-1})	Gauge Profile Width at 75 % of Peak (μs)	Measured Stress (kbar)	Fract. Type
1386	pyrex bar L=63.5 D=12.7	336	pyrex 127mm	0±0.5	-	-	-	D
1387	pyrex bar L=63.5 D=12.7	227	pyrex 127mm+ g+25mm	0±0.5	4900	3.6±0.1	$\sigma=14\pm1$	D
1388	pyrex bar L=63.5 D=12.7	123	pyrex 127mm+ g+25mm	0±0.5	2800	7.5±0.2	$\sigma=9\pm1$	D
1316	steel plate T=15	230	B_4C 127mm	0±0.5	-	-	-	S
1420	TiB_2 plate T=12.7	289	TiB_2 octagon 127mm+g +25.4mm	1±0.5	9000	>0.5	$\sigma=49\pm1$	D
1443	TiB_2 plate T=12.7	300	TiB_2 octagon 127mm+g +25.4mm	-	2800	1.1±0.1	$\sigma=24\pm1$	-
1421	HP Al_2O_3 plate T=14.5	289	HP Al_2O_3 octagon 127mm+g +25.4mm	0±0.5	3300	1.3±0.1	$\sigma=41\pm1$	I
1445	HP Al_2O_3 plate T=14.5	262	HP Al_2O_3 octagon 127mm+g +25.4mm	4±0.5	2100	1.4±0.1	$\sigma=16.5\pm1$	S
1422	SiC plate T=10	292	SiC octagon 127mm+g +25.4mm	2±0.5	2700	1.2±0.1	$\sigma=22\pm1$	D
1444	SiC plate T=10	307	SiC octagon 127mm+g +25.4mm	1±0.5	8700	1.4±0.1	$\sigma=48\pm1$	D

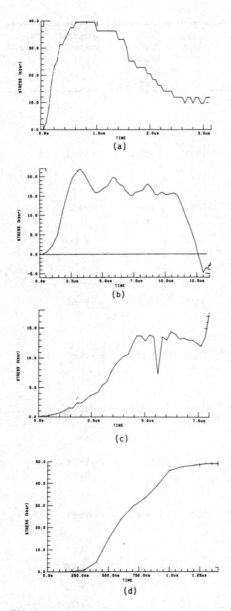

FIG. 3 *Manganin gauge (stress-time) profiles obtained in (a) AD-998 bar/bar experiment 7-1385 at impact velocity of 306 m/s, (b) AD-998 bar/bar experiment 7-1389 at impact velocity of 99 m/s, (c) pyrex bar/bar experiment 7-1387 at impact velocity of 227 m/s, and (d) TiB$_2$ plate/bar experiment 7-1420 at impact velocity of 289 m/s.*

obscured by a cloud of debris, which advances up the bar at the projectile/bar interface velocity, and which extends about one diameter away from the bar/projectile interface. Type S refers to axial splitting, or faulting. Primary axial cracks occur, which branch, leading to breakup of the bar into fragments whose typical dimension is about a quarter of the bar diameter. The fracture front runs out between typically 4 to 6 diameters from the impact plane, and then continues to advance up the bar at the interface velocity. Type D refers to a propagating destruction wave. The material behind the wave is pulverized, and the wave propagates much faster than the interface velocity.

A. ALUMINA

Except at very low velocity, the fracture pattern in alumina was of the S type. There was no evident correlation between the fracture pattern and the duration of the stress or the value of σ_m/σ_i. Nor did hot pressed material differ from sintered material.

Lateral expansion of the AD998 bar/bar impact experiment 7-1385 (shown in Figure 2a) at the impact face and at 1, 2, and 4 diameters away from the impact face is shown in Figure 4. The ratio of σ_m/σ_i in this experiment was 0.7, which indicates that the peak stress was attenuated before reaching the gauge. This is borne out by the stress profile, Figure 3a, which indicates a duration of only 1.8 μs. In contrast, in shot 7-1389 at 99 m/s, the ratio of σ_m/σ_i was 1.1. Here the gauge recorded the peak stress, the stress was sustained, and unloading takes place by the wave reflected back from the rear end of the bar.

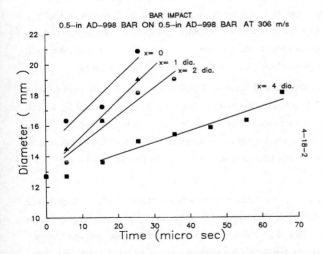

BAR IMPACT
0.5-in AD-998 BAR ON 0.5-in AD-998 BAR AT 306 m/s

FIG. 4 *Lateral expansion of the AD-998 target bar in experiment 7-1385 as a function of the distance from the impact face of the bar.*

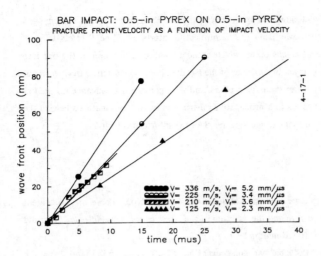

FIG. 5 *Fracture or destruction wave front velocity as a function of the impact velocity in pyrex bar/bar experiments.*

B. PYREX, TiB₂, SiC

Figure 2(b) shows the photograph from shot 7-1386. It is evident that the glass bar was consumed by a propagating damage or destruction wave front. There is a violent radial expansion. The position vs. time of the destruction wave front as a function of the impact velocity is shown in Figure 5. Destruction wave speeds were approximately constant within a given test.   The highest velocity observed was 5.2 mm/μs. This exceeds speeds seen in a previous study [7] of spherically divergent plane-stress damage in soda lime glass, where the fronts propagated at 1.46 to 4.7 mm/μs.

There have only been two shots onto TiB_2, shots 7-1420 and 7-1443, and photographs were only obtained in 7-1420 (Figure 2c). A destruction wave was observed in the material. σ_m/σ_i for this shot was about 0.7. The gauge broke after about 1.5 μs (Figure 3c), thus a good measurement of pulse duration was not obtained. The propagation speed of the destruction wave is difficult to determine because the boundary of the fracture is not clear in opaque bars. Our best estimate is 2.4 mm/μs; this is probably a lower bound.

Two impacts were conducted at SiC bars at essentially the same velocity - shots 7-1422 (Figures 2d, 3d), and 7-1444. There was 2° tilt in 7-1422. The result was a relatively quiescent splitting fracture. The measured pulse had a relatively long rise time, and the peak was only 22 kbar. In contrast, the tilt was 1° in 7-1444. A fast destruction wave was observed (speed > 4.7 mm/μs), the strain rate was much higher, and the peak stress was 48 kbar. In this shot, σ_m/σ_i is about 1.

Our preliminary interpretation is shot 7-1422 is invalid due to excessive target tilt, which caused early failure before the peak stress was obtained. In SiC, this failure involved splitting cracks. In 7-1444, the peak stress was realized in the bar. However, there was still sufficient tilt to cause impact failure. In this case, the failure took place in material under much higher stress and a propagating destruction wave resulted.

IV. CONCLUSIONS

The bar impact test is a useful technique to study failure in brittle materials. Failure in glass and in highly stressed TiB_2 and SiC takes place by a propagating destruction wave. Failure in alumina involves axial splitting leading to faulting.

ACKNOWLEDGMENTS

This work was sponsored by the U.S. Army Research Office under contract DAAL03-88-K-0203. Mark Laber is thanked for his assistance in conducting the experiments.

REFERENCES

1. Z. Rosenberg, M. Mayseless, and Y. Partom, Trans. Am. Soc. Mech. Eng., 51:202 (1984).

2. Z. Rosenberg, and S. J. Bless, Exp. Mech., 26:279 (1986).

3. N. S. Brar, S. J. Bless, and Z. Rosenberg, J. de Physique, Coll. C3, 49:607 (1988).

4. S. J. Bless, N. S. Brar, and Z. Rosenberg, Shock Compression of Condensed Matter, 1989, Eds. S. C. Schmidt, J. N. Johnson, and L. W. Davison, Elsevier Science Pub. p. 939 (1990).

5. W. Janach, Int. J. Rock Mech. Min. Sci., Geomech. Abstr. 13:177 (1976).

6. N. S. Brar and S. J. Bless, Bar Impact Tests on Ceramics and Glass, University of Dayton Research Institute Report, 1990 (in preparation).

7. U. Hornemann and J. Kaltoff, Inst. of Physics Conf. Series No. 70. Oxford, UK (1984).

99

Ballistic Impact Behavior of SIC Reinforced Aluminum Alloy Matrix Composites

S. J. BLESS[1], D. L. JURICK[1], S. P. TIMOTHY[2,3], and M. A. REYNOLDS[2]

[1]University of Dayton Research Institute, Dayton OH 45469-0180, U.S.A.

[2]Alcan International, Ltd., Banbury Laboratory, Banbury UK
[3]present address: CEST, Manchester Science Park, Manchester UK

The ballistic impact performance of three experimental aluminum alloy matrix composites (reinforced with 10% and 20% by volume of SiC particles) was investigated. Projectiles were sintered tungsten at a nominal impact velocity of 1.2 km/s. Target blocks were confined and supported by 6061-T651 aluminum. Performance of the test materials was evaluated using differential efficiency factors, relative to the 6061-T651 alloy. The reinforced aluminum was up to three times more efficient than the unfilled alloy. The superior performance of the composite was due to increased projectile deformation.

I. INTRODUCTION

The evolution of new types of tungsten projectiles threatens to make much lightweight armor technology obsolete. Aluminum plays a major role in armored vehicles, both as a structural armor and as applique armor. However, aluminum alloys are particularly inefficient at stopping modern tungsten bullets. For example, the 22 g penetrator used in these studies will penetrate about 190 mm of 6061-T651 aluminum.

The development of metal matrix composites (MMC) having increased hardness and modulus may provide a means of stepping up the penetration resistance of standard aluminum to meet this category of threats. Therefore, a number of ceramic reinforced aluminum compositions produced by Alcan Banbury Laboratory were evaluated.

Table 1. Materials

Code	Base Alloy	SiC %	ρ (g/cm^3)	K^* (GPa)	υ	Y (GPa)
IZP	2014	20	2.87	90.6	0.305	471
IZQ	2014	10	2.83	78.8	0.325	449
IZR	6061	10	2.84	75.9	0.325	321
	2014	0	2.79	72.5	0.330	448
	6061	0	2.71	66.6	0.330	261

* Calculated from measured values of Young's modulus and Poisson's ratio assuming isotropy of elastic properties.

II. MATERIALS

The MMC materials used in this study were reinforced with SiC particles; 2014-T6 and 6061-T6 alloys were reinforced with 10% and 20% of SiC particles by volume, with a mean particle size of 10-15 μm. Relevant material properties are compared in Table 1 with those of the unreinforced alloys. The symbols in table 1 are ρ (density), K (bulk modulus), v (Poisson's ratio), and Y (flow stress, used in analysis).

Individual target blocks 73.0mm x 73.0mm x 24.1 mm were machined from the host material. The target was made by stacking two blocks together.

III. APPROACH

Penetration resistance or armor performance is usually characterized by mass efficiency, e, defined as the ratio of the areal density of reference armor that will just stop a projectile (W_{REF}) to the areal density of the test armor that will just stop the same projectile. Usually the reference armor is armor steel. However, in the present case, we use as a reference material 6061-T651 aluminum.

Penetration resistance was evaluated by calculating the differential efficiency of the MMC materials using thick backing plate geometry. The MMC blocks were placed in a steel frame to avoid excessive expansion (and hence, easy penetration) due to plastic bulging or radial fracture. The blocks were backed by a thick section of 6061-T651. The Differential Efficiency Factor (e_Δ) [1] of the test element is computed from

$$e_\Delta = \frac{W_{REF} - W_R}{W_A} \qquad (1)$$

where W_A is the areal density of the armor element, W_R is the penetrated areal density under the test armor (density x residual penetration). Residual penetration was measured by radiographing post-impact section of the 6061-T651 substrate.

If e_A = 1, then the performance of the test armor is equivalent to 6061-T651. If e_A = 2, the test armor provides protection at half the weight of 6061-T6.

The projectile mass was 22 g. It had a double conical nose (40° AND 10° half angles) and a major diameter of 7.7 mm. The projectile material was 95% sintered tungsten alloy, and the nominal impact velocity was 1.2 km/s. Except when otherwise indicated, the projectile was launched spin-stabilized. The material properties, shape and velocity of the projectile resemble those of several military tungsten alloy bullets that are used in .50 caliber and larger guns. Thus, the penetration resistance of materials in our experiments is a good indication of their usefulness in armor against this important class of projectiles.

The test matrix included unfilled materials to serve as a baseline. These were also shot in the steel frame since it was possible that the steel frame by itself could enhance the performance of small aluminum blocks. It also included some shots with round nose projectiles. These were added to measure the penetration in the reference material of a blunted projectile, W^B_{REF}. The reason for this is that we might expect $W_R \rightarrow W^B_{REF}$ as $W_A \rightarrow 0$, since the MMC elements blunt the sharp nose of the penetrator.

IV. EXPERIMENTAL RESULTS

The tests that were conducted and resulting data are listed in Table 2. Reference data derived from the tests were as follows: W_{REF} = 52.6 g/cm^2, and W^B_{REF} = 35.1 g/cm^2.

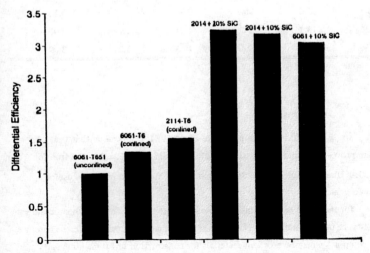

FIG 1 *Differential Efficiency of SiC-Aluminum MMC Materials and Small Aluminum tiles against Tungsten-Alloy Penetrator (referenced to thick 6061-T651).*

Table 2. Tests and Data; 6061-T6 Substrates

Shot No.	Material	W_A (g/cm^2)	Projectile	W_R (g/cm^2)	Velocity (km/s)	e_Δ
2665	6061-T6	13.6	Std.	37.5	0.93	-
2664	6061-T6	13.6	Std.	34.3	1.25	1.34
2669	-	0	Round Nse.	36.2	1.28	-
2668	-	0	Round Nse.	34.6	1.26	-
2667	-	0	Round Nse.	34.6	1.25	-
2578	-	0	Std.	52.4	1.20	-
2577	-	0	Std.	54.3	1.19	-
2579	-	0	Std.	51.0	1.20	-
1700	-	0	Std. unSpun	46.2	1.22	1.14
1701	-	0	Std. unspun	42.4	1.14	1.22
2666	2014-T6	0	Std.	31.2	1.28	1.56
2595	IZP	13.7	Std.	6.2	1.21	3.36
2586	IZP	13.8	Std.	7.1	1.21	3.29
2670	IZP	13.8	Std.	10.0	1.23	3.08
2588	IZQ	13.6	Std.	9.3	1.21	3.18
2590	IZR	13.7	Std.	11.4	1.21	3.00
2594	IZR	13.7	Std.	10.3	1.21	3.09

Values of e_Δ are listed in Table 2 and plotted in Figure 1. It can be seen that the confinement alone improves the penetration resistance of aluminum by 50%. This was true for both 6061-T6 and 2014-T6 aluminum. Thus, combining steel and aluminum may be a useful technique for armored vehicles.

Values of e_Δ for the metal matrix composites were greater than 3. Thus, by this measure the penetration resistance of filled aluminum is twice that of the matrix material alone. Therefore, these metal matrix composites may be excellent materials for construction of future hulls and appliques for armored vehicles.

V. ANALYSIS

Penetration of undeformed projectiles was analyzed using Forrestal's cavity expansion model [2]. This model allows calculation of penetration depth in materials which fail by ductile hole growth. Besides impact geometry, input parameters are the elastic constants and yield strength of the target. There is also a small dependence on the friction coefficient between the penetrator and the target.

Previous applications of this model to unspun steel and tungsten rods penetrating 6061-T6 aluminum found that friction coefficients of $\mu = 0.1$ and 0, respectively, gave the best fits to the data [2,3]. In this work, we found that 0.15 was the best value for the friction coefficient of round-nose projectiles; measured penetrations were predicted with an error of only 2%.

We modified the published version of the spherical cavity expansion model to account for the double conical nose of the projectile. The force on the projectile, F_z, was

$$F_z = C_1 + C_2 V^2 \qquad (2)$$

$$C_1 = \pi a_1^2 (1 + \frac{\mu}{\tan\phi_1}) \alpha_s + \pi a_2^2 (1 + \frac{\mu}{\tan\phi_2}) \alpha_s - \pi a_1^2 (1 + \frac{\mu}{\tan\phi_2}) \alpha_s$$

$$C_2 = \pi a_1^2 (1 + \frac{\mu}{\tan\phi_1}) \beta_s \sin^2\phi_1 + \pi a_2^2 (1 + \frac{\mu}{\tan\phi_2}) \beta_s \sin^2\phi_2 - \pi a_1^2 (1 + \frac{\mu}{\tan\phi_2})$$

$$(3)$$

$$\alpha_s = (2Y/3) [1 + \ln\frac{(1-2\nu) K}{(1-\nu) Y}], \qquad \beta_s = 1.041\rho$$

$$(4)$$

The geometric parameters are a_1, a_2, Φ_1, and Φ_2, which designate, respectively, the radius of the tip cone, the body radius, the half angle of the tip cone, and the half angle of the body cone.

Using this analysis and comparing predictions with experimental results using double-cone projectiles it turned out that for $\mu = 0$, the predicted penetration is 189 mm (which agrees with the data for spun projectiles), whereas for $\mu = 0.15$, the penetration is only 130mm. Thus, the double-cone projectile seems to have a lower friction coefficient than the hemispherical nose projectile.

For the unspun projectile, the observed penetration is predicted by $\mu = 0.04$; a higher friction coefficient is derived when the projectile is not rotating during impact. This analysis also indicates that friction between the projectile tip and the target is an important parameter in this type of penetration.

VI. PENETRATION MECHANICS

The sintered tungsten projectiles did not deform during penetration of the standard 6061-T6 and 2014-T6 targets; both the hemi-spherical-ended and double-nose-cone projectiles remained intact. The targets failed by a conventional "ductile hole growth" mechanism, which is why the Forrestal-type analysis [2] of the penetration mechanics worked satisfactorily. Target cross sections showed smooth cavity walls, resembling closely those shown in [2].

The penetration mechanism in the MMC materials was however very different. The reasons why this class of material worked so well against this type of projectile appear twofold. First, the reinforced alloys were able to deform and fracture the tungsten alloy projectiles which resulted in destabilization of the trajectory of the projectile. Fig. 2 shows a partial metallographic section of a residual penetrator embedded in the 6061-T6 substrate. The original nose has been deformed and fractured. The 6061-T6 substrate shows deformation by adiabatic shear banding [4]. Deformation and fracture of the projectile does not occur symmetrically about the impact axis; the

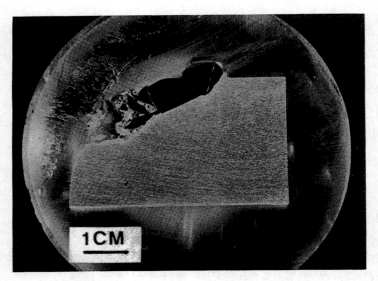

FIG 2 *Projectile After Penetrating MMC Tiles (Shot 2670) (left to right)*.

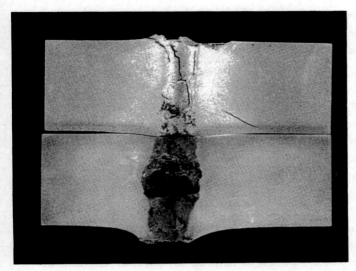

FIG 3 *Penetration Cavity in MMC Tiles (Shot 2590) (top to bottom)*

projectile appears to have become unstable about half-way through the penetration process in Fig. 3. The walls of the craters formed in the MMC's were 'rough', and did not exhibit the features characteristic of the deformation mode observed in the unreinforced alloys.

Secondly, the increased flow strength/hardness of the MMC also plays a role in enhancing the ballistic impact resistance. We can infer this by calculating a differential efficiency factor e^B_Δ based on that for a blunted (i.e. hemi-spherical ended) projectile penetrating 6061-T6 aluminum alloy. If the only effect of the MMC target was to blunt the projectile tip, we would expect values of e^B_Δ to be equal to 1. In fact, we measured this differential efficiency to be $2 < e^B_\Delta < 2.3$.

The Forrestal model [3] as it stands does not predict satisfactorily the penetration of the projectiles into the MMC materials. If we assume initially that the mode of plastic deformation on the MMC targets is the same as for the unreinforced aluminum alloys, i.e. a ductile cavity expansion mechanism, then the combined effect of the deformation of the projectile nose by the MMC and its increased flow strength (using properties in Table 1) in penetration resistance of IZP material, for example, can be calculated as before. We also assume the effective friction factor $\mu = 0.15$, as derived from penetration of hemi-spherically ended projectiles into unreinforced aluminum alloys. The projectile is predicted to emerge from the back surface of the MMC at a velocity of 809 m/s resulting in a calculated residual penetration in the 6061-T6 substrate of 68 mm. Thus, the MMC appears to be harder than predicted using this model. Our initial observations suggest in fact that the MMC did not plastically deform in a straightforward manner by the above mechanism.

Whether our assumptions were too approximate for the analysis of whether the model itself will have to be revised is unclear at the present time. Work is on-going to gain a better understanding of the detailed penetration behavior of these materials.

VII. SUMMARY

The filled SiC-aluminum metal matrix composites tested in this program performed extraordinarily well as armor materials. The outstanding performance relative to conventional aluminum is due to deformation and fracture induced in the penetrator nose and the increased flow strength/hardness of the metal-matrix composite; the former effect may be peculiar to tungsten alloy bullets at velocities 1 to 1.2 km/s. Nevertheless, this is an extremely important class of penetrators for armor design.

The Forrestal cavity expansion model provides a good technique for computing penetration of rigid tungsten bullets in aluminum, and the model has been extended to cover the particular nose shapes used in the present study. However, this model did not work for MMC materials, indicating that ductile cavity expansion is not the primary target deformation.

ACKNOWLEDGEMENTS

The authors are indebted to Dr. N.S. Brar of the UDRI staff who supervised most of the ballistic testing. The work at the University of Dayton was supported by Alcan International.

REFERENCES

1. Z. Rosenberg, S. Bless, Y. Yeshurun, and K. Okajima, 491-498, in Impact Loading and Dynamic Behavior of Materials, ed. C.Y.Chiem, H-D Kunze, L. W. Meyer, 1988.

2. M. Forrestal, K. Okajima, and V. Luk, J. Appl. Mech. 55:755 (1988).

3. M. J. Forrestal, V. K. Luk, and N. S. Brar, Army Solid Mechanics Symp., Newport, RI, 16-18 May 1989.

4. H. C. Rogers, Am. Rev. Mat. Sci. 9:283 (1979).

100

Extent of Damage Induced in Titanium Diboride Under Shock Loading

D.P. DANDEKAR and P.J. GAETA

US Army Materials Technology Laboratory
Watertown, Massachusetts 02172-0001 USA

Shock experiments were performed on hot pressed TiB_2 to test the suggested hypothesis that its spall threshold values assumed vanishingly small magnitude as the compressive stress approaches its HEL due to an incremental loss of cohesiveness of the material through generation of dilatation / microfracture. The results of the present experiments while confirming a decline in the value of spall threshold of TiB_2 as it is shocked to 13.0 GPa does not, however, indicate a concomitant generation of significant dilatation in the material as shown by the measured values of shock parameters and examination of the spalled specimens of TiB_2.

I. INTRODUCTION

Advanced ceramics are increasingly being considered for use under shock loading, i.e., impact loading condition. Limited shock data on some ceramics including TiB_2 have led to a hypothesis that the observed decrease in the values of the spall threshold (tensile strength) as the compressive stress approaches the Hugoniot Elastic Limit (HEL) is due to an incremental loss of cohesiveness of the ceramics under increasing compressive stress [1, 2, 3]. These types of ceramics have zero spall strength at and above the HEL. This incremental loss of cohesiveness is attributed to generation of microfracture and/or dilatation in these ceramics

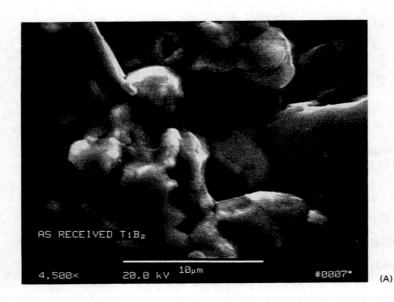

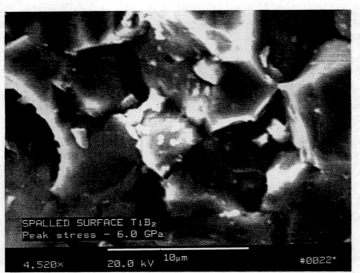

FIG.1 *SEM photographs of (A) as received and (B) spalled TiB₂.*

under plane shock wave loading. However, there neither exists any measurement which quantifies the extent of the loss of cohesiveness in these ceramics as a function of increasing compressive stress nor any determination as to at what stage of the shock process the suggested loss of cohesiveness occurs in the ceramics. In other words, does the loss of cohesiveness occur during compression and/ or during total or partial release of compressive shock? The present work was undertaken to test the above mentioned hypothesis by conducting shock experiments in a hot pressed TiB_2 at peak stresses not exceeding its HEL i.e., 13 GPa. Specifically, these shock experiments were designed to detect whether a loss of cohesiveness in TiB_2 with increasing stress occurs during compression, and/or during partial or total release. The applicable parameters measured during the experiments were shock velocity, particle velocity, impact stress, transmitted stress, release stress states, and in a few cases, reshocked stress state in the material following spallation in spalled specimen, and finally, tensile stress sustained before spallation. The principal idea behind these measurements was that a significant loss in the cohesiveness of TiB_2 will measurably reduce the magnitudes of shock and/ or release wave velocities, and of peak compressive stress attained for a given impact condition in TiB_2.

II. MATERIAL

TiB_2 used in the present investigation was hot pressed by Ceradyne. The elemental composition of this material in weight percentage is as follows: TiB_2 98.89, tungsten 0.83, carbon 0.3, cobalt 0.2, nitrogen 0.37, and oxygen 0.32. The average grain size of the material is 10 mm. Fig.1(A) shows that the as-received material does contain voids. It's density is measured to be 4.49 ± 0.01 Mg/m^3, i.e. slightly less than the single crystal density of TiB_2 i.e. 4.51 mg/m^3. The calculated value of the density of a hot pressed material containing only the elemental impurities given above is 4.53 mg/m^3. And the volume fraction of the voids calculated from the measured density of this material and the calculated density of a void free material is 0.009 ± 0.003. The average values of longitudinal, and shear elastic wave velocities measured by ultrasonic technique are found to be 11.23 ± 0.21 km/s and 7.41 ± 0.13 km/s, respectively. These velocities represent averages of measurements performed on 10 different specimens used in the shock wave experiments.

III. EXPERIMENTAL METHOD

Shock wave experiments were performed on a 10 cm bore diameter single stage gas gun at Materials Technology Laboratory. This gun is capable of accelerating a projectile to a

maximum velocity of 0.8 km/s. The velocity of a projectile at impact is measured by
electrically shorting four charged pins. Similarly charged pins are used to measure planarity of
impact of a projectile on a target and for triggering various devices to measure shock velocity in
TiB_2, and to record shock wave profiles measured by means of either x-cut quartz gauge
and/or manganin (Micromeasurements Inc. Type LM-SS-125CH-048) gauges. The impact
velocities are measured with an uncertainty of 0.5% and the deviation from planarity of impact
amounted to an average of 10^{-3} radians. Stresses measured by x-cut quartz gauges have a
maximum uncertainty of $\pm 3\%$ [4]. Manganin gage records have a similar precision [5].
Temporal measurements by means of manganin gages have larger uncertainty due to
encapsulation of the gage between 10-50mm thick mylar particularly when the gage is
embedded between two discs of TiB_2.

Two types of shock wave experiments were conducted on TiB_2.
Direct impact experiments were conducted to determine shock
state attained at the impact surface of TiB_2 and subsequent release states attained at the surface.
The details of these types of experiments are very well documented in the literature [6].
Transmission experiments were conducted to determine the nature of the transmitted shock
wave profile, nature of release wave, subsequent spall strength of TiB_2 and in a few
experiments reshocked stress state in spalled specimen of TiB_2.

TiB_2 specimens used in the present work were 2 to 12.7 mm thick discs with either
circular or square cross-section with lateral dimension between 25 mm and 50 mm . These
specimens were flat to 5 mm and the opposing faces of the discs were mutually parallel to 2
parts in 10^{-4}. All critical surfaces were lapped and polished.

IV. RESULTS

A. DIRECT IMPACT EXPERIMENTS

The results of direct impact experiments are given in Table 1. The results of these experiments
show that within the range of 2.2 to 13 GPa, the compressive impedance of TiB_2 as measured
at the impact surface varies between 48.6 and 52.0 Gg m^{-2} s^{-1}. The longitudinal elastic
impedance of TiB_2 as obtained from ultrasonic wave velocity and density measurements is
50.4 ± 0.5 Gg m^{-2} s^{-1}. The uncertainty in the measurements of shock impedances of TiB_2 is
on an average 4.2 per cent. Therefore, within the uncertainty of the measurements reported in
Table 1, the shock wave impedance of TiB_2 to 13 GPa is equal to the elastic wave impedance.
Since the measured value of impact surface stress was steady, it is concluded that TiB_2 did not
suffer a measurable degradation during shock compression and remained elastic to 13 GPa.
This table shows that the stepwise release of TiB_2 to 4.8 GPa follows the elastic path in TiB_2.

Table 1. Summary of direct impact experiments

| Experiment | Impact Velocity | Compression | | | First Release | | |
| | | Stress | Particle Velocity | Impedance | Stress | Particle Velocity | Impedance |
	(km/s)	(GPa)	(km/s)	(Gg m^{-2}s^{-1})	(GPa)	(km/s)	(Gg m^{-2}s^{-1})
637*	0.3977	12.950	0.2619	49.4			
822~	0.6071	2.299	0.0458	50.2			
005*	0.2831	9.320	0.1921	48.6	4.793	0.102	49.9
006*	0.3415	11.545	0.222	52.0	5.024	0.0899	49.4

* Tungsten carbide buffer ~ No buffer

In experiment # 005 values of the second release state stress and particle velocity coordinates were 3.566 GPa and 0.077km/s, respectively which also lay on the measured elastic trajectory.

The average values of shock loading impedance and release impedance of TiB$_2$ are found to 50.0±2.1 and 49.9±2.1 Gg m^{-2}s^{-1}, respectively. And these values are indistinguishable from one another. The average shock velocity calculated from the values of the impedance and initial density is 11.14 km/s.

B. TRANSMISSION EXPERIMENTS

Seven transmission experiments were conducted in TiB$_2$ to determine: (i) shock wave profile at various stress levels, (ii) spall threshold as a function of peak stress, and (iii) average velocity of release and shock preceding and following generation of tensile stress. Results of transmission experiments are summarized in Table 2.

This table shows that the the average values of shock wave velocity measured in these experiments is 11.30±0.48 km/s. Transmitted stresses calculated from the measured shock and particle velocities agree well with the predicted stress calculated from the measured particle velocity and the average value of shock impedance of 50 Gg m^{-2} s^{-1} obtained from the direct impact experiments. This agreement is within 3% except in experiment #009 where these two values of stresses differ by 9%. In other words, these experiments show that the impact stress continues to propagate unattenuated in TiB$_2$ to a thickness of 12.7 mm, i.e., there is no measurable loss in the material property of TiB$_2$ while under shock compression. Calculated values of release impedances from the stress-particle velocity pairs listed under compression and release, where the stress is measured by gauges at the TiB$_2$-PMMA or quartz interface,

Table 2. Summary of transmission experiments in TiB₂

				Experiment number			
	925	935	002	009	011	013	018
Impactor*	T	T	T	T	T	C	C
Gage~	M	M	M	M	M	Q	Q
Thickness (mm)							
Impactor	0.999	4.064	1.962	4.038	3.067	5.847	5.811
Target	2.176	8.039	5.995	12.715	2.045	12.696	12.716
Impact velocity (km/s)	0.3539	0.5042	0.5023	0.0879	0.2455	0.4035	0.3018
Compression							
Shock velocity (km/s)	11.48	11.34	11.34	12.18	11.22	10.85	10.70
Stress= (GPa)	9.123	12.836	12.790	2.406	6.186	8.638	6.351
Particle velocity (km/s)	0.1770	0.2521	0.2512	0.0440	0.1228	0.1773	0.1322
Stress# (GPa)	8.850	12.605	12.560	2.198	6.138	8.741	6.465
Release							
Stress+ (GPa)	1.080	1.720	1.700	0.281	5.88	4.118	2.97
Particle velocity (km/s)	0.288	0.444	0.439	0.079	0.1228	0.272	0.196
Impedance (Gg m⁻²s⁻¹)	72.4	57.9	59.0	60.7		47.7	52.8
Spall Strength (GPa)	0.14	0.06	0.1	0.3			
Pullback Stress (GPa)							
Measured	0.947	1.533	1.507	0.245			
Calculated	0.941	1.503	1.50	0.248			

* T: titanium diboride,   C: OFHC copper
~ M: Manganin gauge,   Q: x-cut quartz gauge
= Stress calculated from the measured shock velocity, particle velocity and initial density.
Stress calculated from the average value of shock impedance derived from the results of direct impact experiments.
+ Stress as recorded at titanium diboride - PMMA/or quartz interface, except in experiment 011, where the reported stress is in titanium diboride.

indicate that the material is elastic. A large scatter in these calculated values is due to the effect of cumulative error with which these parameters are measured, including the uncertainty in the Hugoniot of PMMA [7].

Shock wave profiles appear to show dispersion on loading only when the stress exceeds 6 GPa. This is most clearly shown by the wave profiles given for experiments 013

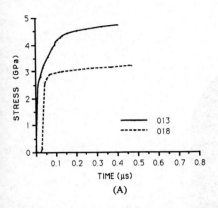

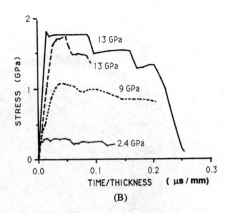

FIG.2 *(A)* *Transmitted stress wave profiles in experiments 013 and 018 as recorded by quartz gages, (B) Transmitted stress wave profiles as recorded by manganin gages.*

and 018, where x-cut quartz gauges were used to record the stress wave profiles [Fig. 2(A)]. The rise times in these experiments were 30 ns or less. The first break observed from the wave profile of experiment 013 occurs at 2.78 GPa, as recorded by the x-cut quartz gauge. This value measured by the quartz gauge, coupled with the observed shock velocity and initial density of TiB2 indicates that this break in the shock profile of TiB2 occurs at 5.9 GPa. Kipp and Grady [8] observed such a break to be located between 4.7 and 5.2 GPa in TiB2.

C. SPALL THRESHOLD

Experiments 925, 935, 002, 009, and 011 were conducted to determine spall threshold in TiB2 as a function of initial compressive stress. Unfortunately spall signal was not obtained in experiment #011. Table 2 lists the values of spall threshold as a function of initial compressive stress. Experiments 935 and 002 were conducted to examine if the spall threshold values were dependent upon the pulse width at a given stress. The values of spall threshold stress observed in these experiments are 0.06 and 0.1 GPa and are indistinguishable from one another. Whereas, the absolute values of the spall threshold are subject to a large error, the observed trend in the diminishing magnitude of spall threshold stress with increasing initial compressive stress is believed to be real. This inference is strengthened by the calculation of the average velocity of release and shock waves preceding and following the generation of tension in these experiments from the measured wave profiles [Fig. 2(B)], and using this value of the velocity to compute pullback stresses and the location of spall plane in the thick specimens of TiB2.

FIG.3   *Recovered spalled specimen of TiB2.*

Spalled materials were recovered in large enough size which were from the central section
where the gages were located in experiments 935 and 009. For example, on the basis of the
calculated value of the above mentioned average velocity for experiment #009, it is predicted
that the spall plane in TiB$_2$ is located at 8.64 mm from the impact surface. The measured
thickness of the pieces recovered from the central region of TiB$_2$ specimen were 8.6 and 4.1
mm, respectively. A photograph of the recovered piece is shown in Fig.3. The calculated
strength of pullback signal is given in Table 2. The measured and the calculated values of
pullback stresses differ by only 2%. Finally, the spall plane in TiB$_2$ is very clean and sharp
i.e., the material appears to spall as soon as the tensile stress reaches the spall threshold value,
unlike an extended spall zone observed in Beryllium oxide [9] .

V. DISCUSSION

The present investigation was initiated to determine extent of damage generated in TiB$_2$ shock
loaded to less than its Hugoniot Elastic Limit value i.e. 13.1 - 13.7 GPa as reported by Kipp
and Grady [8]. The issue at hand was whether the low value of spall threshold and the
vanishing small threshold in ceramics like TiB$_2$ as its HEL is approached were due to an

increase in the induced dilatation in the ceramics with increasing stress. The results of the present work derived from the direct impact and transmission experiments show no evidence of measurable deterioration in the material properties like reduction in shock wave velocity, stress relaxation at the impact surface, attenuation of stress amplitude as the shock wave propagates through TiB_2, and release wave velocity in TiB_2 from its shocked state. In addition, a SEM photograph of the spalled TiB_2 at a magnification of 4500 does not indicate any observable change in the morphology from that of as received material (Fig. 1). This implies that if increasing magnitude of dilatation with increasing stress is responsible for a decreasing value of spall threshold , then the magnitude of dilatation is calculated to be less than 3% in TiB_2. This calculation is based on Mackenzie's work dealing with elastic constants of a solid containing spherical holes [10] and on considerations of errors of measurements in these shock wave experiments on TiB_2. Hence, a more sensitive way of determining the extent of induced dilatation/microfracture in TiB_2 would be to TEM work on the recovered specimens of TiB_2 as done by Vanderwalker [11]. It was shown by her that even at a shock pressure of 1.7 GPa fracture in TiB_2 occurs by nucleation and splitting of b=[001] dislocation loops in the basal plane. The tensile stress of the shock pulse opens the dislocation loops into microcracks. Shocked TiB_2 was found to contain dislocation loops and split loops or voids. Thus, if the TiB_2 samples recovered from varying shocked state upon examination shows increasing density of loops and/or voids, then the vanishingly small spall threshold with increasing stress can be explained in spite of non indication of degradation of shock parameters in TiB_2 when shocked to 13 GPa. Such a work may also clarify the relative roles of voids and impurities present in TiB_2. TEM work on the recovered samples of TiB_2 will be carried out in the near future.

REFERENCES

1. G.I. Kanel, A.M. Molodets, and A.N. Dremen, *Combus. Explos.Shock Waves* 13: 772 (1977).

2. Z. Rosenberg, D. Yaziv, and S. Bless, *J. Appl. Phys.* 58: 3249 (1985).

3. D. Yaziv, and N.S. Brar, *J. De Physique* 49: C3-683 (1988).

4. R.A. Graham, *J. Appl. Phys.* 46: 1901 (1975).

5. Z. Rosenberg, D. Yaziv, and Y. Partom, *J. Appl. Phys.*, *51:* 3702 (1980).

6. P.C. Lysne, C.M. Percival, R.R. Boade, and O.E. Jones, *J. App l. Phys. 40:* 3786 (1969).

7. L.M. Barker, and R.E. Hollenbach, *J. Appl. Phys. 41:* 4208 (1970).

8. M.E. Kipp, and D.E. Grady, "Shock Compression and Release in High Strength Ceramics", in *Shock Compression of Condensed Matter-1989*, S.C. Schmidt, J.N. Johnson and L.W. Davison (eds), Elsevier Publishers, New York, N.Y. 1990.

9. D. Yaziv, S.J. Bless and D.P. Dandekar, "Shock Compression and Spall in Porous Beryllium oxide," in *Metallurgical Applications of Shock-Wave and High-Strain-Rate Phenomena*, L.E. Murr, K.P. Staudhammer and M.A. Meyers (eds.) Marcel Dekker, Inc., New York 1986.

10. J.K. Mackenzie, *Proc. Phys. Soc. London.*, *B63:* 2 (1950).

11. D.M. Vanderwalker, *Phys. Status Solidi.*, *111:* 119 (1989).

101

High-Strain-Rate Characterization of Ceramics in Shear

A. GILAT and M.K. CHENGALVA

Department of Engineering Mechanics
The Ohio State University
Columbus, Ohio 43210, U.S.A.

The torsional split Hopkinson bar technique is used for testing ceramics in shear. Two different geometries, a thin walled tube (spool specimen) and a rectangular prism, are used for specimens. The experiments have been modeled with a three-dimensional finite element analysis to better relate the torque and deformation measured on the Hopkinson bars to the stresses and strains in the specimens. Tests have been conducted on aluminum oxide.

I. INTRODUCTION

Ceramics are currently being used in applications that involve high rate of loadings. Most of the available information with regard to high strain rate response of ceramics has been obtained in shock wave experiments using normal plate impact technique. The yield stress is observed to increase when the Hogoniot Elastic Limit (HEL) is transformed to uniaxial stress and compared with the elastic uniaxial yield stress. An increase in the compressive strength with increasing strain rate of several engineering ceramics has also been noted in uniaxial compressive tests by Lankford [1].

In the present research ceramic specimens made of aluminum oxide have been tested at high rate of loading in shear. The torsional split Hopkinson bar technique is used for the tests. This technique is commonly used for testing ductile materials and its application to a very brittle material is examined.

II.   EXPERIMENTAL TECHNIQUE

In the torsional split Hopkinson bar technique a short specimen is placed between two bars. The specimen is loaded by a torsional wave that is generated in one of the bars. The history of the load and deformation in the specimen is determined by monitoring the stress waves in the bars which remain elastic during the test. A detailed description of the technique, which is a modification of a testing method introduced by Kolsky [2], is given by Hartley and Duffy [3].

Specimens of two different geometries, shown in Fig. 1, are used in the tests. One is the standard short (0.1 in. long) thin walled cylinder made by machining the middle section of a thick walled tube. The other is a small rectangular prism-shaped specimen. In each test two prisms are placed within two circular disc adapters. The assembly is placed between the bars of the split Hopkinson apparatus. The tests with the prism-shaped specimens are an effort to develop a simple inexpensive screening method for ceramics.

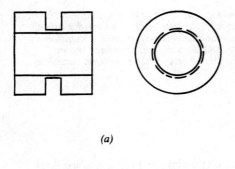

(a)

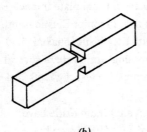

(b)

FIG. 1.   *Specimen geometry, (a) spool specimen, (b) prism specimen.*

III.   RESULTS

Results from tests with spool-shaped specimens are shown in Fig. 2. The figure
shows stress-strain curves obtained in three tests on hot pressed aluminum oxide,
Ebon A made by Cercom Inc. Excellent reproducibility of the results is observed
with maximum shear stress of 246 MPa (average of the three tests). This is also the
maximum tensile stress on planes at 45 degrees to the axial direction. Specimens
recovered after the tests show brittle fracture on these planes. Figure 2 also shows
that following the maximum the stress does not drop immediately to zero. It first
reduces to a level of approximately 140 MPa. The specimens continue to carry load
at this level for sometime (about 50 μsec) and then fail completely.

Results from tests with prism-shaped specimens are shown in Fig. 3. The
figure shows stress-strain curves from two tests on aluminum oxide (the same
material as used for the spool specimens). The stress in this figure is obtained by
assuming a uniform stress distribution in the prisms. The results from the two tests
are almost identical. After initial fracture at 100 MPa the stress continues to
increase to about 200 MPa where total failure occurs.

The stress distribution in the specimens was calculated numerically. A three-
dimensional elastic finite element analysis was done using ABAQUS. For the
experiments with the spool-shaped specimens shown in Fig. 2 the results show a

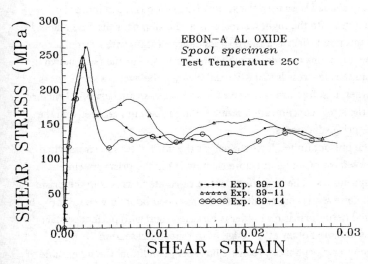

FIG. 2.   *Shear stress-strain curves from tests with spool-shaped specimens.*

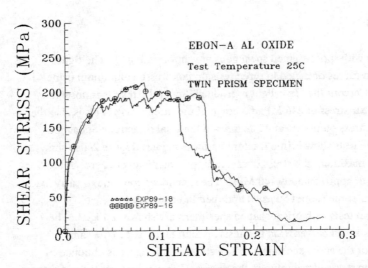

FIG. 3.　*Shear stress-strain curves from tests with prism-shaped specimens.*

stress concentration of 1.7 at the corner between the specimen section and the flanges. If it is assumed that the maximum stress in Fig. 2 corresponds to initial fracture of the material at this corner, then the maximum stress in the material is 418 MPa.

For the prism-shaped specimen the stress distribution is much more complicated. The point with the highest stresses is at the step near the free edge of the specimen. At this point there is a sharp stress concentration with normal tensile stresses in all three directions in addition to the shear stress. For the tests shown in Fig. 3 the maximum tensile stress is 920 MPa at 35° when the average shear stress is 100 MPa (that is when first fracture is observed in the stress-strain curves). Fracture of the material at the stress concentration reduces the stresses in that area and the load carried by the prisms increases.

Based on the present finite element analysis the normal tensile stress when failure is first observed is much higher (more than twice) in the prism specimens than in the spool specimens. The magnitude of the calculated stresses in the prism-shaped specimens, however, is very sensitive to the assumed boundary conditions. In the results presented here the prism's holder is assumed rigid with perfect contact between the prisms and the holder at the edge of the specimen section. If this condition is changed to reflect a more realistic boundary condition, the magnitude of the stress concentration reduces significantly.

The aluminum oxide in room temperature is very brittle and the initial fracture of the specimens occurs during the rise time of the loading wave. The average strain rate in the specimens during this period is in the order of a few hundreds. The maximum tensile stresses in the dynamic tests appear to be higher than stresses measured for this material in static tests (based on information obtained from Cercom Inc. from bending tests). At present, however, no data is available from static tests with specimen geometry and loading configuration similar to the dynamic tests to confirm rate sensitivity.

ACKNOWLEDGMENT

The authors acknowledge support for this research by DARPA.

REFERENCES

1.   J. Lankford, Fracture Mechanics of Ceramics, 5, 625 (1983).

2.   H. Kolsky, Proc. Phys. Soc., London, 62-B, 676 (1949).

3.   K.A. Hartley and J. Duffy, Metals Handbook, 9th Ed., 8, 215 (1984).

102

A Computational Constitutive Model for Brittle Materials Subjected to Large Strains, High Strain Rates, and High Pressures

G.R. JOHNSON AND T.J. HOLMQUIST

Armament Systems Division
Honeywell Incorporated
Brooklyn Park, Minnesota 55428, U.S.A.

This paper presents a computational constitutive model for brittle materials subjected to large strains, high strain rates, and high pressures. The equivalent strength is expressed as a function of the pressure, strain rate, and accumulated damage; and it allows for strength of intact and fractured material. The pressure is primarily expressed as a function of the volumetric strain, but it also includes the effect of bulking for the fractured material. Three examples are presented to illustrate the model.

I. INTRODUCTION

Throughout the past decade there has been much effort directed at developing computational constitutive models for metals subjected to large strains, high strain rates, high pressures, and high temperatures. These conditions are commonly experienced during high velocity impact. The Johnson-Cook strength model [1] and fracture model [2] are typical examples of such models.

Some characteristics of metals are that they generally experience a significant amount of ductility, the strength is not highly dependent on the pressure, and there is little shear strength after fracture.

There are also instances when it is desirable to compute the response of nonmetallic, brittle materials (such as ceramics, glasses, and rocks) to high velocity impact conditions.

Unlike metals, these materials generally experience very little ductility, the strength is highly dependent on the pressure (stronger under compression), and there can be significant shear strength (under compression) after fracture.

This paper presents a computational model for brittle materials, which is analogous to the Johnson-Cook models [1, 2] for metals.

II. DESCRIPTION OF THE MODEL

A general overview of the model is provided in Fig. 1. The initial concept for the model, and a description of the incorporation of such a model into a computer code, are provided in Reference 3.

The available strength (equivalent stress), σ, is dependent on the pressure, P, the dimensionless strain rate, $\dot{\epsilon}^* = \dot{\epsilon}/\dot{\epsilon}_o$ (for $\dot{\epsilon}_o = 1.0$ s^{-1}), and the damage, D. For undamaged material, $D = 0$; for partially damaged material, $0 < D < 1.0$; and for totally damaged (fractured) material, $D = 1.0$. Note that the strength is significantly reduced for fractured material ($D = 1.0$).

T is the maximum tensile hydrostatic pressure the material can experience, and $S1$ and $S2$ are the strengths of the intact material (for $\dot{\epsilon}^* = 1.0$) at compressive pressures $P1$ and $P2$, respectively. After the material has fractured ($D = 1.0$), the slope of the strength is given by $C6$, and the maximum fracture strength is $S3$ (for $\dot{\epsilon}^* = 1.0$).

The strain rate constant is $C3$. If σ_o is the available strength at $\dot{\epsilon}^* = 1.0$, then the strength at the other strain rates is

$$\sigma = \sigma_o (1 + C3 \cdot \ln\dot{\epsilon}^*) \tag{1}$$

It can be seen that the strength increases significantly with pressure, which is consistent with the well-known fact that brittle materials are much stronger in compression than they are in tension. The constants T, $S1$, and $P1$ can generally be determined from quasi-static and/or dynamic (Hopkinson bar) tension, torsion, or compression tests; and the strain rate constant, $C3$, can be determined from comparable quasi-static and dynamic (Hopkinson bar) tests [4].

The higher pressure constants, $S2$ and $P2$, generally require plate impact tests [5, 6]. The interpretation of these tests can be difficult because generally only the net uniaxial stress can be measured. To accurately obtain the constants, it is necessary to determine both the hydrostatic and deviatoric components of stress.

STRENGTH

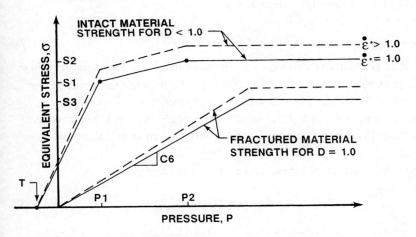

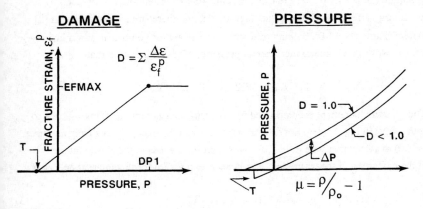

FIG. 1 *Description of the model.*

The post-fracture constant, C6, can be bounded by two different types of tests. A lower bound can be established by axial compression testing of a powdered material which has radial pressure confinement, and an upper bound can be established by axial compression testing of intact material (using displacement control) with radial pressure confinement. For the latter technique, it is the strength after fracture which is of interest. The maximum strength of the fractured material, S3, may sometimes be obtained from these tests, but it usually requires plate impact testing.

The damage for fracture is accumulated in a manner similar to that used in the Johnson-Cook fracture model [2]. It is expressed as

$$D = \Sigma \; \Delta\varepsilon^P \Big/ \varepsilon_f^P \tag{2}$$

where $\Delta\varepsilon^P$ is the plastic strain during a cycle of integration and $\varepsilon_f^P = f(P)$ is the plastic strain to fracture under a constant pressure, P. Referring to Fig. 1, the material cannot undergo any plastic strain at the maximum hydrostatic tension, T, but it increases to ε_f^P = EFMAX at a compressive pressure of P = DP1. This general behavior is consistent with the test results of Heard and Cline [7] where significant plastic strains were obtained under high compressive confinement.

The hydrostatic pressure before fracture (D < 1.0) is simply

$$P = K1 \cdot \mu + K2 \cdot \mu^2 + K3 \cdot \mu^3 \tag{3}$$

where K1, K2, and K3 are constants (K1 is the bulk modulus); and $\mu = \rho/\rho_o - 1$ for current density ρ and initial density ρ_o. For tensile pressures ($\mu < 0$), Equation 3 is replaced by $P = K1 \cdot \mu$. Energy effects probably are not significant, and therefore are not included [8].

After fracture (D = 1.0), bulking (pressure increase and/or volumetric strain increase) can occur [9]. Now an additional incremental pressure, ΔP, is added, such that

$$P = K1 \cdot \mu + K2 \cdot \mu^2 + K3 \cdot \mu^3 + \Delta P \tag{4}$$

The pressure increment is determined from energy considerations. Looking back to the strength model in Fig. 1, there is a drop in strength when the material goes from an intact state (D < 1.0) to a fractured state (D = 1.0). This represents a loss in the elastic internal energy of the deviator and shear stresses. The general expression for this internal energy is

$$U = [s_x^2 + s_y^2 + s_z^2 - 2\nu(s_x s_y + s_y s_z + s_z s_x) \\ + 2(1+\nu) \; (\tau_{xy}^2 + \tau_{yz}^2 + \tau_{zx}^2)] \, / \, (2 \cdot E) \tag{5}$$

where s_x, s_y, s_z are the normal deviator stresses; τ_{xy}, τ_{yz}, τ_{zx} are the shear stresses; ν is Poisson's ratio; and E is the modulus of elasticity.

The loss in this elastic internal energy can be expressed as

$$\Delta U = U_i - U_f \tag{6}$$

where U_i is the elastic energy of the intact material before fracture (D < 1.0) and U_f is the elastic energy immediately after fracture (D = 1.0).

This energy loss (of deviator and shear stresses) can be converted to potential hydrostatic internal energy by adding ΔP. An approximate equation for the energy conservation is

$$\Delta P \cdot \mu_f + \Delta P^2 / (2 \cdot K1) = \beta \cdot \Delta U \qquad (7)$$

where μ_f is μ at fracture and β is the fraction $(0 \leq \beta \leq 1.0)$ of the elastic energy loss converted to potential hydrostatic energy.

The first term $(\Delta P \cdot \mu_f)$ is the approximate potential energy for $\mu > 0$, and the second term $[\Delta P^2 / (2 \cdot K1)]$ is the corresponding potential energy for $\mu < 0$.

Solving for ΔP gives

$$\Delta P = -K1 \cdot \mu_f + \sqrt{(K1 \cdot \mu_f)^2 + 2 \cdot \beta \cdot K1 \cdot \Delta U} \qquad (8)$$

Note that $\Delta P = 0$ for $\beta = 0$, and that ΔP increases as ΔU increases and/or μ_f decreases.

III. EXAMPLES

Various features of this model can be illustrated by the three examples shown in Fig. 2. The tensile strength of the material is $T = 0.2$ GPa, the unconfined compressive strength is $S1 = S2 = 2.0$ GPa, the modulus of elasticity is $E = 220$ GPa and Poisson's ratio is $v = 0.22$. Because the height is $H = 1.0$ m and the area is $A = 1.0$ m^2, the deflection is $\delta = -\varepsilon_z$ and the force is $F = -\sigma_z$. For all three cases, the force, F, is slowly applied until $\delta = 0.02$ m, and then it is slowly released until $F = 0$. The paths are shown for strength versus pressure and for force versus deflection.

For Case A, the material cannot develop any plastic strain $(\varepsilon_f^P = 0)$ or any strength after fracture $(\sigma_f = 0$ for $C6 = 0)$. All of the elastic energy loss (of deviator and shear stresses) is converted to potential hydrostatic energy $(\beta = 1.0)$. Because there is no plastic work, the external work must also vanish. This is clearly shown in the force versus deflection relationship for Case A. The pressure jump at fracture $(\Delta P = 0.56$ GPa$)$ provides for this conservation of energy.

Case B is similar to Case A except that the material is allowed to undergo a small amount of plastic strain $(\varepsilon_f^P = 0.005)$ prior to fracture. Even though the elastic energy loss at fracture, ΔU, is equal to that of Case A, the pressure jump $(\Delta P = 0.37$ GPa$)$ is less because μ_f is greater at fracture.

Case C is similar to Case B except that there is strength after the material has fractured

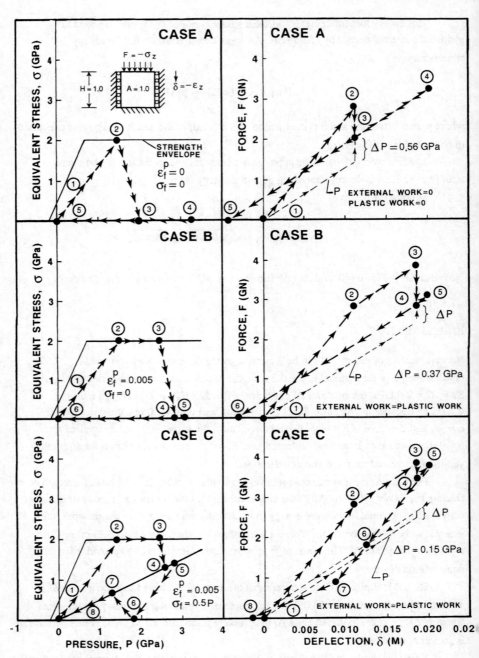

FIG. 2 *Examples of materials responses with the model.*

($\sigma_f = 0.5 \cdot P$ for C6 = 0.5). Here, the pressure jump ($\Delta P = 0.15$ GPa) is reduced from that of Case B because the elastic energy loss at fracture, ΔU, is less. The loading/unloading path is very complex. Of special interest is the elastic unloading between points 5 and 7. Between points 5 and 6, the axial deviator stress, s_z, is in compression; but between points 6 and 7, the same deviator stress is in tension. The net stress, σ_z, which includes the hydrostatic pressure, remains in compression. At point 7, the elastic unloading is complete and the material flows plastically between points 7 and 8.

IV. SUMMARY

A computational constitutive model for brittle materials has been presented and illustrated with examples. It has the capability to include large strains, high strain rates, and high pressures; and it exhibits the major distinguishing features of brittle material behavior.

ACKNOWLEDGEMENTS

This work was funded by the Defense Advanced Research Projects Agency. The many helpful discussions with J. Lankford and C.E. Anderson (at Southwest Research Institute) are also appreciated.

REFERENCES

1. G.R. Johnson and W.H. Cook, "A Constitutive Model and Data for Metals Subjected to Large Strains, High Strain Rates, and High Temperatures," in *Proceedings of the Seventh International Symposium on Ballistics.*, The Hague, The Netherlands, 1983.

2. G.R. Johnson and W.H. Cook, *Eng. Fract. Mech.*, 21:31 (1985).

3. G.R. Johnson, "Implementation of Simplified Constitutive Models in Large Computer Codes," *Dynamic Constitutive/Failure Models*, AFWAL-TR-88-4229, 1988.

4. J. Lankford, *J. Mat. Sci.*, 12:791 (1977).

5. T.J. Ahrens, W.H. Gust, and E.B. Royce, *J. App. Phys.*, 39:4610 (1968).

6. M.E. Kipp and D.E. Grady, *Shock Compression and Release in High Strength Ceramics*, SAND 89-1461 · UC-704, 1989.

7. H.C. Heard and C.F. Cline, *J. Mat. Sci.*, 15:1889 (1980).

8. W.H. Gust and E.B. Royce, *J. App. Phys.*, 42:276 (1971).

9. W.F. Brace, B.W. Paulding, and C. Scholz, *J. Geo. Res.*, 71:3939 (1966).

103

Planar-Shock and Penetration Response of Ceramics

M. E. KIPP, D. E. GRADY, and J. L. WISE

Sandia National Laboratories
Albuquerque, New Mexico 87185, U. S. A.

Two high-strength ceramic materials (boron carbide and titanium diboride) have been subjected to both planar shock and long-rod penetration, and their response has been characterized with time-resolved particle-velocity measurements. In symmetric planar impacts at 1500 and 2200 m/s, shock stresses ranged between 20 and 50 GPa. To examine the ceramic response to divergent waves, targets containing a ceramic layer were impacted at a nominal velocity of 1600 m/s by a 2-mm diameter tungsten penetrator at normal incidence. Penetration targets were built with, and without, a steel cover plate to examine the influence of confinement. Preliminary wavecode analyses of the ceramic response are included.

I.   INTRODUCTION

Ceramics have been repeatedly demonstrated to be effective armor materials against a variety of threats (*e.g.*, Wilkins, *et al.* [1]). Planar impact data for the two ceramics discussed here, boron carbide (B_4C) and titanium diboride (TiB_2), may be found in the collection of Hugoniot data that has been assembled by Gust and Royce [2] and Gust, *et al.* [3], including Hugoniot elastic limits (HEL) and shock Hugoniots for about a dozen ceramic materials. Some shock wave data are also available for titanium diboride [4]. Wave profile data in compression and release in uniaxial strain are presented here for B_4C and TiB_2 (See Ref.[5] for additional details).

In addition to the planar impact experiments, targets containing a ceramic layer were impacted by a long-rod penetrator at normal incidence [6]. The longitudinal particle-velocity history on the rear free surface of the target was monitored on the penetration axis. The diverging waves in the target plate create a velocity record that depends on the impact and penetration process, even though the recording point is spatially removed from the impact point. The one-dimensional (uniaxial strain) data are invaluable in the construction of material models; the divergent data can be directly compared with primary code variables (*e.g.*, particle velocity) to evaluate multidimensional calculations, complementing flash x-ray data (which supply global "snapshots" of the flow field at a few discrete times) and penetration performance (*i.e.*, ballistic limit) data.

II. PLATE IMPACT EXPERIMENTS

Uniaxial strain compressive shock and release waves were produced in the ceramics of interest with a single-stage powder gun (89-mm bore diameter, 2200 m/s maximum impact velocity). The ceramic carried by the projectile was backed by foam, and the target consisted of a disc of similar ceramic backed by an optical-quality single crystal of lithium fluoride. The history of the compression and release wave formed by this impact configuration is measured by monitoring the time-resolved longitudinal motion of the ceramic/lithium fluoride interface using laser velocity interferometry (VISAR) techniques [7] (resolution ∼ 1 ns). The impact velocity and experimental dimensions for each test are provided in Table 1, and the acquired time-resolved velocity profiles are displayed in Figure 1. (The arrival times of the wave profiles were offset to display the records.)

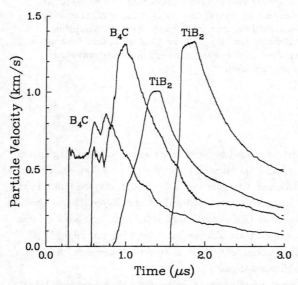

Fig. 1 *Plate-impact particle velocity data for B_4C and TiB_2.*

Table 1: Conditions for Ceramic Plate-Impact Experiments

Material	Impact Velocity (m/s)	Foam Density (kg/m³)	Impactor Thickness (mm)	Target Thickness (mm)
B₄C	1546	320	3.920	9.044
B₄C	2210	640	3.917	9.033
TiB₂	1515	320	3.972	10.804
TiB₂	2113	640	3.337	10.747

Table 2: Elastic Properties

Material	ρ_O kg/m³	C_L m/s	C_S m/s	C_O m/s	ν
B₄C	2516	14040	8900	9570	0.164
TiB₂	4452	10930	7300	6960	0.097

A. MATERIALS

Ultrasonic longitudinal, C_L, and shear, C_S, wave speeds, and reference density, ρ_O, were determined for the ceramic specimens. A summary of these experimental values, accompanied by calculated values for bulk wave speed, C_O, and Poisson's ratio, ν, is provided in Table 2. The titanium diboride was determined to be about 1% porous. Optical microscopy revealed fine-grained, equiaxial grain structures for all samples, with nominal grain sizes of 10 μm for B₄C, and 12 μm for TiB₂.

B. COMPRESSION AND RELEASE PROPERTIES

The wave profiles shown in Figure 1 are distorted somewhat in both amplitude and shape due to the mechanical impedance difference between the lithium fluoride and ceramic. Insight into the ceramic response to shock loading is gained by transforming the particle-velocity history for each experiment into a stress-strain load/release curve. In this way, features in the measured wave profiles associated with wave interactions caused by the sample and window material impedance mismatch can be separated from material response properties of the ceramics (yield, phase transformations, *etc.*). Using the one-dimensional explicit Lagrangian shock-wave propagation code, WONDY [8], neither of these ceramics could be readily represented with traditional elastic/perfectly-plastic material models, (primarily a consequence of the inability to accommodate the very dispersive nature of the unloading wave). To obtain accurate internal stress/strain histories, a parametrized load/unload path was incorporated into WONDY, and exercised in an iterative fashion [9], until the VISAR interface

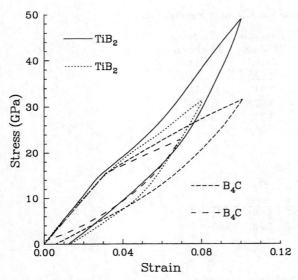

Fig. 2 *Summary of calculated stress/strain load and release paths for B_4C and TiB_2 at the center of the ceramic target.*

particle-velocity history for each experiment was reproduced. (We have assumed that the primary contribution to the stress is the material strain, and any dependence on both strain rate and thermal effects has been neglected.) Ultrasonic data were used to define the initial loading moduli. Load/release paths at the center of the ceramic target are displayed in Figure 2. Note that, although there is almost a factor of two difference in densities, the elastic loading curves of B_4C and TiB_2 are very nearly identical in both low- and high-amplitude cases. The B_4C shows a major loss in strength when compared to TiB_2.

Table 3: Hugoniot Elastic Limits

Test No.	Material	u_M (m/s)	σ_{HEL} (GPa)	P_H (GPa)
1	B_4C	580±30	14.8	22.8
2	B_4C	550±40	14.0	31.4
3	TiB_2	165±15	5.2	31.0
"	"	430±40	13.7*	"
4	TiB_2	150±15	4.7	48.5
"	"	410±20	13.1*	"

* Corresponds to second yield structure in TiB_2

The Hugoniot elastic limits for these ceramics were determined directly from the measured particle-velocity profiles, $u_M(t)$ (Figure 1), accounting for the impedance mismatch between ceramic and window [5]. The Hugoniot elastic limit data (σ_{HEL}) and the corresponding Hugoniot shock pressure, P_H, are tabulated in Table 3. For the titanium diboride, a reasonably well-defined break in both waves at approximately 160 m/s was tentatively selected as a preliminary yield process or phase transformation u_M value. A second break identified at about 420 m/s is a consequence of some structuring feature in the TiB$_2$ material response.

III. ROD IMPACT EXPERIMENTS

The penetration experiments [6] employed a 2-mm diameter, 20-mm long tungsten-alloy rod which impacted a nominally 5-mm or 10-mm thick, 76-mm diameter target plate of ceramic that was backed, in turn, by a 2-mm thick, 51-mm diameter buffer disk of OFHC copper. In some cases, a 2-mm steel cover plate was included to increase confinement. Normal incidence of the penetrator was maintained by mounting it rigidly in the projectile nosepiece. The projectile assembly was accelerated by a single-stage powder gun (89-mm bore diameter) to a nominal impact velocity of 1600 m/s. The particle-velocity history at an observation point, located on the penetration axis on the back surface of the copper buffer, was measured with a velocity interferometer (VISAR [7]).

A. MATERIALS

The target-plate materials were boron carbide and titanium diboride. Ultrasonic longitudinal, C_L, and shear, C_S, wave speeds, and reference density, ρ_O, were determined for the penetrator and the target components in this study. A summary of these properties, accompanied by calculated values for bulk wave speed, C_O, and Poisson's ratio, ν, is provided in Table 4. Minor differences in ceramic properties are visible compared with the data in Table 2.

Table 4: Elastic Properties

Material	ρ_O kg/m^3	C_L m/s	C_S m/s	C_O m/s	ν
B$_4$C	2517	14140	8897	9717	0.1723
TiB$_2$	4509	10850	7450	6610	0.0538
AISI 4340	7810	5884	3202	4578	0.2897
OFHC Cu	8920	4689	2250	3904	0.3506
W-Alloy	17250	5164	2837	3993	0.2840

Table 5: Conditions for Rod Impact Experiments

Target Material	t_{CON} (mm)	t_{TAR} (mm)	t_{BUF} (mm)	V (m/s)	τ (μs)
B_4C		9.69	1.94	1541	1.07
TiB_2 (1)		10.19	2.00	1558	1.33
TiB_2 (2)		10.18	2.00	1567	1.36
TiB_2	1.94	10.20	1.94	1562	1.66
TiB_2	1.94	5.14	2.02	1566	1.23

B. RESULTS

Particle-velocity histories are reported here for one test on a target of B_4C, and four on targets of TiB_2. For each experiment, Table 5 lists the target material, steel confinement thickness, t_{CON}, target thickness, t_{TAR}, copper buffer thickness, t_{BUF}, tungsten-rod impact velocity, V, and transit time, τ, of the leading toe of the wave disturbance through the target assembly. The duplicate experiments (1,2) for TiB_2 established the reproducibility of the VISAR data, and suggested which fine, "non-average" details of the wave profiles were not attributable to special material properties.

The experimental records for the TiB_2 are shown in Figure 3. In all cases, there is an initial peak in particle velocity followed by a temporary drop. The B_4C record (Figure 4) contains some noise during the decay period following the initial peak similar to that observed in uniaxial strain records for B_4C.

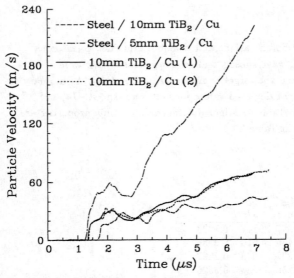

Fig. 3 *Rod impact particle-velocity data for TiB_2.*

Table 6: Computational Parameters

Material	S	Γ	Y_O (GPa)	σ_F (GPa)
B$_4$C	1.0	1.0	12.8	0.4
TiB$_2$	1.0	1.0	15.9	0.3
OFHC Cu	1.489	1.99	0.7	0.3
W-Alloy	1.237	1.54	1.3	1.4

C. CALCULATIONS

The initial calculations corresponding to these penetration experiments were made with standard elastic/perfectly-plastic material models in the Eulerian wave-propagation code CTH [10]. At the stress levels induced by the given impact velocities, the equation of state for each material is governed by the Gruneisen coefficient, Γ, and a linear shock velocity (U_S) vs. particle velocity (U_P) relationship, $U_S = C_O + S\ U_P$, where S is the slope and C_O is the bulk sound speed. An initial yield strength, Y_O, is required, and a fracture strength, σ_F, is assumed for the material. These parameters (summarized in Table 6) coupled with the elastic properties in Table 4 are sufficient to make a first estimate of material response. In the code, mixed-cell yield was set to zero.

The calculated free-surface particle velocities for the unconfined cases are plotted in Figures 4 and 5. As expected from the uniaxial-strain data [5], the calculations for the ceramic targets clearly lack specific agreement with the experimental data, although the general trends are captured. One aspect of ceramic material response not contained by the elastic/perfectly-plastic model is compressive failure.

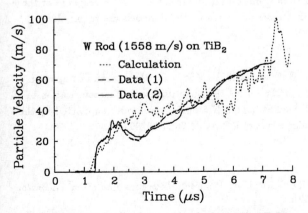

Fig. 4 *Particle-velocity data and calculation for TiB$_2$.*

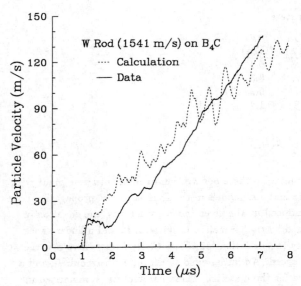

Fig. 5 *Particle-velocity data and calculation for B₄C.*

IV. CONCLUSIONS

The high-resolution data in the present uniaxial and divergent geometries provide ample opportunity to evaluate code and material model behavior in a quantitative manner. The boron carbide and titanium diboride disperse the release waves more widely than should be the case for normal solid response, suggesting that internal damage during compression has altered the state of the material. It is clear that simple material models generally follow the behavioral trends, but do not capture the detailed response of these ceramic materials, suggesting that more advanced material models are required.

ACKNOWLEDGEMENTS

The authors are indebted to D. E. Cox and R. L. Moody for their careful completion of these experiments, and to J. H. Gieske for his thorough ultrasonic characterization of all test materials. This work performed at Sandia National Laboratories supported by the U.S. Department of Energy under contract DE-AC04-76DP00789.

REFERENCES

1. M. L. Wilkins, C. F. Cline, and C. A. Honodel, *Lawrence Radiation Laboratory Report UCRL-71817* (1969).

2. W. H. Gust and E. B. Royce, *J. Appl. Phys.* *42:* (1971) 276.

3. W. H. Gust, A. C. Holt, and E. B. Royce, *J. Appl. Phys.* *44:* (1973) 550.

4. D. Yaziv and N. S. Brar, *J. de Physique, Colloque C3, 49:* (1988) 683.

5. M. E. Kipp and D. E. Grady, *Sandia National Laboratories Report SAND89-1461* (1989).

6. J. L. Wise and M. E. Kipp, *Shock Compression of Condensed Matter - 1989, Ed. by S. C. Schmidt, J. N. Johnson, L. W. Davison,* North-Holland (1990) 943.

7. L. M. Barker and R. E. Hollenbach, *J. Appl. Phys. 43:* (1972) 4669.

8. M. E. Kipp and R. J. Lawrence, *Sandia National Laboratories Report SAND81-0930* (1982).

9. D. E. Grady and M. D. Furnish, *Sandia National Laboratories Report SAND88-1642* (1988).

10. J. M. McGlaun, F. J. Zeigler, S. L. Thompson, and M. G. Elrick, *Sandia National Laboratories Report SAND88-0523* (1988).

104

High-Strain-Rate Compression and Fracture of B$_4$C-Aluminum Cermets

W. R. BLUMENTHAL

Materials Science and Technology Division
Los Alamos National Laboratory
Los Alamos, New Mexico 87545, U.S.A.

The compressive behavior of liquid-metal infiltrated boron carbide-aluminum cermets were studied as a function of strain rate, composition, and microstructure. Hopkinson split pressure bar (HSPB) and quasi-static compression tests were conducted using dumb-bell-shaped specimens. Results showed cermet compressive strength to be independent of loading rate. Strength was also found to be independent of the aluminum alloy used to infiltrate pre-sintered 65 vol% B$_4$C pre-forms. Compositions with the smallest phase size displayed the best strength and ductility.

I. INTRODUCTION

Light-weight cermets (ceramic content > 50 volume %) for armor applications have been of interest for over 20 years [1,2]. These efforts have culminated in the achievement of major breakthroughs in the processing of boron carbide-aluminum and aluminum oxide-aluminum cermets with ceramic contents over 65 vol% within the last five years [3-6]. Characterization of the mechanical response of these cermets is considered to be an important input to their further development and optimization. However, most of the interest in these materials has been devoted to measuring and modelling their fracture toughness and tensile strength [7-9].

Monolithic ceramic tensile failure is typically preceded by extremely localized permanent deformation associated with the initiation and/or propagation of a small number of flaws. Hence, tensile failure will be dependent on the rate of loading only when it is high enough to influence the dominant crack initiation and/or crack propagation mechanism [10]. This is in contrast to the origins of rate dependence in ductile metals which are related to microstructure evolution and the kinetics of deformation mechanisms. Cermets can exhibit measurable permanent deformation prior to failure in both tension and compression which may be associated with microcrack (damage) accumulation. Hence cermet compressive strength may exhibit novel strain rate dependence.

The primary focus of this work was to determine the influence of processing variables (i.e. microstructure and composition) on the compressive behavior of liquid-metal infiltrated boron carbide-aluminum cermets as a function of strain rate.

Recently there have been renewed attempts to correlate the compressive behavior of monolithic ceramics with their ballistic performance [11]. These studies emphasize the necessity for reliable materials property measurements obtained using valid compression and ballistic test methods. Quasi-static compression test methods for brittle materials have continued to evolve up to the present [12,13]. It has been recognized in these studies that tensile stresses can readily develop at specimen interfaces for a variety of reasons and can therefore control the measured failure strength. A proven method for minimizing these interface effects is to use a reduced-gage-section (dumbbell-shaped) specimen [14,15]. The present study extends the use of the dumbbell-shaped specimen to the high strain rate regime.

II. MATERIALS

Four series of cermets were fabricated at the University of Washington by infiltrating liquid aluminum into partially sintered boron carbide pre-forms [4]. Fabrication variables included: phase volume fraction, phase size, and metal phase composition. B_4C volume fractions of 65% and 80% were studied and represent practical boundaries for strong, yet open porosity pre-forms. Average phase sizes of 2.4 and 6.3 microns were measured for the Al and B_4C, respectively, in a "fine-grained" 65 vol% B_4C-pure Al series. A "coarse-grained" version contains average phase sizes of 21 and 47 microns for the Al and B_4C, respectively. Finally, two series of cermets were made by infiltrating either pure aluminum or 7075 aluminum alloy into 65 vol% B_4C pre-forms. All cermets tested were nominally fully dense (less than 2% porosity).

III. EXPERIMENTAL PROCEDURES

A. SPECIMEN GEOMETRY

All tests were conducted using a scaled-down version of the dumbbell-shaped specimen designed by Tracy [14]. Specimens have over-all dimensions of 13 mm x 4.4 mm with nominal gage length of 5 mm and gage diameter of 2.2 mm. The length of this specimen is comparable to the length of right-circular-cylinder specimen (12.5 mm x 6.25 mm) used by Lankford for HSPB studies of aluminum oxide [16]. This specimen length is short enough to permit equilibration of the stresses within the sample prior to failure at strain rates of 10^3 s^{-1} for these high sound speed materials ($C_1 > 9 \times 10^3$ m/s).

Specimen strain is determined in situ using three independent strain gages attached at uniform intervals about the specimen gage circumference. This configuration allows the bending stresses to be determined. Specimen strain gages are required because the "effective" gage length is not known precisely, especially after the gage section yields (permanently deforms).

B. HOPKINSON BAR COMPRESSION

Hopkinson split pressure bar tests were conducted using 12.5 mm diameter, 350 ksi yield strength, maraging steel bars at a nominal strain rate of 10^3 s^{-1}. Specimen stress was calculated from the transmitted bar in the conventional manner as the product of the bar stress and the ratio of bar-to-specimen cross-sectional areas. High band-width (> 3 MHz) strain gage amplifiers and digitizing oscilloscopes were used with a sampling rate of 10^7 points/sec.

Two types of experiments were conducted dynamically: 1) fracture strength and 2) recovery tests. Recovery tests were used to determined yield points and fracture strengths more precisely.

C. QUASI-STATIC COMPRESSION

Quasi-static compression tests were conducted using an Instron Model 1125 testing machine fitted with a precision-machined sub-press. Specimens were aligned in a precision V-block with tungsten carbide loading rams at each end. Testing was performed at a nominal strain rate of 10^{-4} s^{-1}. Specimen strain was obtained in the same manner as with the HSPB, but at a sampling rate of 10 points/sec. and with direct synchronization of the load record.

D. DATA REDUCTION

Quasi-static compression tests were analyzed by calculating the average specimen strain and the bending stress resulting from load eccentricity. The bending stresses were between 5% to 20% of the compressive stress.

In order to analyze the HSPB tests, the bar and specimen records must be synchronized. Recovery tests were used to determine the proper time adjustment between the records to within a few tenths of a microsecond.

IV. RESULTS

Fig. 1 shows a summary of stress versus average strain for the four cermets as a function of strain rate. A comparison of quasi-static and Hopkinson bar results shows that up to strain rates of 10^3 s^{-1}, the peak strength of all the cermets is independent of loading rate. Compression strength is also shown to be independent of the aluminum alloy infiltrate for the fine- grained 65 vol% B$_4$C composition. The fine-grained 65% and 80% B$_4$C compositions (average aluminum phase size = 2.5 microns) all yield at about 1.3% strain. However, the coarse-grained 65 vol% B$_4$C composition (average aluminum phase size = 21 microns) yields at about 0.9% strain. Also stresses measured quasi-statically did not exhibit a decay after peaking and the maximum strains were generally much

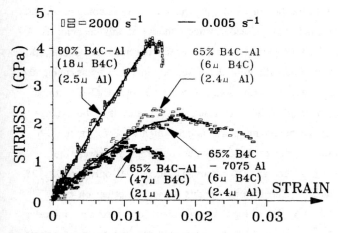

FIG. 1 *Compression stress-strain curves.*

less than those measured dynamically. This illustrates one advantage of the HSPB technique; namely, the sampling rate allows more resolution of the failure event.

The 80 vol% B_4C cermet exhibits twice the ultimate strength compared to the fine-grained 65 vol% B_4C cermets primarily due to its higher elastic modulus. Mechanical moduli (stress/total strain) at an arbitrary strain of 1% is plotted along with the ultrasonic values (zero strain) and the predictions from laminate composite theories (Voigt and Reuss) in Fig. 2. Isotropic composites generally behave as the average of Reuss and Voigt solids as exhibited by the ultrasonic values. However, under compressive strain, the modulus is observed to decay towards the value of a

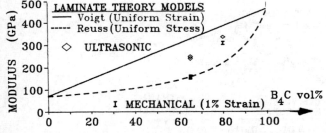

FIG. 2 *Experimental and laminate theory predictions of cermet modulus as a function of composition.*

Reuss solid. In the case of the 65 vol% B_4C cermets the decay is complete to the Reuss value (a bounding condition). Recovery tests show that this behavior is reversible. A simple analysis (Poisson's ratio and stress concentration effects are neglected) of the individual phase strains is demonstrated in Fig. 3 by assuming the 65 vol% B_4C-Al cermets behave as Reuss solids and the 80 vol% B_4C-Al cermet behaves as a Reuss-Voigt-average solid. This analysis indicates that the average elastic strain supported by the aluminum phase is quite high (> 3%) for the 2.5 micron Al phase size cermets prior to the composite yielding. However, the coarse-grained 65 vol% B_4C-Al cermet (21 micron Al phase size) only supports 2% strain in the aluminum phase.   The reasons for this phase size effect on the strength are not clear. Fig. 4, a plot of log strength versus log phase size, indicates that neither phase exhibits Hall-Petch behavior. Undoubtedly the aluminum phase is being constrained by the rigid B_4C structure which significantly increases its flow strength as observed in tension loading [17]. Perhaps this constraint is sensitive to the Al phase size or strength of the B_4C.

Fractography of failed specimens show fracture angles of between 25 to 40 degrees from the compression axis (35 degrees is typical). A previous investigation [18] discovered networks of microcracks intersecting the fracture surface of the fine-grained compositions implying that the yield behavior is related to damage accumulation.   Sliding damage on several planes also suggested multiple sources of failure initiation. However, failure of the lower strength, coarse-grained cermet did not show sliding striations or coalesced microcracks implying little damage accumulation or aluminum flow occurs during failure.

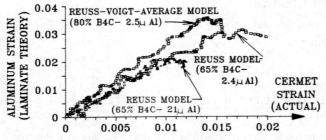

FIG. 3 *Aluminum phase strain analysis assuming laminate theory versus cermet strain.*

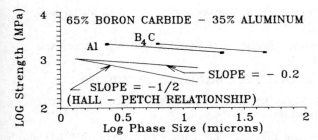

FIG. 4 *Power-law dependence of the compressive strength on constituent phase size.*

V. CONCLUSIONS

The HSPB technique possesses distinct advantages: 1) high time resolution (0.1 microsecond) permits details of the failure to be observed. 2) eccentricity (misalignment) appears to be somewhat lower due to the high rate of load application. 3) The amplitude and duration of applied load can be precisely controlled allowing for iterative (recovery) testing.

The compressive behavior of liquid-metal infiltrated boron carbide-aluminum cermets were studied as a function of strain rate, composition, and microstructure. Results showed cermet compressive strength to be independent of loading rate. Strength was also found to be independent of the aluminum alloy used to infiltrate pre-sintered 65 vol% B_4C pre-forms. Compositions with the smallest phase size (both B_4C and Al) displayed the highest strength and ductility.

REFERENCES

1. M.L. Wilkins, C.F. Cline, and C.A. Honodel, *Light Armor*, UCRL-71817, Lawrence Livermore National Laboratory, Livermore, CA, 1969.

2. M.W. Lindley and G.E. Gazza, *Some New Potential Ceramic-Metal Armor Materials Fabricated by Liquid Metal Infiltration*, AMMRC-TR-73-39, AD-769742, U.S. Army Materials Technology Laboratory, Watertown, MA, 1973.

3. D.C. Halverson, A.J. Pyzik, and I.A. Aksay, *Ceram. Engin. and Sci. Proc.*, 6: 73 (1985).

4. A.J. Pyzik, and I.A. Aksay, *U.S. Patent 4,702,770* (1987).

5. D.C. Halverson and R.L. Landingham, *U.S. Patent 4,718,941* (1988).

6. M.S. Newkirk, A.W. Urquhart, H.R. Zwicker, and E. Brevel, J. Mater. Res., *1*:81 (1986).

7. L.S. Sigl, P.A. Mataga, B.J. Dalgleish, R.M. Meeking, and A.G. Evans, Acta Metall., *36*: 945 (1988).

8. B.D. Flinn, M. Ruhle, and A.G. Evans, Acta Metall., *37*: 3001 (1989).

9. M.K. Aghajanian, N.H. Macmillan, C.R. Kennedy, S.J. Luszcz, and R. Roy, J. Mat. Sci., *24*: 658 (1989).

10. M.E. Kipp, D.E. Grady, and E.P. Chen, Int. Journ. of Fracture, *16*: 471 (1980).

11. Z. Rosenberg and Y Yeshurun, Int. Journ. of Impact Engng., *7*: 357 (1988).

12. S.A. Bortz, and T.B. Wade in *Structural Ceramics and Testing of Brittle Materials*, S.J. Acquaviva and S.A. Bortz (eds.), Gordon and Breach Science, New York, 1968, Chap. III, p. 47.

13. G. Sines and M. Adams in *Fracture Mechanics of Ceramics*, Vol. 3, R.C. Bradt, D.P.H. Hasselman and F.F. Lange (eds.), Plenum Press, New York, 1977, p. 403.

14. C.A. Tracy, Jrnl. of Testing and Evaluation, *15*: 14 (1987).

15. W.F. Brace in *State of Stress in the Earth's Crust*, W.R.Judd (ed.), American Elsevier, New York, 1964, p. 110.

16. J. Lankford, Jrnl. of Material Science, *12*: 791 (1977).

17. M.F. Ashby, F.J. Blunt, and M. Bannister, Acta Metall., *37*: 1847 (1989).

18. W.R. Blumenthal and G.T. Gray III, in *Mechanical Properties of Materials at High Rates of Strain 1989*, J. Harding (ed.), Inst. of Phys. Conf. Series #102, J.W. Arrowsmith Ltd., Bristol, England, 1989, p. 363.

105

Dynamic Response of Magnesia Partially Stabilized Zirconia

S. N. CHANG, S. NEMAT-NASSER, A. NOHARA, and W. P. ROGERS*

Center of Excellence for Advanced Materials
University of California, San Diego
La Jolla, CA 92093, U. S. A.

*Department of Mechanical Engineering
University of Colorado
Boulder, CO 80309, U. S. A.

Dynamic response of magnesia partially stabilized zirconia (Mg-PSZ) is investigated using a split Hopkinson pressure bar. Mechanical properties of Mg-PSZ at high strain rates are compared with those in quasi-static loading. The time-resolved stress was obtained from the transmitter bar signal and the strains (axial and transverse) are measured by strain gauges on the lateral faces of the cubical sample. The observed inelastic behavior is attributed to both phase transformation of precipitates and microcrack development along the loading axis. The axial compressive strain is partially recovered during unloading, probably due to reversible transformation and matrix cracking. Upon unloading, microcracks may form in the vicinity of transformed particles surrounded by the matrix of the cubic phase. Repeated compressive loading-unloading of a sample results in decreasing elastic moduli and increasing Poisson's ratio. The microstructural changes are characterized using scanning and transmission electron microscopy, and X-ray diffractometry.

I. INTRODUCTION

Stress-assisted phase transformation is a main mechanism of imparting high toughness to PSZ; see Hannink and Swain [1] for a review of transformation toughening in zirconia and its alloys. The tetragonal (t) to monoclinic (m) transformation in PSZ produces shear as well as dilatational strains in randomly distributed small particles within the cubic matrix. The applied stress can induce transformation even at levels below the yield stress [2]. The transformation is aided by the mechanical contribution to the thermodynamic driving force. The transformation strain in the individual particles causes the change of macroscopic deformation behavior of PSZ. Since this ceramic is brittle, microcracking is inevitable in later stages of deformation.

The purpose of this paper is to report and discuss inelastic behavior at high strain rates in commercial Mg-PSZ, and compare the results with those obtained in quasi-static compression tests. For more details, see Rogers and Nemat-Nasser [3], and Nohara and Nemat-Nasser [4].

II. EXPERIMENTAL TECHNIQUES

In quasi-static compression tests, axial and transverse strains are measured directly by two strain gauges on two adjacent lateral faces of the cubical sample. This test is not easy to perform since the elastic energy stored in the conventional testing machine drives the sample to catastrophic failure, shortly after microcrack initiation. Also it requires strict alignment of the compression fixture.

The conventional split Hopkinson pressure bar (SHPB) has been modified for application to dynamic recovery testing of ceramics and ceramic composites [5,6]. Recently, the technique has been further developed by a novel mechanism which produces a compression pulse followed by a tensile pulse [7]. The compression pulse duration is controlled by the length of the striker bar. Once the tensile pulse reaches the interface between the specimen and the incident bar, the specimen is automatically recovered, having been subjected to a *single compression pulse* of tailored profile. Two strain gauges are placed on two lateral faces of the sample, in order to measure both axial and transverse strains as functions of time. The average stress in the specimen is obtained from the strain measured in the transmitter bar. In this manner, reliable information about the sample deformation and the load history is obtained.

Normal plate impact experiments have also been performed on Mg-PSZ. Since brittle materials are weak in tension, a new design [8] has been used in order to eliminate in-plane tensile stresses that are developed in the sample when other techniques (e.g., the star-shaped flyer) are used. The design of this improved plate impact configuration is based on 2D and 3D numerical computations. Using this configuration, cracks induced by lateral release waves are essentially eliminated in the specimen for impact velocities up to 66 m/s. However, a specimen impacted at 122 m/s was fractured into four pieces, showing that even with this improved configuration, in-plane tensile waves are generated from the lateral boundaries of the specimen; these experiments are described elsewhere [8].

III. RESULTS AND DISCUSSION

The stress-strain curves for Mg-PSZ tested in SHPB as seen in Figs. 1(a) and (b) are highly reproducible due to the reliability of the new Hopkinson bar measurement techniques [5-7]. The elastic constants are measured from the initial linear portion of both axial and transverse stress-strain curves and are confirmed by ultrasonic measurements: Young's modulus is 205 GPa and Poisson's ratio is 0.26. Upon reloading, these properties are changed: Young's modulus is reduced by 7%, and Poisson's ratio is increased by 8%. The curves exhibit yield phenomena and hardening behavior, similar to ductile metals. The transformation yield stress increases from 0.9 GPa at strain rates of 10^{-4}/s to 1.4 GPa at 270/s, and the fracture stress increases from 1.47 GPa to 2.1 GPa, respectively. In the quasi-static test, the stored elastic energy in the testing machine causes failure of the specimen, once microcracking begins. The rate sensitivity of the yield and the fracture strength has important implications in transformation toughening of PSZ. It can be anticipated that dynamic fracture toughness increases with increasing loading rate.

The slope of the stress-strain curve resulting from the SHPB test reveals the interesting physical-mechanical behavior of this ceramic. The yielding (at point y) is due to the effect of the phase transformation on the overall stress-strain behavior. The constant slope after yielding indicates a stage of constant transformation rate and microcrack sliding rate. This slope increases at point s due to transformation saturation. Probably, the sliding and opening of the cracks are responsible for the continued inelastic deformation in this stage.

Upon unloading, reversed crack sliding may be producing the much lower slope at point c. Then the unloading slope increases to the elastic one. Reversible transformation starts at point r. Additional axial strain recovery can occur during this stage due to unloading cracks normal to the applied compression; Nohara and Nemat-Nasser [4]. It is of interest to note that the actual sample strain (ε_a), directly measured after the recovery, is always much smaller than the final strain recorded by the strain gauges at the end of the Hopkinson bar test. This is possibly due to the transformation and/or microcracking taking place under zero stress, after the loading-unloading event has been completed.

Figure 2 shows transgranular microcracks and inelastically deformed bands on the etched surface of Mg-PSZ tested in SHPB. The deformation bands are considered to be

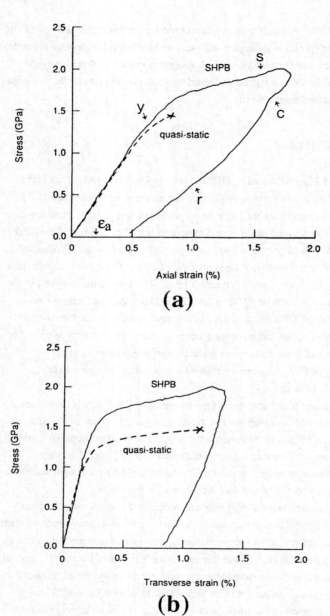

FIG. 1 *Stress-strain curves of Mg-PSZ measured in SHPB and in quasi-static compression; (a) axial ; (b) transverse.*

FIG. 2 *Scanning electron micrograph of Mg–PSZ showing microcracks parallel to the SHPB loading axis and inelastically deformed regions within the individual grains.*

permanent transformation zones which are also observed in quasi-statically compressed samples [2]. The orientation of the bands depends on the corresponding grain orientation. The bands are formed in more than one direction in the same grain. Their orientations do not seem to by only the overall compression axis. This is similar to slip bands in plastically deformed metallic polycrystals.

The transgranular cracks are oriented along the compressive loading axis [9-11]. The cracks propagate from defects (e.g., voids or inclusions). By ultrasonic velocity measurement, it is found that the longitudinal wave velocity in the axial direction is not strongly affected by the compression-induced microcracks. However, the longitudinal wave velocity decreases dramatically in the transverse direction, indicating that penny-shaped cracks are oriented parallel to the loading axis [9,10]. Figure 3 shows two microcracks which have been initiated from a void and have propagated into the matrix. The plane of this TEM micrograph is parallel to that shown in Fig. 2. As the microcrack propagates into the matrix, it skirts the transformed (twinned) monoclinic particles, as is seen in Fig. 4. The reorientation of the crack plane due to this crack-deflection and microcrack bridging by transformed precipitates are two of the important mechanisms of the increased fracture toughness. Microcracking associated with the

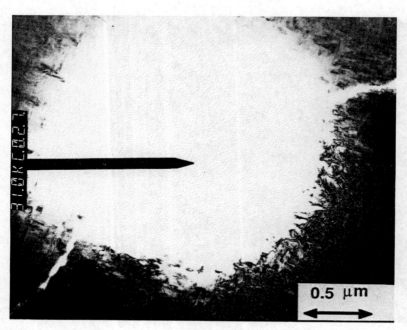

FIG. 3 *TEM micrograph of two cracks initiated from a void.*

transformed monoclinic particles is commonly observed along the interphase boundaries between monoclinic particles and the cubic matrix, whether the transformation has occurred athermally or quasi-statically under the applied stress [12]. Figure 5 shows the tangential (interface) microcracks around the particles in Mg-PSZ, compressed at a strain rate of 10^{-4}/s. The transformation lattice strain can be accommodated by twinning and local decohesion, resulting in the interplanar spacing along the particle boundary. However, this kind of interface microcracking is not observed in samples tested in SHPB at high strain rates. Further observation is needed to fully demonstrate this phenomenon.

Normal plate impact experiments have also been performed to study the dynamic behavior of Mg-PSZ. A new plate impact configuration for soft recovery of brittle materials has been recently developed at the University of California, San Diego [8]. Figure 6(a) shows a recovered Mg-PSZ target assembly impacted at 122 m/s (longitudinal compressive stress is 2.5 GPa). The specimen was surrounded by four lateral momentum traps in this test. It has been fractured into four pieces by in-plane tensile stresses, possibly generated from the interface of the specimen and the lateral momentum traps. No damage is observed in the

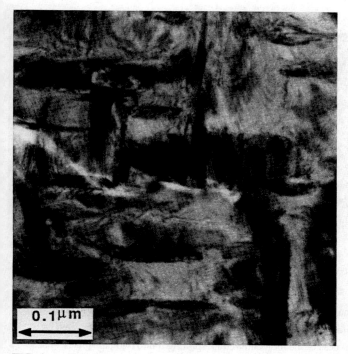

FIG. 4 *A microcrack extending in the matrix by avoiding the transformed monoclinic particles.*

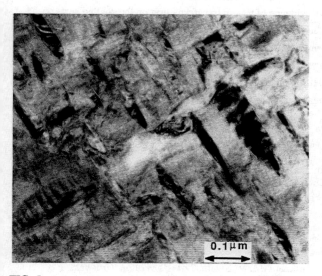

FIG. 5 *Interface microcracks around monoclinic particles transformed at (quasi-static) compressive strain rate of $10^{-4}/s$.*

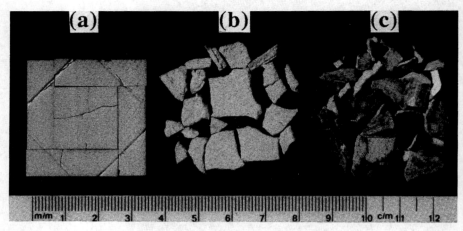

FIG. 6   *Recovered Mg-PSZ target (a) and momentum trap (b and c), impacted by a Mg-PSZ flyer plate at 122 m/s.*

specimen due to the uniaxial compression stress. Figures 6(b) and (c) show a recovered Mg-PSZ momentum trap which originally was a rectangular plate. Spalling has occurred on a single plane at mid-thickness of the momentum trap, splitting it into two half-plates. These half-plates are then fractured by the in-plane tensile stresses. Figure 6(b) shows the fractured half-plate next to the back face of the specimen, and Fig. 6(c) shows the other half-plate. X-ray diffraction analysis shows evidence of t → m transformation on the spalled surfaces. However, it was not possible to detect transformation in the interior of the impacted specimen which did not see any axial tensile stresses. Under the *uniaxial strain* condition which exists in normal plate impact tests, martensitic transformation in iron alloys has been induced by tensile stress only [13,14]. In the SHPB test, a *uniaxial stress* condition exists in the specimen, causing transformation by shear stressing.

ACKNOWLEDGMENTS

The authors thank Mr. J. Isaacs for his assistance with the Hopkinson bar tests. This work was supported by the U.S. Army Research Office under Contract No. DAAL-03-86-K-0169 and DAAL-03-88-K-0118 to the University of California, San Diego.

REFERENCES

1. D. J. Green, R. H. J. Hannink, and M. V. Swain, *Transformation Toughening of Ceramics*, CRC Press, Florida, 1989.

2. A. H. Heuer, M. Ruhle, and D. B. Marshall, *J. Am. Ceram. Soc.* 73: 1084 (1990).

3. W. P. Rogers and S. Nemat-Nasser, *J. Am. Ceram*, Soc. 73; 136 (1990).

4. A. Nohara and S. Nemat-Nasser, *J. Am. Ceram. Soc.*, (submitted).

5. S. Nemat-Nasser, J. B. Isaacs, G. Ravichandran, and J. E. Starrett, "High Strain Rate Testing in the U. S., " in *Proc. of the TTCP TTP-1 Workshop on New Techniques of Small Scale High Strain Rate Studies*, Australia, 1988.

6. G. Ravichandran and S. Nemat-Nasser, "Micromechanics of Dynamic Fracturing of Ceramic Composites: Experiments and Observations," in *Proc. of the 7th Intern. Conf. on Fracture*, Houston, TX, 1989.

7. S. Nemat-Nasser, J. B. Isaacs, and J. E. Starrett, (in preparation).

8. S. N. Chang, Y. F. Li, D-T. Chung and S. Nemat-Nasser, (will be submitted to Experimental Mechanics).

9. S. Nemat-Nasser and H. Horii, *J. Geophys. Res.* 87; 6805 (1982).

10. H. Horii and S. Nemat-Nasser, *J. Geophys. Res.* 90; 3105 (1985).

11. H. Horii and S. Nemat-Nasser, *Phil. Trans. Roy. Soc. Lond.* 319; 337 (1986).

12. B. C. Muddle and R. H. J. Hannink, *J. Am. Ceram. Soc.* 69: 547 (1986).

13. N. N. Thadhani and M. A. Meyers, *Acta Metall.* 34; 1625 (1986).

14. S. N. Chang and M. A. Meyers, *Acta Metall.* 36: 1085 (1988).

Section X
Explosive Welding and Metal Working

106

Equipment for Localization of Explosion

A. A. DERIBAS

Special Design Office for High Rate
Hydrodynamics - Siberian Department
of AS USSR Novosibirsk, 630090, USSR

In this paper a review of investigations about methods of localization of explosion is presented. These investigations were carried out in the Siberian Department of the USSR Academy of Sciences in Novosibirsk. Information on dissipating shields, localisators made from perforated panels, metallic explosive chambers working inside elastic limits of materials and hermetical metallic one-shot localisators is presented.

I. INTRODUCTION

The almost 40-year old history of the development of explosive working of metals shows that these processes occupy much less place then it deserves. I dare to state that the main reasons for this situation are some negative phenomena of explosive, namely air shock wave and its acoustic waves, scattering of debris, seismic phenomena, pollution of atmosphere by the by-products of the explosion. These negative factors including difficulties in regulation with HE have hindered the explosive working of materials in favor of conventional methods of metalworking despite some disadvantages in comparison with explosive metalworking methods.

From this it is clear from this that the methods of localisation of the negative phenomena of explosives have become more and more important.

II. SHOTS IN OPEN AREA

Explosive working of materials in an open area is the most wide spread utillization in the world till now. The maximum mass of HE charges changes depending on the local situation, from several kilogramms to several tons.

In the majority cases the production is carried out in places where explosions proceeded before such as for mining or military needs. The difficulties of organizing new large scale productions increase due to the shortage of open areas especially in heavily populated countries.

III. NON-HERMETIC LOCALISATORS

The main limiting factor in open air blasting is the shock wave and its acoustical after effects. It is necessary to localise the air shock wave the first time around.

The possibility of creating localisators made from perforated panels was developed in Novosibirsk [1]. This device essentially dissipates the air shock wave and prevents the scattering of debris. The scheme of these localisators is shown in fig.1. Rails and bars were used in this construction. The coefficient of perforation is in the order of 0.1. Experiments and calculations showed [2], that the danger zone for a 100-kg charge exploded inside this type of localisator will be in the order 18-20 meters.

It is posible to make this type of localisators for bigger charges of HE. Construction of these localizators is not very expensive because low cost materials are utillized.

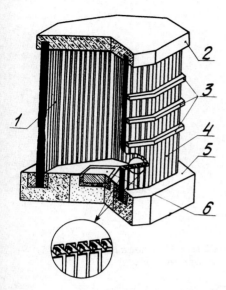

FIG. 1 *Localizator with the completely perforated side surface: 1,4 - perforated pannels, 2 - metallic ceiling, 3 - joining elements, 4 - concrete foundation, 5- work area.*

IV.   UNDERGROUND BLASTING PLACES

Underground blasting places are better for the environment then open places.   Unused tunnels and parts of mining enterprises are used usually for these purposes.

The maximal charge is determined by the stability of the main underground chamber and by the posibility of localizing the shock wave spreading in the transport tunnels.

Pending the strength of the rock, the chamber may not have to be very large.   For example, the blasting chambers in one of the explosive metalworking enterprices in Kazakstan measure 3x3x3 meters, with a maximum charge of 300 kilogramms.

Different doors are usually used for the localisation of air shock waves.   The influence of layers of porous materials on the strength of the doors due to the transformation of the shock-wave impulse was investigated in our Institute [3].

The three-layered construction shown in fig.2 was investigated.

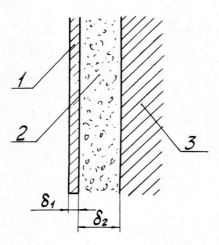

FIG. 2 *Three-layer dissepating shield: 1 - metallic shield, 2 - porous layer, 3 - surface to be protected.*

The experiments and calculations showed that it is possible to transform the dynamic loading of the porous material on the door onto the static frame by the proper selection of the parameters of metallic shield and porous layer.   The strength of the door will increase considerably.   For the first approximation it is possible to assume the porous medium to be gas and to obtain the formula for maximal pressure P max acting on the door:

$$P_{max} = P_o[\iota^2(\gamma-1)/2\rho P_o \delta_1 \delta_2]^{1/(\delta-1)}$$

were P_o - normal pressure

γ - exponent of adiabat of the air

ρ - density of the metallic shield

δ_1 - thickness of the metallic sheild

δ_2 - initial thickness of porous layer

ι - acting impulse of air shock wave

It is possible to optimize the parameters of the three-layered doors and to increase by several orders the capability of the door to withstand the air shock wave.

V. METALLIC EXPLOSIVE CHAMBERS

Metallic explosive chambers are the most popular equipment in the USSR used for the localisation of the effects of explosion in the processes of explosive metalworking.

They are designed and produced for long-term exploitation when all the elements of construction operate below the yield point. The principles of calculation of these chambers were published at the end of 1950 in the USA [5].

In the early 1960's a calculation method for explosive chambers was developed in the Siberian Department of AS USSR [6] and the production of explosive chambers were started. Currently 6 of these explosive chambers have for the last 15 years been utillized for the explosive hardening of railway frogs,and the total amount of these type of chambers under exploitation is more than 30 in the USSR.

The Special Design Office of High-Rate Hydrodynamics owns the rights for producing explosive chambers. Now we produce a series of explosive chambers for a maximum charge from 200 gr to 7 kg HE. The principal scheme of these chambers is shown in fig.3.

Chambers of this type may operate in any industrial building and it is possible to combine their construction with other technologies. These explosive chambers have a unique property which is distinctive from any other methods for the localisation of explosions.

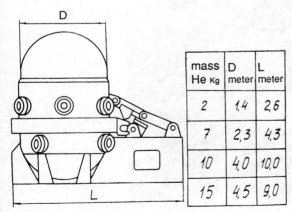

mass He Kg	D meter	L meter
2	1,4	2,6
7	2,3	4,3
10	4,0	10,0
15	4,5	9,0

FIG. 3 *Schematic of metallic explosive chambers.*

In the last two years a new type of explosive chambers-explosive-chemical reactors was created. For Example, these chambers are currently used for the synthesis of diamonds.

VI. HERMETIC ONE-SHOT LOCALISATORS

It is projected that the metallic explosive chamber can safely be used for extended term operations. However, its large weight (ratio of mass of chamber to the mass of charge is of the order of 1000) is a limiting factor in many applications. In many cases it is possible to make a localisator only for one explosive charge of HE for a definite mass. It is possible to allow the plastic deformation or fracturing some elements of the localisator.

Behaviour of different shells under explosive loading was investigated in situations where the different porous media are placed between the explosive and the walls of the shell [7]. Safe localisators were developed with the mass-ratio of the order of 70 instead 1000 without a porous media. A box-shaped localisator after the explosion without porous media is shown in fig.4. The same localisator after the explosion with porous medium is shown in fig.5. It was found from experiments that the porous media are not effective

FIG. 4 *Localizator without porous medium after explosion.*

FIG. 5 *Localizator with the porous medium after explosion.*

if the ratio of characteristic sizes of shell and charge is less than 6.

It follows from this investigation that the ratio of mass of the shell to the mass of the charge may be decreased more than 10 times in comparison with the mass-ratio for explosive chambers.

REFERENCES

1.   Grigoriev G.S., Vershinin V.JU., Klapovskii V.E. *Obrabotka materialov impulsnimi nagruzkami: sb. nauch. tr., - Novosibirsk,* pp. 336-339, (1990).

2.   Grigoriev G.S., Klapovskii V.E. *FGV. -vd. 23, N1,* pp. 104-106, (1987).

3.   Afanasenko S.I., Grigoriev G.S., Klapovskii V.E. *Obrabotka materialov impulsnimi nagruzkami: sb.nauc. tr., Novosibirsk,* pp. 318-324, (1990).

4.   Nesterenko V.F., Afanasenko S.I., Cheskidov P.A., Klapovskii V.E., Grigoriev G.S. *Obrabotka materialov impulsnimi nagruzkami: sb. nauch. tr. - Novosibirsk,* pp.325-335, (1990).

5.   Baker W.E., Allen F.J. *Proc. 3-th Int. Conf. Appl. Mech. ASME.* pp. 79-87. (1958).

6.   Demchuk A.F. *Obrabotka materialov vzrivom: mater. 2-go
      Mezhdunarod. simp. "Ispolzovanie energii vzriva pri obrabotke
      metallicheskikh materialov s novimi svoistvami"*, Marianski Lazni,
      9-12 okt. 1973. - Praga, Vol. 2, pp. 403-410, (1974).

7.   Afanasenko S.I. *Obrabotka materialov impulsnimi nagruzkami: sb.
      nauch. tr. - Novosibirsk*, pp. 291-304, (1990).

107

Effect of Experimental Parameters on the Size of Wavy Interface in Multilayered Material Made By Single-Shot Explosive Bonding Technique

K. HOKAMOTO, M. FUJITA, A. CHIBA and M. YAMAMORI

Graduate School of Science and Technology
Kumamoto University
Kumamoto 860, JAPAN

Department of Mechanical Engineering
Kumamoto University
Kumamoto 860, JAPAN

The method to homogenize the wavy interface in explosively-bonded multilayered plates is investigated under various conditions. In case of the bonding of multilayered copper plates with same thickness, to regulate stand off distance between plates and thickness of driver plate are necessary as to balance acceleration of flying plate by detonating gas pressure and velocity drop by collision. When using multilayered plates with different thickness or dissimilar materials, uniform wave shapes could be obtained by regulating stand off distance between plates.

I. INTRODUCTION

Multilayered explosive bonding is a technique to fabricate multilayered materials with single-shot explosive bonding. This technique is not only possible to fabricate structural transition joint (STJ) [1], but also expected to applicable to fabricate multilayered composites [2,3].

It is well known as a characteristic phenomenon that the explosively bonded interface exhibits wave shape and the formation of wavy interface means a evidence of good bonding. The relationship between parameters in multilayered explosive bonding and the size of waves have already investigated by El-Sobky & Blazynski [4] and Al-Hassani & Salem [5], individually. Al-Hassani & Salem [5] pointed the wavelength in interface is proportional to the kinetic energy lost at each collision of flying plate. In this investigation, we try to obtain the condition to homogenize the wavelength based on the theory proposed by Al-Hassani & Salem [5]. Since the multilayered plates are placed with small stand off distance, the velocity change during this process must be analyzed by considering the acceleration by detonating gas pressure. The authors have already proposed one-dimensional finite differential analysis on the change of flying plate velocity during multilayered explosive bonding process and also shown the possibility to predict the wavelength in interface by utilizing this analysis [6-8]. The effects of some experimental parameters on the change of wavelength are verified in the present investigation.

II. PROCEDURE AND PROCESS OF MULTILAYERED EXPLOSIVE BONDING

Figure 1 shows the schematic illustration of multilayered explosive bonding process. In this investigation, copper plate was used as driver plate and thickness of explosive t_E was fixed 40mm. 0.5mm thick and 9 layers of copper plates were used as multilayered plates in most cases (in FIG.2 and III-A). The parameters to be verified are thickness of driver plate (t_D), thickness of each multilayered plate (t_M), stand off distance between driver and first multilayered plate (SO1) and stand off distance between each multilayered plate (SO2). The following condition is given as a standard condition; t_D=3mm, t_M=0.5mm, SO1=SO2=1mm. Figure 2 shows the analytical change of plate velocity and detonating gas pressure by using our analysis [7] under the standard condition.

The process of multilayered explosive bonding is as follows. At the early stage of detonation, driver plate is immediately accelerated by high detonating gas pressure. When the

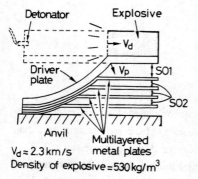

FIG.1   *Schematic illustration of multilayered explosive bonding.*

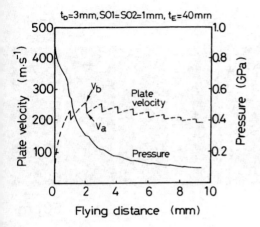

FIG.2   *Analytical change of detonating gas pressure subject to driver plate and plate velocity.*

driver plate collides with first multilayered plates, the flying plate bonded starts to move with a certain flying velocity which depends on the rule of momentum conservation. After the collision, the flying plate is accelerated again by remaining gas pressure and collides with the next plate. Such acceleration by remaining gas pressure and velocity drop by collision are repeated throughout this process.

An example of microstructure of multilayered plates is shown in FIG.3. The experimental condition in FIG.3 is a result of standard condition. In FIG.3, the wavelength from explosive to anvil sides increases at first stage of collision and decreases after several collisions. The change of wavelength shows the same tendency as the change of plate velocity shown in FIG.2. Al-Hassani & Salem reported that the wavelength was almost proportional to the kinetic energy lost by collision ΔKE [5,9]. We also obtained the same result [7] though a little difference is found compared with the result as Al-Hassani & Salem reported [5,9]. Wavelength is possible to predict by empirical relationship between ΔKE and wavelength [7]. The value of ΔKE is expressed as follows [7],

$$\Delta KE = 1/2 \, V_a \, V_b \, \Delta m \qquad [1]$$

where V_a and V_b are the plate velocity just after and just before collision, respectively, and Δm is a mass of collided plate per unit area.

Explosive   Anvil
side   ←   →   side

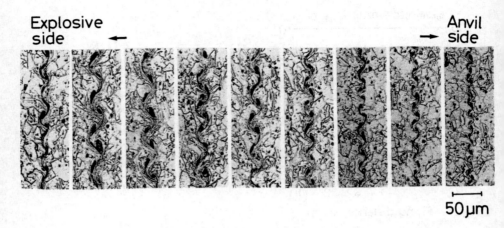

50 μm

FIG.3 *Microstructure of wavy interface in multilayered copper plates by explosive bonding.*

III. RESULTS AND DISCUSSION

A. *Effect of experimental parameters on the wavelength with same thickness of multilayered copper plates*

Effect of thickness of driver plate   Figure 4 shows the effect of thickness of driver plate t_D on the change of plate velocity and ΔKE with flying distance.

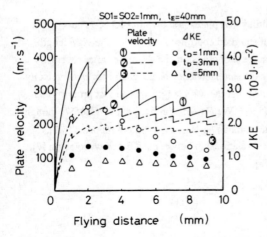

FIG.4 *Analytical change of plate velocity and kinetic energy lost ΔKE for different thickness of driver plate t_D.*

FIG.5 shows the change of predicted and measured wavelength with flying distance. In FIG.5, open marks and dashed lines represent the predicted and solid marks and lines are the measured wavelength, respectively. In FIG.4, it is obvious that the condition of each collision can be homogenized by increasing t_D. The measured wavelength in FIG.5 shows good agreement with the predicted results. The increase of the wavelength at first stage of collision depends on the acceleration of flying plate by detonating gas pressure. Since longer flying distance is required to accelerate the thicker driver plate, the range of acceleration increases with the increase of t_D. By increasing t_D, the measured wavelength in a specimen becomes fine and uniform, but too much increase of t_D induces undesirable decrease of plate velocity. Since the explosive bonding condition [11] is not satisfied, non-welded interface is observed at first interface when t_D is 5mm.

Effect of stand off distance between plates   Figure 6 and FIG.7 shows the effect of stand off distance between driver and first multilayered plate SO1 and stand off distance between multilayered plates SO2 on the change of plate velocity and ΔKE, respectively. When the value of SO1 is small, multilayered plates collide under accelerated condition at early stage of collision. In such a condition, the value of ΔKE increases from explosive to anvil sides for several collisions. When SO2 is small, ΔKE decreases with the progress of collision since the flying plate velocity is not recovered enough before the next

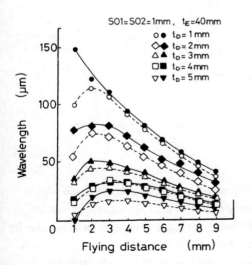

FIG.5   *Change of wavelength as a function of flying distance for different values of t_D.   Open marks are predicted and solid marks are measured wavelength, respectively.*

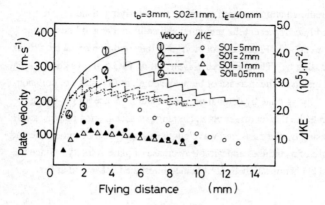

FIG.6 *Analytical change of plate velocity and ΔKE for different distance between driver and first multilayered plate SO1.*

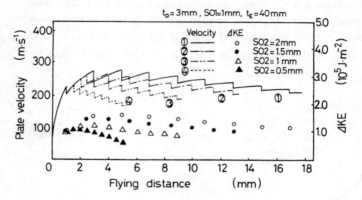

FIG.7 *Analytical change of plate velocity and ΔKE for different distance between multilayered plates SO2.*

collision. On the other hand, when SO1 or SO2 becomes large, the effect of acceleration is disappeared by the increase of flying distance and then the plate velocity decreases with the progress of collision. The acceleration by detonating gas and the velocity drop by collision are almost balanced at SO1=1mm and SO2=1mm, respectively.

Figure 8 and FIG.9 show the effect of SO1 and SO2 on the range of wavelength, respectively.   In these figures, upper and lower lines represent the maximum and minimum wavelength, respectively.   In FIG.8 and FIG.9, the measured wavelength shows good

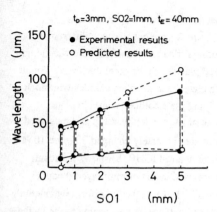

FIG.8 *Change of maximum and minimum wavelength as a function of distance between driver and first multilayered plate SO1.*

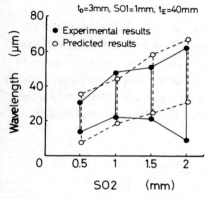

FIG.9 *Change of maximum and minimum wavelength as a function of distance between multilayered plates SO2.*

agreement with the analytical one. It should be noted that the resistance of air gives a effect to decrease the actual flying velocity compared with the analytical plate velocity by increasing flying distance, but the effect of the resistance of air is not considered in this analysis.

B. *Regulation of wavelength when thickness or component of multilayered plates is different*

In case of thickness of multilayered plates is alternately changed In the previous paper [6,7], it was difficult to obtain uniform wave structure when thickness of multilayered plates is different because the wavelength is almost proportional to the mass of collided plate as far as V_a and V_b are not so much different as shown in equation [1]. In this investigation, the authors tried to homogenize the wavelength by regulating stand off distance between multilayered plates with different thickness. Small stand off distance is used not to accelerate the flying plate when the plate collides with thick plate and large stand off distance is used when the flying plate collides with thin plate.

Figure 10 (a), (b) shows the analytical change of plate velocity, ΔKE and measured wavelength when the thickness of multilayered plates are alternately changed. Figure 10 (a) shows the result when the stand off distance is constant, and FIG.10 (b) shows the result when we regulated the stand off distance. In these cases, 6 layers of multilayered copper plates with 0.3 and 0.5mm thick are stacked alternately. In FIG.10 (b), uniform wavelength could be obtained in some extent by regulating stand off distance, but according to the increase of the number of layers, difference of the wavelength becomes large because the recovery of plate velocity by detonating gas pressure is no more expected due to the decrease of detonating gas pressure.

In case of dissimilar metal combinations   For the practical application of explosively bonded multilayered material, the way to fabricate the multilayered material with dissimilar metals should be established. In this section, a combination of mild steel and copper plates with same thickness (0.5mm) is tried to homogenize the wave sizes. In this experiment, three layers of mild steel are placed under copper driver plate with large stand off distance and three copper plates are placed under mild steel plates with small stand off

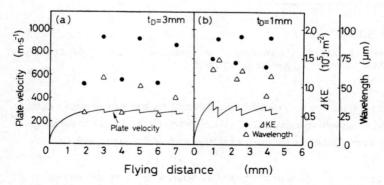

FIG.10 *Analytical change of plate velocity and ΔKE when thickness of multilayered plates are alternately changed ((a); with same stand off distance, (b); with regulated stand off distance)*

Explosive side ← → Anvil side

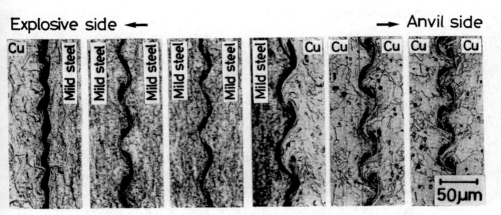

FIG.11 *Microstructure of wavy interface in multilayered plates with dissimilar metals.*

distance. As shown in FIG.11, almost the same wavelength can be obtained by regulating stand off distance. In case of the bonding of harder component, higher value of ΔKE, namely higher collision velocity is required to obtain uniform wave sizes as compared with the case of softer component. This result is explained as a part of energy dissipated by collision is consumed as a deformation to generate a wavy interface [9]. In the bonding of harder component, larger energy is required to form the wavy interface with same size compared with the case of softer component.

IV. CONCLUSIONS

The effects of experimental parameters on the size of wavy interface are investigated in explosively bonded multilayered material. In the bonding of multilayered plates, it is important to balance the acceleration of flying plate by detonating gas pressure and the velocity drop by collision. To regulate thickness of driver plate and stand off distance between plates are required to homogenize the wavelength to satisfy such an experimental condition in the bonding of multilayered plates with same thickness. When thickness or component of multilayered plates is different, uniform wave structure could be obtained only by regulating stand off distance between plates but the number of layers which is possible to homogenize the wavelength is limited to several layers since the recovery of plate velocity by detonating gas is not expected when the flying distance is increased.

ACKNOWLEDGMENTS

This research was supported by Ishihara-Asada Foundation of Iron and Steel Institute of Japan and Foundation of Promotion of Industrial Explosive Technology of Industrial Explosive Society, Japan.

REFERENCES

1. A. Kubota, *J. Japan Soc. Technol. of Plasticity*, 28: 1121(1987). (in Japanese)
2. A. Chiba et al., *J. Japan Inst. Composite Materials*, 9: 108(1983). (in Japanese)
3. A. Chiba et al., *J. Japan Inst. Metals*, 53: 156 (1989). (in Japanese)
4. H. El-Sobky and T. Z. Blazynski, *Proc. 7th Int. Conf. on H.E.R.F.* : 100 (1981).
5. S. T. S. Al-Hassani and S. A. L. Salem, ibid: 208.
6. K. Hokamoto et al., *Proc. 7th Int. Symp. "Use of Explosive Energy in Manufacturing Metallic Materials with New Properties"* : 26 (1988).
7. K. Hokamoto et al., *Mem. Faculty Eng., Kumamoto Univ.*, 33: 95 (1988).
8. K. Hokamoto et al., *Advanced Technology of Plasticity* , 3: 1441 (1990).
9. S. A. L. Salem et al., *Int. J. Impact Engng.*, 2: 85 (1984).
10. L. Lazari and S. T. S. Al-Hassani, *Metallurgical Applications of Shock-Wave and High-Strain-Rate Phenomena*, L.E.Murr, K.P.Staudhammer and M.A.Meyers (eds.), Marcel Dekker Inc., New York, 1986, p.969.
11. F. A. Mckee and B. Crossland, *Proc. 5th Int. Conf. on H.E.R.F.* : 4.11.(1973).

108

Laser-Driven Miniature Plates for One-Dimensional Impacts at 0.5 – ≥ 6 km/s

D. L. PAISLEY

Los Alamos National Laboratory
Dynamic Testing Division
Los Alamos, New Mexico 87545

Miniature (2 - 10-mm thick x ≤600-mm diam) aluminum and copper one-dimensional plates can be launched by a pulsed laser. Accelerations ≥10^10 G's have been recorded with terminal velocities ≥6 km/sec. These thin, high-velocity plates have been used to impart ≥50 GPa for durations ~0.6 - 10 ns. These high-acceleration and -velocity plates and short time duration pulses require improved diagnostics. We have improved the time resolution of VISAR records to ~100 ps per data point. Pulsed laser stereophotography of plates in flight and plate impacts onto PMMA have recorded plate planarity and flight times. These miniature 1-D metal plates have been impacted on targets of like material to evaluate diagnostics for spall studies with shock risetimes of ≤300 ps. These data are recorded primarily to evaluate diagnostic techniques, but spall data at high strain rates has been obtained.

This work was supported by the United States Department of Energy under contract number W-7405-ENG-36.

I. INTRODUCTION

A pulsed laser has been used to launch miniature 1-D plates. Several optical diagnostic techniques have been used to evaluate the velocity, acceleration profiles, planarity, and integrity of plates in flight and on impact. By correlating various techniques, including velocity interferometry (VISAR), stereo photography and streak photography, a complete understanding of the plate performance is possible. Preliminary experiments have been performed with PVDF gauges used in a "thick" gauge mode for pressure and pulse duration The flyer perfomance can be related to the characteristics of the laser pulse that accelerates the plate. Critical laser parameters are: wavelength, temporal profile, spatial profile, and power density imparted to the plate material to be launched.

II EXPERIMENTAL TECHNIQUE FOR MINIATURE 1-D PLATE LAUNCH

Aluminum, copper, or other composites 0.5 - 10 μm thick are physically vapor deposited (PVD) on quartz substrates. A high power (1 - 3 GW/cm^2, typical) 10-ns Nd:YAG laser pulse is directed through the quartz substrate to the quartz/metal interface (Fig. 1) [1-4]. The energy is deposited in the skin depth of the metal at the quartz/metal interface until the metal is converted to a plasma after which the remaining energy in the laser pulse is deposited in the metal plasma. This high temperature and pressure metal plasma accelerates

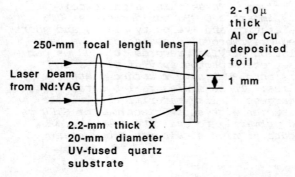

FIG. 1 *The Nd:YAG 10 ns laser pulse is directed through a quartz substrate onto the quartz/metal interface. The laser energy is deposited in the metal forming a plasma that accelerates the tamped remaining solid metal plate.*

the remaining tamped solid metal as a plate the diameter of the laser beam, typically 0.4 - 1 mm diameter. This phenomona has been modeled with LASNEX [5] and Gurney energy models [6] and models are in general agreement with experimental data. Accelerations $\geq 10^{10}$ G's and velocities ≥ 6 km/s have been recorded.

III. DIAGNOSTICS FOR MINIATURE PLATE FREE SURFACE VELOCITY AND CONTOUR AND EXPERIMENTAL RESULTS

A. VISAR (VELOCITY INTERFEROMETRY)

Velocity interferometry (VISAR, Fabry-Perot, or variations of both) is a common technique to measure particle velocity of a material. To date time resolution of recorded velocity interferometry data has been limited to ~1-3 ns/point. Since VISAR data is traditionally recorded on photomultiplier tubes (PMT's) time resolution is limited by the PMT and ancillary electronics. Fabry-Perot (F-P) interferometry is traditionally recorded on an electronic streak camera which has ≤ 1 ns/pt resolution but the "fill time" of the F-P is usually a 2 - 3 ns or greater. By replacing the PMT and ancillary electronic recording method of a VISAR with fiber optics and an electronic streak camera (Fig. 2), the limiting time resolution of the VISAR is reduced to less than or equal to the tau of the VISAR [7-9]. This recording method has resulted in time resolution of ~100 ps/point of recorded data thus permitting detailed structure in "ring up" of plate accelerations (Fig. 3).

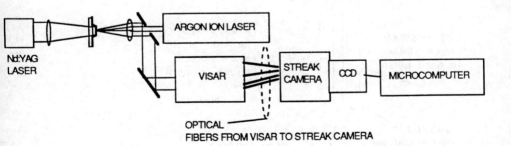

FIG. 2 *The optical output interference signals from the VISAR are directed into optical fibers that are input to an electronic streak camera. The time resolution is determined by the fringe constant (tau) of the VISAR and the resolution of the streak camera.*

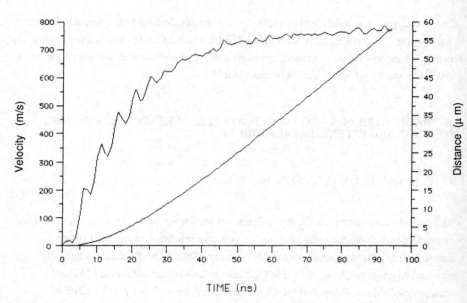

FIG. 3 *The shock "ring-up" of a 11 μm thick copper plate is resolved for numerous reverberations up to a final velocity.*

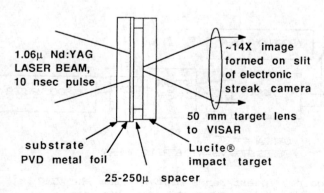

FIG. 4 *A PMMA or fused UV-quartz witness target is placed a known distance from the surface from which the laser-driven flyer plate will be launched.*

B. STREAK CAMERA OF PLATE IMPACTS ON PMMA

Plate planarity and integrity has been recorded by impacting laser-driven plates onto PMMA and time resolving the impact profile over a fixed flight distance (Fig. 4) [10]. These impact records confirm plate integrity and average plate velocity (Fig. 5). By using the experimental set-up in Figure 4 with the VISAR looking through the PMMA and focused on the metal plate to be launched, the plate velocity can be followed as well as the particle velocity of the aluminum/PMMA interface on impact (Figure 6).

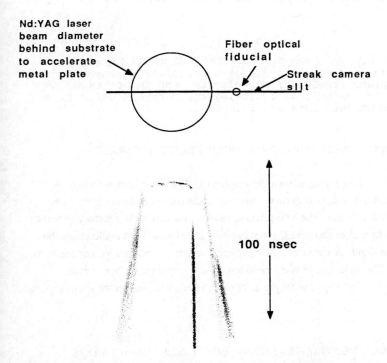

**Nd:YAG laser
beam diameter
behind substrate
to accelerate
metal plate**

**Fiber optical
fiducial**

**Streak camera
slit**

100 nsec

FIG. 5 *The metal flyer plate impacting a PMMA witness target details the profile along a dimeter of the plate. Since the impact time of flight over a fixed distance is known the average plate velocity can be calculated.*

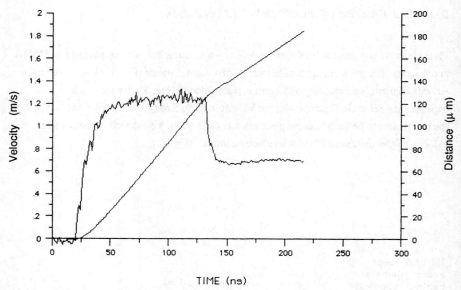

TIME (ns)

FIG. 6 *By using the same experimental set-up as Figure 4 and the VISAR to observe the surface of the plate through the PMMA, the plate acceleration, velocity, and impact time can be determined, as well as following the particle velocity of the metal/PMMA interface after impact.*

C. PULSED LASER STEREOPHOTOGRAPHY OF PLATES IN FLIGHT

Pulsed laser stereo photography of laser-driven plates in flight have been recorded. A portion of the 1.06-μm Nd:YAG pulse is converted to the first harmonic (532 nm) and separated from the 1.06 μm. The 1.06-μm laser beam is directed to the substrate to launch the laser-driven flyer plate and the 532 nm is optically delayed and used to illuminate the 1.06-μm plate in flight. A stereo camera with bandpass filters at 532 nm records the stereo pair of images. The plates can be accelerated toward the camera for surface contour measurements or at right angles to the plane formed by the optical axes of the stereo camera (Fig. 7).

D. PRELIMINARY PVDF GAUGE RECORDS FOR "THICK" GAUGE MODE

Preliminary experiments have been performed using 25-μm thick PVDF gauges [10] in a "thick" gauge mode, that is, the entire pressure pulse duration is less than the time to transit

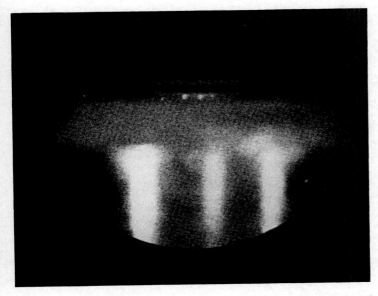

FIG. 7 *The time-delayed second harmonic (532 nm) of the Nd:YAG (1.06 μm) laser used to launch the plate can be used as the pulsed light source to illuminate a pulsed laser photograph of the plate in flight.*

the gauge material. For the PVDF material the transit time is ~8 ns and the pulse duration of our pulses are ≤1 - 6 ns. The shortest FWHM pulse that we have recorded to date is ~8 ns (Fig. 8). However, the frequency response of out recording system was ~ 2 ns/point. The 8 ns FWHM pulse that was recorded could be limited by the gauge, plate planarity, edge effects of the gauge or the recording system. In the near future we will be recording guage signals at ≥6GHz, minimizing edge effects, and further improving planarity. By eliminating or minimizing these experimental uncertainties we should be able to determine the gauge response time.

IV. DIAGNOSTICS FOR PLATE IMPACT AND SPALL

Using the same basic experimental set-up as Fig. 4 and replacing the PMMA with a metal target 25-μm thick, a spall can be generated in the target by laser-driven 10-μm plate impacts. The velocity of the free surface of the spalled sample of the target was recorded by VISAR with ~100 ps/point of recorded data. The spall strength for short duration (~3ns)

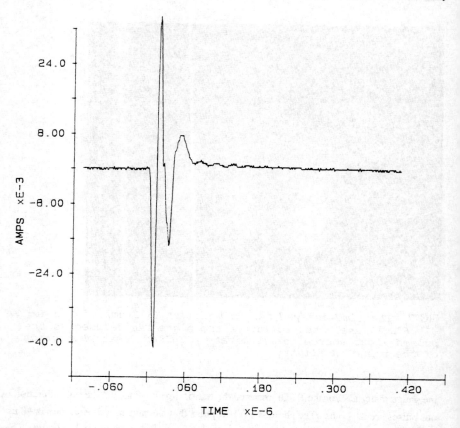

FIG. 8 *Output signal of a PVDF (2 mm x 2 mm) impacted by a 600 μm diameter copper plate 11 μm thick and impact velocity of ~800 m/s. The pulsewidth (~ 8 ns FWHM) of plate impact and three reverbrations at longer pulsewidths and lower pressure are obvious.*

pressure pulses have been recorded where the spall "pull back" velocity minimum is reached in ~1.5 ns after peak free surface velocity before spall. These data (Fig. 9) are similar to Eck and Asay data [11]. The aluminum spall in Fig. 9 rings in the spall whereas the copper sample does not ring. Our pulse durations are appproximately one order of magnitude shorter.

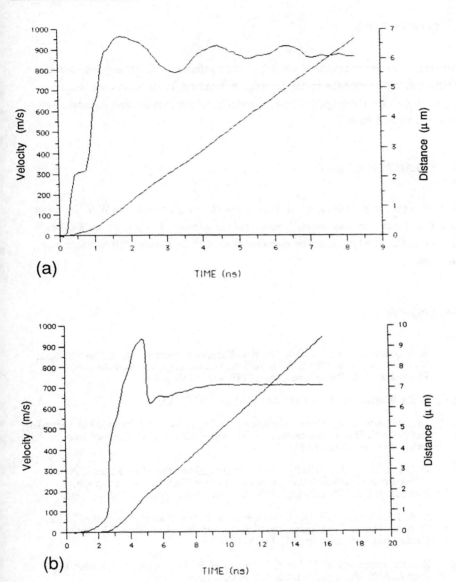

FIG. 9 *Spall records with short(~2-3 ns) pulse durations have been recorded at ~ 100 ps/pt for 6 μm copper impacting 25 μm copper and 10 μm aluminum impacting 25 μm aluminum. Impact spalls are similar with previous data by Eck and Asay for pulse durations of ~60 ns [11].*

V. CONCLUSIONS

Miniature laser-driven one-dimensional flyer plates offer a new opportunity and capability to perform dynamic material property studies with smaller samples, lower cost, possibly higher strain rates, shorter pulse durations, and improved diagnostic time resolution well less than a nanosecond.

VI. ACKNOWLEDGMENTS

N. I. Montoya, D. B. Stahl, and I. A. Garcia conducted experiments and W. F. Hemsing contributed data reduction software and VISAR consultation. C. Gallegos, EG&G/Los Alamos and G. Stradling, Los Alamos provided some of the experimental hardware and software.

REFERENCES

1. S. A. Sheffield and G. A. Fisk, "Particle Velocity Measurements of Laser-Induced Shock Waves using ORVIS", High-Speed Photography, Videography, and Photonics, D. L. Paisley, ed., Proc. SPIE 427, 1983, p 192.

2. S. P. Obenschain, et al., *Phys. Rev. Lett, 50:* 44 (1983).

3. D. L. Paisley, "Laser-Driven Miniature Flyer Plates for Shock Initiation of Secondary Explosives", Shock Compression of Condensed Matter, S. C. Schmidt, et al., (eds.), North-Holland,1989.

4. D. L. Paisley, et al., "Velocity Interferometry of Miniature Flyer Plates with Sub-Nanosecond Time Resolution", Proceeding of the High Speed Photograph, Videography and Photonics Conference, SPIE Proc. 1346, San Diego, 1990.

5. R. A. Kopp and D. L. Paisley, "Acceleration of Flyer Plates by Contained Laser Plasmas - Recent Experiments and Simulations," 20th Anomalous Absorption Conference, Traverse City, MI, July 9-13, 1990.

6. R. J. Lawrence and W. M. Trott, "Analysis of Fully-Tamped Laser-Driven Flyer Plates", APS Meeting, Anaheim, CA, March, 1990.

7. L. M. Barker and R. E. Hollenbach, "Interferometer Technique of Measuring Dynamic Mechanical Properties of Materials," *Rev. Sci. Inst., 36:* 1617(1965).

8. W. F. Hemsing, "VISAR: Interferometer Quadrature Signal Recording by Electronic Streak Camera," Proceeding of the Eighth Symposium (International) on Detonation, July 15-19,1985, p 468.

9. W. F. Hemsing, "VISAR: Displacement Mode Data Reduction", Ibid. 4.

10. L. M. Lee, et al., "Studies of the Bauer Piezoelectric Polymer Gauge (PVF$_2$) Under Impact Loading,"*Shock Waves in Condensed Matter*, Y. M. Gupta, (ed.), Plenum, New York,1986.

11. D. R. Eck and J. R. Asay "The Stress and Strain-Rate Dependence of Spall Strength in Two Aluminum Alloys," Ibid. 10.

109

Vibrational Mechanisms in Explosive Welding

N. NAUMOVICH* and T. NAUMOVICH+

*Byelorussian Powder Metallurgy Association
 Minsk, Byelorussia, USSR

+Byelorussian Polytechnical Institute
 Minsk, Byelorussia, USSR

The existence of rather strict periodicity of the wave
picture in the zone of contact interaction under explosive
welding is observed during high frequency oscillatory
processes in welding plates. The problem about cross
vibrations of two plates under movement with constant
velocity and linear concentrated loading has been considered.
This problem can serve as a mathematical model of plate
deformation processes in explosive welding. A one
dimensional Timoshenko model, taking into consideration shear
and rotation inertia of cross sections, is used for the
description of plate behavior.

I.   INTRODUCTION AND ANALYSIS

The problem of cross vibrations of plates results in a system

of differential equations.

$$E_1 I_1 \frac{\partial^4 y_1}{\partial x^4} + \rho_1 F_1 \frac{\partial^2 y_1}{\partial t^2} - \rho_1 I_1 \left(1 + \frac{E_1}{\lambda\,G_1} \right) \frac{\partial^4 y_1}{\partial x^2\,\partial t^2} +$$

$$+ \frac{\rho_1^{\,2} I_1}{\lambda\,G_1}\,\frac{\partial^4 y_1}{\partial t^4} = \frac{P\ (t+\Delta t)}{\varepsilon_1} - \frac{F(t)}{\varepsilon} \tag{1}$$

$$E_2 I_2 \frac{\partial^4 y_2}{\partial x^4} + \rho_2 F_2 \frac{\partial^2 y_2}{\partial t^2} - \rho_2 I_2 \left(1 + \frac{E_2}{\lambda \, G_2} \right) \frac{\partial^4 y_2}{\partial x^2 \, \partial t^2} +$$

$$+ \frac{\rho_2^2 I_2}{\lambda \, G_2} \frac{\partial^4 y_2}{\partial t^4} + k_o y_2 = \frac{F(t)}{\varepsilon} \qquad (2)$$

where: y_1 and y_2 - cross displacement of upper and lower plates, respectively; E - bulk modulus; G - shear modulus; b,h - width and thickness of plates; F=bh - cross-sectional area; ρ - density; λ=8/9 - coefficient, taking into account the shape of cross-section [1]; I - inertia moment, k_o - stiffness coefficient, $P(t+\Delta t)$ - detonation loading, moving along the x axes with constant velocity V, $P(t+\Delta t)$ is uniformly distributed with pressure p on a certain area $\varepsilon_1 b$. It is supposed that all the detonation parameters are known (Fig. 1).

Material of the plates is considered to be elastic, except the collision zone, which size is equal to $\varepsilon b \varepsilon$, where ε - the width of the interaction zone (small parameter). Material of the collision zone has both elastic and viscous properties. Viscous properties are defined through the coefficient of viscosity η or the relaxation time τ; η = Gτ.

F(t) function defines plates interaction at the zone of their contact. The definition of the F(t) function is one of

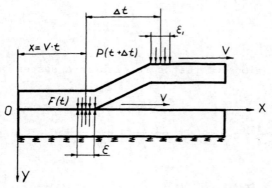

FIG. 1 *Scheme of plates loading at explosive welding*

the major tasks. So when $F(t) = 0$, system (1)-(2) describes the state of two independent dynamic objects, in particular, equation (1) describes the acceleration of the upper plate by detonation loading $P(t+\Delta t)$. It is considered that $F(t)$ function, similar to $P(t+\Delta t)$, is distributed on an area εb, moving along the x axes with constant velocity V. $F(t)$ results from the internal resistance forces of the plates. The Elastic component of plate interaction can be written in the form:

$$F_e = (\varepsilon\, b)\, \frac{\rho_1\, v_{11}}{k_1}\, \left(\frac{d\,z}{d\,t}\right)_{x=vt} \qquad (3)$$

where $\quad k_1 = 1 + \dfrac{\rho_1\, v_{11}}{\rho_2\, v_{12}}\, ; \qquad \left(\dfrac{dz}{dt}\right) = \dfrac{\partial y_1}{\partial t} + v\, \dfrac{\partial y_1}{\partial x} - \dfrac{\partial y_2}{\partial t} - v\, \dfrac{\partial y_2}{\partial x}\, ;$

$v_1 = \sqrt{\dfrac{E}{\rho}}$ .It is supposed that dissipative forces at contact zone are viscous ones and can be written as:

$$F_{1v}(t) = \varepsilon\, \mu_1\, \frac{\partial}{\partial t}\left[\frac{\rho_1{}^2 I_1}{\lambda\, G_1}\, \frac{\partial^4 y_1}{\partial t^4} - \rho_1 I_1\, \frac{E_1}{\lambda\, G_1}\, \frac{\partial^4 y_1}{\partial x^2\, \partial t^2}\right] \qquad (4)$$

$F_{2v}(t)$ expression can be received from (4) substituting all '1' indexes to '2'. Here $\mu = \dfrac{\eta}{G}\left(\dfrac{v_{21}}{r_1}\right)$ – attenuation coefficient , $r_1 = \dfrac{h_1}{\sqrt{12}}$.

$V_2 = \sqrt{\lambda\, \dfrac{G}{\rho}}$ .Boundary and initial conditions have the next form:

$$P(t+\Delta t)=0; \qquad 0 < x < v(t+\Delta t) - \frac{\varepsilon_1}{2}\, ; \qquad v(t+\Delta t) + \frac{\varepsilon_1}{2} < x < 1$$

$$P(t+\Delta t)=p\varepsilon_1 b\, ; \qquad v(t+\Delta t) - \frac{\varepsilon_1}{2} < x < v(t+\Delta t) + \frac{\varepsilon_1}{2}$$

$$F(t) = 0\, ; \qquad 0 < x < vt - \frac{\varepsilon}{2}\, ; \qquad vt + \frac{\varepsilon}{2} < x < 1$$

$$F(t) = F_e(t) - F_v(t); \qquad vt - \frac{\varepsilon}{2} < x < vt + \frac{\varepsilon}{2} \qquad (5)$$

For $x = 0$, $x = 1$ (1- the length of plates) :

$$y \neq 0 ; \qquad \frac{\partial y}{\partial x} \neq 0 ; \qquad \frac{\partial^2 y}{\partial x^2} = 0 ; \qquad \frac{\partial^3 y}{\partial x^3} = 0 ; \qquad (6)$$

For t=0, Y_1=0. Also Y_2 and its partial derivatives are equal to zero.

$\delta y_1/\delta t$ - and higher derivatives are defined from the solution of the problem of acceleration of a plate by detonation loading.

Deflection function has been chosen as an eigenfunction expansion of the problem of bending vibrations of a plate with boundary conditions (6):

$$Y_1 = S_1(t) + \sum_{i=1}^{\infty} N_{1i} \sin (\beta_i t + \gamma_{1i}) f_i(x) ;$$

(7)

$$Y_2 = S_2(t) + \sum_{i=1}^{\infty} N_{2i} \sin (\beta_i t + \gamma_{2i}) f_i(x) ;$$

$f_i(x)$ functions define the shape of the plate bending vibrations [2]. $S(t)$ reflects the possibility of plate displacement with zero frequencies. Boundary conditions equ. (6) allow shear displacement and rotation as a rigid body.

Galerkin's method along the geometrical coordinates and incrementally changeable amplitude method has been used for the investigation of initial system of equations [3]. As a result, the system of ordinary differential equations of the first degree with respect to ($\delta N/\delta t$) has been obtained.

It is assumed that the mechanism of limitation of acting loading amplitude, and dissipative mechanism, in the contact zone, provide an establishment of a stationary motion with constant amplitude N and frequency β. This assumption enables

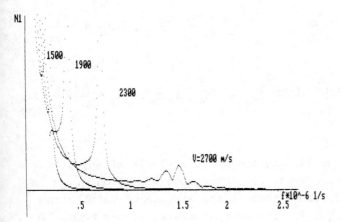

FIG. 2 *Amplitude-frequency characteristic dependence from contact velocity for Al-Al pair.*

the passage from differential equations to the system of linear algebraic equations with respect to N.

The analysis of amplitude-frequency characteristic, (Fig.2) shows that the increase of contact point velocity leads to excitation of higher modes of vibration in the plates. Thus, more terms need to be taken into consideration.

Stability of the stationary motion has been defined with the help of perturbation methods [3]. Characteristic equations, defining the behaviour of the upper plate and serving for definition of the characteristic factor d, has the following form:

$$\delta^2 d^2 - \delta d \, (a_{12} - a_{21}) + a_{11} \, a_{22} - a_{21} a_{12} = 0$$

$$a_{12} = 1.1 \, P_o - 0.94 \, (\tfrac{\varepsilon}{1}) \, \mu_1 \beta_0^4 \, k_2 \; ; \quad a_{21} = - \, P_o + 0.94 \, (\tfrac{\varepsilon}{1}) \, \mu_1 \beta_0^4 k_2$$

$$a_{11} = \beta_0^3 - A_0 \beta_0 + B + 0.5 \, (\tfrac{\varepsilon}{1}) \, \mu_1 \beta_0^4 \, k_2 ; \quad a_{22} = \beta_0^3 - A_0 \beta_0 + B - 0.5 \, (\tfrac{\varepsilon}{1}) \, \mu_1 \beta_0^4 \, k_2 - 0.5 \, P_o$$

$$P_o = \frac{\varepsilon}{1} \frac{1}{\sqrt{12} k_1} \left(\frac{v_{11}}{v_{21}}\right) ; \qquad B = \beta_o^4 \left(\frac{v_{11}}{v} \frac{v_{21}}{v}\right)^2 ; \qquad \beta_o = \beta \left(\frac{r_1}{v_{21}}\right)$$

$$A_o = 1 + \beta_o^2 \left[\left(\frac{v_{11}}{v}\right)^2 + \left(\frac{v_{21}}{v}\right)^2\right] ; \qquad \delta = \beta_o^3 - A_o \beta_o ; \qquad k_2 = \left(\frac{v_{11}}{v}\right)^2 - 1 .$$

Investigated stationary motion is unstable if any of the d values has a positive real part. The analysis of data, presented in Fig.3, shows that in the absence of damping the system is stable, but it becomes unstable at higher modes if there is even a small damping. The boundary of the instability region can be defined if coefficient η or relaxation time τ are known. Unfortunately data given in the literature are approximate. For example soft aluminum has $\eta = (0.03 - 0.27)$ 10^3 Pa. s [4-6]. The instability region boundary calculated for two aluminum plates of $h_1 = 2$ mm, $h_2 = 3.5$ mm according to

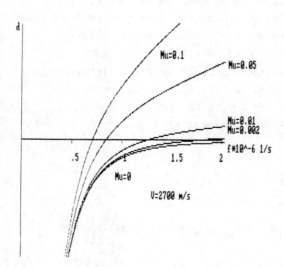

FIG. 3 *Inflúence of attenuation coefficient on stability of the upper plate.*

Fig. 3 and Fig. 4 has:

f_b = 1.25 MHz for η=0.05·10^3 Pa s, μ=0.01 and V=2700 m/s
f_b = 1.12 MHz for η=0.24·10^3 Pa s, μ=0.05 and V=4000 m/s
f_b = 0.78 MHz for η=0.21·10^3 Pa s, μ=0.05 and V=2300 m/s

Thus modes with frequencies higher than f_b are unstable and their amplitudes should exponentially increase with time. So energy, brought to the system, concentrates at higher harmonics, which in turn, having strong damping properties and small values of viscosity coefficient, use this energy for dissipative processes (intensive damping of elastic waves, shear stress relaxation, intensive plastic deformation up to relief of initial crystal structure, material throwen from contact zone). Vibrations caused by dissipative forces are usually called relaxation oscillations. Their frequency are determined by dissipative properties of the material, and energy absorption process, but not by vibrational properties of investigated system. If one supposes that frequency of relaxation oscillation is about stability boundary f_b then

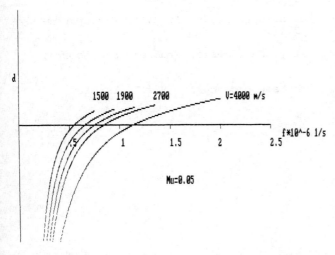

FIG. 4 *Influence of contact velocity on stability of the upper plate.*

calculated space period of contact point or wave length L for
frequency f_b will be L = 1.08, 1.79, 1.47 mm respectively, for
the examples mentioned above.

II. CONCLUSIONS

Mathematical model of joint cross vibrations process of plates
undergoing explosive welding has been speculated. It has been
established, that there is an instability of cross vibrations
of plates at high modes (in MHz frequency region). The cause
of instability is connected with the character of forces acting
at the plate contact zone. Unlimited growth of amplitudes in
the instability region probably causes relaxation vibrations
for conditions of high damping properties of the contact zone.

REFERENCES

1. S. P. Timoshenko, *Statitcheskie i dinamitcheskie problemy
    teorii uprugosti,*. Kiev (1975).

2. N. V. Naumovich, T. M. Naumovich, L. L. Slabodchikova,
    Fizika i technika vysokich davlenij, 31:56 (1989).

3. V. V. Migulin,V. I. Medvedev, E. R. Mustel, V. N. Parygin,
    Osnovy *teoryi kolebanij,* Moskva (1988).

4. *S.* K. Godunov, A. A. Deribas, V. I. Mali, *Fizika goreniya
    i vzryva, N1*:3 (1971).

5. G. V. Stepanov, V. V. Kharchenko, *Problemy prochnosti,
    N8*:59,(1985).

6. V. S. Kozlov, *Problemy prochnosti, N3*:47 (1986).

Author Index

Subject Index

About the Editors

MARC A. MEYERS is Professor of Materials Science, University of California, San Diego, La Jolla. His principal research interests include shock waves, dynamic consolidation, high-strain-rate phenomena, martensitic transformations, and the relationship between structure and mechanical behavior of materials. The author, coauthor, or coeditor of over 120 publications and five books, including (with Lawrence E. Murr and Karl P. Stauhammer) *Metallurgical Applications of Shock-Wave and High-Strain-Rate Phenomena* (Marcel Dekker, Inc.), he is a member of several professional societies and was visiting advisor to the U.S. Army Research Office and a visiting researcher at the National Chemical Laboratory for Industry in Tsukuba, Japan. Dr. Meyers received the B.Sc. degree in mechanical engineering from the Federal University of Minas Gerais, Belo Horizonte, Brazil, and the M.Sc. degree in materials science and the Ph.D. degree in metallurgy from the University of Denver, Colorado.

LAWRENCE E. MURR is Mr. and Mrs. MacIntosh Murchison Professor and Chairman of the Department of Metallurgical and Materials Engineering at the University of Texas at El Paso. The author of over 430 journal articles, he is a consultant, editor, and reviewer. Dr. Murr is the author, editor, or coeditor of over 18 books, including *What Every Engineer Should Know About Material and Component Failure, Failure Analysis, and Litigation; Metallurgical Applications of Shock-Wave and High-Strain-Rate Phenomena* (with Karl P. Staudhammer and Marc A. Meyers); *Industrial Materials Science and Engineering; Solid-State Electronics;* and *Electron and Ion Microscopy and Microanalysis, Second Edition* (all titles, Marcel Dekker, Inc.). He is a Fellow of the ASM (American Society for Metals) International and a member of the Electron Microscopy Society of America; Metallurgical Society of the American Institute of Mining, Metallurgical, and Petroleum Engineers; Institute of Electrical and Electronics Engineers; International Metallographic Society; and Sigma Xi. Dr. Murr received the B.Sc. degree (1963) in physical science from Albright College, Reading, Pennsylvania, and the B.S.E.E. degree (1962) in electronics, the M.S. degree (1964) in engineering mechanics, and the Ph.D. degree (1967) in solid-state science from the Pennsylvania State University, University Park.

KARL P. STAUDHAMMER is Staff Member of the Materials Research and Processing Science group (MST-5) in the Materials Science and Technology Division of the Los Alamos National Laboratory, Los Alamos, New Mexico. The author, coauthor, or coeditor of over 70 publications and three books, including (with Lawrence E. Murr and Marc A. Meyers) *Metallurgical Applications of Shock-Wave and High-Strain-Rate Phenomena* (Marcel Dekker, Inc.), he is a member of the American Society for Metals, the Electron Microscopy Society of America, and the American Institute of Mining and Metallurgical Engineers, among others. He is the recipient of an Alexander von Humboldt U.S. Senior Scientist Award from Germany (1988). Dr. Staudhammer received the B.Sc. degree (1966) in mechanical engineering from California State University, Los Angeles, the M.S. degree (1970) in mechanical engineering and the Engineers degree (1973) in materials science from the University of Southern California, Los Angeles, and the Ph.D. degree (1975) in physical metallurgy from the New Mexico Institute of Mining and Technology, Socorro.